Why Translate Science?

Why Translate Science?

Documents from Antiquity to the 16th Century in the Historical West (Bactria to the Atlantic)

Edited by

Dimitri Gutas

With the assistance of

Charles Burnett
Uwe Vagelpohl

TUTA SUB AEGIDE PALLAS · 1683 ·

BRILL

LEIDEN | BOSTON

Originally published in hardback in 2022 as Volume 160 in the series Handbook of Oriental Studies.

Cover illustration: the beginning of Hippocrates's famous aphorism 'Life is Short, Art is Long', in Syriac and Arabic translation. Paris, BNF, MS arabe 6734, fol. 29v.

The Library of Congress has cataloged the hardcover edition as follows:

Names: Gutas, Dimitri, editor. | Burnett, Charles (Charles S. F.), editor. | Vagelpohl, Uwe, editor.
Title: Why translate science? : documents from antiquity to the 16th century in the historical West (Bactria to the Atlantic) / edited by Dimitri Gutas ; with the assistance of Charles Burnett, Uwe Vagelpohl.
Description: Leiden ; Boston : Brill, [2022] | Series: Handbook of Orientalistik = Handbuch der Orientalistik. Section one, Near and Middle East, 0169-9423 ; volume 160 | Includes bibliographical references and index. |
Identifiers: LCCN 2021061932 (print) | LCCN 2021061933 (ebook) | ISBN 9789004472631 (hardback) | ISBN 9789004472648 (ebook)
Subjects: LCSH: Science–Translating–History. | Scientific literature–Translations–History and criticism. | Communication in science–History. | Translating and interpreting–History.
Classification: LCC Q124 .W48 2022 (print) | LCC Q124 (ebook) | DDC 418/.035–dc23/eng20220218
LC record available at https://lccn.loc.gov/2021061932
LC ebook record available at https://lccn.loc.gov/2021061933

Typeface for the Latin, Greek, and Cyrillic scripts: "Brill". See and download: brill.com/brill-typeface.

ISBN 978-90-04-76089-9 (paperback, 2026)
ISBN 978-90-04-47263-1 (hardback)
ISBN 978-90-04-47264-8 (e-book)

This book is printed on acid-free paper and produced in a sustainable manner.

Contents

Notes on Contributors

Michael Angold
is Professor Emeritus of Byzantine History at the University of Edinburgh. His most recent book is *Nicholas Mesarites: His Life and Works (in Translation)* (Liverpool, 2017).

Pieter Beullens
published (together with Fernand Bossier) the critical edition of Aristotle's *History of Animals* in the Latin translation by William of Moerbeke (two volumes, 2000/2020). His doctoral research focused on the identification of Latin medieval translators of Greek philosophical texts and in particular on the role of Bartholomew of Messina. In his current project as a postdoc of the Flemish Research Fund (FWO-Vlaanderen), he applies tools from computational stylometry to those translations to investigate the development of philosophical concepts.

Charles Burnett
(PhD Cambridge, FBA) is Professor of the History of Arabic/Islamic Influences in Europe, Warburg Institute, University of London. His work centres on the transmission of Arabic science and philosophy to Western Europe, which he has documented by editing and translating several Arabic–Latin texts and by describing the historical and cultural contexts.

David Omar Cohen
studied Classics and German in Amsterdam and Berlin and obtained a doctoral degree on Greek tragedy in 2021. He teaches German, Latin, and ancient Greek in several schools in Amsterdam and is currently working on the publication of his thesis.

Gad Freudenthal
is Senior Research Fellow Emeritus with the CNRS in Paris, France. He has written on the reception of science and philosophy in Jewish cultures, mainly in the Middle Ages and the eighteenth century. His books include: *Aristotle's Theory of Material Substance: Form and Soul, Heat and Pneuma* (Oxford, 1995), *Science in the Medieval Hebrew and Arabic Traditions* (Aldershot, 2005), and the edited volumes: *Science in Medieval Jewish Cultures* (Cambridge, 2011), *Studies on Steinschneider* (with Reimund Leicht, Leiden, 2011), *Latin-into-Hebrew—Studies and Texts*, Volume 1: *Studies* (with Resianne Fontaine, Leiden, 2013). Between 2001 and 2019, he edited *Aleph: Historical Studies in Science and Judaism*.

Dimitri Gutas
is Professor Emeritus of Arabic at Yale University. He has published on the medieval Graeco-Arabic translation movement and its lexicography (*GALex*), the transmission of Greek philosophical texts into Arabic (most recently Aristotle's *Poetics*, Brill, 2012), and Arabic philosophy (*Avicenna and the Aristotelian Tradition*, 2nd ed., Brill, 2014, most recently reprised in U. Rudolph, ed., *Philosophie in der islamischen Welt*, Band 2/1, Basel 2021).

Dag Nikolaus Hasse
is Professor in the History of Philosophy at the Julius-Maximilians-Universität Würzburg, Germany. He was educated at Göttingen and Yale, and received his PhD at the Warburg Institute in London. He is the author of *Avicenna's 'De anima' in the Latin West* (2000) and *Success and Suppression: Arabic Sciences and Philosophy in the Renaissance* (2016), and is the editor of the online *Arabic and Latin Glossary*, a dictionary of the vocabulary of the Arabic–Latin translations of the Middle Ages.

Anthony Kaldellis
is Professor and Chair of Classics at The Ohio State University. He has translated many Byzantine texts into English, especially historians, and has published many books and articles on topics of Byzantine history, literature, and culture. He is currently writing a new history of Byzantium from beginning to end.

Daniel King
specializes in the history of Syriac religion and philosophy, and its contribution to the progress of knowledge. His research is principally concerned with editing Syriac manuscripts containing philosophical material, analysing translation technique and lexicography, and examining how the Syriac tradition adapted to its own environment and idiom the heritage of Greek thought and ideas.

Felix Mundt
PhD (2007) has served as an assistant professor and is currently private lecturer of Classical Philology at the Humboldt-Universität zu Berlin. He is working on a bilingual edition of Gellius's *Noctes Atticae*.

Ignacio Sánchez
is a Senior Research Fellow at the Department of Classics and Ancient History in the University of Warwick. He is the executive editor of the journal *Endowment Studies*, and section-editor of Brill's *Encyclopaedia of Islam Three*

(History of Science). He has published *Epistle On Geography* of the Ikhwān al-Ṣafā' (OUP/IIS, 2014), contributed to *A Literary History of Medicine: Ibn Abī Uṣaybi'ah's 'Uyūn al-Anbā' fī Ṭabaqāt al-Aṭibbā'* (Brill, 2019), and written articles on *adab*, intellectual history, Islamic pious foundations, and medicine in medieval Islam.

Isabel Toral
is Professor of Arabic Studies at the Freie Universität Berlin (faculty member since 2018). She received her PhD from the University of Tübingen (1997) and her habilitation from the Freie Universität Berlin (2008). Between 1997 and 2018, she held numerous research positions in Freiburg, Berlin, London, Göttingen, and Mainz. Her main research fields are the Near East in late antiquity, cultural history, translation in the Islamicate world, and *adab* in classical Islam. Her publications include *Al-Ḥīra* (2014), and 'The Iqd Al-Farīd by Ibn 'Abd Rabbih—the Birth of a Classic', in *Approaches to the Study of Pre-Modern Arabic Anthologies* (2020).

Uwe Vagelpohl
is an assistant professor in Classics and Ancient History at the University of Warwick and, together with Simon Swain, Principal Investigator of the Wellcome Trust project 'Streamlining Galen: Medical Summaries and the Transmission of Medicine in Medieval Islam'. His research focuses on the medieval reception of ancient Greek medicine and philosophy in Arabic. His publications include the editions and translations of the Arabic version of Galen's Commentaries on the Hippocratic *Epidemics*, Books 1 and 2, with the Commentary on Book 6 to appear shortly.

Mohsen Zakeri
(University of Bochum) is a scholar in the field of Near Eastern history and literature, focusing on late antiquity, especially in the area of Iran and Islam. Besides teaching at several American and European universities, he has authored a series of academic works including *Sasanian Soldiers in Early Muslim Society: the Origins of the 'Ayyārān and Futuwwa* (1995), and *Persian Wisdom in Arabic Garb: 'Alī b. 'Ubayda al-Rayḥānī (d. 219/834) and his Jawāhir al-kilam wa-farā'id al-ḥikam* (2007).

CHAPTER 1

Introduction

Dimitri Gutas

Alexander reached Bactria, founding cities named after him along the way, and in subsequent centuries the Romans campaigned in and eventually annexed Iberia, Gaul, and Britain, thus generating a continuous geopolitical area from the west of India to the Atlantic, which constitutes the historical West. In the intellectual history of this vast area no single factor was more instrumental in culturally uniting it and fuelling the motor of civilization—the production and circulation of knowledge—than translation. Translation, of course, is present in multiform ways in all cultures,[1] and was arguably nowhere more integrally enmeshed into the fabric of society than it was already at the dawn of history in ancient Mesopotamia,[2] the civilization antecedent to the one that is our concern; but in the case of the West it evinces a longevous identity of function, nature, and teleology that commands attention, explanation, and, well, admiration.

Throughout the historical period that ranges from Alexander the Great to the sixteenth century, the peoples in the West acquired through translation, studied, and developed the Hellenic scientific and philosophical curriculum of higher learning—in short, science, broadly defined. There were translations, to mention the most prevalent languages, from Greek into Latin, Syriac, Middle Persian, and Arabic; from Syriac into Arabic and from Arabic into Syriac; from Sanskrit into Middle Persian and from Middle Persian into Arabic; from Arabic into Latin, Hebrew, and Greek; from Hebrew into Latin, and from Latin into Hebrew and Greek. Because the scientific knowledge that was transmitted, apart from whatever practical benefits it afforded, was acknowledged by all, tacitly or not, to be a supreme good in itself, it was a source of power and

1 See the gigantic collection of articles on translations from and into almost all languages in H. Kittel et al. (eds), *Übersetzung/Translation/Traduction. Ein internationales Handbuch zur Übersetzungsforschung. An International Encyclopedia of Translation Studies. Encyclopédie internationale de la recherche sur la traduction*, Handbook of Linguistics and Communication Science 26, 3 vols, Berlin: De Gruyter, 2004–2011.

2 See the general survey in the article by A.M. Ulshöfer in the publication cited in the preceding note, 1171–1177, and the analytical studies by F. Briquel-Chatonnet (ed.), *Mosaique des langues, mosaique culturelle: Le bilinguisme dans le Proche-Orient ancient*, Paris: Maisonneuve, 1996, and C. Jay Crisostomo, *Translation as Scholarship: Language, Writing, and Bilingual Education in Ancient Babylonia*, Berlin/Boston MA: De Gruyter, 2019.

 | DOI:10.1163/9789004472648_002

dominance to its possessor. Translation as a social movement was accordingly highly political, and the transmission itself was a deliberate and on occasion aggressive act of acquisition, a conquest, its course related closely to military, political, and social developments throughout this long period of eighteen centuries. At the end of it, all the peoples in this vast area with their various languages attained, by means of translation policies pursued by their elites as a concomitant of self-assertion and political prominence, a similar state of scientific advancement and higher learning. The process was properly the conquest of knowledge.[3] If the religious dimension is also taken into account—the fact that the peoples in the West were all adherents of monotheistic religions in the Biblical tradition—the level of homogeneity that was reached among them, both in the concordance of scientific views and the dialectic between science and religion that affected the process, is remarkable, indeed historic.

In this process, the Graeco-Arabic translation movement of the early ʿAbbāsid period (eighth to tenth centuries) played a pivotal role and constituted the node where translations into Arabic converged and whence they later spread out; and by the same token it brings to light the identity of the historical enterprise and the interconnectedness and continuity of its progression in relay. Recent scholarship has been investigating from outside the historical causes of this and other translation movements from and into the languages mentioned above, but what has not been adequately taken into consideration is the insider evidence provided in contemporary accounts of the motivations and purposes of translation given in the personal statements by the agents in this process, the translators themselves, and the actual or reported eyewitness accounts of the scholars and historians of each society.

This volume aims to collect the documents that present this evidence and accordingly complements and breathes life into the narrative and descriptive accounts of translations in other studies, notably the massive work edited by Kittel and others referred to above. It is intended for all researchers who, in the quest for the historical contextualization of the translations—that is, their causes, function, and significance—wish to study not only the personal status of the translators and the scientists who used their work in their respective societies, but also more broadly the cultural policies of the states and the ruling elites in these societies: their ideological orientations and underpinnings, and their conception of science and its importance in the interplay with traditional beliefs as well as its interface with politics. In short, it is intended that

3 The course of this process is briefly sketched in my article, from which this Introduction is largely taken, 'The Historical and Ideological Dimensions of Graeco-Arabic Studies: The Conquest of Knowledge from Alexander the Great to Meḥmed the Conqueror', *Intellectual History of the Islamicate World*, 3 (2015), 326–350.

this volume present the material for a social history of science. Its aim, as a handbook, is to make this material easily accessible to research, not to conduct the research itself.

The emphasis and the scope of the volume have been accordingly on the motivations for translating science into the languages participating in the relay of translations mentioned above and not on all translations or all languages.[4] The reasons include, most obviously, the absence of such a dedicated focus and comprehensive reach in previous studies. But most significantly the challenge was to extract information that provides insights into the cultural core of a society that is not otherwise explicitly stated, or if it is, is stated for apologetic purposes that conceal the motivating factors beneath—which itself is revealing of them. Thus, at one extreme, among the documents collected for the most extensive translation movement in this history, the Graeco-Arabic one in Baghdad, we are at a loss to find clear and unequivocal statements about the motivations that led to it; while at the other, the Sasanian self-serving and apologetic justification for the Greek to Middle Persian translations, namely the repatriation of knowledge stolen from the ancient Iranians by Alexander the Great, invites further investigation into the motor ideologies in that society. Accordingly, the net has been thrown wide, and the authors of the chapters here have scoured their respective literatures for all documents that might provide some information, or shed some light, however indirectly, on the motivations for translating science. For the period with which we are concerned, science included what we call philosophy as well as 'occult' sciences, notably astrology, divination, and oneiromancy. The authors of these books may be authentic or pseudonymous, and even in one representative case, that of Ibn Waḥšiyya, non-existent, something that encapsulates the obverse of the theme of this book, the translation of authoritative scientific texts: authority vested in a scientific text because it is claimed to be a translation.

It should be noted, however, that the documents selected do not include those that refer to the techniques of translation in particular, although some such passages may have been inserted when deemed helpful for the main objective of motivations for scientific translations. Similarly, religious literature has been excluded, in that apart from the fact that the motivations for translation of religious texts are transparent enough, they refer to a different order of

4 The purpose was not to present a survey of all translations from Greek, secular or religious, even if only in the Caucasus and Africa, which would naturally include languages such as Armenian, Georgian, Coptic, and Ge'ez. For a discussion of some of the issues involved, see Adam McCollum, 'Greek Literature in the Christian East: Translations into Syriac, Georgian, and Armenian', *Intellectual History of the Islamicate World*, 3 (2015), 15–65.

social reality, attitudes, and endeavour,[5] though in this case too some documentation on religious translation has been included where it helps, by contrast or at times by extension, to illuminate scientific translations in a given society. Within these parameters, though, coverage was extensive, if not total, left as it was to the discretion of each author.

The chapters are chronologically arranged in this relay of translations by the language into which the translations were made, ranging from classical Latin to Renaissance Latin. Each chapter is divided into three parts. The introductory essay in Part One provides information about the authors, history, and context of the selected documents, aiming to facilitate, but not offer, their study and analysis. The English translations of the documents that follow in Part Two are either written by the authors themselves, or taken from a satisfactory published version, as indicated. Part Three presents the documents in their original languages following standard editions, with occasional interventions by the author, if deemed necessary, who is responsible for the accuracy of the text in the printed volume. Bibliographical references to primary and secondary literature come at the end.

∴

The essays and documents collected in this volume were first presented and discussed at a conference held in Berlin in November of 2016 in association with the project on 'Aristotle's *Poetics* in the West (of India) from Antiquity to the Renaissance',[6] funded by the Einstein Stiftung Berlin and housed in the Berlin Graduate School Muslim Cultures and Societies (BGSMCS) / Seminar für Semitistik und Arabistik at the Freie Universität, with Beatrice Gründler acting as the academic sponsor and mentor and myself as the Einstein Visiting Fellow. Funds were also kindly provided by the BGSMCS. I gratefully acknowledge this munificent support, especially as it afforded me the opportunity to conduct this research in the city and its institutions where our field has been graced with the contributions of so many scholars. And a big thank you goes to Beatrice

5 These differences become starkly evident when one compares the Graeco-Arabic secular translation movement with the initially contemporary Graeco-Arabic religious translations by Christian Arabs, for a general account of which see Alexander Treiger, 'Christian Graeco-Arabica: Prolegomena to a History of the Arabic Translations of the Greek Church Fathers', *Intellectual History of the Islamicate World*, 3 (2015), 188–227, and the individual studies collected in Barbara Roggema and Alexander Treiger (eds), *Patristic Literature in Arabic Translations*, Leiden: Brill, 2020.

6 https://www.geschkult.fu-berlin.de/e/semiarab/arabistik/Forschung/aristotle_s_poetics/index.html (accessed 11.8.21).

Gründler, their current successor, who made it all possible, with knowledge, skill, grace, and joie de rechercher.

I wish to extend also my heartfelt thanks to the colleagues who participated in the conference and generously offered their expertise in accomplishing this task in a collaborative spirit that emulated the diachronic fraternity of scholars and translators who toiled on the ancient texts. I also thank my colleagues in my home department at Yale, Kevin van Bladel, Benjamin Foster, and Eckart Frahm, who assisted me with their wide knowledge of subjects on Middle Persian and the ancient Mesopotamian languages. In the process of editing the volume, I received unstinting support from all the contributors—particularly from Charles Burnett and Uwe Vagelpohl, without whose assistance the end product would have been much delayed and borne more imperfections than those that may still persist, and for which I am solely responsible.

I would also like to repeat the usual coda in my acknowledgments by thanking Koninklijke Brill NV and our editors, Kathy van Vliet, Joed Elich, and Abdurraouf Oueslati, for accepting and publishing this volume in their *Handbuch der Orientalistik* series. Thanks are also due to Joseph Spooner for accomplishing the Herculean task of copy-editing and proof-reading the English styles of authors of more diverse linguistic backgrounds than those studied in this book; and to Madelon Janse and TAT Zetwerk for their customary expert production of a book with many alphabets and varied layouts. The result has been another testament to Brill's commitment to their motto, 'Humanities Matter'.

CHAPTER 2

Latin Translations of Greek Science and Philosophy: Some Relevant Passages

Felix Mundt and David Cohen

Part One: Essay

The story has been told many times from antiquity to the present day: the translation and adaptation of Greek texts have been crucial in the formation of Latin literature as a whole.[1] By the first century BC, Greek and Roman intellectual cultures were so closely allied that a theory emerged postulating that Latin was in fact a Greek dialect.[2]

Although both this essay and its related texts focus on philosophy and science, it is necessary to preface them with some remarks on the very early history of Latin translation, namely the poetic genres of epic and drama. The history of non-epigrammatic Latin literature begins with a figure deeply rooted in the Greek culture of Magna Graecia: Livius Andronicus, who was probably born in Tarentum between 280 and 260 BC. He translated the *Odyssey* into Latin Saturnian verse and was, according to Cicero (*Brutus*, 72), the first author to stage a literary drama in Latin, probably following a Greek model, in 240 BC. If the theatre was the first place where a broader Roman audience without knowledge of the Greek language came into contact with Greek intellectual culture, it was also the location where snippets of popular philosophy could penetrate the ears and minds of native Latin-speakers.[3] It is therefore not at all unwar-

1 In the most classical terms by Horace, in *Epistulae*, 2.1.156–157: 'Graecia capta ferum uictorem cepit et artes / intulit agresti Latio'. Most recently by Feeney, *Beyond Greek*, 17–91. As a standard reference covering the whole complex of 'Greek into Latin', we recommend Kaimio, *The Romans and the Greek Language*, along with Hutchinson, *Greek to Latin*, for broader cultural context. Cf. Traina, 'Le traduzioni', and Rochette, 'Du grec au latin et du latin au grec'. On Roman theories of translation, see McElduff, *Roman Theories of Translation*, and Robinson, 'Classical Theories of Translation'. On Greek-Roman bilingualism beyond literary translations, see Biville, 'The Graeco-Romans and Graeco-Latin', and Adams, *Bilingualism*.

2 E.g., Dionysius of Halicarnassus, *Antiquitates Romanae*, 1.90.1. Cf. Werner, 'Zur Fremdsprachenproblematik', 16; Stevens, 'Aeolism'; and Rochette, 'Greek and Latin Bilingualism', 285.

3 The influence of Hellenistic philosophy on Greek New Comedy and its Latin successors remains a neglected field; cf. Konstan, 'Crossing Conceptual Worlds'.

 | DOI:10.1163/9789004472648_003

ranted to begin our selection of texts with a quotation from Terence (1.1). In the prologue to his *Adelphoe*, we encounter a few issues that are of the highest importance in the history of Greek–Latin translation.

1. The phrase *verbum de verbo* ('word for word'). Together with its opposite *sensus de/pro sensu* ('meaning for meaning'), this remained a standard expression in Roman translation theory up to and including late antiquity.[4]
2. The moment a Greek genre was for the first time transformed and adapted by a Roman author to the needs of the Latin language and Roman culture signalled the beginning of a Latin tradition that could not be ignored by later writers; here, Terence refers to both Diphilus *and* Plautus. Similarly, any Roman author who adapted or translated Greek philosophy after Cicero could not possibly ignore the Latin terms his great predecessor had already coined; if he deviated, there would have to be good reason to do so.
3. Most Latin translations in any genre, as far as we know, were not word-for-word. Rather, translators gave themselves the freedom to omit or rearrange verses or passages, to translate freely, and to add something of their own. In other words, the line between translation and free adaptation was permeable.
4. Though closely following two earlier poets, Terence claims to stage something 'new'. Thus, a poet who incorporated even large portions of extant works into his own text was by no means considered a thief or plagiarist. Rather, he had internalized the first principles of creativity, in keeping with ancient theories of literature and art, *imitatio/mimēsis* and *aemulatio/zēlōsis*.[5]

In the passage preceding the verses selected from the *Ars poetica* (1.4), Horace tells his readers that a simpler and safer way to become successful as a poet is not to conceive something completely new, but rather to build on extant works and subjects; the new piece should nevertheless not be so close to its model as to resemble a mere translation. We can see also in Cicero how close the *verbum de verbo* topos—which at first glance seems to refer only to translation issues—is related to debates on imitation and originality. In *De optimo genere oratorum*, which served as a preface to his (lost) translations of speeches by Demosthenes

4 The *verbum de verbo/sensus de sensu* opposition is of course also present in other traditions. See Brock, 'Aspects of Translation Technique', for a short comparative survey of the *verbum de verbo* method used in translations of sacred and philosophical texts into Greek, Latin, Arabic, and Syriac.

5 An interesting discussion concerning how the many borrowings from Greek and Latin authors found in the *Aeneid* should be assessed is Macrobius, *Saturnalia*, 6.1–6.

and Aeschines, Cicero positions himself in the dispute between Atticists and Asianists.[6] In reaction to those who accused him of being an Asianist (i.e., of using too much rhetorical *ornatus* compared with the earlier Attic orators such as Lysias), Cicero defends himself by saying that he always oriented himself towards the style of great Athenian orators like Demosthenes and Aeschines, who were different in style but still both fourth-century Athenians. For Cicero, 'Atticism' means simply being as good an orator as the Athenians were, and, above all, as good as Demosthenes. But when translating even he did not follow this greatest of all possible role models slavishly, but rather freely, not 'count[ing] them [the words] out piece by piece to the reader, but rather, as it were, pay[ing] them out by weight' (1.2). In this way, Cicero was able to justify his self-understanding as an independent Roman orator and could at the same time claim originality. This behaviour was typical of him, both as a Roman and personally.[7] And of course, every translator who is proud of his abilities is sure to claim that he is a *sensus de sensu* translator: in the first instance, it demonstrates that he understands the meaning and the cultural context of the original; secondly, it indicates that he has the necessary skills for transferring meaning into his mother tongue, in other words, that he is a master of his own language. It is therefore no coincidence that one of the few pleas in favour of a literal translation concerns a case where individual creativity and adaptation to a specific cultural context were absolutely not required: Holy Scripture (1.5).

The principles and key issues relating to Roman theories of translation and literary transformation outlined so far are also to be found in texts concerned with the translation of philosophy. In book 3 of *De finibus* (1.3), Cato the Younger warns Cicero that Stoic doctrine sometimes seems obscure because of its many unusual or newly coined terms. This brings the discussion to the question of how Romans should deal with the problem when they wish to discuss Stoic philosophy in Latin. In Cicero's opinion, it is of course possible to invent new Latin terms,[8] but one can also draw on an existing Latin word, paraphrase the original by adducing two or more Latin words, or even just leave the Greek word in the

6 There is no clear-cut definition underlying the polemical terms 'Asianism' and 'Atticism' beyond the common notions of 'pompous' and 'plain'. On the opposition in general, cf. Hidber, *Das klassizistische Manifest*, 30–37; and Calboli, 'Atticism'.

7 In the Renaissance, this debate on imitation and originality reappeared as Ciceronianism. On this, see Robert, 'Die Ciceronianismus-Debatte'.

8 What Cicero is thinking of are known as paronyms (*Übersetzungslehnwörter*). These include Livius Andronicus's *versutus* for Greek *polytropos* (*Odyssey*, 1.1). There is a good discussion of the whole passage in Seele, *Römische Übersetzer*, 25–40. Cf. Horace, *Ars poetica*, 52–53: 'New and self-made words will also find acceptance, if they come from a Greek origin and are adapted in a modest way.' The most comprehensive study on Cicero's practice of translation is Poncelet, *Cicéron: Traducteur de Platon*.

text.[9] The use of a Latin word to convey the same meaning as an underlying Greek term, but where the Latin has additional meanings of its own, or possesses very similar but not exactly identical meanings, actually stimulated the development of the Latin language. There are many famous examples: Latin *virtus* (which literally means 'manliness') was used for Greek *aretē* (ἀρετή, which means 'excellence'), while *natura* and *physis* (φύσις), *ratio* and *logos* (λόγος), and *causa* and *aitia* (αἰτία) all found themselves paired.[10]

Because there are clearly many Latin terms that do not have exactly the same meanings as the Greek terms they were intended to render, translation can cause trouble. Cicero is of course perfectly aware of this, but he pushes his argument in a highly polemical passage (II.1) by levelling out any difference in meaning between *hēdonē* (ἡδονή) and *voluptas*. In book 1 of *De finibus*, the politician Lucius Manlius Torquatus argues in favour of Epicurean ethics. In book 2, Cicero refutes Torquatus's views and tackles his opponent by blaming him for the undifferentiated terminology of the Epicurean school. Torquatus insists on defining *hēdonē* (ἡδονή), and not just *hēdonē*, but even the highest form of *hēdonē*, as the absence of pain. Cicero argues that the Greek term *hēdonē* is equivalent to the Latin *voluptas*, and that every Roman would define *voluptas* not as an absence of pain, but as 'joy in the mind, and an agreeable arousing of pleasantness in the body'. Cicero criticises Torquatus for blending together two notions of *voluptas* or *hēdonē*, namely, the usual one and the notion that Epicurus allegedly imposed on the word. In reality, the case is not as simple as Cicero tries to make his readers believe; ironically, even he himself, in *De finibus* 2.23, uses *cupiditas* for *hēdonē*, thus demonstrating that there was no exact one-to-one correlation.[11]

Our next two examples come from Seneca the Younger. Right at the beginning of his dialogue *De tranquillitate animi* (II.2), he explains that he is going to use the term *tranquillitas* for the Greek word *euthymia* (εὐθυμία).[12] He admits that the two words differ in morphemic composition, but claims that both have effectively the same force. If we examine the two words more closely, however, it becomes apparent that the difference between them is not merely a mat-

9 A Greek word appearing in a Latin text has its own kind of expressiveness. Cicero made ample use of this technique; cf. Adams, *Bilingualism*, 339–340.

10 On these examples, see the rich linguistic study by Nicolas, *Utraque lingua*. Glucker, 'Cicero's Remarks', offers a comprehensive list of Greek-Latin word pairs from Cicero's philosophical writings. Cf. also Powell, 'Cicero's Translations', 292–294; and Rosén, 'Greek in Latin, Greek into Latin'.

11 Gordon, *The Invention*, 124–125.

12 This translation goes back to Cicero, *De finibus*, 5.23. Cicero, however, offers *securitas* ('freedom of sorrows') as an alternative translation. Cf. Thomas, 'De la paix', 71.

ter of linguistics: there is difference of meaning. The Greek term indicates that the *thymos* (θυμός, the heart as the seat of emotions) is in good condition, and that it is filled with a positive feeling, *eu-* (εὖ), such as cheerfulness. The Latin word *tranquillitas* is composed of two morphemic elements, *trans* and *quies*. The word's first and oldest meaning suggests the absence of wind on the calm sea.[13] In other words, while the Greek term implies the presence of good emotions, the Latin term implies the absence of any emotions.

In 11.3, Seneca examines the defectiveness of the paradigm of the verb *esse*. For the Greek *ousia* (οὐσία, 'substance', 'being'), Latin authors had already invented the term *essentia*. According to Seneca, this had been a coinage of Cicero's, though the term is not present in the latter's extant works.[14] Seneca himself does not dare to invent a Latin equivalent for the Greek *to on* (τὸ ὄν); instead, he uses the relative clause *quod est*. The present participle *ens* seems to have been coined by Sergius Plautus (see n. 14) or, according to the grammarian Priscianus, by Julius Caesar, probably in his lost treatise *De analogia*.[15] In every case, the mere process of finding an appropriate Latin word or thinking out a new paronym seems to have been seen as an entertaining way of sharpening one's wits. Aulus Gellius (11.4) dedicates a whole chapter of his *Noctes Atticae* to the proper rendering of *polypragmosynē* (πολυπραγμοσύνη). Most interestingly, he does not even mention the Latin *curiositas* as a possible candidate, which was to become the common translation for *polypragmosynē* from late antiquity through to the Renaissance.[16]

Cicero can rightly be called the most influential individual in the development of a basic linguistic toolkit for philosophical discourse in Latin. Quite aside from his own personal motivation to delve into Greek philosophy,[17] one of his most important principles was that whoever wished to excel as either a politician, orator, or philosopher should possess the highest abilities in each of these fields. For Cicero, this was the only way to achieve what was best for the advancement of the Roman state as a whole. In his *Life of Cicero*, Plutarch bestows great honour on Cicero as a philosopher. In 63 BC, the year in which Cicero, as consul, defused the Catilinarian conspiracy, the Roman state was, according to Plutarch, favoured by fortune, since it had at its helm a man who

13 Cf. Plautus, *Mercator*, 877, and *Poenulus*, 753.

14 Calcidius (*ad Tim.*, 35A1–4) and Sidonius Apollinaris (*Carmen* 14, praef.) also assign the invention to Cicero. Quintilian (*Institutio oratoria*, 2.14.2, 3.6.23, 8.3.33) refers to a Stoic named Sergius Plautus. Cf. Reinhardt/Winterbottom, *Quintilian*, 219–220.

15 Quintilian, *Institutio oratoria*, 8.3.33; and Priscian, *Institutiones grammaticae*, 18.8.75. Cf. Reitz/Fuchs, 'Vertere', 54 n. 58.

16 Cf. Leigh, *From Polypragmon to Curiosus*, 57.

17 On this, see Baltussen, 'Cicero's Translation'.

was Plato's ideal of the philosopher-king incarnate.[18] In the passage where he mentions Cicero's activities as a translator of Greek philosophy (III.1), Plutarch represents Cicero not solely as a writer, but as a kind of Roman Socrates, a man who gathered young men of the Roman elite around him, teaching them the principles of philosophy and thus laying the ground for a vital 'national' tradition of philosophy and philosophers. Cicero would have liked this portrait, as it agrees perfectly with his own programme, as outlined especially in the prefaces in the five-book *Tusculanae disputationes* (cf. III.2).[19] As a Roman, he looked to Greek philosophy not as a humble epigone, but as a citizen of the nation that had conquered Greece a hundred years previously. He selected what he deemed to be useful and took the liberty of repurposing philosophical doctrines for his own ends (III.3). Cicero's devotion to the Sceptic Academy provided him with an intra-philosophical foundation for this kind of conscious eclecticism. To take possession of a whole area of knowledge by translating it into one's own language is something to be proud of. In *De natura deorum* (III.4), Cicero makes his interlocutor Balbus claim that the translation of Aratus's *Phaenomena* that Cicero composed in his youth both delights him and has stuck in his memory, precisely because the verses are in Latin. Thus, the translation has the double effect of ennobling the content and of serving as a mnemonic aid.

Whenever scholars compare Cicero's *De re publica* and *De legibus* to the works of Plato that are given the same titles in Latin, they very quickly realise the enormous differences in content and style between the two authors. Both of Cicero's works were influenced by Stoic ideas and make the case for the constitution and legislation of the Roman Republic. They are, moreover, not so much Platonic dialogues, being rather Aristotelian in nature. Nevertheless, Cicero alludes to Plato, and not only through the titles of his works. In IV.1, we can see how he wishes his relationship with Plato to be perceived. At first sight, this passage appears to be another variation on the *verbum de verbo* and *sensus pro sensu* discussion. Yet, there is more to it, since, contrary to what Cicero as an interlocutor says, it is certainly not a simple task to translate Plato's sentences and thoughts (*sententiae*) into Latin. Nonetheless, Cicero is eager to combine the nod to his great predecessor with his claim to be a philosopher in his own right, and not merely a Roman Plato.[20] He wished to be seen as more than a translator. Yet we know of four Greek texts that Cicero translated: in his youth,

---

18 Plutarch, *Cicero*, 52.4.

19 On the programmatic passages in the *Tusculanae disputationes*, see Gildenhard, *Paideia Romana*, and Fögen, *Patrii sermonis egestas*.

20 Quintilian however (*Institutio oratoria*, 10.1.123) calls Cicero a *Platonis aemulus*.

probably as a twenty-year-old, he translated Xenophon's *Oeconomicus* and Aratus's *Phaenomena*, and during his last period of activity as a philosopher, he translated Plato's *Protagoras* and about a quarter of the *Timaeus*. Only a very few fragments of his *Oeconomicus* and *Protagoras* survive; a few more from his translation of Aratus, the *Aratea*, have come down to us (on these see below).[21] However, almost the entirety of the quarter of the *Timaeus* that Cicero translated is extant. Since the Christian Calcidius translated and commented on the first half of the *Timaeus* in the fourth century, there exists a good benchmark for evaluating Cicero's efforts. The translations have been compared several times.[22] Calcidius stays much closer to the original (*verbum de verbo*), whereas Cicero never conceals the fact that he is an orator. For him, it is always a little more important to contribute to the elegance of his mother tongue than to supply a tool for understanding Plato's doctrine. But, as is still the case today, whoever wishes to gain more than a superficial understanding of ancient Greek philosophy must read the original, that is, one must know and use Greek rather than relying solely on Latin translations, adaptations, or summaries. On this topic, in Cicero's *Academica*, the Roman polymath Varro addresses the reasons why he is not concerned with writing (in Latin) on (Greek) philosophy (IV.2). We quote here a passage from McElduff (2013), who notes the irony inherent in Varro's refusal: 'What is a little ironic here is that Varro *did* write a Latin text on philosophy, the now non-extant *De Philosophia*, [...] and he, as a character in a philosophical dialogue, will go on and discuss philosophy in Latin [...]. Still, according to Varro, the pre-existence of Greek texts ensures that they will stay the premier source for philosophy, no matter how sophisticated their replacements.'[23] As in IV.3, Cicero replies with the argument based on notions of pleasure, namely that should Roman readers feel entertained by Latin poems and comedies that have been translated from, or at least closely follow, Greek models, then there is no reason not to be delighted with philosophical texts in Latin that also follow Greek models. In classical times, the main argument of those who were opposed to translating Greek philosophical texts was that anyone who was seriously interested in philosophy should rather read the original texts. In Christian discourse of late antiquity there was another argument against translation: it was claimed that false or dangerous thoughts were better left untranslated and thus kept secret. Generally speak-

21 To be found in Garbarino, *Ciceronis Fragmenta*.

22 Puelma, 'Cicero als Platon-Übersetzer'; Seele, *Römische Übersetzer*, 31, 56–57; Hoenig, 'Εἰκὼς λόγος'; and Reitz/Fuchs, 'Vertere'. Cf. Poncelet, 'Deux aspects'; and Lambardi, *Il 'Timaeus' ciceroniano*.

23 McElduff, *Roman Theories*, 105.

ing, we have not included the many passages from Christian authors that are concerned with translations of Greek theological works into Latin.[24] IV.4 is the exception; this is because it is crucial for the understanding of the problems that may emerge as soon as theological issues arise.

Despite Cicero's sometimes annoying self-awareness as a philosophical writer, his fellow countrymen never rid themselves of the feeling of inferiority associated with the Latin language's (lack of) expressiveness, especially in the field of philosophy.[25] Cicero was the first to mention the alleged poverty of Latin,[26] and his contemporary Lucretius coined the famous phrase *patrii sermonis egestas*, which he used twice (*De rerum natura*, 1.832 and 3.260). In V.1, however, Lucretius the translator and versificator credits himself with possessing a skill similar to that of Epicurus, namely, the ability to expel darkness and unveil truth.[27] Lucretius uses the term *illustrare*, which is interesting: not only does it mean 'to shed light upon something', but, since it is derived from *lustrare*, it is connected with notions of purifying, but also of walking or travelling around, visiting unknown places and making them known, and bringing them into the light. Thus, he presents Greek philosophy as an obscure landscape and the Latin language as a poor candle producing but a meagre flame. There are numerous occurrences of the *egestas* topos in Latin literature. We have included two further examples (V.2–3) in our collection of texts.

Just as with philosophy as an occupation, translating from Greek was seen as a splendid hobby for members of the Roman elite in the imperial period. To them, 'to philosophize' was synonymous with translating Greek philosophical texts, or with expressing Greek thoughts in Roman terms. It did not mean that something new and novel was being produced. The ideals that Cicero typified remained constant during the first and second centuries AD, both in higher education (VI.1) and in the daily life of the literate Roman official (VI.2). Translating Greek philosophy and rhetoric was seen as essential,[28] but even after 200 years of engagement with the issue, translating Plato was still viewed as a personal challenge (VI.3).

Other than the fact that studying and/or translating Greek philosophical texts as a personal exercise was common among Roman intellectuals, we also

---

24 On Latin translations of Christian texts in late antiquity, see Marti, *Übersetzer*, which contains a rich anthology of source texts, as well as Marti, 'Das Übersetzen'.

25 E.g., Pliny, *Epistulae*, 4.18.1. Müller-Wetzel ('Der Topos') offers a short survey of the relevant passages. For a more detailed discussion, see Fögen, *Patrii sermonis egestas*.

26 *Pro Caecina*, 51.2.

27 For a broad discussion of Lucretius's dealing with the difficulties of transforming Greek philosophy into Latin verse, see Sedley, *Lucretius*, 35–61.

28 Cf. Beall, 'Translation'.

know that some ambitious and professional translations were created, especially of Plato and Aristotle. Apuleius (born AD 125 in Madaura in modern Algeria), who is best known for his novel *The Golden Ass*, translated the pseudo-Aristotelian *De mundo*, probably in 170. Apuleius was bilingual and boasted of writing philosophy in both languages. He also declaimed in both, and was proud of composing poems and prose in all main genres *tam Graece quam Latine*.[29] His *De mundo* is a free translation, imitating Cicero in phrasing and attitude.[30] He incorporates portions of more contemporary writers, such as Gellius, and often fails to meet the high standards set by modern philologists for accurate translation.[31] Two medieval Latin translations of *De mundo*, the so-called *Versio Manfrediana* and a translation by Nicolaus Siculus, both of which date from the thirteenth century, are much closer, if less elegant, renditions.

We have already mentioned Calcidius, who translated and commented on the *Timaeus* in the fourth century. His dedicatory letter (VII.1) testifies that friendship and mutual obedience may be motivations for translating philosophy.[32] Also remarkable is his statement that, without a commentary, the translation would be even more obscure than the original. Calcidius's contemporary Marius Victorinus translated the works of several Neoplatonic philosophers, Porphyry and Plotinus among them, as well as Aristotle's *Categoriae* and *De interpretatione*.[33] His translations have all been lost, but quotations in Boethius give us an impression of his importance.

As for the reception of Greek philosophy in Latin in the early Middle Ages, the most important figure covered in our overview is Boethius, certainly as far as Aristotle is concerned. Boethius was born around 480. He was as familiar with the writings of Plato, Aristotle, and the Neoplatonists, as with those of Augustine. His career reached its apex during the reign of Theodoric the Great, but in 524 he was involved in an alleged conspiracy of Roman senators along with the East Roman Emperor Justin I. Boethius was then charged with *crimina maiestatis*, imprisoned, and executed. While in prison, he wrote his famous prosimetric text, *De consolatione philosophiae*, which, generally speaking, is the latest work regularly read and studied by philologists of classical Latin. At the beginning of the sixth century AD, Boethius clearly felt it necessary to trans-

29 Hutchinson, *Greek to Latin*, 139.

30 Harrison, *Apuleius*, 184–187.

31 Because of the many inconsistencies and mistakes in the translation, Apuleius's authorship has frequently been called into doubt. Today, there is a slight majority in favour of authenticity. Cf. Harrison, *Apuleius*, 175–180.

32 Cf. Lucretius in V.1.

33 On Victorinus's translations, see Hadot, *Marius Victorinus*, 179–190.

late the works of Plato and Aristotle into Latin. A Latin-speaker in the early sixth century would have had access to only a few works devoted to Aristotelian logic.[34] These would have included Apuleius's exposition of *De interpretatione*, a work on Aristotle's *Categoriae*, an abbreviated fourth-century translation of the *Analytica priora* and *posteriora* by a certain Vettius Agorius Praetextatus, as well as translations of the *Categoriae* and *De interpretatione* by Marius Victorinus. Dissatisfied with the texts available to him, Boethius originally intended to translate and comment on all of the works of Aristotle and Plato (VII.2). When he died at the age of 44, he was still some way away from accomplishing this ambitious project, but he had managed to translate and comment on the entire *Organon*.[35] In undertaking his translations, Boethius was following a didactic plan, as he wished first to introduce his Latin-speaking audience to the principles of logic and then, step for step, to illuminate all fields of practical and theoretical philosophy for them.[36] In contrast to many of his predecessors,[37] he prefers literal over the literary translation and apologizes for being a *fidus interpres* (VII.3).[38] Boethius's aim was to introduce his readers to Aristotelian and Platonic philosophy, not to make a substantial contribution to Latin literature. A comparison with Jerome's prescription concerning translation of Holy Scripture (I.5) is enlightening here. With Boethius, we have reached the temporal limit of our overview and now turn to Latin translations of technical texts.

Although this essay is almost exclusively concerned with prose, it would be wrong not to mention the text that attracted the most prominent Roman translators, the *Phaenomena*, a didactic poem of about 1,150 verses on stars and constellations, written in the third century BC by Aratus of Soloi. By Cicero's time, it had already become part of the Roman school curriculum.[39] It was translated by Cicero (as noted above), Varro Atacinus, Ovid, Germanicus Caesar (first century AD), and Avienus (fourth century AD), among others.[40]

When Scipio Aemilianus destroyed Carthage in 146 BC, its libraries were plundered and their books given to African client kings. The only work the

34 The following overview is based upon Casey, 'Boethius's Works', 200.

35 Vogel, *Boethius' Übersetzungsprojekt*, 13.

36 Ibid., 14.

37 Including Marius Victorinus; cf. Moreschini, *A Christian in Toga*, 22.

38 Ebbesen, 'The Aristotelian Commentator', 39; and Vogel, *Boethius' Übersetzungsprojekt*, 131–144. On the tradition of the *fidus interpres* topos, see Kelly, *The True Interpreter*, and Kytzler, '*Fidus interpres*'.

39 Hurka, 'Arat und Aratea', 71.

40 For a comparative study of all extant Latin translations of the *Phaenomena*, see Lewis, *From Aratus*; cf. Lewis, 'The Popularity'; Possanza, *Translating the Heavens*; Hübner, *Die Rezeption*; and Hurka, 'Arat und Aratea'.

Romans kept for themselves was a handbook on agriculture in twenty-eight volumes written in Punic by a certain Mago (VIII.1–2). Clearly, and somewhat ironically, it was considered more substantial than the *De agricultura* by Marcus Cato, a man famously known for supporting Carthage's total destruction. Mago's text remained a standard handbook for Romans for at least two hundred years.[41]

The most interesting remarks concerning Latin translations of scientific texts are to be found in the corpus of medical texts. The theory and practice of medicine remained a Greek domain up to the fourth century AD. No Latin work on medicine survives from the Republican period, and there exist only four texts dating from the first three centuries AD. Generally speaking, Latin medical writers imparted Greek medical knowledge to a Latin-speaking audience. The only works based on genuine Roman material were compilations comprising relevant passages taken from Pliny's *Natural History*, known as the *Medicina* and *Physica Plinii*. The bulk of translations of Greek medical texts, however, date from the fifth and sixth centuries AD.[42] Latin medical writers had to deal with methodological issues and difficulties not dissimilar to those faced by their philosophical colleagues (VIII.4).[43] Interestingly, in VIII.3 Pseudo-Celsus explains why he uses impersonal verbs rather than addressing the reader directly in the second person, as was common in Greek texts. In his opinion, direct address 'invokes a bad omen and runs the risk of infecting the reader with the disease in question'.[44] VIII.5 demonstrates that the use of obscure Greek terminology unfamiliar to non-experts was a good method for avoiding Latin words that could be considered indecent,[45] and elsewhere, even if it was not for reasons of tact, it was deemed convenient to maintain the use of foreign medical terms (VIII.6). Indeed, this practice remains prevalent even in our own times.

---

41 On the Latin translation of Mago and its history, see Suerbaum, *Die archaische Literatur*, 576–579; and Fögen, 'The Transformation', 96–97. Cf. Cicero, *De oratore*, 1.249.

42 See the comprehensive catalogue of Latin medical treatises and translations in Langslow, *Medical Latin*, 60–75.

43 Cf. Fögen, 'The Transformation', 94–96. Langslow, *Medical Latin*, is the standard handbook on the Latin medical language. It is supplemented by Langslow, 'Typologies', in which the author sets out a brief theoretical framework for the description of ancient techniques of translation similar to the studies mentioned above in n. 10. On the question of authenticity of VIII.3–4, see Cilliers, 'Forgery'.

44 Fögen, 'The Transformation', 94.

45 Cf. Fögen, 'The Transformation', 105–106.

Part Two: Texts in Translation

Overview

I. *The* verbum de verbo *Topos*

I.1. Publius Terentius Afer, *Adelphoe*, 6–14 (ed. Kauer/Lindsay/Skutsch); translation adapted from Riley, *The Comedies of Terence*, 200.

Publius Terentius Afer (c. 195–c. 158 BC) was a Roman comic playwright, whose dramas were often adapted from Greek examples. The prologue to his *Adelphoe*, performed in 160 BC and based on a play by the Greek playwright Menander (c. 342–c. 291 BC), provides us with a revealing demonstration of this process of adaptation.

The *Synapothnescontes* is a comedy written by Diphilus;[46] Plautus[47] turned it into a play called the *Commorientes*. In the Greek, there is a young man who, in the first part of the play, carries off a courtesan from a procurer; that part Plautus left out entirely. He [i.e., Terence] has adopted this portion in the *Adelphi*, and has adapted it, translating it word for word. We shall now perform this new play. Determine then whether you believe a theft has been committed, or a passage has been restored to notice that before had been passed over in neglect.

I.2. Cicero, *De optimo genere oratorum*, 13–14 (ed. Wilkins); translation adapted from Hubbell, *De inventione*, 365.

Marcus Tullius Cicero (106–43 BC) was a Roman statesman, lawyer, and author of philosophical treatises. His studies in Athens enabled him to convey both the content and

46 Diphilus of Sinope, a comic playwright and contemporary of Menander. Although his work had great influence on Terence and Plautus, only fragments of his plays remain.

47 Titus Maccius Plautus (c. 250–184 BC) was a Roman playwright whose works were, as is also the case with Terence, mostly based on Greek originals.

language of Greek philosophy to a Roman audience; the Latin language—and therefore the whole of Europe—owes many philosophical words to Cicero's linguistic creativity. The dialogue *De optimo genere oratorum*, written in 46 BC, served as an introduction to Cicero's translation of two orations by the Greek orators Aeschines and Demosthenes, yet was never published during Cicero's lifetime.

But since there was a complete misapprehension as to the nature of their style of oratory, I thought it my duty to undertake a task that will be useful to students, though not necessary for myself. That is to say, I translated the most famous orations of the two most eloquent Attic orators, Aeschines[48] and Demosthenes, orations that they delivered against each other. I did not translate them as an interpreter, but rather as an orator, with the same ideas, forms, and, so to speak, shape, and with language fit for our usage. In this process, I did not deem it necessary to render word for word, but instead preserved every category and the force of the words. For I did not think that I should count them out piece by piece to the reader, but rather, as it were, pay them out by weight.

1.3. Cicero, *De finibus bonorum et malorum*, 3.15 (ed. Reynolds); tr. Rackham, *De finibus*, 231.

De finibus bonorum et malorum (*On the Ends of Good and Evil*) was composed in 45 BC and serves as an introduction to Stoic, Epicurean, and Antiochean Platonist philosophy. In book 3, Cato presents Stoic philosophy using Latin terminology.

If when Zeno[49] invented some novel idea he was permitted to denote it by an equally unheard-of word, why should not Cato[50] be permitted to do so too? Though all the same it need not be a hard and fast rule that every word shall be represented by its exact counterpart [*verbum e verbo*], when there is a more familiar word conveying the same meaning. That is the way of a clumsy trans-

48 Aeschines (c. 389–c. 314 BC) was an Athenian statesman and orator. Alexandrian philologists regarded him as one of the greatest Athenian rhetoricians, commonly known as the Ten Attic Orators, a group that included Aeschines's rival Demosthenes (384–322 BC). The works of the Ten Orators were used as class material in Roman higher education.

49 Zeno of Citium (c. 334–262 BC) was the founder of the Stoic school of philosophy and taught in Athens around 300 BC. His work exercised a great influence on Roman thinkers such as Cicero and Seneca.

50 Marcus Porcius Cato Uticensis, also known as Cato the Younger (95–46 BC) to distinguish him from his great-grandfather Cato the Elder, was a Roman statesman and orator, a follower of Stoic philosophy, and a long-time adversary of Julius Caesar. Literary portraits of his life survive in Lucan's epic on the Civil War and in Cicero's dialogue *De finibus*, cited here.

lator. Indeed my own practice is to use several words to give what is expressed in Greek by one, if I cannot convey the sense other. At the same time I hold that we may fairly claim the licence to employ a Greek word when no Latin word is readily forthcoming.

1.4. Horace, *Ars poetica*, 131–134 (ed. Borzsák); tr. Mundt and Cohen.

Quintus Horatius Flaccus (65–8 BC) was a Roman poet who, together with Catullus, was among the first to write shorter poems using the formal heritage of ancient Greek lyric. In his *Letter to the Pisones*, commonly known as the *Ars poetica*, he connects the process of word-for-word translation with the concept of the translator's 'loyalty'.

You will make a subject that is open to the public your own property if you do not linger in easily accessible areas and do not seek to render word for word the way a faithful translator does.

1.5. Jerome, *Epistulae*, 57.5.2 (ed. Hilberg); translation adapted from Fremantle, Lewis, and Martley, *Letters*, 113 f.

Eusebius Sophronius Hieronymus (AD 347–420) was a theologian and translator whose translations of Holy Scripture, commonly known as the Vulgate, count among the most influential in world history. This letter to Pammachius serves as an apologia, as Jerome had previously been charged with having falsified an original work. In order to defend himself, Jerome feels obliged to explain his translation methods.

For I myself not only admit but freely proclaim that in translating from the Greek (except for the Holy Scriptures, where even the order of the words is a mystery) I render sense for sense and not word for word. For this method I have the authority of Cicero, who in such a way has not only translated Plato's *Protagoras* and Xenophon's *Oeconomicus*,[51] but also the two beautiful orations that Aeschines and Demosthenes delivered against each other. This is not the right moment to say what omissions, additions, and alterations he has made, substituting the idioms of his own for those of another tongue. I am satisfied to quote the authority of the translator who, in a prologue prefixed to the orations, has spoken as follows [quotes passage 1.2 above].

---

51 Xenophon (c. 430–354 BC) was an Athenian military strategist, philosopher, and historiographer; several works on Greek history and tractates on philosophical themes by him are extant. His *Oeconomicus*, probably written after 362, is a Socratic dialogue on the principles of household management and agriculture, which, through Cicero's translation, became widely known, at least until late antiquity.

II. *The Limits of* verbum de verbo*: Some Examples*

II.1. Cicero, *De finibus bonorum et malorum*, 2.13, 15 (ed. Reynolds); translation adapted from Yonge, *The Academic Questions*, 130.

Cicero discusses the possibility of using the Latin word *voluptas* as a sufficient equivalent of the Epicurean term *hēdonē* (ἡδονή).

Do they then understand what Epicurus wants to say, and do I not understand it? However, to ensure that you know that I do understand, I will tell you first of all that *voluptas* is the same thing that he calls *hēdonē*. And, indeed, we often have to search for a Latin word equivalent to, and exactly equipollent to a Greek one; but here we had nothing to seek for: for no word can be found that will more precisely express in Latin what *hēdonē* does in Greek than *voluptas*. Now everyone in the world who knows Latin understands by this word two things: joy in the mind, and an agreeable arousing of pleasantness in the body. [...]

Does it not, then, seem to you that I have a sufficient understanding of the meaning of words, or should I at this time of life be taught how to speak Greek, or even Latin? And yet I would ask you to consider, whether—if I, who, as I think, have a very fair knowledge of Greek, still do not understand what Epicurus means—it may not be owing to some fault of his for speaking so as not to be intelligible. And this sometimes happens in two ways, without any blame; either you do so on purpose, as Heraclitus, who was called *skoteinos* [σκοτεινός] for speaking with too much obscurity about natural philosophy; or when the obscurity of the subject itself, not of the language, prevents what is said from being clearly understood, as in the case of Plato's *Timaeus*. But Epicurus, as I imagine, is both willing, if it is in his hands, to speak understandably, and is also speaking not of an obscure subject like the natural philosophers, nor of one that depends on exact rules as the mathematicians do. Moreover, he is discussing a clear and simple matter, which is a subject of common conversation among the common people. You do not deny however that we understand the usual meaning of the word *voluptas*, but only what he means by it: from which it follows, not that we do not understand what the meaning of the word is, but that he follows his own fashion, and leaves our usual one.

II.2. Seneca, *De tranquillitate animi*, 2.3 (ed. Reynolds); translation adapted from Basore, *Seneca: Moral Essays*, II, 213–215.

Lucius Annaeus Seneca (AD 1–65) was a philosopher, playwright, and statesman under Emperor Nero. His dialogue on tranquillity of the soul was written in the early fifties AD.

This stability of mind the Greeks call *euthymia*, 'well-being of the soul', on which there exists an excellent treatise written by Democritus—I call it tranquillity. For it is not necessary to imitate and reproduce words in their Greek form; the thing itself, which is being discussed, must be designated by some name that should have not the shape but the force of the Greek term.

II.3. Seneca, *Epistulae morales*, 58.6–7 (ed. Reynolds); translation adapted from Gummere, *Seneca*, 389–391.

Seneca's *Epistulae morales* are among the most important sources for our understanding of Stoic philosophy among the Romans. These letters, written at the end of Seneca's life and addressed to Lucilius, contain advice on how to become a more devoted Stoic. By means of an example, Seneca discusses the difficulty of finding suitable equivalents to the Greek terms *ousia* (οὐσία) and *to on* (τὸ ὄν), and stresses Cicero's authority regarding the matter.

I suppose you will say: 'What is the purpose and meaning of this preamble?' I shall not keep you in the dark. If possible, I would like to say the word *essentia* to you and obtain an enthusiastic response. If this proves impossible, I shall risk it, even though it put you out of humour. I have Cicero as an authority for the use of this word, and I regard him as a powerful authority. Should you wish a testimony of a later date, I will cite Fabianus,[52] careful of speech, cultivated, and so precise in style that he will suit even our refined tastes. For what can we do, my dear Lucilius? How otherwise may we find a word for that which the Greeks call *ousia* [οὐσία], something that is indispensable, something that is the natural substratum of everything? Accordingly, I ask for your permission to use the word *essentia*. I shall nonetheless take pains to exercise the privilege, which you have granted me, as sparingly as possible; perhaps I shall be happy with the mere right.

Yet what good will your indulgence do me, if I can in no fashion express in Latin the significance of the word that gave me the opportunity to rail at the poverty of our language? And you will condemn our all too narrow Roman boundaries even more, when you find out that there is a word of only one syllable that I am unable to translate. 'What is this?' you ask. It is the word *on* [ὄν]. You think I am lacking in ability, since you believe that the word is ready to hand, that it might be translated by *quod est*. I notice, however, a major difference, since you are forcing me to render a noun by a verb. However, if it needs be so, I shall write *quod est*.

52 Papirius Fabianus, rhetorician and Stoic philosopher of the generation prior to Seneca.

II.4. Aulus Gellius, *Noctes Atticae*, 11.16.1–6 (ed. Marshall); translation adapted from Rolfe, *The Attic Nights*, 337–339.

The commonplace book *Noctes Atticae* by Aulus Gellius (AD c. 130–c. 180) is essentially a compilation of philosophical questions, scientific knowledge, legal debates, and linguistic matters. In the following passage, the author describes some problems he encountered while translating Plutarch into Latin.

We have frequently observed not a few names of things that we cannot express in Latin by single words, as in Greek; and even if we use very many words, those ideas cannot be expressed in Latin so aptly and so clearly as the Greeks express them by single terms. Lately, when a book of Plutarch had been brought to me, and I had read its title, which was περὶ πολυπραγμοσύνης [*Peri polypragmosynēs*], a man who was unacquainted with Greek letters and words asked who the author was and what the book was about. The name of the writer I gave him at once, but I hesitated on the point of naming the subject of the work. At first indeed, since it did not seem to me that it would be a very apt interpretation if I said that it was written *De negotiositate* [*On Busyness*], I began to rack my brains for something else that would render the title word for word, as the saying is. But there was absolutely nothing that I remembered to have read, or even that I could invent, that was not to a degree harsh and absurd, if I fashioned a single word out of *multitudo* and *negotium* in the same way that we say *multiiugus* ['manifold'], *multicolorus* ['multicoloured'], and *multiformius* ['multiform']. But it would be no less unpleasant an expression than if you should try to translate by one word πολυφιλία [*polyphilia*, 'abundance of friends'], πολυτροπία [*polytropia*, 'versatility'], or πολυσαρκία [*polysarkia*, 'fleshiness']. Therefore, after spending a brief time in silent thought, I finally answered that in my opinion the idea could not be expressed by a single word, and accordingly I was preparing to indicate the meaning of the Greek word by a phrase.

III. *Cicero's 'National' Project*

III.1. Plutarch, *Cicero*, 40.1–2 (ed. Lindskog/Ziegler/Gärtner); translation adapted from Perrin, *Plutarch's Lives*, 185.

Plutarch (AD c. 45–c. 125) was the author of a great many works, the most famous of which are his seventy-eight treatises on morals and his parallel biographies of famous Greeks and Romans, such as Alexander the Great and Julius Caesar, or Demosthenes and Cicero. In this last work, he describes Cicero's merits in enriching the Latin language with philosophical terminology.

After this, when the government had been changed to a monarchy, Cicero abstained from public affairs and devoted his time to those of the young men who wished to study philosophy, and mainly from his intimacy with these, since they were of the highest birth and standing, he was once more very influential in the state. He made it his business also to compose and translate philosophical dialogues, and to render into Latin the several terms of dialectics and natural philosophy; for he it was, as they say, who first, or principally, provided Latin names for φαντασία [*phantasia*], ἐποχή [*epokhē*], συγκατάθεσις [*synkatathesis*], and κατάληψις [*katalēpsis*], as well as for ἄτομον [*atomon*], ἀμερές [*ameres*], κενόν [*kenon*], and many others like these, contriving partly by metaphors and partly by new and fitting terms to make them intelligible and familiar.

III.2. Cicero, *Tusculanae disputationes*, 2.5 (ed. Pohlenz); translation adapted from King, *Tusculan Disputations*, 151.

The prefaces of the five books of *Tusculanae disputationes* ('Conversations in Tusculum'), written around 45 BC, are important sources for Cicero's view on cultural transfer from Greece to Rome.

For this reason, I encourage all who have the capacity to wrest from an already languishing Greece the glory that may be won from this field of study and to transfer it to this city, just as our ancestors, by their indefatigable zeal, transferred here all those other glories that indeed could be sought. And in oratory indeed, our fame, from humble beginnings, has reached its zenith, with the result that now, as is the law of nature in almost everything, it is beginning its decline and seems destined in a short while to come to nothing; but Philosophy in the Latin language shall be born in these very days, so let us help her and submit ourselves to contradiction and refutation!

III.3. Cicero, *De officiis*, 1.6 (ed. Winterbottom); tr. Miller, *On Duties*, 9.

De officiis (*On Duties*) was written in 44 BC, shortly before Cicero's death and serves to define the duties of men according to the categories of *honestum* and *utile*.

I shall, therefore, at this time and in this investigation follow chiefly the Stoics, not as a translator, but, as is my custom, I shall at my own option and discretion draw from those sources in such measure and in such manner as shall suit my purpose.

III.4. Cicero, *De natura deorum*, 2.104 (ed. Plasberg/Ax); translation adapted from Brooks, *De natura deorum*, 128.

In *De natura deorum* ('On the Nature of the Gods'), Cicero discusses fundamental theological questions. His discussion partner in book 2 is the Hispanic philosopher Quintus Lucilius Balbus, a specialist on Stoic philosophy. In the following extract, Balbus mentions the translation of Aratus's *Phaenomena* made by Cicero in his youth, claiming that he is still able to remember it, because (*quia*) he was so delighted that it was in Latin.

At this point Balbus looked at me and said: 'I will make use of those verses of Aratus[53] that you translated when you were still quite young, which pleased me so much, because they are in Latin, that I retain many of them in memory.'

IV. *Greek Philosophy in Latin: Is This Possible, Even Desirable?*

IV.1. Cicero, *De legibus*, 2.17 (ed. Powell); translation adapted from Keyes, *De re publica, De legibus*, 391.

Cicero wrote the dialogue *De legibus* (*On Laws*) in his final years. His conversation partners are his younger brother Quintus and their mutual friend Titus Pomponius Atticus. The following extract describes Cicero's relation to Plato, who wrote a dialogue entitled *Nomoi* (*Laws*).

Quintus Nothing could be more unlike his treatment than your previous remarks, and also this preface in regard to the gods. In just one thing you do seem to me to imitate him—in the style of your language. *Marcus* I wish to do so, perhaps; for who can or ever will be able to imitate him in this? It is very easy to translate another man's thoughts, and I might do that, if I did not fully wish to be myself. For what merit is there in presenting the same thoughts rendered in practically the same phrases? *Quintus* I agree with you entirely. I certainly prefer that you be independent, as you have just said.

53 Aratus (c. 310–245 BC) was a Greek didactic poet, whose major work, a poem called *Phaenomena* that describes celestial phenomena, became very popular among the Romans through translations by Cicero and Germanicus Caesar, among others.

IV.2. Cicero, *Academica*, 1.4f., 9f. (ed. Plasberg); translation adapted from Rackham, *De natura deorum, Academica*, 413–415, 421.

In his *Academica*, Cicero treats the Stoic theory of knowledge. With his conversation partners—Atticus (mentioned above) and the grammarian Varro (see VIII.2)—Cicero considers the problem of needing to discuss Greek philosophy in Latin.

Varro I have often deliberated and thought over the topic you are asking me about. So I will not hesitate to reply, but say what comes to me readily, since—as I said before—I have thought for a long time about that very question. When I saw that philosophy had most carefully been expounded in Greek literature, I judged that if any persons of our nation who had been educated in Greek learning felt an interest in the subject, they would rather read Greek writings than ours. On the other hand, if they were holding themselves back from Greek science and literature, they would not care for philosophy, which cannot be understood without Greek learning. So I was not willing to write something that the unlearned would not understand and the learned would not want to read. [...]

Cicero Though it is true that the case you bring forward has some plausibility, as accomplished students on the one hand will prefer to read the Greek writings, and on the other people who do not know those will not read these either, still, tell me now—do you quite prove your point? The truth rather is that both those who cannot read the Greek books will read these and those who can read the Greek will not overlook the works of their own nation. For what reason is there why accomplished Grecians should read Latin poets and not read Latin philosophers? Is it because they get pleasure from Ennius,[54] Pacuvius,[55] Accius,[56] and many others, who have reproduced not the words but the meaning of the Greek poets? How much more pleasure will they get from philosophers, if these imitate Plato, Aristotle, and Theophrastus in the same way as those poets imitated Aeschylus, Sophocles, and Euripides?

---

54 Quintus Ennius (239–169 BC) was a poet who wrote the great epic poem *Annales* on Roman history, of which only fragments survive. He was the first to employ the dactylic hexameter in Latin literature.

55 Marcus Pacuvius (c. 220–130 BC) was an ancient Roman tragic poet and the nephew and pupil of Ennius. His tragedies were founded on Greek subjects. Fragments of his work survive in the works of Cicero, who was a great admirer of his.

56 Lucius Accius (170–c. 86 BC) was a Roman tragic poet, whose work was mostly based on Greek originals. Most of the fragments of his work are preserved in Cicero.

IV.3. Cicero, *De finibus bonorum et malorum*, 1.1 f., 4, 5 f., 7 f. (ed. Reynolds); translation adapted from Yonge, *The Academic Questions*, 95–96, and Rackham, *De finibus*, 7.

I was not ignorant, Brutus, when I was striving to add to Latin literature the same things that philosophers of the most astounding genius and the most profound and accurate learning had previously been writing about in the Greek language, that fault would be found with my endeavours on various grounds. While some men, even those who are not at all unlearned, dislike philosophy, others do not resent a moderate degree of attention being given to it, but do not approve of so much study and labour being devoted to it. There will be others again, educated in Greek literature and despising Latin compositions, who will say that they would rather spend their time reading Greek. [...]

It is, therefore, more difficult to satisfy those men who say that they despise Latin writings. But, first of all, I may express my astonishment at their not being pleased with their mother tongue in matters of the highest importance, when they are fond enough of reading fables in Latin that are translated word for word from the Greek. [...]

If we read 'Would that in the forest ...'[57] just as well as the same passage in the Greek, shall we object to having Plato's discourses on the good and happy life set before the reader in Latin? And, if we for our part we do not act as a mere translator, but, while preserving the doctrines of our chosen authorities, add thereto our own criticism and our own arrangement, what ground have these objectors for ranking the writings of Greece above compositions that are at once brilliant in style and not mere translations from Greek originals? [...]

Although if I were to translate Plato or Aristotle in as bold a fashion as our poets have translated the Greek plays, then I think I should be treating my fellow-countrymen poorly, for having brought those divine geniuses within their reach. However, that is not what I have done so far, nor however do I consider myself forbidden from doing so. Some particular passages, if I deem it desirable, I shall translate, especially from those authors whom I have just named, when there is the possibility of doing so with propriety, just as Ennius often translates passages from Homer, and Afranius[58] from Menander. [...]

57 Cicero quotes the beginning of Ennius's *Medea*, which is modelled after the Euripidean play, but is not a word-for-word translation.

58 Lucius Afranius was a Roman comic poet who lived around the beginning of the first century BC. He was often compared to the Greek playwright Menander, whose work was regarded as a brilliant portrait of everyday life. Only titles and fragments of Afranius's work are extant.

But I have no doubt that the reason why some people take a dislike to Latin literature is that they have happened to meet with certain illiterate and uncouth productions that are bad Greek books in worse Latin versions. I have no quarrel with these persons, provided that they also refuse to read the Greek writers on the same subjects.

IV.4. Jerome, *Epistulae*, 124.1 (ed. Hilberg); translation adapted from Fremantle, Lewis, and Martley, *Letters and Select Works*, 238.

During the last decade of the fourth century AD, Tyrannius Rufinus (344–411) and Jerome were close friends and took pleasure in studying and translating Origen together. While Rufinus translated many of Origen's works—among them the *Peri archōn* (περὶ ἀρχῶν, *De principiis*), which has survived only in his translation—, Jerome came to renounce Origen's teachings. This rift between the friends resulted in a series of pamphlets and mutual reproaches.

About ten years ago, the holy man Pammachius[59] sent me some pages written by a certain man[60] containing the translation of Origen's books *Peri archōn*. Since it was highly deficient, he implored me to produce a Latin translation that preserved the true sense of the Greek, so that a speaker of the Roman language might recognize both the good and the bad utterances of the author, without any bias by the translator.[61] I did what he wished me to do and sent the books back to him. After reading them, he was shocked and locked them up in his bookcase for fear that they might wound the souls of many people once they became publicly known.

V. *The Alleged Poverty of the Latin Language*

V.1. Lucretius, *De rerum natura*, 1.136–145 (ed. Martin); translation adapted from Leonard, *Of the Nature of Things*, 5.

Titus Lucretius Carus (c. 94–c. 55 BC) was a Roman poet, whose epic poem *De rerum natura* (*On the Nature of Things*) is the preeminent source for our knowledge of Epicurian philosophy in Roman times. His translation of Epicurus's ideas was a great feat for the Latin language that thereby demonstrated its ability to treat scientific and philo-

59 Senator and head of the Christian community in the city of Rome, d. AD 409.

60 Rufinus of course.

61 Rufinus had modified some passages in Origen in order to conceal thinking that might have been considered heretical.

sophical themes in translation. In the following extract, the poet explains to his dedicatee Gaius Memmius why he found it necessary to undertake such a difficult task.

I know how difficult it is in Roman verse to bring into the light the dark discoveries of the Greeks, primarily because our poor speech must find strange terms to fit the strangeness of the things. Yet your *virtus* and the hoped-for joy of your sweet friendship do persuade me to bear all toil and wake the calm nights through, while I seek words and verses to, at last, uncloud for you most gloriously the light beyond, with which to view the core of being that is hidden at the centre.

V.2. Cicero, *De finibus bonorum et malorum*, 1.10 (ed. Reynolds); translation adapted from Yonge, *De finibus*, 99.

However, I could never aptly express my wonder as of why this arrogant disdain of everything national arose among us. This is not quite the place for any lecturing on the subject; but my own feelings are, and I have frequently made them known, that the Latin language is not only anything other than deficient—so as to deserve to be wholly disparaged, as commonly supposed—but that it is even more copious than the Greek.

V.3. Seneca, *Epistulae morales*, 58.1 (ed. Reynolds); translation adapted from Gummere, *Seneca*, 387.

How scant of words our language is, nay, how poverty-stricken, I have not completely understood until today. When we happened to be speaking of Plato, a thousand subjects came up for discussion that needed words and yet possessed none; and there were certain others that once possessed words, but have since lost them because we have been too fastidious. Who, however, can endure such fastidiousness in the midst of poverty?

VI. *Translating in the First and Second Centuries AD*

VI.1. Quintilian, *Institutio oratoria*, 10.5.2–3 (ed. Winterbottom); tr. Russell, *Quintilian*, 357.

Marcus Fabius Quintilianus (AD 35–c. 96) was a rhetorician under Emperor Domitianus. He wrote a treatise, the *Institutio oratoria*, in c. 95, on the ways of turning pupils into skilled orators, a process in which translation appears to have been a frequent exercise.

Our earlier orators thought the best exercise was translating Greek into Latin. Crassus[62] in Cicero's *De oratore* says he habitually did this, and Cicero himself in his own person repeatedly advocates it. In fact, he published books of Plato and Xenophon translated in this way. Messalla[63] liked the practice, and we have many speeches of this sort written by him; indeed, he even tried to cope with the delicacy of Hyperides's defence of Phryne,[64] a very difficult thing for a Roman. The purpose of this exercise is plain. The Greek authors are full of varied matter, and they introduced a great deal of art into the practice of eloquence; when we translate them, we are free to use the best possible words, for the words we use will all be our own. As to figures, the main ornament of oratory, we simply cannot help contriving many of these, and of various kinds, because Latin idiom is often different from Greek.

VI.2. Pliny the Younger, *Epistulae*, 7.9.1–5 (ed. Mynors); tr. Radice, *Pliny the Younger: Letters*, 501–504.

Pliny the Younger (AD 62–c. 115) was a lawyer, senator, and epistolographer. His letters provide us with a great impression of Roman cultural life in the late first and early second century AD. He thought of translation as an important exercise for training the critical sense in literature.

You ask me what course of study I think you should follow during your present prolonged holiday. The most useful thing, which is always being suggested, is to translate Greek into Latin and Latin into Greek. This kind of exercise develops in one a precision and richness of vocabulary, a wide range of metaphor and power of exposition, and, moreover, imitation of the best models leads to a like aptitude for original composition. At the same time, any point which might have been overlooked by a reader cannot escape the eye of a translator. All this cultivates perception and critical sense. When you have read a passage sufficiently to remember the subject-matter and line of thought, there is no harm in your trying to compete with it; then compare your efforts with the original and

62 Lucius Licinius Crassus (140–91 BC) was a Roman statesman and rhetorician, who taught Cicero and his younger brother Quintus. A literary portrait survives in Cicero's dialogue on the ideal orator (*De oratore*).

63 Marcus Valerius Messalla Niger (c. 104–c. 50 BC) was a Roman senator and orator. In 80 BC, he was involved in Cicero's defence of Roscius Amerinus.

64 Hyper(e)ides (c. 390–322 BC) was an Athenian orator who was regarded by Alexandrian philologists as one of the Ten Attic Orators, together with Aeschines and Demosthenes (see n. 48). The oration mentioned by Quintilian was written in defence of the courtesan Phryne, who, according to Pseudo-Plutarch, had been accused of impiety.

consider carefully where your version is better or worse. You may well congratulate yourself if yours is sometimes better and feel much ashamed if the other is always superior to yours. You may also sometimes choose a passage you know well and try to improve on it. This is a daring attempt, but does not presume too far when it is made in private; and yet we see many people entering this type of contest with great credit to themselves and, by not lacking confidence, outstripping the authors whom they intended only to follow.

VI.3. Aulus Gellius, *Noctes Atticae*, 17.20.7 f. (ed. Marshall); translation adapted from Rolfe, *The Attic Nights*, 271.

Following the line of thought set out by Quintilian and Pliny, Aulus Gelllius reflects on an attempt he made to translate a passage from Plato's *Symposium*. However, he sees translation not only as a form of exercise, but also as a form of competition with the source text. Here is his response to the philosopher Taurus's warning that it would be extremely difficult to reach Plato's level of stylistic achievement.

Taurus's admonition about Plato's style not only did not keep me back, but it encouraged me to take possession of the elegance of the Greek using Latin words. And in the fashion of little, insignificant animals, which out of wantonness imitate what they have heard or seen, I dared not rival those qualities that I admired in Plato's language, but only made a thin outline of them.

VII. *Translating Plato and Aristotle in Late Antiquity*

VII.1. Calcidius, *Commentarius in Platonis Timaeum*, dedicatory letter to (H)Osius, ed./tr. Magee, 3–5.

Calcidius's translation of and commentary on Plato's *Timaeus* is his only known work. Neither he himself nor the work's dedicatee can be dated exactly. Magee prefers a date slightly after Iamblichus's death in AD 325. Calcidius's text and commentary enjoyed a broad reception in the Middle Ages.

Such, I suppose, is the power of friendship, just to its ability to remove the constraints of nearly impossible conditions, when friends support one another in the performance of an agreed-upon task, one of them by the obligation of commanding and the other by a vow of obeying. You had conceived the hope—one that was worthy of your mind, which blossoms with all forms of liberal education, and of your extraordinary character—of seeing the emergence of a hitherto unattempted work, and had determined that with it Latium should make borrowed usage from the Greeks. And although capable yourself of completing the work both more easily and to better effect, you preferred instead, no doubt because of your admirable sense of modesty, to enjoin it upon one whom you regarded as your other you. Although the matter was a difficult one, could I, please tell me, have refused the obligation when you made such a display of honor in enjoining it upon me? Could I, who had never refused any duty when it came even to customary or routine wishes, have said no to this desire, one so considerable and honorable, when declining this splendid obligation on a plea of ignorance would have proved a calculated simulation of knowledge? And so, feeling certain that it was not without divine instigation that the obligation had been enjoined upon me by you, I complied and consequently not only undertook, with quickened intention and a strengthened sense of hope, to translate the first parts of Plato's *Timaeus* but also produced a commentary on the same on the understanding that without an explanatory interpretation the image of a recondite reality would prove even more obscure than the original itself.

VII.2. Boethius, *Commentarii in librum Aristotelis ΠΕΡΙ ΕΡΜΗΝΕΙΑΣ. Editio secunda*, 2.79–80 (ed. Meiser); tr. Magee, in Moreschini, *A Christian in Toga*, 25.

Anicius Manlius Severinus Boethius (AD c. 480–524) wrote commentaries on works by Plato and Aristotle and epitomized nearly the whole of ancient philosophy in his *De consolatione philosophiae*, which he wrote in the days before his execution under Theodoric. His translations of Aristotle were important for the preservation of this philosopher's works in un-hellenized Western Europe. In this extract, he explains why he is planning to translate every work of Aristotle's he could lay his hands on: through this process, he intended to show that when it came to the most important parts of philosophy, Plato's and Aristotle's thoughts were indeed very much alike.

Assuming divine favour nods in my direction, this is my confirmed plan: although there have been distinguished talents whose studious labour has conferred much upon the Latin language in connection with the subjects which I too am now in the course of treating, nevertheless they have not produced an ordered filament, as it were, or graduated disposition of the disciplines: I will compose commentaries in Latin on every work of Aristotle that comes into my hands, while translating them into the Roman style so that everything stemming from the subtlety of his logical art, from the gravity of his moral understanding, from his insight into the truths of nature—so that I will translate all of it in due order and shine the light of commentary upon it. And I will give Latin form to all of Plato's dialogues by translating and commenting on them as well. Once these tasks have been completed, I will hasten to restore a unified harmony, as it were, to the views of Plato and Aristotle, demonstrating that they, unlike most philosophers, do not disagree on all points but in fact agree on most, the philosophically most important ones included. All of this, assuming life and leisure allow. Given the enormous utility of the undertaking, I will set myself to it despite the labor involved, and those who are free of feelings of petty jealousy really should prove favourable to the matter.

VII.3. Boethius, *Commentarii in Porphyrium*, 1.1 (ed. Brandt); translation adapted from Ebbesen, 'The Aristotelian Commentator', 39.

In the introduction to his translation of Porphyry's *Isagoge*, Boethius makes it clear that it was not his desire to amuse readers with his elegant Latin style, but to make the *veritas* ('truthfulness') to be found in the original work readable for them.

This second exposition will explain the text of our own translation, in which I fear that I have laid myself open to the kind of reproach that people level at any faithful translator, since I have delivered a word-by-word rendition. The thought behind this plan is that in these writings, which are about the search for the knowledge of things, it is not the pleasure of an elegant style, but uncorrupted truth that ought to be expressed. That is why I will seem to have achieved a lot, if these philosophical works composed in Latin may cause, by means of the sincerity of a most faithful translation, there to be nothing more to wish for in the books written by the Greeks.

VIII. *Translations of Scientific Writings*

VIII.1. Plinius, *Naturalis historia*, 18.22–23 (ed. Mayhoff); translation adapted from Rackham, *Natural History*, 303–305.

The *Naturalis historia* is a vast encyclopaedic work by Gaius Plinius Secundus (AD 23–79), usually called Pliny the Elder to distinguish him from his nephew Pliny the Younger (see VI.2). In contrast to the majority of ancient (and pre-modern) technical writers, Pliny is meticulous in naming his sources. In book 1, he gives a full list of the earlier works he used, and on occasion he provides more information on them in the course of his account. As one of his sources in the field of agriculture he names a comprehensive work written by Mago the Carthaginian.

Giving instructions for agriculture was an occupation of the highest dignity even in foreign nations—even kings did it [...], and also Mago the Carthaginian, on whom our senate bestowed so much honour after the taking of Carthage, that when it gave away the city's libraries to the African client kings, it deemed his twenty-eight volumes alone worthy of being translated into Latin. And this despite the fact that Marcus Cato[65] had already written his book of precepts. So they ordered that this task be delegated to people who were acquainted with the Punic language. In this project Decimus Silanus,[66] a man of most distinguished family, surpassed everybody.

VIII.2. Varro, *De re rustica*, 1.1.10 (ed. Keil/Götz); tr. Hooper, *Varro: On Agriculture*, 165–167.

Marcus Terentius Varro (116–27 BC) was a scholar, grammarian, and writer. The greater part of his prolific output is lost, and his *Rerum rusticarum libri* is the only work to have survived intact. The following passage bears witness to the reputation Mago's handbook on agriculture enjoyed among Roman writers.

All these are surpassed in reputation by Mago of Carthage, who gathered into twenty-eight books, written in the Punic tongue, the subjects they had dealt with separately. These Cassius Dionysius of Utica[67] translated into Greek and published in twenty books, dedicated to the praetor Sextilius. In these volumes he added not a little from the Greek writers whom I have named, taking from Mago's writings an amount equivalent to eight books. Diophanes,[68] in Bithynia, further abridged these in convenient form into six books, dedicated to king Deiotarus.[69]

VIII.3. *Epistula [Pseudo-]Celsi ad Pullium Natalem apud Marcellum servata*, 44.5–22 (ed. Liechtenhan); tr. Mundt and Cohen.

Marcellus Empiricus (early fifth century AD) included in his collection of medical preparations a number of letters attributed to great authorities of science and medi-

65 Marcus Porcius Cato (234–149 BC), also known as Cato the Elder to distinguish him from his aforementioned great-grandson, was a Roman senator who wrote *De agricultura*, a treatise on agriculture and the first extant piece of Latin prose.

66 Decimus Junius Silanus was a second-century expert on Punic language and agriculture, whose translation of Mago was to have great influence on Roman agriculture.

67 Cassius Dionysius of Utica was a second-century Greek writer on agriculture, who translated Mago's writings into Greek. His translation is mentioned by the Roman writer Columella, whose work on agriculture survives in its entirety. It is not known whether this translation was produced in collaboration with Silanus.

68 Diophanes of Nicaea was a first-century Greek writer on agriculture. Fragments of his work survive in the *Geoponica*, a tenth-century Byzantine collection of agricultural writings.

69 Deiotarus of Galatia (c. 105–c. 40 BC) was a Celtic/Galatian king and a faithful ally of the Romans.

cine. These included Hippocrates, Pliny, and Celsus. Most of them are pseudepigraphs. Among them is a letter of dedication, attributed to Celsus, to an otherwise unknown Pullius Natalis (see VIII.5).

After having read the two Greek books on medical recipes you sent me, dear Pullius Natalis, wishing that I may translate them into Latin, I willingly gave in to your wish, because they seemed to be a worthy object of my labour, highly beneficial as they are. Otherwise, I would have made an excuse, for it is my firm intention not to adorn anything from the field of medicine with my splendid style and thus with the gift of literary eternity, which is not worthy of the high majesty of the Roman people. Therefore, people now and in the future will possess, with this object of my constant will and my accuracy, a most welcome gift, for there is nothing more pleasant than good health. Nonetheless, I have sought to translate what I found in the Greek documents on medicine literally, to the best of my abilities. I did not deem it necessary thereby to ensure an elegant style, because I thought it more appropriate to render the meanings of the terms and to explain the actual things that found themselves under the names of foreign plants. I must confess that there was one thing I tried to avoid in my translations, contrary to the Greeks, who in their writings on medicine address the reader in the second person singular everywhere. In order to make my work less ominous, however, I preferred to introduce some undefined third person, because I was sure you know very well that I wrote these sorts of good works for the sake of utility rather than for entertainment.

VIII.4. Largius Designatianus, *Ad filios epistula*, 18.4–11 (ed. Liechtenhan); tr. Mundt and Cohen.

Unlike VIII.3, the letter assigned to Largius Designatianus (fourth century AD) also preserved in Marcellus's work is considered to be authentic. In it, he explains to his son why he had translated a Greek letter allegedly written by Hippocrates to a king named Antiochus. This Greek letter, also included in Marcellus's text, is obviously a forgery, since Seleucid kings of this name ruled several centuries after Hippocrates's death.[70]

Some time ago, I read the letter Hippocrates of Cos is said to have written to King Antiochus about the steps necessary to remain in good health. I longed to make his learning and warnings familiar to you, for the sake of your own good health, and that is why I recollected (as far as I could) those parts of his let-

70 See Temkin, *Hippocrates in a World of Pagans and Christians*, 243ff.

ter that I still had in my memory. These I have written down, translating them into the Roman tongue. I must confess that I have avoided translating word for word, as the differences between the two languages often would not allow the text to make sense unless enough attention were given to the actual meaning of the words. That is why I have added explanation where otherwise obscurity would have prevailed.

VIII.5. Celsus, *De medicina*, 6.18.1 (ed. Marx, Corpus Medicorum Latinorum I, 290.12–20); tr. Mundt and Cohen.

Aulius Cornelius Celsus (c. 25 BC–AD c. 50) was a Roman medical writer whose *De medicina*, more than any other work, reveals the Roman world's medical knowledge. In this treatise, he discusses the 'decency' of Greek technical terminology and expresses views on its translation into Latin.

Next up are those words that refer to genitalia, for which the Greeks have words that are both more tolerable and already accepted for use, as they may be found in almost every book or conversation devoted to medicine. We, on the other hand, have words for them that are so much uglier that not even the power of custom has been able to make speakers use them. Hence, for those wishing to preserve both verbal chastity and the learnings of their craft, this explanation will pose some difficulties. Nonetheless, this situation did not discourage me from writing. Thus, first of all, I have described all that I consider to be conducive to bodily health, and afterwards that which everyone will reveal only too unwillingly to someone else, since the path towards healing should be as understandable as possible for unlearned people as well.

VIII.6. Q. Serenus, *Liber medicinalis*, LI, 932–934 (ed. Vollmer, Corpus Medicorum Latinorum II/3, 45); tr. Mundt and Cohen.

Quintus Serenus's *Liber medicinalis* is a didactic poem on medicine, probably dating to the fourth century. Serenus is sometimes identified with Serenus Sammonicus, author of the antiquarian compilation *Res reconditae*. The translation of Greek medical terminology into Latin seems to have been problematic.

More lethal is what the Greeks call *hemitritaeos*. None of our ancestors has been able to name this disease in our language, as I suppose, or was even willing to do so.

Part Three: Original Texts

1. *The* verbum de verbo *Topos*

1.1. Publius Terentius Afer, *Adelphoe*, 6–14 (ed. Kauer/Lindsay/Skutsch).

Synapothnescontes Diphili comoediast:
eam Commorientes Plautu' fecit fabulam.
in Graeca adulescens est, qui lenoni eripit
meretricem in prima fabula: eum Plautus locum
reliquit integrum. eum hic locum sumpsit sibi
in Adelphos, verbum de verbo expressum extulit.
eam nos acturi sumus novam: pernoscite
furtumne factum existumetis an locum
reprehensum, qui praeteritu' neglegentiast.

1.2. Cicero, *De optimo genere oratorum*, 13–14 (ed. Wilkins).

Sed cum in eo magnus error esset, quale esset id dicendi genus, putavi mihi suscipiendum laborem utilem studiosis, mihi quidem ipsi non necessarium. Converti enim ex Atticis duorum eloquentissimorum nobilissimas orationes inter seque contrarias, Aeschini et Demostheni; nec converti ut interpres, sed ut orator, sententiis isdem et earum formis tamquam figuris, verbis ad nostram consuetudinem aptis. In quibus non verbum pro verbo necesse habui reddere, sed genus omne verborum vimque servavi. Non enim ea me adnumerare lectori putavi oportere, sed tamquam appendere.

1.3. Cicero, *De finibus bonorum et malorum*, 3.15 (ed. Reynolds).

Si enim Zenoni licuit, cum rem aliquam invenisset inusitatam, inauditum quoque ei rei nomen inponere, cur non liceat Catoni? Nec tamen exprimi verbum e verbo necesse erit, ut interpretes indiserti solent, cum sit verbum, quod idem declaret, magis usitatum. Equidem soleo etiam quod uno Graeci, si aliter non possum, idem pluribus verbis exponere. Et tamen puto concedi nobis oportere ut Graeco verbo utamur, si quando minus occurret Latinum.

1.4. Horace, *Ars poetica*, 131–134 (ed. Borzsák).

Publica materies privati iuris erit, si
non circa vilem patulumque moraberis orbem,

nec verbo verbum curabis reddere fidus
interpres.

I.5. Jerome, *Epistulae*, 57.5.2 (ed. Hilberg).

Ego enim non solum fateor, sed libera voce profiteor me in interpretatione Graecorum absque scripturis sanctis, ubi et verborum ordo mysterium est, non verbum e verbo, sed sensum exprimere de sensu. Habeoque huius rei magistrum Tullium, qui Protagoram Platonis et Oeconomicum Xenofontis et Aeschini et Demosthenis duas contra se orationes pulcherrimas transtulit. Quanta in illis praetermiserit, quanta addiderit, quanta mutaverit, ut proprietates alterius linguae suis proprietatibus explicaret, non est huius temporis dicere. Sufficit mihi ipsa translatoris auctoritas, qui ita in prologo earundem orationum locutus est: [quotes I.2 above].

II. *The Limits of* verbum de verbo*: Some Examples*

II.1. Cicero, *De finibus bonorum et malorum*, 2.13, 15 (ed. Reynolds).

Ergo illi intellegunt quid Epicurus dicat, ego non intellego? Ut scias me intellegere, primum idem esse dico voluptatem, quod ille ἡδονήν. Et quidem saepe quaerimus verbum Latinum par Graeco et quod idem valeat; hic nihil fuit, quod quaereremus. Nullum inveniri verbum potest quod magis idem declaret Latine, quod Graece, quam declarat voluptas. Huic verbo omnes, qui ubique sunt, qui Latine sciunt, duas res subiciunt, laetitiam in animo, commotionem suavem iucunditatis in corpore. [...]

Satisne igitur videor vim verborum tenere, an sum etiam nunc vel Graece loqui vel Latine docendus? Et tamen vide, ne, si ego non intellegam quid Epicurus loquatur, cum Graece, ut videor, luculenter sciam, sit aliqua culpa eius, qui ita loquatur, ut non intellegatur. Quod duobus modis sine reprehensione fit, si aut de industria facias, ut Heraclitus, 'cognomento qui σκοτεινός perhibetur, quia de natura nimis obscure memoravit', aut cum rerum obscuritas, non verborum, facit ut non intellegatur oratio, qualis est in Timaeo Platonis. Epicurus autem, ut opinor, nec non vult, si possit, plane et aperte loqui, nec de re obscura, ut physici, aut artificiosa, ut mathematici, sed de inlustri et facili et iam in vulgus pervagata loquitur. Quamquam non negatis nos intellegere quid sit voluptas, sed quid ille dicat; e quo efficitur, non ut nos non intellegamus quae vis sit istius verbi, sed ut ille suo more loquatur, nostrum neglegat.

II.2. Seneca, *De tranquillitate animi*, 2.3 (ed. Reynolds).

Hanc stabilem animi sedem Graeci εὐθυμίαν vocant, de qua Democriti volumen egregium est, ego tranquillitatem voco; nec enim imitari et transferre verba ad illorum formam necesse est: res ipsa, de qua agitur, aliquo signanda nomine est, quod appellationis Graecae vim debet habere, non faciem.

II.3. Seneca, *Epistulae morales*, 58.6–7 (ed. Reynolds).

'Quid sibi' inquis 'ista praeparatio vult? quo spectat?' Non celabo te: cupio, si fieri potest, propitiis auribus tuis 'essentiam' dicere; si minus, dicam et iratis. Ciceronem auctorem huius verbi habeo, puto locupletem; si recentiorem quaeris, Fabianum, disertum et elegantem, orationis etiam ad nostrum fastidium nitidae. Quid enim fiet, mi Lucili? quomodo dicetur οὐσία, res necessaria, natura continens fundamentum omnium? Rogo itaque permittas mihi hoc verbo uti. Nihilominus dabo operam ut ius a te datum parcissime exerceam; fortasse contentus ero mihi licere.

Quid proderit facilitas tua, cum ecce id nullo modo Latine exprimere possim propter quod linguae nostrae convicium feci? Magis damnabis angustias Romanas, si scieris unam syllabam esse quam mutare non possum. Quae sit haec quaeris? τὸ ὄν. Duri tibi videor ingenii: in medio positum, posse sic transferri ut dicam 'quod est'. Sed multum interesse video: cogor verbum pro vocabulo ponere; sed si ita necesse est, ponam 'quod est'.

II.4. Aulus Gellius, *Noctes Atticae*, 11.16.1–6 (ed. Marshall).

Adiecimus saepe animum ad vocabula rerum non paucissima, quae neque singulis verbis, ut a Graecis, neque, si maxime pluribus eas res verbis dicamus, tam dilucide tamque apte demonstrari Latina oratione possunt, quam Graeci ea dicunt privis vocibus. Nuper etiam cum adlatus esset ad nos Plutarchi liber et eius libri indicem legissemus, qui erat περὶ πολυπραγμοσύνης, percontanti cuipiam qui et litterarum et vocum Graecarum expers fuit, cuiusnam liber et qua de re scriptus esset, nomen quidem scriptoris statim diximus, rem, de qua scriptum fuit dicturi, haesimus. Ac tum quidem primo, quia non satis commode opinabar interpretaturum me esse, si dicerem librum scriptum 'De negotiositate', aliud institui aput me exquirere, quod, ut dicitur, verbum de verbo expressum esset. Nihil erat prorsus quod aut meminissem legere me aut, si etiam vellem fingere, quod non insigniter asperum absurdumque esset, si ex multitudine et negotio verbum unum compingerem, sicuti 'multiiuga' dicimus et 'multicolora' et 'multiformia'. Sed non minus inlepide ita diceretur, quam si interpretari voce una velis πολυφιλίαν aut πολυτροπίαν aut πολυσαρκίαν. Quamobrem, cum diutule tacitus in cogitando fuissem, respondi tandem non videri mihi significari eam rem posse uno nomine, et idcirco iuncta oratione, quid vellet Graecum id verbum parabam dicere.

III. *Cicero's 'National' Project*

III.1. Plutarch, *Cicero*, 40.1–2 (ed. Lindskog/Ziegler/Gärtner).

ἐκ τούτου Κικέρων, εἰς μοναρχίαν τῆς πολιτείας μεθεστώσης, ἀφέμενος τοῦ τὰ κοινὰ πράττειν ἐσχόλαζε τοῖς βουλομένοις φιλοσοφεῖν τῶν νέων, καὶ σχεδὸν ἐκ τῆς πρὸς τούτους συνηθείας, εὐγενεστάτους καὶ πρώτους ὄντας, αὖθις ἴσχυεν ἐν τῇ πόλει μέγιστον. αὐτῷ δ' ἔργον μὲν ἦν τότε τοὺς φιλοσόφους συντελεῖν διαλόγους καὶ μεταφράζειν, καὶ τῶν διαλεκτικῶν ἢ φυσικῶν ὀνομάτων ἕκαστον εἰς τὴν Ῥωμαϊκὴν μεταβάλλειν διάλεκτον· ἐκεῖνος γάρ ἐστιν, ὥς φασιν, ὁ καὶ τὴν φαντασίαν καὶ τὴν ἐποχὴν καὶ τὴν συγκατάθεσιν καὶ τὴν κατάληψιν, ἔτι δὲ τὸ ἄτομον, τὸ ἀμερές, τὸ κενὸν, καὶ ἄλλα πολλὰ τῶν τοιούτων ἐξονομάσας πρῶτος ἢ μάλιστα Ῥωμαίοις, τὰ μὲν μεταφοραῖς, τὰ δ' οἰκειότησιν ἄλλαις γνώριμα καὶ προσήγορα μηχανησάμενος.

III.2. Cicero, *Tusculanae disputationes*, 2.5 (ed. Pohlenz).

Quam ob rem hortor omnis, qui facere id possunt, ut huius quoque generis laudem iam languenti Graeciae eripiant et transferant in hanc urbem, sicut reliquas omnis, quae quidem erant expetendae, studio atque industria sua maiores nostri transtulerunt. Atque oratorum quidem laus ita ducta ab humili venit ad summum, ut iam, quod natura fert in omnibus fere rebus, senescat brevique tempore ad nihilum ventura videatur, philosophia nascatur Latinis quidem litteris ex his temporibus, eamque nos adiuvemus nosque ipsos redargui refellique patiamur.

III.3. Cicero, *De officiis*, 1.6 (ed. Winterbottom).

Sequimur igitur hoc quidem tempore et hac in quaestione potissimum Stoicos, non ut interpretes, sed, ut solemus, e fontibus eorum iudicio arbitrioque nostro quantum quoque modo videbitur, hauriemus.

III.4. Cicero, *De natura deorum*, 2.104 (ed. Plasberg/Ax).

Atque hoc loco me intuens 'Utar', inquit, 'carminibus Arateis, quae a te admodum adulescentulo conversa ita me delectant, quia Latina sunt, ut multa ex is memoria teneam.'

IV. *Greek Philosophy in Latin: Is This Possible, Even Desirable?*

IV.1. Cicero, *De legibus*, 2.17 (ed. Powell).

Quintus: Nihil enim tam dissimile quam vel ea quae ante dixisti, uel hoc ipsum de deis exordium. Unum illud mihi uideris imitari, orationis genus.—*Marcus:* Velle fortasse: quis enim id potest aut umquam poterit imitari? Nam sententias interpretari perfacile est; quod quidem ego facerem, nisi plane esse uellem meus. Quid enim negoti est eadem prope uerbis eisdem conuersa dicere?—*Quintus:* Prorsus adsentior. Verum ut modo tute dixisti, te esse malo tuum.

IV.2. Cicero, *Academica*, 1.4f., 9f. (ed. Plasberg).

Tum ille: 'Rem a me saepe deliberatam et multum agitatam requiris. itaque non haesitans respondebo, sed ea dicam quae mihi sunt in promptu, quod ista ipsa de re multum ut dixi et diu cogitavi. Nam cum philosophiam viderem diligentissime Graecis litteris explicatam, existimavi si qui de nostris eius studio tenerentur, si essent Graecis doctrinis eruditi, Graeca potius quam nostra lecturos, sin a Graecorum artibus et disciplinis abhorrerent, ne haec quidem curaturos, quae sine eruditione Graeca intellegi non possunt. Itaque ea nolui scribere quae nec indocti intellegere possent nec docti legere curarent. [...]'

Causam autem probabilem tu quidem affers: aut enim Graeca legere malent, qui erunt eruditi, aut ne haec quidem qui illa nescient. Sed eam mihi non sane probas; immo vero et haec qui illa non poterunt et qui Graeca poterunt non contemnent sua. Quid enim causae est, cur poetas Latinos Graecis litteris eruditi legant, philosophos non legant? An quia delectat Ennius Pacuvius Accius multi alii, qui non verba sed vim Graecorum expresserunt poetarum—quanto magis philosophi delectabunt, si ut illi Aeschylum Sophoclem Euripidem sic hi Platonem imitentur Aristotelem Theophrastum.

IV.3. Cicero, *De finibus bonorum et malorum*, 1.1f., 4, 5f., 7f. (ed. Reynolds).

Non eram nescius, Brute, cum, quae summis ingeniis exquisitaque doctrina philosophi Graeco sermone tractavissent, ea Latinis litteris mandaremus, fore ut hic noster labor in varias reprehensiones incurreret. Nam quibusdam, et iis quidem non admodum indoctis, totum hoc displicet, philosophari. Quidam autem non tam id reprehendunt, si remissius agatur, sed tantum studium tamque multam operam ponendam in eo non arbitrantur. Erunt etiam, et ii quidem eruditi Graecis litteris, contemnentes Latinas, qui se dicant in Graecis legendis operam malle consumere. [...]

Iis igitur est difficilius satis facere, qui se Latina scripta dicunt contemnere. In quibus hoc primum est in quo admirer, cur in gravissimis rebus non delectet eos sermo patrius, cum idem fabellas Latinas ad verbum e Graecis expressas non inviti legant. [...]

An 'Utinam ne in nemore ...' nihilo minus legimus quam hoc idem Graecum, quae autem de bene beateque vivendo a Platone disputata sunt, haec explicari non placebit Latine? Quid? Si nos non interpretum fungimur munere, sed tuemur ea, quae dicta sunt ab iis, quos probamus, eisque nostrum iudicium et nostrum scribendi ordinem adiungimus, quid habent, cur Graeca anteponant iis, quae et splendide dicta sint neque sint conversa de Graecis?

Quamquam, si plane sic verterem Platonem aut Aristotelem, ut verterunt nostri poetae fabulas, male, credo, mererer de meis civibus, si ad eorum cognitionem divina illa ingenia transferrem. Sed id neque feci adhuc nec mihi tamen, ne faciam, interdictum puto. Locos quidem quosdam, si videbitur, transferam, et maxime ab iis, quos modo nominavi, cum inciderit, ut id apte fieri possit, ut ab Homero Ennius, Afranius a Menandro solet. [...]

Sed ex eo credo quibusdam usu venire, ut abhorreant a Latinis, quod inciderint in inculta quaedam et horrida, de malis Graecis Latine scripta deterius. Quibus ego adsentior, dum modo de isdem rebus ne Graecos quidem legendos putent.

IV.4. Jerome, *Epistulae*, 124.1 (ed. Hilberg).

Ante annos circiter decem sanctus vir Pammachius ad me cuiusdam schedulas misit, quae Origenis περὶ ἀρχῶν interpretata volumina continerent, imo vitiata, hoc magnopere postulans, ut Graecam veritatem servaret Latina translatio et in utramque partem, seu bene, seu male dixisset ille, qui scripsit, absque interpretis patrocinio, Romana lingua cognosceret. Feci ut voluit, misique ei libros, quos cum legisset, exhorruit, et recludens scrinio, ne prolati in vulgus multorum animos vulnerarent [...].

V. *The Alleged Poverty of the Latin Language*

V.1. Lucretius, *De rerum natura*, 1.136–145 (ed. Martin).

Nec me animi fallit Graiorum obscura reperta
difficile inlustrare Latinis versibus esse,
multa novis verbis praesertim cum sit agendum
propter egestatem linguae et rerum novitatem;
sed tua me virtus tamen et sperata voluptas

suavis amicitiae quemvis efferre laborem
suadet et inducit noctes vigilare serenas
quaerentem dictis quibus et quo carmine demum
clara tuae possim praepandere lumina menti,
res quibus occultas penitus convisere possis.

V.2. Cicero, *De finibus bonorum et malorum*, 1.10 (ed. Reynolds).

Ego autem mirari satis non queo, unde hoc sit tam insolens domesticarum rerum fastidium. Non est omnino hic docendi locus; sed ita sentio et saepe disserui, Latinam linguam non modo non inopem, ut vulgo putarent, sed locupletiorem etiam esse quam Graecam.

V.3. Seneca, *Epistulae morales*, 58.1 (ed. Reynolds).

Quanta verborum nobis paupertas, immo egestas sit, numquam magis quam hodierno die intellexi. Mille res inciderunt, cum forte de Platone loqueremur, quae nomina desiderarent nec haberent, quaedem vero quae cum habuissent fastidio nostro perdidissent. Quis autem ferat in egestate fastidium?

VI. *Translating in the First and Second Centuries AD*

VI.1. Quintilian, *Institutio oratoria*, 10.5.2–3 (ed. Winterbottom).

Vertere Graeca in Latinum veteres nostri oratores optimum iudicabant. Id se L. Crassus in illis Ciceronis de Oratore libris dicit factitasse: id Cicero sua ipse persona frequentissime praecipit, quin etiam libros Platonis atque Xenophontis edidit hoc genere tralatos: id Messalae placuit, multaeque sunt ab eo scriptae ad hunc modum orationes, adeo ut etiam cum illa Hyperidis pro Phryne difficillima Romanis subtilitate contenderet. Et manifesta est exercitationis huiusce ratio. Nam et rerum copia Graeci auctores abundant et plurimum artis in eloquentiam intulerunt et hos transferentibus verbis uti optimis licet: omnibus enim utimur nostris. Figuras vero, quibus maxime ornatur oratio, multas ac varias excogitandi etiam necessitas quaedam est, quia plerumque a Graecis Romana dissentiunt.

VI.2. Pliny the Younger, *Epistulae*, 7.9.1–5 (ed. Mynors).

Quaeris quemadmodum in secessu, quo iam diu frueris, putem te studere oportere. Utile in primis, et multi praecipiunt, vel ex Graeco in Latinum vel ex

Latino vertere in Graecum. Quo genere exercitationis proprietas splendorque verborum, copia figurarum, vis explicandi, praeterea imitatione optimorum similia inveniendi facultas paratur; simul quae legentem fefellissent, transferentem fugere non possunt. Intellegentia ex hoc et iudicium adquiritur. Nihil offuerit quae legeris hactenus, ut rem argumentumque teneas, quasi aemulum scribere lectisque conferre, ac sedulo pensitare, quid tu quid ille commodius. Magna gratulatio si non nulla tu, magnus pudor si cuncta ille melius. Licebit interdum et notissima eligere et certare cum electis. Audax haec, non tamen improba, quia secreta contentio: quamquam multos videmus eius modi certamina sibi cum multa laude sumpsisse, quosque subsequi satis habebant, dum non desperant, antecessisse.

VI.3. Aulus Gellius, *Noctes Atticae*, 17.20.7 f. (ed. Marshall).

Haec admonitio Tauri de orationis Platonicae modulis non modo non repressit, sed instrinxit etiam nos ad elegantiam Graecae orationis verbis Latinis adfectandam; atque uti quaedam animalium parva et vilia ad imitandum sunt, quas res cumque audierint viderintve, petulantia, proinde nos ea, quae in Platonis oratione demiramur, non aemulari quidem, sed lineas umbrasque facere ausi sumus.

VII. *Translating Plato and Aristotle in Late Antiquity*

VII.1. Calcidius, *Commentarius in Platonis Timaeum*, dedicatory letter to (H)Osius (ed. Magee).

Eadem est, opinor, vis amicitiae parque impossibilium paene rerum extricatio, cum alter ex amicis iubendi religione, alter parendi voto complaciti operis adminiculentur effectui. Conceperas animo florente omnibus studiis humanitatis excellentique ingenio tuo dignam spem proventuri operis intemptati ad hoc tempus eiusque usum a Graecis Latio statueras mutuandum. Et quamquam ipse hoc cum facilius tum commodius facere posses, credo propter admirabilem verecundiam, ei potius malueris iniungere quem te esse alterum iudicares. Possemne, oro te, quamvis res esset ardua, tanto honore habito de quo ita senseras iniunctum excusare munus et, qui numquam ne in sollemnibus quidem et usitatis voluntatibus ullum officium recusassem, huic tanto tamque honesto desiderio contradicere, in quo declinatio speciosi muneris excusatione ignorationis callida esset scientiae futura simulatio? Itaque parui certus non sine divino instinctu id mihi a te munus iniungi proptereaque alacriore mente speque confirmatiore primas partes Timaei Platonis aggressus non solum transtuli sed etiam partis eiusdem commentarium feci putans reconditae rei simulacrum sine interpretationis explanatione aliquanto obscurius ipso exemplo futurum.

VII.2. Boethius, *Commentarii in librum Aristotelis ΠΕΡΙ ΕΡΜΗΝΕΙΑΣ. Editio secunda*, 2.79–80 (ed. Meiser).

Mihi autem si potentior divinitatis adnuerit favor, haec fixa sententia est, ut quamquam fuerint praeclara ingenia, quorum labor ac studium multa de his quae nunc quoque tractamus Latinae linguae contulerit, non tamen quendam quodammodo ordinem filumque et dispositione disciplinarum gradus ediderunt, ego omne Aristotelis opus, quodcumque in manus venerit, in Romanum stilum vertens eorum omnium commenta Latina oratione perscribam, ut si quid ex logicae artis subtilitate, ex moralis gravitate peritiae, ex naturalis acumine veritatis ab Aristotele conscriptum sit, id omne ordinatum transferam atque etiam quodam lumine commentationis inlustrem omnesque Platonis dialogos vertendo vel etiam commentando in Latinam redigam formam. His peractis non equidem contempserim Aristotelis Platonisque sententias in unam quodammodo revocare concordiam, eosque non, ut plerique dissentire in omnibus, sed in plerisque, et his in philosophia maximis, consentire demonstrem. Haec si vita otiumque suppetit, cum multa operis huius utilitate nec non etiam labore contenderim, qua in re faveant oportet, quos nulla coquit invidia.

VII.3. Boethius, *Commentarii in Porphyrium*, 1.1 (ed. Brandt).

Secundus hic arreptae expositionis labor nostrae seriem translationis expediet, in qua quidem vereor ne subierim fidi interpretis culpam, cum verbum verbo expressum comparatumque reddiderim.

Cuius incoepti ratio est, quod in his scriptis in quibus rerum cognitio quaeritur, non luculentae orationis lepos, sed incorrupta veritas exprimenda est. Quocirca multum profecisse videbor, si philosophiae libris Latina oratione compositis per integerrimae translationis sinceritatem, nihil in Graecorum litteris amplius desideretur.

VIII. *Translations of Scientific Writings*

VIII.1. Plinius, *Naturalis historia*, 18.22–23 (ed. Mayhoff).

Igitur de cultura agri praecipere principale fuit etiam apud exteros, siquidem et reges fecere, [...] et Poenus etiam Mago, cui quidem tantum honorem senatus noster habuit Carthagine capta ut, cum regulis Africae bibliothecas donaret, unius eius duodetriginta volumina censeret in Latinam linguam transferenda, cum iam M. Cato praecepta condidisset, peritisque Punicae dandum negotium, in quo praecessit omnes vir clarissimae familiae D. Silanus.

VIII.2. Varro, *De re rustica*, 1.1.10 (ed. Keil/Götz).

Hos nobilitate Mago Carthaginiensis praeteriit, Poenica lingua qui res dispersas comprendit libris XXIIX, quos Cassius Dionysius Uticensis vertit libris XX ac Graeca lingua Sextilio praetori misit; in quae volumina de Graecis libris eorum quos dixi adiecit non pauca et de Magonis dempsit instar librorum VIII. Hosce ipsos utiliter ad VI libros redegit Diophanes in Bithynia et misit Deiotaro regi.

VIII.3. *Epistula [Pseudo-]Celsi ad Pullium Natalem apud Marcellum servata*, 44.5–22 (ed. Liechtenhan).

Lectis duobus libris conpositionum Graecis, Pulli Natalis, quos misisti mihi, ut in Latinum sermonem conuerterem, libenter parui tuae uoluntati, dignos eos existimans, quibus operam inpenderem, quoniam plenos utilitatis conperi. Alioquin excusassem me, propositum habens nihil non dignum sublimi Romanorum maiestate de bono medicinae artis splendidissime scribere idque perennitati et litteris mandare. Habebit ergo et praesens tempus et futura posteritas in hoc perpetuae uoluntatis meae curaeque gratissimum munus, siquidem bona ualetudine nihil est pulchrius. Dedi autem operam quantam potui, ut ad uerbum transferrem, quae in Graecis medicinalibus repperi, nec necessariam curam cultus orationis putaui, quia rectius duxi proprietates nominum edere et rem ueram sub herbarum peregrinarum uocabulis explicare. Vnum genus fateor me translationis de industria declinasse, quo fere Graeci utuntur in medicamentorum scripturis, omnia ad secundum personam dirigentes. Ego uero in usum medicamentorum, quae edidi, ominis boni gratia malui quam infinitam inserere personam admoneremque, nisi certum haberem te non ignorare, me huiusmodi bona utilitatis magis quam delectationis causa scripsisse.

VIII.4. Largius Designatianus, *Ad filios epistula*, 18.4–11 (ed. Liechtenhan).

Legi antehac Hippocratis Coi epistulam, medicinae auctoris uel peritissimi, quam de tuenda ualetudine ad Antiochum regem scripsisse fertur. Cuius quia praecepta et monita uobis ad salutis custodiam nota cuperem, collegi, ut potui, quidquid seriem eius continere memineram, eamque ore Romano interpretans scripsi. Caui, fateor, ne ad uerbum uerba transferrem, quo minus pleraque linguarum diuersitate congruerent, si parum ponderis in significationibus ueris fieret; itaque explicandae rei aduersus obscuritatem consului.

VIII.5. Celsus, *De medicina*, 6.18.1 (ed. Marx).

Proxima sunt ea, quae ad partes obscenas pertinent, quarum apud Graecos uocabula et tolerabilius se habent et accepta iam usu sunt, cum in omni fere medicorum uolumine atque sermone iactentur: apud nos foediora uerba ne consuetudine quidem aliqua uerecundius loquentium commendata sunt, ut difficilior haec explanatio sit simul et pudorem et artis praecepta seruantibus. Neque tamen ea res a scribendo me deterrere debuit: primum, ut omnia quae salutaria accepi, conprehenderem; dein, quia in uolgus eorum curatio etiam praecipue cognoscenda est, quae inuitissimus quisque alteri ostendit.

VIII.6. Q. Serenus, *Liber medicinalis*, LI, 932–934 (ed. Vollmer, Corpus Medicorum Latinorum II/3, 45).

mortiferum magis est quod Graecis hemitritaeos
uulgatur uerbis; hoc nostra dicere lingua
non potuere ulli, puto, nec uoluere parentes.

Bibliography

Primary Sources

Boethius, *Commentarii in librum Aristotelis ΠΕΡΙ ΕΡΜΗΝΕΙΑΣ. Editio secunda*, ed. Karl Meiser, Leipzig: Teubner, 1880

Boethius, *Commentarii in Porphyrium*, ed. Samuel Brandt, Paris, 1906

Calcidius, *Commentarius in Platonis Timaeum*, ed. John Magee, Harvard: Harvard University Press, 2016

Celsus, *De medicina*, in Corpus Medicorum Latinorum I, ed. Friedrich Marx, Berlin, 1915, 17–410

Pseudo-Celsus, *Epistula [Pseudo-]Celsi ad Pullium Natalem apud Marcellum servata*, in Corpus Medicorum Latinorum V, ed. Eduard Liechtenhan, Berlin, 1968, 44f.

Cicero, *Tusculanae disputationes*, ed. Max Pohlenz, Stuttgart: Teubner, 1918 (repr. 1965)

Cicero, *Academica*, ed. Otto Plasberg, Leipzig: Teubner, 1922

Cicero, *De natura deorum*, ed. Otto Plasberg and Wilhelm Ax, Stuttgart: Teubner, 1961

Cicero, *De optimo genere oratorum*, ed. Augustus Wilkins, Oxford: Oxford University Press, 1963

Cicero, *De officiis*, ed. Michael Winterbottom, Oxford: Oxford University Press, 1994
Cicero, *De finibus bonorum et malorum*, ed. Leighton Reynolds, Oxford: Oxford University Press, 1998
Cicero, *De legibus*, ed. Jonathan Powell, Oxford: Oxford University Press, 2006
Gellius, *Noctes Atticae*, ed. Peter Marshall, Oxford: Oxford University Press, 1968
Horace, *Ars poetica*, ed. István Borzsák, Leipzig: Teubner, 1984
Jerome, *Epistulae*, ed. Isidorus Hilberg, Vienna/Leipzig, 1918
Largius Designatianus, *Ad filios epistula*, in Corpus Medicorum Latinorum v, ed. Eduard Liechtenhan, Berlin, 1968, 18 f.
Lucretius, *De rerum natura*, ed. Joseph Martin, Leipzig: Teubner, 1959
Pliny the Elder, *Naturalis historia*, ed. Karl Mayhoff, Stuttgart: Teubner, 1967
Pliny the Younger, *Epistulae*, ed. Roger Mynors, Oxford: Oxford University Press, 1963
Plutarch, *Κικέρων*, ed. Claes Lindskog, Konrat Ziegler, and Hans Gärtner, Leipzig: Teubner, 1994
Quintilian, *Institutio oratoria*, ed. Michael Winterbottom, Oxford: Oxford University Press, 1970
Seneca, *Epistulae morales*, ed. Leighton Reynolds, Oxford: Oxford University Press, 1965
Seneca, *De tranquillitate animi*, ed. Leighton Reynolds, Oxford: Oxford University Press, 1977
Serenus, *Liber medicinalis*, in Corpus Medicorum Latinorum II/3, ed. Friedrich Vollmer, Leipzig/Berlin, 1916, 5–52
Terentius, *Adelphoe*, ed. Robert Kauer, Wallace Lindsay, and Otto Skutsch, Oxford: Oxford University Press, 1958
Varro, *De re rustica*, ed. Heinrich Keil and Georg Götz, Leipzig: Teubner, 2nd edn, 1929

Secondary Literature

Adams, James N., *Bilingualism and the Latin Language*, Cambridge, 2003
Baltussen, Han, 'Cicero's Translation of Greek Philosophy', in Siobhán McElduff and Enrica Sciarinno (eds), *Complicating the History of Western Translation: The Ancient Mediterranean in Perspective*, London/New York, 2011, 37–47
Basore, John William, *Seneca: Moral Essays*, II: *De tranquillitate animi. On Tranquillity of Mind*, Cambridge MA, 1928
Beall, Stephen, 'Translation in Aulus Gellius', *Classical Quarterly*, 47 (1997), 215–226
Biville, Frédérique, 'The Graeco-Romans and Graeco-Latin: A Terminological Framework for Cases of Bilingualism', in James N. Adams, M. Janse, and S. Swain (eds), *Bilingualism in Ancient Society: Language Contact and the Written Text*, Oxford, 2002, 77–102
Brock, Sebastian, 'Aspects of Translation Technique in Antiquity', *Greek, Roman, and Byzantine Studies*, 20 (1979), 69–88
Brooks, Francis, *Cicero: De natura deorum*, London, 1896

Calboli, Gualtiero, 'Atticism', in *Brill's New Pauly*, Leiden/Boston, 2003, 324–325

Casey, John Patrick, 'Boethius's Works on Logic in the Middle Ages', in Noel H. Kaylor and Philip E. Phillips (eds), *A Companion to Boethius in the Middle Ages*, Leiden/Boston, 2012, 193–219

Cilliers, Louise, 'Forgery, Fiction and Pseudonym in the Collection of Medical Letters Prefacing Marcellus's *De medicamentis*', *Acta Academica*, 38 (2006), 91–110

Ebbesen, Sten, 'The Aristotelian Commentator', in John Marenbon (ed.), *The Cambridge Companion to Boethius*, Cambridge, 2009, 34–55

Feeney, Denis, *Beyond Greek: The Beginnings of Latin Literature*, Cambridge MA/London, 2016

Fögen, Thorsten, *Patrii sermonis egestas: Einstellungen lateinischer Autoren zu ihrer Muttersprache*, Leipzig, 2000

Fögen, Thorsten, 'The Transformation of Greek Scientific Knowledge by Roman Technical Writers: On the Translating of Technical Texts in Antiquity', *Antike Naturwissenschaft und ihre Rezeption*, 15 (2005), 91–114

Fremantle, William Henry, George Lewis, and William Gibson Martley, *Jerome: Letters and Select Works*, Nicene and Post-Nicene Fathers, ser. 2, vol. 6, New York, 1893

Garbarino, Giovanna (ed.), *M. Tulli Ciceronis fragmenta: ex libris philosophicis; ex aliis libris deperditis; ex scriptis incertis*, Milan, 1984

Gildenhard, Ingo, *Paideia Romana: Cicero's Tusculan Disputations*, Cambridge, 2007

Glucker, John, 'Cicero's Remarks of Translating Philosophical Terms—Some General Problems', in Glucker and Burnett, *Greek into Latin*, 37–96

Glucker, John and Charles Burnett (eds), *Greek into Latin from Antiquity until the Nineteenth Century*, London/Turin, 2012

Gordon, Pamela, *The Invention and Gendering of Epicurus*, Ann Arbor MI, 2012

Gummere, Richard, *Seneca: Ad Lucilium epistulae morales*, Cambridge MA, 1917

Hadot, Pierre, *Marius Victorinus: Recherches sur sa vie et ses œuvres*, Paris, 1971

Harrison, Stephen, *Apuleius: A Latin Sophist*, Oxford, 2000

Hidber, Thomas, *Das klassizistische Manifest des Dionys von Halikarnass: Die praefatio zu De oratoribus veteribus*, Stuttgart/Leipzig, 1996

Hoenig, Christina, 'Εἰκὼς λόγος: Plato in translation(s)', *Methodos*, 13 (2013), unpaginated

Hooper, William Davis, *Varro: On Agriculture*, Cambridge MA, 1934

Hubbell, Henry Mortimer, *Cicero: De inventione, De optimo genere oratorum, Topica*, Cambridge MA, 1949

Hübner, Wolfgang, 'Die Rezeption der Phainomena Arats in der lateinischen Literatur', in M. Horster and Christiane Reitz (eds), *Wissensvermittlung in dichterischer Gestalt*, Stuttgart, 2005, 133–154

Hurka, Florian, 'Arat und Aratea', in Christine Walde (ed.), *Die Rezeption der antiken Literatur*, Der Neue Pauly, Supplemente, Band 7, Stuttgart/Weimar, 2010, 69–75

Hutchinson, G.O., *Greek to Latin: Frameworks and Contexts for Intertextuality*, Oxford, 2013

Kaimio, Jorma, *The Romans and the Greek Language*, Helsinki, 1979

Kelly, Louis G., *The True Interpreter: A History of Translation Theory and Practice in the West*, New York, 1979

Keyes, Clinton Walker, *Cicero: De re publica, De legibus*, Cambridge MA, 1928

King, John, *Cicero: Tusculan Disputations*, Cambridge MA, 1927

Konstan, David, 'Crossing Conceptual Worlds: Greek Comedy and Philosophy', in M. Fontaine and A.C. Scafuro (eds), *The Oxford Handbook of Greek and Roman Comedy*, Oxford, 2014, 278–294

Kytzler, Bernhard, '*Fidus interpres*: The Theory and Practice of Translation in Classical Antiquity', *Antichthon*, 23 (1989), 42–50

Lambardi, Noemi, *Il 'Timaeus' ciceroniano: arte e tecnica del 'vertere'*, Florence, 1982

Langslow, David R., *Medical Latin in the Roman Empire*, Oxford, 2000

Langslow, David R., 'Typologies of Translation Techniques in Greek and Latin: Latin *elticis*: *catelticis* = Greek ἑλκτική: καθεκτική', in Alex Mullen and Patrick James (eds), *Multilingualism in the Graeco-Roman Worlds*, Cambridge, 2012, 141–171

Leigh, Matthew, *From Polypragmon to Curiosus: Ancient Concepts of Curious and Meddlesome Behaviour*, Oxford, 2013

Leonard, William Ellery, *Lucretius: Of the Nature of Things. A Metrical Translation*, New York, 1916

Lewis, Anne-Marie, *From Aratus to the Aratus Latinus: A Comparative Study of Latin Translation*, diss. Ann Arbor MI, 1983

Lewis, Anne-Marie, 'The Popularity of the Phaenomena of Aratus', in Carl Deroux (ed.), *Studies in Latin Literature and Roman History VI*, Brussels, 1992, 94–118

Magee, John (ed.), *Calcidius: On Plato's Timaeus*, Cambridge, 2016

Marti, Heinrich, *Übersetzer der Augustin-Zeit: Interpretation von Selbstzeugnissen*, Munich, 1974

Marti, Heinrich, 'Das Übersetzen philosophischer Texte gegen Ende des 4. Jahrhunderts', in Jacqueline Hamesse and Marta Fattori (eds), *Rencontres de Cultures dans la philosophie médiévale: traductions et traducteurs de l'antiquité tardive au XIVe siècle*, Louvain-la-Neuve/Cassino, 1990, 24–45

McElduff, Siobhán, *Roman Theories of Translation: Surpassing the Source*, New York, 2013

Moreschini, Claudio, *A Christian in Toga. Boethius: Interpreter of Antiquity and Christian Theologian*, Göttingen, 2014

Miller, Walter, *Cicero: De officiis*, London/New York, 1913

Müller-Wetzel, Martin, 'Der Topos von der Spracharmut. Typen seiner Funktionalisierung in der lateinischen Literatur der Antike und Spätantike', in Ulrike-Christine Sander and Paul Fritz (eds), *Muster und Funktionen kultureller Selbst- und Fremd-*

wahrnehmung: Beiträge zur Geschichte der sprachlichen und literarischen Emanzipation, Göttingen, 2000, 120–145

Nicolas, Christian, *Utraque lingua. Le Calque sémantique: domaine gréco-latin*, Louvain/Paris, 1996

Perrin, Bernadotte, *Plutarch's Lives*, Cambridge MA, 1919

Poncelet, Roland, 'Deux aspects du style philosophique latin: Cicéron et Chalcidius, traducteurs du 'Phédre' 245C', *Revue des Études Latines*, 28 (1950), 145–167

Poncelet, Roland, *Cicéron: Traducteur de Platon. L'expression de la pensée complexe en latin classique*, Paris, 1957

Possanza, Mark, *Translating the Heavens: Aratus, Germanicus, and the Poetics of Latin Translation*, New York, 2004

Powell, J.G.F., 'Cicero's Translations from Greek', in id. (ed.), *Cicero the Philosopher: Twelve Papers*, Oxford, 1995, 273–300

Puelma, Mario, 'Cicero als Platon-Übersetzer', *Museum Helveticum*, 37 (1980), 137–178

Rackham, Harris, *Cicero: De finibus bonorum et malorum*, Cambridge MA, 1914

Rackham, Harris, *Cicero: De natura deorum, Academica*, London/New York, 1933

Rackham, Harris, *Pliny the Elder: Natural History*, Cambridge MA, 1950

Radice, Betty, *Pliny the Younger: Letters*, Cambridge MA, 1969

Reinhardt, Tobias and Michael Winterbottom, *Quintilian: Institutio oratoria. Book 2*, Oxford, 2006

Reitz, Christiane and Andreas Fuchs, 'Vertere. Zu Theorie und Praxis des Übersetzens in der Antike', in Albrecht Buschmann (ed.), *Gutes Übersetzen: Neue Perspektiven für Theorie und Praxis des Literaturübersetzens*, Berlin/Boston, 2015, 35–66

Riley, Henry Thomas, *The Comedies of Terence: Literally Translated into English Prose, with Notes*, New York, 1874

Robert, Jörg, 'Die Ciceronianismus-Debatte', in Herbert Jaumann (ed.), *Diskurse der Gelehrtenkultur in der Frühen Neuzeit: Ein Handbuch*, Berlin/New York, 2011, 1–54

Robinson, Douglas, 'Classical Theories of Translation from Cicero to Aulus Gellius', *TextconText*, 7 (1992), 15–55

Rochette, Bruno, 'Du grec au latin et du latin au grec. Les problèmes de la traduction dans l'antiquité gréco-latine', *Latomus*, 54 (1995), 245–261

Rochette, Bruno, 'Greek and Latin Bilingualism', in E. Bakker (ed.), *A Companion to the Ancient Greek Language*, New York, 2010, 281–293

Rolfe, John Carew, *The Attic Nights of Aulus Gellius*, Cambridge MA, 1927

Rosén, Hannah, 'Greek in Latin, Greek into Latin—Reflections on the Passage of Patterns', in Glucker and Burnett, *Greek into Latin*, 1–18

Russell, Donald, *Quintilian: Institutio oratoria. The Orator's Education*, Cambridge MA, 2002

Sedley, David, *Lucretius and the Transformation of Greek Wisdom*, Cambridge, 1998

Seele, Astrid, *Römische Übersetzer: Nöte, Freiheiten, Absichten. Verfahren des literarischen Übersetzens in der griechisch-römischen Antike*, Darmstadt, 1995

Stevens, Benjamin, 'Aeolism: Latin as a Dialect of Greek', *The Classical Journal*, 102 (2006), 115–144

Suerbaum, Werner (ed.), *Die archaische Literatur: Von den Anfängen bis Sullas Tod. Die vorliterarische Periode und die Zeit von 240 bis 78 v. Chr.* Handbuch der Lateinischen Literatur 1, Munich, 2002

Temkin, Owsei, *Hippocrates in a World of Pagans and Christians*, Baltimore, 1991

Thomas, Jean-François, 'De la paix des armes à la tranquillité de l'âme', *Revue des Études Latines*, 89 (2011), 56–75

Traina, Alfonso, 'Le traduzioni', in *Lo spazio letterario di Roma antica Vol. II*, Rome, 1989, 93–123

Vogel, Christian, *Boethius' Übersetzungsprojekt: Philosophische Grundlagen und didaktische Methoden eines spätantiken Wissenstransfers*, Wiesbaden, 2016

Werner, Jürgen, 'Zur Fremdsprachenproblematik in der griechisch-römischen Antike', in Carl Werner Müller, Kurt Sier, and Jürgen Werner (eds), *Zum Umgang mit fremden Sprachen in der griechisch-römischen Antike*, Stuttgart, 1992, 1–20

Yonge, Charles Duke, *The Academic Questions: Treatise De finibus and Tusculan Disputations of M.T. Cicero*, London, 1891

CHAPTER 3

Translations from Greek into Middle Persian as Repatriated Knowledge

Mohsen Zakeri

Part One: Essay

Overview

I. *Introduction: The Sasanians and the Culture of Translation*

The translation of Greek philosophical and scientific literature into Arabic, whether directly or by way of Syriac intermediaries, went hand in hand with translations from the Persian. The history of translation from Greek into Arabic, directly or via Syriac, has been rehearsed frequently in modern research, and its colourful record is easily accessible to everyone.[1] The study of Middle Persian (MP) translations into Arabic, on the other hand, has been neglected.[2] What is missing, or not adequately taken into account, is the fact that a stream of Greek works had been translated into MP under the Sasanians, and that after the collapse of the empire, these Persian adaptations were among the first texts to be translated into Arabic.[3] Thus, it was not without reason that Julius Ruska argued that a large part of the scientific tradition of early Islam was not mediated by the Syrians, but rather through eastern Iranian and Central Asian cultural institutions.[4] Agreeing with Ruska, Franz Rosenthal suggested that the

---

1 For a synoptic treatment of this chapter, see Gutas, *Greek Thought, Arabic Culture*.
2 Kunitzsch, 'Zum "liber hermetis"', 130.
3 Gutas, 'The Presence of Antiquity'.
4 Ruska, *Tabula Smaragdina*, 46–47.

 | DOI:10.1163/9789004472648_004

first information on Aristotle may have reached an Arabic-speaking audience through Pahlavi translations or abridgements:[5] '[…] translations from the Persian and from some Indian scientific material are among the earliest known to us. Even the Muslims' first acquaintance with Aristotle was attributable to the interest of the Sasanians of the sixth century in Greek learning.'[6] In other words, early Muslims first came into contact with Greek science through translations from MP and Syriac, a long time before they turned their attention to the Greek originals.

To study the history of translation in the early Islamic period, a key issue to be tackled at the outset is the circumstances under which the movement for translation into Arabic started. For a long time, the general tendency has been to associate the movement with the rise of the ʿAbbāsids. As a result of this preconception, references to literary and scientific activities under the Umayyads have been systematically ignored or discounted.[7] The exclusion of the century of Umayyad rule from the history of translation in Iran and Mesopotamia has severed the continuous thread that I believe existed in this movement from the Sasanian period to early ʿAbbāsid times. In the course of the present essay, I hope to modify the general perception, which has inadvertently put more weight on the ʿAbbāsids, showing that not only were translations from Greek, Sanskrit, and other languages into MP and Syriac (the language of instruction at western Iranian schools, a fact rarely taken into consideration)[8] carried on in diverse fields before the coming of Islam, but that after a temporary slowdown in pace owing to the havoc caused by the early Arab conquests, the movement continued on its path undeterred: the appearance of new political rulers did not entail the immediate change of ancient institutions. The coming to power of the ʿAbbāsids and their Iranian supporters only accelerated an ongoing practice, creating new centres of scholarship, and the arrival of cheap paper put it in full gear. A new development in the field of translation was the rise of Arabic, which was now exploring the possibility of becoming a universal language and a *lingua scientiae* in the burgeoning new world culture.

Throughout the Sasanian era, interest in foreign contributions to science and in intellectual and technical achievements grew continually, reaching its culmination at the time of the philosopher King Ḫusraw I Anūširwān (Anōšag Rowān, r. AD 531–578). Search for medical, philosophical, and scientific knowledge received political support from Sasanian kings and religious endorsement

5 Rosenthal, *Das Fortleben der Antike im Islam*, 13.
6 Rosenthal, *The Classical Heritage in Islam*, 4.
7 Cf. Ullmann, 'Ḫālid ibn Yazīd und die Alchemie'.
8 See Takahashi, 'Syriac as a Vehicle'.

from a Zoroastrian clerical establishment eager to spread its influence hand in hand with those in power. Religion and politics forged a union that, despite occasional disruptions and turbulent relations, would last to the very end of the dynasty's rule. Intensive propaganda campaigns promulgated the idea that science in all its branches was originally developed in Persia and spread from there to the rest of the world as a result of natural and human disasters (Alexander the Great being the protagonist in the latter), such that seeking and recovering it was both a national and religious duty. That is why we constantly read about Sasanian monarchs commissioning delegations to Greece, India, or Egypt in search of books, and inviting renowned scientists and doctors to the court (see II.1–3). Acquisition of foreign learning, especially Greek and Indian, occupied first position on the imperial agenda.

The inherent benefit in upholding and popularizing an Iran-centric position with respect to the history of science (the idea that all science originated in Persia) is the spread of an open and positive attitude towards the acquisition of other peoples' knowledge. Translation of foreign scientific works is not only a necessary enterprise, but a praiseworthy one: it reclaims and repatriates a lost native heritage. As a result of this line of thinking, translation—presupposing, as it does, the learning of languages and the training of translators—became a kind of 'state-sponsored' endeavour.

When exploring the literature translated under the Sasanians, it must be remembered that theirs was a huge empire: they dominated a vast, multicultural territory comprising many races, religions, and languages. Under these circumstances, translation for the transmission of knowledge was crucial to the survival of the empire. One may recall that several of the scanty samples of the Old Iranian Achaemenid inscriptions are trilingual (written in Old Persian, Elamite, and Neo-Babylonian), and even Šāpūr I (r. AD 241–272), when celebrating his victory over the Romans, created a trilingual inscription at Naqš-e Rostam near Persepolis written in MP, Parthian, and Greek.[9] Literary, religious, and scientific writings in local languages, and translations from one language into another were more often than not encouraged and patronized by kings. Writing a comprehensive history of translation under the Sasanians would require that none of the groups actively participating in this long-term process of cultural transfer and amalgamation be ignored.

Thus, translations done in any language employed in territories under Sasanian rule need to be considered in order to gain a realistic appreciation of the period's colourful intellectual history. As we are limiting our documentation here of necessity to translations from Greek into Zoroastrian Middle Persian,

9 See Huyse, *Royal Inscriptions*.

we are leaving aside several other culturally significant Persian languages and dialects, as well as translation activities in other languages carried out by members of the many religious communities in Iran. To give due consideration to the cultural diversity of the Iranian people, a more comprehensive survey of the field would include the following: the Manichaeans, who are famed for translating their doctrinal texts and commentaries into various languages (e.g., Mānī's Syriac *The Book of the Giants* was translated into MP and subsequently into Arabic, as *al-Ǧabābira*);[10] the Armenians, who, to satisfy the demand of the king of kings for textual evidence of their ancient Iranian nobility, supplied Agatangełos's *History of the Armenians*, which was promptly translated at court into MP;[11] the Christians of the Persian Church, who translated many books from Syriac into MP and vice versa (e.g., Maʿnā of Beth Ardašīr in Persis, who translated the Greek commentaries of Theodore of Mopsuestia, and a number of Persian books into Syriac,[12] and Catholicos Aqāq, bishop of Amida (d. 495/6), who translated an account of Christian beliefs into MP for the Sasanian ruler Qubād (Kawād I), in order to help him understand Christianity);[13] the Jews of Babylonia (e.g., the *Talmud Bavli*, which contains many Iranian features, was composed in Hebrew and Aramaic in a Persian/Sasanian/Zoroastrian cultural milieu, and shows that Babylonian rabbis had access to the Platonic corpus and other Greek texts);[14] and the Buddhists of north-east Iran (e.g., local scholars in Khotan in the fifth–seventh centuries, who were actively engaged in making translations of Buddhist texts from Sanskrit into Khotanese, an eastern Iranian language). Indian medical texts were studied and rendered with enthusiasm into the local languages of Khotan and Kuci (Kucha), as well as into Uigur Turkish and Tibetan (V.1).[15] The late antique Sasanian court was indeed a multilingual, intellectually open one, proud of its cosmopolitanism, hosting as it did Roman philosophers, Indian sages, and Jewish and Christian religious leaders from within and outside the empire.[16]

A formidable problem facing research in the area of pre-Islamic translation in Iran is that not only are the MP originals of the majority of translations lost, so are their early Arabic renderings: our limited knowledge of them is gained from indirect evidence based chiefly on bibliographical references and occasional

10 Henning, 'The Book of the Giants', 55; and Schwartz, 'Qumran, Turfan, Arabic Magic', 231.
11 Payne, *A State of Mixture*, 155.
12 Baumstark, *Geschichte*, 101–102, 105.
13 Sims-Williams, 'Christianity', *EIr*, V, 534.
14 Cf. Boyarin, *Socrates and the Fat Rabbis*, 134.
15 Bailey, 'The Preface to the Siddhasāra-śāstra', 32–33.
16 Payne, *A State of Mixture*, 156.

quotations in secondary and tertiary literature, and the doxographical works of later Muslim biographers, historians, and commentators. Because of this—the scarcity of extant MP texts and the certainty of translation as a cultural force in Iran under the Sasanians—my interpretation of the scant evidence will be maximalist, in an effort to present a narrative that is historically more accurate and without major discontinuities.

II. *Traditional Accounts of the History of Science*

In his *al-Fihrist* (*The Catalogue*), one of our indispensable sources, the tenth-century historian of classical Arabic literature Muḥammad ibn Isḥāq ibn al-Nadīm (d. 380/990) presents a chapter on the history of science and philosophy that is based primarily on the Iranian tradition of the history of science, which he continues where that tradition leaves off. Ibn al-Nadīm begins the seventh chapter of his book with an excursus 'on the rise of science in Islam' and reproduces a long passage from a book by Abū Sahl ibn Nawbaḫt known by the corrupt title *al-Nahmuṭān*[17] (see further below) as a prolegomenon to his comments on the translation into Arabic (*al-Fihrist*, 299–301) (II.3). This is followed by Abū Maʿšar al-Balḫī's account of the discovery of the ancient astrological tables, *Zīǧ al-hazārāt*, in the citadel of Ǧayy in Isfahan (V.6); a brief comment on the history of philosophy in the West and its restriction by Christians; the war of the Roman Emperor Julian (d. 363) against the Sasanian King Šāpūr II, who had a high regard for translations of Greek works into MP (IV.6); the Umayyad prince Ḫālid ibn Yazīd's interest in alchemy and his ordering of books to be translated for him into Arabic; measures taken to translate Greek and Persian administrative-financial *dīwāns*; the caliph al-Maʾmūn's dream, in which he converses with Aristotle, who encourages him to sponsor translations of Greek works; and finally the various delegations sent to Byzantium from Baghdad in search of Greek manuscripts.[18]

After the brief remark on Ḫālid ibn Yazīd, who, in Ibn al-Nadīm's view, was the first to order translations of foreign works on alchemy into Arabic, he continues with the historically more significant conversion of the *dīwān*s, thereby treating the translations of scientific works and the *dīwān*s as related. However, in opening his narrative by excerpting *al-Nahmuṭān*, Ibn al-Nadīm not only showed his acumen in choosing the oldest available history of science in Ara-

17 *Nahmuṭān* is the form in which the title is transmitted in *al-Fihrist*. Kevin van Bladel suggests that it is a mistranscription of the original MP word *tohmagān*, meaning genethlialogy, which is properly rendered in Arabic by Abū Sahl as *mawālīd*, nativities or birth horoscopes ('The Arabic History', 41–42, n. 1). This remains only speculation.

18 Ibn al-Nadīm, *al-Fihrist*, 301–304.

bic, but also signalled the historiographical tradition he was following.[19] His narration, which continues with al-Ma'mūn's dream and conversation with Aristotle, ties the half-Iranian caliph neatly to Greek philosophers and the Persian kings Ardašīr, Šāpūr, and Ḫusraw Anūširwān, who are featured in *al-Nahmuṭān*.[20]

In addition to his introduction to the history of science, Ibn al-Nadīm compiled a list of men who were active as professional or occasional translators from other languages into MP and thence into Arabic. To this list he adjoined the titles of a series of books translated into Persian, the translators of which he could no longer identify. All in all, he named some twenty translators from MP and over one hundred mostly anonymous book titles on diverse subjects, mainly in the fields of history and belles-lettres. Though the single most extensive and valuable list of its kind, Ibn al-Nadīm's inventory is far from being complete, and the data he provided are not always accurate.

The field of translation from MP into Arabic was vast and multifaceted. It began in the early days of Islam and continued unbroken for a long time. For the classical period until the end of tenth century, when translation activity diminished, we know of at least forty translators, and over 300 titles of translated books.[21] A handful of these translator-authors, such as Ibn al-Muqaffaʿ, Abān al-Lāḥiqī, and Sahl ibn Hārūn, are relatively well known, but the majority remain obscure. Not much is known, for example, about Muḥammad ibn Ḫalaf ibn al-Marzbān (b. c. 210/825, d. 309/921), who is said to have translated more

---

19 Ibn al-Nadīm's contribution to the history of science from antiquity to Islam has been well researched in the last two centuries. More recently, it was the focus of the first two chapters of George Saliba's book *Islamic Science* (2007), in which he repositions the ideas presented by Franz Rosenthal, Dimitri Gutas, and others concerning the causes and motivation behind the ground-breaking Arabic translation movement in the early ʿAbbāsid period. Saliba's translation of the relevant passages is fluent and his erudite commentary at times refreshing, but his attempt at rewriting what he terms the 'traditional narrative' of this history—such as his argument that the cultures surrounding those of early Islam had not been participating in the production of scientific and philosophical thought for centuries before Islam (*Islamic Science*, 5, 258 n. 3)—falters by discounting the historical evidence I have presented here. With this and similar blanket generalizations, Saliba tends to disqualify translation as a contributing factor in the progression of science and thus plays down the scientific and intellectual movement in Syriac and Persian schools that flourished under the Sasanians from at least the time of Khusraw Anūširwān in the sixth century.

20 When asked about the men who have had the greatest impact on the course of history, al-Ma'mūn counted five: Alexander, Ardašīr, Bahrām Gor, Ḫusraw Anūširwān, and Abū Muslim; al-Iskāfī, *Luṭf al-tadbīr*, 13–15.

21 Muḥammadī, *al-Mutarǧimūn wa-l-naqalah*, 193–243; de Fouchécour, *Moralia*, 17–120; and Tafażżolī, *Tārīkh-e Adabiyyāt-e Īrān*, 202–314.

than fifty books from MP into Arabic.[22] Here, translators of MP texts into Arabic and the issues related to their work will be mentioned only in so far as they can shed light on the transmission of Greek classics.[23]

Based on textual evidence uncritically compiled by Ibn al-Nadīm and a few other scattered reports, the old Iranian concept of the history of science consists of comments on the beginnings of written culture in pre-historic times, the genesis and accumulation of knowledge in Iran, its dispersal, and its reclamation. It can be outlined as follows: all sciences originated and developed in Iran in antediluvian times; the prophet Zoroaster forecast the coming of a deluge that would devastate the world; measures were taken to preserve the people's ancient heritage, such as the building of impregnable repositories; accumulated knowledge was scattered to different lands (II.2); there was piecemeal recovery of dispersed learning after the deluge; a second 'deluge', in the form of Alexander's invasion of Iran, led to the near total destruction and dispersal of all sacred and profane Iranian wisdom once again; afterwards there were attempts to identify and reclaim the cultural remains of the past, oral and written, from all corners of the inhabited world (II.2); Parthian kings initiated the recuperation process after Alexander (IV.2, IV.1); the Sasanians continued this endeavour enthusiastically; and translation of salvaged wisdom from foreign languages into MP became routine. Translation simply meant repatriating native sciences (IV.3, IV.4).

The sources that tell the above Persian history of science can be differentiated in three narrative traditions, each with its particular emphases: religious, scientific, and historical/literary. The distinctive features in these accounts are fashioned by the tastes and inclinations of those relating the tradition. They share the succession of the events, persons involved, and a few other recurring characteristics, but differ in perspective. A brief outline of these perspectives follows here.

(a) The Religious Tradition: History of the Zoroastrian Holy Scriptures

The indigenous history of the transmission of the Zoroastrian sacred texts is told in several ninth–tenth-century Pahlavi books—the *Dēnkard* (*Acts of Religion*), *Ardā Wīrāz Nāmak*, *Bundahišn*, *Zand ī Wahman Yasn*, *Šahrestānīhā*, *Dēnīg Vizārgird/Vizārgird ī Dēnīg*, etc.—as well as in Arabic (II.7, III.6), and New Persian. Without being consistent, these reports construct a primordial Zoroastrian past; they relate when the books (*Avesta*, *Zand*, *Dēnkard*) were

22 Zakeri, 'Muḥammad b. Khalaf b. al-Marzbān', 343–362.
23 For a brief survey, see Zakeri, 'Translations from Persian'.

first recorded in writing, what they included, places where copies of them were deposited, and what happened to them. This narrative line is more pervasive than those of the other two traditions and sets the historical pattern for them. The longest account is that of the MP *Dēnkard* (II.1). Since the variations and discrepancies in the Zoroastrian tradition are well known in modern scholarship and have been studied frequently,[24] there is no need to reproduce them here. I limit my comments only to the evidence of the *Dēnkard*. Ibn al-Nadīm did not take the contents of the *Dēnkard* into account, perhaps taking Abū Sahl's *al-Nahmuṭān* as sufficient.

When posing the question 'Why translate?' in the context of MP literature, the *Dēnkard* is the principal source to which we are inescapably led. Narrating the longest version of the traditional history of the Mazdayasnian sacred books, the *Dēnkard* cites Indian and Greek works that were translated into Persian at different times and occasionally offers grounds for their having been translated. Although the last redaction of the *Dēnkard* dates to the ninth century, it certainly contains much older material, including a few extracts (410–416) and references that highlight the religious attitudes of several rulers, from the legendary Wištāsp to the Sasanian Ḫusraw I, and how these attitudes contributed to the spread of scientific literature in Persia. This is a particularly illuminating document for the relationship between spiritual and temporal power during the apogee and decline of Sasanian civilization (see further below II.1, II.2).

The essence of the *Dēnkard* extract here (II.1) is this: Ahura Mazdā revealed the contents of the *Avesta* to Zoroaster, who later presented them to the Kayānian King Wištāsp. Wištāsp, or according to another tradition, Dārā, son of Dārā, had two copies of them made, one of which was deposited in the Royal Treasury at Iṣṭaḫr (Staxr), the other in the 'fortress of archives' (*Dež ī nebešt*).[25] At the time of Alexander's conquest, the copy at *Dež ī nebešt* was destroyed by fire, but the other copy was translated into Greek, along with Persian scientific texts (on astronomy, medicine, etc.) that they found beneficial. As a consequence of the conquest, the sacred and profane wisdom and wit of Iran disappeared from the country for a long time in its entirety. The first attempt at restoring the *Avesta* was made under the Arsacids, when King Valaxš (Vologeses) began to reclaim the scattered fragments, supplementing them with current and pertinent oral traditions. This reclamation continued in four successive phases under the Sasanians: Ardašīr I (226–241) ordered the high priest Tōsar to complete the work of collecting the remaining fragments; his son

24 See for example Gignoux, 'La Démonisation d' Alexandre', 87–97.

25 On this, see Shaki, 'Dež ī Nebešt', *EIr*, VII, 348–350.

Šāpūr I (241–272) initiated a search for the scientific documents that had been dispersed by the Greek invaders, from Greece, India, and other countries, and had them added to the *Avesta*; under Šāpūr II (309–379), Ādurbād ī Mahraspandān made a general revision of the canon and ensured its orthodox character against divergent sectarian opinions by submitting himself successfully to an ordeal by fire at the time of a general controversy; finally, a rearrangement of the original texts and a revision of the Pahlavi translations took place under the supervision of the priest Veh-Šāpūr at the time of Ḫusraw I.[26]

Having tied scientific literature closely to the sacred texts, the *Dēnkard* extract given here provides a kind of religiously sanctioned endorsement of acquiring and translating works on the sciences. No limit was set on what was to be translated, or from where. The texts had only to be in harmony with the *Avesta*, or not contradict it. This reverence for the written word and search for knowledge gained momentum on occasion in epochs plagued with socio-political upheavals and religious controversies in which the Zoroastrian faith was deeply involved or under attack. In Sasanian times, the chief priest, following the example of the prophet Zoroaster, at each phase of reform ascends to heaven (*Kartīr*), or is sent to an extra-terrestrial soul journey (*Ardā Wīrāz*), or undergoes an ordeal (*Ādurbād*) in order to ascertain the truth of religion and its ordinances. Such measures are normally undertaken at times when the religious establishment is shaken by some internal or external threat and is in need of reassurance.

(b) The Scientific Tradition: Astronomy/Astrology, Medicine, Alchemy, Geography, Philosophy, Geometry, etc.

The scientific tradition follows that of the religious scriptures, but has its own distinctive features. It is recorded primarily in two complementary Arabic texts of the eighth and ninth centuries, both preserved in Ibn al-Nadīm's *al-Fihrist*.

The first is the so-called *Kitāb al-nahmuṭān fī l-mawālīd* (*Book of the Nahmuṭān*[27] *on Nativities*) by Abū Sahl ibn Nawbaḫt (fl. c. 153–193/770–809) (II.3). The son of Nawbaḫt, Abū Sahl was a Zoroastrian astrologer who converted to Islam at the request of the caliph al-Manṣūr (r. 136–158/754–775) and helped cast the horoscope for the foundation of Baghdad in 145/762. Under Hārūn al-Rašīd (r. 170–193/786–809), Abū Sahl took up a position in the royal library, the *Ḫizānat al-ḥikma* ('Depository of Wisdom'),[28] where he was employed to

26 Adopted from Kellens, 'Avesta. i. Survey of the history and contents of the book', *EIr*, III, 35–44.

27 See n. 17 for the reading of this MP word.

28 Gutas and van Bladel, 'Bayt al-Ḥikma'.

translate Pahlavi books into Arabic. He probably died before the end of Hārūn al-Rašīd's caliphate.[29]

Abū Sahl's account acknowledges the existence of Babylonian science, and that the Egyptians and the Indians borrowed it from the Babylonians, who then lose their wisdom and fall into a period of obscurity. At the time of the Persian King Ǧam, son of Awanǧhān/Awiǧhān (i.e., Ǧamšīd, son of Ṭahmūraṯ, son of Hūšang), the Babylonians became aware of their glorious past and began to reclaim their ancient knowledge on cosmology, the sciences of the stars, drugs, medication, and so forth (from then on Babylonian and Persian culture become indistinguishable from each other). This development continued to the time of Ḍaḥḥāk, son of Qayy. After conquering Iran, Alexander had whatever he needed of the ancient sciences of Persia translated into Greek and sent to Greece (Egypt, or Cyprus), and burnt the originals. The second period of active search for antique science began with the Sasanians Ardašīr I and his son Šāpūr I and continued to the time of Ḫusraw Anūširwān. There is no mention of the Parthians in Abū Sahl's account.

In a detailed study of the contents and sources of this document, Kevin van Bladel identifies parallels between Ibn Nawbaḫt's *Book of the Nahmuṭān* (*Tohmagān*?), as quoted in Ibn al-Nadīm's *Catalogue*, and extant Zoroastrian MP writings, particularly the ninth-century *Dēnkard*. He writes: 'When some sentences of Abū Sahl's Arabic are compared word for word with this passage in the *Dēnkard*, the verbal correspondences are exact enough to indicate that Abū Sahl had before him the same MP text. Abū Sahl was paraphrasing it closely.'[30]

The second Arabic text highlighted here is Abū Ma'šar's *Iḫtilāf al-zīǧāt* (*Book on the Variations among Zīǧes*)[31] on the Persian history of science, from which Ḥamza al-Iṣfahānī quotes an anecdote;[32] the anecdote is repeated by Ibn al-Nadīm without Ḥamzah's name.[33]

As far as Abū Ma'šar Ǧa'far ibn Muḥammad ibn 'Umar al-Balḫī (in Latin Albumasar, b. 171/787, d. c. 272/886) is concerned, recurrent reports on his early study of *ḥadīṯ*, or his first encounter with astrology under the tutorship of the philosopher al-Kindī, belong to the realm of legend. His reliance was clearly on earlier Persian texts and their Arabic translations, including *The Book of*

29 Pingree, 'Abu Sahl Nawbaḵt', *EIr*, I, 369.

30 Van Bladel, 'The Arabic History', 57–58.

31 For further comments on the *Iḫtilāf al-zīǧāt*, see Pingree, *The Thousands*, 128–130.

32 Ḥamza al-Iṣfahānī, *Ta'rīḫ sinī*, 151–153.

33 Ibn al-Nadīm, *al-Fihrist*, 301–302, tr. Dodge, *The Fihrist*, 576–578; and Saliba, *Islamic Science*, 36–37. Šahrazūrī (*Ta'rīḫ al-ḥukamā'*, 50–54) and Iškawarī (*Maḥbūb al-qulūb*, I, 128–129) took over the texts of Abū Sahl and Abū Ma'šar from Ibn al-Nadīm (V.6).

Ǧāmāsp, The Book of Zoroaster on the Paranatellonta of the Various Degrees of the Firmament (*Fī ṣuwar daraǧāt al-falak*), *The Book of the Commentary by the Persians*, and *The Book of the 360 Paranatellonta by the Persians*.[34] Ibn al-Nadīm enumerates thirty-six book titles by Abū Maʿšar; a few have survived. A number of these were translated into New Persian and Byzantine Greek, and by far the largest part into Latin by John of Seville, Hermann of Carinthia, and Hugo of Santalla. As a result, these works have had an intellectual impact the extent of which is only gradually becoming evident.[35]

Abū Maʿšar's history of science, as related by Ḥamza al-Iṣfahānī, was important to Ibn al-Nadīm, for it began where that of Abū Sahl ibn Nawbaḫt left off and brought the ancient history of sciences a step forward, up to the Islamic period. Abū Maʿšar's book *Kitāb iḫtilāf al-zīǧāt* (*Distinction of Zīǧ*es), like Abū Sahl's *Book of the Nahmuṭān*, is lost, except for the short fragment preserved by Ḥamzah and Ibn al-Nadīm. It tells of the Persian kings' continuous preoccupation with the sciences and their successive attempts at securing their preservation. They encouraged writers to write their texts on the most durable of all writing materials, impervious birch bark, and to seek out the places most secure from natural disaster and enemy attack and build depositories there. The legendary citadel of Sārūyah at Ǧayy, Iṣfahān, was erected for this purpose by the no less legendary King Tahmūraṯ, who had been warned of the deluge that would devastate the world. It was in the fortress of Ǧayy that lovers of astrology/astronomy discovered the authoritative ancient Persian book *Hazārāt* (*The Thousands*), the most trusted text on the subject. The topos of the ancient, antediluvian origins of astrology and the secret sciences plays a crucial role here. They are given the same status as the Zoroastrian holy scriptures; see section (a) above.

(c) The Historical/Literary Tradition: The History of Iranian Dynasties and Persian Histories as Sources of Political Ethics

In the introduction to his book *Malāmiḥ yūnānīya fī l-adab al-ʿarabī* (*Greek Features in Arabic Literature*), Iḥsān ʿAbbās discusses the rivalry prevalent among the supporters of Iranian culture and pro-Greek factions in the ninth century.[36] He points out that Iranian supremacy in political philosophy and literature had reached the point where proponents of Greek culture, such as Ibn al-Dāya, were forced to create and present samples of Greek political wis-

34 On these, see Rosenthal, 'x. A list of astrological works', 454–456.

35 For details, see Pingree, *The Thousands*; and Ullmann, *Die Natur- und Geheimwissenschaften*, 316–325.

36 *Malāmiḥ yūnānīyah*, 4–5.

dom. Patricia Crone writes: 'By the tenth century, Persians were associated with governance to such an extent that the Egyptian Ibn al-Dāya felt compelled to defend the Greeks against charges of inferiority in this field.'[37] A reconciliatory gesture was offered by the tenth-century historian al-Mas'ūdī. He speaks of 'seven ancient nations' (the Persians, Chaldeans, Greeks, Egyptians, Turks, Indians, and Chinese) who had a certain body of wisdom in common. Each of these nations excelled in a particular branch of science, so all contributed to the development and flourishing of human culture. The primary contribution of the Persians was in the art of government and statecraft. As Khalidi puts it: 'This was both a theoretical and a practical excellence, for their pre-eminence in government stemmed from the greatness of their empire as well as from the political wisdom which sustained that greatness.'[38] In this period, politics was prerogative of the Iranians, philosophy that of the Greeks, and poetry that of the Arabs.

Acknowledging that the Persians' excellence lay in the field of statecraft neither originated with al-Mas'ūdī nor stopped with him. Earlier Muslim historians and literati had made similar observations: Persians surpass others in kingship (al-Ǧāḥiẓ, *al-Rasā'il*, I, 67), and Ardašīr is a model ruler emulated by other kings (al-Dīnawarī, *Aḫbār*, 47); a similar verdict on Persian monarchy is passed by al-Ṭabarī (I, 148). The historian Abū l-Ḥasan l-Madā'inī (d. 225/839) reports: 'Sulaymān ibn 'Abd al-Malik [Umayyad caliph 96–99/715–717] said: "I am amazed at these Persians. They ruled a thousand years without needing us [i.e., the Arabs] in their politics for a single hour; we have ruled a hundred years, but cannot do without them for an hour."'[39] The eleventh-century historian Ṣā'id al-Andalusī (d. 462/1070) commented: 'The magnificent quality of the Persian kings, for which they became famous, is their superior political and administrative aptitude. This is especially true in the case of the kings of Banū Sāsān. Among them there were kings who surpassed all others in nobility of conduct, in ability to govern and exercise authority, as well as in their widely recognized fame.'[40]

The high regard or admiration for Persian political culture in classical Islamic sources has not escaped the attention of modern researchers. In a brief survey of the relative impact of Greek and Persian culture on the development

37 Crone, *God's Rule*, 151.

38 Khalidi, *Islamic Historiography*, 89, 91, citing al-Mas'ūdī, *Tanbīh*, 92.

39 Al-Zubayr ibn Bakkār, *al-Aḫbār al-Muwaffaqiyāt*, 163. To this al-Rāġib al-Iṣfahānī (*Muḥāḍarāt al-udabā'*, I, 349) adds: 'We are in need of them even in learning our own language!'.

40 Ṣā'id, *Ṭabaqāt*, 16.

of early Arabic literature, Aḥmad Amīn (writing in 1928) makes a pointed observation. He says it is extremely difficult to find any Greek influence on Arabic poetry, or to find an individual of Greek or Roman origin who learned Arabic sufficiently well to compose poetry in it, whereas many native Persians emerged as famous Arabic poets. Moreover, Muslim historians adopted their historiographical method from the Iranians, not from the Greeks. What hampered the impact of Greek culture on Arab-Muslim life, he says, was the lack of sufficient knowledge of Greek literature.[41] In the opinion of Montgomery Watt (1968), 'What is interesting and significant, however, is the contrast between the readiness of the Muslims to accept the Persian tradition and their refusal to borrow—or at least to acknowledge any borrowing—from the classical Greek tradition or the Roman and Byzantine tradition. Even those who accepted Greek philosophy paid relatively little attention to the political thought of Plato and Aristotle. A few works, like *al-Madīna al-fāḍila* of al-Fārābī (d. 950) [...] show Greek influence, but they did not appreciably affect the main stream of Islamic political activity. Likewise the Roman-Byzantine empire received cursory treatment from the Muslim historians. The Muslims regarded this as something alien and inferior.'[42]

Persian know-how in politics and the theoretical underpinnings of the Sasanian concept of kingship were transferred to the Muslims primarily via the medium of historical writings in translation. It is primarily in this literature that the Sasanian kings are portrayed as models of just and benign political rule. The pre-Islamic Iranian tradition of historiography was more concerned with the principles of proper rule and maintenance of social justice than with recording historical events, causes, and effects. This approach to history continued into the early Islamic period. When a prose *Šāhnāma* (*Book of Lords*) was compiled in Persian by the order of Abū Manṣūr, son of ʿAbd al-Razzāq of Ṭūs, in the first half of the tenth century, the reason for it was explained as follows: 'so that men of knowledge may look into it and find in it all about the culture of kings, noblemen, and sages, the royal arrangements, their nature and behavior, good institutions, justice and judicial norms, decisions and administration, the military organization, the art of war, storming expeditions and punitive campaigns, and taking the enemy by surprise as well as their match-making and ways of respecting honor.'[43] The readers of this work will learn from it the art

41 Amīn, *Faǧr al-Islām*, 136.

42 Watt, *Islamic Political Thought*, 80–81. On this subject, see Di Branco, *Storie arabe di Greci e di Romani*.

43 Minorsky, 'The Earlier Preface', 267–268.

of statecraft and 'will be able to get on with anyone in administration'; they will also get pleasure from reading its stories and about matters 'suitable to [their] strivings'.[44]

Iranian historiography follows roughly the same historical periodization, events, and characters as the excerpts given above from Abū Sahl ibn Nawbaḫt and Abū Maʿšar (IV.1). An echo of this history of science can still be heard in Ibn al-Nadīm's account of the origins of story-telling and story-books, where once again, the narrative revolves around the rivalry between the Greeks and ancient Persians for primacy in the creation of the first political fables and entertaining stories (*al-Fihrist*, 363–364). We follow a combination of the above three traditions of history of science in the following section.

III. *Persian Rulers: Preserving and Disseminating Knowledge*

As we have seen, the traditional history of Zoroastrian sacred scriptures and the sciences they endorsed begins with the primordial King Wištāsp and stretches all the way to the Islamic period. From Wištāsp, the first recipient of the faith (in some versions replaced by Dārā, son of Dārā), who is accompanied by the first Zoroastrian sage Ǧāmāsp, the story jumps to one Parthian Walaxš, who is not accompanied by a priest or sage (II.1), then moves directly to a few Sasanian emperors and their accompanying *mobeds*: Ardašīr/Tansar, Šāpūr I/Kartīr, Šāpūr II/Ādurbād-Ardā Wīrāz, and Ḫusraw Anūširwān/Veh-Šāpūr. Other kings feature rarely in this account.

Sasanian historiography generally does not have anything good to say about the Parthians. Thus the fact that an author who lived in the Sasanian period or immediately after it credits the Arsacid King Walaxš with a meritorious role in the history of the holy scriptures (II.1) is probably the result of a genuine tradition,[45] and by inserting this king in their copies of the cherished holy text, the religious scribes show that it is worthy of credence. This juncture in the narrative is important, in that it ascribes the first attempt at recording the *Avesta* or the *Zand* in written form—a matter that has been heavily disputed in modern discourse for a long time—to the Parthian period. These Parthian writings might have been indeed the Zoroastrian books that Mani cites in his *Kephalaia*, 7, 27: 'Zoroaster came to King Hystaspes and preached in Persia, and selected just and righteous disciples [...] However, he did not write any books. But his disciples after his death remembered (his words) and *wrote the books which they read to-day*.'[46]

---

44 Shahbazi, 'Historiography', 329.

45 Cf. Pines, 'A Parallel', 42.

46 Henning, 'Disintegration', 47.

In addition to the Parthian Walaxš, al-Ṯaʿālibī recorded the story of another Arsacid king, who transferred books to Iran after defeating the Greeks (IV.2). In general, al-Ṯaʿālibī depicts the Parthians as just and valiant. Of Aqfūr Šāh (Pakores/Pacorus, son of Orodes)[47] he says that he recovered the ancient Persian *kāwiyānī* banner and took good care of it. Seeking revenge for Dārā's blood, he fought the Romans and brought back to Iran in triumph the sources on medicine, astronomy, and philosophy that Alexander had carried away from Iran.[48] This unique reference, missing from common accounts, is another building block in the history of science tradition, discarded apparently by the pro-Sasanian factions when elimination of traces of the Parthians and their culture was on the agenda.

Under the Seleucids and for a period of some one hundred years, Greek became the dominant language in the former Achaemenid territories, where the Persian aristocracy adopted it for communication and in treaties, inscriptions, and legends on coins.[49] Soon after the conquest of Persia, on the orders of Alexander, the children of the nobility were encouraged to learn Greek. Plutarch probably overstates the case when he affirms that 'Homer was commonly read and the children of the Persians, of the Susanians, and of the Gedrusians [Baluchis] learned to chant the tragedies of Sophocles and Euripides.'[50] Olbrycht observes that the Epigones—the elite Iranian soldiers in Alexander's army—were taught Greek.[51] 'Hence it was Greek, the *lingua franca* of the Mediterranean that became the chief language used in western Asia, alongside Aramaic under Alexander and the Seleucids.'[52] Greek, Parthian, and Aramaic were still used concurrently in Seleucid and Arsacid times.

When the Parthians came to power and replaced the Seleucids, Greek continued to be used as a *lingua franca* until at least the first century AD. Parthians enjoyed the vivacity of Greek culture, and many of them supported it passionately. As an example of this, Olbrycht quotes Flavius Philostratus, who reports the following exchange between Apollonius of Tyana and the Parthian King Vardanes (r. AD c. 40–47):

47 An attribute of Aškān; in some listings he is the first Parthian. Cf. Ṣafā, *Ḥamāsasurāʾī*, 550–551.

48 Yarshater, 'Iranian National History', 475.

49 Klima, 'The Arsacids', 25.

50 Plutarch, *On the Fortune or the Virtue of Alexander*, 1.5, quoted in Boyce and Grenet, *A History of Zoroastrianism*, III, 56.

51 Plutarch, *Alexander*, 47.3, 71.1; and Diodorus, 17.108.1–2.

52 Olbrycht, 'Parthians, Greek Culture', 139–140.

> 'O king, do you know the Greek tongue thoroughly, or have you a smattering of it perhaps, in order to be able to express yourself and appear polite in case any Greek arrives?' 'I know it thoroughly,' replied the king, 'as well as I do my native language.'[53]

Plutarch's remark that the Parthian King Orodes II was 'well acquainted both with the Greek language and literature'[54] is a familiar citation. Orodes's ally, Artavasdes, king of Armenia, wrote tragedies in Greek. The distinguished general Surena, described by Plutarch as a true Parthian with 'Scythian' manners and looks, must have been familiar with Greek literature, since he learned that there were, in the Roman camp at Carrhae, obscene Greek books, which the Parthians discovered when they took the camp; and he had one of these works, the *Milesiaca* by Aristides, displayed during his triumphal march in Seleucia-Ctesiphon on the Tigris.[55] Plutarch made a sarcastic remark that 'many of the royal line of the Arsacidae were sprung from Milesian and Ionian courtesans' [*Crassus*, 32], which was true of Vologases I, who was the son of a Greek concubine.'[56] Plutarch says: 'While this was going on, it happened that Hyrodes [Orodes II] was at last reconciled with Artavasdes the Armenian, and agreed to receive the latter's sister as wife for his son Pacorus, and there were reciprocal banquets and drinking bouts, at which many Greek compositions were introduced. For Hyrodes was well acquainted both with the Greek language and literature, and Artavasdes actually composed tragedies, and wrote orations and histories, some of which are preserved.'[57]

Mithradates VI, king of Pontus (d. 63 BC), was a Persian born to a Greek mother. He was allied with Parthia at an early date and idealized democratic Greece, despising the Romans as uncivilized barbarians. 'In the first book of the *Memorabilia* of Favorinus there is a statement that Mithradates the Persian set up a statue of Plato in the Academy and inscribed upon it these words (in Greek): "Mithradates the Persian, the son of Orontobates, dedicated to the Muses a likeness of Plato made by Silanion."'[58] In her fascinating biographical account *The Poison King*, Adrienne Mayor writes: 'After his death, Mithradates' personal library and papers were taken to Rome, and translated into Latin by

53 Olbrycht, 'Parthians, Greek Culture', 129; cf. Flavius Philostratus, *Vita Apollonii*, 1.32. Apollonius lived in the first century. *The Life of Apollonius of Tyana* was finished in AD 217.

54 *Plutarch's Lives III*, *Crassus*, 33.2.

55 *Plutarch's Lives III*, *Crassus*, 32.

56 Tacitus, *Annals*, 12.44; cited by Olbrycht, 'Parthians, Greek Culture', 132.

57 *Plutarch's Lives III*, *Crassus*, 33.

58 Diogenes Laertius, *Lives*, I, 301.

Pompey's secretary Lenaeus (95–25 BC). Pliny, who studied Mithradates' own handwritten notes, praised his erudition. "We know from direct evidence and by report," wrote Pliny, that Mithradates "was a more accomplished researcher into biology than any man before him. In order to become immune to poison by making his body accustomed to it, he alone devised the plan to drink poison every day, after first taking remedies." At the height of his reign, Mithradates "amassed detailed knowledge from all his subjects, who covered a substantial part of the world." His international library of ethno-botanical and toxicological treatises may have described drugs used by the Druids of Gaul, Mesopotamian doctors, and the works of Hindu Ayurvedic ("long-life") practitioners. The theriac of Sushruta (ca. 550 BC) boasted eighty-five ingredients, and the *Mahagandhahasti* of Charaka (300 BC) had sixty.'[59] Mayor goes on to say: 'Mithradates was reportedly so fluent in the languages of his subjects and soldiers that he never required interpreters. Aulus Gellius remarked that "he was thoroughly conversant in the dialects of the 25 nations that he ruled, and spoke each language as if it were his native tongue." Pliny, who personally studied Mithradates' library and letters, declared, "Mithradates spoke or read the languages of 22 nations; he could address and listen to the petitions of all of his subject peoples without interpreters." Valerius Maximus cited Mithradates' linguistic proficiency as a shining example of "industrious study."'[60]

Under the Sasanians, the royal enterprise of translation began early, with Ardašīr I and his son, Šāpūr I, who were especially inquisitive about Greek and Indian astrological learning. Greek and Indian texts on astrology and astronomy, physics, medicine, and mathematics were translated into MP (II.1, IV.3, IV.5). Scientific knowledge that originated in India, Egypt, and the Hellenistic world was sought out by Persian scholars and then translated into MP or Syriac, refined, synthesized, and augmented at different centres of learning, starting in the early third century at Ǧondēšāpūr (for which see further below). One of the most striking features of Ǧondēšāpūr was its cosmopolitanism, part Persian, part Greek, with a liberal sprinkling of Jews, Christians, Babylonians, and Egyptians, even attracting scholars and traders from as far away as India.

Ardašīr, the son of Pāpak and founder of the Sasanian dynasty, was the first of the kings in this family to patronize the collecting and translating of foreign books (II.1). It is thus not by chance that al-Masʿūdī declares that Tansar/Tōsar, the mysterious and half-legendary high priest of Ardašīr, is said to have been fluent in Platonic doctrine, following chiefly the opinion of Socrates.[61]

59 Mayor, *The Poison King*, 240–241.

60 Ibid., 254.

61 Al-Masʿūdī, *Murūǧ*, I, 289; and *Tanbīh*, 87.

Ardašīr's son Šāpūr I (r. 241–272) followed in his father's footsteps, accelerating the process of acquisition of the sciences. He created or expanded the city that came to be known as Ǧondēšāpūr (Arabic Ǧundīšābūr) and settled a large number of Roman prisoners of war there after defeating Emperor Valerian in 260. The new city was favoured by the Persian king as a royal residence, and he encouraged the sciences there. In 270, two years before his death, Šāpūr I is said to have married a daughter of the Roman Emperor Aurelian (r. 270–275); she was accompanied by various craftsmen and two Greek physicians, who settled in Ǧondēšāpūr and taught Hippocratic medicine to the populace in the city. The story of this imperial marriage is related by Ibn al-Qifṭī, Bar Hebraeus, and, with less detail, several others, on account of its impact on the spread of Greek sciences to Persia. Samuel M. Stern and Ursula Weber have collected the relevant reports and studied them in full.[62] Stern calls this 'the rather tantalizing story about the transmissions of Greek learning to Sasanian Persia'.[63] Šāpūr I also encouraged scholars from Persia and India to reside in Ǧondēšāpūr, and with their help local physicians developed a rich fund of medical knowledge (IV.4, IV.5).[64] Many of the deportees arriving in Iranian territory at this time were Christians. No longer persecuted, they prospered and multiplied in Ḫuzistān (in particular Ǧondēšāpūr), Persis, and eastern Iran; they built churches and monasteries, and even set up bishoprics.[65] Greek and Syriac came into wider use,[66] and various books on the sciences (particularly astronomical works, including Ptolemy's) were translated into Pahlavi.[67] Šāpūr I has been called a Platonic philosopher-king, who used religion as a tool to unite and educate his people.[68]

Šāpūr II (r. 309–379) collected the literature of Persia scattered in India, Rome, and other lands that treated of medicine, stars and their movement,

62 Stern, 'Abū 'Īsā Ibn al-Munajjim's Chronography', 437–466; and Weber, *Prosopographie*, 1–6, who offers the most complete study of this royal marriage and its sources.

63 Stern, 'Abū 'Īsā Ibn al-Munajjim's Chronography', 443. Stern's article has a section on 'The transmission of Greek scientific books into Persia' (447–448), where he says: 'I do not wish to take up here the whole question of Greek science in Sasanid Persia, especially as I hope to publish soon a study on *Greek Astrology in Sasanid Persia and its transmission to the Arabs*, where I propose to deal exhaustively with the subject.' This project was not finished before his premature death in 1969.

64 See Dols, 'Origins of the Islamic Hospital', 367–368. See also Delaini, *Medicina del corpo*, an insightful study with a new perspective.

65 Cf. *Chronicle of Seert*, 1907, 221.

66 Shahbazi, 'Šāpur I: History', *EIr* online: http://www.iranicaonline.org/articles/shapur-i (accessed 4.5.2020), referring to Brock, 'Some Aspects of Greek Words in Syriac', 91–95.

67 Henning, 'An Astronomical Chapter', 245.

68 Shaked, *Dualism in Transformation*, 108.

time and space, substance and accident, creation, becoming and decay, and many crafts and skills. He ordered copies of all of them to be deposited in the Royal Treasury and elsewhere (II.2) and is further credited with the foundation of an academy with faculties of astronomy (with an observatory), theology, and medicine (with a hospital).[69] The historian Ibn al-Qifṭī (d. 1248) described Ǧondēšāpūr as a centre of medical synthesis and innovation. The native physicians readily adopted the scientific methods of other peoples, Indian and Greek, and studied, modified, and expanded medical knowledge through their own discoveries, making rapid progress in the field and developing new treatments along pharmacological lines; in time their therapy was judged superior to that of the Greeks and the Indians. They discussed medical laws on a regular basis and recorded their conclusions in books that were passed on to the Muslims (IV.6, IV.7).

Šāpūr II has gone down in history as a philosopher-king too. In 358, Eustathius, a noble Roman and a pupil of the philosopher Iamblichus, was sent to the Persian court as an ambassador of Constantius II (r. 337–361), son of Constantine, to negotiate 'as a master of persuasion' with the king, as Ammianus (I, 339) puts it. According to Eunapius (fl. fourth–fifth centuries), his impact on Šāpūr II was such that the Persian king was almost ready to exchange his regalia for the philosopher's cloak.[70] The shah was impressed with Eustathius's eloquence and would have given up his own crown to become a philosopher like him, had not members of the Zoroastrian clergy intervened. We have no report about how communication between the king and the philosopher took place, or how the intermediary dragoman(s) worked. According to Pahlavi sources, Šāpūr II held religious disputations.[71] One of his chief physicians was the Persian Christian doctor Theodore, reputed to be skilled in his art. Šāpūr, or, according to others, his grandson Bahrām, built a Christian church for him, which signals his prominence. Theodore wrote a *Compendium* or *kunnāš* (medical pandect, presumably in Syriac), and Ibn al-Nadīm informs us that it was later translated into Arabic (*al-Fihrist*, 360).

By the sixth century, Christians in the Persian Empire were reading and translating Greek works, and engaging in debates informed by Greek philosophy.[72] Teachers and clergy from the Sasanian world repeatedly participated in theological debates in the early Byzantine court. Greek sources discuss only one of these debaters, a certain Paul the Persian, possibly from Nisibis, and

69 Dols, 'Origins of the Islamic Hospital', 368.
70 O'Meara, *Platonopolis*, 17, *fide* Eunapius, *Lives of the Sophists*, 397.
71 See Boyce, *Zoroastrians*, 118; and Shaki, 'Dēnkard Account', 119.
72 Brock, 'Antagonism to Assimilation', 21–22; see also Walker, *The Legend*, 164–205.

his Manichaean opponent. Another Paul the Persian, a Nestorian philosopher active at the court of the most important Sasanian emperor to sponsor translations, Ḫusraw Anūširwān, is better known through his surviving works.[73]

Ḫusraw I Anūširwān, who has gone down in history as the archetype of the just king,[74] had developed an unconditional love for Greek philosophy.[75] In the *Chronicle of Seert*, the anonymous Nestorian author, writing in Arabic perhaps in the ninth century, declares: 'He [Ḫusraw] was very learned in philosophy, which he had studied, it is said, under Mar Bar Samma [Barsauma, Barsuma, Barṣauma], bishop of Qardū, and under Paul the Persian Philosopher, who, being unable to obtain the metropolitan see of Persia, renounced the Christian religion.'[76] Ḫusraw's admirers 'ascribed to him knowledge of Plato and Aristotle down to the last subtleties; he was said to have read and understood the most difficult dialogues, such as *Timaeus* and *Parmenides*' (see IV.9).[77] These reports are not unfounded. When Justinian's edict of 529 prohibited ethnic Hellenes from teaching, it was at Ḫusraw Anūširwān's court that the philosophers sought refuge, according to the well-known story.[78] Philosophical discussions took place in Ctesiphon, as evidence of which we still have, even if only in a later Latin translation, the answers that Priscianus of Lydia gave to ten philosophical questions posed by Ḫusraw Anūširwān. Carlo Nallino affirmed that under Anūširwān philosophical works by Aristotle and some of the Platonic dialogues had been translated into Persian.[79] Wiesehöfer writes: 'Besides philosophy, theology, and statesmanship, Khosrow was also interested in foreign contributions to law and medicine. In his *Res gestae* he refers to his interest in Eastern Roman and Indian law, and in the *Book of a Thousand Judgments* (*Mādayān ī hazār dādestān*) there are several references to the king and his legal advisers. Particularly worthy of note is a decree of the king regarding reforms in legal matters. Another proof of the open intellectual climate of the period is the Christian-

73 On the question of the identity of the three(?) Nestorian Syriac scholars of this name active in the sixth century, see the entries on them by L. van Rompay in *Gorgias Encyclopedic Dictionary of the Syriac Heritage*, S. Brock et al. (eds), Piscataway NJ, 2011, 324–325.

74 Christensen, *L'Iran*, 374–375.

75 For the philosophical activities of Chosroes, see Tardieu, 'Chosroès'.

76 *Chronicle of Seert*, 1911, 147. For more on Paul the Persian, see Teixidor, 'Aristote en syriaque', 723–743.

77 Altheim and Stiehl, 'New Fragments', 4.

78 Given the near uniqueness of this event, it is in fact the second 'brain drain' toward the East, the first being the Dyophysites' settlement at Nisibis in 489. For a discussion, see Delaini, 'La scuola di Gundēšābūr', especially 91–118.

79 Nallino, 'Tracce di opere greche', 345–363.

Syriac book of laws dating from this period.'[80] The king's eagerness to study the works of the Greek philosophers is a remarkable testimony to his broadmindedness, as well as to his desire to learn and make use of new ideas. It is also noteworthy that—according to Agathias—the Sasanian ruler was praised not only in the East but also in the West. John of Ephesus (c. 507–c. 586), a historian who wrote in Syriac in the service of Justinian, is evidently reporting the truth when he describes the king as fond of reading the religious books of all creeds, as well as philosophy.[81]

When Paul the Persian Philosopher compiled a commentary on Aristotle's *De interpretatione* in MP, he dedicated it to Ḫusraw Anūširwān.[82] This work was later translated into Syriac by Severus Sēbōḫt of Nisibis (575–667). Severus was of Persian ancestry and equally well versed in Persian and Greek, a fact demonstrated by the translation;[83] this allows us to suggest that Paul's synopsis (*muḫtaṣar*) of Aristotelian logic may have been a translation from Persian as well.[84] Severus became bishop of the monastery of Qenneshre on the Euphrates and was one of the most learned men of his time in the fields of astronomy and philosophy. Several works by him in these fields survive, notably treatises on the astrolabe and on the constellations; letters on points of logic addressed to Aitalaha of Nineveh and to a Periodeutes Yaunan; and a treatise written in 638, *On the Syllogisms of Aristotle*. He polemicized against the idea that the Greeks had a monopoly on wisdom and science and drew attention to the legendary contributions of the Babylonians, known through Greek sources, identifying the Greeks with the ancestors of the Syrians of his day.[85] Two centuries later, Ibn Waḥšīyah made a similar claim for Nabataean literature.

It is also to be noted that the Christian (Jacobite) physician and scholar Sergius of Rēšʿaynā (d. 536) was the first in the early sixth century to translate Greek philosophical and medical texts into Syriac and write commentaries on

80 Wiesehöfer, *Ancient Persia*, 218.

81 Cited by Cameron, 'Agathias on the Sassanians', 173.

82 See Gutas, 'Paul the Persian', 239; and Teixidor, 'Aristote en syriaque', 733–734.

83 Brock, 'Syriac Attitudes to Greek Learning', 23. See further on this van Hoonacker, 'Le Traité du philosophe syrien', 73.

84 Baumstark, *Geschichte*, 246; Mieli, *La Science arabe*, 66, 67 n. 3; Brock, 'Syriac Attitudes to Greek Learning', 23. Paul's text is edited by Land, *Anecdota Syriaca*, 1–32 (text), 1–30 (trans.); see also Brock, 'Syriac Attitudes to Greek Learning', 32, nn. 48 and 64. For more on these, see Hugonnard-Roche, 'Du commentaire à la reconstruction'; id., 'Un Organon court en syriaque'; id., 'Sur la lecture tardo-antique du Peri Hermeneias'; and id., 'Les traductions du grec au syriaque'.

85 For the Syriac text and translation, see Tannous, 'Syria between Byzantium and Islam', 353–354, with useful additional comments.

Aristotle's treatises on logic, the *Organon*.[86] He probably spent some time at the Persian court: the historian Agathias mentions as his friend and interpreter at the Sasanian court a Sergius who translated into Greek for him a synoptic history of the kings of Persia.[87] Sergius's contemporary Stephen of Edessa, after a period of service as court physician to Anūširwān's father King Qubād I (d. 531), returned to his native city, later (in 543) acting as head of an embassy when it was besieged by Ḫusraw Anūširwān.[88] Sergius of Rēšʿaynā addressed his commentary on the *Categories*, as well as several of his translations of Galenic medical texts, to one Theodore, bishop of Karḫ Ǧuddān, a town located on the River Diyālā in Iran. As Takahashi observes: 'if this is correct, we have an instance from the first part of the sixth century where knowledge of Galenic medicine and Aristotelian logic crossed the border into Persia in a Syriac guise.'[89] The presence of Syriac-speaking scholars at the court opens up the possibility that the Aristotelian books mentioned above may have reached Iran in their Syriac renderings, and that Mārsfand could have used them in his Persian commentary.[90] One might venture that these Aristotelian works on logic in MP were probably among the books that Ibn al-Muqaffaʿ translated into Arabic later.

We learn from Ibn al-Nadīm that 'in olden times the Persians had translated some works on logic and medicine into Persian, and these were later translated further by ʿAbdallāh ibn al-Muqaffaʿ and others into Arabic.'[91] A book on logic survives under Ibn al-Muqaffaʿ's name, but no work on medicine is credited to him. In the sixth chapter of his unique *al-Amṯāl al-ṣādirah ʿan buyūt al-šiʿr*,[92] Ḥamza al-Iṣfahānī reports that a philosopher at the court of King Ḫusraw Anūširwān, Mārsfand, son of Mihrak, wrote a commentary on the three books on logic attributed to Aristotle, which had been translated into MP by a committee of ten philosophers. When the completion of the commentary was brought

86 Gignoux, 'Syriac Language', *EIr* online, http://www.iranicaonline.org/articles/syriac-language-03-translators-greek-ideas (accessed 3.5.2020).

87 Agathias, *The Histories*, 133 = book IV, 3. Cf. Nutton, 'From Galen to Alexander', 13 and n. 117.

88 Nutton, *Ancient Medicine*, 301.

89 Takahashi, 'Syriac as a Vehicle', 36.

90 For an overview of the broader cultural context of transmission of Aristotelian logic and Greek medicine into Syriac during the sixth century, see the third chapter in Walker, *The Legend*, 180–205; Sergius of Rēšʿaynā holds a key place in Walker's presentation. See also Hugonnard-Roche, 'Sur les versions syriaques'; id., 'Aux origines de l'exégèse orientale'; id., 'Les traductions syriaques de l'*Isagoge*'; id., 'Les *Catégories* d'Aristote'; and id., 'Note sur Sergius de Reshʿaynā'.

91 Rosenthal, *The Classical Heritage in Islam*, 45–46.

92 Berlin, Staatsbibliothek, MS Or. Quart. 1215; see Schoeler, *Verzeichnis*, 304–312.

to the attention of Anūširwān, he invited the resident philosophers to royal colloquia to discuss the merits of these texts (V.5).[93] One of the philosophers attending this important gathering could have been Paul the Persian Philosopher.[94]

Al-Ǧāḥiẓ (d. 255/869) places Ibn al-Muqaffaʿ among the early translators of Aristotle (*al-Ḥayawān*, I, 76), and classical Muslim historians of science agree that Ibn al-Muqaffaʿ translated at least parts of the Aristotelian *Organon*. Ibn al-Qifṭī says (*Taʾrīḫ al-ḥukamāʾ*, 220): 'Ibn al-Muqaffaʿ was the first in the Islamic nation to engage in the translation of books on logic for [the caliph] Abū Ǧaʿfar al-Manṣūr [r. 135–158/754–75] [...] he translated the three Aristotelian books on logic, namely, the *Categories*, the *De interpretatione*, and the *Prior Analytics*; and it is said that he translated the *Isagoge* of Porphyry [...]'. Ibn Abī Uṣaybiʿah (*ʿUyūn al-anbāʾ*, 413) writes to the same effect, adding: 'His renderings in translation were fluent, easy to grasp.' From all such renderings, only his general treatise on Aristotelian logic (*al-Manṭiq*) has survived. In some passages Aristotle is followed very closely; in others there are additions clearly obtained from elsewhere. Ibn al-Muqaffaʿ's work may be only the translation of a commentary on Porphyry's *Isagoge*, and Aristotle's *Categories*, *De interpretatione*, and *Prior Analytics*; and since it is stated that this was a translation, it could not have been a commentary by Ibn al-Muqaffaʿ himself.

IV. *Translations of Scientific Works*

Several individual translated works warrant attention here. In a short survey of Greek works that we are sure reached the Muslims by way of the MP translations, Kevin van Bladel identified only six or seven titles of which traces remain today, but there were certainly many more, as we have seen.[95] We learn from a passage of Ibn Nawbaḫt's *Kitāb al-nahmuṭān*, studied in detail by van Bladel, that Šāpūr I persisted in his policy of searching for foreign books to be translated, 'until all those books were copied in Persian, according to the views of Hermes the Babylonian, who had been king over Egypt, and Dorotheus the Syrian, and Fydrws the Greek from the city renowned for learning, Athens, and Ptolemy of Alexandria, and Farmāsb [Parameśvara?] the Indian. They commented on them and taught them to the people just as they had taken from all [these] books, the origin of which was from Babylon. Then, after those two

93 Ḥamza al-Iṣfahānī, *al-Amthāl al-ṣādirah*, 749.

94 Christensen, *L'Iran*, 427.

95 Van Bladel, 'The Arabic History'.

[Ardašīr and Šāpūr] Kisrā Anūširwān collected and harmonized them and put them into use, because of his desire for and love of learning.' (II.3).[96]

With regard to Ibn Nawbaḫt, Pingree writes:

> However, we know [...] that Dorotheus' hexameters were translated into Pahlavī under Ardashīr I, the Sassanian emperor from about 222 to 237, or under his son, Shāpūr I, who ruled from about 237 to 267; and that it was later elaborated during the reign of Khusro Anūshirwān, between 531 and 578. The principal Arabic translation of the lost Pahlavi version of Dorotheus was apparently due to the efforts of 'Umar ibn al-Farrukhān al-Ṭabarī in about the year 800. This translation proves that Ibn Nawbakht was correct; for it contains not only a number of Arabic transliterations of Pahlavī technical terms, but two horoscopes that were added to the text as illustrations by Sassanian scholars. One is datable 20 October 281, the other 26 February 381.[97]

Only brief excerpts survive from the original Greek of the didactic poem on horoscopic astrology by Dorotheus of Sidon (AD c. 50–100), in quotations by later authors, but the Arabic translation made of the whole by 'Umar ibn al-Farruḫān (fl. c. 750) of the lost Pahlavi version is extant.[98] 'Umar also wrote commentaries on other books by Dorotheus, which had probably reached him in Middle Persian as well.[99]

In addition to the Arabic translation of the Pahlavi version of Dorotheus of Sidon, we have the Arabic version of a Sasanian astrological treatise entitled *Kitāb Zarādušt fī l-mawālīd wa-aḥkāmihā* (*Zoroaster's Book of Nativities and their Judgments*), which synthesized Persian science with both Greek and Indian sources.[100] Moreover, the Pahlavi translation of the anonymous *Kitāb asrār al-nuǧūm* (*Book of the Secrets of the Stars*), ascribed to Hermes and used as

96 Ibid., 47.

97 Pingree, 'Classical and Byzantine Astrology', 229–230. The Pahlavi text of Dorotheus is to be found in an Arabic manuscript in the library of the University of Leiden, Oriental 891, which contains on fols 1r–28v a *Kitāb fī bayān al-ifrādāt* ascribed to Dorotheus; another copy of the text exists in Syria, Maktabat al-Asad, no. 47871. For a new reassessment and challenge of Pingree's proposed MP version of Dorotheus' work, see Cottrell and Ross, 'Persian Astrology'.

98 Dorotheus of Sidon, *Carmen Astrologicum*, edited by D. Pingree. The edition is accompanied by a parallel text in Arabic and passages in Greek and Latin.

99 Ibn al-Nadīm, *al-Fihrist*, 328.

100 Pingree, 'Astronomy and astrology in the Sasanian period', *EIr*, II, 859.

a source by Abū Maʿšar (d. 886), must have been made during the sixth century, probably during the long reign of Ḫusraw Anūširwān, since it certainly existed by the beginning of the next century, when it was used by Mahānkard (c. 637) in his Pahlavi version of Zarādušt's *Kitāb al-mawālīd* (v.3)[101] To be added to the list of translated works is the anonymous *Liber hermetis de stellis beibeniis*, a Hermetic astrological book that attributes 'planetary temperaments' to the fixed stars (as opposed to planets), and by doing so brings them into the astrological field.[102]

Eugenio Griffini[103] has brought to wider attention a manuscript on astrology/astronomy at the Biblioteca Ambrosiana in Milan entitled *Kitāb ʿarḍ miftāḥ al-nuǧūm* (*The Latitude of the Stars*), at the end of which it says that the book was translated into Arabic in the year 125/783. It does not say who translated it and from which language, but it was probably from MP: Ibn al-Nadīm and Ṣāʿid al-Andalusī list a book of that title among the writings attributed to Hermes, as does Ibn al-Nadīm, who adds another, *Kitāb ṭūl miftāḥ al-nuǧūm* (*The Longitude of the Stars*).[104] An indication of the antiquity of *Kitāb ʿarḍ miftāḥ al-nuǧūm* is its use of technical terms that were later abandoned or replaced; for example, the book uses *al-kawākib al-sayyāḥa* for planets, whereas in later literature the planets are called *al-kawākib al-mutaḥayyira* or *al-kawākib al-sayyāra*.[105] Moreover, van Bladel has shown that 'Iranian astrologers in the service of al-Manṣūr [r. 136/754–755] and his successors introduced Hermes into Arabic letters from astrological Hermetica available to them in Middle Persian. Some of their translations of Hermetica from Middle Persian into Arabic are extant, such as *Qaḍīb al-ḏahab* [*The Golden Bough*][106] and *Asrār al-nuǧūm fī l-kawākib al-bābāniyya* (Kunitzsch suggested that this work had been translated into Middle Persian from Greek).'[107]

It was the Italian scholar Carlo Nallino in 1922 who directed researchers' attention towards pre-Islamic Iranian written culture and its ties with Greek literature. After reviewing a few references to Persian translations of Greek works by Nöldeke (who maintained that the Syrian translator of the *Alexander Romance* translated it not directly from the Greek, but from a Pahlavi original), and Fraenkel (who agreed with Nöldeke), Nallino showed that a number

101 Pingree, 'Classical and Byzantine Astrology', 234.
102 Kunitzsch, 'Zum "liber hermetis"'.
103 Griffini, 'Lista dei manoscritti arabi', 110–114.
104 *al-Fihrist*, 327 (tr. Dodge, *The Fihrist*, 638); and Ṣāʿid, *Ṭabaqāt*, 19.
105 Ullmann, *Die Natur- und Geheimwissenschaften*, 290.
106 Cf. Schoeler, *Verzeichnis*, 175–177.
107 Van Bladel, 'Hermes and Hermetica', 183. Kunitzsch suggested that the latter work had been translated into Middle Persian from Greek.

of Greek texts had indeed been translated directly into MP. He introduced three cases of Greek scientific works that had passed on to the Arabs through Pahlavi intermediaries: Cassianus Bassus's *Geoponica* (*Agriculture*), the MP *Warz-nāmag*; Vettius Valens's astrological *Anthologiae*, the MP **Wizīdag* (*Selection*[*s*], => the Arabic, *al-Bizīdaj*); and the *Paranatellonta tois dekanois* (*Representations of the Decans*). The latter work, by a Babylonian astrologer active around the first century AD named Teucer (also spelled Teucros, and in Arabic Tankalūšā, among other forms), deals with constellations that rise together at certain degrees of the zodiacal signs.[108]

Theophilus of Edessa (also known as Theophilus ibn Tūmā, AD c. 695–785, a Syrian Maronite Christian author writing in Greek) explains that the Persians, having been lovers of wisdom throughout the East, translated Greek books into their own language. He mentions by name specific authors of astrological works: Dorotheus, Valens, Critodemus, and Timocharis.[109] Theophilus was a court astrologer for the caliphs al-Manṣūr and al-Mahdī and was familiar with MP traditions: he had mastered Persian, Arabic, Syriac, and Greek.[110] A view regarding the sources of Persian wisdom differing from that of Theophilus is expressed in Pseudo-Ǧāḥiẓ's *Bāb al-ʿirāfa*: 'From them [i.e., the Indians] Chosroes [*akāsira*] and the Persians learned all the sciences. They took from them and imitated them. In particular, however, this applies to medicine, astronomy, arithmetic, the *wahm* [a magical art], divination, clairvoyance, physiognomy, and [the science of] the lunar stations and similar matters. As for the Byzantines [*rūm*], they have taken all things from the Greeks.'[111] Clearly the authors writing in the early ʿAbbāsid period were not in agreement as to whence the Persians had obtained their scientific knowledge.

Among the foreign works, Greek and Indian, imported into Iran and translated during the reign of Šāpūr I, was the *Almagest* of Ptolemy, a contemporary of Vettius Valens (both were astrologers living in Alexandria): 'No less a work than Ptolemy's *Syntaxis* was rendered into Pahlavi in the third century AD, less than a hundred years after its original composition.' (II.2).[112] Kunitzsch made a

---

108 Nallino, 'Tracce di opere greche', 345.

109 Pingree, 'Classical and Byzantine Astrology', 237; van Bladel, *The Arabic Hermes*, 34, n. 50; and *Catalogus Codicum Astrologorum*, I, 129–131.

110 Pingree, 'Classical and Byzantine Astrology', 236.

111 This is my translation from *Bāb al-ʿirāfa wa-l-firāsa ʿalā maḏhab al-furs*, a text attributed to al-Ǧāḥiẓ, edited and translated into Russian by Inostranzev in his 'Materialy iz arabskich', Texts, 3–27 (here 4), translation and commentary, 28–120; the passage is also cited by Pines, *Beiträge*, 120.

112 Pingree, *The Thousands*, 7–13.

good case for supposing that the Arabic form of the title, *Almagest*, was derived, not directly from the Greek, but from a MP translation of the book.[113]

The pseudo-Aristotelian lapidary book, translated into Arabic as *Kitāb al-aḥǧār* and dated to the eighth–ninth centuries, belongs to the late Sasanian Persian-Syriac tradition. This book on gemmology differs from the *Mineralogy* by Theophrastus: it is certainly an Eastern compilation and contains many Iranian names for precious stones.[114] Julius Ruska discussed the Syriac-Persian origins of this lapidary and showed that the so-called *Steinbuch des Aristoteles* is a product of Persia in the sixth century.[115] Ruska's observation was later confirmed in several articles by Manfred Ullmann.[116] Another short lapidary text, in Pahlavi, was published by de Menasce;[117] Chunakova suggested that this might have reached Iran via a translation from Syriac.[118] In his *al-Ǧamāhīr*, al-Bīrūnī declares that he acquired a book, written in Damascus during the caliphate of ʿAbd al-Malik ibn Marwān (r. 685–705), that dealt with the qualities of gemstones and their values (*Kitāb al-ǧamāhīr*, 50); it is hard to say whether this could have been related to the book under consideration here.

Levey and Ullmann have shown that a book on substitute drugs by one Pythagoras (Badīġuras), a Greek doctor living in Iran, was written in MP. It is preserved only in the Arabic version passed on to the Muslims, *Abdāl al-adwiya*,[119] and consists of a list of *succedanea*, the drugs an apothecary could use as substitutes when he could not provide or procure the drugs prescribed by a doctor. Such lists were common in antiquity, and one, the *Peri antemballomenōn*, is attributed to Galen. Badīġuras's booklet 'immediately reveals its Iranian origin because it uses almost exclusively ancient Persian names for plants and drugs.'[120]

The author of the MP book *Bundahišn* had some knowledge of the Greek scientific literature that had reached Iran under the Sasanians; for example, the contents of chapter XXVIII resemble the pseudo-Hippocratic tract *Peri hebdomadōn*, and chapter II (on astronomy) shows an acquaintance with Greek and

113 Kunitzsch, *Der Almagest*, 115–125.
114 Nasr, 'Life Sciences', 410.
115 Ruska, *Das Steinbuch*, 43–47; and id., *Tabula Smaragdina*, 171.
116 Ullmann, 'Steinbuch des Xenokrates', 59; and id., 'Neues zum Steinbuch des Xenokrates', 59–76.
117 De Menasce, 'Un lapidaire pehlevi', 180–185. For further discussion of this, see Chunakova, 'The Pahlavi Lapidary', 73–84.
118 Chunakova, 'The Pahlavi Lapidary', 76.
119 See Levey, *Substitute Drugs*; Ullmann, 'Die Schrift des Badīġūras', 230–248; id., *Islamic Medicine*, 17–18; and Kunitzsch, 'Über das Frühstadium', 279–280.
120 Ullmann, *Islamic Medicine*, 17–18.

Indian science.[121] In the *Dēnkard* (428–429), several foreign works are listed by name: the *Almagest* of Ptolemy; books on logic (*tark*, in Sanskrit *tarka*); books on rhetoric (*vyākarana*); books on casting horoscopes (*horā*, in Sanskrit 'hours'); and books on astronomy (*kāla*) (II.2). The text adds that since these works were foreign, scholars reworked and brought them up to date (*nōk ārāst*), and occasionally even kept the original titles. Zaehner showed that among these books was a *Compendium* (*Hangirdīg*) that dealt, among other things, with the origin of the Sasanians' four social classes.[122]

From the same period as the *Dēnkard* come the comments on Šāpūr II recorded by Ibn Qutaybah (d. 276/889). Having become very sick, Šāpūr contemplated abdication, but his advisors convinced him to write to the king of India and ask for an expert physician. The Indian doctor cured him and as his reward was allowed to settle in the city of Susiana in south-east Iran to practise his art; his medical knowledge passed on to the Persians there, who combined it with what they had learned from the Roman settlers in the area (IV.7). In his polemic against the claims Persians made concerning the history of science (III.3), Ibn Qutaybah says that the Persians received from Byzantium the book by Euclid (i.e., the *Elements*); the *Almagest* by Ptolemy; the books on logic by Aristotle; the *Book on Music* (*Kitāb al-mūsīqā*), perhaps by Pythagoras; and, from India, the *Book on Chess* and the *Kalīlah wa-Dimnah*. They then syncretized Indian medicine with that of the Byzantines and produced a *Compendium of Medicine* at Šūš, where Ǧondēšāpūr was located.[123] Apparently this *Compendium* was still available in Arabic to the thirteenth-century author Ibn al-Qifṭī (d. 646/1248), who praises it highly.[124]

Another distinct chapter in the scientific tradition consists of books on equestrian sport, hunting, hippology, hippiatry, ornithology, and veterinary medicine. The unspoken literary competition between supporters of Greek scientist-philosophers and adherents of ancient Persian sages and kings (see above II. (c)) comes to the fore predominantly in Islamic-period texts in these fields. Debates and observations as to who was first in these areas and who borrowed from whom created a lively series of books that, apart from their historical significance, are important for a variety of reasons, including for the history of translation. This class of literature, of interest to modern research on Eastern hunting traditions and their impact on the Mediterranean world, has not received the attention it deserves. Arabic and Persian texts on hunt-

121 McKenzie, 'Bundahišn', *EIr*, IV, 547–551.
122 Zaehner, *Zurvan*, 139–140; cf. de Menasce, *Le Denkart*, 27.
123 Ibn Qutaybah, *Faḍl al-ʿArab*, 205–206.
124 Ibn al-Qifṭī, *Taʾrīḫ al-ḥukamāʾ*, 132.17–133.17.

ing with birds of prey and equestrian sport, which goes hand in hand with the former, are invaluable sources on veterinary medicine for raptors, medicines, and other agents used in treatment and diagnosis, or monitoring progress or response to treatment. Texts on such matters are treasuries of medical procedures and technical vocabulary. In particular, books dedicated to the study of falcons offer precise ornithological information, not merely about falcons, but about their prey as well. As was the case in several reports cited above, Alexander the Great plays a central role in this field too (V.9, V.10).

Writing about the literature available on veterinary surgery and the care of animals in his time, Ibn al-Nadīm (*al-Fihrist*, 377) names a Persian book translated by or for the ʿAbbāsid prince Isḥāq ibn ʿAlī ibn Sulaymān entitled *About the Veterinary Care of Walking Animals, Horses, Mules, Cows, Sheep, and Camels, with Knowledge of Their Cost and Defects*. Another was called *Veterinary Surgery by the Persians* (*al-Bayṭara li-l-Furs*), where the author names, under the rubric 'The books composed about birds of prey, sport with them, and medical care of them by the Persians [...]', *The Book of Falcons by the Persians* (*al-Buzāt li-l-Furs*). In his *Medical History of Persia*, Cyril Elgood writes:

> I have in my private collection of medical manuscripts a pseudo-Aristotelian fragment, written in Persian. The preface states that it is a translation. It is clear that the translation is not a very early one; but it may well be that although the original version was not Greek, it was a Pahlaví collection of Greek veterinary lore. The manuscript is short and is entitled 'The Book of the Horse, a Pamphlet of Aristotle the Sage or Farasnáma'. The last translator has added an interesting preface:
>
> > This book has been collected from the sayings of Aristotle the Sage in praise of horses, the good and the bad, their faults and virtues of breeds and uses, and their proper colour. It was composed for the Two-horned One, as men called Alexander the Great. It was translated by the Imám of These Times, the Sun of the People and the Faith, Moulána Muḥammad bin Ḥusayn (may God have mercy upon his soul) out of the language of the Turks [*sic*] into the language of the Persians, so that anyone who might have difficulty with the Turkish tongue should be able to dip into this book whenever he had need and might learn all that is connected with a horse.[125]

125 Elgood, *Medical History*, 318.

This *Farasnāma* (*On Hippology*) dated 555/1160, a work attributed to Aristotle and supposedly written at the behest of Alexander the Great, has been published in an edition based on half a dozen manuscripts, all of which name the translator from Arabic (not Turkish) into Persian as Šams al-Dīn Muḥammad ibn Ḥasan (or Ḥusayn).[126]

Ipocras, a doctor from Sind in India, is said to have written a book on horse therapy on behalf of the Persian King Ḫusraw I Anūširwān, who ruled in the mid-sixth century. Probably originally composed in Sanskrit and then translated into Pahlavi, it was translated into Arabic at an unknown date and thence into Latin in 1277 by Moses of Palermo at the command of Charles I of Anjou. Although the Arabic text is not preserved, Ipocras's work was well known in Arabic-speaking countries.[127] The most important medieval Persian book on 'the art of venery' is the *Bāznāma* (*On the Art of Falconry*) by Abū l-Ḥasan 'Alī ibn Aḥmad al-Nasawī (b. 393/1003, d. 493/1100), who claimed to have studied with Avicenna.[128] Al-Nasawī was a famous mathematician, astronomer, physician, and veterinarian, and wrote books in all these fields. He refers to his sources in general terms—the *Šikārnāma* (= *Bāznāma*) of the Sāmānids (probably *The Book on Hunting* compiled for 'Abd al-Malik ibn Nūḥ) (V.9), and the *Šikārnāma* of the Soghdians, Indians, Romans (Byzantines), and Persians—adding to these his own experiences of a long life engaged in the study of medicine and its practice (the mixing of drugs and pastes, antidotes, etc.).[129] The only source identified specifically by al-Nasawī is *Bāznāma-e Parvīz Malik Pārsiyān* (*The Book of Falconry of Parvīz, King of the Persians*).[130]

Detlef Möller has dedicated a detailed study to the Arabic literature on falconry in the Middle Ages. For him, this literature must have had its origins in the eighth century, and certainly existed in the ninth, with translations from Persian and Greek swiftly acquiring independent status. Möller attempted to identify the primary sources of treatises current in Arabic that bore general names such as *The Byzantine Writings*, *The Indian Writings*, and *The Persian Writings*. He argues that these writings were translated into Arabic during the Umayyad and (particularly) the early 'Abbāsid periods, and became the basis for all later Arabic works on falconry. He notes that the *Kitāb al-Furs* (*The Book of the Per-*

126 Richard, 'Un traité persan d'hippiatrie', 95–101; edition of the text in Tāğbaḫš, *Tārīḫ-e dāmpiziškī*, 414–428, with a newer edition in Zangī, *Farasnāma-e Qayyim Nahāwandī*.

127 Weidenhöfer, Heide, and Peters, 'Zur Frage der Kontinuität', 63. The sources given for this information are in Sponer, 'Die Pferdeheilkunde', 60, and Froehner, *Tierheilkunde*, 85.

128 Al-Nasawī, *Bāznāma*, 67, 68, 69.

129 Cf. ibid., 76, 87, 104, 111, 114, 121, 124, 126, 150.

130 Ibid., 127.

sians) preserved partially in the manuscript Fātiḥ 3566 (58a–65a; 78b–86a) is about falcons. It was probably translated into Arabic in the second/eighth century and it deals with the treatment of hawk diseases and instructions for the care of the hawks during training.[131]

The study of the impact of foreign sciences on Persian schools of medicine, pharmacy, and alchemy and thence on cultures stretching to the west of Iranian domains, especially on the Arabo-Islamic world of the seventh and eighth centuries, remains an unexplored field.

V. *Conclusion*

No matter what one may make of the Persian approach to the history of science and the reasons for this knowledge's being widely circulated, there should be no doubt that in the long run their approach legitimized and encouraged the acquisition and translation of major foreign works; it removed barriers for acquiring them and paved the way for the accumulation of scientific and literary treasures by Sasanian society. What had been distant and inaccessible to many now became 'our knowledge', open to all.

The phenomenon of appropriating the intellectual accomplishments of foreign cultures through translation, set in full gear in the early decades of the Sasanians, was seen as something natural, and it continued uninterrupted until the end of the tenth century, albeit with a temporary slowdown after the Muslim conquests. In the first, pre-Islamic phase, translations were made from Sanskrit, Greek, and other languages into Syriac and MP, and in the second phase, after Islam, from MP, Syriac, and Greek into Arabic. Neither the broader setting nor the traditional patterns of patronage changed significantly: actors in the transfer of cultural goods characterized their endeavours as a continuation of the old ways. 'True Muslims'—those who thought they knew Islam better than the caliphs, emirs, and scientists, and who distinguished between Islamic sciences (Qurʾān, *ḥadīṯ*, the Arabic language, etc.) and sciences of the ancients (*al-ʿulūm al-qadīma, ʿulūm al-awāʾil, ʿulūm al-ʿaǧam*)—readily accused those engaged in 'foreign sciences' of unbelief and *zandaqa*. But seeking out and translating foreign wisdom was a deep-rooted mindset, and those who challenged it were unable to change it. When the physical means of distributing scientific works became more affordable, with the arrival of paper in Baghdad in the eighth century, the translation of foreign works naturally gained momentum. The support of the early caliphs al-Manṣūr and al-Maʾmūn offered to astronomers, philosophers, and physicians was seen by them as continu-

131 Möller, *Studien*, 47–59. See also Sezgin, *GAS*, III, 355.

ing a long-standing practice, which in turn strengthened the foundations of their rule. One may compare the flourishing of science under al-Ma'mūn in all its aspects to that in the time of Ḫusraw Anūširwān, the differences between the two periods lying mainly in the extent to which scientific endeavour was undertaken and the numbers of documents surviving. We only need look to the scientists—who they were, where they came from, and the nature of their work and studies—to appreciate the continuity that existed in the translation movement.

Part Two: Texts in Translation

Overview

I. Knowledge Dispersed, Knowledge Reassembled: Ancient Precedents
II. Knowledge Dispersed, Knowledge Reassembled: from Alexander to Islam
III. Persian Achievements in the Sciences
IV. Persian Rulers: Preserving and Disseminating Knowledge
   (a) Before the Sasanians
   (b) The Sasanians: Ardašīr I
   (c) Šāpūr I
   (d) Šāpūr II
   (e) Bahrām V
   (f) Ḫusraw I Anūširwān
V. The Translation of Scientific Works

I. *Knowledge Dispersed, Knowledge Reassembled: Ancient Precedents*

Reports about collecting, plundering, or destroying books are a frequent motif in ancient writings about libraries. Persian and Arabic sources are not the only ones that recount a transfer of books from Iran to Egypt and Greece: two sources describe respectively the transportation of foreign books to Iran and their repatriation at later dates. Firstly, the Achaemenid king Xerxes is said to have brought to Iran an immense library that was collected by Peisistratus in Athens and later brought back to Greece by Seleucus Nicator, as Aulus Gellius reports (I.1). Secondly, the libraries of the Egyptian temples were taken by Cambyses (I.2) and returned by Ptolemy Philadelphus, as reported by Ptolemy on the so-called Satrap Stele/Stela (granite, hieroglyphic, Late Middle Egyptian text). The stele was set up by Ptolemy I in 311 BC, after his victory over Demetrius Poliorcetes at Gaza in 312 BC, when he was still just the viceroy of Egypt on behalf of the king of Macedon and had not yet assumed the prerogatives of kingship. The inscription relates how Ptolemy (who had accompanied Alexander to Iran) returned to Egypt from the conquered Persians all the ritual implements and sacred writings of the temples of Upper and Lower Egypt, restoring them to their proper places.[132] Ryholt writes: 'Had these texts not been considered of utmost importance, they would hardly have been mentioned so prominently by Ptolemy. Evidently the Persians had also taken an interest in these texts. If they had merely wished to deprive the Egyp-

132 Ryholt, 'Libraries in Ancient Egypt', 24; and Ritner, 'The Satrap Stela', 393.

tians of these writings, in which their ancient culture and science was codified, they would surely just have set them ablaze, rather than to try and carry them off wholesale to Persia.' Ptolemy's claim to have returned the ancient belongings of the temples confiscated by the Persians—a claim made to bolster his position and legitimize himself as local ruler and rightful successor to Alexander IV—corresponds perfectly with the Zoroastrian claim that Alexander's men sent Persian books to Egypt. The Satrap Stele could then be seen as 'evidence in stone' for this transportation of cultural goods. Later comments to the effect that the Persians had stolen them earlier are in general part and parcel of the endeavour to misrepresent Achaemenid rule in Egypt. Ptolemy's fame for collecting books in every language and from all over the world allows us to conjecture further that he probably also carried off other literary items from Persia.

I.1. Aulus Gellius, *Attic Nights* (*Noctes Atticae*); tr. Rolfe, II, 139.11–32.

In his *Attic Nights*, Aulus Gellius (c. 125–after AD 180) recounts two incidents relating to the loss and gain of books: first the theft of the library of Athens by the Persian king Xerxes I (r. 486–65 BC) after capturing the city during the second invasion of Greece in 480 BC; then the restoration of these books to Athens by Seleucus I (r. 305–281 BC), the founder of the Hellenistic Seleucid empire that came to control most of the eastern parts of the territory conquered by Alexander the Great (r. 336–323).

The book-stealing motif came to play a central role in later Persian narratives about the reclamation of ancient Persian knowledge allegedly dispersed during the campaigns of Alexander the Great, which will be discussed further below (see II.1). Whereas in Gellius's account the Persians plundered Greek written heritage, Zoroastrian authors maintained that the Greeks under Alexander stole Persian religious and scientific writings, and that translating Greek books back into Persian constituted an act of repatriation of ancient lost learning.

Gellius rounds off his account on the theme of the annihilation of books and libraries by also mentioning a more recent episode of wholesale cultural destruction: the fire that consumed the famous library of Alexandria during Julius Caesar's siege of the city in 48 BC.

An older but close parallel to the events described in the present excerpt is the return of scrolls and other artefacts from Persia to Egypt a generation or so after Seleucus by the Hellenic ruler of Egypt, Ptolemy II Philadelphus (r. 285–46 BC) (see I.2).

Who was the first of all to establish a public library; and how many books there were in the public libraries at Athens before the Persian invasions?

The tyrant Pisistratus is said to have been the first to establish at Athens a public library of books relating to the liberal arts. Then the Athenians themselves added to this collection with considerable diligence and care; but later Xerxes, when he got possession of Athens and burned the entire city except the citadel, removed that whole collection of books and carried them off to Persia. Finally, a long time afterwards, king Seleucus, who was surnamed Nicanor, had all those books taken back to Athens.

At a later time an enormous quantity of books, nearly seven hundred thousand volumes, was either acquired or written[133] in Egypt under the kings known as Ptolemies; but these were all burned during the sack of the city in our first war with Alexandria, not intentionally or by anyone's order, but accidentally by the auxiliary soldiers.

I.2. Isidore of Seville, *Etymologies* (*Etymologiae*), VI.iii.1–5, 43.1–45.10; tr. Barney et al., 138–139.

Roughly a millennium after the events described by Aulus Gellius, Isidore (d. 636), archbishop of Seville and the first Christian writer to compose an encyclopaedic digest of the knowledge of his day, included Aulus's comments in his monumental *Etymologies*. In the entry 'De bibliothecis' ('On Libraries', 138–139), we read:

1. A library (*bibliotheca*) takes its name from Greek, because books are deposited there, for βιβλίων means 'of books' (*liber*) and θήκη means 'repository'.
2. After the Law (i.e. Torah) was burned by the Chaldeans, the scribe Ezra, inspired with the divine spirit, restored the library of the Old Testament when the Jews had returned to Jerusalem, and he corrected all the scrolls of the Law and Prophets, which had been corrupted by the gentiles, and he ordered the whole Old Testament into twenty-two books, so that there might be as many books in the Law (i.e. the Old Testament) as they had letters of the alphabet.[134]

---

133 I.e., copied from other manuscripts.

134 The story of the Old Testament recounted here closely resembles that of the compilation of the religious texts of Zoroastrians, the *Avesta*: after being dispersed and partially destroyed, the remaining pieces were assembled and recompiled in twenty-one books (*nasks*). Pines, 'A Parallel', suggested a connection between Jewish stories about losing and reassembling the texts of the Torah on the one hand and attempts by various Persian kings at recovering lost portions of the *Avesta* through collecting and 'collating' foreign wisdom with a primary 'copy' of the *Avesta*, a frequent topos of MP literature; see II.1 below.

3. Among the Greeks, however, the Athenian tyrant Pisistratus is thought to have been the first to establish a library. Afterwards, when Athens was burned, Xerxes brought this library, which had been augmented by the Athenians, to Persia; after a long time Seleucus Nicanor brought it back to Greece.
4. Hence arose, in kings and in other cities, a desire to procure the scrolls of diverse nations and to render them into Greek by translators.[135]
5. Thus Alexander the Great and his successors set their minds to building libraries with every book. Ptolemy in particular, known as Philadelphus, and very perceptive about all kinds of literature, collected for his library not only pagan authors, but even divine literature, because he emulated Pisistratus in his zeal for libraries. In his time seventy thousand books[136] could be found in Alexandria.

II. *Knowledge Dispersed, Knowledge Reassembled: from Alexander to Islam*

II.1. *Acts of Religion* (*Dēnkard*), I, 411.17–414.15 (ed. Madan); tr. Shaki, 'Dēnkard Account', 118–120, with a few superficial changes by Zakeri.

The *Acts of Religion* (*Dēnkard*) is the most important religiously inspired Middle Persian composition that has come down to us.[137] It has been denoted a 'Mazdean encyclopaedia'[138] and is certainly encyclopaedic in scope, though has little in common with a modern encyclopaedia. Of its original nine volumes, only seven survive: the first two books and part of the third are no longer extant. The texts that make up the *Dēnkard* were written by various authors at different times. The *Dēnkard*'s purpose is in part apologetic: as followers of an oppressed religion, its Zoroastrian authors felt obliged to highlight the superiority of their religious teachings over those of the conquering Muslims. Although the need for such a book must have been felt very early in the Mazdean circles of Islamized Iran, the first draft of the compilation that became the *Dēnkard*, the work of a head of the Zoroastrian community by the name of Ādurfarnbag ī Farroḫzādān,[139] was undertaken during the reign of the caliph al-Maʾmūn (r.187/813–

135 The MP *Acts of Religion* (*Dēnkard*), a tenth-century encyclopaedic compendium of Zoroastrian religious beliefs and practices, described the Sasanian ruler Ḫusraw I Anūširwān (r. AD 531–579) in similar terms; see II.1 below.
136 Cf. the number of 700,000 books cited by Aulus Gellius (in I.1 above).
137 For an overview of its contents, see Gignoux, 'Dēnkard', *EIr*, VII, 284–289.
138 E.g., in the title of de Menasce's *Le Denkart: Une encyclopédie mazdéenne*.
139 See Aḥmad Tafażżolī, 'Ādurfarnbag ī Farroxzādān', *EIr*, I, 477–478.

218/833). The surviving redaction was created somewhat later by Ādurbād ī Ēmēdān,[140] one of Ādurfarnbag's successors, whose life spans the second half of the ninth century.

The text of the *Dēnkard* (edited by M.D. Madan in two vols, 1911) is highly problematic, so much so that one observer called it 'undoubtedly the most corrupt text in an Iranian language of any period' and complained about its ambiguous script, terse style, and lack of punctuation.[141] Despite the abstruse and sometimes painfully vague nature of the text, the passages in book 4 that describe the history of the Zoroastrian sacred books and the transmission of knowledge in Iran has been investigated and translated more than twenty times since the early nineteenth century. This part of the *Dēnkard*, composed during the reign of Ḫusraw I (as is indicated in the text), remains one of the most controversial and hotly debated specimens of Pahlavi literature. Notwithstanding the legendary nature and often unverifiable details of the historical account it narrates, it has caught the attention of researchers in a variety of disciplines: it is for example of relevance to the ongoing debates about the date of the prophet Zoroaster, the controversies surrounding the written or oral transmission of the *Avesta*, the role of Alexander the Great in the *Avesta*'s destruction, the existence of libraries and archives in ancient Iran, the contribution of kings and high priests to the recompilation of the sacred books, and the interest of Sasanian society in foreign books and their translation. It is particularly important for our purpose here because of its lengthy account of the dispersal and reclamation of Persian written heritage. Shorter versions of this account with minor differences are told in several other MP and Arabic texts.

Book 4 of the *Dēnkard* is the shortest among the seven extant books and is presented as an extract from Ādurfarnbag Farroxzādān's *Ēwēn-nāmag*, a work that expounded customs, arts, and sciences.[142] The compiler, familiar with a range of Persian, Greek, and Indian scientific writings, also considered meteorological issues, astronomical and medical observations, and geometry. He enumerated translations into MP of several scientific texts from Greece and India, including Ptolemy's *Almagest* and works on logic, rhetoric, astrology, and astronomy (see II.2 below). The following excerpt describes how the Zoroastrian texts were collected, then scattered during Alexander the Great's conquest of Persia and reassembled by subsequent Persian kings. It starts with the figures of Wištasp, a legendary king contemporary to and an early follower of Zoroaster, and King Dārā or Dārāb II, the last king of the Kayanid dynasty, who is often identified with the last Achaemenid ruler Darius III (r. 336–30 BC).[143] The next figure, Walaxš, may be the Parthian monarch Vologases IV (r. 147–191 BC), and the rest belong to the Sasanian

---

140 See Aḥmad Tafażżolī, 'Ādurbād Ēmēdān', *EIr*, I, 477.

141 Zaehner, *Zurvan*, 6.

142 It was part of a whole genre of books with the same title that aroused the interest of later Arabic authors; see Aḥmad Tafażżolī, 'Āīn-nāma', *EIr*, I, 691–692.

143 See Aḥmad Tafażżolī, 'Dārā(b) (1)', *EIr*, VII, 1–2.

dynasty: its founder, Ardašīr I (r. 224–242); his son and successor Šāpūr I (r. 240–270); Šāpūr II (r. 309–379); and Ḫusraw I (r. 531–579). The role of these rulers in reassembling the corpus of Zoroastrian religious writings and spreading religious learning is acknowledged in many sources; see below, sections IV (b), (c), (d), and (f).

Of particular interest is Ardašīr's successor, Šāpūr I (r. 240–270), who is specifically credited with an effort to reassemble the corpus of Zoroastrian religious writings, the *Avesta*, and harmonize Zoroastrian religious doctrine across his realm. Interestingly, some of the assembled texts are said to have dealt with, for example 'movement, time, space, substance, creation, becoming, passing away', and these coincide with the subject matter of Aristotle's *Physics* (*Φυσικὴ ἀκρόασις*). The general tenor of this excerpt and the previous one, which ascribes a similar plan to Šāpūr's father without, however, mentioning the religious dimension, 'suggests very clearly' according to Shaked, 'that in the view of the author of the *Dēnkard* the wisdom contained in those Indian, Greek and other books is part of the original Iranian knowledge, dispersed and then, at the time of Shapur, brought back to Iran and considered together with the Avesta again.'[144]

(1) Dārāy, son of Dārāy, commanded that two written copies of all Avesta and Zand, even as Zardusht had received them from Ohrmazd, be preserved: one in the Royal Treasury and one in the Fortress of Archives.

(2) Walaxš the Aškānian commanded that a memorandum be sent to the provinces (instructing them) to preserve, in the state in which they had come down in (each) province, whatever had survived in purity of the Avesta and Zand as well as every teaching derived from it which, scattered through the land of Iran by the havoc and disruption of Alexander, and by the pillage and plundering of the Macedonians, had remained authoritative, whether written or in oral transmission.

(3) His Majesty Ardašīr, King of kings, son of Pāpag, acting on the just judgment of Tansar,[145] demanded that all those scattered teachings should be brought to the court. Tansar assumed command, and selected those which were trustworthy, and left the rest out of the canon. And thus he decreed: 'Henceforth only those are true compositions which are based on the Mazda-worshipping Religion, for now there is no lack of information and knowledge concerning them.'

144 Shaked, 'Mihr the Judge', 21. Cf. Zaehner, *Zurvan*, 8. On the topos of repatriating ancient knowledge, see further Shaked, 'From Iran to Islam', 58. See also Shaked's partial translation in 'Mihr the Judge', 48. It might be noted that the Achaemenid tradition of the 'Royal Treasury' as a depository of books continues through this time.

145 Tōsar or Tusar/Tansar, a Zoroastrian high priest during the reign of Ardašīr I.

(4) Šābuhr, King of kings, son of Ardašīr, further collected the non-religious writings on medicine, astronomy, movement, time, space, substance, accident, becoming, decay, transformation, logic and other crafts and skills which were dispersed throughout India, Byzantium and other lands, and collated them with the Avesta, and commanded that a copy should be made of all those writings which were flawless, to be deposited in the Royal Treasury [...]

(5) Šābuhr, King of kings, son of Hormizd, induced men from all provinces to orient themselves towards God through disputation, and put forward all oral traditions for consideration and examination. After the triumph of Ādurbād,[146] through his declaration put to trial by ordeal, [in disputation] with all those sectaries and heretics who studied the nasks, he made the following statement: 'Now that we have gained an insight into the Religion in the worldly existence, we shall not tolerate anyone of false religion, and we shall be still more zealous.' And thus did he do.[147]

(6) His present Majesty Xosrow, King of kings, son of Kawād, after he had put down heresy and evil dominion with fullest antagonism, greatly promoted detailed knowledge and investigation in the matter of all heresy within the four estates, according to the revelation of the Religion. And at an assembly (of representatives) of the realm he declared: 'We have recognized the truth of the Mazda-worshipping Religion; and the wise can with confidence establish it in the world by discussion. But to be a bounteous propagator (of the faith) and a foremost sage lies essentially not so much in discussion as in purity of thought, word, act, the guidance of the Good Spirit and the worship of the gods, paid purely in conformity with the Holy Word.'

II.2. *Acts of Religion* (*Dēnkard*), I, 428.15–429.6 (ed. Madan); tr. Zaehner, *Zurvan*, 139.

In the following passage, the unnamed source of book 4 of the *Dēnkard* explicitly mentions Indian and Greek scientific works and describes how they were compared with what the author terms 'the original literature', *bun nibēg*. The *bun nibēg*, the book containing all topics kept at the Royal Treasury (*pad ganj ī šāhīgān*), perhaps an ency-

146 Ādurbād ī Mahrspandān, a Zoroastrian high priest during the reign of Šāpūr II; see Aḥmad Tafażżolī, 'Ādurbād ī Mahrspandān', *EIr*, I, 477.

147 Tōsar seems to have assembled the *mobeds* (priests) who were the repositories of oral knowledge. The version of the *Avesta* he compiled may have differed from that of Vologases, leading to yet further attempts at collecting and unifying Zoroastrian scripture.

clopaedia, was also called *Hangirdīg* (*Collected Works*) or *corpus librorum*, which contained 'every branch of knowledge' (*abar hamāg saxwan*; *Dēnkard*, I, 429.4–5). The attitude this excerpt conveys is clear: whatever their origin and language, Persian scholars freely consulted scientific writings and used them for their own purposes.

> The -----*kośa* of the Indians and the *Megistē* [i.e., Almagest] of the Greeks and other (writings) of that sort were combined with the original literature (the Avesta?) in the Royal Treasury because they offered a reasonable corrective to students of this particular science. Their writings have recently been arranged by scholars, for they were brought from foreign countries: they have been considered and studied, nor have they been neglected or slightly received on account of their inferiority and foreign name. It happened indeed that with the growth of knowledge they became more highly esteemed, and they did not suffer on account of the outlandish names attached to the books. No single book or volume was discovered which contained in its entirety all the science and learning (contained) in (other) books and volumes: but each (school) carried out research on the basis of its own original book or volume. They call the book in the Royal Treasury which deals with all topics the Compendium(?).[148]

II.3. Ibn al-Nadīm, *The Catalogue* (*Kitāb al-fihrist*), 299.9–301.2 (ed. Tağaddud); tr. van Bladel, 'The Arabic History', 44–47, the transliteration adapted.

The famous astrologer Abū Sahl ibn Nawbaḫt (fl. c. 153/770–193/809), a convert from Zoroastrian faith in the service of several early ʿAbbāsid caliphs, wrote a now-lost book on astrology, *Book of the Nahmuṭān on Nativities* (*Kitāb al-nahmuṭān*[149] *fī l-mawālīd*). In the introduction of this book, which Ibn al-Nadīm quotes at length, the author presents a history of science, including the destruction and dispersal of accumulated knowledge during Alexander the Great's campaigns, and the efforts by successive Persian rulers to find and reassemble this learning. This fascinating text, which is quoted by several later Arabic authors, has been translated and commented upon by a number of modern scholars (see, 'II. Traditional Accounts' above).

According to Abū Sahl's narrative, the Babylonians established the sciences, which were then adopted by the Egyptians and the Indians. After a period of neglect, the Babylonian sciences were revived under the mythic Persian King Ǧam, but subsequently dispersed again. The campaigns of Alexander the Great then obliterated and scattered

148 Another translation: Shaki, 'Dēnkard Account', 123 (partial). See also Bailey, *Zoroastrian Problems*, 85–86.

149 On the title of this book, which appears in the manuscripts of Ibn al-Nadīm's *al-Fihrist* as *Kitāb al-nahmuṭān*, see above n. 17.

what was left in Persia. Finally, several Persian rulers made efforts to collect together whatever there was to find of this lost body of learning and compile new texts on the basis of ancient wisdom.

Abū Sahl ibn Nawbaḫt said in the *Kitāb al-nahmuṭān*:

1. The [different] types of sciences and the [various] kinds of books had become numerous, along with the different methods and procedures from which are derived the stars' indications of the existence of things before their causes are manifest and people know them, as the Babylonians described it in their books, and the Egyptians learned from them, and the Indians practiced it in their own country.
2. [This was] just as the first human beings had been before they yielded to sin, committed evil deeds, and fell into the depths of barbarism (*al-ǧahāla*), to the point that their intellects deceived them and led them astray from their wits. For that is what had affected them—in the reports in the books about their affairs and their deeds—to a degree that robbed them of their intellects, confused their wits, and wrecked their religion for them. They became confused and went astray, understanding nothing.
3. They continued like this for some time until they received support from those who followed them, springing up from their progeny, sown from their loins, bringing recollection, comprehension, and understanding of these matters, and knowledge of the past, including the inherent conditions of the world, and regulating its first part, and the beginning of administering its middle part, and the end of its last part, and the condition of its inhabitants, the positions of the celestial circuits of heaven, their paths, their degrees and minutes, their mansions, both the superior and the inferior among them, in their routes and all their courses. That was in the time of King Ǧam, son of Awanǧhān.
4. The scholars understood this and composed books on it, and they explained their compositions. In addition to composing them on that, they described the world, its majesty, starting point of its celestial tethers (*mubtada' asbābihā*) and its foundation, its stars, the status of drugs, remedies, magic incantations, and other instruments of man that he uses for the good and evil things that he desires.
5. That is how they were for a long age until al-Ḍaḥḥāk ibn Qayy became king—(inserted here:) 'From the words of another author: *dah āk* means (in Persian) "ten calamities," and then the Arabs turned it into al-Ḍaḥḥāk; we return to the words of Abū Sahl [...] ibn Qayy in the period, successive stage (*nawba*), and dominance of Jupiter and its authority to govern the years. In the land of al-Sawād [al-'Irāq] he built a city the name

of which was derived from the name of Jupiter. Then he gathered there knowledge and scholars and built twelve palaces (*qaṣr*) according to the number of the zodiacal signs; he named them with their [i.e., the zodiacal signs'] names, stored the books of the scholars (in libraries) [in them], and housed the scholars in them.

(Inserted here:) From the words of another author: He built seven houses according to the number of the seven planets and designated each one of the houses for a man; he designated the house of Mercury for Hermes, the house of Jupiter for Tīnkalūs, and the house of Mars for Ṭīnqarūs; we return to the words of Abū Sahl:

The people submitted to them, submitted to their doctrines, and administered their affairs because they understood their superiority to them in the various sciences and useful mechanisms (*ḥiyal al-manāfiʿ*)[.]

6. [This lasted] until a prophet was sent in that time, but they rejected [him] when he appeared. Their knowledge could not comprehend anything about him. Much of their opinion was confused, their authority was dissolved, and their inclinations and their community became at variance. So every scholar among them went to a country to dwell in it and to be in it and to become chief of its people. One of them was a scholar called Hermes. He was among them one of the most perfect in intellect, most accurate in knowledge, and most subtle in investigation. He went down to the land of Egypt and ruled its people as king, civilized it, improved the conditions of its inhabitants, and revealed his knowledge there.

7. The bulk and the majority of it (the knowledge) remained at Babylon until Alexander, king of the Greeks, went forth, invading the land of Fārs, from a city of the Romaeans called Macedonia (Maqdūniyah), when he refused (to pay) the tribute that was still due to the Babylonians and the kingdom of Fārs, killed the king Dārā, son of Dārā, made himself master over his kingdom, ruined the towers (*al-maǧādil*) built with the demons and giants (*bi-l-šayāṭīn wa-l-ǧabābirah*), and destroyed the varieties of learning that had been inscribed in the different kinds of buildings, written on their stones and timber, tearing them down and burning them, scattering their arrangement.

8. He copied the part of it that was gathered in the offices and the libraries (*fī l-dawāwīn wa-l-ḫazāʾin*) in the city of Iṣṭaḫr. He translated it into the Romaean and Egyptian tongues. Then, after he finished copying what he needed of it, he burned what had been written in Persian (*al-fārisīyah*) and a script called *kaštaǧ*. He took whatever he needed of astrology, medicine, and natural science (*al-ṭabāʾiʿ*), and he sent those books, and

the rest of the sciences, property, treasuries [libraries?] and scholars that he obtained, to the land of Egypt.

9. Some things [of these books] had survived in the region of India and China that the kings of Fārs had copied in the time of their prophet, Zarādušt, and Ǧāmāsb the learned (*al-ʿālim*). They [the kings] preserved them there when their prophet and Ǧāmāsb had warned them about what Alexander would do, that he would conquer their country, destroy what he could of their books and sciences, and transfer it from them to his country.

10. Then learning perished in al-ʿIrāq. The scholars were torn apart, disagreed, and became few. The people became partisan and factious. Every party (*ṭāʾifah*) among them had its own king, so they were called the party kings (*mulūk al-ṭawāʾif*). The kingdom of Rome (by contrast) was united for a single king—after the division, confusion, and warfare that had been among them—after King Alexander, and they became thereby a single power. The kingdom of Babylon remained dispersed, weak, and corrupted.

11. They continued to be subjugated, overcome, unable to protect a sanctuary or to repel an injury, until Ardašīr, son of Bābak, of the lineage of Sāsān, became king. He harmonized their controversy, united their division, conquered their foe, and took mastery over their country. He obtained authority over all of them. He did away with their partisanship and their kingdom became upright for him. Then he sent [messages] to India and China about the books with them, and to Rome, and he copied what had reached them, and he investigated a few remnants that remained in al-ʿIrāq. He gathered those of them that had been scattered and harmonized those of them that were at variance. His son Sābūr did that after him until all those books were copied in Persian, according to the views of Hermes the Babylonian, who had been king over Egypt, and Dorotheus the Syrian, and Fydrws the Greek from the city renowned for learning, Athens, and Ptolemy of Alexandria, and Farmāsb the Indian. They commented on them and taught them to the people just as they had taken from all [these] books, the origin of which was from Babylon.

12. Then, after those two, Kisrā Anūširwān collected and harmonized them and put them into use, because of his desire for and love of learning. The people of every time and age have experiences occurring and science renewed according to the determination of the stars and zodiacal signs, which is in charge of governing time by the command of God—his glory is exalted! End of Abū Sahl's discourse.[150]

150 Other translations: Dodge, *The Fihrist*, 572–575; Saliba, *Islamic Science*, 31–33 (English);

II.4. Al-Bīrūnī, *Vestiges of Past Centuries* (*al-Āṯār al-bāqiyah 'an al-qurūn al-ḫāliyah*), 24.7–12 (ed. Sachau); tr. Sachau, *The Chronology*, 28.11–21, the transliteration adapted.

In his *Vestiges of Past Centuries*, the celebrated polymath al-Bīrūnī (d. c. 440/1048) proposed a division of human history into several eras, starting with the Era of the Creation and the Era of the Deluge, and discussed the expressed opinions of various ethnic and religious communities on these eras.

The significance of the following excerpt, which recounts Persian views on the deluge, does not lie in what it may offer on whether (and why) the deluge was not part of Persian accounts of the history of mankind, but rather in a topos related to the compilation and transmission of knowledge: the preservation of books for posterity. In this particular case, al-Bīrūnī reports that some of his Persian sources credit the legendary third ruler of the world, Tahmūraṯ, with receiving warning of the deluge and taking measures to ensure the survival of the sciences.

It is related that Tahmūrath on receiving the warning of the Deluge—231 years before the Deluge—ordered his people to select a place of good air and soil in his realm. Now they did not find a place that answered better to this description than Ispahān. Thereupon he ordered all scientific books to be preserved for posterity, and to be buried in a part of that place, least exposed to obnoxious influences. In favour of this report we may state that in our time in Jayy, the city of Ispahān, there have been discovered hills, which, on being excavated, disclosed houses filled with many loads of that tree-bark with which arrows and shields are covered, and which is called *Tūz*, bearing inscriptions of which no one was able to say what they are, and what they mean [cf. v.6 below].

II.5. Al-Bīrūnī, *Vestiges of Past Centuries* (*al-Āṯār al-bāqiyah 'an al-qurūn al-ḫāliyah*), 36.2–4 (ed. Sachau); tr. Sachau, *The Chronology*, 42.5–12, slightly adapted.

Appointed Arab governor of Ḫurāsān in 704/705, the soldier and administrator Qutaybah ibn Muslim (d. 96/715–716) consolidated Muslim rule in the area and substantially expanded the territory under Muslim control into Transoxania and Tokharistan. In 711/712, he took advantage of internal conflicts in the neighbouring state of Ḫwārizm,

Cottrell, 'L'Hermès Arabe', cols 394–401 (French). Partial English translations also in Gutas, *Greek Thought*, 39–40; and Pingree, *The Thousands*, 9–12.

which had nominally accepted Muslim rule in the 690s but defected soon after, and reconquered it. Al-Bīrūnī, himself of Ḫwārizmian origin, noted the brutality of Qutaybah's campaign and accused him of extinguishing the flame of Ḫwārizmian culture.[151] Qutaybah's alleged destruction of books and ruthless killing of learned men bears resemblance to reports about atrocious measures taken by the early Muslim conqueror of Egypt, ʿAmr ibn al-ʿĀṣ (d. 42 or 43/662 or 664), and the conqueror of Persia, Saʿd ibn Abī Waqqās (d. betw. 50/670–671 and 58/677–678; see II.7 below).

Qutaybah ibn Muslim had extinguished and ruined in every possible way all those who knew how to write and to read the Ḫwārizmī writing, who knew the history of the country and who studied their sciences. In consequence these things are involved in so much obscurity that it is impossible to obtain an accurate knowledge of the history of the country since the time of Islam (not to speak of pre-Muhammadan times).

II.6. Al-Bīrūnī, *Vestiges of Past Centuries* (*al-Āṯār al-bāqiyah ʿan al-qurūn al-ḫāliyah*), 48.11–15 (ed. Sachau); tr. Sachau, *The Chronology*, 58.5–14, slightly adapted.

When discussing the Ḫwārazmian calendar, al-Bīrūnī observed that the Ḫwārazmians did not have special terms for intercalary days and ascribes this to the extinction of Ḫwārazmian culture during the campaign of Qutaybah ibn Muslim mentioned above (see II.5).

They do not use or even know special names for the *epagomenae* [i.e., intercalary days], but I believe that this fact simply arises from the same confusion regarding these names which prevails among the Persians and Soghdians. For after Qutaybah ibn Muslim al-Bāhilī had killed their [i.e., the Ḫwārazmian] learned men and priests, and had burned their books and writings, they became entirely illiterate (forgot writing and reading), and relied in every knowledge or science which they required solely upon memory. In the long course of time they forgot that on which there had been a divergence of opinion, and kept by memory only that which had been generally agreed upon.

151 According to Bartold, *Turkestan*, 1, no earlier source confirms al-Bīrūnī's dramatic description of the deliberate eradication of Ḫwārazmian culture. Sachau, 'Zur Geschichte', 498–499, suggested that this and the following passage, which refers to the burning of Ḫwārazmian books by Qutaybah, may have been an echo of Alexander the Great's destruction of Persepolis in 330 BC.

II.7. Ibn Ḫaldūn, *The Introduction* (*al-Muqaddima*), in id., *Kitāb al-ʿibar*, I, 631.12–20 (ed. Šaḥādah); tr. Rosenthal, *The Muqaddima*, III, 113–114, slightly adapted.

Arabic sources repeatedly retell the following story, which purports to explain how the Greeks acquired the sciences from the Persians in the aftermath of Alexander the Great's conquests. Ibn Ḫaldūn's version is part of a tradition that stretches back hundreds of years. There are, however, some interesting remarks in his account: Ibn Ḫaldūn links Alexander's purging of Persian intellectual heritage with measures introduced by the Muslim general Saʿd ibn Abī Waqqās after the battle of Qādisīyah in 15/636. The spoils of his campaign are said to have included numerous books that Ibn Abī Waqqās, after requesting instructions from the caliph ʿUmar (r. 13/634–23/644) as to what to do with them, destroyed.[152] The phrasing of the report suggests that Ibn Ḫaldūn may not have believed in the wholesale destruction of Persian books by Alexander: in spite of Alexander's ruinous actions, Muslim troops still found an 'innumerable amount of their books and scholarly records'. We saw this same recurrent motif of cultural devastation in al-Bīrūnī's comments on the culture of his Ḫwārazmian homeland (see II.4–6 above).

Among the Persians, the intellectual sciences played a large and important role, since the Persian dynasties were powerful and ruled without interruption. The intellectual sciences are said to have come to the Greeks from the Persians, (at the time) when Alexander killed Darius and gained control of the Achaemenid empire. At that time, he appropriated the books and sciences of the Persians. However, when the Muslims conquered Persia and came upon an innumerable amount of their books and scholarly records, Saʿd ibn Abī Waqqās wrote to ʿUmar ibn al-Ḫaṭṭāb, asking him for permission to take them and distribute them as booty among the Muslims. On that occasion, ʿUmar wrote him: 'Throw them into the water. If what they contain is right guidance, God has given us better guidance. If it is error, God has protected us against it.' Thus, the (Muslims) threw them into the water or into the fire, and the sciences of the Persians were lost and did not reach us.

152 The episode bears a striking resemblance to the alleged destruction of the library of Alexandria by the Muslim conqueror of Egypt, ʿAmr ibn al-ʿĀṣ. According to Bar Hebraeus's *Compendious History of Dynasties* (*Taʾrīḫ muḫtaṣar al-duwal*, 175.17–176.12), ʿAmr likewise requested directives from the caliph ʿUmar on how to deal with the books in the Alexandrian royal library and was ordered to destroy them on similar grounds.

II.8. Dawlatšāh Samarqandī, *Memoirs of the Poets* (*Taḏkirat al-šuʿarāʾ*), 30.3–11 (ed. Browne); tr. Zakeri.

Roughly two centuries after the events described in the previous example (II.7), the governor of Ḫurāsān, ʿAbdallāh ibn Ṭāhir (d. 230/844–845), was charged with a similar cultural atrocity: according to the much later report by the Persian literary writer Dawlatšāh Samarqandī[153] (d. 900/1495 or 913/1507) quoted below, he ordered that all Persian books, especially those written by Zoroastrian scholars, be destroyed.[154] Needless to say, the claim that ʿAbdallāh ibn Ṭāhir, a highly educated member of the ʿAbbāsid elite and a lover of books and literature,[155] would engage in such an evil act is entirely spurious. Dawlatšāh Samarqandī writes:

According to another report, when the amir ʿAbdallāh ibn Ṭāhir, who was governor of Ḫurāsān in the days of the ʿAbbāsid caliphs, stayed in Nīšāpūr one day, somebody presented him with an old book, either as a gift or because it was customary. (ʿAbdallāh) asked him: 'What book is this?' He said: 'This is the romance of Wāmiq and ʿAḏrāʾ,[156] a pleasing tale which the sages of King Anūširwān compiled.' The amir replied: 'We are the people who read the Qurʾān. We desire nothing besides the Qurʾān and the accounts of the Prophet. To us such books as this are useless, and this book is the work of Zoroastrians [*moġān*], and in our view unacceptable.' He then gave orders that the book be thrown into water and issued a decree commanding that, in his domain, wherever books written by the older Persians and Zoroastrian scholars were kept, they must all be burnt.

III. *Persian Achievements in the Sciences*

III.1. Tatian of Adiabene, *Address to the Greeks* (*Oratio ad Graecos*), 84.1–16; tr. Whittaker, *Oratio ad Graecos and Fragments*, 3, corrected and adapted.

Tatian (d. AD c. 180) was one of the earliest Christian philosophers of the East. Born in Adiabene of Syria, he returned to his homeland after a long sojourn in Rome and opened a Christian school of philosophy. Although thoroughly Hellenized, he was well aware of the achievements of other peoples and the debt owed them by the ancient Greeks. In his polemical *Address to the Greeks*, written about 169, he explicitly reminded

153 See Melvin-Koushki, 'Dawlatshāh Samarqandī'.
154 The episode is also recounted in Browne, *A Literary History*, 12.
155 See Bosworth, 'The Ṭāhirids', 45–79.
156 See B. Utas, 'Wāmiḳ wa ʿAdhrā'', *EI2*, XI, 134.

his readers of this debt. The various combinations of peoples and inventions Tatian listed, including the invention of letter writing and magic by the Persians, were well known to his audience from ancient catalogues of inventions and discoveries.[157]

Do not maintain a totally hostile attitude to foreigners, men of Greece, nor resent their beliefs. For which of your own practices did not have a foreign origin? The most famous of Telmessians[158] invented divination through dreams, Carians foreknowledge through stars, Phrygians and the most ancient Isaurians the lore of bird-flights, Cyprians divination by the entrails of sacrificial animals, Babylonians astronomy, Persians magic, Egyptians geometry, Phoenicians the alphabet. Therefore stop calling your imitations inventions! It was Orpheus who taught you the practice of poetry and singing, also initiation in mysteries; the Etruscans taught sculpture, Egyptian chronological records the composition of history. From Marsyas and Olympus you acquired flute-playing; both were Phrygians, country folk who composed music with a shepherd's pipe. Tyrrhenians taught the trumpet, Cyclopes bronze-work, letter writing the woman who at one time led the Persians, as Hellanicus says; Atossa[159] was her name.[160]

III.2. Al-Ǧāḥiẓ, *Refutation of the Christians* (*al-Muḫtār fī l-radd ʿalā l-naṣārā*), 62.6–63.2 (ed. al-Šarqāwī); tr. Finkel.

Among the numerous works penned by the Baghdad polymath al-Ǧāḥiẓ (d. 869) is his famous *Refutation of the Christians*, which has attracted much attention in recent times. It was written at the time of the caliph al-Mutawakkil (r. 847–861), who, after fighting the Muʿtazilites, was on the move to abolish the considerable freedoms that the 'People of the Book' had gained for themselves since the time of al-Maʾmūn; in particular, the propaganda of Christians (and even Jews) had become widespread. Christian Armenia, perhaps at the instigation of the Byzantine government, had also caused uprisings. Al-Mutawakkil was able to force the Christians into submission by military force and now needed to make a solid intellectual statement against them;

157 Trelenberg, in Tatian, *Oratio ad Graecos*, 84 n. 1 and 85 n. 9.

158 Probably the Lycian town of Telmessus, which housed an oracle of Apollo. Cf. Whittaker in Tatian, *Oratio ad Graecos and Fragments*, 3 n. b.

159 The Atossa here was in all probability not the Persian queen of the same name (d. 475 BC), the wife first of Cambyses II and then Darius I, and the mother of Xerxes I (see I.1, I.2 above), but a legendary figure; see Whittaker in Tatian, *Oratio ad Graecos and Fragments*, 3 n. d.

160 Cf. the translation by B. Pratten in Tatian, *Address to the Greeks*, 5, and Trelenberg's German translation in Tatian, *Oratio ad Graecos*, 85.

this was commissioned from the vizier Fatḥ ibn Ḫāqān to al-Ǧāḥiẓ. In the treatise, he applies his masterly rhetoric to write a scathing polemic against the Christians. Among other things, al-Ǧāḥiẓ rejects the possibility, since ancient Greek writers lived in the distant past, that their books and intellectual heritage could be claimed by contemporary generations of Byzantines and Christians.[161]

If our masses knew that the Christians and Byzantines have no science and rhetoric, and are not people of deep reflection, and possess nothing but the handicrafts of woodwork, carpentry, the art of painting and silk-weaving, they would remove them from the list of men of culture, and would strike their names off the roll of philosophers and scientists. For the books of *Logic*, *De Generatione et Corruptione*, *Meteorologica*, etc., are by Aristotle, and he was neither Byzantine nor Christian. The *Almagest* is by Ptolemy, and he was neither Byzantine nor Christian. The *Book of Euclid* [*Elements*?] is by Euclid, and he was neither Byzantine nor Christian. *The Book of Medicine* is by Galen, and he was neither Byzantine nor Christian. This also holds true for the books of Democritus, Hippocrates, Plato, and others. All these authors belong to a people who became extinct, but whose intellectual reputation has endured. They were the Greeks, whose religion was unlike the religion of the Christians, and their way of life was totally different. The Greeks were savants, and these are manufacturers. It was by chance of geographical proximity that they adopted the Greek books. They either ascribed some of the books to themselves or tampered with their contents so as to make them appear Christian. However, if the work was popular and its contents well known, such that they could not change its title, they then claim that the Greeks were one of the Byzantine tribes, and take pride in the superiority of their religion over that of the Jews, Arabs, and Hindus. They even claim that our wise men are followers of their wise men, and that our philosophers follow theirs. Such is the state of affairs![162]

III.3. Ibn Qutaybah, *The Superiority of the Arabs and Insight into their Sciences* (*Faḍl al-ʿarab wa-l-tanbīh ʿalā ʿulūmihā*), 205.1–206.15 (ed. Ḫāliṣ); tr. Savant and Webb, *The Excellence of the Arabs*, 214.14–216.15.

The theologian and litterateur Ibn Qutaybah (d. 276/889) wrote his *Superiority of the Arabs* to defend the artistic and intellectual achievements of the Arabs at a time of serious political instability: during his lifetime Arab noble families had been sidelined

161 Webb, 'Foreign Books', 42.

162 I have used Finkel, 'A Risāla of al-Jāḥiẓ', 326–327; Fletcher, *Anti-Christian Polemic*, 69; Gibson, 'Closest in Friendship?', 118–120; and Hainthaler, 'Ǧāḥiẓ und seine Schrift', 243–256.

in the caliphate by new military and administrative elites of Turkic and eastern Iranian background, who seized the state for personal gain, and deposed or murdered caliphs at will.[163]

While in general Ibn Qutaybah was respectful of the accomplishments of other peoples, his observations about the Persians are particularly scathing, though himself was of Persian extraction. He seems to have resented them, at least at some point, for their cultural confidence, which he interpreted as a lack of respect for Arabs and the Arabic language, and he may have blamed them for what he saw as the dilution of the Arabness of Islam itself.[164] In the present excerpt, Ibn Qutaybah denies the Persians any original literary and scholarly achievements and claims that they had merely imitated the peoples that preceded them. In this context, he also mentions the Persian Alexander narrative, according to which Alexander the Great stole and then destroyed Persian literature (see IV.1). He argues that this episode is told only by the Persians and lacks corroborating evidence from non-Persian sources.

But everything the Persians know, they learned from other peoples—they walked in the footsteps of other nations.

When we inquire about the forefathers of medicine, we will be led to Hippocrates and Galen; when we inquire about the origins of astronomy and computation, we will be led to Euclid and the *Almagest*; when we inquire about the realm of logic, we will be led to the book of Aristotle; when we inquire about musicology, we will be led to the *Book of Music*. All these works were written by the Romans and Greeks. The Persians have no more of a share in them than any other beneficiary who profits from them.

The Romans developed agronomy; the Indians have chess, *Kalīlah and Dimnah*, and computation using nine sigla, and they possess ancient, accurate medical knowledge derived from discoveries quite different from those of the Greeks.

The Persians themselves admit this in their *Book of Kings*, where they record that when Shapur grew old and his vision dulled and his powers faded, he complained to his subjects that he was too weak to rule and ordered them to find someone proficient to replace him. But this was too much for them to bear, and they rejected this suggestion outright. They asked his permission to find him a doctor, and he agreed. They sent a messenger with great gifts to the king of India, asking him to send the best doctor he had. The doctor tended to Shapur until the king recovered his strength, his skin regained its color, his sight

163 Ibn Qutaybah, *The Excellence*, x–xi.

164 Ibid., xv.

returned, and he could ride on the hunt and cavort with women. Shapur repaid the doctor with great munificence, and asked him to choose the part of the kingdom where he would most like to settle. The doctor chose Susa, settled there, and the people of Susa thereby inherited his medical knowledge.

They also say that a Roman prisoner was settled in Susa, and that the Persians learned medicine from him, too. Thus the people of Susa became the doctors of the Persians. These accounts must be true since we find a combination of Indian and Roman elements in the *Medical Compendium* composed in Susa.

Some of the Bigots[165] argue, on the other hand, that when Alexander invaded Persia, killed and captured its inhabitants, and laid waste, he transported their scientific books to Greece, had them translated into Greek and burned the originals—thus Persian knowledge became Greek. But we can rebut this by saying: 'The first account is corroborated by your very own books, whereas the second account is a claim about the actions of a different people, so you need corroboration from the Romans to prove it.' [See further below, IV.7.]

III.4. Ibn al-Faqīh, *The Book of Countries* (*Kitāb al-buldān*), 319.12–15 (ed. al-Hādī); tr. Zakeri.

Ibn al-Faqīh al-Hamadānī's *The Book of Countries*, a ninth-century geographical work that is more literary in vein than technical, includes a long debate in a caliphal session (*maǧlis*) on the relative superiority of countries and cities to one another in which he quotes the following comment by the Zoroastrian Yazdǧird, son of Mahbandād al-Kasrawī, comparing the merits of Babylon (Baghdad) with those of Egypt:

They [i.e., the Egyptians] cannot boast to us [i.e., the Babylonians] neither of Hermes at the time he lived or of anyone like him in our time; nor of Aġatīmūn at this time. They may or may not know that Babylon in Iraq was the centre of science and scientists, and the hub of wisdom and the wise men. But when the victorious king [i.e., Alexander the Great] defeated the kings of the inhabited world, he transferred the wisdom of the past and of wise men to the land of al-Mīnā [i.e., Egypt?] or thereabouts.

III.5. Ibn Abī Ṭāhir, *The Book of Baghdad* (*Kitāb Baġdād*), I, 157.7–158.5 (ed. Keller); dialogue tr. Toorawa, *Ibn Abī Ṭāhir Ṭayfūr*, 76, slightly adapted here.

Ancient Persian scientific learning was admired and sought after in the early ʿAbbāsid period. Shawkat Toorawa writes that the 'role of Persian culture in the elaboration

---

165 I.e., people who deny the pre-eminence of the Arabs, often in favour of the Persians; see Ibn Qutaybah, *The Excellence*, xvi–xvii.

of Arab-Islamic culture has long been recognized'.[166] This is illustrated in the following anecdote by the Persian-born author and critic Ibn Abī Ṭāhir (d. 280/893),[167] in his multi-volume history of Baghdad, the *Book of Baghdad*. The episode takes place in Raqqah (Syria), at the court of the governor of Ḫurāsān, Muḥammad ibn Ṭāhir ibn al-Ḥusayn, where the litterateur Abū ʿAmr Kulṯūm ibn ʿAmr al-ʿAttābī (d. 208/823 or 220/835) addresses a Persian-speaker named Yaḥyā ibn al-Ḥasan in Persian. The dialogue ensues.

Abū l-Ḥasan Aḥmad ibn Muḥammad al-Muhallabī told me this on the authority of Yaḥyā ibn al-Ḥasan ibn ʿAlī ibn Muʿāḏ ibn Muslim, who said: 'I was at Raqqah with Muḥammad ibn Ṭāhir ibn al-Ḥusayn, (sitting) at a lake, when I called one of my servants and talked to him in Persian. Al-ʿAttābī joined us, witnessed our exchange and addressed me in Persian. I said to him: "Abū ʿAmr, what's with you and this gibberish?" […] He answered me saying, "I've gone to that country of yours three times and copied the Persian books that are in the library [*ḫizāna*] in Merv." The books were brought there and added to the holdings by Yazdajird; they have remained there to this day. Then he said, "I copied what I needed, and then traveled to Nishapur. I had traveled ten farsakhs to a village known as Dhūdar, when I remembered a book I had not yet finished with. So I returned to Merv and spent several months there." […] So I said, "Abū ʿAmr, why did you transcribe Persian books?", and he replied, "Are ideas [*maʿānī*] and eloquence anywhere but in the books of the Persians? We may have the medium [*luġah*, language, i.e., Arabic], but they have the time-honored meanings [*maʿānī*]." After that, we would (often) trade gems of Persian literature and converse at length in Persian.'[168]

III.6. Ibn Ḫaldūn, *The Introduction* (*al-Muqaddima*), I, 631.2–7 (ed. Šaḥādah); tr. Rosenthal, *Muqaddima*, III, 113, slightly adapted.

Ibn Ḫaldūn, the father of modern sociology, believed that the two great nations of the past who left their mark on ancient science were the Persians and the Greeks. However, the Greeks were obliged to the Persians for excellence in philosophical and scientific achievements, since Alexander the Great had plundered and transferred Persian science and learning to the Greeks, providing them with the means whereby their genius might blossom. He writes as follows.

166 Toorawa, *Ibn Abī Ṭāhir Ṭayfūr*, 76.
167 Ibid., 3–5; and Rosenthal, 'Ibn Abī Ṭāhir', *EI2*, III, 682–693.
168 Another translation: Ibn Abī Ṭāhir Ṭayfūr, *Taʾrīḫ Baġdād*, II, 71 (German).

It should be known that as far as our historical information goes, these sciences [i.e., the intellectual sciences, *al-ʿulūm al-ʿaqlīya*] were most extensively cultivated by the two great pre-Islamic nations, the Persians and the Greeks [*rūm*]. According to the information we have, the sciences were greatly in demand among them, because they possessed an abundant civilization and were the ruling nations immediately before Islam and its time. In their regions and cities, the sciences flourished greatly. The Chaldaeans and before them the Syrians, as well as their contemporaries, the Copts, were much concerned with sorcery and astrology and the related subjects of powerful (charms) and talismans. The Persian and Greek nations learned these things from them.

IV. *Persian Rulers: Preserving and Disseminating Knowledge*

(a) *Before the Sasanians*

IV.1. Ḥamza al-Iṣfahānī, *History of the Years of Persian Kings* (*Taʾrīḫ sinī l-mulūk al-ʿaǧam*), 34.16–24 (ed. Maskūnī); tr. Daudpota, *The Annals*, 90.

The historian Ḥamza al-Iṣfahānī (fl. 961), who was well versed in Persian antiquities, attributes the burning of the religious scriptures and obliteration of Persia's scientific legacy and historical records to Alexander, and the partial recovery of the sciences to the Parthians.

Alexander parcelled out the eastern countries among the tribal kings, and took away from their country to the West the sciences of astronomy, medicine, philosophy and agriculture, after he had got them translated into Greek and Coptic. When Alexander perished and the country fell into the hands of the tribal kings, they gave up all warfare and contention among themselves. They tried to excel one another only in abstruse problems. It was in their days that books such as Marwak, Sindbād, Barsinās, Shīmās and the like, which are commonly found in the hands of people, were composed. Such books were about 70 in number.[169]

169 For a discussion of the titles cited in this passage, see Perry, *The Origin*, 28–29; and Zakeri, *Persian Wisdom*, I, 116–118.

IV.2. Al-Ṯaʿālibī, *History of Exalted Lives* (*Taʾrīḫ ġurar al-siyar*), 458.5–459.5 (ed. Zotenberg, *Histoire des Rois des Perses*); tr. Zakeri.

When the eleventh-century historian Abū Manṣūr al-Ṯaʿālibī wrote his universal history, he used Persian sources for his account of the early history of Iran. In the following excerpt, he credits an Aškānian king[170] named Aqfūr with returning the books plundered by Alexander the Great. Al-Ṯāʿālibī's chronology is difficult to reconstruct: he claims that Aqfūr was succeeded by his son Šāpūr, who allegedly lived at the time of Jesus. We know of no Parthian ruler by that name at that time, but do know of several from the following Sasanian dynasty (224–651), some of whom were famed for their learning and their efforts in restoring ancient Persian knowledge (see IV.3, IV.4).

King Aqfūr the Aškānian

Ruler of Madāʾin, most of ʿIrāq, and Fārs. Other princes addressed him as king in their correspondence and offered him gifts, not as a tribute, but out of respect. He recovered the banner of the Kayanids [i.e., Achaemenids, *drafš kāviyān*] from the place where it was hidden and carefully preserved it. He defeated the Greek prince who ruled Mosul and Sawād on behalf of Alexander and drove him out. He then raided Roman territory and, wishing to avenge (the death of) Dārā,[171] he demonstrated the power of his arms to the greater part (of the people there), carrying off the men in ships and drowning them, so that large numbers of them were killed. He destroyed many of their fortresses and brought back the works of medicine, astronomy, and philosophy that Alexander had brought to their (country). He reduced the tax burden of his subjects and conducted himself in the best possible way.

(b) *The Sasanians: Ardašīr I*

IV.3. Al-Ṯaʿālibī, *History of Exalted Lives* (*Taʾrīḫ ġurar al-siyar*), 485.2–4 (ed. and French tr. Zotenberg, *Histoire des Rois des Perses*); tr. by Zakeri.

With al-Ṯaʿālibī's next report, we are on safer historical ground. It ascribes an effort to collect and create new copies of the dispersed Persian literature to Ardašīr I (r. 224–242), the founder of the Sasanian dynasty.

---

170 I.e., a member of the Arsacid dynasty of the Parthian Empire (247 BC–AD 224).

171 I.e., Darius III (r. 336–30 BC), the last Achaemenid ruler of Persia.

(Ardašīr) ordered the collection of copies of the religious, medical, and astronomical books, some of which Alexander had burnt and the largest part of which he had carried off to Rome. He had them newly transcribed and verified, sparing no effort and spending large sums of money on them.[172]

(c) *Šāpūr I*

IV.4. Abū Sulaymān al-Siǧistānī, *Selection from the Depository of Wisdom* (*Muntaḫab ṣiwān al-ḥikma*), 12.3–6 (ed. Dunlop); tr. Stern, 'Abū 'Īsā Ibn al-Munajjim's Chronography', 454.

The next source includes a detail that seems to confirm the cultural exchanges between Persians and Romans during the reign of Šāpūr I mentioned in the *Dēnkard*. As we have seen above, books often changed hands in the course of military campaigns, either through looting (see II.3 above), or as part of a peace deal. In his *Selections*, a history of Greek and Islamic philosophers, the tenth-century philosopher Abū Sulaymān al-Siǧistānī (d. c. 375/985)[173] briefly alludes to the events of the early 240s, when Persian troops invaded Mesopotamia and Syria. In retaliation, the Roman Emperor Gordian III (r. 238–244) led a large army into Persia and successfully pushed back the Persians, but died during the campaign. His successor, Philip I 'the Arab' (r. 244–249), was forced to conclude a humiliating peace and relinquish the territories conquered by Gordian.

The chronology of events in Abū Sulaymān's report is somewhat confused: it is not Šāpūr I, son of Ardašīr, who was nicknamed *Ḏū l-aktāf*, but Šāpūr II (r. 309–379, see IV.7 below). Since Gordian III died campaigning in Persia, the books mentioned in this story may have changed hands as part of the peace concluded in the same year between Šāpūr and Philip.[174]

He [(Aḥmad) ibn 'Alī ibn Yaḥyā] says: The kings of Persia conquered the provinces of the Greeks and the Romaeans, and took several towns which were repositories of books about their philosophy and science, such as Upper Mesopotamia, Syria, Egypt, Constantinople, etc. They took the books of the philosophers, which were to be found there, partly (as booty), partly as presents. The books of astronomy, geometry, arithmetic, music, medicine and mechanics were given as presents by Gordianus king of the Romaeans to Sābūr, son of Ardašīr, called the 'Man of the Shoulders.'

---

172 Another translation: al-Ṯaʿālibī, *Ġurar*, 485 (French).

173 See Stern, 'Abū Sulaymān Muḥammad b. Ṭāhir b. Bahrām al-Sijistānī al-Manṭiqī', *EI2*, I, 151–152.

174 See Stern, 'Abū 'Īsā Ibn al-Munajjim's Chronography', 448; and Göbl, *Der Triumph*, 7–31.

IV.5. Abū l-Fidā', *A Short History of Mankind* (*al-Muḫtaṣar fī aḫbār al-bašar*), I, 47.23–48.1 (Cairo edition); tr. Stern, 'Abū 'Īsā Ibn al-Munajjim's Chronography', 454.

Abū Sulaymān al-Siǧistānī's somewhat confused chronology (IV.4) reappears in an abbreviated form of the same events in a much later source, a history from the Creation to his own days compiled by the Kurdish historian Ismā'īl ibn 'Alī Abū l-Fidā' (d. 732/1331)[175] and entitled *A Short History of Mankind*. Abū l-Fidā' added a point that reveals the importance Greek books seem to have held for Šāpūr: his personal interest in Greek philosophy, which led him to commission translations into Persian.

After him [i.e., Ardašīr] his son, Sābūr, son of Ardašīr, reigned for 31 years and 6 months [...] When eleven years of his reign had passed, he went with his army and took Nisibis from the Romaeans. He then entered the territory of the Romaeans, who were at that time pagans, before their adoption of Christianity; he conquered several towns in Syria by force and killed their inhabitants. Subsequently he went against Rome. The king of the Romaeans, who was at that time Gordianus [...] made peace with Sābūr and showed him obedience. Sābūr was greatly interested in collecting the philosophical books of the Greeks and translated them into the Persian language.

(d) *Šāpūr II*

IV.6. Ibn al-Nadīm, *The Catalogue* (*Kitāb al-fihrist*), 302.15–303.2 (ed. Taǧaddud); tr. Rosenthal, *The Classical Heritage*, 45–47, slightly adapted.

The following report deals with events during the reign of the Sasanian king Šāpūr II (309–379). It describes a key episode in the history of conflict between Rome and Persia, framing it as a chapter in the history of philosophy. It tells of the reintroduction of philosophy by the Roman Emperor Julian (r. 361–363), his disastrous Persian campaign and death, and the end of the philosophical interlude in the wake of the accession of Julian's successor and the peace between Rome and Persia. At the very end, the report briefly mentions the translation of some logical and medical texts into Persian and their subsequent translation into Arabic by the Persian writer and court secretary Ibn al-Muqaffa' (d. c. 139/757).[176]

175 See Talmon-Heller, 'Abū l-Fidā''.

176 See Latham, 'Ebn al-Moqaffa', Abū Moḥammad 'Abd Allāh Rōzbeh', *EIr*, VIII, 39–43.

The report forms part of the series of narrations about the origins and history of philosophy and the sciences that prefaces Book 7 of Ibn al-Nadīm's *Catalogue*.[177] They all refer to the important role played by Persia and Persian rulers, either in the preservation of knowledge or its recovery after it had been dispersed.

According to another report, philosophy was prohibited in olden days, and only sages and those known to be receptive to it by nature were permitted to occupy themselves with it. Philosophers used to study the horoscopes of those who wished to become philosophers. If someone's horoscope indicated that he was destined for philosophy, they accepted him as an attendant and instructed him in philosophy. Otherwise, they did not do so. Philosophy was known among the Greeks and the Romans before Christianity. When the Romans became Christians, it was forbidden to them. They burnt part of the philosophical literature and part of it they preserved in libraries. They were not allowed to discuss anything philosophical that contradicted the prophetic laws.

Later the Romans [i.e., Byzantines] lapsed from Christianity and turned again to the views of the philosophers. The reason was this: the Roman Emperor Julian, whose minister was Themistius, the commentator of Aristotle's works, was staying in Antioch when Sābūr Ḏū l-Aktāf[178] attacked him. Julian defeated him, either in battle or, as is related, because Sābūr had marched to the land of the Byzantines in order to seize control of it. However, his intentions became known, and he was seized—reports differ on this point. Julian penetrated into Persian territory as far as Ǧondēšāpūr—to this day there is a breach (in the wall) there called 'the Byzantine Breach'—and besieged the Persian leaders, their knights, and other guardians of the empire. The siege lasted a long time, and Julian had difficulty in capturing the city.

Meanwhile, Sābūr was kept prisoner in Julian's castle in Byzantine territory. His daughter fell in love with him and let him escape. He succeeded in passing through the country unrecognized, reaching Ǧondēšāpūr and entering it. This gave his friends there fresh courage. Soon they made a sally and attacked the Byzantines, since they regarded the escape of Sābūr as a favorable omen. Julian was taken prisoner and Sābūr had him killed. The Byzantines were left without a ruler. Now Constantine the Elder was in the army, and as the Byzantines could

177 For the first narration, Abū Sahl ibn Nawbaḫt's history of learning, see II.3 above; for the second one by the ʿAbbāsid court astrologer Abū Maʿšar (d. 272/886), which recounts how Persian kings saved ancient learning from the deluge, see IV.5 below.

178 I.e., Šāpūr II. He acquired his nickname, 'he who pierces' or 'pulls out shoulders', during his successful campaign against Arab nomad tribes in 325.

not agree whom they should appoint as ruler and, besides, were too weak to resist him, since Sābūr was interested in him, Sābūr appointed him ruler. Later he also showed himself favorably inclined to the Byzantines for his sake and allowed them to leave his territory after imposing on Constantine the condition that he would plant an olive-tree for every date palm that had been felled in southern Iraq and Persia, and would send him from Byzantine territory first implements and then people in order to rebuild what Julian had destroyed. This he did.

So the country became Christian again, as it had been. Philosophical works were again prohibited, as before, and stored in libraries to the present day. In olden times the Persians had translated some works on logic and medicine into Persian, and these were later translated further by ʿAbdallāh ibn al-Muqaffaʿ and others into Arabic.[179]

IV.7. Ibn Qutaybah, *al-Maʿārif*, 658.17–659.2 (ed. ʿUkāšah); tr. Zakeri.

Shāpūr II [...]: an early decrepitude befell him, and ills increased in him, so he sent to the king of India asking him to send him a doctor. [This doctor] treated him until his gait and his skin gained strength, his eyesight became strong, his desire for women returned, and he became able to ride. Consequently, he granted favours to this doctor and let him choose from his country a city to settle in, and he chose the city of Sūs [i.e., Shūsh/Susa in Khuzistān], where he stayed until he died. His medical knowledge was passed on to the people of Susa, and they became the doctors of Persian people because of that, as well as by what they inherited from the Byzantine captives who settled there. [See above, III.3.]

(e) *Bahrām* V

IV.8. Niẓāmī, *Haft Paykar*, in *Kullīyāt-e ḫamsa-e Niẓāmī*, 66.1–14 (ed. Dastgardī); tr. Meisami, *The Haft Paykar*, 42.14–43.17.

The poet Niẓāmī (d. 598/1209) recites the story of how the prince Bahrām Gūr (Wahrām Gōr, r. 421–439) was sent as a youth to al-Ḥīrah, the capital of the Laḫmids, an Arab vassal kingdom of the Sasanian empire, where a Zoroastrian priest taught him Persian, Arabic, and Greek, and where he also learned geometry and other subjects. According to this account, he became so erudite that by interpreting astrological 'indicators'

179 Other English translations: Dodge, *The Fihrist*, 579–581; and Saliba, *Islamic Science*, 41–42.

(*namūdārs*) taken from astronomical tables (*zīǧ*es) and by employing an astrolabe, he was able to unveil all the mysteries of the universe. Poetic licence aside, a report about Bahrām Gūr's stay at the Laḫmīd court and his wide-ranging education also found its way into the *History of Prophets and Kings* (*Taʾrīḫ al-rusul wa-l-mulūk*) by the tenth-century historian al-Ṭabarī (d. 310/923). The young prince allegedly asked for 'erudite educators, well trained in the methods of teaching, who can instruct me in the art of writing, archery and the knowledge of the law'.[180]

The prince, within that lofty tower,
was raised and taught for several years.
Save study he had no delight;
to knowledge, reason was his guide.
The Persian, Greek, and Arab speech
a Magian master did him teach.
Munzir, that skilled and loving king,
knew well the heavenly reckoning:
The seven planets and twelve signs
disclosed their secrets to his mind.
With geometric lines he worked,
all problems solved, Mijistî-like.
He had surveyed the Nile-blue sphere
in every part. From secret stores
Of far-seeing vision, he had gained
great wisdom through his keenest mind.
Finding the prince in judgement wise,
a student skilled in mysteries,
He gently slate and stylus brought,
and Heaven's secrets to him taught.
Each secret notion that was hidden,
whether it spoke of earth or Heaven,
He joined together, one by one,
and taught the prince, until Bahrâm
Became so wise and learned that
he knew the truth of every art.
With true celestial reckoning
he took the veil from the unseen;

180 Al-Ṭabarī, *Taʾrīḫ*, II, 69.10; translation in al-Ṭabarī, *The History*, V, 83.

Then, setting up his style and slate,
unlocked the scheme of heavenly fate.[181]

(f) *Ḫusraw I Anūšīrwān*

IV.9. Agathias, *Histories* (*Historiae*), 77.4–78.5 (section 28.1–6) (ed. Keydell); tr. Frendo, *The Histories*, 62.19–63.14 (section 28.1–6).

The Sasanian ruler most often cited as an avid supporter of translation, especially of Greek philosophy, is Ḫusraw I, called 'the Immortal Soul' (Anūšīrwān) (r. 531–579). His reputation reached all the way to Constantinople, where the historian Agathias (d. c. 580),[182] who for some of the information on the history of Persian rulers in his *Histories* relied on extracts from Persian sources,[183] felt compelled to comment on it. His account, while entertaining, is mostly hostile. It tries to paint both Ḫusraw as well as those who disseminated and believed in his alleged philosophical expertise as gullible fools and betrays little understanding of Persian culture, religion, and politics.[184]

28.1 After first saying a few words about Chosroes I shall return directly to my earlier narrative. Chosroes has been praised and admired quite beyond his deserts not just by the Persians but even by some Romans. He is in fact credited with being a lover of literature and a profound student of philosophy and somebody is supposed to have translated the works of Greek literature into Persian for him. 2 It is rumored moreover that he has absorbed the whole of the Stagirite [i.e., Aristotle] more thoroughly than the Paeanian orator [i.e., Demosthenes][185] absorbed the works of the son of Olorus [i.e., Thucydides],[186] that his mind is filled with the doctrines of Plato the son of Ariston and that not even the *Timaeus*, bristling as it does with geometrical theorems and scientific speculations, would elude his grasp, nor for that matter the Phaedo or the Gorgias or any other of the polished and more intricate dialogues, as for instance the *Parmenides*. 3 Personally, I could never bring myself to believe that he was so remarkably well-educated and intellectually brilliant. How could

181 See also Charles Wilson's translation in Nīẓāmī, *The Haft Paikar*, I, 46–47.

182 For his life and work, see Cameron, *Agathias*, 1–12.

183 Cf. Cameron, 'Agathias', 74–76, 112–116.

184 Cf. Cameron, 'Agathias', 114–115.

185 I.e., the Athenian orator Demosthenes (d. 322 BC), who lived in the Athenian district of Paiania.

186 I.e., the historian Thucydides (d. c. 400 BC). The idea is that Ḫusraw was allegedly more familiar with the writings of Aristotle than Demosthenes was with those of Thucydides.

the purity and nobility of those time-honoured writings with all their exactitude and felicity of expression be preserved in an uncouth and uncivilized tongue [i.e., Middle Persian]? 4 Moreover one may well ask how a man brought up from childhood in the glamorous atmosphere of the court, surrounded by pomp and adulation, and then succeeding to a utterly barbarous style of life of which battles and maneuvers were a regular feature, could hope to achieve any real competence or proficiency in this branch of learning. 5 Yet if people were to praise him on the score that, in spite of being a Persian and in spite of being weighed down with the cares of empire and the responsibility of governing so many nations, he still showed some interest in acquiring a smattering of literature and liked to be considered something of a dilettante, in that case I should add my own voice to the general chorus and should not hesitate to regard him as superior to the rest of the barbarians. 6 But those who attribute exceptional wisdom to him and call him the rival of all philosophers that have ever lived, claiming that, in the manner of the Peripatetic definition of superior culture, he has mastered every branch of science, thereby disclose the unreality of their pretensions and make it plain to all that they are merely echoing the ill-considered opinions of the crowd.[187]

IV.10. Agathias, *Histories* (*Historiae*), 80.5–81.21 (sections 30.3–7 and 31.1–4) (ed. Keydell), tr. Frendo, *The Histories*, 65.1–66.16 (sections 30.3–7 and 31.1–4).

As part of his almost gleeful demolition of Ḫusraw's reputation as a wise and just ruler, Agathias included a long report about the doings of a philosophical impostor, a Syrian physician named Uranius, who was notorious in Constantinople for his grandiloquent outpourings on philosophical issues. Uranius supposedly managed to insinuate himself into the retinue of the Byzantine ambassador to Persia and presented himself to Ḫusraw, who was promptly taken in by him.[188]

To add emphasis to his portrayal of Ḫusraw as inept, Agathias then launches into the story of a very different visit, to the court of Ḫusraw's father, Kawād I (r. 488–496 and 498–531), by men who actually deserved to be called philosophers. The following, decidedly acerbic account is our only source for this remarkable episode, which captured the imagination of numerous historians of philosophy: the emigration of a number of Neoplatonic philosophers to Persia following the closure of the School of Athens in 529 by Emperor Justinian I (r. 527–565).[189]

187 See also Cameron's translation in 'Agathias', 165.

188 See Agathias, *The Histories*, 64–65; cf. Cameron's translation in 'Agathias', 167.

189 Migrating to Persia to escape religious persecution was not unprecedented: after the closure of the Persian School of Edessa in 489 by Justinian's predecessor Zeno (r. 474–475

Justinian's religious policies, especially his persecution of Monophysites, are a matter of record, as are his attempts to suppress Hellenic paganism for good.[190] That this was not a good time to teach pagan philosophy in Byzantium is obvious. In spite of its hostility, Agathias's account also shows that the Persian authorities not only welcomed the philosophical refugees, but also ensured their safety once they had resolved to return to Byzantium a few years later.

The few years these philosophers spent in Persia must have seen some vigorous exchanges of ideas between them and their Persian hosts. A taste of the wide range of topics discussed is documented in a Greek treatise on key philosophical issues written by one of Ḫusraw's guests, the philosopher Priscian (fl. first half of the sixth century): his *Answers to Ḫusraw* (*Solutiones ad Chosroem*). It survives in a Latin translation possibly produced in the ninth century by a translator who did not understand the book's subject matter.[191] In the preface to this treatise, Priscian writes these revealing dedicatory words: 'When I present to you this book, I have done nothing but to offer you a fruit that I have picked from your own garden. It is in the same fashion that people make sacrifice to Gods from among their own creatures.'[192] Paul the Persian, a contemporary philosopher and convert to Zoroastrianism active at the court of Ḫusraw I in the same period, expressed the same sentiment in the introduction to his book on logic, which he addressed to the king.[193] It is therefore more likely that the philosophers and Ḫusraw parted on good terms, especially since the latter took care to include his guests' well-being and freedom from persecution on their return to Byzantium as one of the terms of the peace treaty he concluded with Justinian in 532.

30.3 [Ḫusraw] swore on many occasions that he had never before seen his [Uranius's] equal, in spite of the fact that he had previously beheld real philosophers of great distinction who had come to his court from these parts.

Not long before Damascius of Syria, Simplicius of Cilicia, Eulamius of Phrygia, Priscian of Lydia, Hermes and Diogenes of Phoenicia and Isidore of Gaza, all of them, to use a poetic turn of phrase, the quintessential flower of the philosophers of our age, had come to the conclusion, since the official religion of the Roman empire was not to their liking, that the Persian state was much superior. So they gave a ready hearing to the stories in general circulation

---

and 476–491), a number of Nestorian physicians and teachers left the city for the safety of Nisibis and Ǧondēšāpūr on Persian territory; cf. Browne, *Revised Translation*, 146.

190 See for example the description in John Malalas, *The Chronicle*, 262–263.

191 See Priscian of Lydia, *Answers*, 1, 5.

192 Ibid., 13.

193 Cf. Gutas, 'Paul the Persian', 246–247.

according to which Persia was the land of 'Plato's philosopher king' in which justice reigned supreme. Apparently the subjects too were models of decency and good behaviour and there was no such thing as theft, brigandage or any other sort of crime. Even if some valuable object were left in no matter how remote a spot nobody who came across it would make off with it, but it would stay put and, without any one's guarding it, would be virtually kept safe for whoever left it until such a time as he should return.

4 Elated therefore by these reports which they accepted as true, and also because they were forbidden by law to take part in public life with impunity owing to the fact that they did not conform to the established religion, they left immediately and set off for a strange land whose ways were completely foreign to their own, determined to make their homes there. 5 But in the first place they discovered that those in authority were overbearing and vainglorious and so had nothing but disgust and opprobrium for them. In the second place they realized that there were large numbers of housebreakers and robbers, some of whom were apprehended while others escaped detection, and that every form of crime was committed. 6 The powerful in fact ill-treated the weak outrageously and displayed considerable cruelty and inhumanity in their dealings with one another. But the most extraordinary thing of all was that even though a man could and did have any number of wives people still had the effrontery to commit adultery. 7 The philosophers were disgusted by all these things and blamed themselves for ever having made the move.

31.1 The opportunity of conversing with the king proved a further disappointment. It was that monarch's proud boast that he was a student of philosophy but his knowledge of the subject was utterly superficial. There was no common ground either in matters of religion since he observed the practices I have already described. Finally the vicious promiscuity which characterized Persian society was more than the philosophers could stand. All these factors, then, combined to send them hurrying back home as fast as they could go. 2 So despite the king's affection for them and despite the fact that he invited them to stay they felt that merely to set foot on Roman territory, even if it meant instant death, was preferable to a life of distinction in Persia. Accordingly they resolved to see the last of barbarian hospitality and all returned home.

3 Nevertheless they derived from their stay abroad a benefit which was neither slight nor negligible, but which was to secure them peace of mind and contentment for the rest of their days. 4 A clause was inserted in fact in the treaty, which at that time was being concluded between the Romans and the Persians, to the effect that the philosophers should be allowed to return to their homes and to live out their lives in peace without being compelled to alter their traditional religious beliefs or to accept any view which did not coincide with

them. Chosroes insisted on the inclusion of this point and made the ratification and continued observance of the truce conditional on its implementation.[194]

IV.11. Miskawayh, *The Experiences of Nations* (*Taǧārib al-umam*), I, 203.6–16 (ed. Abū l-Qāsim Imāmī); tr. Wiesehöfer, *Ancient Persia*, 216–217.

The spirit of seeking and acquiring knowledge irrespective of its origin that so many sources associate with Ḫusraw I Anūširwān is captured in an autobiographical work ascribed to the king himself, the *Book of Deeds* (*Kārnāmag*). According to the extract below, quoted from the Arabic text by the eleventh-century Persian philosopher Miskawayh (d. 421/1030), Ḫusraw claimed that he modelled his conduct in office on his study not just of the lives of his own ancestors and previous Persian kings, but also on examining the wisdom in Greek and Indian writings.

When we had finished studying the conduct of our forefathers—we started with them because they deserved it most, and we did not let up (seeking) the truth but redoubled our efforts, and we came into the closest contact with the truth (?)—, we then studied the conduct of the Greeks and the Indians, selected what was praiseworthy on the basis of our intelligence and discerned it out through our understanding. Among all this we adopted anything we considered an adornment of our rule and instituted it as (our) law and practice, and we did not let ourselves be influenced by our desires. We then informed and apprised them (i.e. the Greeks and Indians) of it and wrote to them (telling them) about those aspects of their conduct that we disapproved of, that we discouraged them from and that we directed them (to do). We did not, however, force anybody to change his religion or his community of believers nor begrudge them what (happened) before our time. In addition, we also did not refuse to learn about their ways, for resolving to find truth and knowledge and following it is the greatest adornment of kings whereas the greatest harm to kings comes from refusing to learn and abstaining from the acquisition of knowledge. A person who does not learn does not become learned.

IV.12. Ṣāʿid al-Andalusī, *The Categories of Nations* (*Ṭabaqāt al-umam*), 49.6–15 (ed. Cheikho); tr. Zakeri.

The Categories of Nations by Ṣāʿid al-Andalusī (419/AD 1029–462/AD 1070) is an eleventh-century history of science written in Muslim Spain (at Toledo) in 1068. It has

194 Cf. Cameron's translation in 'Agathias', 167–169.

chapters on the contributions made to the progress of science by different nations (the Indians, the Persians, the Chaldeans, the Greeks, the Byzantines, the Egyptians, the Israelites, the Arabs, and the Andalusians). The following piece gives an introduction to the beginnings of the translation of scientific books in Islam.

The first of philosophical sciences cultivated by the Muslims were logic and astronomy. The first scholar to distinguish himself in the art of logic under the ʿAbbāsids was the Persian orator ʿAbdallāh ibn al-Muqaffaʿ, secretary of Abū Ǧaʿfar al-Manṣūr. He translated the three Aristotelian treatises on the formal parts of logic, namely the *Categories*, the *Interpretation*, and the *Analytics*. Ibn al-Muqaffaʿ declared that only the first of these three works had been translated in his time. He also translated the introduction to the *Logic*, which is known as the *Isagoge*, written by Porphyry of Tyre. He wrote commentaries on his translations in a simple and accessible way, and translated the Indian book known as *Kalīlah wa-Dimnah* and was thus the first to translate Persian books into Arabic. He is the author of excellent writings, among others, an epistle on *adab* and politics, as well as the booklet *al-Yatīma* (*The Unique Pearl*), on submission to the prince.[195]

v. *The Translation of Scientific Works*

v.1. The Khotanese introduction to the translation of Ravigupta, *The Quintessence* (*Siddhasāra*); ed. and tr. Bailey, 'The Preface to the Siddhasāra-śāstra', 32–33.

Indian medicine was translated not just into Persian and Arabic but also into several other languages, including the eastern Middle Persian language of Khotan, an ancient Iranic Buddhist kingdom in central Asia. One example is the seventh-century Khotanese translation of one of the central texts of Ayurvedic medicine, the *Quintessence* (of Medicine, *Siddhasāra*)[196] by the Indian physician Ravigupta (fl. AD c. 650).[197] The Sanskrit original was first translated into Tibetan and from Tibetan into Khotanese and Uighur Turkish; whereas the Sanskrit and Tibetan texts are extant in full, we only have fragments of the latter two translations. Ibn al-Nadīm and Ibn Abī Uṣaybiʿah attribute the Arabic translation of this 'Indian' book to Ibn Dahn, the director

195 Cf. Blachère's French translation, *Livre des catégories des nations*, 101; and the English translation by Salem and Kumar, *Science in the Medieval World*, 86. See also Ibn al-Qifṭī, *Taʾrīḫ al-ḥukamāʾ*, 220; and Ibn Abī Uṣaybiʿah, *ʿUyūn*, 413.

196 The *Quintessence* was also translated into Arabic; see Emmerick, 'Ravigupta's Siddhasāra'.

197 See Kahl, *Sanskrit, Syriac and Persian Sources*, 22–26, 134–152.

of the Barmakī hospital in Baghdad. Oliver Kahl conjectures a direct Arabic translation from the original Sanskrit; however, Ibn Dahn's affiliation with the Barmakids, and thus Ḫurāsān, makes the Khotanese version also a plausible source.

Particularly interesting is the short verse introduction by the anonymous Khotanese translator, which has survived. Its author described how the Khotanese ruler commissioned the translation of this work, and how it was accomplished with the help of what seems to have been a Chinese scholar in residence trained in Ayurvedic medicine.

1. I come with reverence to the Three Jewels[198] with faith in love, ... that I may be able to translate this *śāstra*-treatise according to its true meaning, 2. the *Siddhasāra-śāstra* which stands among books on medicine like the head complete with all the six faculties, manifest bestower of life, destroyer of ills, a vast ocean of favour for the protectorless. 3. This Indian *śāstra*-treatise here was replete with meaning, being so studied formerly by the *ācārya*-teachers,[199] in foreign language, moreover the style difficult, a text to be acquired for knowledge, so much so that it remained for us unproductive. 4. So far did this go that here there were not two or three (copies?); as to the reader, there was but one. The medical physicians discarded it entirely as a learner's book; they repressed the uses of it in this world; they ceased. 5. With garlands of *yoga* medical practices they practised their treatment in the whole kingdom. The disease was unrecognized because they knew no theory of it. Unsuitable was the *doṣa*-basis, time and season otherwise also. They were applying to them improper medicaments; many beings died. 6 At that time there was the Tsai-siang by name Dpal-brtan. He obtained the treatise, fragmentary, in *ttāgūttaa*-language. But he wrote it with a supplement so that through him it became complete. For the translation he had, however, no comprehension and no ability. 7 He humbly presented this treatise to the Court. The gracious god (i.e. king) of his favour himself ordered to translate this text from the *ttāgūttaa*-language completely, for the sake of future beings who would be ill. 8 As to such medicaments as were comprised in it, the Tai-śi,[200] great physician, acquainted with this treatise, truly helped without grudging, excellent man; he brought it to completion in the month Rarūya.

198 I.e., the three core elements of Buddhism: the Buddha, cosmic order (*dharma*), and religious community (*sangha*).

199 I.e., highly learned or senior teachers.

200 I.e., a Chinese medical teacher.

V.2. *The Book of Ostanes* (*Kitāb Usṭānis*), 85.10–86.9 (ed. Berthelot, *La Chimie*, III); tr. van Bladel, *The Arabic Hermes*, 54–55.

Persian books could occasionally be discovered outside Persia. The anonymous author of the *Book of Ostanes*, an alchemical treatise ascribed to a legendary pre-Islamic founding figure of alchemy who frequently figures in Arabic and Persian alchemical literature,[201] reported one such find. He also notes how keen the Persian priests were to reassemble Persian knowledge including a copy of this book:[202] the theme of reclaiming ancient wisdom was thus not just associated with Persian rulers (see II.1, II.2 above).

Then, after it, there is a Persian inscription full of insight and knowledge. What I tell you now is the knowledge that became manifest to me and that I acquired by reading it. It said: 'Egypt is superior to (all other) cities and districts because God the Exalted has bestowed on its inhabitants wisdom and knowledge in all things. Yet the people of Egypt and those of all far-away countries need the Persians and they cannot succeed in any of their endeavours without (the knowledge) that originates from Persia. Do you not see that all philosophers who have devoted themselves to this science have turned to the Persians, adopted them as brothers and asked them to send them (knowledge) that originates from Persia and is not found anywhere else? Have you not heard (the story that) a certain philosopher [i.e., an Egyptian priest] wrote to a Persian Magian (and said): "I have found one of the books of the ancient sages written in Persian, which I cannot read. Send me one of your wise men who can read for me the book I have found, and fulfil my request without delay before my life runs out, I expire and have no need any more for knowledge. If you do this, I will be greatly obliged and grateful to you as long as I live." The sages of the people of Persia wrote to him (and said): "When your letter reached us, we were delighted with what you wrote to us about and we have hastened to send to you with this (letter) the sage that you have asked for to read to you your book and reveal to you what is concealed from you, for we consider it an obligation on our part towards you. When you have finished your book, as it should be, take care to send to us without delay a copy of that book, because our forefathers were the ones who composed that book. Therefore let us share it with you, which is as it should be. Peace."' This is all I read that was written in Persian on this plaque.[203]

201 See for example Ullmann, *Die Natur- und Geheimwissenschaften*, 184–186, and van Bladel, *The Arabic Hermes*, 48–54.

202 For a discussion of this passage and its background, see van Bladel, *The Arabic Hermes*, 54–57.

203 Cf. Berthelot, *La Chimie*, III, 121–122.

V.3. Zar⟨ā⟩dušt, *Book of Nativities and their Judgments* (*Kitāb fī l-mawālīd wa-aḥkāmihā*), Istanbul, Süleymaniye, MS Nuruosmaniye 2800, fols 230v–243v; ed. and tr. Zakeri.

The valuable introduction of this influential text was first translated by Afnan,[204] and subsequently studied by Sezgin (*GAS*, VII, 81–86), Pingree,[205] and Gutas,[206] among others. I have adopted Afnan's translation with major changes, supplemented with the sections he omitted.

The earliest treatise on genethlialogy in Arabic is the *Book of Nativities and their Judgments* by one Zarādušt, 'the oldest surviving genuine Arabic translation of a Pahlavi astrological work that we possess'.[207] The book starts with an introduction explaining that it was first written in the language of the *dēn-dabīrīh* (i.e., Avestan script), allegedly by the prophet Zoroaster. It was dispersed by Alexander, but subsequently retrieved, and later translated into Middle Persian and commented upon by Māhānkard, son of Mehrziyār/Mehryār, for Māhūyah, son of Māhānāhīḏ the Marzbān of Ḫurāsān, in the year in which Ctesiphon fell to the Arabs (i.e., 16/637). The Pahlavi version was translated into Arabic by Saʿīd Ibn Ḫurāsānḫurrah for the Ispahbadh Sunbāḏ, a commander in the army of Abū Muslim (132/750–137/755).[208]

The *Book of Nativities*, which is the fifth of the five books of astrology attributed to Zoroaster, describes the procedure for interpreting the horoscope of a person cast at or of the time of birth, illustrated by examples of planetary positions that are datable to the Sasanid period. Based on internal evidence, Pingree maintained that the underlying work was a Sasanian astrological treatise translated originally from Greek, authored by a third-century Ḥarrānian astrologer known as Aelius the Wise.[209]

This is a book that Māhānkard translated [for Māhūyah ibn Māhānāhīd].[210] He [i.e., Saʿīd Ibn Ḫurāsānḫurrah, mentioned below] who translated the astro-

204 Afnan, *Philosophical Terminology*, 77–78.

205 Pingree, 'Classical and Byzantine Astrology', 234–235; and id., *From Astral Omens*, 44–46.

206 Gutas, *Greek Thought*, 37–38.

207 Pingree, *From Astral Omens*, 44.

208 Pingree, 'Classical and Byzantine Astrology', 234.

209 Pingree, 'Classical and Byzantine Astrology', 234–235; id., *From Astral Omens*, 44–46; and id., 'The Ṣābians of Ḥarrān', 12–13. See also Kunitzsch, 'The Chapter', 243; Panaino, 'Sopravivvenze del culto iranico', 343–348; Cottrell, 'L'autobiographie de Zoroastre', 178–183; and Cottrell and Ross, 'Persian Astrology', 99–102.

210 As becomes clear further below, due to a copyist's mistake, the name of the governor has moved to the next line, causing confusion in the text. The Tehran manuscript Dānišgāh 347 G, fols 1v–30v, has the correct word order. This is the notorious Ḫurāsānian governor (*marzbān*) Māhūyah Ibn Māhānāhīḏ (not Māfanā), the *ḫudā-dušman*, 'The Enemy of the Lord', who betrayed and killed Yazdgird; see al-Ṭabarī, *The History*, XV, 79–80, 83–87.

nomical books of Zoroaster for the commander Sunbāḏ in the days of Abū Muslim[211] possessor of the rule, said: 'I translated this book from among the books of Zoroaster which consist of five parts [i.e., a Pentateuch]:

1) *Kitāb Hay'at al-falak* (*The Book of Astronomy*);
2) *Kitāb Ṣuwar daraǧ al-falak al-ṯalāṯi mi'ah wa-l-sittīn wa-mā yaṭlu'u fī kull daraǧah minhā min anwā' al-ṣuwar wa-l-tamāṯīl* (*Book of the Images of the 360 Degrees of the Celestial Sphere and That Which Appears in Each of its Degrees of Different Images and Figures*);
3) *Kitāb al-Asrār al-nuǧumīyah wa-mā qad ḫafiya 'an sā'ir ahl hāḏa l-'ilm min al-'awāmm* (*The Book of Astrological Mysteries and That Which is Hidden to other Participants from among the Commoners in this Science*);
4) *Kitāb al-Qirānāt wa-l-duwal wa-l-milal wa-l-mulūk wa-l-anbiyā' wa-mā qad maḍā min ḏālika wa-mā huwa ātin min ba'd ilā an yatalāšā l-'ālam* (*Book of Conjunctions and of States, and of Nations, and of Kings, and of Prophets, and of That Which has Passed and That Which Will Come to Pass in Future, until the World Dissolves*);
5) *Kitāb al-Mawālīd wa-l-kusūfāt wa-taḥāwīl sinī l-'ālam wa-tasyīr qismat al-dunyā wa-taḥāwīl sinī l-mawālīd wa-l-masā'il* (*Book of Nativities and Eclipses and Revolutions of the Years of the World and Prorogation of the Division* [*qismat*] *of the World and Revolutions of the Years of Nativities and Interrogations*).

I did not come across any book by Zoroaster containing the philosophical sciences or including the natures of men and countries, or combining medicine and occult sciences. For when Alexander conquered the kingdom of Darius the king, he had them all translated into the Greek language. Then he burnt the original copies that were kept in the treasure-houses of Darius and killed all those whom he suspected of knowing by heart any of these subjects. Only a small part of the books has survived, through the people who knew them by heart and fled from Alexander to the islands of the seas and the peaks of the mountains. Returning to their homes after the death of Alexander, these were written down from their memories. What they wrote down from memory was fragmentary. Much of it had passed away and little had remained. So Māhānkard translated what still survived by his time—when the rule of the Persians fell to the Arabs [i.e., 16/637]—and the translations that he made from these were from the language of the *dēn-dabīrīh* to the language of Persian *Darī*

211 Abū Muslim Ḫurāsānī (d. 137/755), chief architect of the overthrow of the Umayyad dynasty and its replacement with the 'Abbāsids in 132/749; see Agha, 'Abū Muslim', *EI Three*.

(*al-luġah al-fārisīyah al-darīyah*). Then later Saʿīd ibn Ḫurāsānḫurrah translated them into the Arabic language in order that this science should not fall into desuetude and its outlines be wiped away.

This is the fifth book of the five books of astrology on nativities. Māhānkard Ibn Mehrziyār said: [...] 'This follows the first book in which the description of celestial spheres and what is in them is already given.'

Saʿīd Ibn Ḫurāsānḫurrah the Dragoman said: 'The commandant Sunbād̠,[212] after realizing the merit of this science [i.e., astrology], asked me to translate for him Māhānkard's book *Fī sirr ʿilm Zardušt* (*On the Mystery of Zoroaster's Science*). This 'mystery' [i.e., astrology] had been in the virtuous language that Zoroaster spoke, and Māhānkard translated it for Māhūyah, son of Māhānāhīd̠ in the year in which the worshippers of Satan [i.e., Arabs] invaded the House of Justice; this House of Justice was a council hall constructed by Anūširwān and was only entered by people of virtue and religion at that time; thus the slaves put on the robes of kings, and the Lord Yazdgird [i.e., the last Sasanian monarch of Iran, Yazdgird III (r. 633–651)] was driven out from his capital, and the Ḫāqān crossed the River [Oxus], and with this the kingdom disappeared. Then, as the commandant Sunbād̠ saw that the Persian language had fallen out of use and that the Arab language had surpassed other languages, and God brought forth the rule of the Prince ʿAbd al-Raḥmān b. Muslim [i.e., Abū Muslim], 'the Trustee of the House of Prophet' and their avenger, he wished to spread this mystery with the Arabic language to facilitate its learning, so that the tradition of studying it would not become obliterated and its virtue not forgotten. These two books were circulating in the hands of the treasure-keepers and read in the *dēn-nāmeh* (*The Book of Religion*).'

V.4. *Anūširwān's Book of Falconry* (*Bāznāma-ye Nawšīrwānī*), 12.11–13.2 (ed. Bīniš); tr. Zakeri.

The anonymous treatise entitled *Anūširwān's Book of Falconry* is a medieval Persian book on the art of training and hunting with birds of prey. The introductory chapter of this important text is unfortunately very corrupt and at times exceedingly confusing; many words, particularly foreign personal names and titles, remain illegible; nonetheless, it has some interesting details about the history of falconry. A conclusive inter-

212 Sunbād̠ (d. 137/755), a follower of the ʿAbbāsid general Abū Muslim, fomented a short-lived revolt against the ʿAbbāsid caliph al-Manṣūr (r. 136/754–158/775) after the latter had Abū Muslim killed in 137/755; see Madelung, 'Sunbadh', *EI2*, IX, 874–875. For a discussion of his interest in preserving Pahlavi literature via translations into Arabic, see Anthony, 'Chiliastic Ideology', 652–653.

pretation of its contents will have to await a more reliable edition based on additional manuscripts that have recently come to light. The anonymous author cites a personal motivation for compiling his *Book of Falconry*. He credits every nation (Babylonians, Greeks, Persians, Syrians, Arabs, Turks, etc.) with having tamed and trained at least one bird of prey and describes how they passed on the expertise gained about that particular bird to other nations. With respect to his sources, he says:

Ancient (Persian) masters who toiled and gathered experience in this art [i.e., falconry] did not have access to Greek learning. I was much interested in this field and continuously sought knowledge about it until I heard that there were books on this subject available in Baghdad written by the ancients who lived in the city of Balīnās. In the days of Alexander and Aristotle, the library [*dār al-kutub*] was moved to Alexandria and it remained there until twelve rulers had ruled after Alexander. Then a woman became king and moved the Alexandrian library to Antioch, where it remained until the reign of King Alānī[?][213] and his son Constantine.[214] When his uncle[215] rebelled and wanted to remove him from the throne, Constantine's mother[216] advised him: 'adopt a new creed that will win the support of the people for you.'

I paraphrase the rest of this garbled text:[217]

Constantine's advisors convinced him that as long as the library existed in Antioch, the common people would not follow the new creed [i.e., Christianity]. Constantine thereupon ordered that the library be destroyed and its books burned. A man at that time rescued a few books and brought them to a physician named Šammās, who translated these Greek books into Syriac. Among them there was a text about the history of the kings and the art of hunting.

213 Probably the Roman Emperor Constantius I, who co-ruled the empire as part of Diocletian's tetrarchy (293–c. 313), first as Caesar (junior emperor) with Maximian (293–305), then as Augustus (senior emperor) in the West (305–306).

214 The Roman Emperor Constantine I, called the Great. He started out as Caesar in the West (306–312), then declared himself Augustus in the West and became the senior Augustus (312–324) and finally sole emperor (324–337).

215 Possibly a reference to Maxentius, Augustus of the West (306–312) and rival of the future Emperor Constantine. His mother's advice probably refers to the legend according to which Constantine's adoption of military standards with Christian symbols secured his victory in the decisive Battle of the Milvian Bridge outside Rome in 312.

216 Empress Helena (d. c. 330). According to a legend that arose at the end of the fourth century, she excavated the True Cross of Christ in Jerusalem; she is now considered a saint in all major churches.

217 *Bāznāma-e Nawšīrwānī*, 14–20.

Upon hearing this, I became excited and looked for someone who knew the Syriac language. A man who served in the army of the Turkish general Arlaq who was a native of Pāryāb near Marw and fluent in all languages finally translated these books for me (into Persian). On the basis of these translations, I understood their contents. According to Symmachus [Sūmāḫus]: 'the first person to write about hunting was a Greek king called Demetrius [Dīmiṭriyus]. All hunters were his students in the training of falcons.'[218]

'Īsā ibn 'Abbās reported on the authority of Ḫiradmihr Ḫurdād, who said: 'Bahrām [son of Šāpūr] related on the authority of the Babylonian masters that when the new art of hunting with falcons [*bāz*] arrived in Persia, there lived a king named Šāpūr, son of Salm, who ordered hunters to catch and train hunting falcons.'[219]

Those who hunt with hawks [*čarġ*] are all followers of the Arabs. Ġiṭrīf ibn Qudāmah al-Ġassānī [fl. eighth century] related that when the news about hunting with saker falcons reached the kingdom of the Persians, Bahrām, son of Šāpūr sent (an envoy) to the ruler of Ḥīrah, a man of the Banū Naṣr ibn Ḫidam, and asked for a saker falcon for hunting. Bahrām hunted with this bird for a long time.[220]

A king in Greater Rome [*rūmīyah al-kubrā*] at this time with the name SBĀBWS/BSĀYŪS was the first to train a shaheen [or Barbary] falcon for hunting.

The master hunters of Babylon reportedly possess a book by King Philip, son of Perdiccas [Fīlifūs ibn Bārīfūs].[221] The book was written at the time of Sūqīā[?] and was full of stories about past kings. When Alexander set out on his campaign, he took this book with him. A Persian man called Nīqūlās translated it (into Persian).

Bishop Lūmānis[?] reported that when King Constantine observed the way shaheen falcons hunt, he adopted them as hunting birds.[222]

Ġiṭrīf ibn Qudāmah related from a book he had read that the first person to hunt with eagles [*ālah*] was the king of the West [*maġrib*] called Kālūs [or Kābūs], who originally came from the Nīl of the West [*farang*]. The scholars at the time did not approve of it as a hunting bird, so King Kālūs sent it as a present to King Bahrām. At Bahrām's court the eagle attacked and killed his son. In

218 Ibid., 14–16.

219 Ibid., 16–17.

220 Ibid., 17–18.

221 Apparently a garbled rendering of the name of the Macedonian King Philip II (r. 359–36 BC).

222 *Bāznāma-e Nawšīrwānī*, 18–19.

revenge, Bahrām sent a panther to (Kālūs's court), claiming that it was harmless. There the panther killed a brother of the king.[223] [See further below, v.9, v.10.]

v.5. Ḥamza al-Iṣfahānī, *Proverbs Originating from Poetic Verses* (*al-Amṯāl al-ṣādira ʿan buyūt al-šiʿr*), 749.9–12 (ed. al-Ḍubayb); tr. Zakeri.

The Persian philologist and historian Ḥamza al-Iṣfahānī (d. betw. 350/961 and 360/970–971)[224] wrote several works on proverbs. The present text, taken from his collection of *Proverbs Originating from Poetic Verses*, describes an episode from the reign of the Persian king Ḫusraw I Anūširwān (see IV.9, IV.10 above) that illustrates the translation and appropriation of Greek philosophical works, specifically Aristotelian logical writings. While the details of this story may be entirely fictional, its substance is confirmed by the later translation of Persian writings on Aristotelian logic by Ibn al-Muqaffaʿ (see also IV.7 above).

Mārsfand, son of Mihrak, related that when he finished his commentary on the three books on logic attributed to Aristotle, he informed Kasrā Anūširwān, son of Qubāḏ. Thereupon the king called in the philosophers (*ḥukamāʾ*) who had recommended them and were in charge of translating them; they were ten (in number). He said to them: 'Tell us about the extent to which you have delved into them so that we can compare what you think with the opinion we have formed of them.'

After this introduction the ten philosophers deliver brief hyperbolic statements about their impression of these translated works. Ḫusraw then commends them for having clarified the books' contents from all angles and praises the books.[225]

v.6. Ibn al-Nadīm, *The Catalogue* (*Kitāb al-fihrist*), 301.8–302.5 (ed. Taǧaddud); tr. Saliba, *Islamic Science*, 36–37, slightly adapted.

Abū Maʿšar al-Balḫī (d. 886) was the leading astrologer of the ninth century. He lived and worked at Baghdad, but grew up in the ancient city of Balḫ, once an outpost of Hellenism in central Asia, where he studied before moving into the centre of the caliphate. He represents the Sasanian intellectual heritage of science and philosophy in the early Muslim tradition. Abū Maʿšar was a prolific author. The following piece taken

---

223 Ibid., 19–20.

224 See Rosenthal, 'Ḥamza al-Iṣfahānī', *EI2*, III, 156.

225 Ḥamza al-Iṣfahānī, *al-Amṯāl*, 749.13–750.25.

from his famous *Book of the Differences between Astronomical Tables* (*Kitāb iḫtilāf al-zīǧāt*) describes how the ancient Persian kings, predicting a deluge that threatened the world, chose the best writing materials in the world to preserve their books on the sciences, which they then deposited at the Sārūyah fortress in the city of Jayy in Isfahan. The depository continued to exist until the time of Ibn al-Nadim in the tenth century. Sources following Ibn al-Nadīm push Abū Ma'šar's antediluvian story forward in time and transpose it to Alexander, whose invasion of Iran performs a function similar to that of the universal Flood.

Abū Ma'šar says, in his book *Kitāb iḫtilāf al-zīǧāt*, that the Persian kings' love for the preservation of the sciences, their extreme care in perpetuating them across the ages, and their concern to protect them from natural disasters, both climatic as well as earthly mishaps, has led them to seek for them the most stable of writing material [*makātib*], most durable, and least likely to be affected by decay and effacement, namely, the bark of *ḫadank* [poplar tree], which is called *tūz*. The Indians, the Chinese, and the other neighbouring nations imitated them in that respect. They also chose the same wood for their arrows on account of its rigidity and smoothness and its durability.

Once they have gathered the best writing material they could find, on which their sciences could be saved, they sought for them a building in a place on Earth that had the best soil and clay, least likely to cause decay, and farthest from earthquakes and mudslides. They searched the regions of the kingdom, and did not find any place under the sky that had most of those qualities other than Iṣfahān. Then they searched within that region and could not find better than the encampment [*rustāq*] of Ǧayy. Nor could they find within Ǧayy a place that had more of those qualities than the place where the city of Ǧayy was built much later on.

They came to Quhunduz [i.e., *kuhandiz*, the old fortress], which was inside the city of Ǧayy, and deposited their sciences there. It stands to this day and is now called Sārūyah. And because of this building, people knew who first built it. And that is because many years ago, a section of that construction fell, which revealed a vault constructed of very hard clay [*siftaq*]. There they found a great variety of the sciences of the ancients, preserved on *tūz* bark, and written in ancient Persian.

One of those books was brought to someone who could understand it. He read it and found in it a letter written by one of the ancient kings of Persia. In it he related that when King Ṭahmūraṯ, who loved the sciences and the scientists, had learned about the climatic event that was to take place in the West, where rain would fall continuously, and would become overabundant and bypass the usual limits, and that a period of two hundred and thirty-two years and three hundred days would separate his first day of reign from the said event—according to the astrologers who warned him at the beginning of his reign that this event was going to pass from the West to the East—he ordered the engineers to find the best place in the kingdom in terms of soil and air. It was they who selected for him the structure known as Sārūyah, which is still standing inside the city of Ǧayy. He then ordered that this solid building be erected. When it was finished, he transported to it many of the various scientific texts from his own library, and they were all transferred to *tūz* bark. He placed them in one side of the house in order that they be preserved after the passage of this event.

Among the (treasure books) was a book attributed to one of the ancient sages, which contained specific years and revolutions from which one could extract the mean motions of the planets and the causes of their motions. The people at the time of Ṭahmūraṯ, and the Persians who lived before them, used to call those revolutions the cycles of *hazārāt* [thousands]. Most of the scientists of India, as well as all the kings who ruled over them, and the ancient Persians and Chaldeans, who dwelt in tents in ancient Babylonia, used to extract the mean motions of the seven planets from these years and revolutions. People who lived at the time dated (the mean motions) according to (the *zīǧ*), which they found to be the most correct according to the test and the most concise of all the *zīǧ*es that were known at the time. Astrologers extracted from it then a *zīǧ* that they called the *Šahriyār*, which means the king of *zīǧ*es. This is the end of Abu Maʿšar's statement.

v.7. Al-Ṯaʿālibī, *History of Exalted Lives* (*Taʾrīḫ ġurar al-siyar*), 629.7–633.6 (ed. Zotenberg); tr. Zakeri.

The following account describes how one of the best known works of Persian and Arabic literature, the collection of animal fables entitled *Kalīlah and Dimnah* (*Kalīlah wa-Dimnah*) was brought from India to Persia and translated into Persian and ultimately Arabic (see II.4 and IV.1). The Arabic version then became the source for translations into many other languages, including Syriac, Greek, Hebrew, and Spanish.[226]

The present excerpt, which tells of the Persian court physician Burzōy's voyage to India and his discovery of the texts that would contribute to the formation of the *Kalīlah and Dimnah*, is only one of at least five versions of this story attached as a

226 For the translation history and filiation of the early translations of the book, and an overview of the sources for each of the book's chapters in these early translations, see de Blois, *Burzōy's Voyage*, 1–11 and 62.

preamble to the book in a number of Arabic and Persian manuscripts. They vary substantially in length and the story elements, but they all agree that Burzōy, the physician of King Ḫusraw I Anūširwān, travelled to India, found the *Kalīlah and Dimnah*, and brought it with him to Persia.[227]

Al-Ṯaʿālibī's text, which combines essentials of the shorter, presumably oldest recension of the story and the longer, more detailed but less plausible account, does not form part of the *Kalīlah and Dimnah* itself, but a very similar report can be found in the *Book of Kings* (*Šāhnāma*), the Persian epic poem written around the turn of the eleventh century by the poet Abū l-Qāsim Firdawsī (d. 411/1020). Firdawsī and al-Ṯaʿālibī allude to the Arabic translation by Ibn al-Muqaffaʿ and the Persian verse version created by the poet Abū ʿAbdallāh Ǧaʿfar ibn Muḥammad al-Rūdakī (d. 329/941).[228]

The story of the doctor Burzōy and the *Kalīlah wa-Dimnah*

Anūširwān had one hundred and twenty physicians, Greek, Indian, and Persian. One of the most illustrious Persian doctors and the one most devoted to the study of books was Burzōy. He read in one of those books that there are mountains in India on which marvellous medicinal plants grow that revive the dead. He kept thinking about this and had the ambition to seek out and obtain them; so he eventually informed Anūširwān about his plan and asked permission to set off and seek the object of his desire. The king authorized his departure, paid for the journey, and provided him with a letter addressed to the king of India (asking him) to support (Burzōy).

Burzōy left for the capital of India. When he arrived there and presented Anūširwān's letter to the king, the king received him with generous hospitality and allowed him to do whatever he wanted and encouraged him in his ambition to seek out the medicinal plants where they were supposed to be. (Burzōy) kept striving, working hard, exerting and tirelessly devoting himself to gathering, collecting, classifying, and combining medicinal plants, until after a long time he could say, as the people of Baghdad are used to say: 'We continued doing nothing until we were done.'[229]

He was disheartened and discouraged, because he had not acquired what he wanted and wasted his time, and he imagined the shame he would feel in front of his master if he were to return to his court empty-handed. He then asked for the most skilful doctor and greatest sage in all of India and was told of a very old man. He went to him, told him his story, and related what he had read in a

227 See de Blois, *Burzōy's Voyage*, 40–43.

228 This is de Blois version 3; see *Burzōy's Voyage*, 41, 51–52.

229 I.e., we did all we could, but to no avail.

book, that there were mountains in India covered with medicinal plants capable of reviving the dead. (The old man) said to him: 'O Burzōy, you learned one thing, but (other) things were hidden from you. Do not you understand that this is an allegory of the ancients? By the mountains, they mean the scholars, by plants their salutary and profitable words, and by the dead the ignorant. (The ancients) mean that the scholars who educate the ignorant with their maxims are, as it were, reviving the dead. These maxims are contained in a book entitled *Kalīlah and Dimnah*, which is found only in the treasury of the king.'

Burzōy, relieved of his worries and happy with what he had heard, asked the king to lend him the book and, by doing so, do King Anūširwān a favour worthy of recognition. (The king) answered: 'I will give the order to lend it to you, first out of consideration for your master, but also for your sake, under the condition that you look at it in my presence and do not take a copy for yourself.' Burzōy declared that he would strictly obey his orders. From then on, he attended daily the reception hall of the king, asked for the book and studied it, memorized its meanings and committed them to writing when he returned home until he had finished it completely. He then asked the king for permission to return to his master's court. The king granted it to him, gave him presents and a robe of honour.

When Burzōy arrived before Anūširwān, he related what had happened to him, announced that he was in possession of the book and then presented it to him. Anūširwān admired it, showered (Burzōy) with presents, and ordered Buzurgmihr to translate it into the Pahlavi language. Burzōy humbly beseeched the king to allow his name and his account to be put at the head of the first chapter, which Anūširwān granted. The book remained in the safe-keeping of the kings of the Persians until Ibn al-Muqaffaʿ translated it into Arabic and Rūdakī into Persian verse on the order of the emir Naṣr ibn Aḥmad.[230]

v.8. Ẓahīrī Samarqandī, *The Book of Sindbād* (*Sindbādnāma*), 25.4–11 (ed. Ahmed Ateş); tr. Morley, in Clouston, *The Book of Sindibād*, xliii.

Ẓahīrī Samarqandī's *The Book of Sindbād*, written c. 556/1161 and based on older Persian versions,[231] is a mirror for princes based on the narrative motif of seven wise men who educate a ruler's son and, through story-telling, save him from his stepmother's life-threatening allegations. This is one of the most widespread story collections ever, and after taking shape in its original context it was translated into numerous languages

230 See the French translation in al-Ṯaʿālibī, *Ġurar*, 629–633.

231 For the complex translation history of this work (MP–Arabic–Greek/Syriac/Hebrew/Darī-Persian, etc.), see Clouston, *The Book of Sindibād*, xxxv–liii.

and handed down worldwide. The following piece from its introduction illuminates the background of the book to some extent.

This book was originally in the Pahlavi tongue, and till the time of the honorable Amīr Nāṣir al-Dawlah Abū Muḥammad Nūḥ ibn Manṣūr Sāmānī,[232] it had not been translated by anyone. The just emir Nūḥ ibn Manṣūr decreed that Ḫwājah Amīd Abū l-Fawāris Fanārūzī should translate it into Persian and set right the discrepancies and errors he might find therein. In the year 339/949 Ḫwājah Amīd Abū l-Fawāris undertook the task, and converted it into the Persian Darī language. However, its phrasing was poor; it lacked every ornamentation and embellishment.[233]

v.9. *The Royal Book of Hunting* (*Šikārnāma-e Ḫusrawī* or *Malikšāhī*), xvi,6–xvii,14 (ed. Nafīsī, *Kitāb-e naṣīḥat-nāma*); tr. Zakeri.

Šikārnāma and *Bāznāma* are common titles for books on the training and upbringing of falcons and other birds of prey, hunting with them, and the principles of their medical treatment. *The Royal Book of Hunting* is one among a series of such compilations, the creation of the oldest version of which is commonly attributed to Anūširwān. The book was composed during the time of the Seljuq Sultan Ǧalāl al-Dawlah Malikšāh I (r. 465/1072–485/1092).[234] The introduction to this unpublished book offers a translation history of Persian hunting literature. Sa'īd Nafīsī quoted this introduction in his edition of the *Qābūsnāma*. Written in the eleventh century, it was taken over almost verbatim in the later *Šikārnāma-e Īlḫānī* (of the eighth/fourteenth century), which is, as its author 'Alī ibn Manṣūr Ḥalwāyī says, based on the *Šikārnāma-e Ḫusrawī*. A modified version appears also in Ḫudāyār-ḫān Dāwūd 'Abbāsī's *Ṣayd al-murād fī qawānīn al-ṣayyād* (fifteenth century or later),[235] which in turn is based on the *Šikārnāma-e Īlḫānī*.[236]

This is the book *Šikārnāma* [*The Book of Hunting*], which has been a source for extracting the rules of hunting since the days of Afrīdūn [the mythic Persian

232 The Sāmānid ruler of Ḫurāsān and Transoxania, Nūḥ ibn Naṣr (r. 331/943–343/954); see Bosworth, 'Nūḥ b. Naṣr b. Aḥmad', *EI2*, VIII, 109–110, and id., 'Sāmānids', *EI2*, VIII, 1025–1029.

233 W.H. Morley, in a letter to E.W. Lane, dated August 1841, cited by Clouston, *The Book of Sindibād*, xliii; corrected and adapted.

234 Nafīsī, *Kitāb-e naṣīḥat-nāma*, xvi.

235 There are several manuscripts of the *Ṣayd al-murād* in different libraries. It is now edited in an unpublished Tehran University dissertation.

236 The former was edited by Douglas C. Phillott; see Ḫudāyār-ḫān Dāwūd 'Abbāsī, *Ṣayd al-murād*.

king-hero Frēdōn]. The beginning of the art of venery goes back to those early days, and since then learned men and masters of the art have toiled and exerted themselves with maximum effort to compose books in ancient languages for the libraries of kings, which were then translated from Pahlavi into Persian. These books had been translated from other languages into Pahlavi at the time of Anūširwān the Just. The reason for translating them was that when Anūširwān stepped onto the throne, he became infatuated with hunting with beasts of prey and developed a great passion for this art, to the extent that no other pleasure, fun or play could compete with hunting for him. He spent most of his time pursuing this art. One day a group of chief huntsmen were holding a discussion on the beasts of prey in his presence, basing their arguments on books and personal experience. The king called in his grand minister, the wise Buzurgmihr, and said to him: 'I would like you to dispatch people to all corners of the empire to collect books on hunting written for the bygone kings and kept in their libraries. You should then invite all the presently known masters of the hunt and study these books with them so that you fully learn the good or the obsolete rules, in order to summarize what is useful in them, adding whatever these hunt-masters have learned by experience. In this way you should create a new updated handbook in Pahlavi [i.e., Middle Persian], so that if someone has a question on the subject he may refer to it as a canon of rules in this art.'

Following the king's order, Buzurgmihr translated those books into Pahlavi, supplementing them with whatever he had learned by deliberation and hearing from the hunt-masters about these sciences. The book so produced was entitled *Šikārnāma-e Anūširwān* (*Anūširwān's Book of Hunting*).

Later on, Abū l-Fawāris 'Abd al-Malik ibn Nūḥ [Sāmānid ruler, r. 343/954–350/961] became an expert in the art of hunting and surpassed all others before him. One day, he ordered his vizier to look for *Šikārnāma-e Anūširwān* and find it. When this was done, they invited all the hunt-masters of the day and read that book to them and recorded whatever (improvements) they suggested. Then, on the order of the minister, the book was translated into New Persian, and the new practical learning of the masters was incorporated in it. The king followed the rules of this book to the end of his days. After he passed away, King Ḫalaf of Sīstān (r. 352/963–393/1002, d. 399/1009), who was an expert master in the art of hunting and was infatuated with it, went to Buḫārā in search of knowledge on this art, looking for the book created for 'Abd al-Malik ibn Nūḥ; then he invited many masters including [… a list of falconers, often mangled, follows] to his court and read the book with them. They all acknowledged the accuracy of the principles of the ancestors (in the book) and added whatever they thought was necessary to it. Ḫalaf always followed the rules of this book.

When the time of Sulṭān Malikšāh arrived, he ordered (his vizier) Niẓām al-Mulk to assemble the masters of this art and create a new book. Niẓām al-Mulk sent people to faraway lands and brought in from every country a man who was well-informed in this art; many came. They sat down together for many days and created this *Šikārnāma-e Ḫusrawī* for Malikshāh. [See above, V.4, and further below V.10.]

V.10. Ḫudāyār-ḫān Dāwūd 'Abbāsī, *Hunting the Goal in the Laws of the Hunter* (*Ṣayd al-murād fī qawānīn al-ṣayyād*), 23.8–24.1 (ed. Ṣadīqī, 'Ḥakīm Nasawī'); tr. Zakeri.

The Persian *Hunting the Goal*, which depends on the above text (V.9) for its outline of the history of hunting, also includes two informative sections on the training and medical care of birds of prey and hunting animals. These reports attribute the 'first' book on hunting to mythic antediluvian Iranian kings such as Ǧamšīd or Frēdūn, but as usual, it is the Sasanian King Anūširwān who sends for Greek works written by Galen, Hippocrates, and others. From there on, Anūširwān remains a source of inspiration for successive generations who follow his insight as a tradition worthy of emulation. After assigning the first book on hunting to Ǧamšīd, Ḫudāyār-ḫān says:

This book remained intact until the time of Anūširwān. One day a group of hunt-masters were discussing 'animals' in front of his throne, citing whatever they knew from the books of Galen and Hippocrates. Thereupon he called for Buzurgmihr and told him: 'You should send people to all corners of the world to collect books and treatises that have been written about animals and translate them into Pahlavi, supplementing them with whatever you have learned by your wit and that which the masters of the day recommend, such that it shall remain a memorial in time.' In accordance with this decree, the great minster Buzurgmihr combined the knowledge that he had, verifying its validity with the informed masters (of which the written texts were devoid), into a book and entitled it *Bāznāma-e Nawšīrwānī* (*Anūširwān's Book of Falconry*). This remained in use for a long time and became a reference book for all ages. Later on, King Abū l-Fawāris 'Abd al-Malik ibn Nūḥ [r. 343/954–350/961], the king of Ḫurāsān, became interested in this science and surpassed all previous kings in it. He decided to translate the *Bāznāma-e Nawšīrwānī* from Pahlavi into Persian. He told his vizier: 'I would like you to find the *Bāznāma-e Nawšīrwānī* and order it to be translated from Pahlavi into Persian.' The honourable master Abū l-Buḫtarī translated it into Persian and called it *Ǧawāriḥ-nāma-e Šāhanšāhī* (*The Royal Book on the Beasts of Prey*).

Part Three: Original Texts

1. *Knowledge Dispersed, Knowledge Reassembled: Ancient Precedents*

1.1. Aulus Gellius, *Attic Nights* (*Noctes Atticae*), I, 138.8–22 (ed. Rolfe).

Quis omnium primus libros publice praebuerit legendos; quantusque numerus fuerit Athenis ante clades Persicas librorum in bibliothecis publicorum.

Libros Athenis disciplinarum liberalium publice ad legendum praebendos primus posuisse dicitur Pisistratus tyrannus. Deinceps studiosius accuratiusque ipsi Athenienses auxerunt; sed omnem illam postea librorum copiam Xerxes, Athenarum potitus, urbe ipsa praeter arcem incensa, abstulit asportavitque in Persas. Eos porro libros universos multis post tempestatibus Seleucus rex, qui Nicanor appellatus est, referendos Athenas curavit.

Ingens postea numerus librorum in Aegypto ab Ptolemaeis regibus vel conquisitus vel confectus est ad milia ferme voluminum septingenta; sed ea omnia bello priore Alexandrino, dum diripitur ea civitas, non sponte neque opera consulta, sed a militibus forte auxiliaris incensa sunt.

1.2. Isidore of Seville, *Etymologies* (*Etymologiae*), VI.iii.1–5, 43.1–45.10 (ed. Lindsay).

1. Biblioteca a greco nomen accepit eo quod ibi recondeantur libri, nam biblio librorum, teca repositio interpretantur.
2. Bibliotecam veteris testamenti Esdras scriba post incensam legem a Caldeis, dum Iudei regressi fuissent in Hierusalem, divino afflatus Spiritu reparavit, cunctaque legis ac prophetarum volumina quae fuerant a gentibus corrupta correxit, totumque vetus testamentum in viginti duos libros constituit, ut tot libri essent in lege quot habebantur et litterae.
3. Apud Grecos autem bibliotecam primus instituisse Pisistratus creditur, Atheniensium tirannus, quam deinceps ab Atheniensibus auctam Xerxes, incensis Athenis, evexit in Persas, longoque post tempore Seleucus Nicanor rursus in Greciam rettulit.
4. Hinc studium regibus urbibusque ceteris natum est comparandi volumina diversarum gentium, et per interpretes in grecam linguam vertendi.
5. Dehinc magnus Alexander vel successores eorum instruendis omnium librorum bibliotecis animum intenderunt, maxime Ptolomeus cognomento Filadelfus omnis litteraturae sagacissimus, cum studio bibliotecarum Pisistratum emularetur, non solum gentium scripturas, sed etiam et divinas litteras in bibliotecam suam contulit; nam septuaginta milia librorum huius temporibus Alexandriae inventa sunt.

11. *Knowledge Dispersed, Knowledge Reassembled: from Alexander to Islam*

11.1. *Acts of Religion* (*Dēnkard*), 411.17–414.15 (ed. Madan); transcription Zakeri; cf. Shaki, 'Dēnkard Account', 115–118.

1. Dārāy ī Dārāyān hamāg abestāg zand čiyōn zardušt az Ohrmazd padīrift nibištag 2 pačēn ēk pad ganǰ ī *šāhīgān ēk pad diz ī nibišt dāštan framūd.
2. Walaxš ī aškānān abestāg ud zand čiyōn abēzagīhā andar mad estād hammōg-iz ī aziš harw čē az wizend ud āšuftkārīh ī Alkasandar ud ēwār ud rōb ī Hrōmāyān andar Erānšahr pargandagīhā abar nibištag tā čē uzwān abespārišnīg pad dastwar mānd estād andar šahr čiyōn frāz mad estād nigāh dāštan ō šahrīhā [ī] ayādgār kardan framūd.
3. ōy bay Ardašīr šāhān šāh ī Pābagān pad rāst dastwarīh ⟨ī⟩ tōsar ān-iz hammōg ī pargandag hamāg ō dar xwāst. Tōsar abar mad ān ī *ēwar frāz padīrift ud abārīg az dastwar hišt. ud ēn-iz framān dād kū frāz ō amāh harw nigēzišn ān ē bawēd az dēn mazdēsn čē nūn-iz āgāhīh ud dānišn aziš frōd nēst.
4. šābuhr ī šāhān šāh ī Ardašīrān nibēgīhā-iz ī az dēn bē abar bizišgīh ud star-gōwišnīh ⟨ud⟩ čandišn ud zamān ud gyāg ud gōhr ⟨ud⟩ jahišn ud bawišn ud wināhišn ud jadag-wihērīh ud gōwāgīh ud abārīg kirrōgīh ud abzār andar Hindūgān ud Hrōm ⟨ud⟩ abārīg-iz zamīgīhā pargandag būd abāz ō ham āwurd ud abāg abestāg abāz handāxt ud harw ān ī drust pačēn ō [ō] ganǰ ī *šaāhīgān dād⟨an⟩ framūd ud estēnīdan ī hamāg *arist ⟨ag⟩-ān abar dēn mazdēsn ō uskār kard.
5. šābuhr šāhān šāh ī ohrmazdān hamāg kišwarīgān pad pahikārišn *yazdān āhang kard ud hamāg gōwišn ō uskār ud wizōyišn āwurd. pas az bōxtan ī ādurbād pad gōwišn ī passāxt abāg hamāg awēšān jud-sardagān ud nask-ōšmurdān-iz ī jud-ristagān ēn-iz guft kū nūn ka-mān pad stī be dīd kas-iz ag-dēnīh bē nē hilēm wēš abar tuxšāg tuxšēm ud ham-gōnag kard.
6. im bay husraw šāhān šāh ī kawādān čiyōn-iš ahlamōγīh ud sāstārīh spurr-*hamēstārīhā wānīd pad paydāgīh az dēn andar harw ahlamōγīh ⟨ī⟩ 4 pēšag āgāhīh ud uskārišn ī gōkānīg wasīhā bē abzūd ud ēn-iz pad gēhān hanǰamnīh guft kū rāstīh ī dēn mazdēsn bē dānist ōšyārān pad uskārišn ōstīgīhā tuwān pad gētīg be estēnīd. ud abardar abzōnīg ud pēš-xrad būd⟨an⟩ mādayān nē pad uskār⟨išn⟩ bē pad abēzagīh ī menišn gōwišn ud kunišn ud weh mēnōg hāzišnīh ud mānsrīg abēzagīhā ēzišnīh ī yaz-dān šāyēd.[237]

237 Slightly adapted. For a partial transliteration (lacks par. 1), see Zaehner, *Zurvan*, 31–32.

11.2. *Acts of Religion* (*Dēnkard*), 428.15–429.6 (ed. Zakeri, and Shaki, 'Dēnkard Account', 123).

Kāl kōšāk [kāla koša] ī Hindōg ud Magistīg ī Hrōmāyīg ud abārīg ī az ān šōn abāg bun nibēg ī pad ganj ī šāhīgān [šasabīgān] ōh handāxt ud čē andar-šān wirāy-išn(īg) ud abāg čim ō xʷāstārān ī ham āgāhīh ōh nimūd nibēg-sān ī warōmand nōg ārāst ān ī čē az bē kišwar āwurd ē nigerīd ud ē wizōst ud frāz nē hišt nigō-nīh ud bēgān(ān)ag nāmīh rāy nē kam padīrift ud būd ka waxšišn ī dānāgīh rāy wēš-iz stāyīdan pad čašm bē kišwarīg nibēgīhā nahādan nē wizud ēg šān nibēg ud mādayān pad buwandagīh ī har āgāhīh ud dānišn ī andar nibēgīhā ud mādayānīhā paydāg nē kard bē-šān jud jud az bun nibēg ud mādayān ī xwēš xwāstan ud hangirdīg xwānd nibēg ī pad ganj ī šāhīgān abar hamāg saxwan.

11.3. Ibn al-Nadīm, *The Catalogue* (*Kitāb al-fihrist*), 299.9–301.2 (ed. Tağaddud).

قال أبو سهل بن نوبخت في كتاب النهمطان:

قد كثرت صنوف العلوم وأنواع الكتب ووجوه المسائل والمؤاخذ التي اشتقّ منها ما يدلّ عليه النجوم ممّا هو كائن من الأمور قبل ظهور أسبابها ومعرفة الناس بها على ما وصف أهل بابل في كتبهم وتعلّم أهل مصر منهم وعمل به أهل الهند في بلادهم على مثال ما كان عليه أوائل الخلق قبل مقارفتهم المعاصي وارتكابهم المساوي ووقوعهم في لجج الجهالة إلى أن لبست عليهم عقولهم وأضلّت عنهم أحلامهم فإنّ ذلك قد كان بلغ منهم فيما ذكر في الكتب من أمورهم وأعمالهم مبلغاً دلّه عقولهم وحيّر حلومهم وأهلك عليهم دينهم فصاروا حيارى ضلّالاً لا يعرفون شيئاً فلم يزالوا على ذلك حيناً من الدهر حتّى أيّد من خلف من بعدهم ونشأ من أعقابهم وذرأ من أصلابهم بالتذكّر لتلك الأمور والفطنة لها والمعرفة بها والعلم للماضي من أحوال الدنيا في شأنها وسياسة أوّلها والمؤتنف من تدبير أوسطها وعاقبة آخرها وحال سكّانها ومواضع أفلاك سمائها وطرقها ودرجها ودقائقها ومنازلها العلوى منها والسفلى بمجاريها وجميع أنحائها وذلك على عهد جم بن أونجهان الملك فعرفت العلماء ذلك وضعته في الكتب وأوضحت ما وضعت منه ووصفت مع وضعها ذلك الدنيا وجلالتها ومبتدأ أسبابها وتأسيسها ونجومها وحال العقاقير والأدوية والرقى وغير ذلك ممّا هو آلة للناس يصرفونها فيما هو موافق لأهوائهم من الخير والشرّ فكانوا كذلك برهة وعصراً حتّى ملك الضحّاك بن قيّ—من غير كلام أبي سهل: قال: ده آك معناه عشر آفات فجعلته العرب الضحّاك—رجعنا إلى كلام أبي سهل: ... بن قيّ في حصّة المشتري ونوبته وولايته وسلطانه من تدبير السنين بأرض السواد بنى مدينة اشتقّ اسمها من اسم المشتري فجمع

فيها العلم والعلماء وبنى بها اثني عشر قصراً على عدد بروج السماء وسمّاها بأسمائها وخزن كتب أهل العلم وأسكنها العلماء—من غير كلام أبي سهل: بنى سبعة بيوت على عدد الكواكب السبعة وجعل كلّ بيت منها إلى رجل فجعل بيت عطارد إلى هرمس وبيت المشتري إلى تينكلوس وبيت المرّيخ إلى طينقروس—رجعنا إلى كلام أبي سهل: فانقاد لهم الناس وانقادوا لقولهم ودبّروا أمورهم لمعرفتهم بفضلهم عليهم في أنواع العلم وحيل المنافع إلى أن بعث نبيّ في ذلك الزمان فإنّهم أنكروا عند ظهوره وما بلغهم من أمره علمهم واختلط عليهم كثير من رأيهم فتشتّت أمرهم واختلفت أهواؤهم وجماعتهم فأمّ كلّ عالم منهم بلدة يسكنها ويكون فيها ويترأّس على أهلها وكان فيها عالم يقال له هرمس وكان من أكملهم عقلاً وأصوبهم علماً وألطفهم نظراً فسقط إلى أرض مصر فملك أهلها وعمر أرضها وأصلح أحوال سكّانها وأظهر علمه فيها.

وبقي جلّ ذلك وأكثره ببابل إلى أن خرج الإسكندر ملك اليونانيّين غازياً أرض فارس من مدينة للروم يقال لها مقدونيّة عند الذي كان من إنكاره الفدية التي لم تزل جارية على أهل بابل ومملكة فارس وقتله دارا بن دار الملك واستيلائه على ملكه وهدمه المدائن وإخرابه المجادل المبنية بالشياطين والجبابرة وإهلاكه ما كان في صنوف البناء من أنواع العلم الذي كان منقوشاً مكتوباً في صخور ذلك وخشبه بهدم ذلك وإحراقه وتفريق مؤتلفه ونسخ ما كان مجموعاً من ذلك في الدواوين والخزائن بمدينة إصطخر وقلبه إلى اللسان الروميّ والقبطيّ ثمّ أحرق بعد فراغه من نسخ حاجته منها ما كان مكتوباً بالفارسيّة وكتّاب يقال له الكشتج وأخذ ما كان يحتاج إليه من علم النجوم والطبّ والطبائع فبعث بتلك الكتب وسائر ما أصاب من العلوم والأموال والخزائن والعلماء إلى بلاد مصر.

وقد كانت تبقّت أشياء بناحية الهند والصين كانت ملوك فارس نسختها على عهد نبيّهم زرادشت وجاماسب العالم وأحرزتها هناك لمّا كان نبيّهم زرادشت وجاماسب حذراهم من فعلة الإسكندر وغلبته على بلادهم وإهلاكه ما قدر عليه من كتبهم وعلمهم وتحويله إيّاه عنهم إلى بلاده.

فدرس عند ذلك العلم بالعراق وتمزّق واختلفت العلماء وقلّت وصار الناس أصحاب عصبيّة وفرقة وصار لكلّ طائفة منهم ملك فسمّوا ملوك الطوائف واجتمع ملوك الروم لملك واحد بعد الذي كان فيهم من التفرّق والاختلاط والتجارب قبل ملك الإسكندر فصاروا بذلك يداً واحدة ولم يزل ملك بابل منتشراً ضعيفاً فاسداً ولم يزل أهله مقهورين مغلوبين لا يمنعون حريماً ولا يدفعون ضيماً إلى أن ملك أردشير بن بابك من نسل ساسان فألّف مختلفهم وجمع متفرّقهم وقهر عدوّهم واستولى على بلادهم واجتمع له أمرهم وأذهب عصبيّتهم واستقام له ملكهم فبعث إلى بلاد الهند والصين في

الكتب التي كانت قبلهم وإلى الروم ونسخ ما كان سقط إليهم وتتبع بقايا يسيرة بقيت بالعراق فجمع منها ما كان متفرّقاً وألّف منها ما كان متبايناً وفعل ذلك من بعده ابنه سابور حتّى نسخت تلك الكتب كلّها بالفارسيّة على ما كان هرمس البابليّ الذي كان ملكاً على مصر ودوروسوس السريانيّ وقيدروس اليونانيّ من مدينة أثينس المذكورة بالعلم وبطلميوس الإسكندرانيّ وفرماسب الهنديّ فشرحوها وعلّموها الناس على مثل ما كانوا أخذوا من جميع تلك الكتب التي كان أصلها من بابل.

ثمّ جمعها وألّفها وعمل بها من بعدهما كسرى أنوشروان لنيّته كانت في العلم ومحبّته ولأهل كلّ زمان ودهر تجارب حادثة وعلم مجدّد لهم على قدر الكواكب والبروج الذي هو ولي تدبير الزمان بأمر الله تعالى جدّه.

انقضى كلام أبي سهل.

11.4. Al-Bīrūnī, *Vestiges of Past Centuries* (*al-Āṯār al-bāqiyah ʿan al-qurūn al-ḫāliyah*), 24.7–12 (ed. Sachau).

وقالوا إنّ طهمورث لمّا اتّصل به الإنذارُ وذلك قبل كونه بمائتين و إحدى وثلثين سنة أمر باختيار موضع في مملكته صحيح الهواء والتربة فلم يجدوا أحقّ بهذه الصفة من إصبهان فأمر بتجليد العلوم ودفنها في أسلم المواضع منه.

وقد يشهد لذلك ما وجد في زماننا بجيّ مدينة إصفهان من التلال التي انشقّت عن بيوت مملوءة أعدالاً كثيرة من لحاء الشجرة التي يلبس بها القسيّ والترسة وتسمّى التوز مكتوبة بكتابة لم يدر ما هي وما فيها.

11.5. Al-Bīrūnī, *Vestiges of Past Centuries* (*al-Āṯār al-bāqiyah ʿan al-qurūn al-ḫāliyah*), 36.2–4 (ed. Sachau).

و كان قتيبة أباد من يحسن الخطّ الخوارزميّ ويعلم أخبارهم و يدرّس ما كان عندهم و مزّقهم كلّ ممزّق تخفيت لذلک خفاء لا يتوصّل معه إلى معرفة حقائق ما قبل عهد الإسلام به.

11.6. Al-Bīrūnī, *Vestiges of Past Centuries* (*al-Āṯār al-bāqiyah ʿan al-qurūn al-ḫāliyah*), 48.11–15 (ed. Sachau).

ولا يستعملون فيها أسماء على حدة ولا يعلمون بها وأنا أظنّ أنّ ذلك كان لهم بمثل الاختلاف الواقع فيه الفرس وأهل السغد ثمّ لما كان من إهلاك قتيبة بن مسلم الباهليّ كتبتهم وقتله هرابذتهم وإحراقه

كتبهم وصحفهم بقوا أمّيّين يعوّلون فيما يحتاجون إليه على الحفظ فلمّا طال عليهم الأمد فاتهم ما اختلف فيه وحفظوا ما اتّفق عليه.

II.7. Ibn Ḫaldūn, *The Introduction* (*al-Muqaddima*), in id., *Kitāb al-ʿibar*, I, 631.12–20 (ed. Šaḥādah).[238]

وأمّا الفرس فكان شأن هذه العلوم العقليّة عندهم عظيماً ونطاقها متّسعاً لما كانت عليه دولتهم من الضخامة واتّصال الملك. ولقد يقال إنّ هذه العلوم إنّما وصلت إلى يونان منهم حين قتل الإسكندر دارا وغلب على مملكة الكينيّة فاستولى على كتبهم وعلومهم. إلّا أنّ المسلمين لمّا افتتحوا بلاد فارس وأصابوا من كتبهم وصحائف علومهم ما لا يأخذه الحصر ولمّا فتحت أرض فارس ووجدوا فيها كتباً كثيرة كتب سعد ابن أبي وقّاص إلى عمر بن الخطّاب ليستأذنه في شأنها وتنقيلها للمسلمين. فكتب إليه عمر أن اطرحوها في الماء فإن يكن ما فيها هدىً فقد هدانا الله بأهدى منه وإن يكن ضلالًا فقد كفانا الله. فطرحوها في الماء أو في النار وذهبت علوم الفرس فيها عن أن تصل إلينا.

II.8. Dawlatšāh Samarqandī, *Memoirs of the Poets* (*Taḏkirat al-šuʿarāʾ*), 30.3–11 (ed. Browne).

و نيز حكايت كنند كه امير عبد الله بن طاهر كه بروزگار خلفاى عباسي امير خراسان بود روزى در نيشاپور نشسته بود، شخصي كتابى آورد و بتحفه پيش او بنهاد، پرسيد كه اين چه كتابست؟ گفت اين قصّه وامق وعذراست و خوب حكايتي است كه حكما بنام شاه نوشيروان جمع كرده اند، امير فرمود كه "ما مردم قرآن خوانيم بغير از قرآن و حديث پيغمبر چيزي نمى خواهيم ما را ازين نوع كتاب در كار نيست و اين كتاب تأليف مغانست و پيش ما مردودست"، فرمود تا آن كتاب را در آب انداختند و حكم كرد كه در قلمرو من هر جا كه از تصانيف عجم و مغان كتابى باشد جمله را بسوزانند.

238 Cf. Ḥāǧǧī Ḫalīfah, *Kašf al-ẓunūn*, I, col. 679.11–19.

III. *Persian Achievements in the Sciences*

III.1. Tatian of Adiabene, *Address to the Greeks* (*Oratio ad Graecos*), 84.1–16.[239]

Τατιανοῦ πρὸς Ἕλληνας
Μὴ πάνυ φιλέχθρως διατίθεσθε πρὸς τοὺς βαρβάρους, ὦ ἄνδρες Ἕλληνες, μηδὲ φθονήσητε τοῖς τούτων δόγμασιν. Ποῖον γὰρ ἐπιτήδευμα παρ' ὑμῖν τὴν σύστασιν οὐκ ἀπὸ βαρβάρων ἐκτήσατο; Τελμισσέων γὰρ οἱ δοκιμώτατοι τὴν δι' ὀνείρων ἐξεῦρον μαντικήν, Κᾶρες τὴν διὰ τῶν ἄστρων πρόγνωσιν, πτήσεις ὀρνίθων Φρύγες καὶ Ἰσαύρων οἱ παλαίτατοι, Κυπρίοι θυτικήν, ἀστρονομεῖν Βαβυλώνιοι, μαγεύειν Πέρσαι, γεωμετρεῖν Αἰγύπτιοι, τὴν διὰ γραμμάτων παιδείαν Φοίνικες. Ὅθεν παύσασθε τὰς μιμήσεις εὑρέσεις ἀποκαλοῦντες. Ποίησιν μὲν γὰρ ἀσκεῖν καὶ ᾄδειν Ὀρφεὺς ὑμᾶς ἐδίδαξεν, ὁ δὲ αὐτὸς καὶ μυεῖσθαι·Τουσκανοὶ πλάττειν, ἱστορίας συντάττειν αἱ παρ' Αἰγυπτίοις τῶν χρόνων ἀναγραφαί. Μαρσύου δὲ καὶ Ὀλύμπου τὴν αὐλητικὴν ἀπηνέγκασθε, Φρύγες δέ οἱ ἀμφότεροι, τὴν διὰ σύριγγος ἁρμονίαν ἄγροικοι συνεστήσαντο. Τυρρηνοὶ σάλπιγγα, χαλκεύειν Κύκλωπες, καὶ ἐπιστολὰς συντάττειν ἡ Περσῶν ποτε ἡγησαμένη γυνή, καθά φησιν Ἑλλάνικος Ἄτοσσα δὲ ὄνομα αὐτῇ ἦν.

III.2. Al-Ǧāḥiẓ, *Refutation of the Christians* (*al-Muḫtār fī l-radd ʿalā l-naṣārā*), 62.6–63.2 (ed. al-Šarqāwī).

ولو علمت العوام أن النصارى والروم ليست لهم حكمة ولا بيان، ولا بعد روية، إلا حكمة الكف من الخرط والنجر والتصوير، وحياكة البزيون لأخرجتهم من حدود الأدباء، ولمحتهم من ديوان الفلاسفة والحكماء، لأن كتاب المنطق والكون والفساد، وكتاب العلوي، وغير ذلك، لأرسطاطاليس، وليس برومي ولا نصراني. وكتاب المجسطي لبطليموس، وليس برومي ولا نصراني. وكتاب إقليدس لإقليدس، وليس برومي ولا نصراني. وكتاب الطب لجالينوس، ولم يكن رومياً ولا نصرانياً. وكذلك كتب ديمقراط وبقراط وأفلاطون، وفلان وفلان. وهؤلاء ناس من أمة قد بادوا وبقيت آثار عقولهم، وهم اليونانيون، ودينهم غير دينهم، وأدبهم غير أدبهم، أولئك علماء، وهؤلاء صناع أخذوا كتبهم لقرب الجوار، وتداني الدار، فمنها ما أضافوه إلى أنفسهم، ومنها ما حولوه إلى ملتهم. إلا ما كان من مشهور كتبهم، ومعروف حكمهم، فإنهم حين لم يقدروا على تغيير أسمائها زعموا أن اليونانيين قبيل من قبائل الروم، ففخروا بأديانهم على اليهود، واستطالوا بها على العرب، وبذخوا بها على الهند، حتى زعموا أن حكماءنا أتباع حكمائهم، وأن فلاسفتنا اقتدوا على أمثالهم، فهذا هذا.

239 Cf. the text of Whittaker in Tatian, *Oratio ad Graecos and Fragments*, 2.1–16, and Marcovich in Tatian, *Tatiani Oratio*, 7.1–16.

III.3. Ibn Qutaybah, *The Superiority of the Arabs and Insight into their Sciences* (*Faḍl al-ʿarab wa-l-tanbīh ʿalā ʿulūmihā*), 205.1–206.15 (ed. Ḫāliṣ).

وكلّ ما يعلمه أهلُ فارس فهم له متعلِّمون وفيه لغيرهم متبِّعون ولأعقاب الأمم واطئون.

فإن نحن سألنا عن قدماء الأطبّاء دُللنا على أبقراط وجالينوس وإن سألنا عن أوّل علم النجوم والحساب دُللنا على كتاب إقليدس وكتاب المجسطيّ وإن سألنا عن حدّ المنطق دُللنا على كتاب أرسطوطاليس وإن سألنا عن علم اللحون دُللنا على كتاب الموسيقا وهذا كلّه للروم واليونان وليس لأهل فارس فيه إلّا ما لغيرهم من القابسين المستفيدين.

وللروم الفلاحة وللهند الشطرنج وكتاب كليلة ودمنة والحساب بالحروف التسعة ولهم طبّ قديم صحيح عن استنباط يخالفون في كثير منه اليونانيّين.

ومن الدليل على ذلك ما أقرّ به أهل فارس على أنفسهم في كتاب سير ملوكهم فإنّهم ذكروا أنّ سابور لمّا أسنّ وكَلَّ بصرُه ووهنت قواه شكا إلى أهل مملكته الضعف عن سياستهم وأمرهم بالتماس من يضطلعُ بأمورهم فأكبروا ذلك وقطعوا به وسألوه الإذن لهم في طلب الأطبّاء له فأذن لهم فأرسلوا إلى ملك الهند رسولاً وبعثوا إليه بهديّة عظيمة وسألوه أن يبعث إليهم طبيباً من أفاضل من عنده ففعل فلم يزل يعالجه حتّى اشتدّ عصبه وانبسط جلده وارتدّ بصره وركب للصيد وهشّ للنساء. فأحسن مكافأة الطبيب وأمره أن يتخيّر أحبّ المواضع إليه من مملكته لينزله فاختار السوس فسكنها فورث طبّه أهل السوس.

قالوا: وقد كان أيضاً أسكن السوس سبياً من سبي الروم فتعلّموا منه الطبّ فصار أهل السوس أطبّاء أهل فارس. وهذا خبر صادقٌ لأنّا نجد في جامع الطبّ المعمول بالسوس أخلاطاً هنديّة وأخلاطاً روميّة.

فإن ادُّعي أنّ الأسكندر لمّا دخل أرض فارس وقتل فيها وسبي وأخرب، نقلَ كتب علومهم إلى الروم وترجمها بلسانهم وأحرق أصولها التي كانت عندهم فصارت علومُهم للروم قلنا: خبركم هو إقرار على أنفسكم يُقبلُ فيه قولكم وخبركم الثاني دعوى لما في أيدى غيركم تحتاجون معه إلى إقرار الروم لكم به وإحضار بيّنة وبرهان.

III.4. Ibn al-Faqīh, *The Book of Countries* (*Kitāb al-buldān*), 319.12–15 (ed. al-Hādī).

وليس لهم أن يفاخرونا بهرمس في زمانه أو بمثله في زماننا ولا بأغاتيمون في أواننا هذا وهم يعلمون أو لا يعلمون أنّ بابل العراق كانت مركز العلم والعلماء ومكان الحكمة والحكماء ولكنّ الملك المظفّر لمّا غلب على ملوك العمران نقل الحكمة والحكماء إلى بلاد المينا أو نحوها ممّا قد سلف.

III.5. Ibn Abī Ṭāhir, *The Book of Baghdad* (*Kitāb Baġdād*), I, 157.7–158.5 (ed. Keller).

حدّثني أبو الحسن أحمد بن محمّد المهلّبيّ قال: حدّثني يحيي بن الحسن بن عليّ بن معاذ بن مسلم قال: إنّي بالرقّة بين يدي محمّد بن طاهر بن الحسين على بركة إذ دعَوت بغلام لي فكلّمته بالفارسيّة فدخل العتّابيّ وكان حاضراً في كلامنا فتكلّم معي بالفارسيّة. فقلت له: أبا عمرو ما لك وهذه الرطانة؟ قال: فقال لي: قدمت بلدتكم هذه ثلاث قدمات وكتبت كتب العجم التي في الخزانة بمرو وكانت الكتب سقطت إلى ما هناك مع يزدجرد فهي قائمة إلى الساعة. فقال: كتبت منها حاجتي ثمّ قدمت نيسابور وجزتها بعشر فراسخ إلى قرية يقال لها ذودر فذكرت كتاباً لم أقض حاجتي منه فرجعت إلى مرو فأقمت أشهرا. قال: قلت أبا عمرو: لِمَ كتبت كتب العجم؟ فقال لي: وهل المعاني إلّا في كتب العجم والبلاغة اللّغة لنا والمعاني لهم. ثمّ كان يذاكرني ويحدّثني بالفارسيّة كثيراً.

III.6. Ibn Ḫaldūn, *The Introduction* (*al-Muqaddima*), in id., *Kitāb al-ʿibar*, I, 631.2–7 (ed. Šahādah).[240]

واعلم أنّ أكثر من عني بها في الأجيال الذين عرفنا أخبارهم الأمّتان العظيمتان في الدولة قبل الإسلام وهما فارس والروم فكانت أسواق العلوم نافقة لديهم على ما بلغنا لمّا كان العمران موفوراً فيهم والدولة والسلطان قبل الإسلام وعصره لهم فكان لهذه العلوم بحور زاخرة في آفاقهم وأمصارهم.

وكان للكلدانيّين ومن قبلهم من السريانيّين ومن عاصرهم من القبط عناية بالسحر والنجامة وما يتبعها من الطلاسم وأخذ ذلك عنهم الأمم من فارس ويونان.

240 Cf. Ḥāǧǧī Ḫalīfah, *Kašf al-ẓunūn*, I, col. 679.5–10.

IV. *Persian Rulers: Preserving and Disseminating Knowledge*

(a) *Before the Sasanians*

IV.1. Ḥamza al-Iṣfahānī, *History of the Years of Persian Kings* (*Ta'rīḫ sinī al-mulūk al-ʿağam*), 34.16–24 (ed. Maskūnī).

قسم الاسكندر بلاد المشرق على ملوك الطوائف، ونقل عن بلدانهم علم النجوم والطب والفلسفة والحراثة إلى بلدان المغرب، بعد ان حوّلها إلى اليونانية والقبطية. فلما هلك الاسكندر وحصلت البلاد في أيدي الطوائف رفعوا الحرب والتجاذب فيما بينهم، فكان الواحد منهم إنما يغلب الآخر بالمسائل العويصة. ففي أيامهم وضعت الكتب التي هي في أيدي الناس من مثل: كتاب مروك وكتاب سندباد وكتاب برسناس وكتاب شيماس، وما أشبهها من الكتب التي يبلغ عددها قريبا من سبعين كتابا.

IV.2. Al-Ṯaʿālibī, *History of Exalted Lives* (*Ta'rīḫ ġurar al-siyar*), 458.5–459.5 (ed. Zotenberg).

أقفورشاه الأشكانيّ

ملك المدائن وأكثر العراق وفارس وكاتبه الملوك بالشاهيّة وأهدوا إليه على سبيل المكارمة لا على جهة الضريبة وحصّل درفش كاوابان من بعض مخابئه واحتاط عليه وقهر الروميّ الذي كان على الموصل والسواد من يد الإسكندر وطرده عنهما ثمّ غزا الروم وطلب بثأر دارا ونكأ في معظمهم وكان يحمل رجالهم في السفن فيغرّقهم حتّى أتى على كثير منهم وهدم كثيراً من حصونهم وحوّل ما كان نقل الإسكندر إليهم من كتب الطبّ والنجوم والفلاسفة وخفّف عن الرعيّة وسار أحسن سيرة.

(b) *The Sasanians: Ardašīr I*

IV.3. Al-Ṯaʿālibī, *History of Exalted Lives* (*Ta'rīḫ ġurar al-siyar*), 485.2–4 (ed. Zotenberg).

أمر بتحصيل نسخ〈 الكتب 〈الدينيّة والطبّيّة والنجوميّة التي كان الإسكندر أحرق بعضها وحمل إلى الروم معظمها ورسم بتجديدها وتقييدها وصرف العنايات إليها وأنفق الأموال الكثيرة عليها.

(c) *Šāpūr I*

IV.4. Abū Sulaymān al-Siğistānī, *Selection from the Depository of Wisdom* (*Muntaḫab ṣiwān al-ḥikma*), 12.3–6 (ed. Dunlop).

قال: وقد افتتح ملوك فارس كوراً لليونانيّين والروم وغلبوا على مدائن كانت معادن لكتبهم التي تشتمل على الفلسفة والحكمة كالجزيرة والشام ومصر وقسطنطينيّة وغيرها من البلدان فأخذوا ما كان فيها من كتب الحكماء بعضها وبعضها بالهديّة. وأما كتب النجوم والهندسة والعدد والموسيقى والطبّ والحيل فأهداها غورديانوس ملك الروم لشابور بن أردشير المقلّب بذي الأكتاف.

IV.5. Abū l-Fidā', *A Short History of Mankind* (*al-Muḫtaṣar fī aḫbār al-bašar*), I, 47.23–48.1 (Cairo edition).

ثمّ ملك بعده ابنه سابور ابن أردشير إحدى وثلاثين سنة وستّة أشهر... . ولمّا مضى من ملكه إحدى عشرة سنة سار بعساكره وفتح نصيبين من الروم ثمّ سار وتوغّل في بلاد الروم وهم على عبادة الأصنام وذلك قبل تنصّرهم وافتتح من الشام عدّة مدن عنوة وقتل أهلها ثمّ سار إلى جهة روميّة فصانعه ملك الروم وهو حينئذٍ غرذيانوس ... ودخل تحت طاعة سابور المذكور وكان لسابور المذكور عناية عظيمة بجمع كتب الفلسفة لليونانيّين ونقلها إلى اللغة الفارسيّة.

(d) *Šāpūr II*

IV.6. Ibn al-Nadīm, *The Catalogue* (*Kitāb al-fihrist*), 302.15–303.2 (ed. Tağaddud).

حكاية أخرى

كانت الحكمة في القديم ممنوعاً منها إلّا من كان من أهلها ومن علم أنّه يتقبّلها طبعاً وكانت الفلاسفة تنظر في مواليد من يريد الحكمة والفلسفة فإن علمت منها أنّ صاحب المولد في مولده حصول ذلك له استخدموه وناولوه الحكمة وإلّا فلا. وكانت الفلسفة ظاهرة في اليونانيّين والروم قبل شريعة المسيح عليه السلام فلمّا تنصّرت الروم منعوا منها وأحرقوا بعضها وخزنوا البعض ومنع الناس من الكلام في شيء من الفلسفة إذ كانت بضدّ الشرائع النبويّة.

ثمّ إنّ الروم ارتدّت عائدة إلى مذاهب الفلاسفة وكان السبب فى ذلك أنّ ليوليانس ملك الروم وكان ينزل بأنطاكية وهو الذي وزر له ثامسطيوس مفسّر كتب أرسطاليس لمّا قصده سابور ذو

الأكّاف وظفر به ليوليانس إمّا في حربه له وإمّا لأنّ سابور كما يقال مضى إلى أرض الروم ليقبض أمرها ففطن له وقبض عليه والحكاية في ذلك مختلفة. وإنّ ليوليانس سار إلى أرض العجم حتّى بلغ جنديسابور وبها إلى وقتنا هذا ثلمة يقال لها ثلمة الروم لحصر رؤساء الأعاجم والأساورة وبقايا حفظة الملك وأطال المقام عليها واستصعب عليه فتحها.

وكان سابور محبوساً في بلد الروم في قصر ليوليانس فعشقته ابنته فخلّصته فطوى البلاد مختفياً إلى أن وصل إلى جنديسابور فدخلها وقويت نفوس من بها من أصحابه وخرجوا من فورهم فأوقعوا بالروم تفأولاً بخلاص سابور. فأسر ليوليانس فقتله واختلفت الروم وكان قسطنطين الأكبر في جملة العسكر فاختلفت الروم فيمن يولّونه وضعفوا عن مقاومته وكان لسابور عناية بقسطنطين فولّاه على الروم ومنّ عليهم بسببه وجعل لهم طريقاً إلى الخروج عن بلاده بعد أن شرط على قسطنطين أن يغرس بإزاء كلّ نخلة قطعت من أرض السواد وبلاده شجرة زيتون وأن ينفّذ إليه من بلاد الروم من يبنى ما هدمه ليوليانس بعد أن ينقل الآلة من بلاد الروم فوفى له.

وعادت النصرانيّة إلى حالها فعاد المنع من كتب الفلسفة وخزنها إلى ما كان عليه إلى الآن. وقد كانت الفرس نقلت في القديم شيئاً من كتب المنطق والطبّ إلى اللغة الفارسيّة فنقل ذلك إلى العربيّ عبد الله بن المقفّع وغيره.

IV.7. Ibn Qutaybah, *al-Maʿārif*, 658.17–658.2 (ed. ʿUkāšah).

شابور ...: وعجل عليه الهرم، وكثرت به العلل، فبعث إلى ملك الهند يسأله أن يبعث إليه طبيبا، فعالجه حتى اشتدّ عصبه وجلده، وقوى بصره، وهش للنساء، وأطاق الركوب، فأحسن إلى ذلك الطبيب، وأمره أن يتخير من بلاده بلدا ينزله، فاختار مدينة السوس حتى هلك، فورث طبه أهل السوس، فصاروا أطباء أهل فارس لذلك، ولما ورثوا عمّن سكنها من سبى الروم.

(e) *Bahrām V*

IV.8. Niẓāmī, *Haft Paykar*, in *Kullīyāt-e ḫamsa-e Niẓāmī*, 66.1–14 (ed. Dastgardī).

شاهزاده در آن حصار بلند * پرورش ميگرفت سالى چند
جز به آموختن نبودش راى * بود عقلش به علم راهنماى
تازى و پارسى و يونانى * ياد دادش مغ دبستانى

منذر آن شاه با مهارت و مهر * آیتی بود در شمار سپهر
بود هفت اختر و دوازده برج * پیش او سرگشاده درج به درج
به خط هندسی عمل کرده * چون مجسطی هزار حل کرده
راصد چرخ آبگون بوده * قطره تا قطره قطر پیموده
از نهانخانهای دوراندیش * باز داده خبر به خاطر خویش
چون که شهزاده را به عقل و برای * دانش آموز دید و رمز گشای
تخت و میلش نهاد پیش به مهر * دروی آموخت رازهای سپهر
هر ضمیری که آن نهانی بود * گر زمینی گر آسمانی بود
همه را یک به یک بهم بر دوخت * چون بهم جمله شد درو آموخت
تا چنان بهره مند شد بهرام * کاصل هر علم را شناخت تمام
در نمودار زیج و اصطرلاب * درکشیدی ز روی غیب نقاب.

(f) *Ḫusraw I Anūširwān*

IV.9. Agathias, *Historiae*, 77.4–78.5 (section 28.1–6) (ed. Keydell).

28.1 Ἀλλὰ γὰρ βραχέα ἄττα περὶ Χοσρόου διεξελθὼν αὐτίκα ἔγωγε ἀνὰ τὰ πρότερα καὶ δὴ ἐπανήξω. ὑμνοῦσι γὰρ αὐτὸν καὶ ἄγανται πέρα τῆς ἀξίας, μὴ ὅτι οἱ Πέρσαι, ἀλλὰ καὶ ἔνιοι τῶν Ῥωμαίων, ὡς λόγων ἐραστὴν καὶ φιλοσοφίας τῆς παρ' ἡμῖν ἐς ἄκρον ἐλθόντα, μεταβεβλημένων αὐτῷ ὑπό του ἐς τὴν Περσίδα φωνὴν τῶν Ἑλληνικῶν ξυγγραμμάτων. 2 καὶ τοίνυν φασίν, ὅτι δὴ ὅλον τὸν Σταγειρίτην καταπιὼν εἴη μᾶλλον ἢ ὁ ῥήτωρ ὁ Παιανιεὺς τὸν Ὀλόρου τῶν τε Πλάτωνος τοῦ Ἀρίστωνος ἀναπέπλησται δογμάτων καὶ οὔτε ὁ Τίμαιος αὐτὸν ἀποδράσειεν ἄν, εἰ καὶ σφόδρα γραμμικῇ θεωρίᾳ πεποίκιλται καὶ τὰς τῆς φύσεως ἀνιχνεύει κινήσεις, οὔτε ὁ Φαίδων οὔτε ὁ Γοργίας, οὐ μὲν οὖν οὐδὲ ἄλλος τις τῶν γλαφυρῶν τε καὶ ἀγκυλωτέρων διαλόγων, ὁποῖος, οἶμαι, ὁ Παρμενίδης.

3 ἐγὼ δὲ οὕτως αὐτὸν ἄριστα ἔχειν παιδείας, καὶ ταῦτα τῆς ἀκροτάτης, οὐκ ἄν ποτε οἰηθείην. πῶς μὲν γὰρ οἷόν τε ἦν τὸ ἀκραιφνὲς ἐκεῖνο τῶν παλαιῶν ὀνομάτων καὶ ἐλευθέριον καὶ πρός γε τῇ τῶν πραγμάτων φύσει πρόσφορόν τε καὶ ἐπικαιρότατον ἀγρίᾳ τινὶ γλώττῃ καὶ ἀμουσοτάτῃ ἀποσωθῆναι; 4 πῶς δὲ ἂν ἀνὴρ βασιλείῳ τύφῳ ἐκ παίδων καὶ κολακείᾳ πολλῇ γεγανωμένος δίαιτάν τε λαχὼν ἐς ὅ τι βαρβαρικωτάτην καὶ πρὸς πολέμους ἀεὶ καὶ παρατάξεις ὁρῶσαν, πῶς δὴ οὖν ὧδε βιοὺς ἤμελλε μέγα τι καὶ λόγου ἄξιον ἐν τοῖσδε ἀπόνασθαι τοῖς διδάγμασι καὶ ἐνασκηθῆναι;

5 εἰ μὲν οὖν ἐπαινοίη τις αὐτόν, ὅτι δὴ βασιλεύς γε ὢν καὶ Πέρσης ἐθνῶν τε τοσούτων καὶ πράξεων μέλον αὐτῷ, ὁ δὲ ἐφίετο γοῦν ὅμως ἁμῇ γέ πῃ ἀπογεύεσθαι λόγων

καὶ τῇ περὶ ταῦτα γάνυσθαι δόξῃ, ξυνεπαινέσαιμι ἂν καὶ ἔγωγε τὸν ἄνδρα καὶ μείζονα θείην τῶν ἄλλων βαρβάρων. 6 ὅσοι δὲ λίαν αὐτὸν σοφὸν ἀποκαλοῦσι καὶ μονονουχὶ τοὺς ὅποι ποτὲ πεφιλοσοφηκότας ὑπερβαλλόμενον, ὡς καὶ ἁπάσης τέχνης τε καὶ ἐπιστήμης τὰς ἀρχὰς καὶ αἰτίας διαγιγνώσκειν, ὁποῖον τὸν ἄγαν πεπαιδευμένον οἱ ἐκ τοῦ περιπάτου ὁρίζονται, οἱ δὴ οὖν ταῦτα οἰόμενοι ἐκείνῃ ἂν μάλιστα φωραθεῖεν οὐ τῶν ἀληθῶν ἐστοχασμένοι, μόνῃ δὲ τῇ τῶν πολλῶν ἑπόμενοι φήμῃ.

IV.10. Agathias, *Historiae*, 80.5–81.21 (sections 30.3–7 and 31.1–4) (ed. Keydell).

30.3 καίτοι πρότερον ἀρίστους ὡς ἀληθῶς ἐτεθέατο φιλοσόφους, ἐνθένδε ὡς αὐτὸν ἀφικομένους. οὐ πολλῷ γὰρ ἔμπροσθεν Δαμάσκιος ὁ Σύρος καὶ Σιμπλίκιος ὁ Κίλιξ Εὐλάμιός τε ὁ Φρὺξ καὶ Πρισκιανὸς ὁ Λυδὸς Ἑρμείας τε καὶ Διογένης οἱ ἐκ Φοινίκης καὶ Ἰσίδωρος ὁ Γαζαῖος, οὗτοι δὴ οὖν ἅπαντες τὸ ἄκρον ἄωτον, κατὰ τὴν ποίησιν, τῶν ἐν τῷ καθ' ἡμᾶς χρόνῳ φιλοσοφησάντων, ἐπειδὴ αὐτοὺς ἡ παρὰ Ῥωμαίοις κρατοῦσα ἐπὶ τῷ κρείττονι δόξα οὐκ ἤρεσκεν ᾤοντό τε τὴν Περσικὴν πολιτείαν πολλῷ εἶναι ἀμείνονα, τούτοις δὴ τοῖς ὑπὸ τῶν πολλῶν περιᾳδομένοις ἀναπεπεισμένοι, ὡς εἴη παρ' ἐκείνοις δικαιότατον μὲν τὸ ἄρχον καὶ ὁποῖον εἶναι ὁ Πλάτωνος βούλεται λόγος, φιλοσοφίας τε καὶ βασιλείας ἐς ταὐτὸ ξυνελθούσης, σῶφρον δὲ ἐς τὰ μάλιστα καὶ κόσμιον τὸ κατήκοον, καὶ οὔτε φῶρες χρημάτων οὔτε ἅρπαγες ἀναφύονται, ἀτὰρ οὐδὲ τὴν ἄλλην μετιόντες ἀδικίαν, ἀλλ' εἰ καί τι τῶν τιμίων κτημάτων ἐν ὅτῳ δὴ οὖν χώρῳ ἐρημοτάτῳ καταλειφθείη, ἀφαιρεῖται ὅστις οὐδεὶς τῶν ἐντυγχανόντων, μένει δὲ οὕτω, εἰ καὶ ἀφύλακτον ᾖ, σωζόμενον τῷ λελοιπότι, ἔστ' ἂν ἐπανήκοι·

4 τούτοις δὴ οὖν ὡς ἀληθέσιν ἀρθέντες καὶ πρός γε ἀπειρημένον αὐτοῖς ἐκ τῶν νόμων ἀδεῶς ἐνταῦθα ἐμπολιτεύεσθαι, ὡς τῷ καθεστῶτι οὐχ ἑπομένοις, οἱ δὲ αὐτίκα ἀπιόντες ᾤχοντο ἐς ἀλλοδαπὰ καὶ ἄμικτα ἤθη, ὡς ἐκεῖσε τὸ λοιπὸν βιωσόμενοι. 5 πρῶτα μὲν οὖν τοὺς ἐν τέλει ἀλαζόνας μάλα εὑρόντες καὶ πέρα τοῦ δέοντος ἐξωγκωμένους ἐβδελύττοντό γε αὐτοὺς καὶ ἐκάκιζον· ἔπειτα δὲ ἑώρων, ὡς τοιχωρύχοι τε πολλοὶ καὶ λωποδύται οἱ μὲν ἡλίσκοντο, οἱ δὲ καὶ διελάνθανον, ἅπαν τε εἶδος ἀδικίας ἡμαρτάνετο. 6 καὶ γὰρ οἱ δυνατοὶ τοὺς ἐλάττονας λυμαίνονται ὠμότητί τε πολλῇ χρῶνται κατ' ἀλλήλων καὶ ἀπανθρωπίᾳ. καὶ τὸ δὴ πάντων παραλογώτερον· ἐξὸν γὰρ ἑκάστῳ μυρίας ὅσας ἄγεσθαι γαμετὰς καὶ τοίνυν ἀγομένοις, ἀλλὰ μοιχεῖαί γε ὅμως τολμῶνται. 7 τούτων δὴ οὖν ἁπάντων ἕκατι οἱ φιλόσοφοι ἐδυσφόρουν καὶ σφᾶς αὐτοὺς ᾐτιῶντο τῆς μεταστάσεως.

31.1 Ἐπεὶ δὲ καὶ τῷ βασιλεῖ διαλεχθέντες ἐψεύσθησαν τῆς ἐλπίδος, ἄνδρα εὑρόντες φιλοσοφεῖν μὲν φρυαττόμενον, οὐδὲν δὲ ὅ τι καὶ ἐπαΐοντα τῶν αἰπυτέρων, ὅτι τε αὐτοῖς οὐδὲ τῆς δόξης ἐκοινώνει, ἕτερα δὲ ἄττα ἐνόμιζεν, ὁποῖα ἤδη μοι εἴρηται, τήν τε τῶν μίξεων κακοδαιμονίαν οὐκ ἐνεγκόντες, ὡς τάχιστα ἐπανήεσαν. 2 καίτοι ἔστεργέ τε αὐτοὺς ἐκεῖνος καὶ μένειν ἠξίου, οἱ δὲ ἄμεινον εἶναι σφίσιν ἡγοῦντο ἐπιβάντες μόνον τῶν Ῥωμαϊκῶν ὁρίων αὐτίκα, οὕτω παρασχόν, καὶ τεθνάναι ἢ μένοντες παρὰ Πέρσαις τῶν μεγίστων γερῶν μεταλαγχάνειν. οὕτω τε ἅπαντες οἴκαδε ἀπενόστησαν, χαίρειν εἰπόντες τῇ τοῦ βαρβάρου φιλοξενίᾳ.

3 ἀπώναντο δὲ ὅμως τῆς ἐκδημίας, οὐκ ἐν βραχεῖ τινι καὶ ἠμελημένῳ, ἀλλ' ὅθεν αὐτοῖς ὁ ἐφεξῆς βίος ἐς τὸ θυμῆρές τε καὶ ἥδιστον ἀπετελεύτησεν. 4 ἐπειδὴ γὰρ κατ' ἐκεῖνο τοῦ χρόνου Ῥωμαῖοί τε καὶ Πέρσαι σπονδὰς ἔθεντο καὶ ξυνθήκας, μέρος ὑπῆρχε τῶν κατ' αὐτὰς ἀναγεγραμμένων τὸ δεῖν ἐκείνους τοὺς ἄνδρας ἐς τὰ σφέτερα ἤθη κατιόντας βιοτεύειν ἀδεῶς τὸ λοιπὸν ἐφ' ἑαυτοῖς, οὐδὲν ὁτιοῦν πέρα τῶν δοκούντων φρονεῖν ἢ μεταβάλλειν τὴν πατρῴαν δόξαν ἀναγκαζομένους. οὐ γὰρ ἀνῆκεν ὁ Χοσρόης μὴ οὐχὶ καὶ ἐπὶ τῷδε συστῆναι καὶ κρατεῖν τὴν ἐκεχειρίαν.

IV.11. Miskawayh, *The Experiences of Nations* (*Taǧārib al-umam*), I, 203.6–16 (ed. Abū l-Qāsim Imāmī).

ولمّا فرغنا من النظر في سير آبائنا—وبدأنا بهم وكانوا أحقّ بذلك فلم ندع حقّاً إلّا أكثرناه ووجدنا الحقّ أقرب القرابة—نظرنا في سير أهل الروم والهند فاصطفينا محمودها وجعلنا عيار ذلك عقولنا وميّزناه بأحلامنا فأخذنا من جميع ذلك ما زيّن سلطاننا وجعلناه سنّة وعادة ولم تنازعنا أنفسنا إلى ما تميل إليه أهواؤنا، وأعلمناهم ذلك وأخبرناهم به، وكتبنا إليهم بما كرهنا لهم من السير ونهيناهم عنه، وتقدّمنا إليهم فيه، غير أنّا لم نكره أحداً على غير دينه وملّته ولم نحسدهم ما قبلنا، ولا مع ذلك أنفنا من تعلّم ما عندهم، فإنّ الإقرار بمعرفة الحقّ والعلم والاتّباع له من أعظم ما تزيّنت به الملوك، ومن أعظم المضرّة على الملوك الأنفة من التعلّم، والحمية من طلب العلم، ولا يكون عالماً من لا يتعلّم.

IV.12. Ṣāʿid al-Andalusī, *The Categories of Nations* (*Ṭabaqāt al-umam*), 49.6–15 (ed. Cheikho).

إن أول علم أعتُني به من علوم الفلسفة علم المنطق والنجوم. فأما المنطق فأول من اشتهر به في هذه الدولة (الدولة العباسية) عبدالله بن المقفع، الخطيب الفارسي كاتب أبي جعفر المنصور. فأنه ترجم كتب أرسططاليس المنطقية الثلاثة، التي في صورة المنطق. وهي كتاب قاطيغورياس وكتاب باري أرمنياس وكتاب أنولوطيقا. وذكر أنه لم يتَرجَم منه إلى وقته إلا الكتاب الأول فقط. وترجَم كذلك المدخل إلى كتاب المنطق، المعروف بالإيساغوجي لفرفوريوس الصوري، وعبر عما ترجم من ذلك بعبارة سهلة، قريبة المأخذ. وترجم الكتاب الهندي المعروف بكليلة ودمنه. وهو اول من ترجم من اللغة الفارسية إلى اللغة العربية. وله تآليف حسان منها رسالة في الآداب والسياسة و منها رسالته المعروف باليتيمة في طاعة السلطان.

v. *The Translation of Scientific Works*

v.1. The Khotanese introduction to the translation of Ravigupta, *The Quintessence* (*Siddhasāra*) (ed. Bailey, 'The Preface', 32–33).

1 aurga tsūṃ raṃnāṃ drraina ṣadi jsa brīya
...
mahaiśvarä brahmāṃ: tti ma āysda yināṃde -
khu ttu śāsträ byūttä yinūṃ nva arthä| 1
2 siddhasārä śāsträ cu mara vījye myāña -
kamalä māñaṃdä cu kṣyāṃ iṃdrryau jsa uspurrä -
pichaṣṭu jsīñi hūrākä d⟨u⟩khāṃ jinākä -
anāhaṃ mu'sdī'nai mahāsamudrrä 2
3 ṣi' maṃ śāsträ hīdvaṃ ye pvī'stai arthäna -
tta ttai sājīra udaiśä' pīsāṃ brrāṃbe -
ihīye phara jsa garkha vyaṃjaṃ pātcä -
dasta berä ttidä muhye harye akīrä 3
4 daṃdä tsve khu ri maṃ dva drrayi ni ya -
cu jsaṃ pūśākä ṣi' jsāṃ ttaṃdī śau -
hāysai dīṣṭaṃdä vīja sājaka baiśä -
daiṣṭta-karma pastrīyāṃdä hanaśa tsvāṃda 4
5 yaugamālyo jsa yuḍāṃdä śau-kṣīrä krra -
apaysāṃdä āchai cvai nayä ni bvīra -
viysaṃ 3 dūṣä' kālä u rve haṃdari pātcä -
pijsanīra-ṃ aprrasama arve muḍa phari satva 5
6 ttiña beḍa tcai-syāṃ ye paḍa-ttanä nāma =
tu śāsträ byaudi khaṇḍä ttāgūttau phari jsa =
samīśe pīḍai khve ṣi' hamye uspurrä -
byūttä vaskai ra ni ya bvāma hota 6
7 ṣi' tu śāsträ jsoṇakyi ṣṭaṃ rrvī vī haiṣṭe -
miṣḍi gyastī mu'śdi' udiśä hamaiyi ttu -
ttāgūtto ṣṭāna uspurrä ṣe' pasti byūte
ustimaṃjsyāṃ satvāṃ kiṃṇä cu āchinuḍa 7
8 cu maṃ ye drrāma arve pvīstyedä -
ttayi-śī mistä vījä ttye śāsträ bvākä -
ṣi' hā pā gīsti vinau mātsarä śirkä
rarūyi māśti uspurrä dāśe ṣi' 8

v.2. *The Book of Ostanes* (*Kitāb Usṭānis*) (ed. Berthelot, *La Chimie*, III, 85.10–86.9/Ar.).

ثمّ كان بعده كتّاب فارسيّ في فقه وعلم كثير فهذا الذي أقول لكم الآن هو الذي بدا لي من كتّابه وأحصيت من علمه. إنّه قال:

أمّا مصر فذات فضل على المدائن والكور وذلك لما قد أعطى الله تعالى أهلها من الحكمة والعلم بجميع الأشياء. وأمّا فارس فإليها يحتاج أهل مصر وأهل الآفاق كلّهم ولا يصلح شيء من أعمالهم إلّا بالذي يخرج من فارس. ألا ترى أنّه ليس أحد من الفلاسفة الذين كانت همّتهم في هذا العلم إلّا وقد بعث إلى أهل فارس فاتّخذتهم إخوة وسألهم أن يبعثوا إليه الذي مخرجه من أرضهم ولا يوجد في غير بلادهم؟ ألم تسمع ببعض الفلاسفة إذ كتب إلى مجوس أهل فارس: إنّي أصبت كتّاباً من كتب الحكماء الأوّلين بقلم فارس لا أستطيع القراءة له فابعثوا إليّ من حكمائهم بحكيم يقرأ لي كتّابي الذي أصبت فإنّ لكم عندي إن فعلتم ذلك يداً حسنة وأنا شاكر لكم أبداً ما بقيت فعجّلوا عليّ بالذي سألتكم قبل أن تفنى حياتي فأصير ميّتاً لا أحتاج إلى شيء من العلم. فكتب إليه حكماء أهل فارس: إنّه لمّا وصل إلينا كتّابك كثر فرحنا بالذي كتبت إلينا به وعجّلنا إليك مع ذلك الحكيم الذي سألت ليقرأ لك كتّابك ويظهر لك ما خفيّ عليك منه لأنّا رأينا ذلك لك علينا حقّاً واجباً فانظر إذا أتممت كتّابك كالذي ينبغي فعجّل إلينا نسخة ذلك الكتّاب لأنّ آباءنا الأوّلين هم الذين وضعوا ذلك الكتّاب فليكن لنا فيه معكم نصيب فإنّه كذلك ينبغي والسلام. هذا كلّ الذي قرأت ممّا كان في ذلك اللوح من الكتّاب الفارسيّ.

v.3. Zar⟨ā⟩dušt, *Book of Nativities and their Judgments* (*Kitāb fī l-mawālīd wa-aḥkāmihā*), Istanbul, Süleymaniye, MS Nuruosmaniye 2800, fols 230v–243v; ed. Zakeri.[241]

هذا كتّابُ ترجمه ماهانكرد.

الذي ترجمَ كتب زردشت النجومية في ايام أبي مسلم صاحب الدولة قال ترجمت هذا الكتّاب من كتب زردشت لاصبهبد بن سنباد و ماهويه بن ماهاناهيد وكانت خمسة كتب:

241 I would like to thank Emily Cottrell for kindly making a copy of this manuscript available to me.

1) احدها كتاب هيئة الفلك و وصف ما فيها من الكواكب والصور وعللها وجميع ما يحيطُ به الفلك الأعلى من هذه الأصناف وفنونها.

2) وكتاب صور درج الفلك الثلثماية والستين وما يطلع في كل درجة منها من انواع الصورة والتماثيل.

3) وكتاب الأسرار النجومية وما قد خفي عن سايراهل هذا العلم من العوام.

4) وكتاب القرانات والدول والملل والملوك والأنبياء وما قد مضى من ذلك وما هو آتٍ من بعد إلى أن يتلاشى العالم.

5) وكتاب المواليد والكسوفات وتحاويل سنى العالم و تسيير قسمة الدنيا وتحاويل سنى المواليد والمسائل.

ولم أدرك من كتب زردشت شيئا يحوي العلوم الفلسفية او يضم طبائع الناس والبلدان او يجمع الطبَ والعلوم الغامضة إذ كان الأسكندر لما تغلبَ على مملكة دارا الملك نقل ذلك كله إلى اللغة اليونانية وأحرق الأصول التى كانت في خزائن دارا وقتل من ظنّ أنه يحفظ شيئا منها وانما تخلص ما تخلص من الكتب بحفظ من حفظه منهم وهرب من هرب من الأسكندرإلى جزائر البحار وقلل الجبال. ثم لما عادوا إلى اماكنهم بعد هلاك الإسكندر كُتبَ ذلك من حفظهم وكان ما كتبوه من اوليل (اقاويل؟) الحظفة مبتورا قد مر منه الكبير وبقي القليل فترجم ماهانكرد ما ادركه في زمانه حين انقلبت دولة الفرس إلى العرب وكان ترجمة ما ترجم من ذلك من اللغة الدين دبيرية إلى اللغة الفارسية الدرية ثم ترجمها من بعد سعيد بن خراسانخرة إلى اللغة العربية لأن لا يتدرِسَ هذا العلم ولا ينحي رسمهم.

وهذا هو الكتاب الخامس من كتب النجوم في المواليد. قال ماهانكرد بن مهرزيار: [...]

هذا هو تِلو الكتاب الأول الذي سبق فيه وصف الافلاك بما فيها. قال سعيد بن خراسانخره الترجمان: أن اصبهبد سنباذ لما أيقن بفضل هذا العلم سألنى أن أترجم له كتاب ماهانكرد في سر علم زردشت وكان هذا السر باللسان الفاضل الذي نطق به زردشت فترجمه ماهانكرد لماهويه بن ماهاناهيد المرزبان عام دخلت عبدة الشياطين دار العدل ودار العدل مجلس بناه انوشروان كان لا يدخله إلاَ اهل الفضل والدين في ذلك الزمان وابتزّت للعبيد ثياب الملوك وانزعج خداة يزدجرد عن مستقره، وعبر خاقان النهر وذهب بها الملك. ثم انّ الأصبهبذ سنباد لما رآى لغة الفرس متروكة الإستعمال ولغة العرب قد بذَّت سايرَ اللغات واتى الله بدولة الامير عبدالرحمن بن مسلم امين آل رسول الله وطالب

ثارهم أحبَّ أن يبسُطَ هذا السر باللسان العربي لتسهيل معرفته فلا يندرس رسمه ولا يجهَل فضلُه. وكان هذان الكّابان تتداولهما ايدي الخزنة ويقرآن في الدين نامه.

v.4. *Anūširwān's Book of Falconry* (*Bāznāma-ye Nawšīrwānī*), 12.11–13.2 (ed. Bīniš).

گفت استادان پيشين كه رنجها بردن و تجربها بسيار كردن الا كه بر علم يونانيان نتوانستن مطلع شدن و پيوسته طالب اين علم بودم تا شنيدم در بغداد كتب افتاده است از اوائل كه در شهر بليناس بودن و در روزگار اسكندر و ارسطاطاليس دار الكتب را باسكندريه نقل كردن و بماند آنجا تا دوازده ملك مملكت براند. بعد ذي القرنين و بعد ايشان زني در پادشاهي نشست وآن دار الكتب را نقل كرد بانطاكيه و همچنان بود تا نوبت ملك الاني و پسرش ملك قسطنطين رسيد. چون عم وي ستم كرد و مي خواست كه پادشاهي از وي ستاند پس مادر گفت ملك قسطنطين را كه ترا هيچ بهتر از آن نيست كه ملت بنياد نهي تا خلق بر سر تو جمع شوند.

v.5. Ḥamza al-Iṣfahānī, *Proverbs Originating from Poetic Verses* (*al-Amṯāl al-ṣādira ʿan buyūt al-šiʿr*), 749.9–12 (ed. al-Ḍubayb).

ذكر مارسفند بن مهرك أنّه لمّا فرغ من تفسير الكتب الثلاثة المنسوبة إلى أرسطاطاليس في المنطق رفع خبرها إلى كسرى بن قباذ أنوشروان فعندها دعا الحكماء الذين قرّروها وتولّوا ترجمتها وكانوا عشرة فقال لهم: "أخبرونا عن المقدار الذي وقعت به من قلوبكم لنسبر به مبلغ رأيكم بما عندنا فيها."

v.6. Ibn al-Nadīm, *The Catalogue* (*Kitāb al-fihrist*), 301.8–302.5 (ed. Tağaddud).

حكاية أخرى

قال أبو معشر في كتاب اختلاف الزيجات: إنّ ملوك الفرس بلغ من عنايتهم بصيانة العلوم وحرصهم على بقائها على وجه الدهر وإشفاقهم عليها من أحداث الجوّ وآفات الأرض أن اختاروا لها من المكاتب أصبرها على الأحداث وأبقاها على الدهر وأبعدها من التعفّن والدروس لحاء شجر الخدنك ولحاؤه يسمّى التوز. وبهم اقتدوا أهل الهند والصين ومن يليهم من الأمم في ذلك واختاروها أيضاً لقسيّهم التي يرمون عنها لصلابتها وملاستها وبقائها على القسيّ غابر الأيّام.

فلمّا حصّلوا لمستودع علومهم أجود ما وجدوه في العالم من المكاتب طلبوا لها من بقاع الأرض وبلدان الأقاليم أصحّها تربة وأقلّها عفونة وأبعدها من الزلازل والخسوف وأعلكها طيناً وأبقاها على الدهر بناءً. فانتفضوا بلاد المملكة وبقاعها فلم يجدوا تحت أديم السماء بلداً أجمع لهذه الأوصاف من إصفهان ثمّ فتّشوا عن بقاع هذا البلد فلم يجدوا فيها أفضل من رستاق جيّ ولا وجدوا في رستاق جيّ أجمع لما راموه من الموضع الذي اختطّ من بعد فيه بدهر داهر مدينة جيّ.

فجاؤوا إلى قهندز هو في داخل مدينة جيّ فأودعوه علومهم. وقد بقي إلى زماننا هذا وهو يسمّى ساروية ومن جهة هذه البنية درى الناس من كان بانيها. وذلك أنّه لمّا كان قبل زماننا هذا بسنين كثيرة تهدّمت من هذه المصنعة ناحية فظهروا فيها على أزج معقود من طين السفتق فوجدوا فيه كتباً كثيرة من كتب الأوائل مكتوبة كلّها في لحاء التوز مودعة أصناف علوم الأوائل بالكتابة الفارسيّة القديمة.

فوقع بعض تلك الكتب إلى من عُني به فقرأه فوجد فيه كتاباً لبعض ملوك الفرس المتقدّمين يذكر فيه أنّ طهمورث الملك المحبّ للعلوم وأهلها كان انتهى إليه قبل الحدث المغربيّ الذي كان من جهة الجوّ خبره في تتابع الأمطار هناك وإفراطها في الدوام والغزارة وخروجها عن الحدّ والعادة وأنّه كان من أوّل يوم من سني ملكه إلى أوّل يوم من بدء هذا الحدث المغربيّ مائتان وإحدى وثلاثون سنة وثلاثمائة يوم—وإنّ المنجّمين كانوا يخوّفونه من أوّل ابتداء ملكه تعدّي هذا الحدث من جانب المغرب إلى ما يليه من جانب المشرق—فأمر المهندسين بإيقاع الاختيار على أصحّ البقاع في المملكة تربة وهواءً فاختاروا له موضع البنية المعروفة بساروية وهي قائمة إلى الساعة داخل مدينة جيّ فأمر بابتناء هذه البنية الوثيقة. فلمّا فرغ له منها نقل إليها من خزائنه علوماً كثيرة مختلفة الأجناس فحوّلت له إلى لحاء التوز فجعلها في جانب من ذلك البيت لتبقى للناس بعد احتباس هذا الحدث.

وإنّه كان فيها كتاب منسوب إلى بعض الحكماء المتقدّمين فيه سنون وأدوار معلومة لاستخراج أوساط الكواكب وعلل حركاتها. وإنّ أهل زمان طهمورث وسائر من تقدّمهم من الفرس كانوا يسمّونها أدوار الهزارات. وإنّ أكثر علماء الهند وملوكها الذين كانوا على وجه الأرض وملوك الفرس الأوّلين وقدماء الكلدانيّين وهم سكّان الأحوية من أهل بابل في الزمان الأوّل إنّما كانوا يستخرجون أوساط الكواكب السبعة من هذه السنين والأدوار وإنّه إنّما ادّخره من بين الزيجات التي كانت في زمانه لأنّه وسائر من كان في ذلك الزمان وجدوه أصوبها كلّها عند الامتحان وأشدّها اختصاراً. واستخرج منه المنجّمون في ذلك الزمان زيجاً سمّوه زيج الشهريار ومعناه ملك الزيجات. هذا آخر لفظ أبي معشر.

v.7. Al-Ṯaʿālibī, *History of Exalted Lives* (*Taʾrīḫ ġurar al-siyar*), 629.7–633.6 (ed. Zotenberg).

قصّة برزوية الطبيب وكتاب كليلة ودمنة

كان لأنوشروان مائة وعشرون طبيباً بين روميّ وهنديّ وفارسيّ وكان برزوية من أشرف أطبّاء الفرس وأكثرهم دراسة للكتب. فوجد في بعضها أنّ ببلاد الهند جبالاً فيها من غرائب العقاقير ما يوحي الموتى فما زال ذلك يدور في رأسه ويسمو بهمّته إلى تطلّبه وتحصيله حتّى أخبر أنوشيروان بما في نفسه واستأذنه للنهوض والسعي في الظفر ببغيته فأذن له وأعانه على سفرته وزوّده من الكتّاب إلى ملك الهند ما يكون سبباً لإنجاحه.

واستقلّت به الركاب إلى واسطة الهند فلمّا دخلها وأوصل كتّاب أنوشيروان إلى ملكها أكرمه وحكّمه في مناه وأنهضه لطيّته في تطلّب العقاقير من مظانّها. فما زال يجدّ ويجتهد ويتعب ويدأب في اجتنائها والتقاطها وتأليفها وتركيبها حتّى كان مثله بعد حين من الدهر كما تقول عامّة بغداد: »ما زلنا في لا شيء حتّى فرغنا.«

واستشعر الكأبة والانخزال لما فاته من مراده وضاع من أيّامه وتصوّر الخجل من صاحبه إذا عاد مخفقاً إلى حضرته. فسأل عن أطبّ الأطبّاء وأحكم الحكماء بأرض الهند فدُلّ على شيخ عالى السنّ فأتاه وقصّ عليه قصّته وذكر له ما قرأه في بعض الكتب من حديث جبال الهند واشتمالها من العقاقير على ما يحيي الموتى. فقال له: »يا برزوية حفظت شيئاً وغابت عنك أشياء أما علمت أنّ ذلك رمز للقدماء والمراد بالجبال العلماء وبالعقاقير كلامهم الشافي الكافي وبالموتى الجهّال يعنون أنّ العلماء يؤدّبون الجهّال بحكمهم فكأنّهم يحيون الموتى وهذه الحكم محصورة في كتّاب مترجم بكليلة ودمنة ليس يوجد إلّا في خزانة الملك.«

فسُرّي عن برزوية وسُرّ بما سمع ورغب إلى الملك في إعارته إيّاه الكتّاب وتقليد الملك أنوشيروان بذلك منّة مشكورة فقال: »سآمر بإعارته إيّاك إيجاباً لصاحبك أوّلاً ورعاية لحقّك ثانياً على شريطة أن تعيره بين يديّ لحظك ولا تتّخذ منه نسخة لنفسك.« فأجاب برزوية بالسمع والطاعة وجعل يحضر في كلّ يوم مجلسه ويدعو بالكتّاب فينظر فيه ويتحفّظ معانيه ويقيّدها بالكتابة إذا رجع إلى منزله حتّى أتى على جميعه. واستأذن الملك للعود إلى حضرة صاحبه فأذن له وأهدى إليه وخلع عليه.

وحين وصل إلى أنوشيروان أخبره بقصّته وبشّره بحصول الكتّاب لديه ثمّ عرضه عليه. فأعجب به وأجزل صلته وأمر بزرجمهر بنقله إلى اللغة الفهلويّة. فتلطّف برزوية وتضرّع إلى الملك في الإذن

لافتتاح الباب الأوّل منه باسمه وذكره فأجابه إليه. ولم يزل الكتاب مخزوناً عند ملوك الفرس حتّى نقله ابن المقفّع إلى العربيّة والروذكي بأمر الأمير نصر بن أحمد إلى الشعر بالفارسيّة.

v.8. Ẓahīrī Samarqandī, *The Book of Sindbād* (*Sindbādnāma*), 25.4–11 (ed. Ateş).

این کتاب بلغت پهلوی بوذه است و تا بروزگار امیر اجلّ عالم ناصر الدین ابو محمّد نوح بن منصور السامانی—انار الله برهانه—هیچ کس ترجمه نکرده بوذ امیر عادل نوح بن منصور فرمان داد خواجه عمید ابو الفوارس فناروزی را تا بزبان فارسی ترجمت کنذ وتفاوت و اختلالی کی بذو راه یافته بوذ بر دارذ و درست و راست کنذ بتاریخ سنه تسع وثلثین وثلثمایه خواجه عمید ابو الفوارس رنج برگرفت و خاطر در کارآورده و این کتاب را بعبارت دری پرداخت لکن عبارت عظیم نازل بود و از تزیّن و تحلّی عاری و عاطل.

v.9. *The Royal Book of Hunting* (*Šikārnāma-e Ḫusrawī* or *Malikšāhī*), xvi.6–xvii.14 (ed. Nafīsī, *Kitāb-e naṣīḥat-nāma*).

این کتاب شکارنامه است که از زمان شاه افریدون باز که استخراج این علوم کرده اند و ابتدای جانورداری در آنروزگار بوده و حکما و استادان زحمت ها کشیده اند و بانواع ریاضت و فکر بر زبانهاء پیشینیان جهة کتب خانه سلاطین انار الله براهینهم تصنیفات کرده بودند و بزبان فارسی ترجمه کرده اند از زبان پهلوی که در زمان انوشیروان عادل از زبانهاء دیگر بپهلوی ترجمه کرده بودند و سبب ترجمه کردن و نهادن این کتاب آن بوده که چون نوبت پادشاهی انوشیروان رسید او را هوس جانورداری اتفاق افتاد و درین فن بغایت هوس بکمال داشت چنانك هیچ ذوق و تنعم و تفرج او را چون شکار جانور نبود و بیشتر اوقات بدین شعل مشغول بودی. روزی جمعی از میرشکاران در پیش او در باب جانور از کتب و تجربه بحث می کردند. پادشاه طلب فرمود خواجه بزرجمهر حکیم را و گفت چنان خواهم که باطراف کس فرستی و کتب شکارنامها که در زمان ملوك ماضی جهة کتب خانه ایشان تصنیف کرده اند جمع کنی و استادان که درین روزگار جمع شده بریشان خوانی و صلاح و فساد آن بکلی بدانی و آنچه خلاصه باشد از کتب جمع گردانی و آنچه این جماعت دانند اضافت کنی و بزبان فهلوی تجدید کنی تا دستوری باشد و بر روی روزگار بماند و اگر مشکل افتد بدان رجوع توان کرد. بموجب فرموده خواجه بزرجمهر آن کتب بزبان فهلوی ترجمه کرد

و آنچه خواجه را از روی حکمت در آن باب معلوم شده و استادان دانستند بدان اضافت کردند و آن کتاب را شکارنامه انوشیروان نام نهادند و دستوری شد بروی روزگار. بعد از آن شاهنشاه ابو الفوارس عبد الملك بن نوح رحمة الله علیه را اتفاق این فن افتاد از ملوك ما تقدم بگذشت. روزی وزیر را فرمود که کتاب شکارنامه انوشیروانی طلب کنند و بدست آرند و استادان که در روزگار او جمع بود حاضر کردند و کتاب بر ایشان خوانند و آنچه از ایشان دانند اضافت کنند و بزبان فارسی ترجمه کنند. بموجب فرموده وزیر آن کتاب بزبان فارسی ترجمه کرد و آنچه استادان دانستند اضافت کرد. ملك تا آخر عمر بدان کار کردی و بعد از روزگار او شاه خلف سیستانی که درین فن استادی ماهر بود و هوس ابن فن او را بغایت کمال بود بهوس این فن به بخارا آمد و این کتاب طلب کرد و حاضر گردانید و استادان این فن چون دهقان سکنی و دهقان باتکینی را و ابو الفرج بازدار را و زرگری بازدار را و این کتاب حاضر گردانید و بریشان خواند. جماعت بر قول پیشینیان رضا دادند وآنچه ایشان دانستند اضافت کردند و او همیشه بدین کتاب کار کردی [...] چون زمان شاهی بسلطان ملکشاه رسید نظام الملك را گفت که جمعی از آگاهان بدین فن را گرد آورد و کتابی بپردازند او هم مردمی بهر دیار فرستاد و از هر کشوری یك تن مرد آگاه بدین فن آوردند و ایشان چند روز با هم نشستند و این کتاب "شکارنامه خسروی" را برای ملکشاه پرداختند.

V.10. Ḫudāyār-ḫān Dāwūd ʿAbbāsī, *Hunting the Goal in the Laws of the Hunter* (*Ṣayd al-murād fī qawānīn al-ṣayyād*), 23.8–24.1 (ed. Ṣadīqī, 'Ḥakīm Nasawī').

تا در زمان پادشاهی نوشیروان عادل ... روزی جمعی از میرشکاران در پیش تخت او بحث جانور می کردند از کتاب جالینوس و بقراط آنچه ایشان را معلوم بود. پس نوشیروان بزرجمهر حکیم را طلب فرمود و گفت چنان خواهم که کس باطراف عالم فرستی و کتب و رسائلی که در این باب ساخته باشند جمع گردانی و بزبان پهلوی ترجمه کنی و آنچه خواجه را از روی حکمت معلوم شده باشد و استادان روزگار ما دانند اضافت کنی تا بر روی روزگار یادگاری بماند. بموجب فرموده خواجه بزرجمهر آنچه دانست و از استادان معلوم نمود و اصل نسخ از آن خالی بود جمع کرد و آن را "بازنامه نوشیرانی" نام نهاد و مدتها برآن کار فرمودند و دستوری شد بر روی روزگار. بعد از آن شهنشاه ابو الفوارس عبد الملك بن نوح (345–350 هجری) که پادشاه خراسان بود هوس این علم بهم رسانید چنانکه از ملوك ما تقدم در گذشت و خواست که بازنامه نوشیروانی را از زبان پهلوی بزبان فارسی ترجمه کند وزیر خود را گفت چنان خواهم که بازنامه نوشیروانی را حاضر

کنی و بفرمایی تا از زبان پهلوی بفارسی ترجمه کنند و استاد بزرگوار ابو البحتری بزبان فارسی کرد و جوارح نامه شهنشاهی نام نهاد ...

Bibliography

Primary Sources

Abū l-Fidā', Ismā'īl ibn 'Alī, *Kitāb al-muḫtaṣar fī aḫbār al-bašar*, 2 vols, Cairo, 1325/1907

Agathias, *Agathiae Myrinaei Historiarum libri quinque*, ed. Rudolf Keydell, Berlin, 1967

Agathias, *The Histories*, tr. Joseph D. Frendo, Berlin/New York, 1975

al-'Āmirī, Abū l-Ḥasan, *al-I'lām bi-manāqib al-Islām*, ed. 'Abd al-Ḥamīd Ġurāb, Riyadh, 1988

Aulus Gellius, *Attic Nights*, ed. and tr. John C. Rolfe, 3 vols, Cambridge MA, 1927

Bar Hebraeus, *Gregorii Barhebraei Chroncicon Syriacus*, ed. Paul Bedjan, Paris, 1890

Bar Hebraeus, *Chronography of Gregory Abū 'l-Faraj* [...] *commonly known as Bar Hebraeus*, tr. Ernest A. Wallis Budge, London, 1932

Bar Hebraeus, *Ta'rīḫ muḫtaṣar al-duwal*, ed. Anṭūn Ṣāliḥānī al-Yasū'ī, Beirut, 1992

Bāznāma-ye Nawšīrwānī = Bīniš, Taqī, 'Bāznamā-ye Nawširwānī', *Farhang-e Īrānzamīn*, 30/1 (1384 š./2005–2006), 1–106

al-Bīrūnī, Abū Rayḥān Muḥammad b. Aḥmad, *Chronologie orientalischer Völker von Albêrûnî* [*al-Āṯār al-bāqiyah 'an al-qurūn al-ḫāliyah*], ed. Eduard Sachau, Leipzig, 1878

al-Bīrūnī, Abū Rayḥān Muḥammad b. Aḥmad, *The Chronology of Ancient Nations*, tr. Eduard Sachau, London, 1879

al-Bīrūnī, Abū Rayḥān Muḥammad b. Aḥmad, *Alberuni's India* [*Taḥqīq mā li-l-Hind*], tr. Edward Sachau, 2 vols, London, 1910

Catalogus Codicum Astrologorum Graecorum, ed. Franz V.M. Cumont et al., 12 vols, Brussels, 1898–1953

Chronicle of Seert, ed. Addai Scher in *Patrologia Orientalis*, 4 (1907), 212–313; 7 (1911), 92–203

Dawlatšāh Samarqandī, *The Tadhkiratu 'sh-šu'ará* (*"Memoirs of the Poets"*), ed. Edward G. Browne, Leiden/London, 1901

Dēnkard = *The Complete Text of the Pahlavi Dinkard*, ed. Dhanjishab Medherjibhai Madan, 2 vols, Bombay, 1911

al-Dīnawarī, Abū Bakr Aḥmad b. Marwān b. Mālik al-Qāḍī, *al-Muǧālasah wa-ǧawāhir al-'ilm*, ed. Abū 'Ubayda Mašhūr b. Ḥasan Āl Salmān, 10 vols, Beirut, 2002

al-Dīnawarī, Abū Ḥanīfa Aḥmad b. Dāwūd, *al-Aḫbār al-ṭiwāl*, ed. 'Abd al-Mun'im 'Āmir and Ǧamāl al-Dīn al-Šayyāl, Cairo, 1960

Diogenes Laertius, *Lives of Eminent Philosophers*, ed. and trans. R.D. Hicks, 2 vols, London, 1959

Dorotheus of Sidon, *Carmen Astrologicum = Dorothei Sidonii Carmen Astrologicum*, ed. David Pingree, Leipzig, 1976; repr. as *Carmen Astrologicum; Dorotheus of Sidon*, Ascella, 1993

Eunapius, *Lives of the Sophists*, tr. Wilmer Cave Wright, London/New York, 1922

Flavius Philostratus, *Vita Apollonii*, tr. Frederick C. Conybaere, Cambridge MA, 1912

al-Ǧāḥiẓ, *al-Rasāʾil*, ed. ʿAbd al-Salām Muḥammad Hārūn, 4 vols, Cairo, 1964

al-Ǧāḥiẓ, *al-Bayān wa-l-tabyīn*, ed. ʿAbd al-Salām Muḥammad Hārūn, 4 vols, Cairo, 1968

al-Ǧāḥiẓ, *al-Muḫtār fī l-radd ʿalā l-naṣārā* [*Refutation of the Christians*], ed. Muḥammad b. ʿAbdallāh al-Šarqāwī, Beirut, 1411/1991

Pseudo-al-Ǧāḥiẓ, *Book of Illustrations and Evidence* [*Kitāb al-ʿibar wa-l-iʿtibār*] = Rieu, Charles, *Supplement to the Catalogue of the Arabic Manuscripts in the British Museum*, London, 1894, 466–467, no. 684

Pseudo-al-Ġazālī, *Naṣīḥat al-mulūk*, ed. Ǧalāl Humāʾī, Tehran, 1315–1317 š./1936–1938; with corrections, Tehran, 1972

Ḥāǧǧī Ḫalīfa, *Kašf al-ẓunūn*, ed. Şerefettin Yaltkaya, 2 vols, Istanbul, 1941

Ḥamza al-Iṣfahānī, *The Annals of Hamzah al-Isfahani*, tr. Umar M. Daudpota, Bombay, 1932

Ḥamza al-Iṣfahānī, *Taʾrīḫ sinī mulūk al-arḍ wa-l-anbiyāʾ*, ed. Yūsuf Yaʿqūb Maskūnī, Beirut, 1961

Ḥamza al-Iṣfahānī, *al-Amṯāl al-ṣādira ʿan buyūt al-šiʿr*, ed. Aḥmad b. Muḥammad al-Ḍubayb, Beirut, 2009

Ḫudāyār-ḫān Dāwūd ʿAbbāsī, *Ṣayd al-murād* = Phillott, Douglas C., *The Qawānīnu ʿs-ṣayyād of Khudā Yār Khān*, Calcutta, 1908

Ibn Abī Ṭāhir Ṭayfūr, *Taʾrīḫ Baġdād* = Ibn Abī Ṭāhir Ṭayfūr, *Sechster Band des Kitāb Baġdād von Aḥmad Ibn Abī Ṭāhir Taifūr* [*Taʾrīḫ Baġdād*], ed. Hans Keller, 2 vols, Leipzig, 1908

Ibn Abī Uṣaybiʿah, Muwaffaq al-Dīn Abū ʿAbbās Aḥmad b. al-Qāsim, *ʿUyūn al-anbāʾ fī ṭabaqāt al-aṭibbāʾ*, ed. Nizār Riḍā, Beirut, n.d. [c. 1986]; new edn by Emilie Savage-Smith et al., Leiden, 2020

Ibn al-Faqīh, Abū ʿAbdallāh Aḥmad b. Muḥammad b. Isḥāq al-Hamadānī, *Kitāb al-buldān*, ed. Yūsuf al-Hādī, Beirut, 1416/1996

Ibn al-ʿIbrī *see* Bar Hebraeus

Ibn Ḫaldūn, *The Muqaddima: An Introduction to History*, tr. F. Rosenthal, 3 vols, London, 1958

Ibn Ḫaldūn, *Kitāb al-ʿibar = Taʾrīḫ al-ʿallāmah Ibn Ḫaldūn. Kitāb al-ʿibar wa-dīwān al-mubtadaʾ wa-l-ḫabar fī ayyām al-ʿarab wa-l-ʿaǧam wa-l-barbar wa-man ʿāṣarahum min ḏawī l-sulṭān al-akbar*, Beirut, 1961

Ibn Ḫallikān, Abū l-ʿAbbās Šams al-Dīn Aḥmad b. Muḥammad al-Irbilī, *Wafayāt al-aʿyān wa-anbāʾ abnāʾ al-zamān*, ed. Iḥsān ʿAbbās, 8 vols, Beirut, 1968–1972

Ibn al-Muqaffaʿ, *al-Manṭiq*, ed. M.T. Dānishpazhūh, Tehran, 1978

Ibn al-Nadīm, *The Fihrist of al-Nadīm: A Tenth-Century Survey of Muslim Culture*, tr. Bayard Dodge, 2 vols, New York/London, 1970

Ibn al-Nadīm, *al-Fihrist* = Ibn al-Nadīm, Muḥammad b. Isḥāq, *Kitāb al-fihrist*, ed. Muḥammad Riḍā Tağaddud, Tehran, 1971

Ibn al-Qifṭī, *Taʾrīḫ al-ḥukamāʾ* = Ibn al-Qifṭī, ʿAlī b. Yūsuf, *Ibn al-Qifṭī's Taʾrīḫ al-ḥukamāʾ*, ed. Julius Lippert, Leipzig, 1903

Ibn Qutaybah, *al-Maʿārif*, ed. Ṯarwat ʿUkāšah, Cairo, 1981

Ibn Qutaybah, *Faḍl al-ʿarab wa-l-tanbīh ʿalā ʿulūmihā* [*The Superiority of the Arabs and Insight into their Sciences*], ed. Walīd Maḥmūd Ḫāliṣ, 1998

Ibn Qutaybah, *The Excellence of the Arabs*, ed. James E. Montgomery and Peter Webb, tr. Sarah Bowen Savant and Peter Webb, New York, 2017

Isidore of Seville, *Isidori Hispalensis episcopi etymologiarum sive originum libri XX*, ed. Wallace M. Lindsay, 2 vols, Cambridge, 1911

Isidore of Seville, *The Etymologies of Isidore of Seville*, tr. Stephen A. Barney, W.J. Lewis, J.A. Beach, and Oliver Berghof, Cambridge, 2006

al-Iskāfī al-Ḫaṭīb, Abū ʿAbdallāh Muḥammad b. ʿAbdallāh, *Luṭf al-tadbīr*, ed. Aḥmad ʿAbd al-Bāqī, Cairo, 1964; repr. Beirut, 1979

al-Iškawarī al-Daylamī, Quṭb al-Dīn Muḥammad b. al-Šayḫ ʿAlī, *Maḥbūb al-qulūb*, ed. Ibrāhīm al-Dībāǧī and Ḥāmid Ṣidqī, 2 vols, Tehran, 1999

Malalas, John, *The Chronicle of John Malalas*, tr. Elizabeth Jeffreys, Michael Jeffreys, and Roger Scott, Melbourne, 1986

al-Maqrīzī, *Book of Contention and Strife concerning the Relations between the Banū Umayya and the Banū Hāshim*, tr. Clifford E. Bosworth, Manchester, 1984

al-Masʿūdī, *Kitāb al-Tanbīh wa-l-ishrāf*, ed. ʿAbdallāh Ismāʿīl al-Ṣāwī, Cairo, 1938

al-Masʿūdī, *Murūǧ al-ḏahab wa-maʿādin al-ǧawhar*, ed. Charles Pellat, 7 vols, Beirut, 1965–1979

Miskawayh, Abū ʿAlī al-Rāzī, *Tajārib al-umam*, ed. Abū al-Qāsim Imāmī, 7 vols, Tehran, 1987–2001

Mutġabbī b. Farīʿūn, *Kitāb ǧawāmiʿ al-ʿulūm*, ed. Qays Kāẓim al-Ǧanābī, Cairo, 2007

Nāmakīhā ī Mānūščihar. The Epistles of Mānūshchīhar, ed. Ervad Bamanji Nasaryanji Dhabhar, Bombay, 1912

al-Nasawī, *Bāznāma* [*On the Art of Falconry*], ed. ʿAlī Ġarawī, Tehran, 1354 š./1975

Niẓāmī, *The Haft Paikar* (*The Seven Beauties*), tr. Charles E. Wilson, 2 vols, London, 1924

Niẓāmī, *Kullīyāt-e ḫamsa-e Niẓāmī*, ed. Waḥīd Dastgardī, 2 vols, Tehran, 1372 š./1993–1994

Niẓāmī, *The Haft Paykar: A Medieval Persian Romance*, tr. Julie Scott Meisami, Oxford/New York, 1995

Plutarch's Lives III, ed. and tr. Bernadotte Perrin, London, 1916; repr. 1932

Priscian of Lydia, *Answers to King Khosroes of Persia*, tr. Pamela Huby et al., London, 2016

al-Rāġib al-Iṣfahānī, Ḥusayn b. Muḥammad, *Muḥāḍarāt al-udabā'*, 4 sections in 2 vols, Beirut, 1961

al-Rayḥānī, 'Alī ibn 'Ubaydah, *Persian Wisdom in Arabic Garb: 'Alī b. 'Ubayda al-Rayḥānī (d. 219/834) and his Jawāhir al-kilam wa-farā'id al-ḥikam*, ed. and tr. Mohsen Zakeri, 2 vols, Leiden, 2007

al-Ṣafadī, *al-Wāfī bi-l-wafayāt*, 30 vols, Istanbul/Leipzig/Beirut/Wiesbaden, 1931–2000

Ṣā'id al-Andalusī, *Ṭabaqāt al-umam*, Beirut, 1912

Ṣā'id al-Andalusī, *Livre des catégories des nations*, tr. Régis Blachère, Paris, 1935

Ṣā'id al-Andalusī, *Science in the Medieval World: "Book of the Categories of Nations"*, ed. and tr. Sema'an I. Salem and Alok Kumar, Austin TX, 1991

al-Šahrazūrī, Šams al-Dīn Muḥammad b. Maḥmūd, *Ta'rīḫ al-ḥukamā'. Nuzhat al-arwāḥ wa-rawḍat al-afrāḥ*, ed. 'Abd al-Karīm 'Umar Abū Šuwayrib, Tripoli, 1988

al-Šarīf al-Murtaḍā, *Amālī al-Murtaḍā. Ġurar al-fawā'id wa-durar al-qalā'id*, ed. Muḥammad Abū l-Faḍl Ibrāhīm, 2 vols, Cairo, 1954

al-Siǧistānī al-Manṭiqī, Abū Sulaymān, *Muntaḫab Ṣiwān al-ḥikma of Abū Sulaymān al-Sijistānī*, ed. Douglas M. Dunlop, The Hague, 1979

al-Suyūṭī, Ǧalāl al-Dīn 'Abd al-Raḥmān b. Abī Bakr, *Buġyat al-wu'āt fī ṭabaqāt al-luġawīyin wa-l-nuḥāt*, ed. Muḥammad Abū l-Faḍl Ibrāhīm, 2 vols, Beirut, 1979

al-Ṭabarī, Muḥammad ibn Ǧarīr, *Ta'rīḫ al-rusul wa-l-mulūk*, ed. Muḥammad Abū l-Faḍl Ibrāhīm, 10 vols, Cairo, 1960–1969

al-Ṭabarī, Muḥammad ibn Ǧarīr, *The History of al-Ṭabarī*, XV: *The Crisis of the Early Caliphate: The Reign of 'Uthmān (A.D. 644–656/A.H. 24–35)*, tr. R. Stephen Humphreys, Albany NY, 1990

al-Ṭabarī, Muḥammad ibn Ǧarīr, *The History of al-Ṭabarī*, V: *The Sasanids, the Byzantines, the Lakhmids, and Yemen*, tr. Clifford E. Bosworth, Albany NY, 1999

al-Ẓahīrī al-Samarqandī, Muḥammad ibn 'Alī al-Ẓahīrī, *Sindbāḏ-nāme yazan Muḥammed b. 'Alī aẓ-Ẓāhirī as-Samarqandī*, ed. and tr. Ahmed Ateş, Istanbul, 1948

Tacitus, *The Annals: The Reigns of Tiberius, Claudius, and Nero*, tr. John C. Yardley, Oxford, 2008

Tatian of Adiabene, *Address to the Greeks*, in *The Writings of Tatian and Theophilus; and The Clementine Recognitions*, tr. B. Pratten, Marcus Dods, and Thomas Smith, Edinburgh, 1867, 1–45

Tatian of Adiabene, *Oratio ad Graecos and Fragments*, ed. and tr. Molly Whittaker, Oxford, 1982

Tatian of Adiabene, *Tatiani Oratio ad Graecos*, ed. Miroslav Marcovich, Berlin, 1995

Tatian of Adiabene, *Oratio ad Graecos*, ed. and tr. Jörg Trelenberg, Tübingen, 2012

al-Ṯa'ālibī, *Ġurar: Histoire des Rois des Perses*, tr. Hermann Zotenberg, Paris, 1900

Zangī, Muḥammad b. Mubārak [known as Qayyim Nahāwandī], *Farasnāma-e Qayyim*

Nahāwandī (*kuhantarīn Farasnāma-e šināḫta šuda be-Fārsī*) [*The Oldest Known Book on Hippology in Persian*], ed. Nādir Muṭallibī Kāšānī and Ādharnūš Ādhartāš, Tehran, 1395 š./2016

Vizārgird dēnīg = Daftar-ī vizargird dēnīg, ed. Peshotan Sanjana, Mumbai, 1218/1948

al-Yamanī, Muḥammad b. Ḥusayn b. ʿUmar, *Kitāb muḍāhāt amṯāl kitāb Kalīlah wa-Dimnah bi-mā ašbahahā min ašʿār al-ʿarab*, ed. M. Yūsuf, Beirut, 1961

Yāqūt al-Ḥamawī, *Muʿǧam al-buldān*, 7 vols, Beirut, 1995

al-Yaʿqūbī, Aḥmad b. Abī Yaʿqūb b. Jaʿfar b. Wahb, *Mušākalat al-nās li-zamānihim*, ed. William G. Millward, Beirut, 1962

al-Yaʿqūbī, Aḥmad b. Abī Yaʿqūb b. Jaʿfar b. Wahb, 'The Adaptation of Men to their Time: A Historical Essay by al-Yaʿqūbī', tr. William G. Millward, *Journal of the American Oriental Society*, 84 (1964), 329–344

al-Zubayr b. Bakkār, *al-Aḫbār al-Muwaffaqiyāt*, ed. Sāmī Makkī al-ʿĀnī, Beirut, 1996

Secondary Literature

ʿAbbās, Iḥsān, *Malāmiḥ yūnānīya fī l-adab al-ʿarabī* [*Greek Features in Arabic Literature*], Beirut, 1967

Afnan, Soheil, *Philosophical Terminology in Arabic and Persian*, Leiden, 1964

Agha, Saleh Said, 'Abū Muslim al-Khurāsānī', *EI Three*, 2015, http://dx.doi.org/10.1163/1573-3912_ei3_COM_24741 (accessed 28.7.2021)

Altheim, Franz and Ruth Stiehl, 'New Fragments of Greek Philosophers', *East and West*, 12 (1961), 3–18

Amīn, Aḥmad, *Faǧr al-islām*, Beirut, 1928 (repr. 1969)

Anthony, Sean W., 'Chiliastic Ideology and Nativist Rebellion in the Early ʿAbbasid Period: Sunbādh and the Jāmāsp-Nāmah', *Journal of the American Oriental Society*, 132 (2012), 641–655

Bailey, Harold, *Zoroastrian Problems in the Ninth-Century Books*, Oxford, 1943/1971

Bailey, Harold, 'The Preface to the Siddhasāra-śāstra', in *A Locust's Leg: Studies in Honour of Seyyed Hassan Taqizadeh*, ed. Walter B. Henning and Ehsan Yarshater, London, 1962, 31–36

Bartold, Wassilij, *Turkestan Down to the Mongol Invasion*, 2nd edn, London, 1928

Baumstark, Anton, *Geschichte der syrischen Literatur mit Ausschluss der christlich-palästinensischen Texte*, Bonn, 1922

Berthelot, Marcellin, *La Chimie au moyen âge*, III: *L'Alchimie arabe*, Paris, 1893

van Bladel, Kevin, *The Arabic Hermes*, Oxford, 2009

van Bladel, Kevin, 'The Arabic History of Science of Abū Sahl Ibn Nawbaḫt (fl. ca. 770–809) and its Middle Persian Sources', in *Islamic Philosophy, Science, Culture, and Religion: Studies in Honor of Dimitri Gutas*, ed. Felicitas Opwis and David C. Reisman, Leiden, 2012, 41–62

van Bladel, Kevin, 'Hermes and Hermetica', *EI Three*, http://dx.doi.org/10.1163/1573-3912_ei3_COM_23130 (accessed 3.8.2021) de Blois, François, *Burzōy's Voyage to India and the Origin of The Book of Kalilah wa Dimnah*, London: Royal Asiatic Society, 1990

Bosworth, Clifford E., 'The Ṭāhirids and Arabic Culture', *Journal of Semitic Studies*, 14/1 (1969), 45–79

Bosworth, Clifford E., 'Nūḥ b. Naṣr b. Aḥmad', *EI2*, VIII, 109–110

Bosworth, Clifford E., 'Sāmānids', *EI2*, VIII, 1025–1029

Boyarin, Daniel, *Socrates and the Fat Rabbis*, Chicago, 2009

Boyce, Mary, *Zoroastrians: Their Religious Beliefs and Practices*, London, 1979

Boyce, Mary and Frantz Grenet, *A History of Zoroastrianism*, III: *Zoroastrianism under Macedonian and Roman Rule*, Leiden, 1991

Brock, Sebastian, 'Antagonism to Assimilation. Syriac Attitudes to Greek Learning', in *East of Byzantium: Syria and Armenia in the Formative Period*, ed. Nina Garsoïan, Thomas Mathews, and Robert Thompson, Washington DC, 1982, 17–34

Brock, Sebastian, 'Some Aspects of Greek Words in Syriac', in *Synkretismus im syrisch-persischen Kulturgebiet*, Bericht über ein Symposion in Reinhausen bei Göttingen in der Zeit vom 4. bis 8. Oktober 1971, herausgegeben von Albert Dietrich, Göttingen, 1975, 80–108; =id., *Syriac Perspectives on Late Antiquity*, Variorum Reprints CS 199, London, 1984 [Article IV]

Browne, Edward, *A Literary History of Persia: From the Earliest Times until Firdawsi*, London, 1902

Browne, Edward, *Arabic Medicine*, Cambridge, 1921

Browne, Edward, *Revised Translation of the Chahár Maqála ("Four Discourses") of Niẓámí-i-ʿArúḍí of Samarqand*, Cambridge, 1921

Buscherini, Stefano, 'The translation of *Almagest* in Pahlavi: zero in Persian numerical system', in *Non licet stare caelestibus: Studies on Astronomy and its History Offered to Salvo de Meis*, ed. Antonio Panaino et al., Milan/Udine, 2014, 137–150

Cameron, Averil, 'Agathias on the Sassanians', *Dumbarton Oaks Papers*, 23 (1969–1970), 67–183

Cameron, Averil, *Agathias*, Oxford, 1970

Cereti, Carlo G., 'Prolegomena allo studio del quarto libro del Denkard', *Studi orientali e linguistici*, 5 (1994–1995), 107–129

Chunakova, Olga, 'The Pahlavi Lapidary', *Written Monuments of the Orient*, 1 (2016), 73–84

Christensen, Arthur, *L'Iran sous les Sassanides*, 2nd edn, Copenhagen, 1944

Clouston, William A., *The Book of Sindibād; or, The Story of the King, his Son, the Damsel, and the Seven Vazīrs*, Glasgow, 1884

Cottrell, Emily, 'L'autobiographie de Zoroastre', in *Pensée grecque et sagesse d'Orient: Hommage à Michel Tardieu*, ed. Mohammad Ali Amir-Moezzi, Turnhout, 2009, 178–183

Cottrell, Emily, '"L'Hermès Arabe" de Kevin van Bladel et la question du rôle de la littérature Sassanide dans la présence d'écrits hermétiques et astrologiques en langue arabe', *Biblioteca Orientalis*, 72/3–4 (2015), 336–401

Cottrell, Emily and Micah T. Ross, 'Persian Astrology: Dorotheus and Zoroaster, according to the medieval Arabic sources (8th–13th century)', in *Proceedings of the Eighth European Conference of Iranian Studies*, I: *Studies on Pre-Islamic Iran and on Historical Linguistics*, ed. Pavel B. Lurje, St Petersburg, 2019, 87–105

Crone, Patricia, *God's Rule: Government and Islam*, New York, 2004

Delaini, Paolo, 'La scuola di Gundēšābūr. La conoscenza del corpo umano (anatomia e fisiologia) e la trasmissione delle teorie medico-scientifiche nel mondo Sasanide e post-Sasanide', PhD thesis, Università di Bologna, 2012

Delaini, Paolo, *Medicina del corpo, medicina dell'anima: La circolazione delle conoscenze medico-filosofiche nell'Iran Sasanide*, Milan, 2014

Denison Ross, E., 'Ibn Muqaffa' and the Burzoë Legend', *Journal of the Royal Asiatic Society of Great Britain and Ireland*, 3 (1926), 503–505

Di Branco, Marco, *Storie arabe di Greci e di Romani*, Pisa, 2009

Dols, Michael W., 'The Origins of the Islamic Hospital: Myth and Reality', *Bulletin of the History of Medicine*, 61 (1987), 367–390

Dunlop, Douglas M., 'Bīmārestān', *EIr*, IV, 257–267

EIr = *Encyclopaedia Iranica*, ed. Ehsan Yarshater, London, 1982–

EI2 = *The Encyclopaedia of Islam: Second Edition*, ed. H.A.R. Gibb et al., 13 vols, Leiden, 1954–2009

EI Three = *Encyclopaedia of Islam, Three*, ed. Kate Fleet, Gudrun Krämer, Denis Matringe, John Nawas, and Everett Rowson, Leiden, 2007– [online version]

Elgood, Cyril, *Medical History of Persia and the Eastern Caliphate*, Cambridge, 1951

Emmerick, R.E., 'Ravigupta's Siddhasāra in Arabic', in *Studien zur Geschichte und Kultur des Vorderen Orients: Festschrift für Bertold Spuler zum siebzigsten Geburtstag*, ed. Hand Robert Roemer and Albrecht Noth, Leiden, 1981, 28–31

Finkel, Joshua, 'A Risāla of al-Jāḥiẓ', *Journal of the American Oriental Society*, 47 (1927), 311–334 [a partial translation of al-Ǧāḥiẓ, *al-Radd ʿalā l-naṣārā*]

Fletcher, Charles D., 'Anti-Christian Polemic in Early Islam: A Translation and Analysis of Abū 'Uthmān 'Amr b. Baḥr al-Jāḥiẓ' *Risāla*: *Radd ʿalā al-Naṣārā* (A Reply to the Christians)', MA thesis, McGill University, 2002

de Fouchécour, Charles-Henri, *Moralia: Les notions morales dans la littérature persane du 3e/9e au 7e/13e siècle*, Paris, 1986

Froehner, Reinhard, *Die Tierheilkunde des Abu Bekr Ibn Bedr*, Leipzig, 1931

Ǧawād, Muṣṭafā and Aḥmad Sūsa, *Dalīl ḫāriṭaṭ Baġdād al-mufaṣṣal fī ḫiṭaṭ Baġdād qadīman wa-ḥadīṯan* [*A Detailed Guide to Baghdad's Map in Baghdad's Plans, Old and Modern*], Baghdad, 1958

Gibb, Hamilton A.R., 'The Argument from Design. A Muʿtazilite Treatise Attributed to al-Jāḥiẓ', in *Ignace Goldziher Memorial Volume*, pt 1, ed. Samuel Löwinger and Joseph Somogyi, Budapest, 1948, 150–162

Gibson, Nathan P., 'Closest in Friendship? Al-Jāḥiẓ' Profile of Christians in Abbasid Society in "The Refutation of Christians" (*al-Radd 'alā al-Naṣārā*)', PhD thesis, Catholic University of America, 2015

Gignoux, Philippe, 'La Démonisation d'Alexandre le Grand d'après la littérature pehlevie', in *Iranian Languages and Texts from Iran and Turfan: Ronald E. Emmerick Memorial Volume*, ed. Maria Macuch, Mauro Maggi, and Werner Sundermann, Wiesbaden, 2007, 87–97

Gignoux, Philippe, 'Dēnkard', *EIr*, VII, 284–289

Gignoux, Philippe, 'Syriac Language iii. Syriac Translators as the Medium for Transmission of Greek Ideas to Sasanian Iran', *EIr* online: http://www.iranicaonline.org/articles/syriac-language-03-translators-greek-ideas (accessed 3.5.2020)

Göbl, Robert, *Der Triumph des Sāsāniden Šahpuhr über die Kaiser Gordianus, Philippus und Valerianus*, Vienna, 1974

Griffini, Eugenio, 'Lista dei manoscritti arabi nuovo fondo della Biblioteca Ambrosiana di Milano', *Rivista degli Studi Orientali*, 7 (1916), 51–130

Gutas, Dimitri, 'Paul the Persian on the classification of the parts of Aristotle's philosophy: a milestone between Alexandria and Baġdād', *Der Islam*, 60 (1983), 231–267

Gutas, Dimitri, 'The Starting Point of Philosophical Studies in Alexandrian and Arabic Aristotelianism', in *Theophrastus of Eresus: On His Life and Work*, ed. William W. Fortenbaugh, Pamela M. Huby, and Anthony A. Long, New Brunswick NJ, 1985, 115–123

Gutas, Dimitri, *Greek Thought, Arabic Culture*, London, 1998

Gutas, Dimitri, 'The Presence of Antiquity in the Sasanian and Early 'Abbāsid Empires', in *Die Gegenwart des Altertums: Formen und Funktionen des Altertumsbezugs in den Hochkulturen der Alten Welt*, ed. Dieter Kuhn and Helga Stahl, Heidelberg, 2001, 347–352

Gutas, Dimitri and van Bladel, Kevin, 'Bayt al-Ḥikma', *EI Three*, 2009, http://dx.doi.org/10.1163/1573-3912_ei3_COM_22882 (accessed 28.7.2021)

Hainthaler, Theresia, 'Ǧāḥiẓ und seine Schrift ‚Widerlegung der Christen': Eine Annäherung', in *Christsein in der islamischen Welt: Festschrift für Martin Tamcke zum 60. Geburtstag*, ed. Sidney H. Griffith and Sven Grebenstein, Wiesbaden, 2015, 243–256

Henning, Walter Bruno, 'The Disintegration of the Avestic Studies', *Transactions of the Philological Society*, 41/1 (1941), 40–56

Henning, Walter Bruno, 'An Astronomical Chapter of the Bundahishn', *Journal of the Royal Asiatic Society of Great Britain and Ireland*, 3 (1942), 229–248

Henning, Walter Bruno, 'The Book of the Giants', *Bulletin of the School of Oriental and African Studies*, 11/1 (1943), 52–74

Hertel, Johannes, *Das Pañcatantra: Seine Geschichte und Seine Verbreitung*, Leipzig, 1914

van Hoonacker, Albin, 'Le Traité du philosophe syrien Probus sur les premiers analytiques d'Aristote', *Journal Asiatique*, ser. 9, 16 (1900), 70–166

Hoyland, Robert G. (ed.), *Theophilus of Edessa's Chronicle and the Circulation of Historical Knowledge in Late Antiquity and Early Islam*, Liverpool, 2011

Hugonnard-Roche, Henri, 'Sur les versions syriaques des Catégories d'Aristote', *Journal Asiatique*, 75 (1987), 205–222

Hugonnard-Roche, Henri, 'Aux origines de l'exégèse orientale de la logique d'Aristote: Sergius de Reš'aina (†. 536), médecin et philosophe', *Journal Asiatique*, 277 (1989), 1–17

Hugonnard-Roche, Henri, 'Les traductions syriaques de l'*Isagoge* de Porphyre et la constitution du corpus syriaque de logique', *Revue d'histoire des textes*, 24 (1994), 293–312

Hugonnard-Roche, Henri, 'Les *Catégories* d'Aristote comme introduction à la philosophie dans un commentaire syriaque de Sergius de Resh'aynā (†536)', *Documenti e studi sulla tradizione filosofica medievale*, 8 (1997), 339–363

Hugonnard-Roche, Henri, 'Note sur Sergius de Resh'aynā, traducteur du grec en syriaque et commentateur d'Aristote', in *The Ancient Tradition in Christian and Islamic Hellenism*, ed. Gerhard Endress and Remke Kruk, Leiden, 1997, 121–143

Hugonnard-Roche, Henri, 'Du commentaire à la reconstruction: Paul le Perse interprète d'Aristote (sur une lecture du Peri Hermeneias, à propos des modes et des adverbes selon Paul, Ammonius et Boèce)', in *Interpreting the Bible and Aristotle in Late Antiquity: The Alexandrian Commentary Tradition between Rome and Baghdad*, ed. Josef Lössl and John W. Watt, Farnham, 2011, 207–224

Hugonnard-Roche, Henri, 'Les traductions du grec au syriaque et du syriaque à l'arabe (à propos de l'Organon d'Aristote)', in *Eastern Christianity: A Crossroads of Cultures*, ed. Florence Jullien, Leuven, 2012, 305–328

Hugonnard-Roche, Henri, 'Un Organon court en syriaque: Paul le Perse versus Boèce', in *Ad notitiam ignoti: L' 'Organon' dans la 'translatio studiorum' à l'époque d'Albert le Grand*, Turnhout, 2013, 193–215

Hugonnard-Roche, Henri, 'Sur la lecture tardo-antique du Peri Hermeneias d'Aristote: Paul le Perse et la tradition d'Ammonius', *Studia Graeco-Arabica*, 3 (2013), 37–104

Hugonnard-Roche, Henri and Abdelali Elamrani-Jamal, 'Aristote: l'*Organon*. Tradition syriaque et arabe', in *Dictionnaire des philosophes antiques*, ed. Richard Goulet, Paris, 1989–, I, 502–528

Huyse, *Royal Inscriptions* = Huyse, Philip, *Corpus inscriptionum iranicarum*, Pt. 3: *Pahlavi inscriptions*, Vol. 1: *Royal Inscriptions, with their Parthian and Greek Versions*, Texts 1: *Die dreisprachige Inschrift Šabuhrs I. an der Kaʿba-e Zardūšt* (ŠKZ), London, 1999

Inostranzev, Konstantin, 'Materialy iz arabskich istočnikov dlja kul'turnoj istotorii Sasanidskoj Persii: priměty i pověŕja [Materials from Arabic Sources for the Cultural History of Sasanian Persia]', in *Zapiski vostočnago otdělenija Imp. Russkago Archaeol. Obščestva*, St Petersburg, 1907, XVIII, 3–27 (texts), 28–120 (translation and commentary)

Kahl, Oliver, *The Sanskrit, Syriac and Persian Sources in the Comprehensive Book of Rhazes*, Leiden, 2015

Kellens, Jean, 'Avesta. i. Survey of the history and contents of the book', *EIr*, III, 35–44

Khalidi, Tarīf, *Islamic Historiography: The Histories of Mas'ūdī*, New York, 1975

Klima, Otakar, 'The Arsacids', in *History of Iranian Literature*, ed. Jan Rypka, Dordrecht, 1968, 25–30

Kunitzsch, Paul, 'Zum "liber hermetis de stellis beibeniis"', *Zeitschrift der Deutschen Morgenländischen Gesellschaft*, 118 (1968), 62–74; 120 (1970), 126–130

Kunitzsch, Paul, *Der Almagest: Die Syntaxis Mathematica des Claudius Ptolemäus in arabisch-lateinischer Überlieferung*, Wiesbaden, 1974

Kunitzsch, Paul, 'Über das Frühstadium der arabischen Aneignung antiken Gutes', *Saeculum*, 26 (1975), 268–282

Kunitzsch, Paul, 'Stelle beibenie—*al-kawākib al-biyābānīya*. Ein Nachtrag', *Zeitschrift der Deutschen Morgenländischen Gesellschaft*, 131 (1981), 263–267

Kunitzsch, Paul, 'The Chapter on the Fixed Stars in Zarādusht's *Kitāb al-Mawālīd*', *Zeitschrift für Geschichte der Arabisch-Islamischen Wissenschaften*, 8 (1993), 241–249

Land, J.P.N., *Anecdota Syriaca* 4, Leiden, 1895

Latham, Derek, 'Ebn al-Moqaffa', Abū Moḥammad 'Abd Allāh Rōzbeh', *EIr*, VIII, 39–43

Leclerc, Lucien, *Histoire de la médicine arabe: Exposé complet des traductions du grec, les sciences en Orient, leur transmission à l'Occident par les traductions latines*, 2 vols, Paris, 1876

Levey, Martin, *Substitute Drugs in Early Arabic Medicine*, Stuttgart, 1971

Madelung, Wilferd, 'Sunbadh', *EI2*, IX, 874–875

Mayor, Adrienne, *The Poison King: The Life and Legend of Mithradates, Rome's Deadliest Enemy*, Princeton NJ, 2010

McCants, William, *Founding Gods, Inventing Nations: Conquest and Culture Myths from Antiquity to Islam*, Princeton NJ, 2012

McKenzie, D. Neil, 'Bundahišn', *EIr*, IV, 547–551

Melvin-Koushki, Matthew, 'Dawlatshāh Samarqandī', *EI Three*, 2017, http://dx.doi.org/10.1163/1573-3912_ei3_COM_25930 (accessed 28.7.2012)

de Menasce, Jean Pierre, 'Un lapidaire pehlevi', *Anthropos*, 37–40 (1942–1945), 180–185

de Menasce, Jean Pierre, *Le Denkart: Une encyclopédie mazdéenne*, Paris, 1958

Mieli, Aldo, *La Science arabe et son rôle dans l'évolution scientifique mondiale*, Leiden, 1966

Minorsky, Vladimir, 'The Earlier Preface to the *Shāh-nāmeh*', in *Studi orientalistici in onore di Giorgio Levi della Vida*, Pubblicazioni del'Istituto per l'Oriente 52, 2 vols, Rome, 1956, II, 159–179; repr. in id., *Iranica: Twenty Articles* [*Bīst maqāla-ye Mīnorskī*], Tehran, 1964, 260–273

Möller, Detlef, *Studien zur mittelalterlichen arabischen Falknereiliteratur*, Berlin, 1965

Muḥammadī, Muḥammad, *al-Mutarǧimūn wa-l-naqalah ʿan al-fārisīyah fī l-qurūn al-islāmīyah al-ulā*, Beirut, 1964

Muḥammadī, Muḥammad, *Adab wa aḫlāq dar Īrān-e pīš az Islām wa chand namūna az āṯār-e ān dar adabiyyāt-e ʿArabī-Islāmī* [*Manners and Morals in pre-Islamic Iran with a few samples from Arab-Islamic Literature*], Tehran, 1352 š./1973

Nafīsī, Saʿīd, *Kitāb-e naṣīḥat-nāma maʿrūf bi Qābūs-nāma ta'līf Amīr ʿUnṣur al-Maʿālī Kaykāwūs ibn Iskandar ibn Qābūs ibn Wušmgīr ibn Ziyār*, Tehran, 1312 š./1933

Nallino, Carlo Alfonso, 'Tracce di opere greche giunte agli arabi per trafila pehlevica', in *A Volume of Oriental Studies Presented to Professor E.G. Browne*, ed. Thomas W. Arnold and Reynold A. Nicholson, Cambridge, 1922, 345–363; repr. in id., *Raccolta di scritti*, 6 vols, Rome, 1939–1948, VI, 285–303

Nasr, Seyyed Hossein, 'Life Sciences, Alchemy and Medicine', in *The Cambridge History of Iran*, IV: *The Period from the Arab Invasion to the Saljuqs*, ed. Richard N. Frye, Cambridge, 1975, 396–418

Nutton, Vivian, 'From Galen to Alexander: Aspects of Medicine and Medical Practice in Late Antiquity', in *Dumbarton Oaks Symposium on Byzantine Medicine*, ed. John Scarborough, *Dumbarton Oaks Papers*, 38 (1984), 1–14

Nutton, Vivian, *Ancient Medicine*, London, 2004

Olbrycht, Marek Jan, 'Parthians, Greek Culture and Beyond', in *Within the Circle of Ancient Ideas and Virtues: Studies in Honour of Professor Maria Dzielska*, ed. Kamilla Twardowska et al., Cracow, 2014, 129–142

O'Meara, Dominic J., *Platonopolis: Platonic Political Philosophy in Late Antiquity*, Oxford, 2003

Panaino, Antonio, 'Sopravvivenze del culto iranico della stella Sirio nel *Kitāb al-mawālīd* di Zarādušt ed altre questioni di uranografia sasanide', in *Alle soglie della classicità: il Mediterraneo tra tradizione e innovazione. Studi in onore di Sabatino Moscati*, I: *Storia e culture*, ed. Enrico Acquaro, Pisa/Rome, 1996, 343–354

Payne, Richard E., *A State of Mixture: Christians, Zoroastrians, and Iranian Political Culture in Late Antiquity*, Oakland CF, 2015

Perry, Ben Edwin, *The Origin of the Book of Sindbad*, Berlin, 1960; first published in *Fabula*, 3/1 (1960), 7–94

Peters, Francis E., 'The Greek and Syriac background', in *History of Islamic Philosophy*, ed. Seyyed Hossein Nasr and Oliver Leaman, London/New York, 2008

Pines, Shlomo, *Beiträge zur islamischen Atomenlehre*, Berlin, 1936

Pines, Shlomo, 'Aḥmad Miskawayh and Paul the Persian', *Irān-šināsī*, 2/2 (1971), 121–129

Pines, Shlomo, 'A Parallel between Two Iranian and Jewish Themes', *Irano-Judaica*, 2 (1990), 41–51

Pingree, David, *The Thousands of Abū-Maʿshar*, London, 1968

Pingree, David, 'Māshā'allāh; some Sasanian and Syriac sources', in *Essays on Islamic Philosophy and Science*, ed. George F. Hourani, Albany NY, 1975, 5–14

Pingree, David, 'Abu Sahl Nawbaḵt', *EIr*, I, 369

Pingree, David, 'Astrology and Astronomy in Iran. ii. Astronomy and Astrology in the Sasanian period', *EIr*, II, 858–871

Pingree, David, 'Classical and Byzantine Astrology in Sassanian Persia', *Dumbarton Oaks Papers*, 43 (1989), 227–239

Pingree, David, 'The Teaching of the *Almagest* in Late Antiquity', in *The Sciences in Greco-Roman Society*, ed. Timothy Barnes, Edmonton, 1994, 73–98

Pingree, David, *From Astral Omens to Astrology: From Babylon to Bīkāner*, Rome, 1997

Pingree, David, 'The Ṣābians of Ḥarrān and the Classical Tradition', *International Journal of the Classical Tradition*, 9 (2002), 8–35

Richard, Francis, 'Un traité persan d'hippiatrie portant la date de 555 H. (= 1160) de la Bibliothèque Nationale', *Studia Iranica*, 17 (1988), 95–101

Ritner, Robert K., 'The Satrap Stela', in *The Literature of Ancient Egypt: An Anthology of Stories, Instructions, Stelae, Autobiographies, and Poetry*, ed. William Kelly Simpson, New Haven CT/London, 2003, 392–397

Rosenthal, Franz, 'x. A list of astrological works from the 'Discussions' of Abū Maʿshar and Shādhān', in id., 'From Arabic Books and Manuscripts', *Journal of the American Oriental Society*, 83 (1963), 454–456

Rosenthal, Franz, *Das Fortleben der Antike im Islam*, Zürich, 1965

Rosenthal, Franz, *A History of Muslim Historiography*, Leiden, 1968

Rosenthal, Franz, *The Classical Heritage in Islam*, tr. Emile and Jenny Marmorstein, Berkeley CF, 1975

Rosenthal, Franz, 'Ibn Abī Ṭāhir', *EI2*, III, 682–693

Rosenthal, Franz, 'Ḥamza al-Iṣfahānī', *EI2*, III, 156

Ruska, Julius, *Das Steinbuch des Aristoteles*, Heidelberg, 1912

Ruska, Julius, *Tabula Smaragdina*, Heidelberg, 1926

Ryholt, Kim, 'Libraries in Ancient Egypt', in *Ancient Libraries*, ed. Jason König, Katerina Oikonomopoulou, and Greg Woolf, Cambridge, 2013, 23–37

Sachau, Eduard, 'Zur Geschichte und Chronologie von Khwârizm', *Sitzungsberichte der kaiserlichen Akademie der Wissenschaften, philosophisch-historische Classe*, 73/1 (1873), 471–506

Ṣadīqī, Ghulām Ḥusayn, 'Ḥakīm Nasawī', *Mağalla-e Dāniškada-e Adabiyyāt-e Tehran*, 6/1 (1337 š./1958), 12–28

Ṣafā, Dhabīh Allāh, *Ḥamāsasurā'ī dar Īrān*, Tehran, 1333 š./1954

Saliba, George, *Islamic Science and the Making of the European Renaissance*, Cambridge MA, 2007

Schoeler, Gregor, *Verzeichnis der orientalischen Handschriften in Deutschland*, XVII: *Arabische Handschriften. Teil II*, Stuttgart, 1990

Schwartz, Martin, 'Qumran, Turfan, Arabic Magic, and Noah's Name', *Res Orientales*, 15 (2002), 231–238

Sezgin, *GAS* = Sezgin, Fuat, *Geschichte des Arabischen Schrifttums*, 17 vols, Leiden, 1967–2015

Shaked, Shaul, 'Mihr the Judge', *Jerusalem Studies in Arabic and Islam*, 2 (1980), 1–31

Shaked, Shaul, 'From Iran to Islam. Notes on Some Themes in Transmission', *Jerusalem Studies in Arabic and Islam*, 4 (1984), 31–67

Shaked, Shaul, *Dualism in Transformation: Varieties of Religion in Sasanian Iran*, Jerusalem, 1994

Shaked, Shaul, *From Zoroastrian Iran to Islam: Studies in Religious History and Intercultural Contacts*, Aldershot, 1995

Shahbazi, Alireza Shapour, 'Historiography in Pre-Islamic Iran', *EIr*, XII, 325–330

Shaki, Mansour, 'The Dēnkard Account of the History of the Zoroastrian Scriptures', *Archív Orientální*, 49 (1981), 114–125

Shaki, Mansour, 'Dež ī Nebešt', *EIr*, VII, 348–350

Sims-Williams, Nicholas, 'Christianity iv. Christian Literature in Middle Iranian Languages', *EIr*, V, 534

Sponer, Gisbert, 'Die Pferdeheilkunde des Ipocras indicus', diss., Tierärztliche Hochschule Hannover, 1966

Soravia, Bruna, 'Les Manuels à l'usage des fonctionnaires de l'administration (Adab al-Kātib) dans l'islam classique', *Arabica*, 52/3 (2017), 417–436

Stein, Ernest, *Histoire du Bas-Empire II*, Paris, 1949

Stern, Samuel M., 'Abū Sulaymān Muḥammad b. Ṭāhir b. Bahrām al-Sijistānī al-Manṭiqī', *EI2*, I, 151–152

Stern, Samuel M., 'Abū ʿĪsā Ibn al-Munajjim's Chronography', in *Islamic Philosophy and the Classical Tradition: Essays Presented to Richard Walzer*, ed. Samuel M. Stern, Albert Hourani, and Vivian Brown, Oxford, 1972, 437–466

Strohmaier, Gotthard, '*ʿUlūm al-awāʾil* and Orthodoxy: a Famous Monograph of Goldziher Revisited', in *Goldziher Memorial Volume*, ed. Éva Apor and István Ormos, Budapest, 2000, 251–255

Sulaymān-Ḫamīs, Sulaymān, 'Ḥawl muʾallafāt al-Ǧāḥiẓ: al-Dalāʾil wa-l-iʿtibār, al-ʿIbar wa-l-iʿtibār, al-Tafakkur wa-l-iʿtibār: asmāʾ ṯalāṯah li-musammā wāḥid' ['About the Books of al-Ǧāḥiẓ: al-Dalāʾil wa-l-iʿtibār, al-ʿIbar wa-l-iʿtibār, al-Tafakkur wa-l-iʿtibār: Three Titles for One Book'], *al-Azhar*, 29/1 (1367/1947), 183–187

Tafażżolī, Aḥmad, 'Ādurbād Ēmēdān', *EIr*, I, 477

Tafażżolī, Aḥmad, 'Ādurbād ī Mahrspandān', *EIr*, I, 477

Tafażżolī, Aḥmad, 'Ādurfarnbag ī Farroxzādān', *EIr*, I, 477–478

Tafażżolī, Aḥmad, 'Āīn-nāma', *EIr*, I, 691–692

Tafażżolī, Aḥmad, 'Dārā(b) (1)', *EIr*, VII, 1–2

Tafażżolī, Aḥmad, *Tārīḫ-e Adabiyyāt-e Īrān Pīš az Islām* [*The Literary History of Iran in the Pre-Islamic Era*], Tehran, 1376/1997

Tāğbaḫš, Ḥasan, *Tārīḫ-e dāmpiziškī wa piziškī-e Īrān* [*A Medical and Veterinary History of Iran*], Tehran, 1372 š./1993

Takahashi, Hidemi, 'Syriac as a Vehicle for Transmission of Knowledge across Borders of Empires', *Horizons*, 5 (2014), 29–52

Talmon-Heller, Daniella, 'Abū l-Fidā'', *EI Three*, 2008, http://dx.doi.org/10.1163/1573-3912_ei3_SIM_0286 (accessed 28.7.2012)

Tannous, Jack Boulos Victor, 'Syria between Byzantium and Islam: Making Incommensurables Speak. Volume 1', PhD dissertation, Princeton University, 2010

Tardieu, Michel, 'Chosroès', in *Dictionnaire des philosophes antiques*, ed. Richard Goulet, Paris, 1989–, II, 309–318

Taylor, Gail, 'The Physicians of Jundishapur', *e-Sasanika*, 11 (2010), 1–18

Teixidor, Javier, 'Aristote en syriaque: les philosophes de la Haute Mésopotamie au VIe siècle', *Annuaire du Collège de France*, 97 (1996/1997), 723–743

Toorawa, Shawkat M., *Ibn Abī Ṭāhir Ṭayfūr and Arabic Writerly Culture: A Ninth-Century Bookman in Baghdad*, London/New York, 2005

Ullmann, Manfred, *Die Natur- und Geheimwissenschaften im Islam*, Leiden, 1972

Ullmann, Manfred, 'Steinbuch des Xenokrates von Ephesos', *Medizinhistorisches Journal*, 7 (1972), 49–64

Ullmann, Manfred, 'Neues zum Steinbuch des Xenokrates', *Medizinhistorisches Journal*, 8 (1973), 59–76

Ullmann, Manfred, 'Die Schrift des Badīġūras über die Ersatzdrogen', *Der Islam*, 50/2 (1973), 230–248

Ullmann, Manfred, 'Ḫālid ibn Yazīd und die Alchemie: Eine Legende', *Der Islam*, 55 (1978), 181–218

Ullmann, Manfred, *Islamic Medicine*, Edinburgh, 1978

Utas, B., 'Wāmiḳ wa 'Adhrā'', *EI2*, XI, 134

Walbridge, John, *The Wisdom of the Mystic East: Suhrawardī and Platonic Orientalism*, New York, 2001

Walker, Joel, *The Legend of Mar Qardagh: Narrative and Christian Heroism in Late Antique Iraq*, Berkeley CF, 2006

Watt, Montgomery W., *Islamic Political Thought: The Basic Concepts*, Chicago, 1968

Watts, Edward, 'Where to Live the Philosophical Life in the Sixth Century? Damascius, Simplicius, and the Return from Persia', *Greek, Roman, and Byzantine Studies*, 45 (2005), 285–315

Webb, Peter, '"Foreign Books" in Arabic Literature: Discourses on Books, Knowledge and Ethnicity in the Writings of al-Jāḥiẓ', *Journal of Arabic and Islamic Studies*, 12/2 (2012), 16–55

Weber, Ursula, 'Prosopographie des Sāsānidenreiches im 3. Jahrhundert n. Chr.', PhD thesis, University of Kiel, 2017; http://www.uni-kiel.de:8080/klassalt/prosopographie/index.html (accessed 3.5.2020)

Weidenhöfer, Veronika, Martin Heide, and Joris Peters, 'Zur Frage der Kontinuität des hippiatrischen Erbes der Antike: Die Behandlung von Erkrankungen des Bewegungsapparates im Kitāb al-furūsīya wa-l-bayṭara von Muḥammad ibn Ya'qūb ibn aḫī Ḥizām al-Ḫuttulī', *Sudhoffs Archiv*, 89 (2005), 58–95

Wiesehöfer, Josef, *Ancient Persia from 550 BC to 650 AD*, tr. Azizeh Azodi, London/New York, 2001

Yarshater, Ehsan, 'Iranian National History', in *The Cambridge History of Iran*, III.1: *The Seleucid, Parthian and Sasanian Periods*, ed. Ehsan Yarshater, Cambridge, 1983, 359–478

Zaehner, Robert Charles, *Zurvan: A Zoroastrian Dilemma*, Oxford, 1955

Zakeri, Mohsen, 'Das Pahlavi Buch *Kārwand* und seine Rolle bei der Entstehung der arabischen Rhetorik', in *Sprachen, Mythen, Mythizismen: Festschrift für Walter Beltz zum 65. Geburtstag am 25. April 2000*, ed. Armenuhi Drost-Abgarjan and Jürgen Tubach with Mohsen Zakeri, Halle (Saale), 2004, 839–858

Zakeri, Mohsen, 'Translation from Middle Persian (Pahlavi) into Arabic to the early Abbasid period', in *Übersetzung: Ein internationales Handbuch zur Übersetzungsforschung*, ed. Harald Kittel et al., Berlin, 2007, II, 1199–1206

Zakeri, Mohsen, 'Qubādiyān/Kawādiyān. Balkhī-Istakhrī Atlas and its Middle Persian Ancestry', in *Ancient and Middle Iranian Studies: Proceedings of the 6th European Conference of Iranian Studies*, ed. Maria Macuch, Vienna, 2011, 213–224

Zakeri, Mohsen, 'Muḥammad b. Khalaf b. al-Marzbān (d. 309/921) and his Role in Translations from the Middle Persian', in *The Place to Go: Contexts of Learning in Baghdad, 750–1000*, ed. Jens Scheiner and Damien Janos, Princeton NJ, 2014, 343–362

CHAPTER 4

Why the Syrians Translated Greek Philosophy and Science

Daniel King

Part One: Essay

From at least the early ʿAbbāsid era, when the East Syrian patriarch Timothy I (727/8–823) was negotiating power relations between his Christian community and the learned circles at court, the Syriophone Christians of the Near East retained a strong sense of their own heritage, and nowhere more so than in the field of the Greek sciences. Barhebraeus was still claiming four centuries later that all the learning of classical Greece possessed by the Arabs had passed first through the hands of Syriac-speaking Christians (Text 25).

However the nature and long-term significance of their contribution may be adjudged, there can be no doubt that the volume of translation work carried out by the Syrians of late antiquity was very substantial in its own right, certainly of far greater extent than its extant manuscript remains initially suggest. At a time when the ancient world was undergoing transformations so fundamental as to give birth to new cultures and literatures, the Syrian churches showed an eagerness to appropriate the Greek classical heritage that was rivalled by few amongst the Western European successors of the Latin rhetors. When the most famous of ʿAbbāsid-era translators, Ḥunayn ibn Isḥāq, claimed to have translated into Syriac no fewer than ninety-five works by Galen, he was merely presenting himself as the high point of a long tradition already more than three centuries old.

The Syriac obsession with translating into their own language revered works of the Greek masters, a preoccupation not always uncontested even among their own people, began naturally enough with the translation of scriptural and para-scriptural texts. Many of the pseudepigraphical texts that arose in Jewish and Christian circles in the first three centuries of the Christian era were eagerly consumed and then regurgitated in multifarious ways by the devoted populations of the Roman and Persian empires in their own vernaculars, of which the still unliterary Syriac dialect of Edessa was but one instance. We know next to nothing about the processes and contexts in which such translations were produced besides what may be gleaned from philological analyses of the texts

 | DOI:10.1163/9789004472648_005

themselves. Syriac-speaking Jewish communities had translated the Hebrew Scriptures already in the second century, and Tatian's *Harmony of the Gospels* (*Diatessaron*) belongs to a similar age. Attempts to manage the process by the ecclesiastical authorities came later. Fourth-century translations of the 'Separated Gospels' challenged the popularity of Tatian's *Harmony*, and in the early fifth century Bishop Rabbula of Edessa (bishop 411/2–435/6) had gained sufficient central authority over his diocese to try to stamp it out and replace it with a new standardized translation.

Translations of more 'literary' texts were not slow in coming. The earliest dated manuscript written in the Edessan dialect, swiftly becoming a literary standard, was produced in AD 411, and contains carefully crafted translations of Titus of Bostra's *Against the Manichees*, and works by Eusebius of Caesarea, written in a confident orthography and style that testifies to an already well-established technique. From this time onwards, the manuscript remains open a window onto what was a veritable flood of translations of the renowned fourth- and fifth-century doctors of the Greek Church. These translations were in the main sponsored by ecclesiastical authorities and were usually carried out in the monasteries, though we generally lack details of how this process worked. A native system of higher education, built on the foundation of the old Greek schools of rhetoric, from which it branched off in ways inspired by more local concerns, flourished among the Syriac churches from the fifth century onwards. Within the Persian realm, the world-famous school at Nisibis, founded in 489 following the expulsion of some chief leaders and educators from Edessa on the Roman side of the border, became a centre of learning whose fame extended thousands of miles east and west. Although its curriculum remained focused upon the Scriptures, the exegetical techniques and traditions there propagated were those of the Antiochene 'school' of Theodore of Mopsuestia (bishop 392–428), which co-opted those branches of secular learning that might contribute to the exposition of the Holy Books.

In the West Syrian realm (meaning generally the easternmost provinces of the Roman Empire, in which the Syrian Church followed a Miaphysite theology in opposition to the Chalcedonian theology of the imperial church of Constantinople), a monastery was founded between 528 and 531 on the Euphrates at Qenneshre, where Greek studies were especially treasured and promoted and a large bilingual library was preserved and increased. Other monastic centres developed other specialties; for example, the still-flourishing monastery of Mar Mattai near Mosul became renowned for the study and imitation of Greek rhetorical theory; Kashkar's school produced Theodore bar Koni, who brought together scriptural exegesis and Aristotelian philosophy for the Church of the East; while the relatively little-known monastery of Dayr Qunna was able to produce a philosopher of the standing of Abū Bishr Mattā. Many of these schools were also training centers for medicine as well as theology and church leadership. The close relations between medicine and philosophy in the Syriac milieu will emerge as a theme of the texts gathered here.[1]

It was at least partly in the context of schools of this sort that the Syrians began to patronize the translation of secular, as well as of sacred, texts. But they were selective. Medicine and Aristotelian logic hog the lion's share of the extant translations, with astronomy, rhetoric, and grammar not far behind. Mathematics, alchemy, and psychology, the latter so pertinent to the Christian concern for the destiny of the soul, were also popular, but the extent of their penetration is hard to assess accurately on the basis of the extant remains. Technical subjects are not absent either, as witness the Syriac recensions of the Byzantine manuals on agriculture, botany, and pharmaceutics; and in the practical sciences such as architecture, the buildings that crop up all over the sphere of Syriac culture are a monument to the continuity of technical knowledge in the region, quite apart from matters of language change.[2] Other fields, however, are largely absent—metaphysics and ethics were the province of Christian theology, although the ontological frameworks that Greek theologians had already assimilated to Christian doctrine were not unwillingly absorbed into the infrastructures of Syriac theological thought. The speculations in physics so variously pursued by the philosophers of Alexandria are not to be found in Syriac. Those who considered themselves to be philosophers among the Syrians had not yet succeeded in doing what some of their Arabic counterparts did in later ages, namely to pursue metaphysics and ethics outside the environment of strictly spiritual genres. Sergius of Rēšʿayna hints that he intended to author expositions of all Aristotelian subjects, but he seems never to have progressed beyond logic, and nobody else in the Syriac tradition seems to have fulfilled his ambition.

We need also to bear in mind that from the very beginning, and at least until the end of the ninth century, Greek was still a language in which the better-educated Syrians could comfortably work. It is certainly true for philosophy, perhaps rather less for medicine, that Syriac-speaking students were expected to read many of their texts in Greek. Indeed, Sergius's commentary on the *Categories* appears to have been written in the absence of any translation of the text being commented upon. The conceptual framework within which these trans-

1 Debié, 'Sciences et savants syriaques', underlines with plenty of evidence the point that medical schools are central loci for the production of Syriac scientific work.

2 For many of these fields, the best introductions may now be found together in Villey, *Les sciences en syriaque*. For psychology, see Hugonnard-Roche, 'La question de l'âme'; for grammar, King, 'Elements of the Syriac Grammatical Tradition'; for logic, Brock, 'The Syriac Commentary Tradition'.

lations ought to be read is that of a Graeco-Syriac tradition oriented toward and in emulation of a still ongoing Alexandrian tradition of scientific learning.

These last points raise a further significant question, namely whether our overall picture of Syriac science and philosophy has been unrepresentatively skewed by the nature of the remaining evidence. Since such a great quantity of the evidence has been lost to the ravages of time and war, we are left wondering about the rest. By the very nature of the matter, there can be no certain answer; nonetheless a few useful points can be made. First, the rhetoric of the discourse of the 'Greek sciences in Syriac' often outstripped the reality of what was studied and what was translated. Sergius, as we shall see, planned a full complement of Aristotelian translations, of which he accomplished only a fraction; yet this was a fault of late ancient philosophy in general rather than of Syriac in particular—Boethius too planned a comprehensive Latin curriculum that was never accomplished (condemning Western Europe of the next centuries to having only the *Logica vetus*), and the Alexandrian commentators spent so much energy on the *Organon* that they were rather thin on the other parts of philosophy. Nonetheless, there are frequent indications that translations were produced that have left no trace in the manuscripts: a Syriac version of Aristotle's *Rhetoric* was identified from quotations in the work of Barhebraeus;[3] arguments from Philoponus's *Physics Commentary* were incorporated into the seventh-century legend of the holy man Mar Qardagh;[4] Syriac astronomical works suggest that at Qenneshre there were translations at least of Ptolemy's *Handy Tables* and his *Geography*;[5] and by the ninth century there appears to have been a Syriac version of (at least parts of) Alexander of Aphrodisias, *Commentary on Physics*,[6] as well as of Archimedes, *On the Sphere and Cylinder* and *On Triangles*.[7]

These types of data can offer only a partial, and often rather confused, picture of what was translated over the course of late antiquity and in what order and for what purpose. As we emphasized above, Greek continued to be read and used as a medium of study into the ninth century. Hence evidence of the knowledge of a text does not imply evidence of a Syriac version, so long as many had no need of translations. Of mathematics, it would appear that much has

3 Watt, *Aristotelian Rhetoric in Syriac*.

4 Walker, *The Legend of Mar Qardagh*, 189–197.

5 Villey, 'Qennešre et l'astronomie aux VIe et VIIe siècles', 153–154.

6 Arzhanov and Arnzen, 'Die Glossen in Ms. Leyden Or. 583'. As was the case with the pseudo-Aristotelian *Problemata Physica*, this translation seems to have been made by Ḥunayn for a purely Syriac audience, and was later used alongside the Arabic tradition in the teaching of the Baghdad school in the tenth century.

7 Hugonnard-Roche, 'Mathématiques en syriaque', 75ff.

been lost, as we can tell from the sophistication of Qenneshre astronomy and from the Syriac works of Thābit ibn Qurra; but it seems very unlikely, for example, that there were ever any translations of Aristotle's *Physics* or *Metaphysics*, or of Greek commentaries on them. Moreover, it is well known that no works of or commentaries on Plato appear to have been preserved in the Syriac environment, and suggestions have been made why that may have been so.[8]

While we must thus keep constantly in mind the bilingual nature of Syrian society throughout late antiquity, that translations were only ever required for the non-Greek-literate community, and that much scientific and philosophical learning could be both taught and assimilated without any need for translations at all, the actual process of translation nonetheless retains a certain significance and interest for historians of ideas, not least insofar as it offers evidence for the depth and extent of the penetration of Greek learning. For heuristic convenience we may divide the history of Syriac translations of Greek secular works into six stages:

1. An early phase, in which translators were unsystematically translating texts that seemed to be required in various circumstances. Their techniques were still very underdeveloped. Some of the very earliest translations include those of Porphyry's *Isagoge*, possibly Aristotle's *Categories*, and the *Synagoge* of Anatolius of Beirut (the foundation of the *Geoponicon*). These early translations have no extant prologues or paratextual information that would help us better locate their aims and objectives.
2. A more systematic phase, in which especially Sergius of Rēšʿayna made what looks like an attempt to transfer into the Syriac realm the philosophical and medical curricula that were taught in his day at Alexandria.[9] He may also have been responsible for some astronomical, astrological, and alchemical translations. In addition to the foundational subjects of medicine and logic, from this time most likely also dates the Syriac interest in rhetoric and grammar, and possibly astronomy.
3. An institutionalized phase, in which at least some translations were patronized and commissioned by leading educational establishments. At Qenneshre, translators put the *Organon* into Syriac (*Isagoge* to *Sophistical Refutations*), and earlier translations were revised and improved. Astronomy and its basis in mathematics was also a key concern of learned Syri-

8 Bettiolo, 'Scuole e ambienti intelletuali nelle chiese di Siria'; Watt, 'From Sergius to Matta'; and Arzhanov, 'Plato in Syriac'.

9 There is an increasingly wide-ranging bibliography on Sergius. The most important recent work is summarized in Hugonnard-Roche, 'Sergius de Resh'aina', and Watt, 'Sergios (Sargīs) von Rešʿaynā'. Important also is Fiori, 'Un intellectual alexandrin'.

ans in this period. There is a certain homogeneity of technique, which is shared also with translations of Scripture and theological works, strongly suggesting a school setting for the training of translators.

4. There was a flourishing of Syriac translations in the early ʿAbbāsid era. The best-known practitioner is Ḥunayn ibn Isḥāq and his school, but there were plenty of others: Job and Theophilus of Edessa, to name but two. Many of these were doctors, and many are known only by their names and the titles of their lost works. The topics covered at this stage do not appear fundamentally to have changed much, but the numbers of texts covered was growing, as witnessed by, e.g., translations of works by Archimedes and Aristotle's *Rhetoric* and *Poetics*, which probably belong to this period.
5. Among the tenth-century school of logicians at Baghdad were many Syriac philosophers. When needed, these men translated into Arabic preexisting Syriac versions (often those made during phase 4), e.g., Abū Bishr Mattā's version of the *Posterior Analytics* and *Poetics*, and Yaḥyā ibn ʿAdī's of the *Topics*. The knowledge of Greek had largely been lost even among well-educated Syrians by this time, a change from phase 4.
6. Much later, in the twelfth and thirteenth centuries, a number of leading churchmen tried, ultimately unsuccessfully, to resurrect Syriac as a language of learning and science to rival the supremacy of Arabic. This is no longer a period of translation from the Greek, but did involve to some degree the resurrection of the older translations that were still available. Barhebraeus also translated from Arabic where he felt this would enrich the Syriac tradition, and was much under the influence of Ibn Sīnā's works.

This division into phases should not be taken to imply any strict boundaries, and even dating is notoriously difficult in the case of many texts. Broadly speaking, the following chronological sequence could be attached to the foregoing division into phases: 1) fifth century into the early sixth; 2) early sixth century until early seventh century; 3) c. 650–750; 4) c. 750–900; 5) tenth century; 6) twelfth–thirteenth centuries. However, the 'phases' describe styles and attitudes rather than chronological divisions. The *ad hoc* translations of 'phase one' could equally be produced during a later age.

Unfortunately, paratextual information about the contexts of translation is rare in all phases. Many translations are now to be found in manuscripts emanating from monastic contexts that comprise anthologies or collections, often rather eclectic, and usually for the most part ecclesiastical or theological in nature. If there were originally prefaces to these translations, they have been lost in the process of being extracted from their original context and inserted into these anthologies. Our best means of getting at the motivations of the translators and their social contexts is closely to analyze the texts themselves and their contents and to look for clues as to just why it was that their subject matters were of sufficient concern to the intended audience to motivate the work of translation. For the period post-800 we are in possession of only a minute fraction of the Syriac texts that were produced, by the school of Ḥunayn ibn Isḥāq or others; instead we are reliant on what fragments of echoes we can pick up from the Arabic biographers and bibliographers and other indirect sources, such as Patriarch Timothy's letters.

The majority of texts included in this collection come from the second and third phases described above. The first phase includes some important early translations, but no prologues or other paratextual data. For the fourth to the sixth phases, as just mentioned, the textual witnesses we possess are very few indeed.

The Second Phase

Sergius's prologue to his commentary (Text 1) is probably the most important single item from the second phase. It confirms that from the very beginning, medical and philosophical translations and commentaries were intimately related, something that was already a given in the Alexandrian system.[10] It also offers us the first hint that the translation of Greek science was a contested and potentially inflammatory issue, that there were those who disapproved of the texts concerned, and that those who created and copied these texts felt a need to defend their actions.

Text 1 also shows us that Sergius's plan to translate the Galenic corpus both chronologically and logically preceded Theodore's request for further explanations of logic. The deep concern among Syriac philosophers to elucidate for their readers the underlying theoretical foundations of the practical sciences is a recurring theme to be noted in many of the texts here collected. In Text 2 Sergius develops this theme: the lower reaches of philosophy that are located in the *Organon* are the prerequisite foundations for the superstructures of higher philosophy, that which is called 'divine', which concerns God. Like many other bilingual, bicultural, and well-educated persons of his day, Sergius drank heavily of the Neoplatonist cursus of contemporary Greek philosophy, and he had a vision for offering its fruits to those of his countrymen who had not yet partaken of them, and who also stood in need of a mother-tongue version of this curriculum. At this stage, however, translation is a hit-and-miss process, not governed by any very systematic plan. Sergius himself translated *On the Cos-*

---

10 Westerink, 'Medicine and Philosophy in Late Antiquity'.

mos and *On the Principles of the World* (both thought to be genuine Aristotelian texts), but for the *Organon* he planned only an explanatory commentary. He appears loosely to have wanted to expound all Aristotelian science, but if his ambition truly stretched so far, the time available to him sufficed for only a fraction, and we cannot say whether his ability was a match for the rest.

The same themes recur many times in Sergius's reflections. Text 3 briefly recapitulates in aphoristic form the belief that philosophy is the infrastructure that is the *sine qua non* of the practical sciences, especially medicine, and also for higher theoretical sciences, such as that involved in divine knowledge (theology). The contested nature of these aims is again alluded to in an ironic remark that one who 'receives divine ability through the exalted nature of his living [...] has no need of human instruction'—that is, those who see their asceticism as a passport to 'true' philosophy via divine revelation may freely dispense with reading Aristotle. This was no doubt what his opponents argued in what was only the latest incarnation of the already hackneyed contest between Greek wisdom and Christian revelation. Text 4 is the prologue to a shorter commentary that repeats many of the same themes and defensively prays for the reader's tolerance if he should find something at first sight to be distasteful, urging that he should persevere.

The only extant prologue to a translation of a philosophical text by Sergius (Text 5) belongs to the *De mundo*. A foundational text for late ancient cosmology, the *De mundo* naturally lent itself to a monotheistic or Christian Neoplatonist worldview, with which Sergius was clearly in sympathy. He does not make any substantial alterations to the text of the pseudo-Aristotelian treatise, unlike his translation of *On the Principles of the Universe* (Text 6). With the latter we have no prologue and are left to work solely from the actual translation itself. In this case he did make efforts to render the text more acceptable to a Christian audience: he eliminates references to the eternity of the world, and emphasizes that theology is the ultimate final goal of studying cosmology.[11]

Three more texts from Sergius's pen (Texts 7–9) are all prologues to medical, rather than philosophical, works. Sergius had made a translation of Galen's *On Critical Days*, a rather distinctive treatise in which the Greek doctor had laid out the astrological knowledge that all doctors ought to attain as part of their craft.[12] The translation piqued the interest of Sergius's colleague Theodore, for whom the former was subsequently required to author a further expository

---

11 McCollum, 'Sergius of Reshaina as Translator'; and King, 'Alexander of Aphrodisias' *On the Principles of the Universe*'.

12 King, 'Origenism in Sixth Century Syria', 199–204.

treatise outlining key themes in astronomy. The existence of this work testifies to the breadth of the curriculum that Sergius valued and wanted to import into the Syriac world. Text 8 reminds us that the theme of opposition, of which Sergius made much in the philosophical texts, was alive also in the medical field. Its allusion to the envy of those who are seeking to amass riches strongly suggests that translation and knowledge-transfer was not necessarily a *gratis* service offered by well-meaning altruists, although that is the light in which Sergius would like himself to be viewed. While we can hardly draw any details from this scant allusion, we do know from a hostile historical notice about Sergius that he was accused of corrupt financial practice,[13] and it has been recently argued that financial motives may have played as big a part in sixth-century translations, as they are known to have done in the ʿAbbāsid era of patronage.[14] See also Text 16 below, for a case where a patron most likely paid for a commissioned translation.

In Text 9, we see that Sergius was especially concerned in his medical translations to ensure that both the practical and the theoretical sides of the subject were transmitted, in accordance with the pedagogical technique of Galen himself. In adopting this approach, Sergius appears even to have been attempting to improve the Alexandrian canon,[15] a fact that demonstrates that he was not thinking of himself simply as transmitting 'old' authoritative knowledge from a revered past but rather perceived himself to be part of the ongoing development of late ancient *paideia*.[16] For example, although the Alexandrian Galenic medical canon contained only the practical part of *On the Therapeutic Method*, Sergius nonetheless translated the whole work and thereby ensured that the Syriac curriculum would be broader than the Alexandrian by including the theoretical as well as the practical aspects of the subject. This relates closely to Sergius's comment, found in Text 1, to the effect that it had been a question concerning the underlying principles of Galenic medicine that had led him in the first place to the task of introducing Aristotle to a Syriac audience. It was increasingly common in late ancient education to pass quickly over the theoretical basis of a subject in order the more quickly to move onto the practical aspects.[17] Sergius appears to have been swimming against this stream, not only in ensuring a full translation of Galen, but in going deeper, back to Aristotle.

13 Zacharias Rhetor, *Historia ecclesiastica* IX,19. Greatrex, *The Chronicle of Pseudo-Zachariah Rhetor*, 368–371.
14 Bhayro, 'Galen in Syriac'.
15 Bhayro and Brock, 'The Syriac Galen Palimpsest'.
16 Bhayro, 'Sergius of Reš ʿAyna's Syriac Translations of Galen'.
17 Nutton, 'From Galen to Alexander'.

Sergius's endeavours in philosophy were flanked by those of two near contemporaries, Proba and Paul the Persian (Texts 10–12). Their approaches and premises are remarkably similar. Proba's translational and commentarial work on Aristotle was as extensive as that of Sergius, and just as dependent on the traditions of the Alexandrian commentators. The fact that he uses the same metaphor (Text 10) as Sergius did to explain the utility of the study of logic indicates the common sources upon which they both drew. Paul's context, however, is different from that of Sergius or Proba. He worked at the Persian court and appears to have written in Pahlavi. His works constitute evidence for the state of learning at Chosroes's court, but also for the state of Syriac learning in the next phase (seventh century), when two of his treatises were translated into Syriac. Paul seems even freer than was Sergius to emphasize the complementarity of faith and reason, and even to elevate the latter at the former's expense (Text 11).

By the late sixth century, the Syriac world had become especially noted by the more learned of Greek and Latin Europe for its renowned educational institution, the School of Nisibis. We must remember however that, although actual translation of Greek science appears not to have predated c. 500, ever since the second century Edessa had been the locus of a burgeoning native Syriac learning, of which figures such as Bardaiṣan and rhetorically learned texts such as the *Letter of Mara bar Serapion* are but the extant vestiges. According to a note in Eusebius (*EH*, IV,30), the works of the Syriac philosopher Bardaiṣan were even translated into Greek. From the time when Ephrem taught Biblical exegesis and theology and battled the competing sects of Edessa in the 360s, the school became a focus for the incorporation into the Syriac tradition of Greek forms of theological thought, with Theodore of Mopsuestia's exegetical method at the forefront of the movement. Theodore's works were being translated already in the fifth century, along with the work of a range of other Greek theological authors—Athanasius, Gregory Nazianzenus, John Chrysostom. By at least by the middle of the fifth century, we may reasonably label these circles a 'school' in a meaningful sense, one that was drawn gradually into the fold of 'orthodox' Nicene theology as it simultaneously absorbed Hellenic sources for its learning and teaching. After the Third Ecumenical Council (431), however, the theology of Theodore was no longer considered part of orthodox tradition, and the learnèd of Edessa were finally expelled by order of the Emperor Zeno in 489. Some of its exiles established a fresh institution in Nisibis, in the relative safety of the Persian realm.

Following a false step by the pioneering Syriacist Assemani, older scholarship believed that the translations of Greek philosophy had begun already at the schools of Edessa/Nisibis in the fifth century. There is no longer any evidence that this was the case, and even evidence that Greek science or philosophy was read at the school in the following (sixth) century is thin and circumstantial.[18] Nonetheless, Theodoran theology is, in a certain way, built upon an Aristotelian (logical) framework, and Theodore's early readers and followers were more sympathetic to the Stagirite than to the Platonism of the Alexandrian theologians. If this fact was recognized in the schools of Edessa or Nisibis, then it is arguable that the reading of Theodore may have acted as an instigator for the reading (and then translation) of Aristotle as well.

Sometime around the late sixth century, Barḥadbešabba wrote the *Cause of the Foundations of the Schools*, which includes a long, rather tortuous justification of the School of Nisibis's *raison d'être* (Text 13). In the passage quoted, the author concedes (rather as Sergius did, see Text 3) that while special ascetics have no need of 'human' modes of knowledge acquisition, the rest of us require a solid foundation in logic if we are to distinguish the true from the false. This logic Barḥadbešabba specifically refers to as an instrument (*organon*) by which one may come to knowledge of corporeal things, thence to those created principles that reside between heaven and earth, which are the foundation of our knowledge of the divine. This imagined cursus is already familiar in its Syriac guise from the works of Sergius, especially his treatise on the spiritual life,[19] and indicates to us a general commonality of thought between them.

A number of texts written in East Syrian circles in the sixth/early seventh century appear to have adopted into their schemata the particular form of prolegomena that had been made popular in the Alexandrian schools. The impetus for influence of this kind has been attributed to learnèd churchmen such as Mar Aba, who spent time in Alexandria and acquired books and learning before becoming the head of the Church of the East in 540. The evidence, however, does not take us much further than this, and hardly constitutes direct evidence that Aristotelian studies were part of the curriculum of the School of Nisibis in the sixth, or even in the following, century.[20]

Thus while there is a certain amount of evidence that a framework of logical analysis, together with a body of specialist terminology, was adopted into the School of Nisibis from the Neoplatonist commentary tradition on Aristo-

18 Becker, *Fear of God*, 127–129, offers a list of potential connections between Nisibis and Alexandria, but none of it is very concrete.

19 Watt, 'From Sergius to Matta'.

20 See Becker, *Fear of God*, ch. 7, and Brock, 'From Antagonism to Assimilation'. If the garbled use of terminology drawn from the early Aristotle translations (composite, universal, particular: Becker, *Fear of God*, 134–139) is evidence for the teaching of Aristotle in the curriculum, then certainly he was taught rather badly.

tle,[21] nonetheless there is no clear evidence that the School was a locus for the translation of Greek science and philosophy in any way comparable to what occurred at Qenneshre (below). To the extent that Aristotelian philosophy was pursued at Nisibis, its primary motivating factor may well have been the medical curriculum that was also being taught there, in which (as we saw for Sergius) it was widely recognized that Galen had himself started from an Aristotelian mode of logical reasoning.[22]

The Third Phase

In 537, just a year after the death of Sergius of Rēšʿayna, John bar Aphtonia, a lawyer-turned-ascetic who had studied Greek rhetoric at the monastery of St Thomas at Seleuceia near Antioch, founded a new monastery by the Euphrates in his native Syriac territory. John had become abbot of St Thomas's before being expelled, together with his monks, in Justinian's persecutions against the Miaphysite doctrine between 528 and 531. As his panegyricist informs us, John sought out a new location further east where his establishment could remain safe from the heretical Byzantines, and this he found at the so-called Eagle's Nest (Syriac: Qenneshre).[23]

This monastery at Qenneshre became the West Syrian equivalent of the Church of the East's School of Nisibis, at least insofar as it developed into the most renowned centre of learning available, provided the Church with much of its leadership, and was thought of specifically as a focus for the transmission of Hellenic language and knowledge. In a satirical text about the monastery, a demon insults the monks who are 'Hellenists' for having introduced the relics of the anti-Hellenist Ephrem into their institution.[24]

Severus Sebokht (d. 666/7) was abbot-bishop of Qenneshre, and hence also head of the school, during a critical period in its history. He was both a teacher of others and a translator-scholar in his own right who translated into Syriac works of philosophy and astronomy from both Persian and Greek. Neugebauer had a very high opinion of his work in the field of mathematical astronomy.[25] Severus himself doggedly defended the competence of his own culture to enter into scientific fields.

---

21 As suggested in Becker, *Fear of God*, ch. 7.

22 For placing medicine back in the centre of the Nisibis curriculum, see Debié, 'Sciences et savants', 43–45.

23 For the broader significance of Qenneshre, see Watt, 'From Alexandria to Baghdad'. There is also an excellent contextualization of its importance by Tannous, 'Making of the Medieval Middle East', 160–180.

24 Michael Penn, 'Demons Gone Wild'.

25 E.g., Neugebauer, 'The Early History of the Astrolabe'.

In the brief preface to his *Syllogisms of Aristotle* (Text 14), Severus recapitulates themes already familiar—that logic is the instrument by which the true is distinguished from the false, and how one book of the *Organon* functions as the necessary foundation for the next, the *Posterior Analytics* being the summit of that particular epistemological ladder. In the second part of the text, the whole of logic becomes in turn the first part of a ladder of knowledge that leads to assimilation with the divine—again a topos that we have seen already in Sergius and Proba. There is thus a certain continuity between the second and third phases, insofar as the motivations for translating and teaching science and philosophy remained the same for the Qenneshre masters as it had been for their sixth-century predecessors.

In one of his letters (Text 15), Severus offers some polemical reflections on the process of the transmission of the sciences. He returns to a theme familiar both in ancient Greek and in later Arabic accounts, in which the Babylonians were the originators, the Greeks merely the borrowers, of the sciences. Drawing on an old tradition in Jewish and Christian *Apologiae*, we are reminded that Plato himself admitted that Greek learning was derivative and that its true origins lay further east. The astronomical and mathematical achievements of the Indians are also mentioned in this connection. However, Severus's twist on the theme comes in his claim that the Syrians of his day were themselves the heirs and successors of the ancient Babylonians. Hence in developing these scientific achievements, the learnèd of Qenneshre were simply fulfilling their ancestral rights rather than adopting the learning of a foreign culture. Anyone, he insists, who works hard at these subjects, is able to master them—one does not need to be enslaved to the Greek language to do so.

There is a great deal of rhetorical hyperbole in Severus's argument, however, and he was in fact committed to the introduction of the Greek sciences to his compatriots. In his treatises, the astronomical knowledge and calculations are entirely based on Greek sources, as is his work on the construction and uses of the astrolabe, and in his extant works he never makes use of the 'Indian signs' that he so praises in this letter.[26] In his introduction to the *Letter on the Ascending and Descending Nodes* (Text 16), Severus explains how Theon of Alexandria's commentary to Ptolemy's *Handy Tables* offers the very best technique of which he is aware for calculating the ascendant and descendant. Hence he offers a close translation of the text to his interlocutor. At this point we gain a tantalizingly small window onto the political motivation behind at least some of these

26 Hugonnard-Roche, 'Mathématiques en syriaque', 73–74. Cf., however, the argument about the eastward orientation of the astronomy opposed by Bardaiṣan, in Bhayro, 'On the Problem of Syriac "Influence"'.

translations. The request for this work on the ascendant and descendant, written in 662, appears to have originated with a certain Stephanus, referred to as the 'chartularius of all Gezirta', i.e., the Jezira, or Upper Mesopotamia. Nor was this the first work that Severus had written for him, for he refers to two other letters, both of which are extant in the same manuscript, one concerning the date of Easter, the other the conjunctions of the planets.[27] Since these texts appear to have been authored around the time of the beginning of the hegemony of the Umayyad caliphate in Damascus, this Stephanus, who holds senatorial rank in the Byzantine bureaucracy, may have been an official working under the early Umayyads. Since astronomical calculations were used in the planning of military campaigns and for a whole variety of civic activities, we may not have to look far for Stephanus's motivations in asking for these materials from the most eminent astronomer-mathematician in the region. We cannot be sure whether Stephanus read Syriac or whether the texts were translated for him.

Severus Sebokht's students and successors at Qenneshre included those who would continue his work, especially in the philosophical field. There were Jacob of Edessa (d. 708), whose revision of the old translation of the *Categories* was of a piece with his revisions of other (non-scientific) translations, all designed to improve the learning of his people and the closeness of their interaction with Greek traditions; Athanasius of Balad (d. 686), later the chief cleric of his denomination, who translated an introduction to philosophy from a source also used by John of Damascus, in addition to personally translating the *Isagoge, Prior and Posterior Analytics, Topics*, and *Sophistical Refutations*; and George, bishop of the Arab tribes (d. 724), who went systematically through the first three books of the *Organon*, translating and writing commentaries, and probably intended to go further.

George's introductions provide us with some clues as to what he saw was the point of reading and interacting with these texts and why he should bother himself to translate them, improving on existing prior Syriac translations.

Text 17 is an extract from George's general introduction to Aristotelian logic, found at the very beginning of the large manuscript of his works that remains unedited in the British Library. A comparison with the introduction of Philoponus's *Commentary on the Categories* will immediately indicate the close dependence of George upon him.[28] That is, just as Philoponus in his work on Aristotle adopted the Neoplatonist approach that was taught at Alexandria,

27 These, and all the astronomical texts attributed to Severus, may now be found with full discussion in Claude-Villey, *Les Textes astronomiques syriaques*.

28 E.g., compare Philoponus, *In Cat.* 5,34–6,3: 'Now, what is the goal of Aristotelian philosophy? We say that it is the knowledge of the principle of everything, i.e., the creative cause

so George nearly two centuries later saw the virtue in adopting just the same approach for the benefit of non-Graecophone Syrians. The Neoplatonist references to the One, as found here, could be read in a Christianizing way, and it is striking that even in the eighth century, under Umayyad rule, a Christian bishop in Mesopotamia should seek to follow the same path as his Alexandrian predecessors.

The next extract (Text 18), taken from the introduction to the *Prior Analytics*, shows George in more obviously 'theistic' mood, openly identifying the One with the Christian God. The main burden of the passage is to defend 'pagan' philosophy, just as Sergius had done, against the many opponents it attracted. In his arguments one may hear echoes of his mentor Severus Sebokht's discussion about the Babylonians. In any case, George clearly sees a real and immediate value in Aristotelian logic and disdains those who would throw out all pagan wisdom simply because it is not Scripture. He does not immediately tell us, however, just what areas of knowledge or 'utility' he is speaking of. In Text 17 (point 5), we see that he envisioned a cursus leading from logic, through the corporeal and then incorporeal sciences, towards knowledge of the One; and then, in the body of his commentary (Text 19), we receive further confirmation that the utility George places on this type of knowledge is connected to its structural position within an overarching epistemology—for the trinitarian nature of God has been imprinted in nature in manifold ways, one of which is the mysterious threefold nature of the syllogistic system, in which all logic may be reduced to a threefold pattern. For George, philosophy was a gift from God for humankind, not merely as an instrument for distinguishing the true from the false, but as one of the means by which we may come to a full knowledge of our creator.[29]

The Fourth and Fifth Phases

It is only from the 'Abbāsid era that we begin to be able to take a clearer view of the state of philosophical and scientific learning among the East Syrians. The *Cause* (Text 13) offered a glimpse of what was being taught at Nisibis in the pre-Islamic era. Besides this we hear only hints and echoes until around the turn of the ninth century, when the extensive *Scholion* of Theodore bar Koni offers a summary of how East Syrian exegetes had assimilated what they had learned of Aristotelian logic into the task of interpreting Scripture, and

of all things, the one which is always just the same as it is. For he demonstrates that the principle of all things is one and incorporeal, and all things come forth out of it.'

29 For further elucidations of George's doctrines in this area, see Miller, 'George, Bishop of the Arab Tribes'.

how they were starting to move into the field of dialogue with Islam, a subject that was fated to become very significant over the next two centuries.[30] Nonetheless, it is again the continuity of the philosophical tradition that stands out. The definitions of philosophy given by Theodore's continuator Sylvanus of Qardu (Text 20) gives one to think that we still remain firmly implanted in the world of Sergius, of Proba, and of Severus and George. For still the old prolegomena material of the Alexandrian commentaries is being rolled out for the benefit of new generations of students. Philosophy is given pride of place as queen over all other sciences and as the study of being *qua* being. It is considered the means of attaining likeness to God—the same aim that others sought to achieve by ascetical self-denial. Note the ease with which the old 'pagan' Neoplatonist notion of unity with the divine is assimilated into Syrian Christianity, either by considering the pagan philosophers as proto-Christians, or by allowing, as George did, that pagan philosophers could sometimes see the truth despite their handicaps. Both opinions have a strong pedigree in the Christian discourse of late antiquity.

At the same time, at the new focal-point of politics and culture in Baghdad, the long-lived patriarch of the Church of the East, Timothy I (r. 780–823), found the challenge of Islam already a matter of some urgency for the survival of his Christian community.[31] Timothy left behind a series of letters that shed light on his world, and especially on his attempts to negotiate the new power relations between caliphate and Church that obtained after the foundation of Baghdad and the transference there of the patriarchal seat. While many of his extant letters concern matters of Church organization (recall that at this time the Church of the East's reach extended to China and India), others report on discussions he had had with learnèd Muslims at court about philosophical and theological matters, and famously even with the caliph himself.[32] Others offer a window onto his search for Greek and Syriac philosophical sources in the old Syriac monasteries of northern Mesopotamia. As Text 21 demonstrates, Timothy was in part motivated by a feeling new among Syrians, namely a sense of envy of what the Arabs had already achieved in their relatively short history, a sense that there was a need to compete in these fields of endeavour if the Syrian her-

30 For a summary discussion of the contents of Theodore's work, which includes important testimonia to the early engagement with Islam as well as summary philosophical materials, see Griffith, 'Theodore Bar Koni's Scholion'.

31 For Timothy's life and work, see Berti, *Vita e studi di Timoteo I*.

32 E.g., Mingana, 'The Apology of Timothy the Patriarch'. Cf. Suermann, 'Der nestorianische Patriarch Timotheos I'; Griffith, 'The Syriac Letters of Patriarch Timothy I'; and id., 'Patriarch Timothy I and an Aristotelian at the Caliph's Court'.

itage was to survive at all. In this letter, the patriarch points to the richness of the grammatical tradition of language study in Greek and Arabic sources as compared against what was available in Syriac (we do not know if he was aware of the extensive work in this field achieved by the West Syrian bishop Jacob of Edessa). His evident concern further to elucidate the early books of the *Organon* betrays an awareness of the opinion held by many others in the Graeco-Syriac tradition, that the *De interpretatione* was fundamentally a text about language structures.[33]

In his *Ep.*43 (Text 22), Timothy again reverts to the theme of competitiveness in the field of translation, when he boasts about the quality of his own work on the *Topics* compared with the work of others. To a Syrian trained in the Alexandrian prolegomena that appear to have been the staple of all the Syrian schools, the idea of reading the *Topics* before, or independently, of the first four books of the *Organon* would have appeared confused.[34] Hence Timothy feels justified in a degree of cultural snobbishness when he considers the relative achievements of Arabs and Syrians in the field of logic. Here he offers evidence of his interest in gathering the fruits of previous generations of Graeco-Syriac scholarship regarding the various parts of the *Organon*. We do not know whether he was successful in his search, but we can imagine that his aims were closely related to the challenges he faced at the caliphal court.

Ḥunayn ibn Isḥāq's *Risāla* (Text 23), in which the renowned Arabic translator enumerates the books of the Galenic corpus and narrates the translation background of each one, offers useful insights into the social and political context of Syriac translations as well as Arabic ones. A recent study has emphasized the fact that most of the numerous Greek–Syriac translations of Galen that Ḥunayn mentions were translated under the patronage of well-known Syriophone physicians, men such as Yūḥannā b. Māsawayh, Salmawayh, Bakhtīshūʿ b. Jibrīl, and Jibrīl b. Bakhtīshūʿ.[35] The conclusion to be drawn is that the Syriac portion of the work of Ḥunayn and his school was not carried out merely as a stepping stone towards translating into Arabic, but that still in the ninth century there was a substantial constituency of Syriophone physicians and others willing to pay for Greek–Syriac translations for their own use or for the promotion of their profession in the ʿAbbāsid realm. Even the *Risāla* itself (as

33 In BL Add. 14658, a Syriac translation of Dionysius Thrax's grammar has been inserted by a scribe in the place where we would expect a treatise on the *De interpretatione*. Even the manuscript's first cataloguer was fooled.

34 Watt, 'Greek Thought, Syriac Controversies', 169.

35 Watt, 'The Syriac Translations of Ḥunayn ibn Isḥaq and their Precursors', and 'Why Did Hunayn, the Master Translator into Arabic, Make Translations into Syriac?'

suggested by the quoted extract, §3) is fundamentally a translation from an earlier Syriac catalogue that Ḥunayn had prepared for a Syriac patron he mentions. All this brings us back once again to the very practical motivations we noted at the time of Sergius, where the practice of medicine in the community led at least some doctors to seek deeper learning from the roots of their discipline, Galenic medicine and Aristotelian logic. These were the two major areas of concern of Ḥunayn's school.

In his famous *Fihrist*, or index of Arabic literature, Ibn al-Nadīm offers what appears to be plentiful information about the Syriac texts that were still extant in the ninth and tenth centuries, especially among members of the Baghdad school of logic. There was, however, a great deal of which he was unaware. In the field of mathematics, for example, he mentions Syriac texts of which we have no other source of knowledge, but at the same time makes no mention of important texts that we do know existed, translations of Nicomachus of Gerasa, Menelaus, and Archimedes.[36] Thus while we may be grateful for what we can glean from the *Fihrist*, Ibn al-Nadīm's lacunae and imprecise descriptions of texts that he had seen often leave us none the wiser.

The quoted extracts (Text 24) are merely a taster of the types of data that may be drawn from this source, though often we are left guessing about the specific motivations of the translators and about the social and political contexts in which they carried out their work. From these extracts it is at least clear that in Ḥunayn's day it was still very common for significant texts to be rendered only into Syriac and not into Arabic, as was the case with Aristotle's *Posterior Analytics*, an Arabic version of which was not made until the East Syrian Abū Bišr Mattā took the matter in hand in the tenth century. Why Ḥunayn translated the *Posterior Analytics* into Syriac we can only guess, but we may assume, following the pattern of the Galen texts, that they had a patron who needed this text for some reason, but that there was no Arabic patron with a comparable requirement, at least not until the following century. The earlier Syriac tradition had considered this book to be the very summit of Aristotelian logic, and Athanasius of Balad had translated it, not for the whim of an *ad hoc* patron, but as part of a systematic educational curriculum for a readership that was beginning at that time (seventh century) to lose its connection with the Greek language. The second extract confirms that the Syriac language was indeed Ḥunayn's principal focus, and again we may conclude that this was so on account of the particular commissions that he received.

---

36 Hugonnard-Roche, 'Mathématiques en syriaque'; and Takahashi, 'The Mathematical Sciences in Syriac'.

We close this collection of texts with Barhebraeus's whimsical look back at the literary history of his people. We may note in particular the sense that the bishop has of fighting a losing battle against the march of Arabic, a sense of which we saw the first stirrings in Timothy's letters. This competitive streak is still a major factor in Barhebraeus's considerations. He himself relied heavily on Arabic philosophy in his own work, especially on Avicenna, whom he even translated directly. Yet at the same time, he claims a nationalistic priority for the Syrians in the translation and transmission of science. Barhebraeus and his fellows, especially his predecessor Jacob bar Shakko, made a brave attempt to reformulate known philosophy and science in a renovated Syriac diction that did not long outlive them.

Part Two: Texts in Translation

Text 1. Sergius of Rēšʿayna, *On the Aim of All the Works of Aristotle*, Prologue

Renowned translator, doctor, and priest, Sergius of Rēšʿayna was active during the first third of the sixth century (d. 536). He studied philosophy and medicine at Alexandria at a time when Ammonius was head of the philosophical school there and Philoponus a student. His extant writings are all in Syriac, although Greek may in fact have been the language in which he was intellectually more comfortable. In the prologue to his commentary on the *Organon*, entitled *On the Aim of All the Works of Aristotle*, Sergius mentions the translation process that he was accustomed to use when translating Galen, depending as he did on an oral translation, the style and diction of which would then be improved by his assistant Theodore.

Sergius also here explains why the logical works had to be translated, so that students of medicine could return *ad fontes* and understand the foundations of Galenic theory. In Sergius's bilingual Graeco-Syriac context, however, translation has a limited utility. It is clear from this text that Sergius expects his readers to engage directly with the Greek text of Aristotle after reading his (Syriac) introduction.

He also mentions a subject that will become a topos of Syriac philosophy, namely the opposition encountered in the attempt to transfer Aristotelian philosophy into a Syriac context, most likely from defenders of a monastic way of life that depended solely upon Christian writings and rejected all else as of no value. Sergius's own motivation, however, in introducing Syrians to Greek philosophy was in no way to undermine the foundations of their faith, but rather to uphold and strengthen that faith on a truly rational basis.

Date Early sixth century.
Text There is no published version of this work; the principal manuscript witnesses are Paris, BNF, syr. 354, and Birmingham, Mingana 606. French translation in Hugonnard-Roche, *Logique d'Aristote*, 167–170.[37]

1

According to a saying of the Ancients, O our brother Theodore, the bird called the stork finds happiness and gains strength whenever it leaves the inhabited world and finds a deserted place,[38] and there it establishes its retreat for the

37 The following English translation is taken from the Syriac text but makes use of Hugonnard-Roche's interpretations.

38 There is a pun here: in Syriac, the words for 'stork' and 'deserted place' are the same.

rest of its life. In the same way, it seems to me, nobody is able to understand the teachings of the Ancients, nor to penetrate the secrets of science in their books if he does not leave the world and its business, and if he does not distance himself from the flesh—not in terms of location, but in his mind—and leave behind him all of his desires. It is in just this way that the spirit is free to turn in on itself, to contemplate itself, to see clearly what they [the Ancients] wrote, and accurately to judge what has been correctly said and what has not been so stated, for within such a way of life there is nothing to get in the way [of making such judgements], [viz.,] none of those things that oppose its lightness with the weight of the flesh.

II

Therefore, when we were translating certain of the writings of Galen the doctor from the Greek language into that of the Syrians, I myself was translating, and you were writing after me and correcting the Syriac words as the style of that language requires. Also when you saw the wonderful logical divisions [i.e., the logical system of dividing up the material] that were to be found in the writings of [this] man, and the definitions and demonstrations that are frequently and in an excellent manner to be found therein, you asked me where this man had been able to find the principle and the starting-point of knowledge, and whether he had derived such a profit from himself or from some other author before him. For myself, I answered to this, for the sake of the love of learning that is in you, that the origin and the beginning and the principle of all knowledge was Aristotle, not only for Galen and the other doctors like him, but also for all authors who are called philosophers who lived after him. In truth, up until that time when this man lived in the world, all the parts of philosophy and of knowledge in general had been scattered about, rather in the manner of simple drugs, and distributed among the various authors without any order or [system of] knowledge. It was [Aristotle] alone who, like a learned doctor, gathered together all the parts that had been scattered about, and assembled them by his skill and knowledge, and prepared from these the perfect remedy of his teaching, which roots out and removes the grave maladies and diseases of ignorance from those who apply themselves seriously to his writings. Similarly, just as those who carve statues first mould each part of the figure separately, and then assemble them all together as their art requires and hence create the final statue, thus also did Aristotle assemble, adjust, and distribute each of the parts of philosophy into the required order and from them he moulded, in all his books, the perfect and beautiful form of the knowledge of all things.

III

Having heard me say all this, O our brother Theodore, you wanted also to know what was the aim of this man's teachings, the order of his writings, and the sequence of his ideas. As I had tried to explain to you a little of what I was able to remember on this subject, you again asked me to put in writing for you that which I had already explained orally. Whilst apologizing for myself before the importance of the task, I said that I had composed a brief treatise on the aim of Aristotle's philosophy that was enough to instruct readers, so far as possible, about his teachings. You, however, were not quite satisfied, but you amicably pressed me further, not merely for a general study on the whole teaching of this learned man, which is what we had already done before, but rather to explain in brief what we thought about each one of his writings in turn.

IV

Therefore, because I have not been able to escape your request, before going into the heart of the matter, I would ask you, and all those who read this work, not to fall at once into reproofs or sterile criticisms after a single reading of what is here said; but rather to persist in reading and studying it once and twice, or even three or four times if the subject requires it. If something still remains obscure, the reader should not hesitate to refer to someone who can instruct him and explain to him what he does not understand. Thus will he truly avoid the difficulty that is produced in the minds of those who do not understand what they read. And he protects himself also from making pointless slanders and criticisms against the author of the treatise. There are actually many people who, led by the force of their jealousy as if by powerful gusts, the moment they start to read these writings, reprove rather than reflect, for they believe in increasing their own glory by taking others down. But they do not realize that everything, whatsoever it is, is proven by its own power, not by the weakness of something else. Truly the weakness of one power compared to another does not affect the stability of that other any more than the smallness of a natural object increases the greatness of that to which it is compared. If [these people] consider these matters appropriately, then they must examine and accept what is said by the author that is of good consequence and justice, whilst in those matters that they feel are in need of further precision or correction, they ought without envy to correct them from other things, or else just pass over them without reproof, not detracting human nature and not pouring slanders upon them, because [human nature] can hardly succeed in everything. So far as I am concerned, in these matters I would say that whosoever reads them will understand them, even if they do not at the same time share the man's opinion or agree with his teachings.

V

For now, I want to write about the aim of these writings, not only because I have been pressed by your friendship, as I said before, but also because I think that those who read them can draw therefrom an excellent training and preparation for science. Let us therefore get back to our explanation about the aim, on which subject we propose to write.

Text 2. Sergius of Rēšʿayna, *On the Aim of All the Works of Aristotle*, Book 2

The first book of Sergius's treatise had been concerned with the divisions of all knowledge and how the different parts of science may be joined together. At the start of the second book, he reiterates the close association of the various parts of Aristotelian science, especially underlining the foundational significance of logic as the necessary preparation for the elucidation of scientific demonstrations. In the following text, he goes on to adopt Ammonius's metaphor of the building of a house to describe this cursus (cf. Philoponus, *In Cat.*, 10,21–11,33 Busse).

Date Early sixth century.
Text There is no published version of this work; the principal manuscript witnesses are London, BL, Add. 14658; Paris, BNF, syr. 354; and Birmingham, Mingana 606. Translation taken from Watt, 'Sergius of Reshaina on the Prolegomena to Aristotle's Logic: The Commentary on the Categories, Chapter Two', 74–75. Note that the translation assumes certain emendations to the Syriac text that are explained in Watt's commentary (80–91).

In any art whatsoever the end of *theoria*[39] is the beginning of action, and the end of action coincides with the beginning of *theoria*. What I mean is [something] like the following. If a builder is commissioned to build a house, he considers [the matter] in his mind, saying, 'I have been commissioned to make a canopy giving protection against wind, rain, and other destructive [forces] like these. However, I cannot make the canopy unless I first raise load-bearing walls for it, and I cannot build these unless I first dig out and lay the foundations.' Accordingly he first lays down the foundations, then builds the walls, and finally puts over them the roof, which is the end of building. Therefore in these the beginning of *theoria*, namely the consideration in his mind, begins from the canopy and ends with the foundations, but the action, namely the

39 I.e., the theoretical, as opposed to the practical, part of philosophy—this being the most basic of all the divisions of philosophy.

labour of [his] hands, begins from the foundations and ends with the roof. Therefore, as we said just above, the beginning of *theoria* is the end of action, and the beginning of action the end of *theoria*. In the same way Aristotle dealt with the art of logic. First he considered [the matter] in his mind, [saying], 'I wish to create an instrument of discrimination which defines for me in *praxis* good from evil and separates for me in the knowledge of things truth from falsehood. But because this instrument[40] is an art which generates all demonstrations[41] constituted by discourse, it is evident that I must first create this art of demonstrations. And because this art of demonstrations comes from the linkage of discourse which is composed with art, I must first teach about this. And because there is a prior combination of terms, from which this linkage of discourse comes, I must first write about this. Furthermore, because simple namings precede this, I must first teach about these.' So in his thinking he began from the art of demonstrations and proceeded in sequence to the simple namings; but he made the simple namings the beginning of the [act of] teaching about all these, after them taught about the first combination of terms, similarly after this wrote about the correctly and properly composed linkage of discourse, and then taught about the art of demonstrations itself. Subsequently [he taught] about everything useful to it in all [kinds of] ways. So the end of his action in [writing] these things coincides with the beginning of *theoria* on them, as the end of *theoria* on them coincides with the beginning of writing about them.

The book written by him about simple namings is called *Categories*, that which he wrote about their first combination *On Interpretation*, that about the linkage of discourse is named *Prior Analytics*, and that about the art of demonstrations itself is named *Apodeictics*. Together with this there is that called *Topics*, and that about the refutation of sophists which he named *Sophistical Refutations*. With these, therefore, this philosopher completed the whole art of logic, which is, as we have said, an instrument of philosophy and not a part of it. Some people say that the *Art of Rhetoric* which was composed by him is also part of the same [art] of logic. However, let us turn now to the subject itself and start to speak as [well as] we can about the aim of each one of these treatises,

40 The term here translated as 'instrument' represents the Greek *organon*, which was the name for the logical books of Aristotle treated as a corpus. In the late antique philosophical tradition, the logical texts of Aristotle were considered to be not a part of philosophy itself, but rather an instrument for the carrying out of philosophical thinking, hence the term *Organon*, which was applied to the logical works as a group, even though these were not originally designed to be such by Aristotle.

41 The term 'demonstrations' represents the Greek *apodeiktika*, which is also the name given to the subject matter of the book of the *Posterior Analytics*.

beginning the sequence with that on *Categories*, which is about simple namings, and similarly treating each of them one by one in the same way. Then we will go on to his other treatises, those on the parts of *praxis*, and on all natures, teachings, and the other ones which are called 'divine'.

Text 3. Sergius of Rēšʿayna, *On the Aim of All the Works of Aristotle*, Book 7

Towards the end of the treatise, Sergius summarizes the opinion that he has dwelt upon many times in other places, viz. that the study of Aristotelian logic is foundational for all the sciences, and this includes the 'divine sciences' and the understanding of the Scriptures. His understanding of this interconnectedness of knowledge was his principal motivation for translating.

Date Early sixth century.
Text London, BL, Add. 14658, f.60vb.

Without the discipline of logic it is not possible to grasp the power of medical treatises or understand the thought of philosophers; nor, furthermore, can the true meaning be revealed of the divine Scriptures in which is the hope of our salvation—unless someone receives divine ability through the exalted nature of his living so that he has no need of human instruction.

Text 4. Sergius of Rēšʿayna, *Introduction to Aristotle and His Categories, to Philotheos*, Prologue

This text is most likely an abridgment of the longer treatise from which Texts 1–3 were drawn. There is, however, some different material included in it. In the prologue, Sergius again alludes to the opposition that he encounters in his programme of bringing Greek philosophy to the Syrians. They recognize the truth of these matters, he says, but nonetheless oppose them out of envy.

Date Early sixth century.
Text Aydin, *Introduction to Aristotle and his Categories*, 94,3–16; tr. Aydin, ibid., 95 (slightly emended).

It seems to us, O our brother Philotheos, that [some] good comes from these things that are useful and particularly helpful for the knowledge and learning of those who are concerned with the study of the philosophical sciences that constitute the art of logic, [i.e.,] that which relates to fundamental concepts in philosophy, together with the rules, divisions, and arrangements of this art [of logic]. We will [therefore] gather for you as briefly as possible [material] concerning this subject.

Now, whenever we fulfil requests from you, or from others like yourself, we are worried lest you should supply[42] what we are lacking and correct something we say that is out of place, not because you rush to complain about things in the treatise, but rather because you care about what is helpful in it. We have written this for you at the beginning of our treatise because we are convinced that Satan, the enemy of the common good, whenever something is done that is to the general benefit, always causes some people to get stirred up with envy and oppose the arguments, even though they recognize the truth of them. Then, instead of edifying one another in love they destroy one another with envy.

Text 5. Sergius of Rēšʿayna, Syriac version of Pseudo-Aristotle, *De mundo*, Prologue

Sergius may have felt that the *De mundo* was a text that would readily lend itself to monotheistic interpretation and that in any case it offered an excellent introduction to cosmology, a role it performed as part of the Alexandrian curriculum that he had himself studied. There is, however, no trace of 'Christianization' such as we find in the translation of the work of Alexander of Aphrodisias (see Text 6).[43] Again, Sergius alludes to the challenges posed by those who oppose his programme, here hinting that his texts were sometimes tampered with, perhaps with the aim of making them even more objectionable to orthodox Christian theology than they really were.

Date Early sixth century.
Text Lagarde, *Analecta Syriaca*, 134,13–27 = London, BL, Add. 14658, ff.107v–108r. English translation taken from: McCollum, 'Sergius of Reshaina as Translator: The Case of De Mundo', 167–168, which is based on some better readings than Lagarde's text.

The letter composed by Aristotle the philosopher for Alexander the King on the knowledge of the things that exist, which you, the elect, have sent me and commanded that I translate according to my ability from Greek speech to the language of the Syrians, I have received whence you sent it. I have, though, been hindered from the work until now for various reasons, which it is not now time to mention. But now, since it is necessary, I have decided to fulfil your com-

42 A minor emendation. The edition reads 'you should speak'.
43 For further discussion, see McCollum, 'Sergius of Reshaina as Translator'.

mands. Even though many other important things have been preventing me, I have let them all go for the sake of your command for me and have taken pains to accomplish your will. But I urge you, dear sir, that if another copy of this letter is found, in which is anything more or less, please, elect one, do not blame our weakness: that which I have found in the copy that was sent from you, dear sir, I have taken care to preserve completely, neither adding anything to those things written here by the philosopher, nor on the other hand taking away from them, according to my ability.

Text 6. Sergius of Rēšʿayna, Syriac version of Alexander of Aphrodisias, *On the Principles of the Universe*, § 35

Alexander of Aphrodisias's *On the Principles of the Universe* is a text concerned primarily with the motions of the heavenly spheres and how these are caused, together with various ancillary questions about the relationship between these motions and the motions of the sublunar bodies, as well as how both of these relate to the First, or Unmoved, Mover. It was Alexander's aim to harmonize various passages in Aristotle, such as *Physics*, VIII; *Metaphysics*, Λ; and *De anima*, III,10. Alexander was particularly concerned to demonstrate how the heavenly spheres could influence sublunar bodies, thereby grounding those theories of providence and fate that he expounded in his *De fato* and elsewhere.

Sergius's work is a loose adaptation of Alexander's work. In particular, he suppresses the theory of the eternal cosmos and elaborates Alexander's epistemology into a Christianized Neoplatonist form. For Sergius, the understanding of the physical processes by which heavenly bodies may affect earthly ones is a stage on the ladder towards the perfect knowledge of ultimate being. This is the repetition in a new context of a theme seen in the logical works, that all these forms of knowledge are in the final analysis ancillary to the goal of union with the divine. Sergius saw the works of Aristotle and his school (it is unclear whether Sergius believed the present work to be Aristotle's or not) as being in need of translation in order to benefit the Christian religion of his countrymen. The extract below is an addition to the text vis à vis Alexander's original, insofar as this can be reconstructed from the Arabic version.[44]

Date Early sixth century.
Text Fiori, 'L'Épitomé syriaque', § 35 (with French translation).

44 Further discussion in King, 'Alexander of Aphrodisias' *On the Principles of the Universe*'.

Everything desires the Being. It seems that also among men knowledge of It/Him, that which we call *theoria*, is their fulfilment/culmination, because knowledge is the highest blessedness and fulfilment/culmination for men, as we demonstrated in the discourse *On Theoria*, and the *theoria* of It/Him is the high point of all knowledge.

Text 7. Sergius of Rēšʿayna, *The Action of the Moon*

Galen's treatise *On Critical Days* was of particular interest to Sergius, on account of the relationship it seeks to establish between therapeutics and astrology. Having translated it, however, Sergius was faced with a request for further explanations and summaries of its contents. The following text is his prologue to that exposition, in which Sergius again points towards the need to return to the foundations of logic before aspiring to the deep understanding of more particular branches of science.

Date Early sixth century.
Text Sachau, *Inedita Syriaca*, 101,3–102,8; better edition, with French translation, Claude-Villey, *Les textes astronomiques syriaques* (*vie et viie siècles*), 195–196.

After reading our translation of this third treatise, which is the final one of the work *On Critical Days*, our brother Theodore, because you noted the complexity of the thought that the author had placed in this book, you persuaded us to expound for you as briefly as possible the theory to which Galen had recourse, so that by reading it, it might be found agreeable and one may receive some faculty to direct one towards the knowledge of the matters that he deals with. However, I shall speak of the real point of this treatise, not in the way of medical science nor of a theory of natural philosophy, but rather following the path taken by all those who concern themselves with the celestial sphere and the stars therein, and to what, as they say, can occur in this [sublunar] region on account of the numerous variations of the moon and the varying movements of everything that has circular motion. That is why if someone wants to know about some particular subject, such as the critical days that Galen discussed in this third treatise of his, he must first understand the broader subject [*melta d-gawwa*] concerning the opinions of astronomers relative to all the causes of events in this human region, as Aristotle teaches.

Text 8. Sergius of Rēšʿayna, Syriac version of Galen, *Commentary on Hippocrates's On Nutriment*, Prologue

Very few of Sergius's voluminous translations of the Galenic corpus are extant.[45] Although he was better known in his own and subsequent ages as a medical translator, we have far more from his pen of philosophy than of medicine. The current text was only unearthed recently as part of an Arabic translation written in Hebrew characters of Galen's commentary on Hippocrates's work *On Nutriment*.[46] It repeats themes already familiar from the prologues given above. The rest of Sergius's introduction concerns the standard questions that students of scientific texts were required to investigate before beginning the study of any text. These *schemata isagogica* were especially associated with the Alexandrian curriculum, and Sergius's concern to transmit them in the Syriac tradition is testimony to his wider project of sharing the fruits of Greek science with the Syrians in order to deepen their appreciation for and the benefits to be gleaned from the certain knowledge that it offers.

Date Early sixth century.
Text Bos and Langermann, 'The Introduction of Sergius of Rēshʿainā to Galen's Commentary on Hippocrates' On Nutriment', 192. Translation ibid., 198–199, from which the following is taken.

This is Hippocrates's book *On Nutriment* with the commentary of Galen, translated by the translator Yusuf ibn Khaj [*licet* al-Khuri?]. Sergius said: Let us begin, with the help of God, to write what Galen explained concerning Hippocrates' book *On Nutriment*. He [Sergius] said: Dear brother, since you asked me and requested of me many times that I translate for you some of the books of the Greeks into Syriac, but especially the treatise that Hippocrates composed about foods, and this was on account of your love for all knowledge that the soul requires, I assented and will not deny you that which you asked for and requested; for it is one of the things that are required.

I did not want to do any of this, because I know the censure and rebuke that will come upon me due to the envy of those who are not satisfied by anything other than amassing money. This is so because they do not seek the same gain as those who are in search of knowledge [and] who do not have material possessions that might be a substitute for it [i.e., seeking knowledge]. I also beseech you not to neglect seeking beneficial knowledge […]

45 Hugonnard-Roche, 'Note sur Sergius de Reshʿaina'; and Kessel, 'Inventory of Galen's Extant Works in Syriac'.

46 Bos and Langermann, 'The Introduction of Sergius of Rēshʿainā to Galen's Commentary on Hippocrates' On Nutriment'.

Text 9. Sergius of Rēšʿayna, Syriac version of Galen, *On the Powers and Mixtures of Simple Drugs*, Prologue to Book 6

The following is another of Sergius's prologues to a translation of Galen. Although the existence of this translation has been known to scholars ever since the arrival in the British Library of the Dayr al-Suryan MSS in the mid-nineteenth century, the discovery in 2009 of another copy of the text in a privately owned manuscript that has become known as the 'Syriac Galen Palimpsest' has fired an incentive for further research. Although Ḥunayn's *Risāla* informs us that Sergius translated only the second, practical half (books 6–10) of this work and eschewed the more theoretical first half, recent considerations have suggested that this was unlikely.[47] Indeed the contents of this prologue indicate quite clearly that Sergius had already translated the first half, and the whole tenor of his argument favours the importance of the theoretical as a propaedeutic to the practical. This is by no means the first time we have seen this theme emerge from Sergius's prologues.[48] The same basic stance engaged his motivation to move on from Galen to Aristotle.

Date Early sixth century.
Text Merx, 'Proben der syrischen Übersetzung', 244–248, and Bhayro and Brock, 'Syriac Galen Palimpsest', 38. Translation taken from Bhayro and Brock, 'Syriac Galen Palimpsest', 39–40; there is also a German translation in Merx, op. cit.

In many passages, it is said by the writer Galen, O our brother Theodore, that the only true teaching can be the one that first establishes the rules that pertain in general [i.e., theoretical] to the art in the subjects with which the teacher is concerned, and in this way it follows on to the instruction that is in detail [i.e., particular, or practical]. For truly the art of [knowing] the rules is not able to perfect a person who is learning without [any accompanying] instruction on individual details. Nor, again, can instruction on details train a student without [the accompanying] knowledge of general matters. But in the case of whosoever is familiar with only one of these, it occurred to me many times that I should liken him to one who wants to walk with one of his feet, or to labor with one of his hands, or to work with one of the parts of the body which are not able, without the mutual participation of the others, to carry some action or other to completion.

---

47 Bhayro and Brock, 'The Syriac Galen Palimpsest', 40; and Kessel, 'Membra Disjecta Sinaitica I'.

48 Bhayro, 'Sergius of Reš 'Ayna's Syriac Translations of Galen'.

Now, therefore, if a person is going to be with an exact knowledge of the matters [in hand], for him it is by all means a necessary thing, that he should know the general rules, those things which are designated as pertaining to the art of all of the parts of the art, and that he should also be trained carefully and instructed in the specific knowledge of those things that are included in every one of the parts of medicine. Therefore, as one who knows this way accurately, the teacher in every case first teaches about the rules that are general, and he prepares the students, by means of these things, in order that they shall acquire [an idea of] the system of the art. And thus he expounds an exposition to the pupils and he trains their [general] knowledge for a grasp of those things that occur in detail, his aim being to make them also by means of these things perfect in knowledge in all of those things which are right; and also in this book, which is about the power of simple medicines, O our brother, he clearly employed this method of teaching.

For in all of those five treatises that are before this [i.e., Books 1–5], he set down general rules for us concerning all the powers and kinds of herbs, and concerning all the rest of *materia medica*, from where and how it is compounded from the four elements, while he showed clearly which power or which kind or which taste naturally is attached to each one of the medicines [as a result] of mixture in which the heat overwhelms, also that which has cold in it abundantly, and again that [results] from moisture and from dryness, in cases where one of these shall be found overwhelming its fellow in the mixture. But he reproved in these treatises which are before this one, as was right at the time of writing, also those who think that the medicines and herbs of all the things that are here are not compounded from the four elements, but from small atoms that are absorbed into the air that is about them. With regard to the subject of the whole of this book, we have spoken at length because of the love of learning that is in you.

Therefore, in these five treatises which are before this, the author taught concerning the general art, as we said, of the strength and the mixtures of the medicines. But in these last six, of which this is the first of them that is now our intention to translate, he who is instructing the hearers about specific instruction is teaching about things in detail. He begins with the herbs in sequence and describes clearly the strength and efficacy of each and all of them, placing them in the language of the Greeks in the order of the letters from alpha and until o. And when he had completed all the instruction concerning herbs, he then begins again, in the order of the letters, and teaches concerning every other material that is taken for a remedy; from stones and from within the ground, and from flesh and from dung, and from blood and from the fur of beasts. And with these, he completes the discourse of these six latter treatises. In this discourse, then, which we now intend to translate from the Greek language into the Syriac tongue, it being the sixth out of this entire work, but the first of the remaining ones, the writer commences on the discussion, as we said, by setting down in it all the plants whose name begins with the first letter, namely alpha, up to the ninth, that is, iota. And with this, he completes the sixth discourse.

Text 10. Proba of Antioch, *Commentary on the De interpretatione*

A misinformed comment in 'Abdišo''s thirteenth-century catalogue of Syriac authors led to the belief that the Syriac philosopher Proba of Antioch lived in the first half of the fifth century. For many years, this led scholars to argue that his commentaries were evidence for the advanced state that Syriac philosophy had already achieved at that time. But Proba references Proclus (d. 485), a fact that immediately shows 'Abdišo''s comment to be misleading. Moreover, Proba utilizes the prolegomena system of Ammonius of Alexandria, together with a commentarial format that is known also, and was perhaps initiated by, Olympiodorus of Alexandria.[49] His works thus most likely date from the second half of sixth century, sometime after at least the first four books of the logical curriculum (*Isagoge, Categories, De interpretatione, Prior Analytics*) had already been translated into Syriac. He may himself have been responsible for the last of these translations, and he wrote commentaries on the first, third, and fourth of them. We have no work of his on the *Categories*, either because Sergius had already covered that ground, or else his work was not preserved.

The text here cited comes from the introduction to the commentary on *De interpretatione* and explains how the methodology of philosophical enquiry works, building up gradually from a foundation in logic. The metaphor employed is readily recognizable from the similar passage from Sergius's commentary (Text 2 above) and was itself a staple of the Alexandrian approach to studying philosophy. Proba drew from the same sources as Sergius, and both represent a Graeco-Syriac branch of the Alexandrian commentarial tradition. They share its assumptions and also its notions of the utility of philosophy, and hence the utility of translating philosophy for new audiences.

Date Mid- to late sixth century.
Text Hoffmann, *De Hermeneuticis*, 64,1–10, 15–20, with Latin translation. A new edition with French translation and notes is in preparation by H. Hugonnard-Roche.

49 For the dating debate, and issues of whether the philosopher may be identified with the churchman of the same name and period, see Brock, 'The Commentator Probus'.

Therefore, this is the aim: the culmination of forethought. For example, when men come to build a house, first they consider the roof, but because it is not possible for this to stand up by itself, they then give thought to the walls. But because walls are unshakeable only when based upon foundations, they must then give thought to those foundations. This is [what we mean by] the culmination of forethought, which is also the beginning of action. For it is from that point where thought ends that action begins. And the culmination of action is right there at the point where the thinking starts. It is thus appropriate that the aim [of logic] is the culmination of forethought [...] This aim is required because our reasoning is defined towards what we want to do, since it is necessary for the very person who is trying to do something to have some knowledge of what he wants to do. For Plato says, 'O boy, those who debate have but one [method of] beginning: either they must know the very thing that they are set upon doing, or else they will inevitably fall short of it altogether' [*Phaedrus* 237c].

Text 11. Paul the Persian, *Treatise on Logic*, ?as translated by Severus Sebokht

The sixth-century philosopher Paul the Persian[50] wrote at least three introductory works, probably in Pahlavi: 1) a general introduction, of which only the part on the classification of Aristotelian science is extant, via Arabic quotation; 2) a *Treatise on Logic* dedicated to the Persian shah, likely written between 567 and 580; 3) an *Explanation of De interpretatione*. The last of these is stated in the manuscript as being a translation from Pahlavi into Syriac by Severus Sebokht (see Text 13 below), and it is possible that all his works were written in Pahlavi. Although these three texts by Paul are his own work, they are heavily dependent on the Alexandrian commentarial tradition and may be little more than translations of portions and summaries that were available at the Persian court or at Nisibis, or some other significant centre.

A number of features of Paul's *Treatise on Logic* are revealing of the attitudes to philosophy held at the Persian court and of the reasons that some may have had for translating it into their own environments. The current text explains Paul's position on the relation of faith and knowledge. The arguments in favour of the claims of the latter as against the former as a path to knowledge, as they are clearly presented in this text, appear not to have been in any way a stumbling block to Severus Sebokht, abbot of one of the most prestigious monasteries among the West Syrians. He translated it into Syriac and, we may suppose, made use of it in the philosophical curriculum that was pursued at his monastery (see below, Text 14).

50 For further references regarding the question of identity of the various individuals called 'Paul the Persian', see Gutas, 'Paul the Persian'.

Date c. 550–579.
Text Land, *Anecdota Syriaca*, IV, 2,22–3,7, with Latin translation; English tr. by D. King, forthcoming.

We cannot hold on to one of [these teachings] while abandoning the other, nor choose one while rejecting the other. So what we require is definite knowledge, on the basis of which we might accept one [of these various dogmas] and leave the rest aside. Yet there is no definite demonstration of any such [knowledge]. Hence the subject matter of these teachings is to be analysed through faith as well as through [scientific] knowledge. Knowledge is concerned with things that are close, distinct, and knowable, while faith is concerned with all those subjects that are distant, and which cannot be seen or precisely known. The latter involves doubt, while the former does not. All such doubt is divisive, while its absence leads to unanimity. Knowledge is therefore [better] than faith, and choosing the former is superior to [choosing] the latter, for even believers, when questioned about their faith, make their defence in relation to knowledge, asserting as they do that what we have faith in at the present time we shall later have knowledge of, quoting 'for the moment we see things in pictures, like in a mirror, but at that time it will be face to face' [1Cor 13:12].

Text 12. Paul the Persian, *On the Explanation of De interpretatione*, translated by Severus Sebokht

Another statement from Paul on the epistemic centrality of logic. See above (Text 11) for details on Paul and his work. The present text was written in Pahlavi and translated into Syriac in the seventh century. Paul includes only a very short prologue to his work.

Date Second half of sixth century.
Text Hugonnard-Roche, 'Sur la lecture tardo-antique du *Peri Hermeneias*', 48–49 (with French translation); English tr. by D. King, forthcoming.

It is in the use of logic that humans are superior to irrational beings [*aloga*]. Everything that human beings do not know, they come to know by means of [the faculty of] logic, so long as it is knowable and so long as they wish to know it. For logic is the instrument [*organon*] by which everything is known, and that is why one must first of all learn logic.

Text 13. Barḥadbešabba ʿArbaya, *The Cause of the Foundation of the Schools*

The School of Nisibis was the primary locus of learning in the Church of the East and lay within the Persian empire until the Islamic invasions. It was renowned even in the West as a seat of knowledge and culture. Although its main purpose was the religious and theological instruction of the future leaders of the Church, some philosophical studies were carried on there, perhaps initiated by the Catholicos Mar Aba in the middle of the sixth century.[51] The present text belongs to the Syriac *Cause* genre of aetiologies; formally, it is presented as a speech given to a new incoming class of students at the school, in which the speaker looks back to the beginning of Creation and describes cosmic history as a succession of 'schools', of which that of Nisibis is the latest.

The speaker presumes a quite high degree of Graecophone knowledge amongst his audience; he liberally intersperses his text with Greek loanwords and assumes that the audience is familiar with at least the rudimentary foundations of the Alexandrian curriculum, dividing philosophy into the theoretical and the practical, and, just as we have seen with Sergius and Proba, defining logic as the *sine qua non* of not just philosophy but of all knowledge (except, he carefully and self-defensively notes, for those who speak by the Spirit of God).

Date Second half of sixth century.
Text Scher, *Mār Barḥadbšabba ʿArabaya: Cause de la fondation des écoles*, 342–345. English translation: Becker, *Sources for the History of the School of Nisibis*, 111–114 (here lightly emended).

For because all things doubly established in learning are divided into two kinds, theory and practice, it is right to know that the perfection of theory is the exact comprehension of the knowledge of all existents [*hawayye*], while the perfection of action is the excellence of good things.

Therefore, because there is a contrariety [Gk, *enantiotēs*] attached to each one of these, as colour to a body, and accident to essence, I mean the perfection of both intelligence [i.e., theory] and action; on this account, logic was sought as an intermediary, that it might distinguish for us this contrariety from the true perfection of each one of the parts of the soul. For if perfect intelligence is the exact knowledge of all things that exist, it is certain that its contrary is ignorance. And it is because of this that we stand in need of logic, that by it we may distinguish the true from the false. For what is revealed to be true we can grasp by sensible conviction about our knowledge of things. Yet whatever

51 Brock, 'From Antagonism to Assimilation', 22.

is borne witness to by a true demonstration to be a lie, this thing we leave out of all remembrance of truth. It is certain then that without logic, [truth] is not distinguished correctly or known by those who judge these things in a human manner. For the teaching of a person who does not speak by God's Spirit stands in need of logical arguments to be believed by those who hear it.

[…] Therefore by this wonderful instrument [*organon*] of logic,[52] the mind paints all the adorned images of exact knowledge and by it casts a glorious statue of that prototype, so that the intelligence and the logic of this mind be neither idle nor useless. Since it has no alphabet, by which it might compose names and read them, for teaching itself about that essence […] it was necessary for the exercise of his faculties and as a sign of his freedom that the creator establish this corporeality, and adorn it with faculties and types [or: properties], dividing it up into genera and species, and distinguishing it by figures and operations [or: images], and conferring upon it individual properties, and placing it in the gap between heaven and earth. He disposed and wrote, as it were on a tablet, all the visible bodies, so that the intellect could read them there and would be able, by means of knowing them, to have knowledge of the creator of this knowledge.

Text 14. Severus Sebokht, *On the Syllogisms of Aristotle*

One of the leading lights of Syriac philosophy, Severus was abbot-bishop of Qenneshre, the Graecophile monastery on the Euphrates that trained many future leaders of the West Syrian Church. He was active in the middle years of the seventh century (d. 666/7) and wrote numerous treatises, especially in the fields of logic and astronomy. He received high praise from Otto Neugebauer for his work in mathematical astronomy, where he built upon the work of Ptolemy's *Syntaxis* and Theon's *Smaller Commentary*. In logic, he translated the works of Paul the Persian and himself wrote short treatises and letters on various points of the *Organon*. It seems likely, from what we know of extant texts and of the system at Qenneshre, that much of the philosophical study that was pursued there was carried on directly in Greek, and that complete translations of texts were only done as needed, increasingly in the next generation after Severus. However, we can imagine that much *ad hoc* translation was done in the classroom and in the discussion of the learnèd sons of the monastery.

In the prologue (first text below) and exordium (second text) to his brief treatise on the *Syllogisms of Aristotle*, Severus, in the tradition of Sergius and Paul, lays out the cursus of philosophy as he understood it to be. It is especially important to note his

52 See note in Text 2 above regarding the term *organon*.

teaching of the relationship of the *Prior* to the *Posterior Analytics* since, although some teachers in the Greek and Syriac schools taught a shortened *Organon* that ended at *Prior Analytics* I,7, it seems that Severus's pupils at Qenneshre were keen to translate all of the logical works, in accordance with their master's conception of the interrelationships between them.

Date Mid-seventh century.
Text London, BL, Add. 14660, Add. 17156.

I—BL, Add. 14660, f.47b
Our goal in this treatise is briefly to demonstrate, or to detail, the modes of the categorical deductions in the book of *Prior Analytics* of the philosopher Aristotle, namely their composition and analysis, and also how many there are and which they are. Also [to be discussed] are their figures, how many there are and which they are, as being the logical and deductive technique for this. This is useful for us and of great benefit for [gaining] a complete knowledge of the logical and demonstrative theories that are discussed in the book of *Apodeictics*, in which may be found accurate demonstration and the distinguishing of the true from the false by way of logical technique. So, if it is beneficial to know the true and to be far from the false, then this may be discovered by way of deductions.

II—BL, Add. 14660, ff.53v–54r; Add. 17156, f.5v
[Translation from Watt, 'Prolegomena to Aristotelian Philosophy', 155–156]

The student should first know that this book of the *Analytics* is not for itself (*law meṭṭul yāteh*). On the contrary, as the book of *Categories*, which teaches (us) about simple namings, (leads us up) to the *Peri hermeneias*, which (teaches us) about the first combination of simple namings, (which in turn) leads us up (*masseq lan*) to this book of the *Analytics*, so also this book of the *Analytics*, which teaches us about the construction together with the reduction (*šrāyā*) again of categorical syllogisms, leads us up to the use (*ḥšaḥtā*) of the logical treatise (*pragmateia*) of the book of the *Apodeictics*, which is the aim and fulfilment of the whole logical art, which (in turn) is the instrument (*organon*) of the whole of philosophy, which (in turn), according to a fine Platonic word or definition, is assimilation to God according to what is possible for man.

Text 15. ?Severus Sebokht, Letter to ?Basil

The following is the first known text that mentions what we now universally know as the base ten numerals, whose origin lay with the Indian astronomers. It has traditionally been ascribed to Severus, although it is equally possible that Severus was the recipient rather than the author. The identification of the priest Basil as the recipient is also no more than an educated guess.[53] We can, however, date the letter rather precisely on account of the astronomical data it contains.

Severus's much-laboured point is that scientific knowledge is no Greek monopoly, and that in some branches of knowledge, e.g., astronomy, the easterners have the edge. The rhetoric buys in to a broader stream of tradition in which non-Greek philosophers view the translation movement as the *re*-appropriation of their own ancient knowledge, rather than as a secondary or servile imitation of Greek learning. It may be imagined that Severus was here defending the programme of his monastery against those who sneered at 'pagan Greek learning' as an unworthy pursuit for a man of God. A programme of translation may often be perceived as a tacit admission of the inferior cultural status of the target language. Sergius too found himself ever defending philosophy against this sort of opposition.

Date 661/2.
Text Reich, 'Ein Brief des Severus Sebokt' (with German translation).[54] Partial English translation in Takahashi, 'Between Greek and Arabic: The Sciences in Syriac from Severus Sebokht to Barhebraeus', 22–23, from which the following translation is partly drawn.

Concerning the fact that some of the Greeks who are with you, as you [lit.: your brotherhood] wrote, say that the Syrians can know nothing at all of such things, I mean, the computation of the stars and the eclipse of the sun and the moon, believing that all knowledge belongs to the Greeks alone because they speak Greek, they ought to know, since there are the wise Babylonians, that the Babylonians were the first inventors of knowledge, and not the Greeks, as all the writings of the Greeks themselves testify; and after the Babylonians, there were the Egyptians, and then the Greeks. And I do not think anyone will dispute that the Syrians are Babylonians. Those who say, therefore, that the Syrians can know nothing at all of this sort are in great error, seeing that the Syrians were the first inventors and teachers of these things.

Ptolemy, too, testifies that this is so in the *Syntaxis*. For when he sets down there the beginning of the computation of the sun and the moon and the five planets, he does not begin with the years of the Greek kings, but with those of

53 Villey, 'Qennešre et l'astronomie aux VIe et VIIe siècles', 168–170. It was Reich, in his edition of the text, who ventured upon Basil as the recipient.

54 Reich's text contains many typographical errors. Original text, Paris, BNF, syr. 346, ff.168v–171r.

the Babylonian kings, that is, with King Nebuchadnezzar (by Nebuchadnezzar, I do not mean the one in whose time the Prophet Daniel was, but the other who was much earlier than him). Ptolemy writes in the *Syntaxis* that the years that passed from this Nebuchadnezzar, the first of the Babylonian and Persian kings, until [the time of] Philip the Macedonian [i.e., Philip Arrhidaeus], who came after Alexander, the founder of Alexandria, were 424 years, indicating thereby that he found that the origin and basis of those computations that he had made was among the Babylonians, and not from among the Greeks. Hence it was upon this basis, which he had found among [the Babylonians], that he added to and enlarged upon the computations that he made.

Further, the fact that the Babylonians, that is the Syrians, are said to be wiser than the Greeks, is in turn something that the Greeks themselves taught, for when, as they themselves say, someone was questioning the oracle that they have about which people had always been God-fearers, this is what they report that the oracle said: 'Only the Chaldeans received wisdom, and also the Hebrews, those who worship the king, who is of himself, namely, God.'[55] Now the Chaldeans are, according to current knowledge, the Babylonians, and the Babylonians are Syrians. Therefore, even according to the testimony of their own oracle, 'only the Syrians received wisdom'.

We hear about the lateness of learning, i.e., the ignorance, of the Greeks also from the Greeks themselves, in particular from Plato, the foremost celebrated philosopher among them, as he wrote in the treatise *Timaeus*, that 'when Solon, the wisest of the wise, returned from Egypt he told Critias what he had heard from an Egyptian priest, one who was particularly old, for he had said to him, "Solon, Solon, you Greeks always remain children, there is no Greek who is an old man!" Then he had said, "You Greeks are all so childish in your souls, for among all your ideas there is not one that is ancient, not one piece of learning that is ancient in time, but rather for many generations writing was lacking while you were dying without a word." '[56]

So then, according to this testimony, the Greeks did not even know what writing was for many generations, but instead they all died without a word, like dumb and irrational creatures. [Yet] some of them behave arrogantly as if they were the first inventors of the mathematical science of astronomy. This is not so, for science does not belong to *lexis*, or [a set of] words, but the words [i.e., the *lexis*] belong to science; and *lexis* is not the cause of wisdom, but the latter

55 This account was well known in Christian apologetic circles in late antiquity, e.g., *Cohortatio ad Graecos*, 11–12.

56 Cf. *Timaeus*, 20d–e, 22b, 23c; the summary at *Cohortatio ad Graecos*, 12 (Marcovich, 39) is, however, closer to Severus's direct source.

of the former. Those wise Greeks too, who wisely defined philosophy, did not define it [and say] that philosophy is nouns and verbs, *lexeis*, i.e., words, and the Greek language, but rather that philosophy is the 'knowledge of things that are insofar as they are', 'knowledge of divine and human actions', 'art of arts', 'science of sciences', 'meditation on death', 'becoming similar to God insofar as is possible for a human being', and 'love of wisdom'. But this [sc. philosophy] does not belong to the Greeks alone, but can be acquired by all who are diligent, whether they be Greeks or barbarians.

I shall remain silent for now about the knowledge of the Indians, who are not even Syrians, and about their subtle inventions in this teaching of astronomy, which are more artful than those of the Greeks and of the Babylonians, and the rational *methodoi* of their computations, and the system of counting that surpasses [the use of] words, I mean, that with the nine signs.[57] If those who think to themselves that they alone have reached the utmost level of wisdom just because they speak Greek were to take notice of these things, they might perhaps be persuaded, if only too late, that there are others too who know something, not the Greeks alone, but also some among the nations with other languages.

I have said these things not because I despise the wisdom of the Greeks in such matters as these and other similar matters, for I am not totally errant [i.e., I am not altogether unfamiliar with these sciences], but because I want to show that knowledge is common property and belongs to anyone who wishes to be diligent, regardless of whether he is Greek or barbarian.

[The letter ends with a series of rhetorical questions on why Ptolemy made errors in certain matters, as if to demonstrate the secondary and comparatively meagre nature of Greek astronomical learning by comparison with the founts of knowledge from the East].

Text 16. Severus Sebokht, *Letter on the Ascending and Descending Nodes*

The prologue to this short astronomical treatise is of interest primarily for containing a rarity, viz. mention of a patron who seems to have contracted for the present work, itself based on the *Handy Tables* and Theon's *Commentary*. The patron in question, a certain Stephanos holding the senatorial rank of a *vir illustris* in the Byzantine

57 What Severus probably means is that there are two methods in use, one using words (or perhaps letters) and the other using symbols. But there is some uncertainty regarding what Severus means by 'surpasses words'. See the discussion in Hugonnard-Roche, 'Mathématiques en syriaque', 73.

court system, held the post of *chartularius* of the Jezira, i.e., Upper Mesopotamia, in the years around the establishment of the Umayyad caliphate in Syria. A *chartularius* was ordinarily the chief administrative officer attached to a civil service department in the provincial bureaucracy. Did he, like later Arabophone civil service officials in the same region, feel that a close knowledge of astronomy, which had a number of practical applications in governmental planning, would further his civil service career?

Date c. 662.
Text Claude-Villey, *Les Textes astronomiques syriaques* (*VIe et VIIe siècles*), 251–253, with French translation and notes.

[I shall write] about how one may calculate the ascending [*anābibazon*] and descending [*qaṭābibazon*] nodes in accordance with the *Handy Tables* [*procheiros*], or without the *Handy Tables*, and how to predict whether or not there will be a solar or lunar eclipse of sun or moon. Then also [I shall write] about what the friend of Christ Stephanos Illustris, secretary [*chartularius*] of all Gazirta, wrote, or rather recalled, to whom I shall humbly present, O my brother, the following subjects, namely whether the Nisan 14 [i.e., Paschal moon] that will take place in the 19th year of the lunar cycle, in the 8th indiction [*epinemesis*], that is to say next year, should be placed on the 5th or the 6th of Nisan. Attention will also be paid to whether the seven planets may or may not form a conjunction [*sunodos*] with one another, and how often they do it. I have already spoken above about these subjects. We will therefore now speak as far as possible about the calculation of the ascending and descending nodes. As for being able to calculate quickly and without effort the ascending and descending nodes, I was not myself able, and I reckon nobody else was either, to find a calculation of the ascending and descending nodes that is quicker, easier, and less labour intensive than the calculation [made] in accordance with the rule of the *Handy Tables*, as taught by Theon of Alexandria in his *Commentary* [*Scholia*] *to the Handy Tables*.

Text 17. George of the Arabs, *Prolegomena to Aristotelian Philosophy*

George was a learned Syriac speaker, probably originating near Antioch, who in 687 was ordained as bishop of the Arab tribes.[58] He had been a pupil and co-worker of the West Syrian patriarch Athanasius of Balad, himself a son of Qenneshre who translated into

58 For a general introduction to George, see Tannous, 'Between Christology and Kalam?'; and Miller, 'George, Bishop of the Arab Tribes'.

Syriac the *Prior* and *Posterior Analytics, Topics*, and *Sophistical Refutations*. Whether George studied at Qenneshre we have no certain knowledge, but given the nature of his written work and of his connections, it seems more than likely that he did. He was a contemporary and close collaborator of Jacob of Edessa, perhaps Syria's most famous philosopher-bishop, who had also been a protégé of Athanasius of Balad. According to the chronicles, George died in 724. Besides his extensive and varied homilies, letters, and poems, some of which deal with philosophical themes, and which indicate his extensive learning and close engagement with both Greek and Syriac literary works, George's principal legacy is a substantial volume of translation and commentary on the *Organon*, a fortunate survival of the ravages that have decimated so much of the Syriac heritage. The single extant manuscript contains his translations of the *Categories, De interpretatione*, and both books of the *Prior Analytics*, together with extensive commentaries on the *Categories* and both books of the *Prior Analytics*, and an introduction to *De interpretatione*. The whole is preceded by prolegomena whose beginning is sadly lost.

In this extract from the prolegomena, George demonstrates the continuing attachment of the Syrian teachers to the Alexandrian tradition that Sergius and Proba had endeavoured to import into the Semitic milieu. He follows closely the prolegomena pattern to be found in the Alexandrian commentaries and shares their assumptions about the aims and utility of logic. He also has a clear vision of how philosophy leads to the true and full knowledge of God that every ascetic seeks by means of an ascending epistemological ladder. Here logic is not an optional extra to the learning required of a man of God, nor is it merely an instrument for fighting theological battles.[59] A comparable attitude may be found in Alexander of Aphrodisias (CAG, II,1, 6,8–12) and indeed almost any of the Neoplatonist Alexandrians of the fifth and sixth centuries, Christian or otherwise.

Date Early eighth century.
Text Watt, 'Prolegomena to Aristotelian Philosophy'. Translation here quoted, ibid., 146–147.

Point Four. What is the end of Aristotelian philosophy? We say (it is) that we may know the one principal, cause, and creator of all. For the Philosopher demonstrates in the treatise called *Metaphysics* [literally: After the natural things] that the principal and cause is one, bodiless, from which everything has come into being.

59 For a similar ladder of learning in Sergius, together with an analysis of its foundation in Evagrian epistemology, see Watt, 'From Sergius to Matta'.

Point Five. What are the things which lead us to the end? We say that (it is) the doctrine of the things which are in time and change. For from these, by the intermediation of mathematics, we may ascend to those which are in a state always in like manner, and thus after bodiless substances (ascend) to the first cause of all [...]

Point Eight. What must the reader be like? Just, sharp of intellect, [careful] with words, and measured in behavior, so that he may be adorned in all of them.

[Point] Nine. It is necessary [also for] the commentator to be like that, so that he does [not] through virtuosity of intellect presume to exalt and demonstrate that which is badly said, like him who accepts things (merely because they are) *ex cathedra*,[60] nor (conversely) on account of hatred oppose good things in a bad way. Rather, he must be a dispassionate judge of what is said, and first clarify the thought of the ancients and explain what is appropriate to them, and then come to his own judgement.

Text 18. George of the Arabs, *Introduction to the Prior Analytics*

Introducing the syllogistic of Aristotle, George dwells upon the beauties of the gift of philosophy, of its indispensability for human wisdom, and of its perfection among Christians, who alone have perceived its true end in a knowledge of the true God. Like Sergius and Paul, however, he vigorously defends what truth the pagan philosophers, above all Aristotle, did successfully arrive at.

Date Early eighth century.
Text London, BL, Add. 14659, ff.83v–84v, here quoted from the extracts translated in Miller, 'George, Bishop of the Arab Tribes', 312–314 (emended).

And this also should be considered wisely and ungrudgingly, namely that there is no good [more] primary and exalted among all things bestowed on the race of man by God, the Cause of All—for in this good and from this good our rational nature is especially imprinted and it is properly distinguished from all irrational natures—than the gift of true philosophy, which comes from God, the overflowing and immeasurable wellspring of all wisdom and knowledge. From the beginning [this gift] was wisely bestowed and sown in all rational and intellectual nature, even if some, or rather almost all the pagans, having fallen away from exact knowledge of the First Cause, were unable truly to preserve

60 Lit. 'from the tripod', an expression taken from Philoponus, *In Cat.* 6,29 ff., most likely referring to the Delphic oracle. Ammonius, *In Cat.* 8,11 ff., makes the same point without mentioning the tripod.

among themselves accurately the accurate exemplars (*eikones*) and inexpressible beauties of this same gift [...] [This is so] even though some of them conceived of notions that are more like ours [i.e., we Christians], 'being constrained by the truth itself', as St Gregory [of Nyssa] has taught us,[61] in fact nearly everything that is ours. Men are not badly placed with respect to the truth of the teaching of philosophy compared to their fellows [...] Therefore let nobody find fault with philosophy, but with those who make use of it wrongly! And as for us, to the extent that we are able, it is not dishonorable to make use of the evident and clear and uncorrupted exemplars of this knowledge for what may be useful for us [...] Let us not stop from gaining profit from terms and names and ideas and inquiries by question and answer, and the art of ordered speech, and rightly judging all the queries that anyone might raise. Such things are to be found and gathered especially from the writings of Aristotle the philosopher, more than all others.

Text 19. George of the Arabs, *Commentary on the Prior Analytics*

Elsewhere in his commentary, George even sees a mystical truth behind the threefold nature of the syllogism (major premise, minor premise, conclusion) and the divine Trinity.

Date Early eighth century.
Text London, BL, Add. 14659, f.94r–v. English translation quoted in Miller, 'George, Bishop of the Arab Tribes', 314–315.

These are all the premises that are interwoven in the three figures of syllogisms [...] Now there are three figures of arguments, no more and no less, in the entire art ⟨of reason⟩ [...] and it is not possible that three not be found, even if we have ⟨only⟩ one in mind. Now this I take to be a great mystery, that God imprinted the intellect in a threefold and mysterious manner, so that by a trinitarian thought process (*methḥašbanuta*) the intellect might see the whole truth, and by the trinity that is in the intellect might see the Trinity which exists eternally in unity, that which is properly the Truth. ⟨The intellect does so⟩ when it distances itself from falsehood and joins itself to the truth by means of true and reasoned demonstration. And see how this philosopher teaches that there are only three figures, for he says that the middle term has three variations—as I also indicated above ⟨in discussing⟩ other ⟨points⟩, for it passes straight

61 Derived from Aristotle, *Physics*, 188b30.

through ⟨i.e., the first figure⟩, or it descends from above to below ⟨i.e., the second figure⟩, or it turns back from below to above ⟨i.e., the third figure⟩. Lo, in this way he describes the three figures. This, again, is a great mystery: firstly Φ makes plain ⟨or: signifies⟩ in many ways the unlimitedness of Him who holds fast all the ends of the worlds, and further, with ⟨Φ⟩ the entirely holy name of God begins ⟨i.e., φιλάνθρωπος⟩, and it is like the holy cross which leads back to one those who are afar off (cf. Eph. 2:17), and it is like a guardian and a provider through the Existent One, Φ, when one takes it in the Greek ⟨sense⟩.

Text 20. Sylvanus of Qardu, *On the Definitions and Aims of Philosophy*

Sylvanus of Qardu is a little-known East Syrian philosopher who wrote additions to Theodore bar Koni's *Scholion*. He appears to have been more or less contemporary with Theodore, who taught at the School of Kashkar probably around the turn of the ninth century. The school at Kashkar was founded as a branch of the more famous School of Nisibis, and just as in the mother school, the curriculum there would have been focused on the exegesis of Scripture according to the mode of Theodore of Mopsuestia, for which a basic grounding of some sort in Aristotelian logic was expected. The *Scholion* takes the form of a question-and-answer commentary on all the books of the Bible, though in fact its contents form something more like an extensive introduction to the whole East Syrian intellectual tradition (Griffith, 'Theodore Bar Koni's Scholion') and includes a whole book on defining philosophical terms.

The addition from the pen of Sylvanus printed below is of interest insofar as it indicates the types of sources from which the East Syrians drew their Greek philosophy. The definitions of philosophy quoted in this extract are the same as those given by Ammonius in fifth-century Alexandria.[62] Sylvanus may have taken his definitions from a compendium or perhaps directly from Paul the Persian, who puts them in a very similar way. Even well into the Islamic era, the principal teachers of Syriac philosophy still conceived their aims as being the same as that of their Greek forebears.

Date Late eighth / early ninth century.
Text Hespel, *Theodore bar Koni* [...] *Sylvain de Qardu*, 42–46, with French translation.[63]

What is philosophy and what are the definitions of it?

62 Busse, *Ammonius in Porphyrii Isagogen*, 2f.

63 Hein, *Definition und Einteilung der Philosophie*, 86–130, provides an overview of the reception of this question of the definitions of philosophy in the Syro-Arabic philosophical literature.

Philosophy is the science [lit: knowledge] of all beings in the very fact of their being beings.[64] They call it the knowledge of all beings because every discipline that there is has a single subject-matter and a single goal. Medicine, for instance, whose subject-matter is human bodies, has for its goal, the health of [those bodies]; or such as astronomy, whose matter is the stars and whose goal is the knowledge of their movements and actions. It is thus with all disciplines. Each one of them has a single subject-matter and a single goal. But philosophy does not have its own particular subject-matter, but rather it concerns itself with all sensible and intelligible beings universally. Thus also its goal is not one particular thing, but is the science of all things. And one should add also [to this definition] 'in the very fact of their being beings', because the knowledge of things varies among individual people. There are those who know a thing in its nature, in its faculty, and in its activity; there are those who know all about the faculty while knowing nothing of the nature; and there are those who know only its [general] shape. For instance, if there were someone who knew the human body, of what it is constituted, what is the measure of each of the humours within it, what are its faculties, and who knew everything about it, that person is indeed an expert in the body. For even someone who knows about just one of [the body's] faculties may also be called an expert in the body, and even someone who only knows about its exterior shape. And the knowledge of things is not equally [distributed among] all people, so people have spoken of philosophy 'insofar as there are existent beings'. It is evident that they mean that it is a true science.

[Another definition of] philosophy. Philosophy is the precise science of things divine and human.[65] 'Precise science of things' is equivalent to what was said in the first definition. 'Divine and human' [are in the definition] because everything that exists is either intelligible and comprehensible, or else visible and sensible, and 'divine' comprehends all intelligible substances, while 'human' covers all sensible substances, since inner words [*melle d-ruḥa*] habitually refer by these terms to the whole world, both the visible and the invisible, the temporal and the eternal.

[Another definition of] philosophy. Philosophy is a meditation on death,[66] and is defined as a meditation on death because man is composed of two parts, a visible body and an intelligible, invisible soul. And just as he has two

64 As Ammonius, *In Porphyrii Isagogen*, 2,22f. For Paul the Persian's expression of a similar set of definitions to those here offered by Sylvanus, cf. Land, *Anecdota Syriaca*, IV, 4,4–5.

65 Ammonius, *In Porphyrii Isagogen*, 3,2–3.

66 Ibid. 4,15–5,31. Sylvanus's discussion of the bonds between soul and body closely follows Ammonius's exposition at 5,9–31.

parts, so also he has two connections, one natural and the other voluntary. The natural one is that by which the body is united to the soul, while submitting to it and fulfilling all of its impulses in accordance with the commands of the Creator; the voluntary is that by which the soul is united and tied to the body in all its desires, because these are not in accordance with God's command. And it is because this subjection of the soul to the body only occurs by the will and free choice of the soul that it is called 'voluntary', just as the former was called natural because it related to nature. And just as people possess two connections, so they also have two disconnections, one natural and one voluntary, because every connection has a disconnection. The disconnection is natural when the body is disconnected from the soul in accordance with God's limit. This is what all people call natural death. The voluntary disconnection is when the soul is disconnected from the body and becomes dead to all desires, and it is this distance of the soul from the body's desires, which is called voluntary death. It is not [called so] by everyone, but only by experts. Philosophy can therefore be defined as a meditation on death, not the natural death, because philosophers also censure those who suffocate and do away with their own souls before God's limit, but that voluntary death which causes one to destroy all the body's desires for the sake of true knowledge.

[Another definition of] philosophy. Philosophy is a likeness to God to the extent that humans are capable of being like Him.[67] It is defined as 'likeness to God' because God possesses two faculties, one intellectual, by which he acts wisely in all matters, and the other providential, by which he takes care of all beings. Philosophy is like divinity in that it too has these two faculties, one by which it knows beings, the other by which it is concerned with providing for all the needs of any who approach it. The words 'to the extent that humans are capable of being likened to Him' show that it is impossible for men to be wholly like God, neither in their knowledge nor their providence, yet [they can do so] in a figurative and obscure way, like a shadow is to a body, and insofar as it is given to created things to be like their creator.

[Another definition of] philosophy. Philosophy is the discipline of disciplines and the science of sciences. It is defined as the 'discipline of disciplines' because it is the mother and progenitor of all sciences and disciplines, which all draw from it as rivers from sources. Medicine has taken from it the four elements: cold, warm, dry, and wet, and the mixtures [of these that occur] in the body. Geometricians have taken from it the point, the line, the surface, and all

67 Ibid. 3,8–4,14.

the figures of triangles, tetragons, and hexagons. Astronomers have taken from it the course of the stars and their movements. Even if a physician makes use of the four elements and their faculties, yet it does not belong to him to undertake a defence when questioned about what each of them is like and about how they are discovered; rather this is for the philosopher alone, for he is their inventor. And even if the astrologer (*kaldaya*) makes use of the sphere of the zodiac and the course of the heavenly bodies, yet he is quite ignorant of the actual nature of the sun and the moon and each of the heavenly bodies. It is only the philosopher who really has knowledge of this, since it is he who handed these [discoveries] over to [the astrologer]. And it is not only intellectual disciplines that are constituted on the basis [of philosophy], but it is the practical ones as well. The architect has taken from [philosophy] the rule and the plumb line; and the carpenter [has taken from it] the measuring-thread, the straight-rule, and the compasses. Since, then, it is the inventor of all the disciplines, that is why they defined it as the 'discipline of disciplines and the science of sciences', giving us thereby to understand its superiority and excellence compared with all [other] disciplines.

Text 21. Catholicos Timothy I, Epistle 19

One of the most prominent leaders and administrators of the Church of the East, Timothy, was made patriarch in 780 and ruled the Church for around forty years. Among his many concerns was the intellectual vitality of the Church, and it was he who engaged with al-Mahdī on the *Topics* and with another *mutakallim* of the court on Aristotelian philosophy.[68]

Around 785, Timothy was engaged in teaching at one of his Church's principal schools, at Mosul. While there, he seems to have been thinking hard about the intellectual foundations of Syriac learning, both seeking books from the monasteries (probably to copy for his patriarchal centre in Baghdad) and planning how to put that learning on a firmer foundation. In this letter, he notes that the Greeks and Arabs are well ahead of the Syrians in the matter of the linguistic study of their own tongues, and he proposes that Syriac follow a similar method of reflective grammatical research. On the one hand, Timothy's requests from the monastic libraries of northern Mesopotamia indicate the wealth of learning still deposited there, in both Greek and Syriac;[69] on the

68 Griffith, 'The Syriac Letters of Patriarch Timothy I'. On the life and works of Timothy, see Berti, *Vita e studi di Timoteo I*.

69 Brock, 'Two Letters of the Patriarch Timothy'; Watt, 'Syriac Translators and Greek Philosophy in Early Abbasid Iraq'; and id., 'Commentary and Translation in Syriac Aristotelian Scholarship'.

other, he is clearly motivated by a jealousy for the achievements of Arabic sciences, and that desire to emulate becomes from c. 800 onwards one of the prime movers of Syriac translation.

Date c. 785.
Text Braun, *Timothei patriarchae I: Epistulae I*, 127,1–22 and 129,10–18, with Latin translation. An edition and German translation of Timothy's Letters 3–29 has been published recently (Heimgartner, *Die Briefe 3–29 des Ostsyrischen Patriarchen Timotheos I*).[70]

The book you have proposed is quite desirable. Indeed, not only desirable, but rather very necessary. When I have from time to time observed and examined, whether it be on my own or in company with many [other] researchers, now the wealth of our Syriac language, now its poverty, sometimes its material abundance, at other times its lack of a distinctive form, I found that on [the Syriac] side there was nothing at all, or only very little, whereas on the side of other peoples [I did find] some small things, though not a great deal.[71] I thus reflected and considered how I myself might develop a method of this type, although I am rather foolish and lack understanding. For I was excited as a possessor, i.e., a mother-tongue speaker, of the language, both in what I called its abundance and in its poverty. I was also interested in other foreign languages, which some have said, following a well-known ancient tradition of our ancestors, derive from this one of ours. I am speaking of Greek and Arabic, both of which have abundant and rich lexicons and distinctive forms, something that is possible only in a grammar [i.e., when there is a grammar] and not otherwise.

If possible, a memorandum of [the monastery of] Mar Zina's books should be sent to me. Maybe among these there will be discovered some that we have not [yet] found. You have to examine these books personally. Go through them in accordance with all their subject matter and all their [different] forms. It may be that you will come across those two treatises *On Poets*, for I have one.[72] Or perhaps among them you will come across Olympiodorus's commentary on the books of logic [i.e., *Organon*], or [the commentaries] of Stephanos, or Sergius, or Alexander [...]

70 I am indebted to Vittorio Berti for many of the interpretations of Timothy's words in this translation.

71 In light of what Timothy says further on, this probably refers to Greeks and Arabs.

72 It may be that Timothy had heard stories about the famous lost second book of the *Poetics*. See the discussion in Tarán and Gutas, *Aristotle Poetics*, 80–88.

Text 22. Catholicos Timothy I, Epistle 43

In this letter, written at a similar period or a little later than the foregoing, Timothy discusses the quality of his work on the *Topics* vis à vis the work of others. Competition with Arabic scholars again seems to have become a primary motivation in the Syriac sciences. As we saw above, the letters also show us how Timothy sought to plumb the depths of the Graeco-Syriac translation tradition in order to find commentaries and scholia on philosophical books. He expected to find Greek as well as Syriac books in the monastic libraries.[73]

Date 785–800(?).
Text Pognon, *Une version syriaque des aphorismes d'Hippocrate*, XVI–XVIII, with French translation; Heimgartner, *Die Briefe 42–58 des Ostsyrischen Patriarchen Timotheos I*, 65–68, with German translation, 47–52. English translation below taken from Brock, 'Two Letters', 235–237.

1. To the God-loving priest and teacher, Rabban Mar Pethion: Timothy the sinner greets you and hopes to see you.
2. The royal command required of us to translate the *Topika* of the philosopher Aristotle from Syriac into the Arabic tongue. This was achieved, with God's help, through the agency of the teacher Abū Nūḥ. A small part was done by us as far as the Syriac was concerned, whereas he did it in its entirety, both Syriac and Arabic; the work has already reached a conclusion and has been completed. And although there were some others who were translating this from Greek into Arabic—we have written to inform you how and in what way it happened that all this took place—nevertheless (the king)[74] did not consider it worth even looking at the labours of those other people on the grounds that they were barbaric, not only in phraseology, but also in sense, whether because of the natural difficulty of the subject (*hypothesis*)—for you are aware of the style (*eidos*) of the Philosopher in matters of logic, and how and to what extent he infuses obscurity into the beauty of (his) meaning and sense—or as a result of the lack of training of those who approached such things. For you know the extent and magnitude of the toils (*agônes*) and labours such a task requires. But (the king) entirely approved of our labours, all the more so when from time to time he compared the versions with each other.

73 There is plentiful discussion on the details of this letter in Brock, 'Two Letters of the Patriarch Timothy'.

74 Al-Mahdī.

3. Let your Eminence sagely ask and enquire whether there is some commentary or scholia by anyone, whether in Syriac or not, to this book, the *Topika*, or to the *Refutation of the Sophists*, or to the *Rhetorika*, or to the *Poetika*; and if there is, find out by whom and for whom (it was made), and where it is. Enquiries on this should be directed to the Monastery of Mar Mattai—but the enquiries should not be made too eagerly, lest the information, (the purpose of the enquiry) being perceived, be kept hidden, rather than disclosed.
4. Job the Chalcedonian told me that he has seen a small (number) of scholia on the *Topika*, but only, he said, on certain chapters. But let your Chastity doubly enquire about scholia or a commentary on these books.
5. Send us the other volume of Athanasius, so that we can copy it out. We have the first. I think the translation is by Paulā, for on the title of the book the following is inscribed: 'First volume of the holy God-clothed Gregory the Theologian, which the Abbas Mar Paulā translated from Greek into Syriac in the island of Cyprus'. The revision, so it says, is by Athanasius. So much for this.
6. Search out, too, for the treatises on the natural principles (lit. heads) of bodies, written by someone of the Platonic school (*dogma*); it begins: 'Concerning the natural principle of bodies some have said …'. The first treatise gives the opinion of all the earlier philosophers and sets out the Ideas (*ideai*) and Platonic Forms. The second treatise begins by speaking of matter (*hylē*), species (*eidos*) and negation, following Aristotelian teaching (*dogma*). (The author) deals with it in five sections, but the treatise is incomplete. Search out to see if these treatises can be found—both what remains from the second treatise, and the rest of what follows on from these.[75]
7. Search out for a work by a certain philosopher called Nemesius, on the structure of man, which begins: 'Man is excellently constructed as a rational soul and body […]'.[76] He brings the subject to an end in roughly five sections; at the end he promises to deal with the soul, but this second part is missing.
8. Please search out and copy for us Dionysius in the translation of Athanasius or that of Phokas.
9. Peace to you and to all the brethren.

---

75 This text has recently been discovered in a Syriac manuscript and attributed to Porphyry; see Arzhanov, *Porphyry: On Principles and Matter*.

76 This is Nemesius of Emesa's *De natura hominis*. A Syriac version existed, of which fragments may be retrieved from quotations on Syriac authors; see Zonta, 'Nemesiana Syriaca'.

Text 23. Ḥunayn, *Risāla*

Ḥunayn ibn Isḥāq's famous text needs little introduction. The preface to the *Risāla*, however, reminds us that it is itself a translation from Syriac, in which language Ḥunayn had first prepared his overview of the tradition history of the Galenic corpus. Recent research has shown that even in Ḥunayn's day, the majority of the medical translations being carried out in and around Baghdad were translations into Syriac, for Syriophone patrons, for whom their mother-tongue was still the surest basis for sound scientific knowledge.[77]

Date Mid-ninth century.
Text Bergsträsser, *Hunain ibn Ishaq*, 1,4–2,4 (German translation, 1).

[1]
You have mentioned—Allah give you honour!—the need for a book that would gather together all the books of the ancients on medicine that one needs [nowadays] and in which the goal pursued in each of them is set forth, also the arrangement of the parts of each book and the individual parts of knowledge that are found in each, so that it would become an easier task for anyone seeking one of these chapters, when the need arises, to find it, and so he can determine in which book it is found, in which part of it, and where in the part; and you have asked me to undertake this task for you.

[2]
So I told you, God help me, that my memory is not capable of remembering all these books after I had lost my entire collection. And that, after I lost my books, a certain Syriac-speaker had made a similar request, especially concerning the books of Galen and asked me to explain to him which of these books I and others had translated into Syriac and other languages, and [I told you] that I then wrote him a book in Syriac, based on what he asked for in his request for me to write.

[3]
You then asked me—Allah give you honour!—to translate this book for you now, until Allah in his kindness grants the return of these books through your help, and [you asked] that I add to the books of Galen listed in this book any-

77 Watt, 'Why Did Hunayn, the Master Translator into Arabic, Make Translations into Syriac?'.

thing that I had missed, as well as an account of the other books of the ancients on medicine that we have come across. I turn now, Allah willing, to the execution of this request.

Text 24. Ibn al-Nadīm, *Fihrist*

The *Fihrist*, a catalogue of Arabic scientific writings including a vast quantity of information about earlier translators, was written more than a century after the *Risāla*, at a time when Syriac translations were certainly on the wane. As a result, many of them escaped Ibn al-Nadīm's notice. However famous Ḥunayn may have become as an Arabic translator, Ibn al-Nadīm nonetheless points out that most of his work was done in Syriac and was for Syriac patrons. Elsewhere in his catalogue, Ibn al-Nadīm notes that Ḥunayn translated into Syriac a part, and his son Isḥāq the whole, of the *Posterior Analytics*, and that only much later was the whole rendered into Arabic by Abū Bishr Mattā. This is significant for a comparison of motivations. For the Syrians, as we have seen in many texts cited above, the logical sequence of study demanded that the *Posterior Analytics* be the end-point and goal of the preceding books of the *Organon*, which on their own make no sense without it. Ibn al-Nadīm was not aware that a Syriac version of the *Posterior Analytics* had in fact existed since the seventh century. In the early years of the ʿAbbāsid era, the Arabic translators were not motivated in the same way—for instance al-Mahdī commissioned a translation of the *Topics* without having any prior knowledge of the other books, perhaps because it fulfilled his more immediate need for a practical handbook of dialectic.

Date 987/8.
Text Flügel, *Ibn al-Nadīm: Kitāb al-Fihrist*, 309, 348. English translation: Dodge, *The Fihrist of al-Nadīm: A Tenth-century Survey of Muslim Culture*, 600, 682.

Account of the *Apodeiktikos*, which is the *Analytica posteriora*, in two sections. Ḥunayn translated part of it into Syriac and Isḥāq translated it in complete form into Syriac. Mattā translated the version of Isḥāq into Arabic [...]

[...] It was the good fortune of Ḥunayn that the things which Ḥubayš ibn al-Ḥasan al-Aʿsam, ʿIsā ibn Yaḥyā, and others translated into Arabic were attributed to him, Ḥunayn. If we have recourse to the catalogue of Galen's books which Ḥunayn made for ʿAlī ibn Yaḥyā, we learn that most of the things which Ḥunayn translated were [translated] into Syriac, although he may also have corrected and examined the Arabic of other people's translations.

Text 25. Barhebraeus, *Chronicle*

The Syrians' sense that, following the rise of 'Abbāsid-era science, they had been overshadowed by a rival in the theatre of learning persisted long after Timothy's day. Barhebraeus, the great representative of the thirteenth-century Syriac renaissance, himself attempted to combine the best of both traditions, drawing on the old Syriac translations and interpretations of Aristotle, while at the same time trying to introduce into Syriac the advances represented in the work of Ibn Sīnā.[78] Here he takes pride in the history of his own peoples, yet acknowledges that things are very different in his own day. Barhebraeus was even motivated to translate parts of Ibn Sīnā directly into Syriac.[79]

Date c. 1276.
Text Bedjan, *Gregorii Barhebraei Chronicon syriacum*, 98. English translation: Budge, *The Chronography of Gregory Abû'l Faraj*, 92.

There arose among (the Arabs) philosophers, mathematicians, and physicians, who surpassed the ancients in subtlety of understanding. While they built on no foundation other than those of the Greeks, they constructed greater scientific edifices by means of a more elegant style and more studious researches, with the result that although they had received the wisdom from us through translators, all of whom were Syrians, now we find it necessary to seek wisdom from them.

---

78 Much, but by no means the whole of his philosophical œuvre, has been edited. Especially important for an understanding of his sources are studies such as Takahashi, *Aristotelian Meteorology in Syriac*, and Watt, *Aristotelian Rhetoric in Syriac*. For an overview of his work on logic, see Hugonnard-Roche, 'L'œuvre logique de Barhebraeus'.

79 Takahashi, 'Reception of Ibn Sīnā in Syriac', 257, 261ff.

Part Three: Original Texts

Text 1. Sergius of Rēšʿayna, *On the Aim of All the Works of Aristotle*, Prologue[80]

Paris, BNF, syr. 354, ff.1v–3v (13th cent) = P
Birmingham, Mingana 606, ff.52r–53v (20th century) = B

I

ܥܠܬܐ ܐܡܝܪܐ ܡܢ ܩܕ̈ܡܝܐ ܐܘ ܐܚܘܢ ܬܐܕܘܪܐ. ܕܒܣܘܦܐ ܗܘ̇ ܕܡܫܬܥܝܬܗ ܣܘܪܟܐ. ܗܕܝܢ ܫܕܐ ܘܡܬܚܟܡ. ܟܕ ܢܦܘܪܫ ܢܦܫܗ ܡܢ ܥܘܠܐ. ܘܢܣܩܐ ܠܐܬܪܐ ܣܘܪܟܐ. ܘܢܗܪ ܒܡܕܒܪܢܘܬܗ ܡܕܝܢܝܬܐ. ܟܕܡܟܐ ܠܢܒܠܐ ܕܫܘܠܡ ܣܘ̈ܦܗܝ. ܒܗ̇ ܕܝܢ ܒܕܡܘܬܐ ܡܣܬܒܪܐ ܠܝ. ܕܠܐ ܐܢܫ ܡܨܝܐ ܠܡܣܬܟܠܘ ܬܪ̈ܝܨܘܬܗܘܢ ܕܩܕ̈ܡܝܐ. ܘܠܡܗܘܐ ܠܗܘ ܡܢ ܪܐܙܐ[81] ܕܝܕܥܬܐ ܕܡܠܦ̈ܢܘܬܗܘܢ. ܐܠܐ ܐܢ ܦܪܫ ܢܦܫܗ ܡܢ ܥܠܡܐ[82] ܘܡܦܦܩ̈ܘܗܝ ܘܐܪܝܡ ܐܦ ܡܢ ܦܓܪܐ. ܠܘ ܒܐܬܪܐ ܐܠܐ ܒܬܪܥܝܬܐ. ܘܥܒܕܐ ܟܠܗܘܢ ܕ̈ܚܫܚܬܗ ܠܡܫܬܪܗ.[83] ܗܕܝܢ ܗܘ ܒܢ ܪܢ ܡܫܬܡܫ ܗܘܢܐ ܠܦܘܩܕܢ ܠܘܬ ܗܘܐ. ܘܠܡܫܢܪ ܒܟܠܘܡܗ. ܘܠܡܚܙܐ ܢܗܝܪܐܝܬ ܐܝܟ ܕܡܢ ܫܘܢ ܐܬܚܫܒܘ.[84] ܘܠܡܕܢ ܥܦܪܐܝܬ ܕܐܝܟܢ ܐܢܝܢ ܕܬܪܝܨܐܝܬ ܐܬܐܡܪ. ܘܐܝܟܢ ܕܠܗ ܗܟܢܐ ܐܬܬܩܢ. ܡܟܐ ܕܠܐ ܢܬܒܥ ܡܕܡ ܕܡܬܚܙܐ ܠܗ ܒܪܗܛܐ ܕܐܘܪܚܐ[85] ܕܐܝܟ ܗܕܐ. ܡܢ ܐܝܠܝܢ ܕܕܠܡܘܟܠܐ ܕܡܠܠܘܬܗ ܐܬܝܗܝܢ ܒܟܬܒܐ ܦܓܪܐ.

II

ܟܕ ܗܟܝܠ ܡܕܡ ܡܕܡ ܡܢ ܡܟܬ̈ܒܢܘܬܗ ܕܚܠܝܡܘܗ ܐܣܝܐ ܡܫܟܚܝܢ ܗܘܝܢ ܡܢ ܡܡܠܠܐ ܕܝܘ̈ܢܝܐ ܠܒܪܬ ܡܠܐ ܕܗܘܪܛܝܐ. ܐܢܐ[86] ܟܕ ܡܦܫܩ ܗܘܝܬ. ܐܝܬ ܕܝܢ ܟܬܒ ܗܘܝܬ ܒܬܪܝ. ܟܕ ܡܬܦܢܐ ܗܘܝܬ ܬܠܐ ܡܦܫܩܢܘܬܐ. ܐܝܟ ܕܬܟܬܒ ܡܡܠܠܐ ܕܠܗ ܕܠܥܠ ܗܢܐ. ܘܟܕ ܣܓܝܬ ܦܘܠܚܢܐ ܥܡܝܪܐ ܕܡܠܬܐ ܕܐܝܬ ܒܡܟܬ̈ܒܢܘܬܗ ܕܓܒܪܐ. ܘܬܘܣܦ̈ܬܐ ܘܬܘܣܦܬܐ ܕܫܘܚܠܦ ܒܗ̇، ܡܫܚܠܦܢܘܬܐ. ܘܡܟܠܐܢܘܬܐ. ܫܐܠܬܢ. ܕܡܢ ܐܡܝܢܐ[87] ܟܕ ܢܦܩܬ ܠܗ ܠܟܬܒܐ[88] ܗܢܐ ܥܠܬܐ ܘܥܝܪܐ ܕܙܪܝܙܘܬܐ. ܘܦܝܫܐ ܡܠܐܐ ܕܐܝܟ ܗܢܐ ܡܢ ܢܦܫܗ ܐܘ ܡܢ ܐܢܫ ܐܚܪܝܢ ܡܢ ܡܟܬ̈ܒܢܘܬܐ ܕܡܕܒܪܢܘܬܗ. ܐܢܐ ܕܝܢ ܠܘܬ ܗܠܝܢ ܦܫܛܬ ܦܬܓܡܐ ܠܐܣܝܘܬ

80 The text and apparatus for Texts 1–2 have been kindly prepared and made available for this publication by Yury Arzhanov, who is preparing an edition of the whole text. Only variants affecting the translation are included.

81 ܪܐܙܐ P : ܪ̈ܐܙܐ M

82 ܥܠܡܐ P : ܥܠܡܐ ܟܠܗ M

83 ܠܫܬܪܗ P : ܠܡܫܬܪܗ M

84 ܐܬܚܫܒܘ M : ܐܬܚܫܒ P

85 ܕܐܘܪܚܐ M : ܕܐܘܪܚܗ P

86 ܐܢܐ P : ܐܠܐ M

87 ܐܡܝܢܐ P : ܐܡܢܐ M

88 ܠܟܬܒܐ M : ܟܬܒܐ P

ܘܠܦܢܐ ܕܟܢ. ܕܪ̈ܫܐ ܘܫܘܪܝܐ ܘܓܠܝܬܐ ܕܟܠܗ̇ ܪ̈ܕܘܬܐ ܐܪܣܛܛܠܝܣ ܗܘܐ. ܠܘ ܒܠܚܘܕ ܠܟܬܒܘܗܝ ܘܠܐܣܪ̈ܝܐ ܣܕܪ̈ܘܗܝ ܐܣܟܡ̈ܘܬܐ. ܐܠܐ ܘܐܦ ܠܟܠܗܘܢ ܡܟܬܒ̈ܢܐ ܘܦܝܠܣܘܦ̈ܐ ܡܫ̈ܡܗܐ ܕܗܘܘ ܡܢ ܩܕܡܘܗܝ. ܒܕܡܐ ܐܝܟ ܠܪ̈ܒܢܐ ܕܟܕ ܗܢܐ ܚܝܠܐ ܐܝܬܘܗܝ ܠܒܪܢܫܐ ܗܢܐ ܠܡܣܩܪܐ ܕܡܫܬܥܐ. ܒܠܚܘܕ ܡܟܬܒ̈ܢܘܬܐ ܕܦܝܠܣܘܦܘܬܐ ܘܕܪ̈ܫܘܬܐ ܟܠܗ̇. ܒܕܡܘܬ ܚܡܪ̈ܐ ܦܬ̈ܓܡܐ ܡܕܒܪ̈ܢܐ ܗ̈ܢܘܢ ܘܪ̈ܕܝܢ ܒܥܠܬܐܝܬ. ܘܠܐ ܡܫܟܚܐܝܬ ܠܘܬ ܟܠܗܘܢ ܡܟܬܒ̈ܢܐ. ܗܢܐ ܕܝܢ ܗܘ ܠܣܘܟܠܗܘܢ ܒܕܡܘܬ ܐܣܝܐ ܚܟܝܡܐ. ܟܕ ܐܝܟ ܠܟܠܗܘܢ ܡܢ̈ܘܬܐ[89] ܕܡܕܒܪ̈ܢܐ ܗ̈ܘ. ܘܢܓܕ ܐܢܝܢ ܐܘܡܢܐܝܬ ܘܚܟܝܡܐܝܬ. ܘܡܬ̈ܒܝܢ ܡܚܣܢܝܢ[90] ܚܕ ܒܐܘܪܚܐ ܫܠܡܬܐ ܕܝܘܠܦܢܗ. ܕܟܕ ܘܡܫܟܠܠ ܡܢ ܢܦ̈ܩܬܐ ܕܐܝܠܝܢ ܕܡܬܦܪ̈ܫܝܢ ܐܬܩܢܝܬ ܠܘܬ ܡܫܡ̈ܠܝܘܬܗ. ܒܣܘܪ̈ܗܢܐ ܢܡܘܣܝܐ[91] ܕܠܐ ܡܚܒܬܐ. ܐܝܟܢܐ ܐܝܟ ܕܦܠܘܢ ܕܟܬܒܝܢ ܐܢܕܪܘܢܝܩܐ. ܒܠܚܘܕ ܚܕܐ ܡܢ ܡܬ̈ܘܬܗ. ܕܝܠܦܬܐ ܣܟܠܝܢ ܠܗ̇. ܡܢܗ̇ ܘܠܟܐ.[92] ܘܟܕ[93] ܡܬܕܒܪܝܢ ܠܗܘܢ ܚܕܐ ܒܬܪ ܚܕܐ. ܐܝܟ ܕܬܠܡܝܕܐ ܐܘܡܢܘܬܐ. ܘܟܕܝܢ ܐܢܕܪ̈ܘܢܝܩܐ ܒܪ̈ܒܢܐ. ܗܟܢܐ ܐܦ ܗܘ ܪܕܐ ܘܠܣܡ ܘܣܡ. ܒܠܚܘܕ ܚܕܐ[94] ܡܢ ܡ̈ܟܬܘܬܗ ܕܦܝܠܣܘܦܘܬܐ ܒܠܚܘܕܝܐ ܕܬܟܒܪ ܒܢܦܫܗ. ܘܫܒܠ ܡܢܗܘܢ ܒܟܠܗܘܢ ܟܬ̈ܒܘܗܝ، ܚܕ ܒܠܚܘܕ ܐܝܟܢܐ ܘܬܡܝܡܐ ܕܝܕܥܬܐ ܕܟܠܗܘܢ ܗܘܝܐ.[95]

III

ܟܕ ܕܝܢ ܗܠܝܢ ܫܡܥܬ ܡܢܟ ܐܘ̄ ܐܚܘܢ ܬܐܘܕܘܪܐ. ܐܬܬܥܝܪܬ ܒܪ ܫܒܬܗ ܠܡܕܥ. ܕܐܝܟܢܐ ܗܘ ܢܝܫܐ ܕܡܠܦܢܘܬܗ ܕܓܒܪܐ ܗܢܐ. ܘܐܝܟܢܐ ܗܘ ܠܦܘܫܩܐ ܕܡܟܬ̈ܒܘܬܗ ܘܢܣܒܘܬܐ ܕܦܠ̈ܣܘܦܘܗܝ. ܘܟܕ ܡܠܠ[96] ܡܢ ܡܕܡ ܕܚܫܝܚ ܗܘܝܬ ܡܛܠܬܐ ܗܘܝܬ ܠܡܐܡܪ ܡܕܡܝܢ ܥܠ ܗܠܝܢ. ܬܘܒ ܐܦܝܣܬܢܝ[97] ܣܘܟܝ ܕܡܬܬܢ̈ܐ ܐܬܬܘܒ[98] ܠܝ. ܗ̇ܘ[99] ܡܕܡ ܕܡܠܬܐ ܡ̈ܠܬܐ ܗܘܝܬ ܡܕܡܝܢ. ܐܢܐ ܕܝܢ ܟܕ ܡܫܬܐܠ ܗܘܝܬ ܡܢ ܗܕܐ ܡܛܠ ܡܫܠܡܘܬܗ ܕܝܕܥܬܐ ܐܡܪܬ. ܕܟܡܐ ܗܘ ܠܝ ܡܬܐܡܪܐ ܚܕ ܒܡܦܩܢܬܐ. ܥܠ ܢܝܫܐ ܕܦܝܠܣܘܦܘܬܗ[100] ܕܐܪܣܛܛܠܝܣ. ܘܦܣܩ ܠܐܝܠܝܢ ܕܦܓܥܝܢ ܒܗ̇. ܕܢܫܡܥ ܐܢܘܢ ܐܝܟ ܒܡܟܐ[101] ܕܡܫܟܚܐ ܥܠ ܬܪܥܝܬܗ ܕܓܒܪܐ. ܐܢܬ ܕܝܢ ܘܠܐ ܗܟܢܐ ܐܬܦܠܣܦܬ. ܐܠܐ

89 ܡܢ̈ܘܬܐ P : ܡܬܒܘ̈ܬܬܐ M
90 ܡܚܣܢܝܢ P : ܐܝܠܝܢ M
91 ܒܣܘܪ̈ܗܢܐ ܢܡܘܣܝܐ P : ܒܣܘܪ̈ܗܢܐ ܫܦܝ̈ܪܐ ܘܢܡܘ̈ܣܝܐ M
92 ܘܠܟܐ P : ܒܠܗ̇ M
93 ܘܟܕ P : ܘܟܡ M
94 ܚܕܐ M : om. P
95 ܕܝܕܥܬܐ ܕܟܠܗܘܢ ܗܘ̈ܝܐ M : ܕܗܘ̈ܝܐ ܟܠܗܘܢ P
96 ܡܠܠ M : ܡܠܠ ܡܠܠ P
97 ܐܦܝܣܬܢܝ P : ܐܦܝܣܟ M
98 ܐܬܬܘܒ P : ܐܬܥܕܡ M
99 ܘܗܘ P : ܡܢ ܗܘ M
100 ܕܦܝܠܣܘܦܘܬܗ P : ܕܡܟܬܒܢܘܬܗ M
101 ܒܡܟܐ P : ܡܟܐ M

ܬܬܪܐܝܬ ܐܠܗܝܬ ܠܝ ܣܘܟܠܐܝܬ. ܕܢܒܥܐ ܠܗ ܗܟܢܐ ܟܐܢܐܝܬ ܥܠ ܟܠܗ
ܡܠܦܢܘܬܗ ܕܫܠܝܚܐ ܗܘ. ܕܡܛܠ ܥܠܠܬܗ[102] ܕܗܘܢܐ ܥܠ ܐܝܟ ܕܟܬܒܬ ܠܝ ܠܥܠ.
ܐܠܐ ܡܫܪܐܝܬ ܥܠ ܟܠ ܚܕܐ ܚܕܐ ܡܢ ܡܬܒܥܝܢܘܬܗ ܢܦܪܫ ܒܡܫܩܠܬܐ ܗ̇ܘ
ܡܢܐ ܕܡܬܒܥܐ ܠܝ.

IV

ܡܛܠ ܗܟܝܠ ܕܠܐ ܐܬܬܚܝܒܬ ܠܡܦܢܝܘ ܡܢ ܫܐܠܬܟ. ܡܕܡ ܕܐܬܐ ܠܘܬ ܫܘܪܝܐ
ܕܗܠܝܢ. ܗܕܐ ܡܦܝܣ ܐܢܐ ܠܟ. ܘܠܟܠܗܘܢ ܐܝܠܝܢ ܕܦܓܥܝܢ ܒܡܟܬܒܢܘܬܐ
ܗܕܐ. ܕܠܘ ܡܢ ܡܢܝܢܐ ܕܫܪܪܐ ܐܦ ܒܠܚܘܕ ܐܝܠܝܢ ܕܡܬܐܡܪ̈ܢ ܬܢܢ. ܢܪܓܫ
ܐܢܫ ܒܪ ܫܒܬܗ ܠܘܬ ܒܪ̈ܝܬܐ ܘܡܫܪܪܬܐ ܕܠܐ ܡܫܘ̈ܚܢ. ܐܠܐ ܢܒܚܪ ܒܗ ܡܢܐ
ܘܡܫܬܟܚ ܚܕܐ ܕܝܢ ܘܬܪ̈ܬܝܢ. ܘܬܠܬ ܬܘܒ ܘܐܪ̈ܒܥ. ܐܝܟܢܐ ܕܗܟܢܐ ܬܬܒܥ
ܝܒܘܬܐ. ܐܝܢܝܢ ܘܐܦ ܗܟܢܐ ܢܬܒܝܢ ܡܕܡ ܕܒܡܘܬ. ܘܐܦܠܐ ܗܕܝܢ ܢܟܣܢ
ܡܢ ܕܠܡܐܙܠ ܠܘܬ ܐܝܟ ܕܡܫܟܚ ܡܪܬܐ ܠܗ[103] ܘܡܣܘܝܐ ܠܗ ܗܘ ܡܢܐ ܕܠܐ
ܢܕܥ. ܗܟܢܐ ܒܪ ܡܦܘܪ ܢܦܫܗ ܡܢ ܫܥܒܕܘܬܐ ܕܗܘܝܐ ܒܬܪܥܝܬܐ ܕܐܝܠܝܢ
ܕܠܐ ܡܫܬܟܚܝܢ ܗܘ ܡܢܐ ܕܦܪܝܢ. ܣܟܝܢ ܬܘܒ ܢܦܫܗ[104] ܡܢ ܡܐܟܠܘܢܝܐ
ܘܡܫܪܫܘܬܐ ܕܠܐ ܡܕܝܪ̈ܢ ܠܡܬܬܟܒܢܘ ܕܒܠܬܐ. ܐܝܟ ܒܪ ܩ̈ܛܝܢܐ.[105]
ܕܗܟܢܐ ܐܝܟ ܕܡܢ ܢܒܥܐ[106] ܒܪܢܫܐ ܡܬܕܒܪܝܢ ܡܦܪܐܝܬ ܡܢ ܣܡܝܟܐ. ܐܝܟ
ܕܡܢ ܫܘܪ ܡܢܝܢܝܢ ܕܬܬܟܬܒܬܐ. ܒܪ̈ܝܬܐ ܚܠܦ ܩܘܒܠܐ ܢܟܠܘܢ. ܒܡܗܕܪܝܢ ܠܗܘܢ
ܕܒܩ̈ܘܢܐ ܕܐܚ̈ܪܢܐ ܬܫܒܘܚܬܐ ܗܘ ܕܢܫܒܚܘܢ ܚܟܝܡܝܢ. ܠܐ ܕܝܢ ܣܟܠܝܢ
ܗܠܝܢ. ܕܟܠ ܝܒܘܬܐ ܐܝܕܐ ܕܗܝ. ܡܢ ܚܝܠܐ ܗܘ ܕܢܫܒܘܗ̇ ܡܬܟܬܒܐ. ܘܠܘ ܡܢ
ܡܫܝܠܘܬܐ ܕܬܪ̈ܝܬܐ. ܦܣܩܐ ܒܪ ܕܡܫܝܠܘܬܗ ܕܚܝܠܐ. ܫܠܡܘܬܗ ܕܐܚܪܢܐ
ܠܐ ܒܚܕ. ܘܠܐ ܬܘܒ ܒܝܪܘܬܗ ܕܚܝܠܐ ܡܕܡ. ܕܒܘܬܐ ܡܫܟܠܐ ܥܠ ܗܘ
ܡܕܡ[107] ܕܡܬܦܫܩ[108] ܠܗ ܙܕܩ ܗܘܐ ܠܗܘܢ ܒܪ ܠܗܠܝܢ. ܐܝܟܢܐ ܕܒܬܪܝܨܘܬܐ
ܚܙܝܢ ܗܘܘ. ܕܐܝܠܝܢ ܕܡܬܐܡܪ̈ܢ ܢܦܩܐܝܬ ܘܫܪܝܪܐܝܬ[109] ܡܢ ܐܝܟ ܢܟܡܟܘܢ
ܘܢܡܟܠܘܢ. ܘܐܝܠܝܢ ܕܡܬܚܙܝܢ ܕܗܝܟܝܢ ܥܠ ܬܫܡܫܬܐ ܘܬܘܪܬܐ ܕܡܢ ܐܚܪ̈ܢܐ.
ܢܬܪ̈ܝܢ ܕܠܐ ܣܡܝܟܐ. ܐܘ ܢܒܟܪܘܢ ܐܢܫ ܕܠܐ ܒܕܠܝ. ܒܪ ܠܐ ܒܟܣܝܢ ܠܗ ܠܒܢܝܐ
ܐܢܫܐ. ܘܐܫܬܡܝܢ ܒܠܗܘܢ ܡܐܟܠܘܢܝܐ. ܒܗܝ ܕܠܐ ܡܫܬܚܫ ܡܚܝܠ ܒܟܠ. ܗܠܝܢ
ܕܝܢ ܐܡܪܢ ܐܢܐ ܕܢܫܬܟܚ ܐܢܫܐ ܕܗܘܘ[110] ܕܦܪܐ. ܠܘ ܐܝܟ ܡܢ ܕܐܢܫ ܐܢܐ
ܒܬܫܒܘܚܬܗ ܕܒܪܢܐ. ܐܘ ܐܦܝ ܒܪ ܬܪܒܝܬܗ ܗܘ.

---

102 ܥܠܠܬܗ M : ܥܠ̈ܠܬܗ P
103 ܠܗ M : om. P
104 ܢܦܫܗ M : om. P
105 ܩ̈ܛܝܢܐ P : ܩ̈ܛܝܢܘܬܐ M
106 ܢܒܥܐ P : ܡܒܥܐ M
107 ܡܕܡ M : om. P
108 ܕܡܬܦܫܩ M : ܕܡܬܬܫܩ P
109 ܘܫܪܝܪܐܝܬ M : ܘܫܦܝܪܐܝܬ P
110 ܕܗܘܘ M : om. P

V

ܗܟܢܐ ܫܠܡܬ ܐܝܟܐ ܥܠ ܢܦܫܐ ܕܫܡܠܝܘܬܗ.[111] ܐܠܐ ܡܛܠ ܕܐܬܬܚܝܒܬ ܡܢ ܚܘܒܝܢ. ܐܝܟܢܐ ܕܐܬܬܪܝܬ ܡܢ ܠܥܠ. ܘܡܛܠ ܕܡܦܩܕܢ ܐܢܐ ܕܕܘܪܫܐ ܪܒܐ ܗܘܐ ܡܢ ܗܠܝܢ. ܘܟܘܬܪܐ ܪܒܐ ܠܘܬ ܡܕܒܪܬܐ. ܠܐܝܠܝܢ ܕܡܫܬܠܡܢܘܬ[112] ܩܐܡ ܒܗܘܢ. ܡܕܝܢ ܢܦܫܢ ܠܡܠܬܐ ܕܝܠܢ. ܥܠܘܗܝ ܥܠ ܢܦܫܐ ܕܫܦܝܪ ܠܝ ܠܡܟܬܒ.

Text 2. Sergius of Rēšʿayna, *On the Aim of All the Works of Aristotle*, Book 2[113]

London, BL, Add. 14658, ff.2rb–3rb = L
Paris, BNF, syr. 354, ff.16v–17v (13th century) = P
Birmingham, Mingana 606, ff.63v–64v (20th century) = M
Baghdad, Chaldean Monastery, 169 (*olim* Notre Dames des Semences, 51), ff.73r–74v = B

ܒܟܠ ܐܘܡܢܘܬܐ ܐܝܢܐ ܕܗܘ. ܫܘܠܡܐ ܕܬܐܘܪܝܐ ܫܪܝܪܐ ܐܝܬܘܗܝ ܕܡܕܒܪܢܘܬܐ. ܘܫܘܠܡܐ ܬܘܒ ܕܡܕܒܪܢܘܬܐ[114] ܠܫܘܪܪܗ[115] ܥܒܪ ܕܬܐܘܪܝܐ. ܐܡܪ ܐܢܐ ܕܝܢ ܕܐܝܟܢ ܗܟܢܐ. ܕܐܦ ܗܘ ܕܡܬܚܦܛ ܐܪܕܟܠܐ ܕܢܒܢܐ ܒܝܬܐ. ܡܬܚܫܒ ܒܪܥܝܢܗ ܟܕ ܐܡܪ. ܕܐܬܚܦܛ ܠܝ[116] ܠܡ ܠܡܥܒܕ ܡܛܠܠܐ ܕܡܩܘܪܐ. ܡܢ ܙܘܝܬܐ ܘܡܢ ܫܬܐܣܐ ܘܡܢ ܥܪܒܐ ܕܠܒܢܐ ܕܐܝܟ ܗܠܝܢ. ܐܠܐ ܡܛܠܠܐ ܠܡ ܠܐ ܡܥܒܕܢ ܐܢܐ ܕܐܒܢܝܢ. ܐܠܐ ܐܦ ܡܕܒܪܢܘܬ ܐܘܡܢܘܬ ܐܦܣܐ ܠܫܘܡܠܝܘܬܗ. ܠܗܠܝܢ ܕܝܢ ܠܐ ܡܥܒܕܢ ܐܢܐ ܕܐܒܢܐ.[117] ܐܠܐ ܐܦ ܡܕܒܪܢܘܬ ܫܠܡܬ ܘܪܦܝܬ ܫܬܐܣܐ. ܘܗܟܢܬ ܦܐܝ ܡܕܒܪܢܘܬ ܫܬܐܣܐ. ܘܡܢ ܒܬܪ ܗܢܐ ܐܦܣܐ. ܘܗܟܢܐ[118] ܠܫܘܪܬܐ ܦܐܝ ܠܟܠ ܡܢܗܘܢ ܬܛܠܝܠܐ. ܗܘ ܕܗܘܝܘ ܫܘܡܠܝܗ ܕܒܝܬܐ. ܒܗܠܝܢ ܗܟܝܠ ܫܘܪܝܐ ܕܬܐܘܪܝܐ. ܗܘܢܐ ܕܝܢ ܕܫܘܡܠܝܐ ܕܟܠܗ ܪܥܝܢܗ. ܡܢ ܡܛܠܠܐ ܫܪܝ. ܘܥܒܪ ܠܘܬ ܫܬܐܣܐ. ܡܫܡܠܝܘܬܐ ܕܝܢ ܕܗܝ ܗܝ ܦܘܠܚܢܐ ܕܥܒܕܐ. ܡܢ ܫܬܐܣܐ ܫܪܝܬ ܘܥܒܪܬ ܠܘܬ ܬܛܠܝܠܐ. ܘܗܟܢܐ ܐܝܟ ܕܐܡܪܢ ܡܠܠ ܡܢ ܠܥܠ. ܗܘܐ ܫܘܪܝܐ ܕܬܐܘܪܝܐ ܫܘܠܡܐ ܕܡܕܒܪܢܘܬܐ. ܘܫܘܪܝܐ ܕܡܕܒܪܢܘܬܐ ܫܘܠܡܐ ܕܬܐܘܪܝܐ. ܡܕܝܢ ܘܐܦ ܐܪܝܣܛܛܠܝܣ ܒܗ ܒܗܢܐ[119] ܐܬܚܫܒ ܒܗ ܒܐܘܡܢܘܬܐ ܡܠܠܘܬܐ. ܡܕܒܪܢܘܬ ܒܬܪ ܐܬܚܫܒ ܒܪܥܝܢܗ. ܕܢܟܬܘܒ ܐܢܐ ܠܡ ܠܡܥܒܕ ܐܘܪܓܢܘܢ ܦܘܪܫܐ. ܕܡܬܚܫܒ ܠܝ[120] ܒܡܕܒܪܢܘܬܐ ܠܬܚܬܐ ܡܢ

---

111 ܕܫܡܠܝܘܬܗ M : ܕܫܘܠܡܘܬܗ P
112 ܕܡܫܬܠܡܢܐܝܬ P : ܕܡܫܬܠܡܢܘܬܐ M
113 Text and apparatus prepared by Yury Arzhanov.
114 ܘܫܘܠܡܐ ܬܘܒ ܕܡܕܒܪܢܘܬܐ om. P
115 ܬܘܒ add. M, corr. add. B
116 ܠܝ L P : om. B M
117 ܕܐܒܢܐ B M L : ܕܐܒܢܝܢ P
118 ܘܗܟܢܐ L P : ܘܗܟܢܬ B M
119 ܒܗܢܐ B M L : ܒܗܢܐ ܗܢܐ P
120 ܠܝ B M L : ܠܗ P

ܒܥܠܬܐ. ܘܦܩܕ ܠܝ[121] ܒܐܝܕܥܬܗܘܢ ܕܝ̈ܒܘܬܐ[122] ܫܪܝܪܐ ܡܢ ܕܚܠܬܐ. ܐܠܐ
ܡܛܠ ܕܗ̣ܘ ܗܘܐ ܐܘܪܓܢܘܢ܆ ܐܘܡܢܘܬܐ ܐܝܬܘܗܝ ܕܡܠܟܐ ܥܠ ܬܚܘ̈ܡ
ܕܟܝܢ. ܡܛܠܬܐ ܡܬܦܝܣ̈ܝܢ. ܝܕܥܐ ܗ̣ܝ ܕܠܗ̇ ܠܗܕܐ ܐܘܡܢܘܬܐ ܘܬܫܡ̈ܫܬܐ
ܙܕܩ ܠܝ ܡܕܡܝܢܬ ܠܡܥܒܕ.[123] ܘܡܛܠ ܕܗ̇ܝ ܗܕܐ ܐܘܡܢܘܬܐ ܕܬܫܡ̈ܫܬܐ܆ ܡܢ
ܡܘܣܝܩܝ ܡܡܠܠܐ ܕܗܝܟ ܐܘܡܢܝܐܝܬ ܗܘܝܐ. ܙܕܩ ܠܝ ܡܕܡܝܢܬ ܥܠ ܗܘܢܐ ܠܡܟܠܝܘ.
ܘܡܛܠ ܬܘܒ ܕܐܝܬ ܪܘܒܒܐ ܡܕܡܝܐ ܕܫ̈ܡܗܐ ܕܡܢܝܢܗ ܗ̣ܘ[124] ܗܢܐ
ܡܘܣܝܩܝ ܡܡܠܠܐ. ܡܕܡ ܠܡ ܙܕܩ ܠܝ. ܕܟܠ ܗܢܐ ܐܟܬܒ ܡܕܡܝܬ. ܘܡܛܠ ܬܘܒ
ܕܐܦ ܠܗܢܐ ܡܕܡܝܢ ܗ̣ܢܘܢ ܒܡܢ̈ܝܢܐ ܦܫ̈ܝܛܐ. ܫܒܩܐ ܠܝ ܕܟܠ ܗܠܝܢ ܐܠܦ
ܡܕܡܝܬ. ܥܕ̇ܝ ܗܟܝܠ ܒܡܥܒܕܗ ܡܢ ܐܘܡܢܘܬܐ ܕܬܫܡ̈ܫܬܐ. ܘܢܣܒ ܢܩܦܐܝܬ
ܒܡܕܡܐ ܠܡܢ̈ܝܢܐ ܦܫ̈ܝܛܐ. ܡܕܡ ܪܒܐ ܕܡܬܠܒܟܘܬܐ ܕܟܠ ܗܠܝܢ ܒܠܗܝܢ. ܡܢ
ܒܡܢ̈ܝܢܐ ܦܫ̈ܝܛܐ ܚܕܐ ܠܗ. ܘܒܬܪܗܘܢ ܡܠܦ ܥܠ ܪܘܒܒܗܘܢ ܡܕܡܝܐ ܕܫ̈ܡܗܐ.
ܘܗܟܘܬ ܒܬܪ ܗܢܐ ܡܟܬܒ ܥܠ ܡܘܣܝܩܝ ܡܡܠܠܐ ܕܡܬܬܩܢ ܬܪܝܓܐܝܬ
ܘܙܕܩܐܝܬ. ܘܡܢ ܒܬܪܬܐ ܟܠܗ̇ ܕܐܘܡܢܘܬܐ ܕܬܫܡ̈ܫܬܐ. ܘܒܬܪܗ ܥܠ ܒܠܗܝܢ
ܐܠܝܢ ܕܣܟ̈ܝܡܢ ܠܘܬ ܗܕܐ ܒܟܠ ܙܢܝܢ.[125] ܘܫܪܝܢ ܫܘܠܡܐ ܕܡܟܘܪܬܗ ܕܒܗܠܝܢ.
ܠܘܬ ܫܘܪܝܐ ܕܬܐܘܪܝܐ ܕܒܠܗܝܢ. ܐܝܟ ܕܐܦ ܫܪܝܢ ܫܘܠܡܐ ܕܬܐܘܪܝܐ
ܕܒܠܗܝܢ[126]: ܠܘܬ ܫܘܪܝܐ ܕܡܟܬܒܢܘܬܐ ܕܒܠܗܝܢ.

ܒܬܪܟܐ ܗܟܝܠ ܗ̇ܘ ܕܥܒܕ ܠܗ ܥܠ ܡܢ̈ܝܢܐ ܦܫ̈ܝܛܐ. ܡܬܩܪܐ ܕܦܝܬܐܓܘܪܘܣ.
ܗ̇ܘ ܕܝܢ ܕܥܠ ܪܘܒܒܗܘܢ ܡܕܡܝܐ. ܪܫܡ[127] ܕܦܐܪܐܪܟܝܣ. ܘܗ̇ܘ ܕܥܠ
ܡܘܣܝܩܝ ܡܡܠܠܐ ܕܡܫܬܡܗ ܐܢܠܘܓܝܩܐ ܡܕܡ̈ܝܐ ܘܐܣܝ̈ܪܐ. ܘܗ̇ܘ ܕܥܠܝܗ̇
ܕܐܘܡܢܘܬܐ ܕܬܫܡ̈ܫܬܐ. ܐܦܘܕܝܩܛܝܩܐ ܡܬܟܢܝܐ. ܕܟܡܗ ܕܗܢܐ ܐܝܬܘܗܝ.
ܗ̇ܘ ܕܡܬܩܪܐ ܛܘ̈ܦܣܐ. ܘܗ̇ܘ ܕܥܠ ܡܫܘܚܬܐ ܕܣܘ̈ܦܣܛܐ ܕܡܫܘܚܬܗ
ܠܗ[128] ܣܘܦܣܛܝܩܘ ܐ̈ܠܝܛܘ. ܒܗܠܝܢ ܗܟܝܠ ܡܫܠܡ ܠܗ̇ ܦܝܠܘܣܘܦܐ ܗܢܐ܆
ܠܟܠܗ̇ ܐܘܡܢܘܬܐ ܡܫܠܡܬܐ. ܗ̇ܝ ܕܐܘܪܓܢܘܢ[129] ܐܝܬܝܗ̇ ܐܝܟ ܕܐܡܪܝܢ
ܕܦܝܠܘܣܘܦܘܬܐ. ܘܠܐ ܡܢܬܐ ܡܢܗ̇. ܐ̈ܢܫܝܢ ܕܝܢ ܐܡ̈ܪܝܢ. ܕܐܦ ܐܘܡܢܘܬܐ
ܕܪܗܛܘܪܘܬܐ ܗ̇ܝ ܕܥܒܕܐ ܠܗ. ܘܐܦ ܗ̇ܝ ܡܢܗ̇ ܕܠܗ̇ ܕܡܠܝܠܘܬܐ ܐܝܬܝܗ̇.
ܐܠܐ ܣܡ ܢܦܫܐ ܠܝ ܡܛܠ ܟܠܗܘܢ ܕܥܒܪܬܐ. ܘܢܥܒܪ ܠܡܐܡܪ ܐܝܟ[130] ܕܡܟܝܠܢ:
ܥܠ ܢܦܫܐ ܕܟܠ ܚܕܐ[131] ܡܢ ܡܬ̈ܟܬܒܢܘܬܐ ܗܠܝܢ. ܟܕ ܡܫܪܝܢ ܢܩܦܐܝܬ ܡܢ
ܗ̇ܝ ܕܥܠ ܦܝܬܐܓܘܪܘܣ. ܕܐܝܬܝܗ̇ ܥܠ ܡܢ̈ܝܢܐ ܦܫ̈ܝܛܐ. ܘܗܟܘܬ ܒܐܝܕܐ

121 ܠܝ L P : om. B M
122 ܕܝ̈ܒܘܬܐ L P : om. B M
123 ܠܡܥܒܕ L P : ܠܡܥܒܪ B M; dittography in P probably due to homoioarchon (ܘܡܛܠ ܕܗ̇ܝ [...] ܘܡܛܠ ܕܗ̇ܝ)
124 ܗ̣ܘ om. P
125 ܘܒܠܗܝܢ add. P
126 ܐܝܟ [...] ܕܒܠܗܝܢ om. M
127 ܪܫܡ L : ܪܫܝܡ P : ܗ̣ܘ ܪܫܝܡ B M
128 ܠܗ L P : om. B M
129 ܕܐܘܪܓܢܘܢ L P : ܕܒܐܘܪܓܢܘܢ B M
130 ܐܝܟ L P : ܐܝܟ ܡܐ B M
131 ܕܟܠ ܚܕܐ L P : ܕܟܠ ܚܕܐ ܚܕܐ B M

ܒܐܝܕܐ ܥܠ ܟܠ ܚܕܐ[132] ܡܢܗܘܢ ܢܥܒܕܝܢ ܒܗ ܒܝܕܥܐ. ܘܟܕ ܐܬܝܢ ܠܘܬ ܬܪ̈ܟܘܒܬܗ ܐܣܛܘܟ̈ܣܝܐ.[133] ܗܠܝܢ ܕܥܠ ܬܪ̈ܟܘܒܬܗ ܕܡܫܘܪܝܬܐ[134] ܘܥܠ ܫ̈ܘܢܝܐ ܟܠܗܘܢ ܘܡ̈ܫܠܡܢܐ. ܘܗܠܝܢ[135] ܐܣܛܘܟ̈ܣܝܐ ܕܡܬܩܪ̈ܝܢ ܐܠܗܝܬܐ.

Text 3. Sergius of Rēšʿayna, *On the Aim of All the Works of Aristotle*, Book 7

London, BL, Add. 14658, f.60vb.

ܢܦܩܐ ܕܡܡܠܠܘܬܐ [...]ܟܠܚܕ ܡܢ ܗܠܝܢ: ܠܐ ܣܢܝܩܐ ܕܬܪ̈ܟܘܒܬܐ ܕܐܝܟܢܘܬܐ ܡܬܕܪܟܢ: ܘܠܐ ܬܪܝܨܘܬܐ ܕܦܝ̈ܠܣܘܦܐ ܡܬܕܪܟܐ: ܘܠܐ ܬܘܒ ܙܒܢܐ ܥܘܪܐ ܕܟ̈ܬܒܐ ܐܠܗ̈ܝܐ: ܕܒܗܘܢ ܐܝܬܘܗܝ ܗܕܪܐ ܕܫ̈ܡܝܐ ܡܬܓܠܐ: ܐܠܐ ܐܢ ܐܢܫ ܢܦܠ ܒܝܕ ܡܡܠܠܘܬܐ ܕܕܘܒܪ̈ܘܗܝ، ܣܓܝ ܐܠܗܝܐ: ܐܝܟ ܕܠܘ ܢܬܦܢܐ ܥܠ ܕܘܪܫܐ ܐܚܪܢܐ

Text 4. Sergius of Rēšʿayna, *Introduction to Aristotle and His Categories, to Philotheos*, Prologue

Aydin, *Introduction to Aristotle and His Categories*, 94,3–16 = Berlin, Staatsbibliothek, Syr. 88, f.83v.

ܡܛܠ ܗܟܝܠ ܕܐܬܚܫܒܬ ܠܝ ܐܘ̈ ܐܚܘܢ ܦܝܠܘܬܐܘܣ: ܕܐܟܬܘܒ ܐܠܘܬܟ ܡܢ ܗܠܝܢ ܕܣܓ̈ܝܣ ܘܡܬܐܪܐܬ ܡܕܒܪ̈ܝܢ ܠܡܕܒܪܢܘܬܐ ܘܡܫܠܡܢܐ ܕܗ̇ܢܘܢ ܕܒܝܠܝܢ ܠܗܘܢ ܥܠ ܕܘܪܫܐ ܕܡܠܦ̈ܢܐ ܕܦܝܠܘܣܘܦܘܬܐ ܕܒܐܘܡܢܘܬܐ ܡܠܠܝܬܐ. ܗ̇ܝ ܕܠܗܘܢ ܠܗܘ̈ܢܐ ܘܣܘ̈ܟܠܐ ܕܦܝܠܘܣܘܦܘܬܐ: ܒܗ ܡܢ ܦܘܩ̈ܕܢܐ ܘܦܘܠ̈ܚܢܐ ܘܦܠܓܘܬܐ ܕܐܘܡܢܘܬܐ. ܢܬܒܥ ܠܝ ܒܥܘ̈ܬܐ ܐܝܟ ܕܡܫܟܚܐ ܒܠܚܘܕ ܥܠ ܗܕܐ.

ܗܟܝܠ ܗܐ ܐܬܬܘܣܚܢ ܒܗ ܟܐܠܝܢ ܡܢܝ ܘܡܢ ܐܝܠܝܢ ܕܐܟܘܬܟ. ܕܗܘ ܘܗ̇ܘ ܡܟܪܙ ܕܫܠܡ ܡܣܬܪܩܝܢ ܐܢܘܢ ܬܡܠܘܢ.[136] ܘܐܦ ܡܟܪܙ ܠܒܪ ܡܢ ܠܦܘܡܐ ܐܡܪܝܢ ܐܢܘܢ ܬܬܪ̈ܝܢ. ܒܗ ܠܗ ܠܘܬ ܗ̇ܝ ܕܬܬܚܠܘܢ ܒܡܠܬܐ ܕܦܝ̈ܠܣܘܦܐ. ܐܠܐ ܬܬܪܐܝܬ ܠܘܬ ܗ̇ܝ ܕܡܬܩܪܝܐ ܠܓܘܢܐ ܡܬܦܠܣܦܘܬܐ. ܘܗܠܝܢ ܡܢܗ ܕܥܘܪܝܐ ܕܡܠܬܐ ܠܘܬܢ ܒܬܚܘܡ ܒܕܡܫܚܝܢ ܕܡܪܐ ܗܘ ܦܠܢܐ: ܗ̇ܘ ܗܘܢܐ ܕܗܘܬܪܢܐ ܕܟܘܢܐ: ܕܐܡܪܬ ܕܡܕܡ ܕܐܝܬܘܗܝ ܕܟܘܢܝܐ ܟܝܢܐ ܡܫܠܡܢ. ܘܦܬ ܠܐܠܗ̈ܝܢ ܕܬܬܒܟܘܢ ܒܡܫܡܫܢܘܬܐ: ܘܒܗ ܢܕܥܝܢ ܠܗ ܠܓܘܢܗ ܕܡܠܬܐ ܠܗܘܢ ܡܢܗܘܢ. ܘܣܓܝ ܕܒܝܠܘܢ ܘܬܘܒܗܘܢ ܒܣܘܟܐ ܫܦܝܩ ܘܡܫܬܘܕܥܝܢ ܒܫܡܫܘܬܐ.

132 ܟܠ ܚܕܐ L P : ܟܠ ܚܕܐ ܚܕܐ B M
133 ܐܣܛܘܟ̈ܣܝܐ B L M : ܫܬܣ̈ܬܐ P
134 ܡܬܟܪ̈ܟܝܢ ܗ̇ܢܘ ܕܝܢ ܟ̈ܠܗܘܢ add. B M
135 ܘܗܠܝܢ L : ܘܟܠܗܘܢ ܗܠܝܢ P : ܘܗܢܝܢ B M
136 ܬܡܠܠܘܢ Aydin

Text 5. Sergius of Rēšʿayna, Syriac version of Pseudo-Aristotle, *De mundo*, Prologue

Lagarde, *Analecta Syriaca*, 134,13–27 = London, BL, Add. 14658, ff.107vb–108ra.

ܐܓܪܬܐ ܕܟܬܒܬ ܠܝ ܒܥܘܬܟ ܐܚܐ ܕܟܒܝܪܐ ܠܐܪܝܣܛܘܛܠܝܣ ܦܝܠܣܘܦܐ܆
ܠܘܬ ܐܠܟܣܢܕܪܘܣ ܡܠܟܐ ܥܠ ܡܕܡ ܕܗܘܝܐ: ܘܦܩܕܬ ܠܝ ܠܡܦܫܩܘܬܗ ܐܚܝ
ܚܠ܆ ܡܢ ܡܡܠܠܐ ܝܘܢܝܐ ܠܠܫܢܐ ܕܣܘܪ̈ܝܝܐ܆ ܡܠܠܬܗ ܡܢ ܒܪ ܕܟܬܒܬܗ.
ܐܬܟܢܫܬ ܕܝܢ ܡܢ ܒܪܗ܆ ܒܪܡܐ ܠܒܝܬ ܗܢܐ ܡܛܠ ܥܠ̈ܠܬܐ ܡܫ̈ܚܠܦܬܐ.
ܐܠܝܡ ܕܠܘ ܙܒܢܐ ܗܘ ܗܫܐ ܠܡܬܘܐ ܐܢܫ. ܗܫܐ ܕܝܢ ܡܛܠ ܕܗܡ ܐܠܨ̈ܘܬܐ:
ܫܓܚܬ ܠܡܫܠܡܘ ܦܘܩܕܢܟ܆ ܐܦܢ ܡܟܡܟܡ ܗܘ̈ܝ ܠܝ ܫܓܝ̈ܫܘܬܐ ܐܫܬܢ̈ܝܬܐ
ܕܝܠܟ ܐܠܝ̈ܨܬܐ. ܒܗܠܝܢ ܐܪܥܠܬ ܡܛܠ ܦܘܩܕܢܟ ܕܝܠܟ. ܘܐܬܚܫܒܬ ܠܡܫܠܡܘ
ܝܨܝܦܘܬܝ. ܡܛܠ ܐܢܐ ܕܝܢ ܠܗܘܟܝ܆ ܕܐܦ ܗܘ ܕܡܬܥܬܕ ܡܫܝܚܐ ܐܬܪܐ ܕܐܠܗ
ܕܐܓܪܬܐ ܗܕܐ ܕܐܡܪܬ ܒܗ ܡܕܡ ܕܝܬܪ ܐܘ ܒܨܝܪ. ܠܐ ܬܗܘܐ ܒܥܘܬܟ
ܒܥܠܝܐ ܒܬܪ ܡܫܠܡܘܬܗ. ܐܢܐ ܓܝܪ ܗܘ ܡܕܡ ܕܐܫܟܚܬ ܒܝܘܢܝܐ ܕܐܟܬܒܬ܀
ܡܢ ܠܘܬ ܣܘܟܠܝ: ܠܗ ܢܩܘܫ ܠܡܠܬܐ ܒܥܠܡܘܬܐ ܒܡܬܪܬܐ. ܒܪ ܠܐ ܡܘܣܦ
ܐܢܐ ܡܕܡ ܥܠ ܗܠܝܢ ܕܡܬܒܥܝܢ ܠܗ ܠܦܝܠܣܘܦܐ ܗܘܝܢ. ܘܠܐ ܬܘܒ ܐܠܐ ܡܒܨܪ ܐܢܐ
ܡܟܝܠ ܐܚܝ ܚܠܐ܀

Text 6. Sergius of Rēšʿayna, Syriac version of Alexander of Aphrodisias, *On the Principles of the Universe*, § 35

Fiori, *L'Épitomé syriaque*, § 35 = London, BL, Add. 14658, f.107rb–va.

ܒܠܚܘܕ ܠܐܠܗܘܬܐ ܗܘ ܪܐܙܐ. ܕܥܠܡܐ ܕܝܢ ܕܐܝܟ ܡ̈ܕܒܪܢܘܬܐ ܡܕܒܪܢܘܬܗ ܕܝܠܗ.
ܗܝ ܕܐܠܗܘܬܐ ܡܫܡܫܢܝܬܐ ܠܗ. ܐܠܗܘܬܗ، ܥܡܕܠܝܘܬܗ ܡܛܠ ܟܠ ܕܡܫܬܘܬܦܘܬܐ
ܘܥܡܠܝܐ ܡܫܬܪܝܐ ܕܟܠ ܐܢܫܐ. ܐܝܟ ܕܡܫܪܝܐ ܗܘ ܕܟܠ ܬܐܘܪܝܐ ܡܫܘܝ ܠܗ.
ܡܚܘܬܐ ܐܠܗܘܬܗ ܡܫܠܛܐ ܕܝܢ ܕܥܠܡܐ ܡܚܘܬܐ ܐܠܗܘܬܗ ܕܬܐܘܪܝܐ ܕܗܝ.

Text 7. Sergius of Rēšʿayna, *The Action of the Moon*

Sachau, *Inedita Syriaca*, 101,3–102,8; newer edition, with French translation, Claude-Villey, *Les Textes astronomiques syriaques* (VIe et VIIe siècles), 195–196.

ܡܢ ܒܬܪ ܕܦܫܩܢ ܡܐܡܪܐ ܗܘ ܕܬܠܬܐ: ܕܐܝܬܘܗܝ ܐܚܪܝܐ ܕܡܬܒܪܢܘܬܐ
ܗܝ ܕܥܠ ܡܫ̈ܡܫܬܐ ܒܣܘ̈ܪܐ ܐܘ ܐܚܘܢ ܬܐܘܕܘܪܐ: ܡܛܠ ܕܣܝܡܬ ܒܡܩܘܡܘܬܐ
ܕܪ̈ܚܡܐ ܕܗܘܐ ܒܗ ܒܡܐܡܪܐ ܗܘ ܡܬܬܒܥܢܐ. ܐܦܝܣܬ ܠܝ ܠܡܟܬܒ ܠܟ
ܚܠܦܝܗ: ܬܐܘܕܘܪܐ ܗܝ ܕܐܬܚܙܝ ܒܗ ܠܟܠܗܘܢ ܐܝܟ ܕܡܬܚܫܚܐ ܦ̈ܘܫܩܘܬܐ. ܗܘ
ܡܢܐ ܕܒܗ ܩܪܐ ܠܗ ܐܢܫ ܡܬܟܢܫ ܕܢܘܗܪܐ. ܘܢܣܒ ܡܢܗ ܚܝܠܐ ܡܕܡ ܕܡܥܒܕ
ܠܗ ܠܘܬ ܡܕܒܪܬܐ. ܕܗܠܝܢ ܕܡܢ ܗܘ ܐܬܐܡܪ̈ܝ.

ܐܢܐ ܕܝܢ ܐܡܪܬ܆ ܕܡܛܠܬܗ ܟܠܗ̇ ܕܡܐܡܪܐ ܗܘ̇. ܠܐ ܡܢ ܫܘܠܛܢܐ
ܕܐܣܘܬܐ ܐܬܬܚܝܗ̇܆ ܘܠܐ ܬܘܒ ܡܢ ܬܐܘܪܝܐ ܕܦܝܠܘܣܘܦܘܬܐ ܚܫܝܚܬܐ. ܐܠܐ
ܢܗܝܪܐ ܠܗ ܡܢ ܐܘܪܚܐ ܗܝ̇ ܕܙܕܩ ܒܗ. ܟܠܗܘܢ ܐܝܠܝܢ ܕܗܘܘ ܠܗܘܢ ܒܝܘܠܦܬܐ
ܥܠ ܡܘܙܠܬܐ ܘܥܠ ܢܡܘܣܐ ܕܒܗ̇܆ ܘܥܠ ܐܝܠܝܢ ܕܒܐܬܪܐ ܗܢܐ ܐܝܟ ܕܐܡܪܝܢ
ܡܬܗܦܟܝܢ̈. ܡܢ ܫܘܚܠܦܐ ܡܫ̈ܚܠܦܐ ܕܐܐܪ܆ ܘܡܢ ܙܘܥܐ ܡ̈ܫܚܠܦܐ ܕܟܠܗܘܢ
ܗ̇ܢܘܢ ܕܒܣܘܕܪܐ ܡܬܬܠܬܠܝܢ. ܘܒܕܡܘܢ ܐܝܠܝܢ ܠܐܝܢܐ ܕܐܝܟܢܐ ܠܗ ܠܡܕܒܪܘ܆
ܗܘ̇ ܡܢܐ ܕܥܠ ܝܩܘܬܐ ܚܕܐ ܡܫܬܡܠܝܬܐ܆ ܐܡܪܢܐ ܕܥܠ ܡܘܬܪܬܐ ܒܣܘܒܪܐ
ܐܝܟ ܕܐܡܪܢ ܓܠܝܢܘܣ ܒܡܐܡܪܐ ܗܘ̇ ܕܬܠܬܐ܆ ܕܒܗ̇ ܡܠܬܐ ܡܠܬܐ ܕܓܒܐ
ܥܠ ܟܠܗ̇ ܡܫܡܫܢܘܬܐ ܕܗܘܬ ܠܐܦܘܩܪܛܝܣ܆ ܥܠ ܬ̈ܠܬܐ ܟܠܗܘܢ ܐܝܠܝܢ
ܕܡܬܗܦܟܝܢ̈ ܒܐܬܪܐ ܗܢܐ ܐܢܫܐ. ܐܝܟ ܕܡܬܪܬܐ ܠܘܬ ܦܝܠܘܣܘܦܘܬܐ
ܕܐܪܣܛܘܛܠܝܣ܆

Text 8. Sergius of Rēšʿayna, Syriac version of Galen, *Commentary on Hippocrates's On Nutriment*, Prologue

Bos and Langermann, 'The Introduction of Sergius of Rēshʿainā to Galen's Commentary on Hippocrates' On Nutriment', 192,1a.

הדא כתאב אלגדא
לאבקראט שרח גאלינוס נקל יוסף אבן כאג(?)
אלמתרגם קאל סרכס נבדא בעון
אללה פנכתב מא פסרה גאלינוס
מן כתאב אבקראט פי אלגדא. קאל איהא אלאך
אלחביב אנך למא סאלתני וטלבת אלי מרארא
כתירה אן אתרגם לך בעץ כתב אליונאניין
אלי אלסריאניה וכאצה אלמקאלה אלדי וצעה
אבקראט פי אלגדא ודלך לחבך לגמיע מא
יחתאג אליה אלנפס מן אלעלם אעתנית
[ו]לא אחרמך מא סאלתני וטלבת אלי פיה
לאנה ממא יחתאג אליה ולם אכן אחב אן
אפעל שיא מן דלך למערפתי במא ילזמני
מן אלחסדה אלדין לא יגניהם אלא גמע אל
מאל מן אלעתב ואלמלאמה ודלך אנהם
לא יטלבון מנפעה טאלבי אלעלם אלדין ליסת
ענדהם אמואל תגניהם עלי [ען] דלך ואנה איצי
אטלב אליך אלא תגהל ען טלב נפעה עלם

Text 9. Sergius of Rēšʿayna, Syriac version of Galen, *On the Powers and Mixtures of Simple Drugs*, Prologue to Book 6

London, BL, Add. 14,661, ff.1v:1–2v:13; text quoted at Bhayro and Brock, 'Syriac Galen Palimpsest', 38; previous edition: Merx, 'Proben der syrischen Übersetzung', 244–248.

ܒܕܘܟ̈ܬܐ ܡܩܕ̈ܡܬܐ ܐܡܪܢ ܠܗ ܠܡܬܬܒܢܢܐ ܟܠܗܘܢ ܐܘ ܐܚܘܢ ܬܐܕܘܪܐ
ܕܡܠܦܢܘܬܐ ܬܪܝܨܬܐ ܗܕܐ ܟܠܗܘܢ ܡܫܡܥܢܐ ܠܡܗܘܐ ܐܝܟܐ ܕܡܗܕܪܐ
ܗܟܢܐ ܡܬ̈ܘܢܐ ܐܘܣ̈ܝܐ ܠܡܐܡܪ ܕܟܠܗܘܢ ܝܨܦ̈ܬܐ ܕܟܠܗܘܢ ܗܢܘ
ܠܡܠܦܢܐ ܘܡܗܝܢܐ ܐܝܬܝܗ ܢܦܩܐܝܬ ܥܠ ܕܘܪܫܐ ܕܒܐܝܕܐ ܒܐܝܕܐ: ܫܪܝܪܐܝܬ
ܟܕ ܠܐ ܐܘܡܢܘܬܐ ܕܡܬܘܡܐ ܡܫܡܥܢܐ ܕܬܟܪܘܝܗܝ، ܠܗܘ ܕܠܐ ܗܦܪ ܡܢ
ܕܘܪܫܐ ܕܒܝܫܘܬܐ: ܘܠܐ ܬܘܒ ܕܘܪܫܐ ܕܒܐܝܕܐ ܒܐܝܕܐ ܐܝܟ ܡܢܝܢ ܡܠܦܢ
ܠܗ ܠܬܠܡܝܕܐ ܟܠܗ ܡܕܡܬܐ ܕܗܘܝܢ ܕܟܘܐ: ܐܠܐ ܥܠ ܐܝܟܢܐ ܐܝܟܢܐ ܕܚܝܕܐ ܡܢ
ܗܠܝܢ ܟܠܗܘܢ ܡܬܚܫܒ ܕܬܚܝܢ ܩܕܝܡܢ ܐܬܬ ܠܝ ܐܡܪܬܗܘܢ، ܠܗܘ ܕܚܝܕܐ
ܢ̈ܟܠܗ ܚܝܒܐ ܠܡܗܠܟܘ ܐܘ ܒܚܝܕܐ ܐܝܢܐ ܠܡܦܠܚ ܐܘ ܗܝܟ ܒܚܕ ܡܢ ܗܕ̈ܡܐ
ܠܡܕܒܪܘ ܐܝܠܝܢ ܕܠܐ ܡܚܝܢ ܒܟܠܗ ܡܢ ܥܘܬܕܬܐ ܕܟܡ ܫܪܪܐ ܠܡܗܕܪ
ܡܕܒܪܢܘܬܐ ܐܝܕܐ ܗܝ، ܠܡܬܒܪܐܝܬ: ܐܝܢܘ ܗܟܝܠ ܕܟܠܬܗ ܐܢܫ ܠܡܗܘܐ
ܒܢܝܬܗܘܢ ܚܝܠܬܬܐ ܕܝܨܦ̈ܬܐ ܠܗܢܐ ܐܠܝܨܐ ܠܗ ܡܢ ܥܠ ܦܪܘܣ
ܕܗܘܐ ܚܕ ܡܬܘܡܐ ܕܟܘܐ ܗܢܘ ܕܡܬܬܚܡܝܢ ܐܘܣ̈ܝܐ ܕܟܠܗܘܢ ܡܬ̈ܘܬܗ
ܕܐܘܡܢܘܬܐ ܘܕܗܘܐ ܬܘܒ ܡܕܪܫ ܐܝܟܦܐܝܬ ܘܡܬܘܡ ܒܚܝܠܬܐ ܚܫܝܬܐ
ܕܗܠܝܢ ܕܫܟܝܢ ܒܟܠ ܚܕܐ ܡܢ ܡܬ̈ܘܬܗ ܕܐܘܣܝܬܐ: ܐܝܟ ܡܢ ܗܟܝܠ ܕܡܢܗ
ܡܠܦܢܐ ܠܐܘܪܚܐ ܗܕܐ ܚܝܠܬܐܝܬ ܒܟܠ ܕܘܟܐ ܥܠ ܡܬܘܡܐ ܗܢܘܢ ܕܟܘܐ
ܡܠܦ ܡܡܟܐܝܬ ܘܟܕܒ ܠܗܘܢ ܠܬܠܡܝ̈ܕܐ ܒܗ ܗܠܝܢ ܕܗܘܘܢ ܡܢܝܢ ܡܬܡܟܢ
ܕܐܘܡܢܘܬܐ: ܘܡܗܝܢܐ ܡܬܪܬܐ ܡܕܝܢ ܠܗܘܢ ܠܩܢܘܡܐ ܘܡܬܝܩܡ ܡܟܬܗܘܢ
ܠܬܪܝܨܘܬܐ ܕܗܠܝܢ ܕܒܐܝܕܐ ܒܐܝܕܐ ܠܦܘܡ ܐܝܟ ܕܠܚܕܐ ܐܝܠܝܢ ܐܦ ܒܗ
ܗܠܝܢ ܡ̈ܬܘܬܐ ܠܫܘܬܐ ܒܗܠܝܢ ܐܝܠܝܢ ܕܕܘܡ ܘܐܦ ܒܡܬܬܒܢܘܬܐ ܗܕܐ
ܕܐܝܬܝܗ ܥܠ ܚܝܠܐ ܕܡܬܬܡܢܐ ܦܪܝܫܐ ܐܘ ܐܚܘܢ ܒܗ ܒܡܠܦܢܘܬܐ ܗܕܐ
ܐܬܚܫܒ ܠܡܐܡܪܬ: ܒܡܠܗܘܢ ܟܕ ܗܘܝܢ ܚܡܝܢܐ ܡܡܠܐܬܐ ܕܘܪܡ ܗܢܐ ܡܬܘܡܐ
ܕܟܘܐ ܗܘܡ ܠܝ ܥܠ ܟܠܗܘܢ ܚܝܠܐ ܘܪ̈ܗܛܐ ܕܡܫܝܢܐ ܘܥܠ ܟܠܗ ܗܘܠܐ ܐܫܪܬܐ
ܕܐܣܘܬܐ ܕܡܢ ܐܬܡܟܐ ܘܐܬܡܟܐ ܡܬܪ̈ܟܐ ܡܢ ܐܪܒܟܐ ܐܣܛܘܟܣܐ ܒܗ
ܗܘ، ܢܗܝܪܐܝܬ ܕܐܝܢܐ ܚܝܠܐ ܐܘ ܐܝܢܐ ܘܐܝ ܐܘ ܐܝܢܐ ܠܡܟܡܐ ܢܦܩ ܠܥܠ
ܚܕ ܡܢ ܗܡ̈ܬܡܬܐ ܒܝܠܝܬ ܡܢ ܡܘܙܓܐ ܕܟܠܟܐ ܒܗ ܚܡܝܡܘܬܐ ܘܐܝܢܐ
ܬܘܒ ܡܢ ܗܘ ܕܗܟܝܐ ܒܗ ܡܪܝܪܘܬܐ ܘܐܝܢܐ ܬܘܒ ܡܢ ܙܟܘܬܐ ܘܡܢ
ܡܥܫܘܬܐ ܡܟܐ ܕܬܫܬܥܝ ܚܕܐ ܡܢ ܗܠܝܢ ܟܠܟܐ ܠܫܪܝܪܘܬܗ ܒܗ ܒܡܘܙܓܐ:
ܐܟܡܐ ܕܐܦ ܒܗܘܢ ܒܗܠܝܢ ܡܐܡܪ̈ܐ ܕܡܢ ܩܕܡ ܗܢܐ ܐܝܟ ܕܡܟܡܐ ܗܘܐ ܒܪܝܗ
ܕܡܬܬܒܢܘܬܐ ܐܦ ܠܗܘܢ ܕܪܝܫ ܕܠܗ ܡܢ ܐܪܒܟܐ ܐܣܛܘܟܣܐ ܡܡܙ̈ܓܝܢ
ܗܡ̈ܬܡܬܐ ܘܚܡܝܪܐ ܕܟܠܗܘܢ ܝܨ̈ܦܬܐ ܕܬܠܝܢ ܐܠܐ ܡܢ ܦܪ̈ܕܐ ܘܚܡܪ̈ܬܐ ܕܦܘܡܐ
ܒܐܪ ܕܟܠܗܘܢ ܕܗܠܝܢ: ܒܪܝܟܐ ܗܘ ܕܥܠ ܥܠܬܐ ܕܟܠܗ ܡܬܬܒܢܘܬܐ ܗܕܐ
ܐܡܪܢ ܠܝ ܒܗܡܝܢ̈ܬܐ ܠܗܬ ܢܣܒܬ ܡܠܦܢܐ ܕܐܝܟ ܒܗܠܝܢ ܗܟܝܠ ܫܡܝܢܐ
ܡܐܡܪ̈ܐ ܕܡܢ ܩܕܡ ܗܢܐ ܟܠܗ ܕܐܘܡܢܘܬܐ ܕܟܘܐ ܐܝܟ ܕܐܡܪܢ

ܕܣܘܟܠܗܘܢ ܘܡܦ̈ܪܫܘܬܗܘܢ ܕܦܝܠܘܣ̈ܘܦܐ ܐܠܦ ܡܬܬܠܡܕܢܐ: ܒܗܠܝܢ ܕܝܢ ܐܢܬܬܐ
ܐܣ̈ܝܐ ܕܐܝܬܝܗܘܢ܆ ܡܕܡܝܢܗܘܢ ܗܢܐ ܕܗܘܢܐ ܦܝܫ ܠܝ ܠܡܦܫܩܘ ܥܠ ܕܘܪܫܐ
ܣܘ̈ܕܪܐ ܗܘ ܕܡܬܒܥܐ ܠܗܘܢ ܠܥܒ̈ܕܐ ܒܗܝܢ ܒܝ̈ܕܥܬܐ ܕܒܐܝܕܐ ܐܝܕܐ
ܡܠܦ ܓܝܪ ܡܕܝܢܐ ܡܢ ܒܬܪܐ ܢܣܒܐܝܬ ܘܐܡܪ ܕܠܐܝܬ ܫܠܡܗ ܘܡܕܒܪܢܘܬܗ
ܕܟܠ ܚܕ ܚܕ ܡܢܗܘܢ ܓܝܪ ܦܫܛ ܠܗܘܢ ܒܥܡܠܐ ܕܗܘܝܐ ܒܦܠܚܘܬܐ ܕܐܬ̈ܘܬܐ
ܡܢ ܐܠܦܐ ܘܡܕܒܪܢܐ ܠܗ ܘܡܢ ܕܥܠܡ ܬܘܒ ܒܠܗ ܕܘܪܫܐ ܕܟܠ ܒܣܪܐ: ܗܘܝܢ
ܡܕܝܢܐ ܬܘܒ ܒܗ ܒܦܠܚܘܬܐ ܕܐܬ̈ܘܬܐ ܘܡܠܦ ܥܠ ܟܠܗ ܗܘܠܐ ܐܣܝܘܬܐ
ܕܡܬܢܣܒܐ ܠܐܘܣܝܘܬܐ ܡܢ ܒܣ̈ܦܐ ܘܡܢ ܠܗ ܐܪܒܐ ܘܡܢ ܒܣܪܐ ܘܡܢ
ܘܟܠܐ ܘܡܢ ܕܡܐ ܘܡܢ ܡܢܬܐ ܕܣܝܡܬܐ ܘܒܗܠܝܢ ܡܫܠܡ ܠܗ ܠܡܠܬܐ
ܕܫܬܬܩܢܘܗ̈ܝ ܗܠܝܢ ܡܐܡ̈ܪܐ ܐܣܝܐ: ܒܬܪܟܢ ܗܟܝܠ ܗܢܐ ܕܗܘܢܐ ܦܝܫ ܠܝ
ܠܡܦܫܩܘ ܡܢ ܡܡܠܠܐ ܢܦܫܢܝܐ ܠܠܥܠܐ ܦܘܪܫܢܐ ܕܐܝܬܝܗܘܢ ܫܬܝܬܝܐ ܡܢ
ܕܟܠܗ ܡܬܬܠܡܕܢܘܬܐ ܗܕܐ ܘܡܕܡܝܢܐ ܕܝܢ ܕܗܘܠܐ ܕܫܪܝܐ ܡܫܪܐ ܒܗ ܡܬܬܠܡܕܢܐ
ܒܕܘܪܫܐ ܐܝܟ ܕܐܡܪܢܢ ܓܝܪ ܡܫܟܚ ܒܗ ܒܠܚܘܕܘܗܝ ܒܣܪܐ ܕܫܠܡ ܝܫܪܐ ܕܒܝܫܘܬܗܘܢ
ܡܢ ܐܝܬܘܬܐ ܡܬܕܡܝܢܘܬܐ ܕܐܝܬܘܗܝ ܐܠܦܐ ܒܕܡܘܬܐ ܠܗܝ ܕܬܐܒ ܕܐܝܬܘܗܝ
ܘܦܠܐ: ܘܒܗܕܐ ܒܒܝܐ ܫܘܠܡܗ ܕܡܐܡܪܐ ܗܢܐ ܕܫܬܐ:

Text 10. Proba of Antioch, *Commentary on the De interpretatione*

Hoffmann, *De Hermeneuticis*, 64,1–10, 15–20.

ܢܦܩܐ ܗܟܝܠ ܐܝܬܘܗܝ܆ ܫܘܠܡܐ ܕܦܘܩ ܦܪ̈ܘܬܗܘܢ. ܕܐܝܟ ܐܡܝܪ. ܓܝܪ ܐܬܘ
ܒܫܢ̈ܝܐ ܠܗܘܬ ܗ̇ܝ ܕܒܥܠܘܗܝ ܒܬܪܟܢ. ܘܡܕܡܝܢܘܬ ܬܠܝܬܝܐ ܐܬܚܫܒܬ. ܡܛܠ ܕܝܢ ܕܠܐ
ܡܫܟܚ ܗܘ̇ܐ ܗܢܐ ܕܢܦܘܩ ܡܢܗ ܘܠܗ܆ ܗܘܝܢ ܐܦ̈ܝܐ ܐܬܚܫܒܬ. ܕܐܦ̈ܝܐ ܕܝܢ
ܗ̈ܘܝܢ ܠܗ ܡܕܕܒܢܝ̈ܬܐ. ܐܡܪܝ ܕܟܠ ܫܬܐܦ̈ܐ ܡܬܬܗܦ̈ܟܝܢ. ܡܛܠ ܗܢܐ
ܫܬܐܦ̈ܐ ܒܬܪܝܢ ܐܬܚܫܒܘ. ܘܗܢܐ ܐܝܬܘܗܝ ܫܘܠܡܗ ܕܚܘܫܒܐ. ܗ̇ܘ ܡܢ
ܕܐܝܬܘܗܝ ܫܘܪܝܐ ܕܗܦܘܟܘܬܐ. ܡܢ ܗܘܪܝܐ ܒܪ ܡܕܝܢܐ ܗܦܘܟܘܬܐ. ܡܢ
ܐܝܟܢܐ ܕܡܫܠܡ ܚܘܫܒܐ. ܘܗܘܪܝܐ ܡܫܠܡܢܐ ܗܦܘܟܘܬܐ. ܐܝܟܢܐ ܕܒܪ ܚܘܫܒܐ.
ܘܐܝܬ ܗܟܝܠ ܐܬܬܘܣܦ ܢܦܩܐ܆ ܫܘܠܡܐ ܕܦܘܩ ܦܪ̈ܘܬܗܘܢ [...] ܢܦܩܐ ܦ̈ܪܬܒܐ.
ܡܛܠ ܕܡܬܦܫܛܬܐ ܬܗܘܐ ܬܪܝܨܬܐ܆ ܠܗܘܬ ܗ̇ܘ ܡܢܐ ܕܒܚܝܢ ܕܢܦܩܘܪ. ܡܛܠ
ܕܗܘ ܠܗ ܠܗܘ ܕܠܡܟܒܪ ܡܕܡ ܡܚܪ܆ ܕܢܦܘܩ ܐܚ. ܡܢܐ ܟܕܐ ܠܡܟܒܪ. ܐܡܪ
ܟܢ ܦܠܦܠܐ. ܐܘ ܦܠܦܠܐ. ܚܕܐ ܗ̇ܝ ܦܪܝܫܬܐ ܕܗܠܝܢ ܕܡܬܦܫܛܝܢ. ܐܘ ܕܒܪܒܝܢ ܗ̇ܘ
ܡܕܡ ܕܡܠܠܬܗ ܒܬܠܝܢ ܠܦܘܪܩܢܐ. ܐܘ ܕܡܢ ܥܠ ܡܕܡ ܢܦܘܩܘܢ ܡܢ ܐܢܫ̈ܐ.

Text 11. Paul the Persian, *Treatise on Logic*, ?as translated by Severus Sebokht

Land, *Anecdota Syriaca*, IV, 2,22–3,7.

ܐܦܠܐ ܡܚܝܠ ܐܢܐ: ܕܠܚܝ ܢܐܫܐ ܘܠܐܣܝܪܐ ܢܦܩܘܢ. ܐܘ ܠܚܝ ܝܟܦܐ
ܘܠܐܣܝܪܐ ܢܦܠܐ. ܡܬܬܒܥܝܢܐ ܠܝ ܗܟܝܠ ܡܬܬܦܪܫܢܘܬܐ ܟܠܦܬܐ ܕܡܛܠܠܬܗ.
ܒܥܒܘ ܠܥܪܒܐ. ܘܠܚܝ ܢܦܩܝܢ. ܐܠܐ ܠܗܬ ܬܘܣܦܬܐ ܟܠܦܬܐ ܕܗܕܐ. ܡܕܡ

ܝܕܥܬܐ ܕܕܘܒܪ̈ܐ ܚܘܝܢ. ܠܘܬ ܗܝܡܢܘܬܐ ܘܠܘܬ ܝܕܥܬܐ ܡܬܕܒܪܢܐ.
ܝܕܥܬܐ ܓܝܪ ܥܠ ܡܕܡ ܕܩܪܝܒ ܘܓܠܐ ܘܝܕܝܥ ܗܘܐ. ܘܗܝܡܢܘܬܐ ܥܠ ܟܠ
ܝܕ̈ܥܬܐ ܕܪ̈ܚܝܩܢ ܘܠܐ ܡܬܚܙܝܢ ܘܠܐ ܡܬܒ̈ܥܝܢ ܒܫܘܬܦܘܬܐ ܗܘܝܐ. ܘܗܝ ܟܝ
ܒܦܘܠܓܐ. ܗܕܐ ܕܝܢ ܕܠܐ ܦܘܠܓܐ. ܥܠ ܦܘܠܓܐ ܡܚܝܒ ܗܕܡܐ ܡܓܠ. ܠܐ
ܦܘܠܓܐ ܕܝܢ ܥܒܘܕܬܐ. ܝܕܥܬܐ ܐܪܐ ܡܢ ܗܝܡܢܘܬܐ. ܘܠܗܕܐ ܐܪܥܐ ܠܡܟܬܒ.
ܬܘܒ ܡܢ ܗܘ. ܐܦ ܡܗܝ̈ܡܢܐ ܓܝܪ ܒܗ ܡܬܬܕܒܪܝܢ ܥܠ ܐܦ̈ܝ ܗܝܡܢܘܬܐ. ܡܢ
ܝܕܥܬܐ ܚܕܝܢ ܡܚܡ ܒܪܘܚܐ. ܒܗܝ ܕܐܡܪܝܢ ܕܒܗ̇ܘ ܡܕܡ ܕܗܫܐ ܡܗܝܡܢܝܢ.
ܟܬܪܝܢ ܠܗ ܢܝܚܝܢ. ܗܫܐ ܠܟ ܐܝܟ ܕܒܡܚܙܝܬܐ ܚܙܝܢ ܒܦܠܐܬܐ. ܗܝܕܝܢ ܓܝܪ
ܐܦ̈ܝܢ ܠܘܩܒܠ ܐܦ̈ܝܢ.

Text 12. Paul the Persian, *Explanation of the* De interpretatione, translated by Severus Sebokht

Hugonnard-Roche, 'Sur la lecture tardo-antique du Peri Hermeneias', 48–49.

ܒܪܢܫܐ ܡܢ ܗܠܝܢ ܕܠܐ ܡܠܬܐ ܒܡܠܠܘܬܐ ܡܚܘܐ ܘܡܠܦܐ ܕܠܐ ܝܕܥ ܘܡܚܝܠܐ
ܕܬܘܕܥ ܘܡ̈ܢܐ ܕܝܕܥ ܒܗ̇ ܡܠܠܘܬܐ ܝܕܥ ܠܗ. ܡܛܠ ܕܐܘܪܓܢܘܢ ܕܡܠܡܕܢ
ܠܝܕܥ ܡܠܠܘܬܐ ܐܝܬܝܗ̇ ܘܡܠܦܢܘܬܐ ܡܕܡܝܐ ܠܡܠܠܘܬܐ ܗܢܘ ܠܝܕܥ.

Text 13. Barḥadbešabba, *The Cause of the Foundation of the Schools*

Scher, *Mār Barḥadbšabba 'Arabaya: Cause de la fondation des écoles*, 342,12–345,3.

Scher's text is based on a seventeenth-century MS, Seert 109 (C), with variants noted from sixteenth-century MS, Seert 82 (T). Both were destroyed in the genocide of 1915. Most other MSS seem to be copies of these, hence Scher's published text of 1907 remains the key witness to this important treatise.[137] I have noted only a selection of T's variants.

ܡܛܠ ܓܝܪ ܕܟܠܗܘܢ ܐܝܠܝܢ ܕܒܡܠܘܐܐ ܬܪܝܨܐܝܬ ܡܫܩܝܢ ܠܬܪܝܢ ܐܝܬ ܡܬܦܠܓܝܢ:
ܠܝܕܥܬܐ ܘܠܡܣܒܪܢܘܬܐ: ܙܕܩ ܠܡܐܡܪ ܕܐܣܟܘܠܐ ܕܝܕܥܬܐ ܐܝܬܘܗܝ
ܡܕܪܫܘܬܐ ܫܬܐܬܐ ܕܝܕܥܬܗܘܢ ܕܟܠܗܘܢ ܗܢܘܢ. ܕܣܟܘܠܘܬܐ ܕܝܢ
ܐܝܬܘܗܝ ܫܘܡܠܝܗ̇ ܡܣܒܪܢܘܬܐ ܕܛܠ̈ܬܐ܀

ܡܛܠ ܗܢܐ ܕܐܝܬ ܫܘܫܠܡܘܬܐ ܕܢܦܫܐ ܠܚܢܟܐ ܡܢ ܗܠܝܢ: ܐܝܟ ܓܘܢܐ[138]
ܠܟܬܒܐ: ܘܟܪ̈ܝܐ ܠܐܝܬܘܬܐ: ܗܘ ܕܝܢ ܠܣܘܟܠܗ̇ ܕܣܘܟܠܬܢܘܬܐ[139]

137 For full details, see Becker, *Sources for the History of the School of Nisibis*, 65–71.
138 ܛܠܠܐ T (shadow)
139 ܕܡܝܬܪܘܬܐ ܗܘ ܕܝܢ ܕܣܘܟܠܬܢܘܬܐ T (of excellence, that is, of intelligence)

ܘܡܫܒܪܢܘܬܐ: ܥܠ ܗܕܐ ܐܬܟܬܒܬ ܡܠܝܠܘܬܐ ܡܢ̈ܝܚܘܬܐ ܕܬܘܪܨܐ ܠܝ
ܗܕܐ ܡܫܡܠܝܘܬܐ ܡܢ ܥܡܠܝܐ ܥܙܝܙܐ: ܕܥܠ ܚܕܐ ܡܢ ܡܢܘ̈ܬܗ ܕܠܦܬܐ:
ܐܝܟܢܐ ܥܡܠܝܗ̇ ܕܡܕܒܪܢܘܬܐ ܐܝܬܘܗܝ، ܝܕܥܬܐ ܫܪܝܪܬܐ ܕܟܠܗܘܢ
ܐܝܠܝܢ ܕܐܝܬܝܗܘܢ: ܝܕܥܐ ܗܝ ܕܡܫܡܠܝܘܬܗ ܠܐ ܝܕܥܬܐ ܐܝܬܝܗ̇. ܘܡܛܠ
ܗܕܐ ܡܫܡܫܝܢ ܥܠ ܡܠܝܠܘܬܐ: ܕܟܕ ܦܪܫܝܢ ܥܙܝܙܐ ܡܢ ܕܟܠܗܬܐ. ܗ̇ܘ
ܓܒܪ ܕܡܟܠܟܠܐ ܕܐܝܬܘܗܝ ܥܙܝܙܐ: ܠܗ ܠܗܢܐ ܐܫܝܢ ܒܬܘܕܝܬܐ ܫܠܝܡܬܐ
ܕܐܝܬܘܗ̇ ܝܕܥܬܗܘܢ ܕܢܓ̈ܘܬܐ. ܘܗ̇ܘ ܬܘܒ ܡܢܐ ܕܡܬܚܫܒܐ ܟܕ ܬܘܩܦܬܐ
ܥܙܝܙܬܐ ܕܐܝܬܘܗܝ ܟܠܝܘܬܐ: ܠܗܢܐ ܥܒܝܕܝܢ ܠܒܪ ܡܢ ܟܠܗ ܒܡܗܕܝܢܐ ܕܥܙܝܙܐ
ܝܕܥܐ ܗܝ ܕܒܠܚܘܕ ܡܠܝܠܘܬܐ ܠܐ ܡܬܦܪܫܐ ܕܐܝܬܘܗܝ ܘܡܬܝܕܥܐ ܠܐܝܠܝܢ
ܕܐܢܫܐܝܬ ܕܝܠܝܢ ܠܗܘܢ ܠܡܦܩܘܬܐ: ܗ̇ܘ ܓܒܪ ܕܪܘܚܢܐ ܐܠܗܝܐ ܠܐ ܡܟܠܠ: ܥܠ
ܬܫܘ̈ܝܬܐ ܟ̈ܠܝܠܬܐ ܡܫܡܫܐ ܡܠܦܢܘܬܗ ܕܬܬܗܘܝܢ ܠܢܦ̈ܫܬܐ܀

(344,8) [...] ܒܗܢܐ ܡܫܝܠ ܐܘܟܝܢ ܬܡܝܣܐ ܕܡܠܝܠܘܬܐ. ܐܝܟ ܕܐܡܪ ܡܕܒܪܢܐ
ܠܟܠܗܘܢ ܝܘܠܦ̈ܢܐ ܗܕܝܪܐ ܕܡܕܒܪܢܘܬܐ ܫܪܝܪܬܐ. ܘܒܗ ܢܣܒ ܐܪܝܣܛܘܛܠܝܣ ܥܒܝܫܐ
ܕܗ̇ܘ ܬܦܣܢܐ.[140] ܕܠܐ ܡܫܝܠ ܗܘܐ ܠܗ ܗ̇ܝ ܕܡܕܒܪܢܘܬܗ ܘܡܠܝܠܘܬܗ
ܕܗܢܐ ܡܕܒܪܢܐ ܒܡܠܝܠܐ ܘܕܠܐ ܫܢܩܐ: ܒܗ̇ ܕܠܝܬ ܗܘܐ ܠܗ ܐܠܐ ܒܝܬ ܕܒܗ̇
ܢܒܒ ܥܒ̈ܗܝܐ ܘܢܗܘܐ ܘܢܣܟܠ ܡܠܦܢܐ ܕܥܠ ܗܝ ܐܝܬܘܬܐ: [...] ܐܠܝܨܐܝܬ
ܐܝܟ ܕܠܗܦܟܘܗ ܘܠܐܬܐ ܕܢܫܪܘܬܗ ܐܬܘܬܗ ܒܚܘܕܪܐ ܠܗܕܐ ܠܡܫܥܝܘܬܐ.
ܘܦܬܟܗ ܒܫܬܠܐ ܘܒܦܘ̈ܢܐ: ܘܦܠܓܗ ܒܦܢܝ̈ܡܐ ܘܐܕ̈ܫܐ: ܘܦܪܫܗ ܒܐܣܟ̈ܡܐ[141]
ܘܒܡܚܒ̈ܪܢܘܬܐ: ܘܐܘܩܡܗ ܕܬܠܝܬܐ ܡܫܪܝܬܐ: ܘܐܟܠ ܫܡܗ ܒܗܢܐ ܟܘܢܐ
ܪܘܫܡܐ ܕܒܝܬ ܥܡܘܕܐ ܠܐܪܒܥܐ. ܘܐܝܟ ܕܥܠ ܠܘܚܐ ܡܕܡ ܟܬܒ ܘܪܒܒ ܠܗ
ܠܟܠܗܘܢ ܠܒܢ̈ܝܢܫܐ ܡܬܚܫ̈ܒܢܐ ܕܟܠܗܘܢ ܢܦܪܐ: ܘܡܚܘܝܢܘܢ ܢܕܥ ܠܗ̇ ܘܗ̇ܘ ܕܗܘܐ ܠܗ
ܟܠܗ̇ ܕܗܢܐ ܡܠܦܢܐ:

Text 14. Severus Sebokht, *On the Syllogisms of Aristotle*

I—London, BL, Add. 14660, f.47v.
II—London, BL, Add. 14660, ff.53v–54r (A); Add. 17156, f.5v (B).

I

ܢܫܝܐ ܐܝܬ ܠܝ ܒܡܐܡܪܐ ܗܢܐ ܕܠܫܘܐ ܐܝܟ ܕܒܝܬܘܢ̈ܬܐ ܐܘ ܒܝܬ ܢܙܒܘܡ.
ܡܛܠ ܙܢ̈ܝܐ ܕܣܘܠܘܓܝܣܡܘ ܡܠܝܓܘܪ̈ܝܩܐ: ܗܠܝܢ ܕܒܟܬܒܐ ܕܐܢܠܘܛܝ̈ܩܐ
ܡܕ̈ܒܪܢܐ ܕܐܪܣܛܛܠܝܣ ܦܝܠܣܘܦܐ. ܐܡܪ̈ ܐܢܐ ܕܝܢ ܡܛܠ ܪܘܟܒܐ ܘܫܪܟܐ
ܕܠܝܬܗܘܢ܆ ܐܟܣܢܐ ܕܝܢ ܘܪܟܒܐ ܘܐܝܠܝܢ ܐܝܬܝܗܘܢ ܬܘܒ ܕܝܢ ܘܪܟܒܐ
ܘܐܝܠܝܢ ܐܝܟܢ ܐܣܟ̈ܡܐ ܕܠܝܬܗܘܢ܆ ܐܝܟ ܐܣܟܡܘܬܐ ܡܠܝܠܬܐ
ܘܣܘܠܘܓܝܣܛܝܩܘܬܐ ܕܗܘ. ܢܫܝܐ ܠܝ ܒܪ ܗܕܐ܆ ܘܡܛܠ ܡܬܘܪܐ ܠܘܬ

140 ܬܦܣܢܐ ܣܡܘܟܐ T (sustaining prototype)
141 ܒܐܣܟܡܐ ܘܒܕܡܘܬܐ T (by a figure and a likeness)

ܝܕܥܬܐ ܡܫܡܠܝܬܐ ܕܬܐܘܪ̈ܝܐ ܡܠܠܬܐ ܘܡܫܘܚܬܐ ܕܗܠܝܢ ܕܟܘܟܒܐ
ܕܐܣܛܪܘܢܘܡܝܐ ܐܡܝ̈ܢܐ܆ ܗܘ ܕܒܗ ܫܬܐܬܐܝܬ ܡܬܚܙܝܐ ܘܡܬܦܪ̈ܫܐ ܟܠ܇
ܐܘܡܢܘܬܐ ܡܠܠܬܐ ܫܪܪܐ ܡܢ ܕܓܠܝܘܬܐ. ܐܦ ܡܟܝܠ ܗܝ ܕܝܕܥ ܫܪܪܐ
ܘܢܛܪ ܡܢ ܕܓܠܝܘܬܐ ܡܬܪܥܢܝܬܐ ܐܝܬܝܗ̇܇ ܗܕܐ ܕܝܢ ܟܕ ܣܘܠܘܓܝܣܡܘܣ
ܡܚܘܝܐ ܠܡܕܒܪ܆

II

ܙܕܩ ܠܗ ܠܝܠܘܦܐ ܠܡܕܒܪ ܡܕܡܐܝܬ[142] ܐܦ ܗܕܐ ܕܟܬܒܐ ܗܢܐ
ܕܐܣܛܪܘܠܘܓܝܐ܆ ܠܘ ܡܛܠ ܦܬܗ̇[143] ܐܬܘܗܝ. ܐܠܐ ܐܡܝܢܐ ܕܟܬܒܐ
ܕܦܝܬܓܘܪܘܣ ܗܘ ܕܐܦ ܟܠ ܟܘܢ̈ܢܐ[144] ܦܬܓ̈ܡܐ ܟܬܒ. ܠܘܬ ܗܘ
ܕܦܪ̈ܣܝܐ ܙܡܐܢܘܣ. ܗܘ ܕܐܦ ܟܠ ܙܘܕܝܩܘܣ ܡܕܡܝܐ. ܕܒܢ̈ܝܢܐ ܦܬܓ̈ܡܐ
ܠܘܬ ܟܬܒܐ ܗܢܐ ܕܐܣܛܪܘܠܘܓܝܐ ܚܫܚ ܠܝ܆ ܗܟܢܐ ܐܦ ܗܘ ܟܬܒܐ
ܗܢܐ ܕܐܣܛܪܘܠܘܓܝܐ ܗܘ ܕܐܦ ܕܟܠ[145] ܙܘܕܝܐ ܐܡܫܪܐ ܘܫܪܪܐ ܕܡܫܪܪ
ܕܣܘܠܘܓܝܣܡܘܣ ܦܝܬܓܘܪܝܘܢܐ ܟܬܒ ܠܝ܆ ܠܘܬ ܫܥܬܐ ܕܦܪܓܡܛܝܐ
ܡܠܠܬܐ ܕܟܬܒܐ ܐܣܛܪܘܢܘܡܝܐ܆ ܡܫܡܗ ܠܝ܆ ܗܘ ܕܐܬܘܗܝ ܢܝܫܐ ܘܫܘܡܠܝܐ
ܕܟܠܗ ܐܘܡܢܘܬܐ ܡܠܠܬܐ܆ ܗ̇ܝ ܕܐܬܝܗ̇ ܐܘܪܓܢܘܢ ܕܟܠܗ̇ ܦܝܠܘܣܘܦܘܬܐ܆
ܗ̇ܝ ܕܐܝܟ ܫܘܪܪܬ ܡܠܠܬܐ ܦܠܝܓܘܬܐ܆ ܐܘ ܚܢܝܬ ܬܫܡܫܬܐ ܕܡܫܘܚܬܐ
ܐܝܬܝܗ̇ ܕܟܠܗܘܢ ܐܝܟ ܕܡܚܘܝܐ[146] ܠܒܪܢܫܐ ❖

Text 15. ?Severus Sebokht, Letter to ?Basil

Paris, BNF, syr. 346, ff.168v–170v. A previous edition (Reich, *Ein Brief des Severus Sebokt*) contains numerous typographical errors. The following has been re-transcribed from the manuscript.

ܡܛܠ ܕܝܢ ܗ̇ܝ ܕܐܡܪܝܢ ܐܢܫܝܢ ܡܢ ܝܘܢܝܐ ܠܘܬܟܘܢ ܐܡܝܢܐ ܕܐܢܫܘܬܟ ܒܬܒܬ:
ܕܠܐ ܠܡ ܡܚܘܝܐ ܦܝܣ ܕܣܘܪܝܝܐ ܝܕܥܝܢ ܡܕܡ ܕܐܝܟ ܗܘܝܢ: ܐܡܪܬ ܐܢܐ
ܕܝܢ ܫܘܒܚܐ ܕܩܘܒܒܠܐ ܕܦܝܠܘܣܘܦܘ ܕܝܘܢܝܐ ܘܕܦܪܣܝܐ: ܟܕ ܡܫܡܫܝܢ ܕܕܝܘܢܝܐ
ܠܣܘܪܝܐ ܐܝܬܝܗ̇ ܟܠܗ̇ ܝܕܥܬܐ ܡܛܠ ܗ̇ܝ ܒܐܡܪܬ ܕܡܠܟܝܢ ܝܘܢܝܐ: ܙܕܩ ܗܘܐ
ܠܗܘܢ ܡܕܡܝܬ ܠܗܘܢ ܟܕ ܫܡܝܥܐ ܬ̈ܒܠܝܐ ܐܝܬܝܗܘܢ ܠܡܕܒܪ܆ ܕܕܝܕܥܬܐ
ܗܕܐ ܬ̈ܒܠܝܐ ܡܕܡܝܬ ܗܘܘ ܡܫܡܫܝܢܗܘܢ ܘܠܐ ܝܘܢܝܐ܆ ܐܝܟ ܕܠܗܘܢ ܬܡܬܟܢܐ
ܕܠܗܘܢ ܕܝܘܢܝܐ ܗܘܕܝܢ. ܡܢ ܟܠܗ̇ ܕܝܢ ܬ̈ܒܠܝܐ ܐܦܘܦܠܝܐ ܘܗܝܕܝܢ ܝܘܢܝܐ.
ܕܬ̈ܒܠܝܐ ܕܝܢ ܗܘܪܝܢܐ ܐܝܬܝܗܘܢ: ܠܐ ܦܟܪ ܕܐܢܫ ܡܬܚܙܐ. ܡܕܡ ܗ̇ܢܘܢ
ܕܐܡܪܝܢ ܕܠܐ ܡܚܘܝܐ ܦܝܣ ܕܣܘܪܝܝܐ ܝܕܥܝܢ ܡܕܡ ܕܐܝܟ ܗܘܝܢ: ܟܕ ܗܢܘܢ

142 ܡܕܡܝܬ B
143 ܝܬܗ̇ B
144 ܕܟܠ ܟܘܢܢܐ B
145 ܗ̇ܘ ܗܘ ܕܟܠ B
146 ܡܚܘܐ A.

ܣܘܪ̈ܝܝܐ ܗܘܘ ܡ̈ܫܡܫܢܐ ܘܡ̈ܠܦܢܐ ܩ̈ܕܡܝܐ ܕܗܠܝܢ ܦܟܪ ܙܒܢܝܢ. ܘܗܕܐ ܗܟܢܐ ܐܫܬܟܚ: ܦܘܩܕ ܐܝܟ ܚܕܐ: ܐܦ ܗܘ ܦܝܠܘܠܐܘܣ ܟܣܢܘܩܪܛܝܣ. ܗܢܐ ܓܝܪ ܒܗ ܟܬܒܗ ܬܘܒ ܫܘܪܝܐ ܕܚܘܫܒܢܐ ܕܐܣܛܪܘܢܘܡܝܐ ܘܕܗܢܕܣܐ ܘܕܗ̄ܘ ܒܩ̈ܘܒܒܐ ܓ̈ܒܝܐ: ܠܘ ܡܢ ܝ̈ܘܢܝܐ ܕܡ̈ܠܦܢܐ ܡܩܒܠܐ ܗ̇ܢܘ. ܐܠܐ ܡܢ ܗܠܝܢ ܕܡ̈ܠܦܢܐ ܟ̈ܠܕܝܐ. ܐܡܪ ܐܢܐ ܕܝܢ ܡܢ ܠܡܘܡܕܢܝܢ ܡܠܦܐ ܕܐܬܘܪ̈ܝܐ. ܠܡܘܡܕܢܝܢ ܕܝܢ ܐܡܪ ܐܢܐ: ܠܘ ܗ̇ܘ ܕܒܩܕܡܝܬܗ ܐܝܬܘܗ̄ ܗܘܐ ܕܢܝܐܝܠ ܢܒܝܐ. ܐܠܐ ܐܚܪܢܐ ܕܐܝܬܝ ܡܕܡ ܡܢ ܗܢܐ. ܦܫܩ ܡܚܝܠ ܗ̇ܘ ܦܝܠܘܠܐܘܣ ܟܣܢܘܩܪܛܝܣ ܕܗ̇ܘܝܢ ܝ̈ܘܢܝܐ ܕܡܢ ܠܡܘܡܕܢܝܢ ܗܢܐ ܡܕܡܝܐ: ܕܡ̈ܠܦܢܐ ܚܠܦ ܕܝ̈ܘܢܝܐ ܘܕܦܪ̈ܣܝܐ: ܒܕܡܝܐ ܠܦܝܠܝܦܘܣ ܡܩܕܘܢܝܐ ܗ̇ܘ ܕܐܒܘܗ̄ ܐܠܟܣܢܕܪܘܣ ܒܢܝܐ ܕܐܠܟܣܢܕܪܝܐ. ܐ̄ܠܦܐ ܫ̈ܢܝܢ. ܒܪ ܙܒܢ ܒܢ ܗܕܐ: ܕܐܦ ܫܘܪܝܐ ܘܫܬܐܣܬܐ ܕܝܘܠܦܢܐ ܗ̇ܠܝܢ ܕܟܠܗ ܡܢ ܟ̈ܠܕܝܐ ܐܫܟܚ: ܘܠܘ ܡܢ ܝ̈ܘܢܝܐ. ܘܟܕ ܚܙܘ ܟܠ ܫܬܐܣܬܐ ܗ̇ܝ ܕܐܫܟܚܘ ܡܢ ܗ̇ܠܝܢ. ܟܠܗ̇ ܐܘܣܦ ܘܪܒܝ ܚܘ̈ܫܒܢܐ ܗ̇ܠܝܢ ܕܟܬܒܐ.܀

ܕܟ̈ܠܕܝܐ ܕܝܢ ܕܐܝܬܝܗܘܢ ܣܘܪ̈ܝܝܐ: ܕܐܝܬܝ ܢ̇ܣܒܝܢ ܡܢ ܝ̈ܘܢܝܐ ܡܫܬܘܕܝܢ. ܢܐܠܦܘܢ ܛܘܒ ܘܗܕܐ ܡ̈ܠܝܗܘܢ ܕܝ̈ܘܢܝܐ. ܒܪ ܓܝܪ ܡܢ ܡܬܚܫܠ ܗܘܐ ܐܝܟ ܠܡܣܝܡܐ ܕܠܘܬܗܘܢ ܐܝܟ ܡܢܐ ܕܗܘܢ̄ ܐܡܪܝܢ: ܕܐܡܠܝܢ ܐ̈ܢܫܝܢ ܒܪܫܐ ܡܚܬܘܡܐ ܘܗܘܘ ܕܫܠܝ ܠܐܠܗܐ: ܗܟܢܐ ܐܡܪܝܢ ܕܐܡܪ ܡܪܝܐ: ܕܠܟ̈ܠܕܝܐ ܗܘ ܒܠܚܘܕ ܡܠܠܬ ܚܟܡܬܐ ܘܬܘܒ ܠܒܪ̈ܝܐ. ܗ̇ܢܘܢ ܕܠܡܠܟܐ ܕܗ̇ܘܐ ܡܢ ܢܦܫ ܦܓܪܗ ܕܐܝܬܘܗܝ ܐܠܗܐ: ܟ̈ܠܕܝܐ ܕܝܢ ܐܝܟ ܕܡܕܡܝܐ ܢܕܡܐ ܟ̈ܠܕܝܐ ܐܢܘܢ. ܟ̈ܠܕܝܐ ܕܝܢ ܣܘܪ̈ܝܝܐ. ܡܕܝܢ ܘܐܦ ܐܝܟ ܣܗܕܘܬܗ ܕܡܪܝܐ ܗ̇ܘ ܕܠܘܬܗܘܢ: ܠܣܘܪ̈ܝܝܐ ܗܘ ܒܠܚܘܕ ܡܠܠܬ ܚܟܡܬܐ:܀

ܡܛܠ ܕܝܢ ܡܫܡܫܢܘܬ ܡܠܦܢܐ ܐܘܚܕܬ ܠܐ ܢܠܒܫܘܬܐ ܕܝ̈ܘܢܝܐ ܢܥܒܕܘܢ ܬܘܒ ܡܢܗܘܢ ܕܝ̈ܘܢܝܐ. ܐܡܪ ܐܢܐ ܕܝܢ ܡܢ ܦܠܛܘܢ ܦܝܠܣܘܦܐ ܡܕܡܝܐ ܘܡܫܡܫܢܐ ܕܠܘܬܗܘܢ: ܕܡܢܐ ܡܟܬܒ ܒܛܝܡܐܘܣ ܡܬܐܡܪܐ:܀ ܟܕ ܦܢܐ ܗܘܐ ܠܗ ܡܢ ܡܨܪܝܢ ܦܘܠܘܢ ܢܒܝܫ ܣ̈ܒܝܬܐ: ܗܠܝܢ ܐܡܪ ܠܡܪܝܛܐܘܣ.[147] ܐܝܠܝܢ ܕܗܘܘ ܫܒܝܒ ܡܢ ܟܗܢܐ ܡܨܪ̈ܝܐ: ܐܝܢܐ ܕܡܛܠ ܚܟܝܡ ܗܘܐ. ܐܡܪ ܓܝܪ ܠܘܬܗ: ܐܘ̄ ܣܘܠܘܢ ܣܘܠܘܢ: ܐܢܬܘܢ ܝ̈ܘܢܝܐ: ܛܠܝ̈ܐ ܐܢܬܘܢ ܒܟܠ ܙܒܢ. ܣܒܐ ܕܝܢ ܝ̈ܘܢܝܐ ܠܝܬ: ܒܬܪܝܢ ܛܘܒ ܐܡܪ ܕܐܢܬܘܢ ܝ̈ܘܢܝܐ ܫܒܪ̈ܐ ܐܢܬܘܢ ܒܢܦ̈ܫܬܟܘܢ ܟܠܟܘܢ. ܐܦܠܐ ܓܝܪ ܚܕܐ ܬܫܒܘܚܬܐ ܥܬܝܩܬܐ ܐܝܬ ܠܟܘܢ ܒܗܘܢ: ܐܦ ܠܐ ܝܘܠܦܢܐ ܕܥܬܝܩ ܒܙ̈ܒܢܐ. ܐܠܐ ܠܟܘܢ ܠܟܠܟܘܢ ܣܦܪܐ: ܒܫܪ̈ܒܬܐ ܩ̈ܠܝܠܬܐ: ܒܪ ܡܫܬܥܝܬܘܢ ܕܠܐ ܡܠܐ܀

ܡܕܝܢ ܡܫܬܘܕܥܐ ܗܕܐ: ܐܦܠܐ ܗܘ ܣܦܪܐ ܢ̇ܣܒܝܢ ܗܘܘ ܕܕ̈ܪܐ ܡܫܝ̈ܚܝܐ ܝ̈ܘܢܝܐ: ܐܠܐ ܡܫܬܘܗ ܒܠܚܘܕ ܕܠܐ ܡܠܐ. ܗ̇ܘ ܕܝܢ ܫܪܝܪܘܬ ܘܐܝܟ ܠܐ ܬ̈ܠܝܠܐ: ܘܐܡܠܝܢܐ ܡܫܡܫܢܐ ܡܕ̈ܡܝܐ ܕܡܨܪܝܬܐ ܗ̇ܝ ܡܬܐܡܪܐ ܦܝܠܘܣܘܦܘܬܐ ܕܐܣܛܪܘܢܘܡܝܐ ܡܬܐܡܪܝܢ ܐ̈ܢܫܝܢ ܡܢܗܘܢ ܐܢܘܢ. ܠܐ ܐܝܬܝܗ̇ ܗܕܐ ܠܐ

147 ܠܐܪܝܛܐܘܣ P

ܐܝܬܝܗ̇. ܠܘ ܓܝܪ ܠܚܡܫܬܗܘܢ ܐܘܟܝܬ ܕܒܪܬ ܡܠܐ ܐܝܬܝܗ̇ ܝܚܝܕܝܬܐ. ܐܠܐ ܕܝܚܝܕܝܬܐ
ܐܝܬܝܗ̇ ܗ̇ܝ ܕܒܪܬ ܡܠܐ. ܘܠܘ ܠܚܡܫܬܗܘܢ ܐܝܬܝܗ̇ ܟܠܬܐ ܕܫܡܝܢܝܬܐ: ܐܠܐ ܗܕܐ
ܕܗ̇ܝ. ܐܦ ܓܝܪ ܡܘܬܐ ܫܡܝܢܝܐ. ܗ̇ܠܝܢ ܕܫܡܝܢܝܐܝܬ ܬܫܡܫܬܗ ܠܦܝܠܘܣܘܦܘܬܐ. ܠܘ
ܗܕܐ ܬܫܡܫܬܐ ܕܐܝܬܝܗ̇ ܦܝܠܘܣܘܦܝܐ: ܥ̈ܠܡܐ ܘܡ̈ܠܟܐ: ܠܚ̈ܡܫܬܗܘܢ ܐܘܟܝܬ[148] ܘܩ̈ܢܝܬ
ܩ̈ܠܐ ܘܠܚܝܐ ܢܦܫܐ. ܐܠܐ ܕܦܝܠܘܣܘܦܝܐ ܐܝܬܝܗ̇. ܝܚܝܕܝܬܐ ܕܗܠܝܢ ܕܐܝܬܝܗܘܢ
ܒܗ̇ܝ ܕܐܝܬܝܗܘܢ: ܘܝܚܝܕܝܬܐ ܕܗܘܦܪ̈ܝܐ ܐܠܗ̈ܝܐ ܘܐܢܫ̈ܝܐ: ܘܐܘܡܢܘܬܐ
ܕܐܘܡܢܘ̈ܬܐ ܘܝܚܝܕܝܬܐ ܕܝ̈ܚܝܕܝܬܐ: ܘܗܘ ܓܝܪ ܕܡܢܘܬܐ: ܘܕܘܡܢܝܐ
ܕܐܠܗܘܬܐ[149] ܒܡܢܐ ܕܡܬܝܕܥܐ ܠܒܪ̈ܢܫܐ: ܘܪܘܚܢܝܬܐ ܕܫܡܝܢܝܬܐ:[150] ܗܕܐ ܕܝܢ ܠܘ
ܕܒ̈ܢܝܢܫܐ ܒܠܚܘܕ: ܐܠܐ ܘܕܟܠ ܫܡܝܢ̈ܝܐ ܡܬܝܕܥܐ ܕܐܘܦ̈ܝܐ. ܘܐܝܢ ܡܘܬܐ ܒܝܬ ܐܝܬܘ̄ܗ̇
ܗܢܐ ܘܐܝܢ ܒܪܒܝܬܐ.

ܥܬܝܕ ܐܢܐ ܕܝܢ ܗܫܐ ܠܡܐܡܪ: ܡܛܠ ܝܚܝܕܝܬܐ ܕܗܕ̈ܡܘܬܐ. ܗ̇ܠܝܢ ܕܐܦܠܐ
ܗܕ̈ܡܐ ܐܝܬܝܗܘܢ. ܘܥܠ ܥܒܝܕܬܐ ܦ̈ܠܝܛܬܐ ܕܠܗܘܢ ܕܒܡܠܦܢܐ ܗܘܐ
ܕܐܣܛܪܘܢܘܡܝܐ: ܗ̇ܠܝܢ ܕܐܬܡܪ ܐܘܚ̈ܕܢܝܢ ܡܢ ܕܗܘܝܐ ܒܝܬ ܘܡܢ ܕܒܟ̈ܠܝܐ:
ܘܡܬܘܕܘ ܬܠܝܠܬܐ ܕܢܘ̈ܥܒܝܫܘܬܗܘܢ. ܘܢܘܥܒܘܬܐ ܕܟܬܒܐ ܠܚܠܬܐ: ܐܡܪ ܐܢܐ
ܕܝܢ ܗ̇ܝ ܕܡܬ̈ܩܪܒܐ ܐܬ̈ܘܬܐ. ܗܠܝܢ ܕܐܝܠܘ ܐܪܒܥܐ ܒܗܘܢ ܗ̇ܠܝܢ ܕܡܬܩܪܝܢ
ܒܫܡܗܘܢ ܕܗܘܝܢ ܒܠܣܘܕ̈ܝܗܘܢ ܡܛܠ ܠܫܪܪܗ̇ ܕܫܡܝܢܝܬܐ. ܡܛܠ ܗ̇ܝ ܒܠܣܘܕ
ܕܡܠܟܝܢ ܗܘܕܥܢܘܬܐ. ܡܬܚܫܚܝܢ ܗܘܘ ܒܒܪ ܐܦ ܐܢ ܡܥܘܪܢܘܬܐ: ܕܐܝܬ ܗܘ̣
ܐܦ ܐܫ̈ܬܐ ܕܢܕܒܝܢ ܡܕܡ. ܠܘ ܡܘܬܐ ܒܠܚܘܕ: ܐܠܐ ܐܦ ܐܢ̈ܫܝܢ ܡܢ ܒܟ̈ܡܟܐ
ܐܣܝ̈ܘ ܠܒܢܐ ❖

ܘܗܠܝܢ ܐܡܪܬ: ܠܘ ܒܪ ܡܫܠܐ ܐܢܐ ܡܫܡܫܬܐ ܕܡܘܬܐ ܕܒܪܐܫܝܬ ܗܠܝܢ
ܘܒܐܣܛܘܟܝܬܐ ܕܢܕܥܝܢ ܠܗܘܢ ܠܐ ܬܗܘܦ. ܠܘ ܓܝܪ ܒܠܚܘ ܗܘܐ ܝܠܦܐ ܐܢܐ. ܐܠܐ
ܒܪ ܚܒܪܐ ܐܢܐ ܕܐܝܟܘܐ ܕܪ̈ܟܘܐ ܐܝܬܝܗ̇ ܝܚܝܕܝܬܐ: ܘܕܠܐ ܡܢ ܕܢ̈ܝܫܐ ܕܗܘܦܐ
ܫܡܝܢܐ: ܐܝܢ ܡܘܬܐ ܒܝܬ ܢܫܘܦܐ ܐܝܬܘ̄ܗ̇ ܗܢܐ ܘܐܝܢ ܒܪܒܝܬܐ.❖

Text 16. Severus Sebokht, *Letter on the Ascending and Descending Nodes*

Claude-Villey, *Les Textes astronomiques syriaques* (*VIe et VIIe siècles*), 251–253.

ܡܛܠ ܗ̇ܝ ܕܐܬܡܠܟܐ ܣܓܝ ܐܢܬ ܡܛܠ ܐܢܐܒܝܒܙܘܢ ܘܩܛܐܒܝܒܙܘܢ. ܐܝܟ
ܦܪ̈ܘܫܝܗܘܢ ܘܣܘܦܝ ܡܢ ܦܘܫܩܝܗܘܢ ܘܕܐܬܡܠܟܐ ܬܘܒ ܢܦܩܬ ܐܢܬ ܒܝܕ ܕܐܝܢ
ܗ̇ܘܝܐ ܐܘܛܘܠܡܝܘܣ ܕܡܫܬܥ̈ܝܐ ܘܕܗܘܪܐ ܐܘ ܠܐ ܗ̇ܘܝܐ. ܡܛܠ ܕܝܢ ܗܠܝܢ ܕܐܒܝܕ
ܐܘܟܝܬ ܒܝܫ ܕܐܪ ܪܘܫܡ ܠܡܚܝܒܐ ܡܪܝ ܗܦܟܘܗܝ ܐܦܝܩܘܩܠܘܣ ܘܡܪܘܦܘܪܝܐ
ܕܠܒܗ̇ ܚܘܫܒܬܐ. ܕܐܬܩܪ̈ܝܢ ܠܗ ܡܢ ܟܘܪܣܘܬܗ ܒܐܪ̈ܝ ܐܣܘܬܝܟ. ܡܛܠ
ܡܢ ܕܢ̇ ܕܗܘܣܪܐ ܕܢܦܩ ܗ̇ܝ ܕܗܘܝܐ ܒܥܒܕܬܐ ܕܐܠܦ ܕܣܕܪܐ ܕܠܡ ܕܗܘܣܪܐ

148 ܟܝܬ P
149 om P
150 ܗܢܘ ܕܝܢ ܕܐܠܗܘܬܐ (that is, divinity) add. P, *in errore*

ܒܐܦܝܠܘܓܝܣܡܘܣ ܕܝܢ ܗܕܐ ܕܒܬܪܐ ܐܘܟܝܬ ܕܐܬܬܪ ܡܪܒܥܐ: ܕܐܝܟ ܒܫ̈ܡܥܐ
ܠܡ ܒܢܡܘܣ ܙܕܩ ܠܡ ܠܡܫܬܡܥܘ ܐܘ ܒܩ̈ܛܠܐ: ܬܘܒ ܕܝܢ ܘܡܛܠ ܗܘ̇ ܕܐܝܟ ܚ̈ܒܪܝܢ
ܠܡ ܐܚܪܢܐ ܗܠܝܢ ܥ̈ܒܕܐ ܠܛ̈ܒܝܢ ܣܘܥܪܢܐ ܠܘܬ ܫ̈ܪܪܐ ܐܘ ܠܐ. ܘܕܠܒܫܐ
ܬܓ̈ܪܐ ܚ̈ܒܪܝܢ. ܗܘܐ ܡܢ ܒܪܘ: ܒܗܠܝܢ ܕܡܢ ܠܒܪ ܡܕܡܝܐ ܐܡܝܢܐ. ܡܛܠ ܕܝܢ
ܫܘܚܠܦܐ ܕܐܝܬܝܩܝܘ ܘܦܘܠܝܛܝܩܝܘ ܗܫܐ ܐܝܟ ܕܡܬܚܙܝܐ ܐܡܪܝܢ ܘܡܛܠ
ܡܢ ܗܘ̇ ܕܡܬܚܙܝܐ ܕܡܠܠܝܘܬܐ ܘܕܠܐ ܒܬܠܐ ܢܦܫܐ ܕܠܫܒܥ ܠܐܝܬܝܩܝܘ ܒܬܪ
ܘܠܦܘܠܝܛܝܩܝܘ: ܐܢܐ ܟܡ ܠܐ ܒܕܘܟ ܐܫܟܚܬ. ܦܩܕ ܐܒܐ ܕܝܢ ܕܐܦܠܐ ܐܢܫ
ܐܫܟܝܡ: ܫܘܚܠܦܐ ܕܐܝܬܝܩܝܘ ܘܕܦܘܠܝܛܝܩܝܘ: ܕܡܠܠ ܘܦܫܩ ܘܕܦܣܝ ܠܘܬ
ܒܗ ܒܡܠܬܐ: ܠܒܪ ܡܢ ܫܘܚܠܦܐ ܗܘ̇ ܕܐܝܟ ܡܠܘܐ ܕܦܪܘܒܠܝܡܘܣ: ܐܡܝܢܐ
ܕܟܠܗ ܬܐܘܪܝܐ ܗܘ̇ ܐܠܟܣܢܕܪܘܣ ܒܡܡܠܠܐ ܕܠܗ ܕܦܪܘܒܠܝܡܘܣ܀

Text 17. George of the Arabs, *Prolegomena to Aristotelian Philosophy*

Watt, 'Prolegomena to Aristotelian Philosophy', 144.

ܡܛܠܘ ܕܐܪ̈ܒܥܐ ܕܡܢܝܢܐ ܐܝܬܘܗܝ ܫܘܠܡܐ ܕܦܝܠܣܘܦܘܬܐ
ܐܪܣܛܘܛܠܝܬܐ. ܘܐܡܪܝܢ ܕܢܕܒ ܠܗ ܢܦܫܐ ܟܠܬܐ ܘܒܒܘܕܐ ܕܟܠ. ܡܚܘܝܐ
ܓܝܪ ܗܘ ܦܝܠܣܘܦܐ ܒܡܬܠܐ ܗܘ ܕܡܬܦܪܬܐ ܒܬܪ ܒܢ̈ܝܬܐ ܕܢܕ ܢܦܫܐ ܘܟܠܬܐ
ܐܝܬܘܗܝ ܕܠܐ ܓܘܫܡ. ܕܡܟܝܢ ܥܠ ܐܬܐ ܠܗܘܢܐ܀
ܡܛܠܘ ܕܫܡܥܐ ܕܡܢܝܢܐ ܐܝܬܘܗܝ ܗܠܝܢ ܕܡ̈ܬܟܠܝܢ ܠܗ ܠܘܬ ܫܘܠܡܐ.
ܘܐܡܪܝܢ: ܕܡܠܦܢܘܬܐ ܕܗܠܝܢ ܕܒܟܝܢܐ ܘܒܫܘܠܡܐ ܐܝܬܝܗܝܢ. ܡܢ ܗܠܝܢ ܓܝܪ
ܒܝܢ ܡܬܝ̈ܕܥܢܘܬܐ ܕܫ̈ܠܝܦܢܐ ܡܬܬܟܠܝܢ ܠܘܬ ܗܠܝܢ ܕܒܟܝܢܐ ܗܘ̣ ܗܟܘܬ ܐܝܬ
ܠܗܘܢ: ܘܗܟܢܐ ܡܢ ܒܬܪ ܐܘ̈ܣܝܐ ܕܠܐ ܓܘܫܡ ܠܘܬ ܥܠܬܐ ܡܕܒܪܢܝܬܐ ܕܟܠ܀
ܡܛܠܘ ܕܬܠܬܐ ܕܡܢܝܢܐ ܐܝܬܘܗܝ ܐܕܫܐ ܕܡܬܟܬܒܢܘܬܗ. ܒܟܠ ܓܘܢ ܕܗܘ
ܫܬܝܬܐ ܘܢܗܝܪܐ ܒܦܘܫܩܐ ܕܡܠܬܐ. ܒܟܠ ܙܢ ܓܝܪ ܚܙܘ ܡܢ ܡܬ̈ܟܬܒܢܘܬܐ
ܕܡܠܬܐ ܪܗܛܘܪܝܬܐ [...] ܚܢܢܐ ܕ[ܗܘ]ܚ̈ܪܢܐ [...] ܘܠܐ ܢܗܝܪܐܝܬ ܡܠܒܫ: ܠܘ
ܡܛܠ ܚܢܢܐ ܕܡ[ܟܬܒ]ܘܬܐ ܐܠܐ [ܒ]ܚܢܢܐܝܬ ܚܒܪ ܗܕܐ܀ [...]
ܡܛܠܘ ܕܬܡܢܝܐ. ܕܐܝܟ ܐܝܢܐ ܙܕܩ ܠܩܪܘܝܐ ܕܢܗܘܐ. ܒܐܢܐ: ܘܫܪܟܐ
ܬܪܬܝܢܬܐ: [ܘܫ]ܦܝܠܐ ܠܘܬ ܩ̈ܠܐ: ܘܡܬܚܫܒܢܐ ܒܟ̈ܝܢܐ ܐܝܟܢܐ ܕܒܟܠܗܘܢ
ܢܗܘܐ ܡܬܚܫܒܬ.
ܕܬܫܥܐ. [ܘܠ]ܡܦܫܩܢܐ ܗܟܢܐ ܙܕܩ ܠܗ ܕܢܗܘܐ. ܐܝܟܢܐ ܕ[ܠ]ܐܒܫܦܪܘܬ
ܬܪܬܝܢܬܐ ܢܣܒܐ ܕܢܣܒ ܘܢܣܒܐ ܗܠܝܢ ܕܒܡܫܬܐܝܬ ܐܡܪ̈ܝܢ: ܐܝܟ ܗܘ ܕܡܢ
ܦܪܘܕܐ ܠܗܠܝܢ ܢܣܒ. ܘܠܐ ܠܗܠܝܢ ܦܠ̈ܓܘܬܐ ܒܡܫܬܘܬ ܘܠܐ ܢܣܒ ܡܛܠ
ܗܟܢܐ. ܐܠܐ ܕܢܣܒ ܠܗ ܣܘܥܪܢܐ ܙܕܩ ܠܗ ܕܢܗܘܐ ܗܠܝܢ ܕܡܬܐܡܪ̈ܢ.
ܘܡܕܡܝܐܝܬ ܡܢ ܠܬܪܝܢܬܐ ܕܒܬܪ̈ܝܐ ܢܗܘܐ: ܘܢܦܫܘܩ ܗܠܝܢ ܕܠܗܘܢ ܥܩ̈ܪܐ.
ܘܗܕܝܢ ܠܕܝܢܐ ܕܡܢܝܢܗ ܢܩܬܐ܀

Text 18. George of the Arabs, *Introduction to the Prior Analytics*

London, BL, Add. 14659, ff.83v–84r

ܟܕ ܐܦ ܗܕܐ ܙܢܐ ܣܟܝܡܐܝܬ ܘܠܐ ܣܘܦܣܛܝܩܐܝܬ ܕܢܬܪܟܐ܆ ܕܠܝܬ ܠܓܡܪܐ
ܪܫܝܬܐ ܘܡܬܚܠܦܘܬܐ܆ ܟܠܗܘܢ ܗܠܝܢ ܕܐܫܬܠܡ ܠܚܘܫܒܐ ܕܒܢܝܢܫܐ ܡܢ ܐܠܗܐ
ܚܠܦܬܐ ܕܟܠ܆ ܕܒܗ ܘܡܢܗ ܡܬܝܪܐܝܬ ܡܬܘܡ ܒܢܝܐ ܡܠܝܠܐ ܗܢܐ ܕܝܠܗ.
ܘܡܬܦܪܫ ܕܡܠܝܠܐܝܬ ܡܢ ܟܠܗܘܢ ܗܠܝܢ ܠܐ ܡܠܝܠܬܐ܆ ܐܝܟ ܒܡܘܣܝܩܘܬܐ
ܕܦܝܠܣܘܦܘܬܐ ܫܪܝܪܬܐ܆ ܗܝ ܕܡܢܗ ܕܐܠܗܐ܆ ܗܘ ܡܒܘܥܐ ܥܡܘܩܐ ܘܠܐ
ܡܬܬܚܡܢܐ܆ ܕܟܠܗܘܢ ܣܟܡܬܐ ܘܡܕܒܪ̈ܢܬܐ܆ ܐܫܬܦܛܬ ܘܐܙܕܪܥܬ ܣܢܝܩܐܝܬ
ܡܢ ܫܘܪܝܐ܆ ܒܟܠܗ ܒܢܝܐ ܡܠܝܠܐ ܘܡܕܒܪܢܐ. ܘܐܦ ܢ̈ܒܝܐ ܡܢ ܗܠܝܢ܆ ܕܠܗܘܬ
ܫܦܝܐ. ܡܠܘܗܝ ܕܝܢ ܐܝܟ ܕܒܡܘܫܠܡܐ ܟܠܗܘܢ܆ ܟܕ ܡܢ ܡܕܒܪܬܐ ܣܘܟܠܝܬܐ
ܕܚܠܬܐ ܡܕܒܪܢܝܬܐ ܢܦܠܘ܆ ܠܐ ܗܘܘ ܕܢܥܘܪܘܢ ܒܗܘܢ ܣܘܟܠܐܝܬ ܘܡܩ̈ܢܐ
ܣܘܟܠܬܐ ܘܥܘܦܪ̈ܐ ܠܐ ܡܬܬܠܠܢܐ ܕܠܘܬܗ ܕܒܗܘܡܘܬܐ [...] ܐܝܟ ܢ̈ܒܝܐ
ܒܢܒܘܗܝ ܗܠܝܢ ܕܠܘܬܢ ܦܬܚܝܢ ܠܢ܆ ܐܝܟ ܕܡܫܝܚܐ ܠܒܪ̈ܢܫܐ ܗܘܐ ܐܠܦ ܠܢ܆ ܟܕ
ܒܢܗ ܕܐܪ̈ܙܐ ܡܬܐܡܪܝܢ܆ ܐܝܟ ܕܒܡܬܠܐ ܟܠܗܘܢ ܗܠܝܢ ܕܝܠܗ ܐܬܘܗܝ ܠܒܪ܆
ܘܠܘ ܦܓܪ ܒܫܢܝܐܝܬ ܐܬܬܬܡܫܘ ܚܝܝ ܪܐܙܐ ܕܡܠܦܢܐ ܕܦܝܠܣܘܦܘܬܐ܆ ܟܕ
ܚܝܝ ܡܢ̈ܘܗܝ ܡܬܦܫܩܝܢ܆ [...] ܠܐ ܐܢܫ ܢܟܘܠ ܠܦܝܠܣܘܦܘܬܐ܆ ܐܠܐ
ܠܗܠܝܢ ܕܡܫܬܐܠܝܬ ܒܗ ܐܬܘܚܝܕܘ. ܘܣܡܝ ܬܘܒ ܐܝܟܢܐ ܕܡܟܫܟܠܝܢ܆ ܕܢܒ̈ܝܐ
ܘܢܗܘܘܢ ܘܠܐ ܡܫܬܚܠܝܢ: ܡܘܩܢܐ ܕܠܗܘܢ ܕܢܚܬܐ ܗܕܐ܆ ܠܐ ܥܒܕܢܐ ܕܢܬܝܚܕ
ܒܗܘܢ ܠܗܬ ܗܝ ܕܠܝܬ ܚܫܝܫܐ [...] ܠܐ ܢܗܘܐ ܡܢ ܗܝ ܕܐܢܬܬ̈ܐ ܘܡܢ ܒܢ̈ܝ ܩ̈ܠܐ.
ܘܡܢ ܥܡ̈ܗܐ ܘܦ̈ܘܠܚܐ܆ ܘܒܩ̈ܘܪܝܐ ܕܥܘܡܪ̈ܐ ܘܒܣܘܦܬܐ ܕܒܘܟܐ܆ ܘܐܘܡܢܘܬ
ܡܠܬܐ ܕܡܬܠܦܢܐ܆ ܘܡܥܒܕܢܘܬ ܕܢܗܐ ܠܟܠ ܒܝܬܐ ܗܝ ܕܡܬܬܬܡܝܐ ܡܢ
ܐܢܫܐ ܕܡܘ. ܕܗܠܝܢ ܕܬܪܥܝܬ ܥܩܒܝܢ: ܘܡܬܬܪܥܝܬ ܡܬܬܟܢܝܢ. ܡܢ ܡܬܬܟܢܫܢܘܬܗ
ܕܐܪܣܛܘܛܠܝܣ ܦܝܠܣܘܦܐ. ܡܬܪ ܡܢ ܟܠܗܘܢ.

Text 19. George of the Arabs, *Commentary on the Prior Analytics*

London, BL, Add. 14659, f.94r–v

ܗܠܝܢ ܐ̈ܢܫܝܢ ܦܪ̈ܘܛܝܣܝܣ ܟܠܗܘܢ ܕܒܗܠܝܢ ܬܠܬܐ ܐܣܟ̈ܡܐ ܕܣܘܠܘܓܝܣܡ̈ܘܣ
ܡܬܚ̈ܙܝܢ [...] ܗܘ ܕܝܢ ܬܠܬܐ ܐܣܟ̈ܡܐ ܕܣܘ̈ܥܪܢܐ܆ ܘܠܐ ܡܬܝܪܬܐ ܘܠܐ
ܣܝܡ̈ܬܐ܆ ܒܟܘܠ ܐܘܪܚܐܘܢܬܐ܆ [...] ܘܠܐ ܡܣܬܟܐ ܕܠܘ ܬܠܬܐ ܢܬܚܣܘܢ܆
ܘܐܦ ܡܬܗܘܝܢ ܣܝ ܗܝ܆ ܕܐܢܐ ܪܐܙܐ ܪܟܐ ܓܡܠ ܐܢܐ ܠܗ. ܕܐܡܪܢܐ ܟܪ
ܥܘܦܪ̈ܐ ܐܢܫ̈ܝܐ ܠܐ ܡܬܬܠܠܢܐ ܕܐܪ̈ܟܢ ܐܠܗܐ ܟܠ ܝܠܕܗ܆ ܘܒܗܕܐ
ܬܘܒ ܬܠܝܬܝܬ ܘܙܘܢܝܬ ܠܓܒܝܗ ܠܗܘܢܐ܆ ܐܡܪܢܐ ܕܒܡܬܣܥܪܢܘܬܐ
ܕܬܠܬܝܘܬܐ ܠܟܠ ܥܙܝܙܐ ܢܣܒܐ. ܘܡܬܠܬܝܘܬܐ ܕܒܗ܆ ܠܬܠܬܝܘܬܐ ܗܝ
ܐܝܬܝܬܐ ܕܟܠܢܘܗܘܬܐ܆ ܕܗܝ ܐܝܬܝܗ ܥܙܝܙܐ ܡܬܚܒܪܝܬ ܢܣܒܐ. ܟܕ ܡܢ ܢܣܘ
ܡܢ ܕܟܠܬܐ ܘܡܬܬܘܦ ܠܥܙܝܙܐ܆ ܒܝ ܡܢܣܝܘܬܐ ܥܙܝܪܬܐ ܘܡܠܠܬܐ. ܘܣܝ
ܐܝܟ ܡܠܦ ܗܘ ܦܝܠܣܘܦܐ܆ ܡܛܠ ܗܝ ܕܬܠܬܐ ܐܣܟ̈ܡܐ. ܒܠܝܘܕ ܐܬܝܗܘܢ.

ܐܡ̇ܪ ܓܝܪ ܕܬܘܣܦܬܐ ܡܫܟܚܢܐ܆ ܬܠܬܐ ܦܘ̈ܠܫܐ ܐܝܬ ܠܗ̇܇ ܐܝܟܢܐ ܕܐܦ
ܠܟܠ ܟܠܗܝܢ ܐܣܛܘܟܣܐ ܢܟܝܢܬ. ܐܘ ܓܝܪ ܐܘܪܬܝܐܝܬ ܟܬܒ܆ ܐܘ ܡܢ ܠܟܠ ܠܘܬܗ
ܢܣܒܬ܆ ܐܘ ܡܢ ܠܘܬܗ ܠܟܠ ܟܠܗ̇܀ ܗܐ ܒܗܢܐ ܐܣܟܡܐ[151] ܐܡܪ ܠܗܘܢ
ܕܐܬܘܬܗܘܢ ܠܬܠܬܐ ܐܣܟܡ̈ܐ. ܗܕܐ ܗܘܝܐ܆ ܕܙܪܐ ܪܝܫܐ ܐܬܘܬܗ̇. ܡܕܡܟܐܝܬ
ܡܢ Φ ܕܠܐ ܠܡܫܡܫܘܬܐ ܕܝܠܗ̇ ܕܠܟܠܗܘܢ ܩܘ̈ܒܐ ܩܠܐ ܕܟܠܡܐ ܐܝܟ ܕܡܬ ܡܫܡܫܐ܆
ܒܪ̈ܝܫܐ ܡܬܟܬܒܢܐ. ܘܐܝܟ ܕܗܘܝܐ ܡܟܪܙܐ ܥܘܠܐ ܡܢ ܒܬܪ ܟܠ ܕܐܠܦܐ܆ ܘܐܝܟ
ܢܠܒܫܐ ܡܫܡܫܐ. ܗܘ ܕܠܗܠܝܢ ܕܬܪܝܢ ܐܝܟ ܠܣ܆ ܘܐܝܟ ܢܦܘܪܐ ܘܡܕܝܢܝܐ.
ܟܕ ܐܝܬܘܬܐ Φ ܒܪ ܡܫܠܡܬܐ ܗܘܝܢ ܢܫܩܒ ܐܝܟ܀

Text 20. Sylvanus of Qardu, *On the Definitions and Aims of Philosophy*

Hespel, *Theodore bar Koni* [...] *Sylvain de Qardu*, 42–46.[152]

ܡܢܐ ܗ̱ܝ ܦܝܠܘܣܘܦܘܬܐ ܘܐܝܠܝܢ ܐܝܢܘܢ ܬܚܘܡ̈ܝܗ̇

ܦܝܠܘܣܘܦܘܬܐ ܐܝܬܝܗ̇ ܝܕܥܬܐ ܕܟܠܗܘܢ ܗܘ̈ܝܐ ܒܗ̇ ܕܐܝܬܝܗܘܢ ܗܘ̈ܝܐ. ܐܡܪܝܢ ܠܗ̇ ܓܝܪ ܝܕܥܬܐ ܕܟܠܗܘܢ ܗܘ̈ܝܐ܆ ܡܛܠ ܕܟܠ ܐܘܡܢܘܬܐ ܕܐܝܬ܆ ܚܕ ܡܠܘܐܐ ܘܚܕ ܥܒܘܕܠܝܐ ܡܢܝܢܐ܆ ܐܝܟܢܐ ܕܐܣܝܘܬܐ ܕܐܝܬ ܠܗ̇ ܡܠܘܐܐ ܦܠ̈ܓܐ ܐܢܫ̈ܝܐ ܘܥܒܘܕܠܝܗ̇ ܣܘܟܠܡܝܢܘܢ܆ ܐܘ ܐܝܟ ܐܣܛܪܘܢܘܡܝܐ ܕܗܠܐ ܕܝܠܗ̇ ܐܝܬܝܗܘܢ ܒܟܘ̈ܟܒܐ܆ ܘܥܒܘܕܠܝܗ̇ ܝܕܥܬܐ ܕܙܘܓ̈ܝܗܘܢ ܘܡܥܒܕܢܘܬܗܘܢ. ܗܟܢ ܘܟܠܗܝܢ ܐܘܡ̈ܢܘܬܐ ܟܠ ܚܕ ܡܢܗܝܢ ܚܕ ܡܠܘܐܐ ܘܚܕ ܥܒܘܕܠܝܐ ܡܢܝܢܐ. ܦܝܠܘܣܘܦܘܬܐ ܕܝܢ ܠܘ ܡܕܡ ܚܕܢܝܐ ܡܢܝܢܐ ܗܟܢܐ ܕܝܠܗ̇܆ ܐܠܐ ܟܠܗܘܢ ܗܘ̈ܝܐ ܡܬܦܪ̈ܫܢܐ ܒܝܕܥܐܝܬ ܘܡܬܝܕ̈ܥܢܐ܆ ܗܟܢ ܐܦ ܥܒܘܕܠܝܗ̇ ܠܘ ܡܣܝܒܪܢܘܬܐ ܗ̱ܝ ܐܠܐ ܝܕܥܬܐ ܕܟܠܗܘܢ܆ ܐܘܣܝܐ ܕܝܢ ܟܠܗ̇ ܕܐܝܬܝܗܘܢ ܗܘ̈ܝܐ܆ ܡܛܠ ܕܝܕܥܬܐ ܕܡܟܢܘܬܐ ܐܫܪܝܪܐܝܬ ܘܐܫܪܝܪܬܐ ܐܝܬܝܗ̇ ܠܟܝ̈ܢܐ. ܐܝܬ ܓܝܪ ܕܩܪܐ ܠܗ̇ ܠܝܕܥܬܐ ܒܫܡܗ̇ ܘܚܟܡܬܗ̇ ܘܡܫܡܫܢܘܬܗ܆ ܘܐܝܬ ܚܠܝܗ̇ ܡܫܡܫ ܒܗ̇ ܠܚܝܠܗ̇ ܠܐ ܝܕܥ܆ ܘܐܝܬ ܕܒܐܣܟܡܝܗ̇ ܟܠܗܘܢ. ܐܡܪܝܢܐ ܕܐܝܟ ܐܝܬ ܐܢܫ ܕܢܕܥ ܠܦܓܪܐ ܐܢܫܝܐ ܕܡܢ ܡܢ ܦܐܡ. ܘܡܟܢܐ ܐܝܬܝܗ̇ ܡܫܟܚܬܐ ܕܚܕ ܚܕ ܡܢ ܒܐܦܘ̈ܗܝ ܕܟܡ܆ ܘܡܟܢܐ ܐܠܐ ܣܠܩܘ̈ܗܝ܆ ܘܢܕܥ ܠܟܠܗܘܢ ܕܠܗ̇܆ ܗܟܢܐ ܢܕܥܝܗ ܗ̱ܘ ܒܦܪܕܐ ܕܦܓܪܐ. ܡܬܐܡܪ ܓܝܪ ܢܕܥܝܗ ܕܦܓܪܐ܆ ܐܦ ܗ̱ܘ ܕܒܚܕ ܡܢ ܣܠܩܘ̈ܗܝ ܡܟܝܠ܆ ܐܦ ܗ̇ܘ ܕܠܐܣܟܡܝܗܘܢ ܟܠܗܘܢ ܕܦܓܪܐ ܢܕܥ. ܘܡܛܠ ܕܠܐ ܥܒܘܕܐ ܢܕܥܐ ܕܦܓܪܐ ܠܟܠ ܐܝܟ܆ ܐܡܪܝܢܢ ܠܦܝܠܘܣܘܦܘܬܐ ܕܟܠܗ̇ ܕܐܝܬܝܗܘܢ ܗܘ̈ܝܐ܆ ܝܕܥܐ ܠܗܘܢ ܕܐܝܬܝܗܘܢ ܟܠ ܝܕܥܬܗ ܫܪܝܪܬܐ.

ܦܝܠܘܣܘܦܘܬܐ. ܦܝܠܘܣܘܦܘܬܐ ܐܝܬܝܗ̇ ܝܕܥܬܐ ܫܪܝܪܬܐ ܕܡܬܝܕ̈ܥܢܬܐ ܐܠܗܝ̈ܬܐ ܘܐܢܫܝ̈ܬܐ. ܗ̇ܝ ܕܝܢ ܕܝܕܥܬܐ ܫܪܝܪܬܐ ܕܡܬܝܕ̈ܥܢܬܐ܆ ܥܒܘܕܐ ܒܗ̇ ܗ̇ܝ ܕܐܬܐܡܪܬ ܒܬܘܣܦܬܐ ܡܕܡܐ. ܗ̇ܝ ܕܐܠܗܝ̈ܬܐ ܘܐܢܫܝ̈ܬܐ܆ ܡܛܠ

151 ܕܢܐ marg

152 Note that the lacunae in Hespel's text need to be filled out from bar Bahlul's quotation of Sylvanus's definition of philosophy (Payne-Smith, *Thesaurus Syriacus*, 3106), as per his apparatus.

ܕܥܠ ܡܕܡ ܕܐܝܬ ܐܘ ܡܬܝܕܥܢܐ ܘܡܣܬܟܠܢܐ ܐܝܬܘܗܝ܆ ܐܘ ܡܬܚܙܝܢܐ ܘܡܬܪܓܡܢܐ ܒܗ̇ܝ ܕܐܠܗܘܬܐ ܣܥܪ ܥܠܝܗܘܢ ܐܘܣܝܐܣ ܡܬܝܕܥܢܝܬܐ ܘܒܗ̇ܝ ܕܐܢܫܘܬܐ ܣܥܪ ܥܠܝܗܘܢ ܐܘܣܝ̈ܣ ܡܬܪ̈ܓܡܢܝܬܐ. ܐܝܟ ܕܐܦ ܡ̈ܠܐ ܕܪܘܚܐ ܠܥܠܡܐ ܟܠܗ ܒܗܠܝܢ ܥܒܕ̈ܘܗܝ ܡܚܘܝܢ ܦܘܡ ܠܗ̇܆ ܕܡܬܚܙܝܐ ܘܕܠܐ ܡܬܚܙܝܐ ܘܕܒܙܒܢܐ ܘܕܠܥܠܡ.

ܦܝܠܣܘܦܘܬܐ. ܦܝܠܣܘܦܘܬܐ ܐܝܬܝܗ̇ ܗܘܓܝܐ ܕܒܡܘܬܐ ܘܡܬܬܕܡܝܢܘܬܐ ܕܝܢ ܗܘܓܝܐ ܕܒܡܘܬܐ ܡܛܠ ܕܒܪܢܫܐ ܡܢ ܬܪܬܝܢ ܡܪ̈ܟܒ ܡܢ ܦܓܪܐ ܡܬܚܙܝܢܐ ܘܡܢ ܢܦܫܐ ܡܬܝܕܥܢܝܬܐ ܘܠܐ ܡܬܚܙܝܢܝܬܐ. ܘܐܝܟܢܐ ܕܡܢ̈ܘܬܗ ܬܪܬܝܢ ܐܝܠܝܢ܆ ܐܦ ܐܣܘܪ̈ܐ ܬܪܝܢ ܐܝܬ ܠܗ܆ ܚܕ ܚܝܝܐ ܘܚܕ ܡܝܬܘܬܐ. ܚܝܝܐ ܐܝܬܘܗܝ܆ ܗ̇ܘ ܕܒܗ ܡܬܐܣܪ ܦܓܪܐ ܥܡ ܢܦܫܐ܆ ܒܗ ܡܬܐܣܪ̈ܢ ܘܥܒܕ ܠܗ̇ ܒܟܠܗܘܢ ܘܚܝ̈ܝܢ ܐܝܟ ܦܘܩܕܢܗ ܕܒܪܘܝܐ܆ ܘܡܝܬܘܬܐ ܐܝܬܘܗܝ܆ ܗ̇ܘ ܕܒܗ ܡܬܦܪܫܐ ܘܡܬܐܣܪܐ ܢܦܫܐ ܥܡ ܦܓܪܐ ܒܟܠܗܘܢ ܕܒܪ̈ܝܬܗ ܡܛܠ ܕܠܘ ܐܝܟ ܦܘܩܕܢܗ ܐܠܗܝܐ ܐܝܬܘܗܝ. ܗܕܐ ܕܬܫܬܡܫ ܢܦܫܐ ܠܦܓܪܐ ܐܠܐ ܡܢ ܚ̈ܝܠܐ ܘܐܢܪܓܝܐ ܢܦܫܐ. ܥܠ ܗܕܐ ܐܫܬܡܫܗ ܚ̈ܝܠܝܗ̇ ܐܝܟ ܗ̇ܘ ܘܡܕܒܪܢܐ܆ ܡܛܠ ܕܚܝܠܐ ܐܝܬܘܗܝ܆ ܐܝܬܝܪ ܚܝ̈ܠܐ. ܘܐܝܟܢܐ ܕܐܣܘܪ̈ܐ ܬܪܝܢ ܡܢܐ ܒܪܢܫܐ܆ ܗܟܢܐ ܐܦ ܡܘ̈ܬܐ ܬܪܝܢ܆ ܚܕ ܟܝܢܝܐ ܘܚܕ ܨܒܝܢܝܐ܆ ܡܛܠ ܕܥܠ ܐܣܘܪܐ ܡ̈ܘܬܐ ܐܝܬ ܠܗ. ܡܘܬܐ ܓܝܪ ܟܝܢܝܐ ܐܝܬܘܗܝ܆ ܒܗ ܡܬܦܪܫܐ ܦܓܪܐ ܡܢ ܢܦܫܐ ܐܝܟ ܦܘܩܕܢܐ ܕܡܢ ܐܠܗܐ܆ ܗܢܐ ܕܡܫܬܡܗ ܡܘܬܐ ܟܝܢܝܐ ܡܢ ܟܠܗܘܢ ܒܢ̈ܝܢܫܐ. ܐܚܪܢܐ ܕܝܢ ܕܨܒܝܢܝܐ ܐܝܬܘܗܝ܆ ܒܗ ܡܬܦܪܫܐ ܢܦܫܐ ܡܢ ܦܓܪܐ ܘܗܘܝܐ ܡܬܬܚܬܐ ܠܟܠܗܘܢ ܪ̈ܓܝܓܬܗ ܘܗ̇ܝ܆ ܗܕܐ ܙܕܩܘܬܗ ܕܢܦܫܐ ܡܢ ܪ̈ܓܝܓܬܗ ܕܦܓܪܐ ܡܫܬܡܗ ܡܘܬܐ ܨܒܝܢܝܐ. ܠܐ ܗܘܐ ܡܢ ܟܠ ܐܢܫ܆ ܐܠܐ ܡܢ ܝܚ̈ܝܕܝܐ ܒܠܚܘܕ. ܗܘܓܝܐ ܡܟܝܠ ܕܒܡܘܬܐ܆ ܡܬܚܫܒܐ ܦܝܠܣܘܦܘܬܐ܆ ܠܐ ܗ̇ܘܐ ܕܒܡܘܬܐ ܟܝܢܝܐ܆ ܡܛܠ ܕܐܦ ܒܗܠܝܢ ܦܝܠܘܣܘܦܐ ܠܐܠܝܢ ܕܚܢܩܝܢ ܢܦܫܗܘܢ ܡܕܡ ܦܘܩܕܢܐ ܕܡܢ ܐܠܗܐ ܐܠܐ ܗ̇ܘ ܡܘܬܐ ܨܒܝܢܝܐ ܕܡܛܠ ܐܢܫ ܡܢ ܟܠܗܘܢ ܪ̈ܓܝܓܬܗ ܕܦܓܪܐ ܡܢ ܩܕܡ ܥܪܝܪܬܐ.

ܦܝܠܣܘܦܘܬܐ. ܦܝܠܣܘܦܘܬܐ ܐܝܬܝܗ̇ ܕܘܡܝܐ ܕܐܠܗܐ[153] ܐܝܟ ܕܡ̈ܨܝܐ ܕܒܪܢܫܐ ܒܗ ܒܢ̈ܝܢܫܐ. ܡܬܬܕܡܝܢܘܬܐ ܕܝܢ ܕܠܐܠܗܐ[154] ܡܛܠ ܕܐܠܗܐ ܬܪܝܢ ܚ̈ܝܠܐ ܡܢܗ܆ ܚܕ ܝܕܘܥܬܢܐ ܕܒܗ ܚܟܡ ܟܠ ܡܕܡ ܘܐܚܪܢܐ ܡܦܪܢܣܢܐ ܕܒܗ ܚܝܦ ܕܟܠܗܘܢ ܗܘ̈ܝܐ. ܕܡܝܐ ܕܝܢ ܦܝܠܣܘܦܘܬܐ ܠܐܠܗܘܬܐ܆ ܡܛܠ ܗ̇ܝ ܬܪܝܢ ܚ̈ܝܠܐ ܡܢܗ ܚܕ ܕܝܕܥܐ ܒܗ ܠܟܠܗܘܢ܆ ܘܐܚܪܢܐ ܕܡܦܪܢܣܐ ܒܗ ܠܡܦܪܢܣܘ ܥܠ ܣܘܢܩܢܝܢ[155] ܠܐܝܢܐ ܕܡܬܒܥܐ ܠܗ. ܗ̇ܝ ܕܝܢ ܐܝܟ ܕܡ̈ܨܝܐ ܕܒܪܢܫܐ ܒܗ ܒܢ̈ܝܢܫܐ ܡܣܘܝܐ ܕܠܘ ܒܠܚܘܕ ܡܨܝܐ ܕܒܪܢܫܐ ܒܢ̈ܝܢܫܐ ܒܐܠܗܐ܆ ܠܐ ܒܝܕܥܬܗܘܢ ܘܠܐ ܒܡܦܪܢܣܘܬܗܘܢ ܐܠܐ ܠܦܘܩܕܢܗ ܘܡܦܪܢܣܘܬܗ ܘܐܝܟ ܚܝܠܐ ܠܡܕܒܪܢܐ ܘܐܝܟ ܡܐ ܕܡܨܝܐ ܠܡܕܒܪܘ ܕܒܪܢܫܐ ܒܥܒܕܝܗܘܢ.

153 ܐܠܗܘܬܐ bar Bahlul
154 ܐܠܗܘܬܐ bar Bahlul
155 ܒܗ ܥܠ ܣܘܢܩܢܝܢ ܠܡܦܪܢܣܘ Hespel

ܦܝܠܘܣܘܦܘܬܐ. ܦܝܠܘܣܘܦܘܬܐ ܐܝܬܝܗ̇ ܐܘܡܢܘܬ ܐܘܡܢܘ̈ܬܐ܆ ܘܚܟܡܬ
ܚ̈ܟܡܬܐ. ܐܘܡܢܘܬ ܐܘܡܢܘ̈ܬܐ ܒܘܬ ܡܬܬܚܡܐ܇ ܡܛܠ ܕܐܡܪܐ ܐܝܬܝܗ̇
ܘܠܘܦܬܐ ܕܟܠܗܝܢ ܝ̈ܕܥܬܐ ܕܐܘܡܢܘ̈ܬܐ܆ ܘܟܠܗܝܢ ܡܢܗ̇ ܢܦܩܝܢ ܐܝܟ
ܢܗܪ̈ܘܬܐ ܡܢ ܡܒܘܥܐ. ܐܘܣܝܘܬܐ ܓܝܪ ܡܢܗ̇ ܩܒܠܬ ܐܪܒܥܐ ܐܣܛܘܟܣܐ܇
ܢܘܪܐ܇ ܫܡܝܐ܇ ܘܪܓܠܕ܆ ܘܡܢܝܢܐ ܘܡܫ̈ܚܬܐ ܕܐܦܝܪ̈ܐ ܘܡܘܣܝܩܐ ܡܢܗ̇
ܩܒܠܘ[156] ܢܘܡܘܣܐ ܘܣܘܪܝܓܐ ܘܥܠܝܡܘܬܐ܆ ܘܐܘܡܢ̈ܝܐ ܟܠܗܘܢ ܕܐܪ̈ܓܢܘܢ
ܘܕܠ̈ܠܐ ܠܗܘܢ܇ ܘܕܐܣܟܝܡ̈ܐ ܠܗܘܢ܆ ܘܕܐܣܛܪܘܢܘܡܝܘܣ ܡܢܗ̇ ܩܒܠܘ[157] ܡܕܪ̈ܫܬܐ
ܕܒ̈ܘܟܪܐ ܘܦܘ̈ܚܡܐ܇ ܐܦ ܓܝܪ ܡܬܚܫܒ ܐܣܝܐ ܒܐܪ̈ܟܐ ܐܣܛܘܟܣܐ
ܘܣܠܩܘ. ܐܠܐ ܕܡܕܢܐ ܐܝܬܝܗܘܢ, ܚܝܠܐ[158] ܕܚܕ ܚܕ ܡܢܗܘܢ܆ ܘܗܝ ܐܝܟܢܐ ܗܘܬ
ܥܒܝܕܘܬܗܘܢ܆ ܠܐ ܕܝܠܗ̇ ܐܝܬܝܗ̇ ܠܡܕܒܪ ܡܟܐ ܒܪܘܚܐ ܒܝ ܡܬܚܫܠ. ܐܠܐ
ܕܦܝܠܘܣܘܦܐ ܒܠܚܘܕ ܡܛܠ ܕܗ̇ܝ ܗܘ ܐܝܬܝܗܘܢ, ܡܫܬܡܥܝܢ. ܘܐܦ ܡܬܚܫܒ ܒܠܒܐ
ܒܪܘܚܐ ܕܡܕܝܠܬܐ ܘܒܪܘܗܡܐ ܕܢܦܫܗ̇܆ ܐܠܐ ܕܡܕܢܐ ܐܝܬܝܗܘܢ, ܚܝܠܐ
ܕܫܡܫܐ ܘܕܗܡܝܪܐ ܘܕܝܢ ܝܢ ܡܢ ܢܗܝܪܐ ܠܐ ܢܦܩ ܗܕܐ. ܐܠܐ ܦܝܠܘܣܘܦܐ
ܒܠܚܘܕ ܘܢܕܥ܆ ܡܛܠ ܕܗ̇ܝ ܗܘ ܐܫܟܚ ܐܝܟ ܠܗ. ܘܠܐ ܒܠܚܘܕ ܗܠܝܢ ܐܘ̈ܡܢܘܬܐ
ܡܫܬܟܠܠܢܘܬܐ ܡܢܗ̇ ܢܦܩܝܢ ܘܡܬܚܫܒܝܢ ܐܠܐ ܘܐܦ ܗܠܝܢ ܒܠܬܫܬܐ. ܐܪ̈ܕܟܠܐ
ܓܝܪ ܡܢܗ̇ ܥܒܠ ܒܢܝܢܘܬܐ ܘܦܠܓܐ܆ ܘܢܓܪ̈ܐ ܣܦܠܐ ܘܬܪܥܘܬܐ ܘܕܒܝܠܐ.
ܘܕܒܚ̈ܐ ܡܛܠ ܕܗ̇ܝ, ܐܝܬܝܗ̇ ܡܫܬܡܫܢܘܬܐ ܕܟܠܗܝܢ ܐܘ̈ܡܢܘܬܐ܆ ܒܠ ܗܕܐ
ܬܫܡܫܬܗ̇ ܐܘܡܢܘܬ ܐܘܡܢܘ̈ܬܐ ܘܚܟܡܬ ܚ̈ܟܡܬܐ܆ ܒܝ ܡܗܕܝܢ ܒܗܕܐ
ܡܫܬܟܠܠܘܬܗ̇ ܘܡܫܬܪܪܘܬܗ̇ ܕܗܢܝܢ ܟܠܗܝܢ ܐܘ̈ܡܢܘܬܐ.

Text 21. Catholicos Timothy I, Epistle 19

Braun, *Timothei patriarchae I: Epistulae I*, 127,1–22 and 129,10–18. A new edition and German translation of Timothy's Letters 3–29 has now been published, albeit too late for its text to be used in this volume. M. Heimgartner (ed.) *Die Briefe 3–29 des Ostsyrischen Patriarchen Timotheos I*, Corpus Scriptorum Christianorum Orientalium 700, Leuven: Peeters, 2021.

ܡܬܒܥܝܢܐ ܗܘ̣ ܓܝܪ ܠܬܒܐ ܗ̇ܘ ܕܐܡܪܬ. ܘܠܘ ܒܠܚܘܕ ܡܬܟܒܝܢܐ ܐܠܐ
ܘܐܦ ܡܠܟ ܐܢܢܩܝܐ. ܒܝ ܓܝܪ ܫܪܪ ܗܘܝܬ ܒܝܢܝ ܘܒܝܢ ܡܬܟܒܡܐ܆ ܐܝܟ ܐܡܪܬ,
ܡܢ ܕܒܥܝܬ ܠܝ ܘܠܢܦܫܝ܆ ܘܐܝܟ ܐܡܪܬ, ܕܝܢ ܕܒܥܝܬ ܡܠܟܘ̈ܬܐ ܡܒܩܒܝܢܐ܆ ܗܢܫܐ
ܡܢ ܒܡܠܟܘܬܐ ܕܠܫܢܐ ܕܝܠܢ ܗܘܪܝܢܐ܆ ܗܫܐ ܕܝܢ ܒܝܬ ܥܝܢܘܬܐ܆ ܘܒܝܢ ܡܢ
ܒܡܕܝܢܬܐ ܗܝܕܐ ܕܝܠܢ ܗܘܠܢܝܬܐ܆ ܒܝܢ ܕܝܢ ܒܝܬ ܠܗܘܢܐ ܕܐܕܫܐ ܡܫܝܚܝܐ ܕܝܠܗ܆
ܡܫܡܫ ܗܘܝܬ ܡܢ ܕܝܠܗ ܡܢ ܘܠܐ ܡܕܡ ܐܘ ܡܠܟ ܒܬܘܪܝܬܐ܆ ܡܢ ܕܐܢܫܝ̈ܐ
ܕܝܢ ܠܗ ܡܠܟܘܬܐ ܡܢ܆ ܒܬܘܪܝܬܐ ܕܝܢ ܘܐܦ ܗܘܝܢ܆ ܓܒܠ ܗܘܝܬ ܡܢ ܬܪܒܝܬܐ
ܘܡܫܬܥܝܬܐ܆ ܐܝܟܢܐ ܕܐܢܐ ܡܠܘܟ ܒܠ ܡܬܟܕܘܒ ܕܐܝܟ ܗܕܐ ܐܣܟܡܐ.
ܒܝ ܓܠܝܐ ܐܢܐ ܗܘܝܬ ܘܒܫܢܝܘܬ ܙܒܢܐ ܡܬܒܠܠ ܐܢܐ. ܓܠܝܢ ܗܘܝܬ ܡܢ

156 ܘܡܫܪܐ ܡܢܗ̇ ܩܒܠ Hespel
157 ܘܕܐܣܛܪܘܢܘܡܝܐ ܡܢܗ̇ ܩܒܠ Hespel
158 ܗܘ ܝܢܐ ܕܟܠ bar Bahlul

ܒܪ ܐܝܟ ܕܫܠܝܚܐ ܥܡܬ ܘܒܥܘܬܐ ܕܠܒܢܝܐ ܘܡܟܬܒܢܘܬܗ ܥܡܬ ܗܿܝ ܕܐܡܪܬ
ܘܒܪܢܫܘܬܗ. ܠܐܢ ܗܘܘܬ ܥܡܬ ܘܒܠܥܕ ܐܣܝܘܬܐ ܘܠܘܒܪܐ. ܗܿܠܝܢ ܕܐܝܟ
ܕܝܢ ܥܡܟܐ ܘܝܢ ܗܘܢܐ ܕܠܝܢ ܐܬܬܝܬܘ. ܐܝܟ ܡܥܠܡܢܘܬܐ ܗܿܝ ܒܬܝܡܬܐ
ܥܡܬ ܘܐܟܡܢܬܐ ܕܪܕܝܐ. ܘܢܝܢܐ ܥܡܬ ܐܘܪܚ ܐܝܢܐ ܘܗܢܝܐ ܐܥܡܟܠܝܐ: ܗܿܠܝܢ
ܕܒܬܪܬܝܗܘܢ ܡܥܪܬܚܝܢ ܘܡܣܝܡܝܢ. ܘܟܠܗܝܢ ܐܘܪܚ ܐܝܢܐ ܘܒܐܕܥܐ ܡܢܝܢܐ
ܕܗܠܝܢ: ܗܿܘ ܡܟܐ ܕܒܪܢܫܐ ܠܝܗ ܠܢܘܕ ܐܠܐ ܠܘ ܐܣܝܢܐܝܬ ܡܥܡܝܢܐ ܕܢܡܘܐ.
ܘܐܢ ܡܢܝܢܐ ܢܬܕܪ ܠܝ ܒܗܘܕܝܐ ܕܒܬܒܘܗܝ ܕܡܪܝ ܘܠܝܐ. ܒܕܥܐ ܒܪ ܕܐܝܬ
ܒܗܘܢ ܡܕܡ ܕܠܐ ܢܒܥܝܢ ܒܗ. ܘܐܝܬ ܣܝ ܐܝܠܝܢ ܠܒܬܒܘ̈ܗܝ ܒܡܝܡܝܢ.
ܕܬܒܥܐ ܐܝܠܝܢ ܒܟܠ ܥܕ̈ܡܝܢ ܘܒܟܠ ܐܣܟܡ̈ܝܢ. ܕܠܡܟܐ ܬܥܒܝܕ ܒܗܘܢ ܗܿܝ ܬܪܝܢ
ܡܐܡ̈ܪܝܢ ܕܒܠ ܦܐܘܬ̈ܠܐ. ܣܝ ܒܪ ܐܝܬ ܠܝ. ܐܘ ܕܠܡܟܐ ܬܥܒܝܕ ܒܗܘܢ
ܦܣܥܐ ܕܐܠܡܦܣܝܪܘܣ ܕܗܠܝܢ ܒܬܒ̈ܐ ܕܡܠܠܘܬܐ. ܐܘ ܕܐܘܣܛܦܘܣ ܐܘ
ܕܗܡ ܒܗܡ ܐܘ ܕܐܠܟܣܢܕܪܘܣ [...]

Text 22. Catholicos Timothy I, Epistle 43

Pognon, *Une version syriaque des aphorismes d'Hippocrate*, XVI–XVIII; Heimgartner, *Die Briefe 42–58 des Ostsyrischen Patriarchen Timotheos I*, 65–68.

ܠܪܚܡ ܐܠܗܐ ܪܝܡ ܦܪ̈ ܦܬܝܘܢ ܡܥܝܢܐ ܘܡܠܦܢܐ. ܛܝܡܬܐܘܣ ܚܛܝܐ ܫܐܠ
ܫܠܡܟܘܢ ܘܡܒܣܡܐ ܠܚܘܒܟܘܢ.

ܟܬܒܐ ܕܛܘܦܝܩܐ ܕܐܪܣܛܘܛܠܝܣ ܦܝܠܣܘܦܐ: ܬܒܥܢ ܒܐܝܕ̈ܝܢ ܦܘܣܕܢܐ
ܡܠܦܢܐ. ܐܝܡܢܐ ܕܢܦܫܟܘܢ، ܡܢ ܗܘܪܝܢܐ ܠܒܪܬ ܩܠܐ ܕܛܒ̈ܝܐ ܗܘ ܕܐܦ ܗܘܬ
ܒܡܕܪ̈ܫܐ ܕܐܠܗܐ ܒܡܢܝܢܘܬ ܕܪܒܐ ܐܒܘ ܢܘܚ. ܡܢܠ ܗ̈ܢ ܒܪ ܩܠܝܠ ܡܢ
ܠܥܠܐ ܗܘܪܝܢܐ. ܡܢܗ ܕܝܢ ܒܠܐܝܬ. ܘܡܢ ܗܘܪܝܢܐ ܐܚܪ̈ܢܐ ܘܡܢ ܛܝܒܐ. ܢܦܫ
ܡܢ ܒܪܘ ܗܘܒܪܐ ܗܒܐ ܘܐܬܬܡܠܝ. ܘܒܪ ܛܒ ܐܝܬ ܗܘܘ ܐܢܫ̈ܝܢ ܐܣܝ̈ܐ
ܕܠܗܢܐ ܡܬܪܓܡܝܢ ܡܢ ܠܥܒܐ ܢܘܝܐ ܠܠܥܒܐ ܛܝܒܐ: ܡܕܡܝܢ ܒܪ ܒܬܒܝܢ
ܠܝ ܗܠܝܢ ܘܐܘܕܒܢܝ. ܕܐܝܡܢܐ ܥܡܬ ܘܒܐܝܢܐ ܙܝܢ ܒܪ ܕܒ ܕܢܩܘܡ ܐܠܝܢ
ܕܗܘ̈ܝ: ܐܠܐ ܒܒܬ̈ܒܐ ܡܢ ܗܿܠܝܢ ܐܣܝ̈ܐ ܘܐܦܠܐ ܕܢܣܘܪ ܒܗܘܢ ܐܥܘܝ.
ܒ̈ܪܒܝܢ ܗܘܘ ܒܪ ܒܐܡܬ. ܠܘ ܒܠܚܘܕ ܒܡܠܬܐ ܐܠܐ ܘܒܗܘܢ ܒܗܘ̈ܢܐ. ܘܐܢ
ܡܛܠ ܒܡܣܬܘܬܗ ܕܡܦܬܘܬܗܘܢ: ܢܕܒ ܐܝܬ ܗܘ ܒܪ ܠܐܕܥܐ
ܕܠܗ ܕܦܝܠܘܣܘܦܐ ܒܡܠܝܢ ܬܠܝܬܝܐ: ܕܐܝܡܢܐ ܥܡܬ ܘܕܒܡܐ ܒܡܛܠܝܐ ܐܥܕ
ܒܠ ܐܦ̈ ܥܘܦܪܐ ܕܣܘܩܠܐ ܘܕܗܘܝܐ: ܘܐܢ ܡܛܠ ܠܐ ܡܕܪܥܘܬܐ ܕܗܿܠܝܢ
ܕܠܐܝܟ ܗܠܝܢ ܐܬܦܢܒܘ. ܡܦܣܬ ܗܿܘ ܒܪ: ܕܒܠ ܒܡܒܐ ܘܕܐܝܟ ܐܝܠܝܢ
ܐܟ̈ܬܒܐ ܘܡܬ̈ܠܐ ܣܝܡܐ ܝܒܘܬܐ ܕܕܐܝܟ ܗܕܐ. ܒܠܐܝܬ ܕܝܢ ܪܡܐ ܒܠ
ܒܡ̈ܠܐ ܗܿܠܝܢ ܕܗܠܝܢ. ܡܛܪ ܕܝܢ ܩܡ ܒܪ ܐܥܘܝ، ܐܝܠܝܢ ܒܪܡ ܙܡ ܐܝܟ ܕܒܦܘܣܩܢܐ
ܒܠ ܣܒ̈ܕܐ.

ܛܥܐܠ ܗܒܝܠ ܣܝܡܘܬܟ ܒܝܪܚܬܐ ܘܬܒܥܒ. ܐܝܬ ܗܿܘ ܒܪ ܐܪܐ ܠܬܒܥܐ
ܗܢܐ ܛܘܦܝܩܐ: ܐܘ̇ ܠܗܘ ܡܒܣܢܐ ܕܣܘܦܣܛܝܩܐ: ܐܘ ܠܗܘ ܕܪܗܛܘܪܝܩܐ ܐܘ ܠܗܘ
ܕܦܘܐܝܛܝܩܐ: ܦܘܥܣܐ ܒܝܬ ܡܕܡ ܐܘ ܣܒ̈ܬܢܐ ܕܐܢܫ̈ܝܢ ܐܝܟ ܕܒܣܘܪܝܝܐ ܐܘ
ܠܐ. ܘܐܢ ܩܡ ܐܝܬ: ܒܡܕ ܕܒܬܒܘ ܘܠܡܒܘ ܘܐܝܬܐ. ܥܒܐܠܐ ܕܝܢ ܕܒܠ ܗܕܐ

ܢܩܘܫܘܢ ܠܬܪ̈ܥܐ ܒܡܡܪܐ ܕܡܪܝ، ܡܬܠ. ܒܪܡ ܕܝܢ ܠܐ ܗܘܐ ܒܠܝܠܐܝܬ ܢܩܘܫܘܢ ܫܘܐ̈ܠܐ. ܕܝܢ ܟܕ ܬܬܪܝܟ ܒܥܘܬܬܐ ܬܬܦܫܩ ܘܠܐ ܬܬܟܠܠ.

ܐܡܪ ܠܝ ܐܚܘܢ ܗ̇ܘ ܒܠܥܕܘܢܝܐ: ܕܝܕܥܐ ܡܕܡ ܪ̈ܚܘܩܐ ܡܢ ܡܫܩ̈ܠܐ ܕܦܝܠܣܘܦܐ. ܕܫܪ̈ܒܐ ܠܗ ܫܪ̈ܒܐ ܡܢܗ. ܐܠܐ ܢܬܦܫܩܢ ܬܒܥ ܐܝܟܢܐܝܬ ܒܠ ܡܫܘܠ̈ܬܐ ܐܘ ܦܘܫܩܐ ܕܗܠܝܢ ܒܬܪ̈ܐ.

ܫܪܪ ܠܝ ܦܘܫܩܬܐ ܐܣܛܪܘܢܘܡܝܐ ܕܐܝܬܘܗܝ ܕܒܛܠܡܝܘܣ. ܡܬܡܪܝܬܐ ܟܕ ܐܝܬ ܠܝ. ܘܡܣܬܒܪܢܐ ܠܝ ܕܡܬܚܙܝܢܐ ܡܢ ܕܦܘܠܐ ܗܝ. ܗܟܢܐ ܟܕ ܐܡܪ ܘܐܦ ܒܡܦܠܐܘܣ ܕܠܗ ܕܬܠܬܐ ܦܪܝܫ. ܦܘܫܩܬܐ ܠܗ ܡܬܚܫܒܬܐ ܕܡܕܪܫܐ ܘܠܚܕ ܠܐܝܠܗܝܢ ܒܐܓܪ̈ܬܗܘܢ ܬܐܘܠܘܓܘܣ. ܕܦܫܩܗ ܡܢ ܗܘܢܐ ܠܗܘܢܐ: ܐܟܡܐ ܕܐܡܪ، ܦܘܠܐ ܒܡܦܩܬܗ ܬܪܬܐ. ܬܘܪܓܡܐ ܕܝܢ ܐܝܟ ܡܠܬܐ ܕܐܬܘܗܝ ܐܝܬܘܗܝ، ܘܗܕܐ ܡܢ ܗܘܠܐ.

ܒܗܕ ܕܝܢ ܘܡܛܠ ܡܐܡܪ̈ܐ ܕܟܠ ܙܪ̈ܥܐ ܒܬ̈ܠܝܬܐ ܕܠܓܘܫ̈ܝܐ. ܕܡܬܚܬܝܢ ܠܐܢܫ ܡܢ ܕܐܓܪܐ ܦܠܐܛܘܢܝܩܘܣ. ܕܐܬܘܗܝ ܢܝܫܘܗܝ ܗܘܠܐ. ܟܠ ܙܢܐ ܡܢܐ ܕܠܓܘܫ̈ܝܐ ܐܝܬ ܕܐܡܪܘ.. ܘܡܚܝܐ ܬܪܝܨܘܬܐ ܕܟܠܗܘܢ ܦܝܠܣܘ̈ܦܐ ܡܕ̈ܢܚܝܐ ܒܡܐܡܪ̈ܐ ܡܕܒܪܢܐ. ܘܡܚܘܝܢ ܠܐܢܫܘܬܐ ܘܠܬܘܩ̈ܢܐ ܦܠܛܝ̈ܢܘܣ. ܒܡܐܡܪ̈ܐ ܕܝܢ ܕܬܪܝܢ ܡܙܪܐ ܕܐܢܫܝ ܡܛܠ ܗܘܠܐ ܘܐܕܫܐ ܘܟܠܝܘܬܐ. ܐܝܟ ܕܒܓܢܣܐ ܐܪܝܣܛܛܠܝܣ ܘܐܡܪ ܒܗ ܐܝܟ ܣܓܝܐܐ ܪ̈ܫܢܐ. ܘܠܐ ܚܕܐ ܠܗ ܥܘܠܡܐ ܠܡܐܡܪܐ. ܒܗܕ ܕܟܕ ܡܬܚܒܝܢ ܡܐܡܪ̈ܐ ܗܠܝܢ. ܘܡܛܠ ܗ̇ܘ ܡܐ ܕܢܦܩ ܡܢ ܡܐܡܪܐ ܕܬܪܝܢ: ܘܡܛܠ ܗܘܢܐ ܕܫܪ̈ܝܐ ܕܒܬܪܗܘܢ..

ܒܗܕ ܬܘܒ ܘܟܠ ܗܢܝܐ ܠܐ ܝܗܒܐ ܕܐܝܟ ܦܝܠܣܘܦܐ ܕܡܬܘܪܐ ܢܦܫܗܘܢ ܕܟܠ ܬܘܩܢܗ ܕܒܢܝܢܐ ܘܐܝܬܘܗܝ، ܪܢܝܢ ܗܘܐ.. ܒܪܢܫܐ ܡܢ ܢܦܫܐ ܡܬܚܫܒܬܐ ܘܦܓܪܐ ܠܒܪ ܥܦܪܐ ܡܬܘܡ. ܘܢܗܒ ܡܢ ܥܘܠܡܐ ܠܗܢܐ ܒܒܚܝܢܘܬܐ ܕܣܒܪܐ ܪ̈ܘܚܢܐ ܬܘܒ ܣܓܝܐ. ܡܬܥܘܕܐ ܕܝܢ ܒܥܘܠܡܗ ܟܠ ܢܦܫܐ ܠܒܢܝܢ. ܘܠܬܘܒܗܝ ܗܢܐ ܬܘܪܓܡܐ.

ܦܘܫ ܘܚܘܒ ܘܬܘܒ ܠܝ ܕܢܫܬܘܫܘܢ ܡܫܘܬܐ ܕܐܬܘܗܝ ܐܘ ܕܦܘܫܩܐ. ܥܠܡܐ ܒܚܝܝ ܘܒܡ ܐܢܫܝܢ ܒܠܗܘܢ.

Text 23. Ḥunayn, *Risāla*

Bergsträsser, *Hunain ibn Ishaq*, 1,4–2,4 (also, with some variants, Lamoreaux, *Ḥunayn ibn Isḥāq on His Galen Translations*, 2).

[1]

ذكرت أكرمك الله الحاجة الى كتاب يجمع فيه ثبت ما يحتاج اليه من كتب القدماء فى الطب ويتبين الغرض فى كل واحد منها وتعديد المقالات من كل كتاب وما فى مقالة مقالة منها من أبواب العلم لتخف به المؤونة على الطالب لباب باب من تلك الأبواب عند الحاجة تعرض الى النظر فيه ويفهم فى أى كتاب يوجد وفى أى مقالة منه وفى أى موضع من المقالة وسألت أن أتكلف ذلك لك

[2]

فأعلمتك أيدك الله أن حفظى يقصر عن الإحاطة بجميع تلك الكتب إذ كنت قد فقدت جميع ما كنت جمعته منها وأن رجلا من السريانيين قد كان سألنى بعد أن فقدت كتبى شبيها بهذا فى كتب جالينوس خاصةً وطلب منى أن أبين له ما ترجمته انا وغيرى من تلك الكتب الى السريانية والى غيرها فكتبت له كتابا بالسريانية نحوت فيه النحو الذى قصد اليه فى مسئلته إياى وضعه

[3]

فسألت أكرمك الله أن أترجم لك ذلك الكتاب فى العاجل الى أن يتفضل الله بما هو اهله من رد تلك الكتب على يدك فأضيف الى ما ذكرته فى ذلك الكتاب من كتب جالينوس شيئا إن كان شذ عنى منها وذكر سائر ما وجدناه من كتب القدماء فى طب وأنا صائر الى ما سألت من ذلك إن شاء الله

Text 24. Ibn al-Nadīm, *Fihrist*

Flügel, *Ibn al-Nadīm: Kitāb al-Fihrist*, 309, 348.

الكلام على ابو ديقطيقا وهو انا لوطيقا الثانى. مقالتين نقل حنين بعضه الى السريانى. ونقل اسحق الكل الى السريانى. ونقل متى نقل اسحق الى العربى

من سعادات حنين ان ما نقله حبيش بن الحسن الاعسم، وعيسى بن يحي، وغيرهما الى العربى، ينحل الى حنين. واذا رجعنا الى فهرست كتب جالينوس الذى عمله حنين الى على بن يحي، علمنا ان الذى نقل حنين اكثره الى السريانى، وربما اصلح العربى من نقل غيره، او تصفحه.

Text 25. Barhebraeus, *Chronicle*

Bedjan, *Gregorii Barhebraei Chronicon syriacum*, 98.

ܘܡܛܘ ܒܗܘܢ ܦܝܠܣܘܦܐ ܘܡܬܡܛܝܩܘ ܘܐܣܘܬܐ ܕܟܠܗ ܠܥܬܝܩܐ
ܒܝܘܢܝܐܝܬ ܗܘܝܐ. ܠܗ ܕܝܢ ܟܠ ܫܬܐܣܬܐ ܐܣܝܘܬܐ ܗܘܝܐ: ܐܠܐ ܟܠ
ܕܐܬܪܟܡܐ ܗܘܝܐ: ܒܒܢܝܐ ܫܬܠܠܗ ܣܬܟܬܘܢܐ ܕܙܘܙܒܝܢ ܒܪ ܠܒܣܝܘ ܝܬܪ
ܫܡܠܬܐ: ܘܒܘܚܝܐ ܕܐܦܪ ܡܠܦܢܝܢ. ܐܢܫܐ ܕܝܢ ܗܘܘ ܕܒܪܝܢ ܢܫܡܥܘܢ
ܠܫܡܥܬܐ ܒܪ ܡܬܪܓܡܢܐ: ܕܟܠܗܘܢ ܫܘܪܝܐ: ܗܐ ܡܢܗܘܢ ܐܫܬܘܡܝܢ
ܠܡܬܐܠ ܫܡܥܬܐ.

Bibliography

Primary Sources

Manuscripts

Baghdad, Chaldean Monastery, 169 (*olim* Notre Dames des Semences, 51) (Text 2)
Berlin, Staatsbibliothek, Syr. 88, f.83v (Text 4)
Birmingham, Mingana, 606 (Texts 1, 2)
London, BL, Add. 14658 (Texts 2, 3, 5, 6)
London, BL, Add. 14659 (Texts 18, 19)
London, BL, Add. 14660 (Text 14)
London, BL, Add. 14661 (Text 9)
London, BL, Add. 17156 (Text 14)
Paris, BNF, syr. 346 (Text 15)
Paris, BNF, syr. 354 (Texts 1, 2)
Seert 82 (Text 13)
Seert 109 (Text 13)

Printed Books

Aydin, Sami, *Sergius of Reshaina: Introduction to Aristotle and his Categories, Addressed to Philotheos. Syriac Text, with Introduction, Translation, and Commentary*, Leiden/ Boston MA: Brill, 2016

Becker, Adam H., *Sources for the History of the School of Nisibis*, Translated Texts for Historians 50, Liverpool: Liverpool University Press, 2008

Bedjan, (ed.), *Gregorii Barhebraei Chronicon syriacum*, Paris: Maisonneuve, 1890

Bergsträsser, Gotthelf (ed.), *Hunain ibn Ishaq über die syrischen und arabischen Galen-Übersetzungen*, Abhandlungen für die Kunde des Morgenlandes 17.2, Leipzig: Brockhaus, 1925

Bhayro, Siam, and Sebastian Brock, 'The Syriac Galen Palimpsest and the Role of Syriac in the Transmission of Greek Medicine in the Orient', *Bulletin of the John Rylands Library*, 89/1 (2012), 25–43

Bos, Gerrit, and Y. Tzvi Langermann, 'The Introduction of Sergius of Rēsh'ainā to Galen's Commentary on Hippocrates' On Nutriment', *Journal of Semitic Studies*, 54/1 (2009), 179–204

Braun, Oskar (ed.), *Timothei patriarchae I: Epistulae I*, Corpus Scriptorum Christianorum Orientalium 74/75, Leuven, 1914

Brock, Sebastian P., 'Two Letters of the Patriarch Timothy from the Late Eighth Century on Translations from Greek', *Arabic Sciences and Philosophy*, 9 (1999), 233–246

Brooks, E.W., *Historia Ecclesiastica Zachariae Rhetori Vulgo Adscripta II*, Corpus Scriptorum Chistianorum Orientalium 84, Leuven: Imprimerie Orientaliste, 1953

Budge, E.A. Wallis, *The Chronography of Gregory Abû'l Faraj, the Son of Aaron, the Hebrew Physician, Commonly known as Bar Hebraeus, Being the First Part of His Political History of the World*, London: Oxford University Press, 1932

Busse, A. (ed.), *Ammonius in Porphyrii Isagogen*, Commentaria in Aristotelem Graeca 4.3, Berlin, 1891

Claude-Villey, Émilie, *Les Textes astronomiques syriaques* (*VIe et VIIe siècles*), thesis, L'Université de Caen Basse-Normandie, 2011/12

Dodge, Bayard, *The Fihrist of al-Nadīm: A Tenth-century Survey of Muslim Culture*, London/New York: Columbia University Press, 1970

Fiori, Emiliano, 'L'Épitomé syriaque du *Traité sur les causes du tout* d'Alexandre d'Aphrodise attribué à Serge de Res'ayna', *Le Muséon*, 123 (2010), 127–158

Flügel, Gustav, *Ibn al-Nadīm: Kitāb al-fihrist*, Leipzig: F.C.W. Vogel, 1871

Greatrex, Geoffrey, *The Chronicle of Pseudo-Zachariah Rhetor: Church and War in Late Antiquity*, Translated Texts for Historians 55, Liverpool: Liverpool University Press, 2011

Heimgartner, Martin (ed.), *Die Briefe 42–58 des Ostsyrischen Patriarchen Timotheos I*, Corpus Scriptorum Christianorum Orientalium 644/645, Leuven: Peeters, 2012

Heimgartner, Martin (ed.), *Die Briefe 3–29 des Ostsyrischen Patriarchen Timotheos I*, Corpus Scriptorum Christianorum Orientalium 700, Leuven: Peeters, 2021

Hespel, Robert (ed.), *Theodore bar Koni, Livre des Scolies* (*recension d'Urmiah*): *Les collections annexeés par Sylvain de Qardu*, Corpus Scriptorum Christianorum Orientalium 464, Leuven: Peeters, 1984

Hoffmann, J.G.E., *De Hermeneuticis apud Syros Aristoteleis*, Leipzig, 1873

Hugonnard-Roche, H., 'Sur la lecture tardo-antique du *Peri Hermeneias* d'Aristote: Paul le Perse et la tradition d'Ammonius: Édition du texte syriaque, traduction française et commentaire de l'*Élucidation du Peri Hermeneias* de Paul le Perse', *Studia Graeco-Arabica*, 3 (2013), 37–104

Kessel, Grigory, 'Membra Disjecta Sinaitica I: A Reconstitution of the Syriac Galen Palimpsest', in *Manuscripta Graeca et Orientalia: Mélanges monastiques et patristiques en l'honneur de Paul Géhin*, ed. A. Binggeli, A. Boud'hors, and M. Cassin, Orientalia Lovaniensia Analecta 243, Leuven: Peeters, 2016, 469–496

Lagarde, A. de, *Analecta Syriaca*, Leipzig, 1858

Lamoreaux, John C., *Ḥunayn ibn Isḥāq on His Galen Translations*, Eastern Christian Texts, Provo UT: Brigham Young University Press, 2016

Land, J.N., *Anecdota Syriaca*, 4 vols, Leiden: Brill, 1862–1875

Marcovich, Miroslav (ed.). *Cohortatio ad Graecos / De monarchia / Oratio ad Graecos* [attributed to Pseudo-Iustinus], Patristische Texte und Studien 32, Berlin/New York: W. de Gruyter, 1990

Merx, A., 'Proben der syrischen Übersetzung von Galenus' Schrift über die einfachen Heilmittel', *Zeitschrift der Deutschen Morgenländischen Gesellschaft*, 39 (1885), 237–305

Penn, Michael, 'Demons Gone Wild: An Introduction, and Translation of the Syriac Qenneshre Fragment', *Orientalia Christiana Periodica*, 79 (2013), 367–399

Pognon, H., *Une version syriaque des aphorismes d'Hippocrate*, Leipzig: J.C. Hinrichs, 1903

Reich, Edgar, 'Ein Brief des Severus Sebokt', in *Sic Itur ad Astra: Studien zur Geschichte der Mathematik und Naturwissenschaften*, ed. Menso Folkerts and Richard Lorch, Wiesbaden: Harrassowitz, 2000, 478–489

Sachau, E., *Inedita Syriaca*, Vienna, 1870

Scher, A., *Mār Barḥadbšabba ʿArabaya: Cause de la fondation des écoles*, Patrologia Orientalis 4:4, Paris, 1907; republished, Turnhout: Brepols, 1981

Takahashi, Hidemi, *Aristotelian Meteorology in Syriac: Barhebraeus, Butyrum Sapientiae, Books of Mineralogy and Meteorology*, Leiden: Brill, 2004

Tarán, Leonardo, and Dimitri Gutas (eds), *Aristotle Poetics: Editio Maior of the Greek Text*, Mnemosyne Supplements 338, Leiden: Brill, 2012

Watt, John W., *Aristotelian Rhetoric in Syriac: Barhebraeus, Butyrum Sapientiae, Book of Rhetoric*, Aristoteles Semitico-Latinus 18, Leiden: Brill, 2005

Watt, John W., 'Sergius of Reshaina on the Prolegomena to Aristotle's Logic: The Commentary on the Categories, Chapter Two', in *De l'Antiquité Tardive au Moyen Âge: Études de logique aristotélicienne et de philosophie grecque, syriaque, arabe et latine offertes à Henri Hugonnard-Roche*, ed. Elisa Coda and Martini Bonadeo, Études Musulmanes XLIV, Paris: Vrin, 2014, 31–57

Watt, John W., 'The Prolegomena to Aristotelian Philosophy of George, Bishop of the Arabs', in *Christsein in der islamischen Welt: Festschrift fur Martin Tamcke zum 60. Geburtstag*, ed. Sidney H. Griffith and Sven Grebenstein, Wiesbaden: Harrassowitz, 2015, 141–163

Secondary Literature

Arzhanov, Y., 'Plato in Syriac Literature', *Le Muséon*, 132 (2019), 1–36

Arzhanov, Y., *Porphyry: On Principles and Matter*, Berlin: De Gruyter, 2021

Arzhanov, Y., and Rüdiger Arnzen, 'Die Glossen in Ms. Leyden Or. 583 und die syrische Rezeption der aristotelischen Physik', in *De l'Antiquité tardive au Moyen Âge: Études de logique aristotélicienne et de philosophie grecque, syriaque, arabe et latine offertes à Henri Hugonnard-Roche*, ed. Elisa Coda and Cecilia Martini Bonadeo, Études Musulmanes XLIV, Paris: Vrin, 2014, 415–463

Becker, Adam H., *The Fear of God and the Beginning of Wisdom: The School of Nisibis and the Development of Scholastic Culture in Late Antique Mesopotamia*, Philadelphia PA: University of Pennsylvania Press, 2006

Berti, Vittorio, *Vita e studi di Timoteo I patriarca cristiano di Baghdad*, Studia Iranica 41, Paris, 2009

Bettiolo, P., 'Scuole e ambienti intelletuali nelle chiese di Siria', in *Storia della filosofia nell' Islam medievale*, ed. Christina D'Ancona Costa, Turin: Einaudi, 2005, 48–100

Bhayro, Siam, 'Sergius of Reš 'Ayna's Syriac Translations of Galen: Their Scope, Motivation, and Influence', *Bulletin of the Asia Institute*, 26 (2016), 121–128

Bhayro, Siam, 'Galen in Syriac: Rethinking Old Assumptions', *Aramaic Studies*, 15/2 (2017), 132–154

Bhayro, Siam, 'On the Problem of Syriac "Influence" in the Transmission of Greek Science to the Arabs: The Cases of Astronomy, Philosophy, and Medicine', *Intellectual History of the Islamicate World*, 5 (2017), 211–227

Brock, Sebastian P., 'From Antagonism to Assimilation: Syriac Attitudes to Greek Learning', in *East of Byzantium: Syria and Armenia in the Formative Period*, ed. N. Garsoïan, T. Matthews, and R. Thomson, Washington DC: Dumbarton Oaks Center for Byzantine Studies, 1982, 17–32

Brock, Sebastian P., 'The Syriac Commentary Tradition', in *Glosses and Commentaries on Aristotelian Logical Texts: The Syriac, Arabic, and Latin Traditions*, ed. C. Burnett, Warburg Institute Surveys and Texts 23, London, 1993, 3–18; reprint, *From Ephrem to Romanos: Interactions between Syriac and Greek in Late Antiquity*, Aldershot: Ashgate, 1999, ch. XIII

Brock, Sebastian P., 'The Commentator Probus: Problems of Date and Identity', in *Interpreting the Bible and Aristotle in Late Antiquity*, ed. John W. Watt and J. Lössl, Ashgate, 2011, 195–206

Debié, Muriel, 'Sciences et savants syriaques: Une histoire multiculturelle', in *Les sciences en syriaque*, ed. Émilie Villey, Paris: Guenther, 2014, 9–66

Fiori, Emiliano, 'Un intellectuel alexandrin en Mésopotamie. Essai d'une interprétation d'ensemble de l'œuvre de Sergius de Rēš'aynā', in *De l'Antiquité tardive au Moyen Âge: Études de logique aristotélicienne et de philosophie grecque, syriaque, arabe et latine offertes à Henri Hugonnard-Roche*, ed. Elisa Coda and Cecilia Martini Bonadeo, Études Musulmanes XLIV, Paris: Vrin, 2014, 59–90

Griffith, Sidney H., 'Theodore Bar Koni's Scholion: A Nestorian Summa Contra Gentiles from the First Abbasid Century', in *East of Byzantium: Syria and Armenia in the Formative Period*, ed. N. Garsoïan, T. Matthews, and R. Thomson, Washington DC: Dumbarton Oaks Center for Byzantine Studies, 1982, 53–72

Griffith, Sidney H., 'The Syriac Letters of Patriarch Timothy I and the Birth of Christian Kalām in the Mu'tazilite milieu of Baghdad and Basrah in early Islamic times', in *Syriac Polemics: Studies in Honour of Gerrit Jan Reinink*, ed. Wout Jac. van Bekkum, J.W. Drijvers, and A.C. Klugkist, Leuven: Peeters, 2007, 103–132

Griffith, Sidney H., 'Patriarch Timothy I and an Aristotelian at the Caliph's Court', in *The Christian Heritage of Iraq: Collected Papers from the Christianity of Iraq I–V Seminar Days*, ed. Erica C.D. Hunter, Piscataway NJ: Gorgias Press, 2009, 38–53

Gutas, Dimitri, 'Paul the Persian on the classification of the parts of Aristotle's philosophy: a milestone between Alexandria and Baġdâd', *Der Islam*, 60 (1983), 231–267

Hein, C., *Definition und Einteilung der Philosophie: Von der spätantiken Einleitungsliteratur zur arabischen Enzyklopädie*, Frankfurt a. M.: Peter Lang, 1985

Hugonnard-Roche, H., 'Aux origines de l'exégèse orientale de la logique d'Aristote: Sergius de Res'aina (†536), médecin et philosophe', *Journal Asiatique*, 277 (1989), 1–17

Hugonnard-Roche, H., 'Note sur Sergius de Resh'aina, traducteur du grec en syriaque et commentateur d'Aristote', in *The Ancient Tradition in Christian and Islamic Hellenism: Studies on the Transmission of Greek Philosophy and Sciences, dedicated to H.J. Drossaart Lulofs*, ed. G. Endress and R. Kruk, Leiden: Brill, 1997, 121–143

Hugonnard-Roche, H., *La Logique d'Aristote du grec au syriaque: Études sur la transmission des textes de l'* Organon *et leur interpretation philosophique*, Paris: Vrin, 2004

Hugonnard-Roche, H., 'L'œuvre logique de Barhebraeus', *Parole de l'Orient*, 33 (2008), 129–143

Hugonnard-Roche, H., 'Mathématiques en syriaque', in *Les sciences en syriaque*, ed. Émilie Villey, Paris: Guenther, 2014, 67–106

Hugonnard-Roche, H., 'La question de l'âme dans la tradition philosophique syriaque (VIe–IXe siècle)', *Studia Graeco-Arabica*, 4 (2014), 17–64

Kessel, Grigory, 'Inventory of Galen's Extant Works in Syriac', in *Ḥunayn ibn Isḥāq on His Galen Translations*, ed. John C. Lamoreaux, Eastern Christian Texts, Provo UT: Brigham Young University Press, 2016, 168–193

King, Daniel, 'Alexander of Aphrodisias' *On the Principles of the Universe* in a Syriac Adaptation', *Le Muséon*, 123 (2010), 159–191

King, Daniel, 'Origenism in Sixth Century Syria. The Case of a Syriac Manuscript of the Pagan Philosophers', in *Origenes und seine Bedeutung für die Theologie- und Geistesgeschichte Europas und des Vorderen Orients*, ed. Alfons Fürst, Adamantiana: Texte und Studien zu Origenes und seinem Erbe 1, Münster: Aschendorff, 2010, 179–212

King, Daniel, 'Elements of the Syriac Grammatical Tradition as these Relate to the Origins of Arabic Grammar', in *The Foundations of Arabic Linguistics: Sibawayhi and the Earliest Arabic Grammatical Theory*, ed. Amal Marogy, Leiden: Brill, 2012, 189–209

McCollum, Adam, 'Sergius of Reshaina as Translator: The Case of De Mundo', in *Interpreting the Bible and Aristotle in Late Antiquity*, ed. John W. Watt and J. Lössl, Ashgate, 2011, 165–178

Miller, D.R., 'George, Bishop of the Arab Tribes, on True Philosophy', *Aram*, 5 (1993), 303–320

Mingana, A., 'The Apology of Timothy the Patriarch before the Caliph al-Mahdi', *Bulletin of the John Rylands Library*, 12 (1928), 137–226

Neugebauer, O., 'The Early History of the Astrolabe', *Isis*, 40/3 (1949), 240–256

Nutton, V., 'From Galen to Alexander, Aspects of Medicine and Medical Practice in Late Antiquity', *Dumbarton Oaks Papers*, 38 (1984), 1–14

Suermann, H., 'Der nestorianische Patriarch Timotheos I und seine theologischen Briefe im Kontext des Islam', in *Syriaca: Zur Geschichte, Theologie, Liturgie, und Gegenwartslage der syrischen Kirchen. Ausgewählte Vorträge des deutschen Syrologen-Symposiums vom 2.–4. Oktober 1998 in Hermannsburg*, ed. M. Tamcke, Münster, 2000, 17–30

Takahashi, Hidemi, 'Reception of Ibn Sīnā in Syriac—The Case of Barhebraeus', in *Before and after Avicenna: Proceedings of the First Conference of the Avicenna Study Group*, ed. D. Reisman, Islamic Philosophy, Theology and Science 52, Leiden: Brill, 2003, 249–281

Takahashi, Hidemi, 'Between Greek and Arabic: The Sciences in Syriac from Severus Sebokht to Barhebraeus', in *Transmission of Sciences: Greek, Syriac, Arabic and Latin*, ed. Haruo Kobayashi and Mizue Kato, Tokyo: Organization for Islamic Area Studies, Waseda University, 2010, 16–39

Takahashi, Hidemi, 'The Mathematical Sciences in Syriac: From Sergius of Resh-'Aina and Severus Sebokht to Barhebraeus and Patriarch Ni'matallah', *Annals of Science*, 68/4 (2011), 477–491

Tannous, Jack, 'Between Christology and Kalam? The Life and Letters of George, Bishop of the Arab Tribes', in *Malphono w-Rabo d-Malphone: Studies in Honor of Sebastian Brock*, ed. G. Kiraz, Gorgias Eastern Christian Studies 3, Piscataway NJ: Gorgias Press, 2008, 643–688

Villey, Émilie (ed.), *Les sciences en syriaque*, Études syriaqes 11, Paris: Guenther, 2014

Villey, Émilie, 'Qennešre et l'astronomie aux VIe et VIIe siècles', in *Les sciences en syriaque*, ed. Émilie Villey, Paris: Guenther, 2014, 149–190

Walker, Joel, *The Legend of Mar Qardagh: Narrative and Christian Heroism in Late Antique Iraq*, Berkeley CF: University of California Press, 2006

Watt, John W., 'Syriac Translators and Greek Philosophy in Early Abbasid Iraq', *Journal of the Canadian Society for Syriac Studies*, 4 (2004), 15–26

Watt, John W., 'Commentary and Translation in Syriac Aristotelian Scholarship: Sergius to Baghdad', *Journal of Late Antique Religion and Culture*, 4 (2010), 28–42

Watt, John W., 'From Sergius to Matta. Commentary and Translation in Syriac Aristotelian and Monastic Tradition', in *Interpreting the Bible and Aristotle in Late Antiquity*, ed. John W. Watt and J. Lössl, Ashgate, 2011, 239–258

Watt, John W., 'The Syriac Translations of Hunayn ibn Ishaq and their Precursors', in *Geschichte, Theologie und Kultur des syrischen Christentums: Beiträge zum 7. Deutschen Syrologie-Symposium in Göttingen, Dezember 2011*, ed. Martin Tamcke and Sven Grebenstein, Göttinger Orientforschungen: Syriaca 46, Wiesbaden: Harrassowitz, 2014, 423–445

Watt, John W., 'Why Did Hunayn, the Master Translator into Arabic, Make Translations into Syriac? On the Purpose of the Syriac Translations of Hunayn and his Circle', in *The Place to Go to: Circles of Learning in Baghdad in the Ninth and Tenth Centuries*, ed.

Jens Scheiner and Damien Janos, Studies in Late Antiquity and Early Islam, Princeton NJ: Darwin Press, 2014, 353–377

Watt, John W., 'Sergios von Reš'ayna', in *Grundriss der Geschichte der Philosophie, begründet von Friedrich Ueberweg, völlig neubearbeitete Ausgabe: Die Philosophie der Antike, 5. Die Philosophie der Kaiserzeit und der Spätantike, Teilband 3*, ed. C. Horn, C. Riedweg, and D. Wyrwa, Basel: Schwabe, 2018, § 194

Westerink, L.G., 'Medicine and Philosophy in Late Antiquity', *Janus*, 51 (1964), 169–177

Zonta, M., 'Nemesiana Syriaca: New Fragments from the Missing Syriac Version of the *De Natura Hominis*', *Journal of Semitic Studies*, 36 (1991), 223–258

CHAPTER 5

Why Do We Translate? Arabic Sources on Translation

Uwe Vagelpohl and Ignacio Sánchez

Part One: Essay[1]

The Greek–Arabic translation movement and translations into Arabic from other languages, such as Syriac, Persian, and Sanskrit, were immensely influential during the formative period of classical Arabic literature, between the second/eighth and the fourth/tenth centuries. Contemporary Arabic sources are full of information about ancient philosophy and sciences, translations and translators, and sponsors and scholars who eagerly read and argued about these texts. In spite of this wealth of information, however, it is difficult to find statements that give a clear-cut answer to the question as to why we translate.

To understand why contemporary observers were apparently not overly concerned with this question and to assess exactly what their writings tell us about translation, it is useful first to describe the nature and extent of the relevant sources.

1 *Types of Material on Translation in the Arabic Literary Tradition*

Extant sources that provide information about translation, translators, sponsors, and other relevant subjects fall into two categories. The first are the translations themselves. Translators and scribes often recorded information about these at the beginnings, the ends, or in the margins of the texts. With few exceptions, their notes tell us little beyond what was translated, when, and by whom. In some cases however, we also learn who paid for the translation and why it was undertaken. Translation notes in the body of translated texts constitute a related source. Almost all such notes in Arabic translations are from the pen of one man, the physician and master translator Ḥunayn ibn Isḥāq (d. 260/873).[2] Ḥunayn annotated the translations he produced with explana-

1 This research for this chapter was funded by the Wellcome Trust (grant no. 212385/Z/18/Z). The transliteration of Arabic names and terms follows the standard established by the Deutsche Morgenländische Gesellschaft.

2 See below, 255–256, and 1.5 with n. 52.

DOI:10.1163/9789004472648_006

tory notes, glosses, and clarifications. He also frequently explained his procedure, remarked on philological problems, and added material to fill gaps in the text or aid comprehension. None of these notes tells us how Ḥunayn would have explained why he translated. Rather, they offer a window into the practice of translation and the mindset of a translator who confidently reconstructed the text, added missing material, and sometimes even contradicted the original author.

The second category of Arabic material about translation comes from classical Arabic literature, which in our case means mainly contemporary bibliographical and biographical compilations, but can also mean other writings, for example historical, philosophical or religious works. In addition to basic information about specific translations—who translated what, for whom, and when—these works offer a substantial amount of detail on the biographies of individual translators, the activities of sponsors, and the translation history of specific texts or entire genres of texts; alternatively, they can criticise translations and translators and sometimes even question the value of translation itself. It is this second category of material from which the sample texts below are drawn. We hope they will be a representative selection from what is a much larger body of material.

The most informative sources for an understanding of the translation movement are contemporary bibliographical works. Two of them stand out because they do not just offer a wealth of information but became essential sources for all later writers who grappled with the subject: Ḥunayn ibn Isḥāq's so-called *Epistle* (*al-Risālah*) and the *Catalogue* (*Kitāb al-fihrist*) by Ibn al-Nadīm (d. 385/995 or 388/998).

Ḥunayn ibn Isḥāq was the most important and prolific translator in Islamic history.[3] He was also the only active participant in the translation movement who wrote a detailed account of his own translation activities and that of his associates. This comes in the form of a letter addressed to one of his sponsors, the courtier ʿAlī ibn Yaḥyā al-Munaǧǧim (d. 275/888–889),[4] who had asked Ḥunayn for a bibliographical inventory of Syriac and Arabic translations of Galen. Ḥunayn replied with the *Epistle*,[5] which contains precious bibliographical information about previous translations, prosopographical details

3 See Gotthard Strohmaier, 'Ḥunayn b. Isḥāq', *EI Three*.

4 He is also mentioned in Ibn Abī Uṣaybiʿah's list of translation sponsors quoted below, II.2 with n. 267.

5 See Ḥunayn ibn Isḥāq, *al-Risālah*. Gotthelf Bergsträßer's edition and German translation of the *Epistle*, with additions and revisions in Bergsträßer, *Neue Materialien*, are preferable to John C. Lamoreaux's recent *Ḥunayn ibn Isḥāq on His Galen Translations*, a re-edition and English translation of only one of the recensions of the *Epistle*; see the review by Gutas, 'A

about translators and patrons, and technical information concerning translation practices (III.1–4). He probably composed it in the year 241/855–856 and updated it eight years later; other additions were made shortly after his death.[6] Ḥunayn's *Epistle* became a crucial source of information about medical translations for all subsequent historians of Arabic scientific and medical literature, including the next author, Ibn al-Nadīm.

Ibn al-Nadīm was a bookseller and copyist of manuscripts, an occupation he inherited from his father. We know little about his life, except that he lived and worked in Baghdad, was taught by some of the most eminent scholars of his day, and moved in the circle of ʿĪsā ibn ʿAlī (d. 391/1001), the son of the 'good vizier' ʿAlī ibn ʿĪsā ibn al-Ǧarrāḥ (d. 334/946) and an important sponsor of translations.[7] His *Catalogue*, finished in 377/988, was meant to be a comprehensive index of all books in Arabic by Arab and non-Arab authors.[8] Of its ten chapters the seventh in particular, which deals with books on philosophy and the 'ancient sciences', is an invaluable source on Arabic philosophical, medical, and scientific literature, as well as translations into Arabic. Ibn al-Nadīm included not only extensive lists of books, translators, and patrons (II.1, II.3, and II.6), but also provided information about the authorship of translations (II.10).

Another important scholarly genre, which developed a little later and relied heavily on the aforementioned sources, is collections of biographies focused on science, especially medicine, the so-called *ṭabaqāt* (classes). The most important works of this genre are *The Classes of Nations* (*Ṭabaqāt al-umam*, II.5) by Ṣāʿid al-Andalusī (d. 462/1070); *The Classes of the Physicians and Sages* (*Ṭabaqāt al-aṭibbāʾ wa-l-ḥukamāʾ*, I.2, I.5, I.11, II.7) by Ibn Ǧulǧul (d. after 384/994–995); the *History of Learned Men* (*Taʾrīḫ al-ḥukamāʾ*, I.10, IV.3) by Ibn al-Qifṭī (d. 646/1248); and *The Best Accounts of the Classes of Physicians* (*ʿUyūn al-anbāʾ fī ṭabaqāt al-aṭibbāʾ*, I.8, I.12, II.2, II.8, II.9, II.11, III.5) by Ibn Abī Uṣaybiʿah (d. 668/1270). The biographies of scholars included in these works contain important bibliographical information, often in the form of book lists, and also abound in anecdotes that provide useful details about patronage networks and the relationship of these scholars with Muslim rulers, courtiers, and officials.

New "Edition"'. An abridged version of the *Epistle* that only records titles, length in books, and translator of each work was edited and translated by Käs in 'Eine neue Handschrift'.

6 See Ullmann, *Wörterbuch*, 30–31.

7 See Harold Bowen, "ʿAlī b. ʿĪsā', *EI2*.

8 See Johann W. Fück, 'Ibn al-Nadīm', *EI2*. The *Catalogue* has been edited several times. The sample texts below quote the text of Ibn al-Nadīm, *Kitāb al-fihrist*. On occasion we refer to readings in two other editions, by Riḍā Taǧaddud (*Kitāb al-fihrist li-l-Nadīm*) and Ayman Fuʾād Sayyid (*al-Fihrist li-l-Nadīm*). The text has been translated into English by Bayard Dodge as *The Fihrist of al-Nadīm*.

The numerous quotations from Greek sources in Arabic translation bear testimony to the popularity of these works and their reception by Muslim authors.

II *What Do These Sources Tell Us about Translation?*

As noted, Muslim scholars left numerous statements about translation. The bulk of these statements deals with the Greek–Arabic translation movement, a concerted effort to translate Greek philosophy, science, and medicine into Arabic that was sponsored by caliphs, high-ranking officials, wealthy individuals, and scholars, and which lasted from roughly the late second/eighth to the middle of the fourth/tenth century. Even after the end of systematic translation activities, individual scholars continued to undertake translations, and read and comment on them; translation remained a frequent subject of discussion in scholarly writing up to and during the Mamluk period (648/1250–922/1517).

There are few references in our sources to translation in the period prior to the accession of the ʿAbbāsid dynasty (132/749–656/1258), especially for the first few decades of Islamic history until 41/661 (I.1, I.3, I.4). What little there is can be found in narratives on the life of the Prophet Muḥammad and the Muslim conquests and mostly reflects the diplomatic and administrative needs of the new Muslim state. There is as yet no mention of science and philosophy. With few exceptions, the sources are also silent about any attempt to translate scientific works under the predecessors of the ʿAbbāsids, the Umayyad dynasty (41/661–132/749). The first exception is the fictitious scholarly activity of the Umayyad prince Ḫālid ibn Yazīd ibn Muʿāwiyah (d. 85/704), who is said to have sponsored translations of Greek and Coptic alchemical works (e.g., II.1).[9]

Also doubtful is the historicity of reports about the Umayyad rulers' patronage of medical works. Some sources (e.g., I.2) identify the Jewish physician Māsarǧawayh as one of the earliest translators and credit him with the translation of a medical handbook, originally written in Syriac, which was then officially endorsed by the caliph ʿUmar II ibn ʿAbd al-ʿAzīz (r. 99/717–101/720). Among the translations from Middle Persian listed in Ibn al-Nadīm's *Catalogue*, we also find mention of astrological material, some of which may have been translated in support of the ʿAbbāsid political agenda during the decades-long conflict that would lead to the overthrow of the Umayyad dynasty.[10]

In contrast to this earlier period, the systematic effort to translate the available corpus of Greek philosophical, scientific, and medical literature into Ara-

---

9 II.1 records him as the patron of the first translator listed by Ibn al-Nadīm, Iṣṭifan the Elder. On Ḫālid's alleged interest in alchemy, see Ullmann, 'Ḫālid ibn al-Yazīd', esp. 193–194, 211–218.

10 See Gutas, *Greek Thought*, 25–27, esp. 27.

bic during the heyday of the ʿAbbāsid dynasty figures very prominently in historiographical writings of the time and was eagerly seized upon by both pro-ʿAbbāsid and anti-ʿAbbāsid scholars. The acquisition of Greek wisdom through procuring ancient Greek books from the Byzantine authorities and translating them into Arabic, often presented in mythical terms, became a central element of ʿAbbāsid propaganda.[11] Some narratives traced the acquisition of Greek books to the caliphate of Hārūn al-Rašīd (r. 170/786–193/809), particularly his campaigns against the Byzantines in Cilicia (I.5), or to missions dispatched by this and other caliphs or wealthy patrons to gather manuscripts in Byzantine territory (II.6). One of the most popular narratives that sought to explain the impetus behind the Greek–Arabic translation movement revolves around a dream in which the son of Hārūn al-Rašīd, al-Maʾmūn (r. 198/813–217/833), encountered Aristotle. According to this widely reported episode, the philosopher instructed the caliph about the nature of the good. Inspired by Aristotle's wisdom, the caliph resolved to gather ancient Greek books and sponsor their translation (I.9).[12]

At the same time, the acquisition and translation of Greek books became a talking point among religious critics and polemicists, who blamed the ʿAbbāsids for the corrosive effect these ancient sciences had on religion. One polemical motif that reappears in various guises involved hidden libraries: Byzantine rulers had allegedly locked away ancient Greek manuscripts to protect their subjects from their pernicious influence, but gladly handed them over to ʿAbbāsid diplomats sent in search of ancient knowledge, in the hope that they would corrupt Islam (IV.4).

Translation was also discussed in practical terms both by the translators themselves and by the scholars who read them. The translators' claim to be able to convey the exact meaning of a text from one language to another was frequently undermined by the problematic nature of the translated texts with their often convoluted style and obscure terminology. Terminological problems were a common concern in medical works, for example, especially when it came to botanical and pharmacological terms (e.g., IV.6), where a mistranslation could make the difference between a drug and a poison. The basis for the medieval pharmacopoeia was *On Medical Material* (*De materia medica*) by Dioscorides (fl. AD c. 40–90). As Ibn Ǧulǧul explained in the introduction to one of his works on simple drugs, the first translation of Dioscorides was deficient, and its improvement in al-Andalus by a team of scholars was a regarded

11 Ibid., 28–60.
12 Ibid., 95–104.

as a major scientific achievement (I.12). Another discipline in which the proper translation of technical terms was often discussed was philosophy, especially metaphysics and logic, and philosophers such as al-Fārābī devoted many pages to dealing with the problems involved in rendering the nuances of Aristotle's terminology into other languages (IV.7).

The dangers of misinterpretation—or outright manipulation—were also considered a major problem in translations of religious works. Even though Muslim tradition accepted the divine mission of previous prophets and the truth of the message revealed to them, the rendition of this message in Hebrew and Greek, and the translation of the Torah and the Gospels into Arabic, raised many concerns and suspicions (III.6). Translators were also accused of adulterating literary works, such as the fables of Kalīlah and Dimnah, in order to convey Manichaean doctrines and other heresies (IV.1).

The translation of any work of a technical nature required such a wide range of skills that some third-/ninth-century observers did not believe that anyone, least of all the translators active at that time, possessed them. Al-Ǧāḥiẓ (d. 255/869), the famous Muʿtazilite author and polymath from Basra, argued both sides in his writings (IV.5d, IV.5e). His particular obsession was the role of the translator as an active agent rather than a passive intermediary, but he was not the only scholar who wrote about or warned against the shortcomings of translators and the material problems involved in the transmission of texts.[13]

Needless to say, meaning does not only reside in words. Scholars were especially concerned about the relationship between content and form, between the meaning of utterances (*maʿnā*) and the way in which this meaning is conveyed (*lafẓ*). This affected the translation of poetry in particular. According to medieval authors, poetry was an almost innate art form among the Arabs. They boasted of the Arabs' ability to declaim verses spontaneously and considered this art the quintessence of Arabness. Unsurprisingly, some scholars argued that it was impossible to translate poetry because the meaning of verses would inevitably be distorted without their poetic form (IV.5a–c).

Finally, some translators wrote down first-person accounts of their own experiences that offer extraordinary insights into the practicalities of their work. The exceptional testimony of Ḥunayn ibn Isḥāq, mentioned above, and of other scholars allows us better to understand the intermediary role of Syriac translations, the nature and extent of collaboration among translators, the

13 See al-Ǧāḥiẓ's remarks in IV.5c, IV.5f, and IV.5g. Abū Ḥayyān al-Tawḥīdī (d. 414/1023), a great admirer of al-Ǧāḥiẓ, also discussed this problem; see IV.8a–c.

process of revising translations, and the efforts scholars and translators made to acquire Greek manuscripts (III.1–4, III.6–7).

III Context: Translation, Religion, and Ideology

The attitude contemporary observers took toward translation was largely determined by their approach. One such approach was through religion, which also became relevant for the reception of scientific works in general. The Qur'ān was revealed in Arabic and, with the conversion of non-Arab peoples to Islam, Muslim scholars and authorities had to face the problem of making the word of God accessible to the new members of the religious community.[14]

This problem was complicated by a core Islamic principle that was first referred to in the third/ninth century, but possibly formulated even earlier: the dogma of the inimitability of the Qur'ān (*i'ǧāz al-Qur'ān*), which postulates that the language of the Book of God is beyond human capabilities and therefore impossible to imitate. By virtue of the divine nature of the Qur'ān, its revelation and subsequent recitation by the Prophet Muḥammad are considered a miracle. Muslim scholars also provided a historical explanation for the miraculous nature of the Qur'ān based on a categorization that assigns to each prophet the most meaningful miracle for their people: Moses turned his staff into a snake because magic enjoyed the highest regard among Jews; Jesus brought Lazarus back to life and cured lepers because medicine was the most important discipline for Christians; and Muḥammad recited the Qur'ān because language and poetry were at the heart of Arabic culture. These ideas can be found as early as the third/ninth century in the works of al-Ǧāḥiẓ and the theologian Ibn Qutaybah (d. 276/889) and have been used in apologetics ever since.[15]

This categorization of miracles is a perfect example illustrating the complexities of the issue of translation in classical Islam. Al-Ǧāḥiẓ was also the author of one of the first refutations of Christian beliefs, in which he raised doubts about divergences in Christian and Jewish religious sources and the translation of the Torah and the Gospels.[16] His scepticism concerning the mere possibility of translation stems to a great extent from his understanding of the Arabic language as the language chosen by God to communicate His revelation (on al-Ǧāḥiẓ's attitude towards translation, see IV.5a–g). But the implications of this dogma went further. Christians were regarded as the people of medicine, and a mere glance at the texts collected in this article demonstrates the prepon-

14 On this problem, see Zadeh, *The Vernacular Qur'an*, 214–250.

15 See al-Ǧāḥiẓ, 'Ḥuǧaǧ al-nubuwwah', 278.8–280.15, and Ibn Qutaybah, *Ta'wīl muškil al-Qur'ān*, 12.7–13.

16 Al-Ǧāḥiẓ, 'al-Radd 'alā l-naṣārā', 329.18–331.3.

derance of Christian translators who were more often than not also physicians (see II.1–2). Does this mean that these Christians were automatically the heirs of Greek wisdom? According to al-Ǧāḥiẓ, they certainly claimed so, and he was categorical in denying them this status. First of all, al-Ǧāḥiz argued, Galen, like other ancient Greek authors, was not a Christian and, more importantly, Christian Arabs and Byzantines were not Greek.[17]

Religion, ethnicity, and an ideology of cultural supremacy form the backdrop to all early discussions of translation, and it is here that an implicit answer to the question as to why we translate may be sought. As Dimitri Gutas has shown, these concerns were very significant for the Muslim caliphs who sponsored the translation movement.[18] The ʿAbbāsid conception of history was based on teleological and messianic principles, and with their accession, they considered themselves to be the successors of previous dynasties, religions, and civilizations: they were the heirs not just of the Prophets, but also of the sages of antiquity.

IV *Instead of a Conclusion*

The final section of the samples assembled below is entitled, perhaps somewhat presumptuously, 'Why Do We Translate?'. It contains a single text, a quotation from a treatise by the philosopher al-Kindī (d. 252/866), a contemporary of Ḥunayn ibn Isḥāq and sponsor of a number of earlier translations of philosophical works (V.1). Al-Kindī's remarks—that the accumulation of knowledge is a process that spans generations, and that we owe a debt of gratitude to our forebears, whatever their religious and ethnic background—are echoed by other authors at the time.

This noble sentiment offers an answer to our opening question that is generic at best. It also illustrates the kind of response that seems to emerge from the flood of pronouncements on translation that can be found in the Arabic literary tradition. As stated at the beginning, not a single one of our sources can be said to give a direct and unequivocal answer to the question 'why do we translate?' The picture our sources sketch out is that of a society in which translation and translated texts were very much part of the intellectual every-

17 Al-Ǧāḥiẓ, 'al-Radd ʿalā l-naṣārā', 314.9–315.10. Al-Ǧāḥiẓ also wrote an epistle against medicine and physicians that was refuted by Abū Zakariyā' al-Rāzī (d. 313/925 or 323/935; see below, I.8 with n. 80). Both works are lost; see Pellat, 'Nouvel essai', 160, n. 221. A century later, the Muʿtazilite theologian ʿAbd al-Ǧabbār (d. 415/1025) devoted a substantial part of his polemic against the Christians to a refutation of the scientific value of medicine; see his *Taṯbīt*, esp. 612–641.

18 See Gutas, *Greek Thought*, esp. 28–60 and 75–106.

day life during roughly the first two centuries of ʿAbbāsid rule. Not even the one translator who, thanks to his own writings, is more than a name and the sum of a few anecdotes, saw fit to comment on his and his contemporaries' motivations: there is no evidence Ḥunayn asked himself that question, neither in its most general form—why do we translate at all?—nor in its more specific forms: why translate medicine or philosophy? why translate this particular work?

The debates on translation that the sources record invariably touch on particular aspects and problems of translation, for example the process itself; obstacles to correct translation; and the impact of translation on people's religious beliefs. The arguments put forward in the third/ninth century in some of the earliest surviving literary and scholarly works seem to document an already mature stage of an ongoing debate about translation. They illustrate a situation in which writers felt confident discussing technical details and assessing the impact of translation on society. This suggests that they and their predecessors had had an opportunity to observe this phenomenon for some time, form their opinions, and hone their arguments.

On grounds of plausibility, one could argue that if the question 'why do we translate?' was posed explicitly, it must have been before the start or at the very beginning of the Greek–Arabic translation movement, but if there was an answer, we do not know it. What we do know is that the realities of a multilingual state and the practical necessities of administration led to translation activities of various sorts from early on, such as the use of interpreters, or the widely reported translation of administrative records, the 'Arabization' of the tax registers (*dīwāns*, I.3, I.4). At the same time, there must have been constant and extensive exchanges between travellers, traders, scholars, and all sorts of people in the expanding Islamic state. In such an environment, the question as to why one translates seems moot. This, together with the lack of reliable documentation for events that precede the ʿAbbāsid era, may explain why we do not hear of any sustained debate about the advisability of or the justification for translation.

What the authors on whom we rely seem to deem much more important were questions of cultural transfer: who took knowledge from whom. The reports of this mostly legendary transfer—which range from mythical accounts of Alexander the Great's pillage and translation of Persian literature into Greek, to the more contemporary but no less mythical accounts of the transfer of philosophical and medical teaching from Alexandria to Baghdad[19]—either mention translation, but do so baldly, as something that took place but was

19 See Gutas, 'The "Alexandria to Baghdad" Complex of Narratives', 155–193.

not in question, or do not mention it at all: for the authors and compilers of these reports, it was clearly an uncontroversial activity.

The same holds true for the various stories that seek to explain how the vast numbers of Greek works that were translated fell into Muslim hands in the first place. We hear of books given as spoils of war or tributes to Muslim rulers, or of books recovered through the efforts of scholars and their sponsors, who collected them inside and outside Muslim territories. Translation is always mentioned as one step in the transfer of knowledge: manuscripts are acquired, translations commissioned, and then, signally, the public is encouraged to read these books. But again, the question as to why we translate is not asked.

The transfer of knowledge was for our sources as much a fact of life as translation, but it seems to have been more under scrutiny than translation itself. This may explain the continued popularity of origin stories, which seek to give certain authors, works, or scholarly fields the sanction of ancient or contemporary authorities by claiming, for example, that the knowledge involved can be traced back to the sages of antiquity; that it has been sought out through the ages by emblematic rulers of other cultured nations, such as the Greeks and the Persians; that after the founding of the Islamic state, its pursuit had been sanctioned by caliphs and scholars; and that given the defeat or current weakness of the other carriers of this knowledge, especially the Greeks and the Persians, it is incumbent on the Muslim community to find, translate, and save this knowledge from oblivion.[20]

This is where we come back to al-Kindī and his paean to his predecessors. By looking at translation through the lens of cultural transfer, transferred knowledge itself takes centre stage rather than the process of transfer and translation. Translation, an everyday activity in a multilingual society, was just one of several steps to get to this knowledge. Perhaps this is another reason why the question 'why do we translate?' does not figure in our sources: the obvious value of knowledge, the end result of the process of transmission and translation, may have obviated the need to justify the individual steps of the process.

20 See Gutas, *Greek Thought*, 83–95.

Part Two: Texts in Translation

Overview

I. History: Origins, Genres, Works
II. People: Translators and Sponsors
III. Methods: Procedures and Approaches
IV. Problems: Difficulties and Criticisms
V. Why Do We Translate?

I. *History: Origins, Genres, Works*

I.1. Al-Masʿūdī, *The Book of Admonition and Revision* (*Kitāb al-tanbīh wa-l-išrāf*), 282.12, 283.2–5 (ed. de Goeje); tr. Vagelpohl and Sánchez.

Shortly before his death in 345/956, the historian and geographer al-Masʿūdī wrote an abridgment of his best-known work, a massive chronicle of world history from Creation to his days entitled *The Meadows of Gold and Mines of Gems* (*Murūǧ al-ḏahab wa-maʿādin al-ǧawhar*).[21] The abridgment, the *Book of Admonition and Revision* (*Kitāb al-tanbīh wa-l-išrāf*), amounts to less than a quarter of its source, but adds some material not contained in it.[22] The present passage illustrates the importance of translation in the political and diplomatic sphere from the very beginning of Islamic history. Zayd ibn Ṯābit (d. betw. 42/662 and 56/676), one of the scribes of the Prophet Muḥammad who wrote down the revelations he received, is better known for his role in the compilation of the Qurʾān under the caliph ʿUṯmān (r. 23/644–35/656) and as a transmitter of *ḥadīṯ*.[23]

Among the Prophet's secretaries was [...] Zayd ibn Ṯābit al-Anṣārī, a member of the Ġanm ibn Mālik ibn al-Naǧǧār branch of the Ḫazraǧ tribe. He wrote (diplomatic letters) to the kings and responded at the behest of the Prophet, and he interpreted for him in Persian, Greek, Coptic, and Ethiopic. He learned (interpreting) in Medina from speakers of these languages.[24]

---

21 See below, II.4.
22 See Charles Pellat, 'al-Masʿūdī', *EI2*.
23 See Michael Lecker, 'Zayd b. Thābit', *EI2*.
24 Another translation: al-Masʿūdī, *Le Livre de l'avertissement*, 371–372 (French).

I.2. Ibn Ǧulǧul, *The Classes of the Physicians and Sages* (*Ṭabaqāt al-aṭibbāʾ wa-l-ḥukamāʾ*), 61.2–5 (ed. Fuʾād Sayyid); tr. Vagelpohl and Sánchez.[25]

Ibn Ǧulǧul (d. after 384/994–995) was a Cordovan physician and historian of medicine, famous for his works on *materia medica* and his collection of medical biographies entitled *The Classes of Physicians and Sages* (*Ṭabaqāt al-aṭibbāʾ wa-l-ḥukamāʾ*). This compilation is the first extant history of medicine after the *Chronology of the Physicians* (*Taʾrīḫ al-aṭibbāʾ*) by Isḥāq ibn Ḥunayn (d. 289/911)[26] and the first of its kind in Western Islam. Ibn Ǧulǧul is the first Arab historian who quotes from Latin sources, as he proudly states in his introduction, in which he mentions the Latin translation, by Jerome (d. 420), of the *Chronicle* (*Chronica*) by Eusebius of Caesarea (d. 339–340), and the *Histories Against the Pagans* (*Historiae adversus paganos*) by Orosius (d. after 418).[27] His medical biographies also include fragments from the *Etymologies* (*Etymologiae*) by Isidore of Seville (d. 636), most likely the result of excerpts translated *ex professo* by Mozarabs, since no full translation of this work is known.[28]

If the following report from Ibn Ǧulǧul's *Classes of Physicians* is to be credited, the translation into Arabic by the Jewish physician and translator Māsarǧawayh[29] of a medical handbook by the Christian Syriac physician Ahrun[30] was one of the earliest translations of a medical text. Doubts about its historicity notwithstanding, the text suggests that translations of scientific and medical works into Arabic may have already begun during the Umayyad era (41/661–132/750).[31]

The politics of this report are delicate: a translator, explicitly identified as non-Muslim, embarked on his work without any prompting from the Umayyad authorities. His work is then endorsed, after a suitable period of reflection and prayer, by ʿUmar II, the only Umayyad caliph renowned for his piety. The Umayyads, including their relatives who ruled al-Andalus during the lifetime of Ibn Ǧulǧul, are therefore entirely absolved from any responsibility for the translation. On the contrary, the precedent set by ʿUmar II can now be used to exonerate both the practice of translation and engaging

---

25 Parallels: Ibn al-Qifṭī, *Taʾrīḫ al-ḥukamāʾ*, 324.19–325.4; and Ibn Abī Uṣaybiʿah, *ʿUyūn al-anbāʾ*, II/1, 421.8–11.

26 See below, I.6 with n. 63.

27 On the Arabic translations of Orosius's *Histories*, see below, I.12 with n. 122.

28 On Ibn Ǧulǧul, see Ignacio Sánchez, 'Ibn Juljul, Abū Daʾūd Sulaymān b. Ḥassān', *EI Three*.

29 See Meyerhof, 'Mediaeval Jewish Physicians', 435–437.

30 The medical *Handbook* (*Kunnāš*) of Ahrun ibn Aʿyān al-Qass (fl. first/seventh century), originally written in Greek, was translated early on into Arabic through a Syriac intermediary. The translation itself is lost, but the work was frequently quoted by later authors. See Albert Dietrich, 'Ahrun', *EI2*; and Ullmann, *Die Medizin*, 87–89.

31 See Richter-Bernburg, 'Islamischer Hellenismus', 48–49; and Gutas, *Greek Thought*, 23–24.

with the ancient sciences in general, which had been rejected previously because they were associated with the ʿAbbāsid dynasty.[32]

(Māsarǧawayh) was a Syriac-speaker of the Jewish faith, and during (the rule of) the (early) Marwānids he undertook to translate the book of the Christian priest Ahrun ibn Aʿyān into Arabic. ʿUmar ibn ʿAbd al-ʿAzīz[33] found it [i.e., the translated book] in the (caliphal) library and ordered it to be brought out and deposited in his oratory. ʿUmar asked God for guidance as to whether to publish it among the Muslims so that they could benefit from it, and when he had done so for forty days, he made it public to the people and disseminated it among them.

1.3. Al-Balāḏurī, *The Conquests of the Regions* (*Futūḥ al-buldān*), 193.1–9 (ed. de Goeje);[34] tr. Hitti, *The Origins*, I, 301, adapted by Vagelpohl and Sánchez.

The third-/ninth-century historian al-Balāḏurī (d. c. 278/892) chronicled the expansion of the Muslim state from the wars of the Prophet Muḥammad to the conquest of Iraq and Persia in his *Conquests of the Regions* (*Futūḥ al-buldān*), which is a shortened version of a more comprehensive but lost historical work about the same period. In addition to political and military history, the work contains invaluable information about the social and cultural history of the early Islamic era.[35]

Until some time after the initial conquest, the administrative structures of the former Byzantine and Persian provinces of the Islamic state were largely kept intact, which also meant that administrative records continued to be kept in their original languages, Greek and Persian. The episodes described in the following two texts mark the transition to an administrative system that relied exclusively on the Arabic language. The translation of Greek administrative records into Arabic in the western provinces was either part of the reforms instituted by the Umayyad caliph ʿAbd al-Malik (r. 65/685–86/705), as the following text suggests, or was commissioned by his son Hišām (r. 105/724–125/743).[36]

---

32 In another report quoted below (I.12), Ibn Ǧulǧul mentions the role the Umayyad ruler of al-Andalus played in the translation of an important Greek pharmacological work.

33 The Umayyad caliph ʿUmar II.

34 Parallels: Ibn al-Nadīm, *Kitāb al-fihrist*, I, 242.25–30, translated in Rosenthal, *The Classical Heritage*, 48; and Ibn al-Nadīm, *The Fihrist*, II, 583 (English).

35 See Franz Rosenthal and Carl Heinrich Becker, 'Balādhurī', *EI2*; and Charles Bosworth, 'Balāḏorī', *Encyclopaedia Iranica*.

36 The shorter and less colourful account of this event in Ibn al-Nadīm (*Kitāb al-fihrist*, I, 242.25–30) credits either Hišām or ʿAbd al-Malik; see Gutas, *Greek Thought*, 17, 23.

The translation of the state registers of the Greek (territories)

They said: Greek remained the language of the state registers of Syria until the reign of ʿAbd al-Malik ibn Marwān, who in the year 81/700–701 ordered them to be translated. The reason was that a Greek chancery secretary desiring to write something and finding no water (to prepare ink) urinated in the inkstand. Hearing this, ʿAbd al-Malik punished the man and gave orders to Sulaymān ibn Saʿd[37] to translate the registers. Sulaymān requested ʿAbd al-Malik to assign to him the land tax of the Jordan province for one year. ʿAbd al-Malik granted his request and assigned him to the governorship of the Jordan. No sooner had the year ended, than the translation was finished and Sulaymān brought the registers to ʿAbd al-Malik. The latter called his secretary Sarǧūn [Sergius][38] and showed it to him. This distressed Sarǧūn and he left ʿAbd al-Malik sorrowful. A group of Greek secretaries ran into him and he said: 'Seek your livelihood in any other profession than this, for God has cut it off from you.'

I.4. Al-Balāḏurī, *The Conquests of the Regions* (*Futūḥ al-buldān*), 300.8–301.9 (ed. de Goeje);[39] tr. Hitti, *The Origins*, I, 465–466, adapted by Vagelpohl and Sánchez.

The translation of the Persian administrative records described below took place somewhat later than that of the Greek records. It was instigated by the Umayyad governor of Iraq, al-Ḥaǧǧāǧ ibn Yūsuf (d. 95/714), allegedly so that he would not have to depend on his Persian bureaucrats to inspect the tax register.[40]

Translation of the state register of the Persian (territories)

Al-Madāʾinī ʿAlī ibn Muḥammad ibn Abī Sayf[41] reported to me from his teachers: Persian remained the language of the register of the land tax of al-Sawād

37 According to Ibn al-Nadīm (*Kitāb al-fihrist*, I, 242.27–28), Abū Ṯābit Sulaymān ibn Saʿd (fl. late first/seventh century) was a client of the Prophet's nephew Ḥusayn and directed the caliphal chancery during the reign of ʿAbd al-Malik.

38 Sarǧūn ibn Manṣūr (fl. first/seventh century), a member of a Damascene Melkite family who already had administrative responsibilities under the Byzantines, headed the financial administration (*dīwān*) for the Umayyad caliphs from Muʿāwiyah I (r. 41/661–60/680) to ʿAbd al-Malik; see Sprengling, 'From Persian to Arabic', 182.

39 Parallels: Ibn al-Nadīm, *Kitāb al-fihrist*, I, 242.8–25, translated in Rosenthal, *The Classical Heritage*, 47–48; and Ibn al-Nadīm, *The Fihrist*, 582–583 (English).

40 See Albert Dietrich, 'al-Ḥadjdjādj b. Yūsuf', *EI2*; Gutas, *Greek Thought*, 26. The episode is discussed in detail by Sprengling, 'From Persian to Arabic', 194–197.

41 The early Arab historian al-Madāʾinī (d. 228/843); see Ursula Sezgin, 'al-Madāʾinī', *EI2*.

and the rest of Iraq. When al-Ḥaǧǧāǧ became governor of Iraq, he made Zādān Farrūḫ ibn Bīrī his secretary,[42] and the latter was assisted by Ṣāliḥ ibn ʿAbd al-Raḥmān,[43] a client of the Banū Tamīm, who knew both Arabic and Persian. Ṣāliḥ's father was one of the captives of Siǧistān. Through Zādān Farrūḫ, Ṣāliḥ had access to al-Ḥaǧǧāǧ, who enjoyed his company. One day (Ṣāliḥ) said to (Zādān): 'You are the reason why I (have access) to the governor, and I see that he has taken a liking to me. I am not sure if he prefers me over you and you will fall.' He replied: 'Do not believe that: he needs me more than you because he will not find anyone other than me whom he trusts to keep his books.' (Ṣālih) said: 'By God, if you wanted me to convert the accounts into Arabic, I would do so.' (Zādān) replied: 'Convert a sample and I will see.' (Ṣāliḥ) did so. (Zādān) told him to claim to be sick, which he did. Al-Ḥaǧǧāǧ then sent his own physician, but he found nothing wrong with him. This reached Zādān Farrūḫ's ears and he ordered him to appear.

In the days of ʿAbd al-Raḥmān ibn Muḥammad ibn al-Ašʿaṯ al-Kindī,[44] Zādān Farrūḫ was killed on his way from some house to his own or some other man's home. Al-Ḥaǧǧāǧ then made Ṣāliḥ secretary in his place. Ṣāliḥ reported to al-Ḥaǧǧāǧ the conversation that took place between him and Zādān Farrūḫ about translating the register. Al-Ḥaǧǧāǧ decided to render the register into Arabic and charged Ṣāliḥ with the task. Mardānšāh ibn Zādān Farrūḫ[45] asked Ṣāliḥ: 'What do you do with *dahūyah* [tenths] and *bīstūyah* [twentieths]?' He replied: 'I shall write "*ʿušr*" [tenth] and "*nuṣf ʿušr*" [half-tenth] (instead).' (Mardānšāh) said: 'And what about *wīd* [excess]?' He replied: 'I shall write "*ayḍan*" [also], "*al-wīd al-nayyif*" [the surplus excess], and "*al-ziyādah tuzād*" [the surplus is added].' (Mardānšāh) then said: 'May God efface your trace from the world as you have effaced the trace of the Persian language!' Ṣāliḥ was later offered 100,000 dirhams in order to show that it was impossible to translate the register and to refrain from doing it, but he refused and trans-

42 Zādān Farrūḫ ibn Bīrī (d. 82/701) was a Persian tax official in the employ of the governors of Iraq from Ziyād ibn Abīhi (d. 53/673) onward; see Sprengling, 'From Persian to Arabic', 185–190.

43 Abū l-Walīd Ṣāliḥ ibn ʿAbd al-Raḥmān (d. shortly after 101/720), a Persian freedman and Muslim convert, the successor of Zādān Farrūḫ as the administrator of taxes in Iraq and later the province's chief treasurer under the caliph Sulaymān (r. 96/715–99/717); see Sprengling, 'From Persian to Arabic', 191–208.

44 ʿAbd al-Raḥmān ibn Muḥammad ibn al-Ašʿaṯ al-Kindī (d. 85/704), an Arab general during the early Umayyad caliphate and the leader of an unsuccessful rebellion against the governor of Iraq, al-Ḥaǧǧāǧ ibn Yūsuf; see Laura Veccia Vaglieri, 'Ibn al-Ashʿath', *EI2*.

45 The son of Zādān Farrūḫ and, like his father, a tax official in the service of the governors of Iraq.

lated it. ʿAbd al-Ḥamīd ibn Yaḥyā,[46] the secretary of Marwān ibn Muḥammad,[47] used to say: 'By God, Ṣāliḥ has bestowed a great blessing upon the secretaries!'

In the account he gave to me ʿUmar ibn Šabbah[48] said: Abū ʿĀṣim al-Nabīl[49] told me that Sahl ibn Abī l-Ṣalt[50] reported: Al-Ḥağğāğ gave Ṣāliḥ ibn ʿAbd al-Raḥmān a deadline for translating the register.

I.5. Ibn Ğulğul, *The Classes of the Physicians and Sages* (*Ṭabaqāt al-aṭibbāʾ wa-l-ḥukamāʾ*), 65.2–5 (ed. Fuʾād Sayyid); tr. Vagelpohl and Sánchez.[51]

One of the more important translators and physicians of the generation immediately preceding the master translator Ḥunayn ibn Isḥāq[52] was Yūḥannā ibn Māsawayh.[53] In addition to attending to several caliphs from Hārūn al-Rašīd to al-Mutawakkil (r. 232/847–247/861) and producing translations at their behest, he also taught medicine in a lecture circle attended by Ḥunayn ibn Isḥāq himself; the story of their falling out and subsequent reconciliation is recounted below (II.8). The details of the text are in all

46 ʿAbd al-Ḥamīd ibn Yaḥyā al-Kātib (d. 132/750), secretary to several Umayyad caliphs, epistolographer, and one of the earliest masters of Arabic literary prose. His style was largely inspired by Persian secretarial practices, as the philologist Abū Hilāl al-ʿAskarī pointed out in his *Dīwān al-maʿānī*, II, 89.10–12. See Wadād al-Qāḍī, "Abd al-Ḥamīd b. Yaḥyā al-Kātib', *EI Three*.

47 The Umayyad caliph Marwān II ibn Muḥammad ibn Marwān (r. 127/744–132/750).

48 ʿUmar ibn Šabbah (d. 262/878) was a scholar of history and poetry whose works, now mostly lost, with the exception of his *History of Medina* (*Taʾrīḫ al-Madīnah al-munawwarah*), became important sources for later historians; see Stefan Leder, "Umar b. Shabba', *EI2*.

49 Abū ʿĀṣim al-Nabīl (d. 212/828), who was born in Mecca but later taught in Basra, was a transmitter of Prophetic traditions.

50 Sahl ibn Abī l-Ṣalt al-Sarrā (fl. late second/eighth century), originally from Basra, was a renowned transmitter of Prophetic traditions; see Ibn Ḥağar al-ʿAsqalānī, *Tahḏīb al-tahḏīb*, II, 125.7–36.

51 Parallels: Ṣāʿid al-Andalusī, *Kitāb ṭabaqāt al-umam*, 36.11–14; Ibn al-Qifṭī, *Taʾrīḫ al-ḥukamāʾ*, 380.10–14; and Ibn Abī Uṣaybiʿah, *ʿUyūn al-anbāʾ*, II/1, 446.5–7.

52 Ḥunayn ibn Isḥāq was the most accomplished and productive translator of the ʿAbbāsid translation movement. He was also a practising physician and composed numerous original works on medicine that enjoyed great popularity. The most famous, finished posthumously by his nephew Ḥubayš, is the *Introduction to Medicine* (*al-Madḫal fī l-ṭibb*), also known as *Questions on Medicine* (*Masāʾil fī l-ṭibb*). This work was widely used as a teaching text for medicine. A version of Ḥunayn's *Introduction* was later translated into Latin under the title *Isagoge Joannitii ad tegni Galieni*. See above, 255–256; and Gotthard Strohmaier, 'Ḥunayn b. Isḥāq', *EI Three*.

53 See Jean-Claude Vadet, 'Ibn Māsawayh', *EI2*; Meyerhof, 'New Light', 717; Micheau, 'Mécènes et médecins', 152–155; and Gutas, *Greek Thought*, 118–119.

probability apocryphal; while Hārūn al-Rašīd attacked Ancyra twice during his reign, the two cities were not conquered until 223/838, during the reign of his son, the caliph al-Muʿtaṣim (r. 218/833–227/842).[54]

Yūḥannā ibn Māsawayh: a Syriac Christian. (Hārūn) al-Rašīd appointed him to translate the ancient medical books that were found in Ancyra, Amorium, and the Byzantine territory after the Muslims had conquered them. He made him the chief translator and put him in charge of skilled scribes who drafted (texts for him). (Yūḥannā) served (the caliphs) Hārūn (al-Rašīd), al-Amīn,[55] and al-Maʾmūn,[56] and kept this (office) until the days of (the caliph) al-Mutawakkil [i.e., between 232 and 247/847 and 861].[57]

1.6. Ibn al-Nadīm, *The Catalogue* (*Kitāb al-fihrist*), I, 265.20–24 (ed. Flügel); tr. Vagelpohl and Sánchez.[58]

The following quotation from Ibn al-Nadīm's *Catalogue* outlines the complex translation history of the *Elements* (*Stoicheia*)—the foundational text of Greek mathematics and geometry—by Euclid (fl. third century BC). According to Ibn al-Nadīm, the book was translated and revised several times at different stages of the Greek–Arabic translation movement, in part supported by the caliphal authorities: a testament to its importance and the interest it aroused.[59]

Account of (Euclid's) book on the *Principles of Geometry*. Its name is al-Isṭrūšiyā,[60] which means 'the principles of geometry'; al-Ḥaǧǧāǧ ibn Yūsuf ibn Maṭar translated it twice.[61] One version, the first, was known as al-Hārūnī,[62] the second was the one for (the caliph) al-Maʾmūn. It was known as al-Maʾmūnī and people relied on it.

54 See Gutas, *Greek Thought*, 123; and Micheau, 'Mécènes et médecins', 155.
55 The ʿAbbāsid caliph al-Amīn (r. 193/809–198/813).
56 The ʿAbbāsid caliph al-Maʾmūn.
57 Another translation: Micheau, 'Mécènes et médecins', 154–155 (French, partial).
58 Parallel: Ibn al-Qifṭī, *Taʾrīḫ al-ḥukamāʾ*, 64.2–6.
59 On the complex textual tradition of the various translations of Euclid's *Elements*, see Brentjes, 'Textzeugen und Hypothesen'.
60 A somewhat garbled Arabic transliteration of the original Greek title, *Στοιχεῖα*.
61 Al-Ḥaǧǧāǧ ibn Yūsuf ibn Maṭar (fl. 169/786–218/833) translated Euclid's *Elements* at the request of Hārūn al-Rašīd and revised it during the reign of al-Maʾmūn. For the latter he also translated Ptolemy's astronomical tables; see Gregg De Young and Sonja Brentjes, 'al-Ḥajjāj b. Yūsuf b. Maṭar', *EI Three*. Ibn Abī Uṣaybiʿāh (*ʿUyūn al-anbāʾ*, II/1, 468.14–18) reported that he was part of a mission dispatched by al-Maʾmūn to Byzantium in search for books; see Gutas, *Greek Thought*, 148.
62 I.e., it was translated for the caliph Hārūn al-Rašīd.

Isḥāq ibn Ḥunayn[63] translated it and Ṯābit ibn Qurrah al-Ḥarrānī[64] corrected it. Abū ʿUṯmān al-Dimašqī translated several volumes of it. I saw the tenth of them[65] in Mosul, in the library of ʿAlī ibn Aḥmad al-ʿImrānī.[66] One of his pages was Abū l-Ṣaqr al-Qabīṣī. He studied the *Almagest* under his supervision during our own time.[67]

I.7. Ibn al-Nadīm, *The Catalogue* (*Kitāb al-fihrist*), I, 267.29–268.4 (ed. Flügel); tr. Vagelpohl and Sánchez.[68]

Another foundational scientific work, the *Mathematical Treatise* (*Syntaxis mathematica*) or *Almagest* on the motion of planets and stars by Ptolemy (d. AD c. 170), also went through several stages of translation and revision. The earliest stage was again supported by the state through a commission by Yaḥyā ibn Ḫālid ibn Barmak (d. 190/805), the vizier of the caliph Hārūn al-Rašīd and himself a generous patron of scholars and translators.[69]

Account of the *Almagest*. This work is in thirteen books. The first person to see to its translation and publication in Arabic was Yaḥyā ibn Ḫālid ibn Barmak. A group of people translated it for him, but did not have the skill (to do it), and he

63 Isḥāq ibn Ḥunayn, son of the physician and master translator Ḥunayn ibn Isḥāq, followed in his father's footsteps and translated a number of medical works, but is mainly known as a translator of philosophical texts; see below, II.9, and Gotthard Strohmaier, 'Isḥāq b. Ḥunayn', *EI2*.

64 Ṯābit ibn Qurrah (d. 288/901) was a mathematician and astronomer from the northern Mesopotamian city of Ḥarrān who worked as a translator in Baghdad under the patronage of the Banū Mūsā family; see Roshdi Rashed and Régis Morelon, 'Thābit b. Ḳurra', *EI2*. He belonged to the Ṣābians, a late antique religious group concentrated in Ḥarrān, whose cult centered around the worship of the moon and other astral bodies; see Strohmaier, 'Ḥarrān'.

65 The Christian physician Abū ʿUṯmān Saʿīd ibn Yaʿqūb al-Dimašqī (fl. fourth/tenth century) was a translator mainly of philosophical and medical texts in the service of the 'good vizier' ʿAlī ibn ʿĪsā ibn al-Ǧarrāḥ. See Ibn al-Nadīm, *Kitāb al-fihrist*, I, 298.23–25; Ibn al-Qifṭī, *Taʾrīḫ al-ḥukamāʾ*, 409.14–16; and Gerhard Endress, 'Abū ʿUthmān al-Dimashqī', *EI Three*.

66 ʿAlī ibn Aḥmad al-ʿImrānī (d. 344/955–956), born in Mosul, was a mathematician, expert in arithmetic and geometry, and a renowned collector of books. His private library was open to anyone willing to learn. See Ibn al-Qifṭī, *Taʾrīḫ al-ḥukamāʾ*, 233.13–19, and Ibn al-Nadīm, *Kitāb al-fihrist*, I, 283.1–4.

67 Other translations: Ibn al-Nadīm, *The Fihrist*, II, 634–635 (English); and Kapp, 'Arabische Übersetzer', 164 (German).

68 Parallel: Ibn al-Qifṭī, *Taʾrīḫ al-ḥukamāʾ*, 97.20–98.2.

69 On Yaḥyā ibn Ḫālid and the Barmakid family, see Gutas, *Greek Thought*, 128–129, and Kevin van Bladel, 'Barmakids', *EI Three*.

was not satisfied with it. He then commissioned Abu Ḥassān[70] and Salm, the head of the House of Wisdom,[71] to translate it. They had the (requisite) skill and worked hard to revise it after having sent for the best translators, examining their translation, and selecting what was most eloquent and accurate.

People say that al-Ḥaǧǧāǧ ibn Maṭar[72] also translated it. The person who worked on it was al-Nayrīzī.[73] Ṯābit (ibn Qurrah)[74] corrected the entire work in the old translation. Isḥāq (ibn Ḥunayn)[75] translated this book and Ṯābit corrected his work, (but it was) an unsatisfactory translation because his correction of the former was better.[76]

I.8. Ibn Abī Uṣaybiʿah, *The Best Accounts of the Classes of Physicians* (*ʿUyūn al-anbāʾ fī ṭabaqāt al-aṭibbāʾ*), II/2, 886.3–888.2 (ed. Savage-Smith et al.); tr. Savage-Smith et al., III/2, 986–987, adapted by Vagelpohl and Sánchez.[77]

The Damascene medical scholar Ibn Abī Uṣaybiʿah (d. 668/1270) is the author of the most comprehensive prosopographical source about medieval medicine that has come down to us, *The Best Accounts on the Classes of Physicians* (*ʿUyūn al-anbāʾ fī ṭabaqāt al-aṭibbāʾ*). This work covers the history of medicine from its inception until the time of the author and contains the biographies of more than 450 physicians and philosophers from antiquity to the author's lifetime.[78]

Ibn Abī Uṣaybiʿah devotes a short chapter in this work to Indian medicine. It includes several biographies of ancient Indian authors of medical works and of two Indian physicians who worked at the ʿAbbāsid court. The author largely relies on information from Ibn al-Nadīm's *Catalogue*, which has a section on Indian medical works translated into Arabic. Since the exact reading of the Arabic transliterations of Indian names is not always clear and Ibn Abī Uṣaybiʿah and Ibn al-Nadīm (or their editors) do not always agree, alternative readings are provided in the notes here.

---

70 Unidentified.

71 Salm (fl. third/ninth century), a translator from Persian into Arabic, served as the head of the caliphal library, the so-called House of Wisdom (*bayt al-ḥikmah*); see Dimitri Gutas and Kevin van Bladel, 'Bayt al-Ḥikma', *EI Three*.

72 See above, I.6 with n. 61.

73 Abū l-ʿAbbās al-Faḍl ibn Ḥātim al-Nayrīzī (fl. c. 285/900) was a Persian astronomer and geometer. He wrote a commentary on al-Ḥaǧǧāǧ ibn Maṭar's Arabic translation of Euclid's *Elements*. See Jan Hogendijk, 'al-Nayrīzī', *EI2*.

74 See above, I.6 with n. 64.

75 See above, I.6 with n. 63.

76 Another translation: Ibn al-Nadīm, *The Fihrist*, II, 639 (English).

77 Parallels: cf. the list of Indian medical works translated into Arabic in Ibn al-Nadīm, *Kitāb al-fihrist*, I, 303.5–13, translated in Ibn al-Nadīm, *The Fihrist*, II, 710 (English).

78 See Juan Vernet, 'Ibn Abī Uṣaybiʿah', *EI2*.

(The physician and astrologer) Ṣanǧahal the Indian[79] was followed in India by a group of scholars, including Bākahr, Rāḥah, Ṣakih, Dāhir, Ankar, Zankal, Ǧabhar, Andī, and Ǧārī, all of whom composed many well-known works on the art of medicine and other sciences. They rank amongst the great learned men of India: they were skilled in the art of medicine, and they elucidated the rules governing the science of astrology. The scholars of India study and imitate their works, transmitting them from generation to generation. Many of those works have been translated into Arabic:

I have found that al-Rāzī, in his *Comprehensive Book* [*on Medicine*][80] and other works, drew upon a number of works by Indian scholars, such as the *Book of Šarak* the Indian,[81] a work that ʿAbd Allāh ibn ʿAlī translated from Persian into Arabic after it had first been translated from Sanskrit into Persian;[82] the *Book of Susrud*,[83] which discusses the symptoms of diseases, methods of treatment, and remedies—it is divided into ten chapters, and was translated into

---

79 Nothing is known about this author, and the Arabic transliteration probably distorted the original name beyond recognition. Sanǧahal is also mentioned by Ibn al-Nadīm as the author of a work entitled *The Secrets of Questions* or *Problems* (*Kitāb asrār al-masāʾil*); see *Kitāb al-fihrist*, I, 271.2.

80 Abū Bakr Muḥammad ibn Zakariyāʾ al-Rāzī, a towering figure of Islamic intellectual history, was one of the most distinguished Muslim medical scholars and practitioners, and also an independent-minded philosopher and freethinker; see Lenn E. Goodman, 'al-Rāzī', *EI2*, and Ullmann, *Die Medizin*, 128–136. Among his many medical works, pride of place belongs to the *Comprehensive Book* (*al-Kitāb al-ḥāwī*), a vast compilation of al-Rāzī's extensive notes and excerpts on diseases and their treatment published posthumously by his students. On the contents and structure of the *Comprehensive Book*, see Kahl, *The Sanskrit, Syriac and Persian Sources*, 2–7. On al-Rāzī's massive medical œuvre, see Richter-Bernburg, 'Abū Bakr Muḥammad al-Rāzī's (Rhazes) Medical Works'.

81 This is the *Caraka-Saṃhitā*, a medical compendium written by the Indian physician Caraka (second century BC). Its author's name was transliterated in Arabic variously as Šarak, Sīrk or Ǧarak. See Kahl, *The Sanskrit, Syriac and Persian Sources*, 18–20, 86–129.

82 This ʿAbd Allāh ibn ʿAlī is unknown. A similar translation process involving an intermediate Persian version has been documented in the introduction to the Arabic version of another well-known Indian book, the *Book of Poisons* (*Kitāb al-sumūm*) by Pseudo-Cāṇakya (Šānāq in Arabic sources). This book was first translated into Persian by the Indian physician Mankah (Sanskrit Māṇikya; see below, II.1 with n. 214) with the help of a certain Abū Ḥātim al-Balḫī, and later into Arabic by al-ʿAbbās ibn Saʿīd al-Ǧawharī; see Strauß, 'Giftbuch des Šānāq', 2.7–10 (Arabic).

83 This is the *Suśruta-Saṃhitā* (*Kitāb Susrud* in Arabic), the medical compendium of the Indian physician Suśruta (sixth century BC), considered the oldest attempt to systematize Ayurvedic medicine. This work was translated by Mankah (see below, II.1 with n. 214) for the Barmakid Yaḥyā ibn Ḫālid (Ibn al-Nadīm, *Kitāb al-fihrist*, I, 303.6–7). See Sezgin, *GAS*, III, 197–198, 200–201; Ullmann, *Die Medizin*, 106; and Kahl, *The Sanskrit, Syriac and Persian Sources*, 14–18, 72–85. For further references, see Ibn Abī Uṣaybiʿah, *ʿUyūn al-anbāʾ*, III/2, 987, with n. 7.

Arabic at the order of Yaḥyā ibn Ḫālid;[84] the *Book of Nidān*, which discusses the symptoms and diagnosis of four hundred and four ailments, but does not deal with treatments;[85] the *Book of Sindhašār*, whose title means *The Book of the Quintessence of Success*;[86] a book on the differences in the views held by the Indians and the Greeks concerning heat and cold, the effectiveness of various medicines and the division of the year;[87] a book on the interpretation of the names of drugs, in ten (different languages);[88] the *Asānkar*,[89] a compendium; a book on the treatments of pregnant women in India;[90] a compendium on drugs used by the Indians;[91] the *Book of Nūfašal*, which discusses one hundred ailments and one hundred remedies;[92] a book on the treatment of women, by an Indian woman named Rūsā;[93] a book on sugar in India; a book on the view of Nāqil the Indian on various species of snakes and their venom; and a book on delusions concerning diseases and maladies by the Indian author Abū Qubayl.[94]

---

84 Yaḥyā ibn Ḫālid ibn Barmak, vizier of Hārūn al-Rašīd; see above, I.7 with n. 69.

85 The *Nidāna* (i.e., pathology), also known as *Rogaviniścaya* (i.e., diagnosing disease), is a compilation of other medical treatises written by Mādhava, an Indian physician who lived AD c. 700. This work was probably translated during the caliphate of Hārūn al-Rašīd; see Ullmann, *Die Medizin*, 105. It is quoted by ʿAlī ibn Rabban al-Ṭabarī, *Firdaws al-ḥikmah*, 563.4–5 and 578.12–23, and al-Rāzī, *al-Kitāb al-ḥāwī*, XXIIIb, 148.5; see Kahl, *The Sanskrit, Syriac and Persian Sources*, 26–27, 152–153.

86 This is the *Siddhasāra*, 'Quintessence (of Medicine)', written by Ravigupta (fl. AD c. 650). Arabic sources rendered this title as *Kitāb Sindistār* and glossed it as 'quintessence of success' (*ṣafwat al-nuǧḥ*). It was translated by Ibn Dahn the Indian, the director of the hospital of Ǧondēšāpūr (Ibn al-Nadīm, *Kitāb al-fihrist*, I, 303.8–9; see below, II.1 with n. 216). See Sezgin, *GAS*, III, 198–199; Ullmann, *Die Medizin*, 105; and Kahl, *The Sanskrit, Syriac and Persian Sources*, 22–26, 134–152. On the title *Siddhasāra*, see Emmerick, 'Ravigupta's Siddhasāra'.

87 Unidentified.

88 This is probably the work entitled *Names of Indian Medicines* (*Asmāʾ ʿaqāqīr al-Hind*) that Mankah translated for Isḥāq ibn Sulaymān ibn ʿAlī (Ibn al-Nadīm, *Kitāb al-fihrist*, I, 303.11–12; see below, II.1 with n. 214).

89 Vāgbhata's *Aṣṭāṇgahṛdaya-Saṃhitā*, rendered into Arabic as *Istankār al-ǧāmiʿ*. This work was translated by Ibn Dahn, the director of the hospital of Ǧondēšāpūr (Ibn al-Nadīm, *Kitāb al-fihrist*, I, 303.7; see below, II.1 with n. 216). See Sezgin, *GAS*, III, 198–199, and Ullmann, *Die Medizin*, 105.

90 Unidentified.

91 This work, similarly unidentified, is also listed by Ibn al-Nadīm, *Kitāb al-fihrist*, I, 303.9–10.

92 Unidentified.

93 ʿAlī ibn Rabban al-Ṭabarī quotes from the books of 'an Indian woman' on the treatment of women's diseases; see *Firdaws al-ḥikmah*, 591.9–594.13.

94 Unidentified. Ibn al-Nadīm attributes this work to *t-w-q-š-t-l*, probably the same person Ibn Abī Uṣaybiʿah calls Nūfašal above.

1.9. Ibn al-Nadīm, *The Catalogue* (*Kitāb al-fihrist*), I, 243.3–14 (ed. Flügel); tr. Vagelpohl and Sánchez.[95]

The following undoubtedly apocryphal episode, recorded not just in Ibn al-Nadīm's *Catalogue* but quoted in a number of other biographical works, purports to explain why so many philosophical and scientific works were translated into Arabic. The surge in translation activity is ascribed to the ʿAbbāsid caliph al-Maʾmūn, who, inspired by a dream in which he saw and conversed with Aristotle, sent scholars to the Byzantine empire to collect ancient Greek works and translate them. Translations of ancient Greek, Persian, and Sanskrit philosophy, science, and medicine obviously long preceded al-Maʾmūn's reign and may in fact already have started as early as the first quarter of the second/eighth century. In addition, several of al-Maʾmūn's predecessors vigorously supported the translation movement. Rather than providing an incentive for the translation movement, the dream may illustrate the effect it had on intellectual attitudes at the time and the high esteem in which ancient Greek philosophers such as Aristotle were held.[96]

Why there are so many books in these parts on philosophy and other ancient sciences.

One reason for this is that al-Maʾmūn saw in a dream of his a man of reddish-white complexion with a high forehead, joined eyebrows, bald head, dark-blue eyes, and handsome features, sitting on his chair. Al-Maʾmūn said: 'I had the impression that I was standing in front of him, filled with awe. I asked him who he was. He replied: "I am Aristotle." I was happy to be with him and said: "O wise man, may I address a question to you?" He replied: "Ask!" I said: "What is good?" He replied: "Whatever is good according to reason." I asked: "What else?" He replied: "Whatever is good according to religious law." I asked: "And what else?" He replied: "Whatever society considers good." I asked: "What else?" And he replied: "Nothing else."'

According to another tradition: 'I [i.e., al-Maʾmūn] said: "Give me further (instructions)." Aristotle replied: "He who gives you good advice for gold, consider him to be like gold. It is your duty to believe in the oneness of God."'

This dream was one of the most important causes for the publication of books (on philosophy etc.). Al-Maʾmūn and the Byzantine king, whom al-Maʾmūn had just defeated, exchanged several letters. He wrote to the Byzantine

95 Parallel: Ibn al-Qifṭī, *Taʾrīḫ al-ḥukamāʾ*, 29.8–15. Alternative versions can also be found in Ibn Abī Uṣaybiʿa, *ʿUyūn al-anbāʾ*, II/1, 468.5–8, and Ibn Nubātah, *Sarḥ al-ʿuyūn*, 213.

96 See Gutas, *Greek Thought*, 95–104.

king and asked for permission to send people to select books on the ancient sciences from those stored in the libraries on Byzantine territory.

(The Byzantine king) at first would not allow it, but later gave his consent. Then al-Ma'mūn sent a group of people, among them al-Ḥaǧǧāǧ ibn Maṭar,[97] Ibn al-Biṭrīq,[98] and Salm, the head of the House of Wisdom,[99] and others, and they made their choice among the material that they found there. When they brought it to (al-Ma'mūn), he ordered them to translate it, and this was done. Yūḥannā ibn Māsawayh[100] is also said to have been among those sent (at that time) to Byzantine territory.[101]

I.10. Ibn al-Qifṭī, *History of Learned Men* (*Ta'rīḫ al-ḥukamā'*), 61.4–13 (ed. Lippert); tr. Vagelpohl and Sánchez.

Ibn al-Qifṭī (d. 646/1248) was an Egyptian bureaucrat in the service of the Atabegs of Aleppo. In addition to his administrative duties, he was an accomplished scholar; medieval sources credit him with twenty-six books, only two of which have survived. One of these works is the *History of Learned Men*, a compilation of biographies of Muslim scholars. It is preserved in an epitome written by Muḥammad ibn 'Alī al-Zawzanī in 647/1249 that contains 414 biographies of philosophers, physicians, mathematicians, astronomers, and other scholars.[102]

The following passage deals once more with the contribution of the caliph al-Ma'mūn to the Greek–Arabic translation movement. Although he may not have been the first caliphal patron of scholars and translators, his active support of the translation movement led to a flood of new translations. Among them was a standard text of Greek geometry, the *Conics* (*Conica*) by Apollonius of Perga (d. c. 190 BC), parts of which had been lost before it could be translated into Arabic. It had a profound impact on the development of mathematics in the Islamic world.[103]

When the (Greek) works were brought to (the caliph) al-Ma'mūn from the Byzantine empire, no more than the first book of this work [i.e., the *Conics*] was brought, which consisted of seven chapters. When the work was translated, its introduction indicated that it (actually) consists of eight chapters and

97 See above, I.6 with n. 61.
98 See below, I.11 with n. 106.
99 See above, I.7 with n. 71.
100 See above, I.5 with n. 53.
101 Rosenthal, *The Classical Heritage*, 48–50, adapted. Another translation: Ibn al-Nadīm, *The Fihrist*, 583–584 (English).
102 See Albert Dietrich, 'Ibn al-Ḳifṭī', *EI2*.
103 See Hélène Bellosta, 'Apollonius of Perge', *EI Three*.

that the eighth chapter covers the contents of the (preceding) seven chapters and some extra material, and that it stipulates useful (mathematical) terms and (has other) desirable benefits. From that time until this day, people in this field have been looking for this chapter, but have not found any information about it. Without a doubt, it is a treasure (worthy of) kings, because the kings of Greece held these disciplines in high regard. I had talked about the fate of this chapter with someone in our time who had some interest in this field, or claimed to do so. He told me that it had been found and started to describe it, but in his description he mentioned things that do not match the words of (the work's) author. I then knew that he knew nothing at all, and I turned away from him and left him in his ignorance.

I.11. Ibn Ǧulǧul, *The Classes of the Physicians and Sages* (*Ṭabaqāt al-aṭibbāʾ wa-l-ḥukamāʾ*), 67.6–11 (ed. Fuʾād Sayyid); tr. Vagelpohl and Sánchez.[104]

Yaḥyā (or Yūḥannā) ibn al-Biṭrīq (fl. first half of the third/ninth century) was, according to Ibn Ǧulǧul, the client (*mawlā*) of the caliph al-Maʾmūn, a status that usually implied a conversion to Islam. He was the son of another translator, al-Biṭrīq, who was apparently active during the reign of the caliph al-Manṣūr (r. 136/754–158/775). We know little about the life of either father or son, and their similar names meant that there was frequent confusion about which of them produced which translation. In general, however, the father seems to have been more interested in medical texts, while the son may have been one of the first translators of philosophical works into Arabic.[105]

The following report about the discovery of *The Secret of Secrets* (*Sirr al-asrār*), probably the most widely known book associated with Ibn al-Biṭrīq, is clearly apocryphal and was perhaps meant to impress readers and pique their interest in the book.[106] It is a pseudo-Aristotelian work that purports to be a collection of letters by Aristotle to Alexander the Great and covers a wide range of topics. It enjoyed great popularity in the Middle Ages and was translated into several languages. The preface to the Arabic version, parts of which are quoted below, claims that Ibn al-Biṭrīq translated the Syriac version of the original Greek into Arabic. Modern scholarship, however, considers it to be an Arabic work written in the fourth/tenth century.[107]

104 Parallel: the introduction of the *Secret of Secrets* (see below) in Badawī, *al-Uṣūl al-yūnānīyah*, 69.5–11, tells a similar story with some variants, e.g., associating the temple with Asclepius instead of Hermes.

105 See Dunlop, 'The Translations of al-Biṭrīq'; and Micheau, 'Yaḥyā (or Yūḥannā) b. al-Biṭrīḳ', *EI2*.

106 See Dunlop, 'The Translations of al-Biṭrīq', 147–148; and Forster, *Das Geheimnis*, 50–54.

107 See Forster, *Das Geheimnis*, 111–112.

The translator Yūḥannā ibn al-Biṭrīq, a client of the Commander of the Believers al-Ma'mūn, was a faithful translator, excellent at rendering ideas but unable to speak Arabic properly. He translated many works of the ancients, and he is the translator of the book Aristotle wrote for Alexander known as *The Secret of Secrets*, that is, *The Book of Government on the Conduct of Statecraft*.

Yūḥannā (ibn al-Biṭrīq) reports that he went out to find it. He headed for (ancient) temples in his search for it until he arrived at the temple of ʿAbd al-Šams,[108] which Hermes the Great[109] had built for himself to glorify God in it. He said: 'In it, I encountered a pious monk of outstanding knowledge and keen intelligence. I made myself agreeable and employed finesse so that he gave me access to the books deposited at the temple. Among them, I found written in gold what I was looking for and what the Commander of the Faithful had ordered me to find. Having accomplished my mission, I then returned to his victorious presence.'[110]

I.12. Ibn Abī Uṣaybiʿah, *The Best Accounts of the Classes of Physicians* (*ʿUyūn al-anbā' fī ṭabaqāt al-aṭibbā'*), II/2, 928.4–930.9 (ed. Savage-Smith et al.); tr. Savage-Smith et al., III/2, 1037–1038, adapted by Vagelpohl and Sánchez.[111]

The following passage was part of Ibn Ǧulǧul's introduction to his *Explanation of the Names of the Simple Drugs in the Book of Dioscorides* (*Kitāb tafsīr asmā' al-adwiyah al-mufradah min kitāb Dīsqūrīdis*). Only fragments from this treatise have come down to us.[112] This first-person account of the Arabic translation of Dioscorides's *On Medical Material* (*De materia medica*)[113] in al-Andalus has survived because it is quoted by Ibn Abī Uṣaybiʿah in his *Best Accounts of the Classes of Physicians*.[114]

108 The use of this name as a toponym is strange and probably the result of scribal error. It probably refers to ʿAyn Šams (Heliopolis), the site of a temple devoted to Hermes according to Arabic sources; see al-Maqrīzī, *Kitāb al-mawāʿiẓ*, I, 617.12.

109 On Hermes and the Hermetic tradition in Arabic, see Kevin van Bladel, 'Hermes and Hermetica', *EI Three*.

110 Another translation: Dunlop, 'The Translations of al-Biṭrīq', 147–148 (English).

111 Parallel: the text is quoted in the appendices of ʿAbd al-Laṭīf al-Baġdādī, *Kitāb al-ifādah*, 549.7–550.21.

112 A reconstruction of Ibn Ǧulǧul's treatise on the basis of the Madrid manuscript Biblioteca Nacional MSS/498 and quotations in various sources has been published by Garijo, *Ibn Ŷulŷul*.

113 Pedanius Dioscorides (AD c. 40–90) was a Greek physician, botanist, and pharmacologist, mainly known for his *On Medical Material*, the most influencial book in medieval pharmacopoeia. See Alain Touwaide, 'Pedanius [1]', *New Pauly*; César E. Dubler, 'Diyuskūridīs', *EI2*. For the Arabic transmission and the various translations, see Ullmann, *Untersuchungen*.

114 See Meyerhof, 'Die Materia Medica', 72–74; and Ullmann, *Untersuchungen*, 61–63. On Orosius's *History*, see Levi Della Vida, 'La traduzione araba', and Penelas, 'A Possible Author'.

At the beginning of his book[115] (Ibn Ǧulǧul) states: The book of Dioscorides was translated in Baghdad during the ʿAbbāsid caliphate in the days of Ǧaʿfar al-Mutawakkil.[116] It was translated from Greek into Arabic by Iṣṭifan ibn Basīl the translator,[117] and his work was examined by Ḥunayn ibn Isḥāq the translator,[118] who corrected and certified the translation.[119] Those Greek names that Iṣṭifan knew how to translate, he translated into Arabic; and those for which he did not know of an Arabic equivalent, he left in Greek. He trusted that God Almighty would send someone after him who would know (the names) and render them into Arabic, for nomenclature is but the agreement of the people from each land to name the different classes of drugs as they consider appropriate, sometimes by (etymological) derivation, sometimes by other ways upon which they agree. Iṣṭifan was confident that among those who were to come after him some would know those classes of drugs for which he was unable to find a name in his time, and would name them according to what they had learned in their own time, so that the (drug names) would finally be known.

Ibn Ǧulǧul says: This book came to al-Andalus in the translated version made by Iṣṭifan, containing those drug names that he knew how to translate into Arabic and those he did not know. The people profited from all that could be understood from it, both in the East and in al-Andalus, until the days of al-Nāṣir ʿAbd al-Raḥmān ibn Muḥammad, who at that time was the ruler of al-Andalus.[120] Romanos, the emperor of Constantinople, presented him with splendid gifts, in the year 337/948–949, I think.[121] Among those

115 I.e., Ibn Ǧulǧul's *Explanation of the Names of the Simple Drugs*.

116 I.e., the caliph al-Mutawakkil.

117 See below, II.1 with n. 165.

118 See below, I.5 with n. 52.

119 This translation was apparently not completely superseded by the one produced in al-Andalus; see Touwaide, 'Translation and Transliteration'. In addition to the previously known Arabic versions of the *Materia medica*, namely Iṣṭifan ibn Basīl's translation from the Greek (Ullmann, *Untersuchungen*, 21–24, 48–49) and two extant seventh-/twelfth-century translations from Syriac by Abū Sālim al-Malaṭī and Mihrān ibn Manṣūr (Ullmann, *Untersuchungen*, 18–19, 339–340, 341–355), Manfred Ullmann has identified an older translation from the Greek, produced around the year 183/800 (*Untersuchungen*, 9–10, 69–78).

120 ʿAbd al-Raḥmān III al-Nāṣir ruled as governor from 300/912 to 316/929, and as caliph until 350/961.

121 The date does not fit the reign of the Byzantine emperor Romanos II (r. 959–963). In 948–949, his father Constantine VII was emperor, but Romanos II might have been associated with his father: on some imperial seals they are shown together as co-regents. There is also evidence for a box with a portrait of Constantine VII that was sent to the Umayyad caliph in 949; see Walker, *The Emperor*, 90. On this embassy and the practice of presenting books as gifts, see Signes Codoñer, 'La diplomacia del libro'.

presents there was a copy of the book of Dioscorides illuminated with marvellous Byzantine illustrations of plants; the book was written in Greek [*iġrīqī*], i.e., ancient Greek [*yūnānī*]. Together with it, he sent the book of Orosius, the author of stories, which contains a wonderful history of the Romans, with information about the past, stories about the first kings, and many profitable things.[122]

Romanos wrote to al-Nāṣir: You will not profit from the book of Dioscorides unless you have someone with knowledge of the Greek language, who will recognize the characteristics of those drugs. If there is someone able to do this in your land, then you will enjoy, O King, the benefits of the book. As for the book of Orosius, you have in your land, among the Latin Christians, some who can read it in Latin, and if you allow them, they will translate it for you from Latin into Arabic.

Ibn Ǧulǧul says: In that time, none of the Christians of Cordova was able to read Greek [*iġrīqī*], i.e., ancient Greek [*yūnānī*], and the book of Dioscorides was kept in the library of ʿAbd al-Raḥmān al-Nāṣir, in Greek, without being translated into Arabic. Thus, the book remained in al-Andalus while Iṣṭifan's translation from Baghdad circulated among the people. In his answer to Romanos, al-Nāṣir asked him to send someone able to speak Greek and Latin to teach some slaves so that they would become translators, and the emperor Romanos sent al-Nāṣir a monk called Niqūlā,[123] who arrived in al-Andalus in 340/951–952.

At that time there were a number of physicians in Cordova who were interested in seeking, investigating, and inquiring about the names of drugs in the book of Dioscorides for which the Arabic terms were still unknown. The physician who was most eager to investigate the matter was Ḥasdāy ibn Šaprūṭ al-Isrāʾīlī,[124] who sought to be close to the king, ʿAbd al-Raḥmān al-Nāṣir. The monk Niqūlā won (Ibn Šaprūṭ's) favour, preference, and high regard, and he explained to him the names of the drugs in the book of Dioscorides that had previously been unknown. (Ibn Šaprūṭ) was the first physician in Cordova to

122 This is Paulus Orosius's *History Against the Pagans* (*Historiae adversus paganos*), which was also translated into Arabic in al-Andalus in the fourth/tenth century; see Penelas, 'A Possible Author', and Sahner, 'From Augustine to Islam'. In the introduction to his *Classes of Physicians*, Ibn Ǧulǧul listed this work as one of his sources; see *Ṭabaqāt al-aṭibbāʾ*, 2.1. For an edition of the Arabic Orosius, see Orosius, *Kitāb Hurūšiyūš*.

123 Nothing is known about this monk, apart from his role in the translation of this work.

124 Ḥasday ibn Šaprūṭ (d. c. 350/970–971) was a Jewish physician at the courts of ʿAbd al-Raḥmān III and al-Ḥakam II. As a member of one of the most important Jewish families in al-Andalus, he was also a patron of sciences and a diplomat. See Eliyahu Ashtor, 'Ḥisdai (Ḥasdai) Ibn Shapruṭ', *Encyclopaedia Judaica*.

prepare the great theriac [*tiryāq al-fārūq*] following the precise explanation of the botanical information contained (in the book). Other physicians of that time who embarked on the task of investigating the question of the names of the drugs in the book of Dioscorides and identifying their classes were: Muḥammad, known as al-Šaǧǧār ['the botanist'];[125] someone known as al-Šabānisī;[126] Abū 'Uṯmān al-Ǧazzār, who had the nickname of al-Yābisah;[127] Muḥammad ibn Sa'īd al-Ṭabīb;[128] 'Abd al-Raḥmān ibn Isḥāq ibn al-Hayṯam;[129] and Abū 'Abd Allāh al-Ṣiqillī,[130] who spoke Greek and recognized the characteristics of the drugs.

Ibn Ǧulǧul says: This group (of physicians) and Niqūlā the Monk lived at the same time. I was able to meet (Ibn Šaprūṭ) and Niqūlā the Monk in the days of al-Mustanṣir, and fraternized with them at that time.[131] Niqūlā the Monk died at the beginning of the reign (of al-Mustanṣir al-Ḥakam). With the research of this group who investigated the names of drugs of the book of Dioscorides, he had made it possible for their characteristics to be known in Cordova, especially regarding (the plants) of al-Andalus, dispelling every doubt from our hearts; he provided knowledge of their characteristics, and explained the meaning of their names almost without mistakes, apart from a few of them—about ten drugs—with which he was not familiar and about which he had no knowledge.[132]

1.13. Ibn Suwār, annotation to the translation of Aristotle's *Sophistical Refutations* in Aristotle, *Manṭiq Arisṭū*, III, 1017.5–1018.13 (ed. Badawī), and Aristotle, *Les Catégories*, 198–199 (ed. Georr); tr. Rosenthal, *The Classical Heritage*, 22–23, adapted by Vagelpohl and Sánchez.

---

125 Unidentified.

126 All manuscripts of Ibn Abī Uṣaybi'ah's *Best Accounts* read al-Basbāsī, but this is most certainly a misspelling for al-Šabānsī. Qāsim ibn Muḥammad al-Qurašī al-Marwānī al-Šabānsī (fl. second half of fourth/tenth century) was a physician and a poet. Ibn al-Abbār reported that he was a learned man (*min al-udabā'*) and one of the teachers of Sa'īd ibn Fatḥūn al-Saraqusṭī al-Ḥammār; see Ibn al-Abbār, *al-Takmilah*, II, 194.16 (n. 518).

127 Unidentified. The *nisbah* is probably a scribal error for al-Yābisī, 'native of Ibiza'.

128 Unidentified.

129 Ibn al-Hayṯam (fl. second half of fourth/tenth century) was a Cordovan physician; see Sezgin, GAS, II, 309–310; Ullmann, *Die Medizin*, 26; and Cabo González, 'Ibn al-Hayṯam'.

130 Unidentified.

131 I.e., the Umayyad caliph al-Ḥakam II (r. 350/961–366/967).

132 Other translations: Rosenthal, *The Classical Heritage*, 194–197 (English); and 'Abd al-Laṭīf al-Baġdādī, *Kitāb al-ifādah*, 495–497 (French).

Some Arabic translations of Greek works were not made directly from the Greek originals, but from Syriac intermediaries. This process is best attested in the case of medicine, where we have a comprehensive overview of the translations in both languages from the pen of the translator himself, Ḥunayn ibn Isḥāq.[133] Syriac translations also formed the basis of a large number of philosophical translations, including that of Aristotle's *Sophistical Refutations* (*De sophisticis elenchis*) discussed in the next text.[134] This text illustrates the problem Arabic translators had when their Syriac source was inferior: if they were lucky, there were translated commentaries to help them understand the text; if not, as in this case, they were on their own. The following notes are the annotations with which the editor of a compilation of Aristotle translations,[135] the philosopher Ibn Suwār,[136] introduced Aristotle's *Sophistical Refutations*. Confronted with three equally unsatisfactory translations,[137] he took the unusual step of including all three in his compilation in the hope that the reader would pick the correct interpretation whenever the text was problematic.

I have copied this translation from a manuscript written by the venerable Abū l-Ḫayr al-Ḥasan ibn Suwār, may God be pleased with him. It ends with the following remarks: I have copied this translation from a manuscript which appears to me to have been written by Abū Naṣr al-Fārābī.[138] The first part of that manuscript was carefully written and correct, but the second was faulty.

The venerable Abū l-Ḫayr ibn Suwār, may God be pleased with him, says: In order to understand the ideas expressed in the language from which he is translating, the translator must first grasp them in the sense of the original author and must be familiar with the usage of the language from which he is translating, as well as that into which he is translating. However, the monk

133 See above, I.5 with n. 52.

134 On the complex Arabic translation and commentary history of this work, see Peters, *Aristoteles Arabus*, 23–25; and Ian R. Netton, 'al-Sūfisṭā'iyyūn', *EI2*.

135 This important source, extant in a single manuscript, contains Arabic translations of Aristotle's logical works, the *Organon*, and the introduction to logic, the *Isagoge*, by Porphyry (d. AD c. 305); see Hugonnard-Roche, 'Un manuscrit savant'.

136 Abū l-Ḫayr al-Ḥasan ibn Suwār, also known as Ibn al-Ḫammār (d. c. 421/1030), was a Syriac Christian philosopher and student of the great Aristotelian philosopher Yaḥyā ibn 'Adī (d. 363/974), himself a student of one of the greatest philosophers of Islam, al-Fārābī (d. 339/950); see Daiber, 'The Meteorology', 220–221.

137 See Walzer, *Greek into Arabic*, 83.

138 Abū Naṣr Muḥammad ibn Muḥammad al-Fārābī (d. 339/950–951), known as the 'second teacher' (Aristotle being the first) was the second major Muslim philosopher after the pioneering efforts of the 'philosopher of the Arabs' Abū Yūsuf Ya'qūb ibn Isḥāq al-Kindī; see Jean Jolivet and Roshdi Rashed, 'al-Kindī', *EI2*. See further Richard Walzer, 'al-Fārābī', *EI2*, and below, IV.7.

Athanasius[139] was not familiar with Aristotle's ideas in this work. Hence it was inevitable that mistakes crept into his (Syriac) translation. Since the translators mentioned above, who rendered this book into Arabic on the basis of the Syriac translation of Athanasius, had no commentary at their disposal, they relied on their own insight in order to understand the ideas of the work. Everyone endeavoured to arrive at the truth and to understand the purpose intended by the philosopher [i.e., Aristotle], and they (each) altered what they translated into Arabic of Athanasius's (Syriac) translation. As we desire to become acquainted with the views of each one of them, we have reproduced all the translations we were able to obtain, so that all of them may be studied and jointly used for the comprehension of (Aristotle's) meaning.

The excellent Yaḥyā ibn ʿAdī[140] composed a Syriac and Arabic commentary on this book. I have seen most of it and estimate (that it comprises) about two thirds of the work. I presume that he completed it, but after his death it could not be found among his books. My opinion about this fluctuates. Sometimes I think he may have destroyed it because he was dissatisfied with it, while at other times I suspect that it was stolen, which I consider more likely. He produced the said translation of the book before compiling his commentary on it, hence it is a little obscure, for he did not always grasp the meaning correctly and based his translation on the Syriac text.

Nowadays there also exists a manuscript of the Greek commentary by Alexander of Aphrodisias.[141] At the beginning a quire is missing. Very little of it has so far been made accessible (through translation).

From what I have heard, Abū Isḥāq Ibrāhīm ibn Bakkūš[142] translated this work from Syriac into Arabic, and, in collaboration with the Greek priest and

139 Athanasius of Balad (d. AD 686), a West Syrian scholar and Greek–Syriac translator, mainly of philosophical works. See Adam H. Becker, 'Athanasius of Balad', *EI Three*; and Michael Penn, 'Athanasios II of Balad', *GEDSH*.

140 The Christian Arab philosopher Yaḥyā ibn ʿAdī, the teacher of the author of the present text, is credited with a number of translations or revisions of existing translations of Aristotle's work. His library became an important source of information about Greek philosophy and translations of philosophical works for the *Catalogue* of his contemporary Ibn al-Nadīm. See Gerhard Endress, 'Yaḥyā b. ʿAdī', *EI2*.

141 Alexander of Aphrodisias (fl. AD c. 200) was a philosopher and celebrated commentator of Aristotle's works. A number of his commentaries were translated into Arabic; see Goulet and Aouad, 'Alexander d'Aphrodisias'.

142 Ibrāhīm ibn Baks (or Bakkūš) al-ʿAššārī (fl. fourth/tenth century) was a Christian physician active in Baghdad and is credited with four Arabic translations: Aristotle's *Sophistical Refutations* (Ibn al-Nadīm, *Kitāb al-fihrist*, I, 249.27–29; Ibn al-Qifṭī, *Taʾrīḫ al-ḥukamāʾ*, 37.15–16); Aristotle's *On Generation and Corruption* (Ibn al-Nadīm, *Kitāb al-fihrist*, I, 251.4;

geometrician, Yūḥannā, who is known by the name of Ibn Fatīlah,[143] worked on the correction of a number of passages on the basis of the Greek text, but I have not seen (his work).

Abū Bišr,[144] may God have mercy on him, is also said to have corrected the first translation or produced another translation, but I have not seen (his work).

I have noted all this here in order that those who use this book should be fully informed about the situation and (understand) why I have reproduced all translations in the manner described.[145]

II. *People: Translators and Sponsors*

II.1. Ibn al-Nadīm, *The Catalogue* (*Kitāb al-fihrist*), I, 244.1–245.10 (ed. Flügel); tr. Vagelpohl and Sánchez.

The scale and extent of Arabic translations from various languages is best illustrated by the long list of translators Ibn al-Nadīm compiled for his *Catalogue*, based on the

Ibn al-Qifṭī, *Taʾrīḫ al-ḥukamāʾ*, 40.18); and Theophrastus's *The Causes of Plants* and *On Sense Perception* (Ibn al-Nadīm, *Kitāb al-fihrist*, I, 252.8–10; Ibn al-Qifṭī, *Taʾrīḫ al-ḥukamāʾ*, 107.5–6; Bar Hebraeus, *Taʾrīḫ muḫtaṣar al-duwal*, 93.18–19 and 94.1). Two sources suggest that he was already active during the first half of the fourth/tenth century: according to Ibn Abī Uṣaybiʿah (*ʿUyūn al-anbāʾ*, II/1, 301.14–15), a physician named Abū l-Ḥasan ibn Nafīs commissioned the translation of Ibn Sarābiyūn's medical handbook, *al-Kunnāš al-ṣaġīr*, in 312/924–925, and according to ʿAbd al-Ǧabbār (*Taṯbīt*, 619.2–3, 7–12), Ibn Nafīs was the teacher of Ibn Baks, one of the physicians who struggled to fight an epidemic that spread in Baghdad in 330/941–942. See Ullmann, *Die Natur- und Geheimwissenschaften*, 73–74; and Gutas, *Greek Thought*, 151.

143 Ibn al-Nadīm (*Kitāb al-fihrist*, I, 282.20–24; cf. Ibn al-Qifṭī, *Taʾrīḫ al-ḥukamāʾ*, 380.2–4) reported that Yūḥannā ibn Yūsuf ibn al-Ḥāriṯ ibn al-Biṭrīq (fl. fourth/tenth century), nicknamed al-Ḫūrī or al-Qass ('the priest') and also known as Ibn Fatīlah, wrote on mathematics and geometry and also translated from Greek. Since there are two translators nicknamed al-Qass, one of which (Yūsuf; see below, II.2 with n. 249) was a physician rather than a mathematician, Yūḥannā may also be identical to the Yūsuf al-Qass credited by Ibn Abī Uṣaybiʿah (*ʿUyūn al-anbāʾ*, II/1, 574.3–4) with a Syriac book on triangles, or translation of a work on triangles ascribed to Archimedes.

144 Abū Bišr Mattā ibn Yūnus (d. 328/940) was a Syriac–Arabic translator and Aristotle commentator. Of his translations, the Arabic version of Aristotle's *Poetics* (*Poetica*) and *Posterior Analytics* (*Analytica posteriora*) formed part of Ibn Suwār's Aristotle compilation (see above, I.13, with nn. 136 and 137). See Gerhard Endress, 'Mattā b. Yūnus', *EI2*. Abū Bišr was one of the participants in the debate between a logician and a grammarian reported in the *Book of Enjoyment and Conviviality* (*Kitāb al-imtāʿ wa-l-muʾānasah*) by al-Tawḥīdī (d. 414/1023), samples of which are quoted below (IV.8a–d).

145 Another translation: *Les Catégories* (ed. Georr), 199–200 (French).

accounts of his informants and the written sources he assembled.[146] Ibn Abī Uṣaybiʿah took Ibn al-Nadīm's list, amplified it with additional contextual information, and added a section on the sponsors of translations (see II.2 below). Since Ibn al-Nadīm's *Catalogue* and Ibn Abī Uṣaybiʿah's *Best Accounts* (or their editors) sometimes disagree about the reading of individual names, the following texts are annotated with corrections and relevant alternative readings.

The Names of the Translators from (Foreign) Languages into Arabic

Iṣṭifan al-Qadīm: he translated books on alchemy and other subjects for Ḫālid ibn Yazīd ibn Muʿāwiyah.[147]
Al-Biṭrīq: he lived in the days of (the caliph) al-Manṣūr, who ordered him to translate some ancient books.[148]
His son Abū Zakariyāʾ Yaḥyā ibn al-Biṭrīq, who belonged to the group of al-Ḥasan ibn Sahl.[149]
Al-Ḥaǧǧāǧ (ibn Yūsuf) ibn Maṭar: he translated for al-Maʾmūn and was the person who translated the *Almagest* and Euclid.[150]
Ibn Nāʿimah, whose name was ʿAbd al-Masīḥ ibn ʿAbd Allāh al-Ḥimṣī al-Nāʿimī.[151]
Sallām al-Abraš:[152] he was one of the early translators in the days of the Barmak family,[153] and among his translations is (Aristotle's) *Physics*. Thus

146 See Gutas, *Greek Thought*, 136, with n. 45.

147 Iṣṭifan, nicknamed 'the Elder' probably to differentiate him from Iṣṭifan ibn Basīl, seems to be a legendary figure, as are other translators who allegedly worked for Ḫālid ibn Yazīd (see below, II.3), the grandson of the Umayyad caliph Muʿāwiyah; see Ullmann, 'Ḫālid ibn Yazīd', esp. 193–194 and 211–218.

148 Little is known about al-Biṭrīq, a Christian scholar whose translations, produced mainly during the caliphate of al-Manṣūr, are easily confused with those of his son Yaḥyā below. See above, I.11 with n. 105, and Dunlop, 'The Translations of al-Biṭrīq'.

149 See above, I.11 with n. 105.

150 See above, I.6 with n. 61.

151 ʿAbd al-Masīḥ ibn ʿAbd Allāh al-Ḥimṣī al-Nāʿimī was a Syriac Orthodox scholar mainly known for his translation of the so-called 'Theology of Aristotle', a paraphrastic selection from the last three *Enneads* by Plotinus revised by the philosopher al-Kindī (see below, V.1), and Aristotle's *Sophistial Refutations*. See Zimmermann, 'The Origins'.

152 Little is known about Sallām al-Abraš, apart from this brief note. See Peters, *Aristoteles Arabus*, 32, and Gutas, *Greek Thought*, 72–73. Historical sources mention a certain Sallām al-Abraš al-Ḫaṣī (the eunuch), probably the same person, in the service of several ʿAbbāsid caliphs from the time of al-Manṣūr until al-Maʾmūn; see Muḥammad ibn Ǧarīr al-Ṭabarī, *Taʾrīḫ*, VIII, 63.23 and 595.16 = al-Ṭabarī, *Annales*, III, 293 and 1065. Sallām also figures as a reporter of anecdotes in the major *adab* works of the ʿAbbāsid period, such as the *Book of Songs* (*Kitāb al-aġānī*) by Abū l-Faraǧ al-Iṣbahānī (d. 356/967), or the *Relief after Hard-*

reported our lord Abū l-Qāsim ʿĪsā ibn ʿAlī ibn ʿĪsā,[154] may God support him.

Ḥabīb ibn Bahrīz, the metropolitan of Mosul: he translated a number of books for al-Maʾmūn.[155]

Zarūyā ibn Mā Ḥawah/Māǧūh[?] al-Nāʿimī al-Ḥimṣī.[156]

Hilāl ibn Abī Hilāl al-Ḥimṣī.[157]

Theodore.[158]

Fiṯiyūn [Pethion].[159]

Abū Naṣr ibn Āwā[?] ibn Ayyūb.[160]

---

ship (*al-Faraḥ baʿd al-šiddah*) by al-Tanūḫī (d. 384/994); none of these sources refers to him as a translator. Being a eunuch, he was probably of Byzantine descent, and he was a prominent figure at court with his own clients, such as Aḥmad ibn Rašīd al-Kātib (Ibn Abī Uṣaybiʿah, *ʿUyūn al-anbāʾ*, II/2, 891.9–10). Rather than being author of translations himself, he may have commissioned others.

153 The Barmakid family of administrators and ministers played an important role in the political and cultural life of early ʿAbbāsid Baghdad until their downfall in the year 187/803 under Hārūn al-Rašīd. Their most powerful member was the vizier of Hārūn al-Rašīd, Yaḥyā ibn Ḫālid ibn Barmak. See Gutas, *Greek Thought*, 128–129, and Kevin van Bladel, 'Barmakids', *EI Three*.

154 The son of the vizier ʿAlī ibn ʿĪsā ibn al-Ǧarrāḥ; see above, 256 with n. 7.

155 ʿAbdīšūʿ bar Bahrīz was the metropolitan of Ḥarrān, Mosul, and Ḥazzah. According to Ḥamzah al-Iṣfahānī, he was the translator of a *History of the Greeks* (*Aḫbār al-yunānīyīn*); see Ibn al-ʿAdīm, *Buġyat al-ṭalab*, IV, 1599; and Barbara H. Roggema, 'Abdisho bar Bahrīz', *GEDSH*.

156 Unidentified. Ibn Abī Uṣaybiʿah (see below, II.2) claimed that his skills were limited.

157 Hilāl ibn Abī Hilāl al-Ḥimṣī (d. 218/833) was a mathematician associated with the Banū Mūsā and one of the translators of Apollonius's *Conic Sections*; see Apollonius, *Conics*, xviii, 620, 628.

158 A comparison with the list of translators transmitted by Ibn Abī Uṣaybiʿah (see below, II.2) suggests that this Theodore is most likely the bishop of Karḫ Ǧuddān of the same name, a patron of the sciences and translator closely associated with the scholar and translator Sergius of Rēšʿaynā (see below, II.2), who is also mentioned in Ḥunayn ibn Isḥāq's *Epistle* (Ḥunayn ibn Isḥāq, *al-Risālah*, 12.21–22/Ar. and 10/Ger.). See Aydin, *Sergius of Reshaina*, 10–11, n. 1.

159 Pethion (fl. second half of third/ninth century) was an East Syrian Christian and author of a lost ecclesiastical history. He is one of Ibn Abī Uṣaybiʿah's and Bar Hebraeus's main sources of information on the physicians of Ǧondēšāpūr, probably via quotations in the *Chronography* (*Maktbānut zabne*) by Elijah of Nisibis (d. 1046). Although nothing is known about his life, it has been suggested that he is the Pethion of Elam to whom the Catholicos Timothy addressed his ninth letter (see Berti, *L'Au-delà de l'âme*, 26–32) and identical with Pethion ibn Ayyūb al-Sahhār al-Tarǧumān, who translated the biblical books of Job, Jeremiah, Isaiah, Ezekiel, and Ecclesiastes into Arabic; see Vaccari, 'Le versioni arabe', 413–416, and Frank, 'The Jeremias'.

160 Unidentified. In Ibn Abī Uṣaybiʿah's list, this translator is named Abū Naṣr ibn Nārī ibn Ayyūb; see below, II.2.

Basīl al-Muṭrān ['the metropolitan'].[161]
Abū Nūḥ (Ibrāhīm) ibn al-Ṣalt.[162]
Usṭāṯ [Eustathius].[163]
Ḥayrūn [Heron?].[164]
Iṣṭifan ibn Basīl.[165]
Ibn Rābiṭah.[166]
Tiyūfīlā [Theophilus].
Šamlī.[167]
ʿĪsā ibn Nūḥ.[168]
Quwayrī, whose name was Ibrāhīm, with the patronym Abū Isḥāq.[169]

---

161 According to Ibn al-Nadīm (*Kitāb al-fihrist*, I, 250.21–22) and Ibn al-Qifṭī (*Taʾrīḫ al-ḥukamāʾ*, 39.6–7), Basīl translated a commentary by Porphyry on Aristotle's *Physics*.

162 Ibrāhīm ibn al-Ṣalt was a physician who belonged to the Church of the East. Almost nothing is known about his life. Ḥunayn ibn Isḥāq mentions him three times as the author of translations into Arabic and Syriac, one sponsored by the Banū Mūsā; see Ḥunayn ibn Isḥāq, *al-Risālah*, 31.8, 34.18, and 35.13–14/Ar., and 25, 28, and 29/Ger.

163 Usṭāṯ (Eustathius) was one of the scholars of the circle of the philosopher al-Kindī, for whom he translated Aristotle's *Metaphysics*; he was also the translator of the *Nicomachean Ethics*. See Endress, 'The Circle of al-Kindī', 52; and Ullmann, *Die Nikomachische Ethik*, 16–19. It has been suggested that Usṭāṯ could have translated Aristotle's zoological works; see Kruk, *The Arabic Version of Aristotle's Parts of Animals*, 18–23.

164 Unidentified.

165 Little is known about the life of Iṣṭifan ibn Basīl (d. c. 298/910), presumably the son of the above-mentioned Basīl. He was a member of the Church of the East and, according to Ḥunayn ibn Isḥāq's *Epistle*, the author of nine Galen translations commissioned by Muḥammad ibn Mūsā; he also translated Oribasius's *Medical Collections* (*Collectiones medicae*). Ḥunayn and Iṣṭifān collaborated on the translation of Galen's *Causes of Breathing* (*De causis respirationis*) (Ḥunayn ibn Isḥāq, *al-Risālah*, 24.8–11/Ar. and 20/Ger.) and Ḥunayn corrected Iṣṭifan's translation of Dioscorides's *On Medical Material* (Ibn Abī Uṣaybiʿah, *ʿUyūn al-anbāʾ*, II/2, 928.4–6). See Ignacio Sánchez, 'Iṣṭifān ibn Basīl', *EI Three*; D'Ancona, 'Greek into Arabic', *EI Three*; Georg Graf, *Geschichte*, II, 131; and Ullmann, *Untersuchungen*, 21–24.

166 In Ibn Abī Uṣaybiʿah's list (see below, II.2), this name is combined with the above-mentioned Ḥayrūn (or Ḥīrūn) to Ḥayrūn ibn Rābiṭah.

167 Unidentified. Šamlī translated Galen's *On Juice* (*De bonis malisque sucis*) (see Ḥunayn ibn Isḥāq, *al-Risālah*, 36.9/Ar. and 29/Ger.; Ibn al-Nadīm, *Kitāb al-fihrist*, I, 290.25–26; Ibn al-Qifṭī, *Taʾrīḫ al-ḥukamāʾ*, 131.6); and Aristotle's *Metaphysics Lambda* (see Ibn al-Nadīm, *Kitāb al-fihrist*, I, 251.30; Ibn al-Qifṭī, *Taʾrīḫ al-ḥukamāʾ*, 42.5).

168 Unidentified.

169 According to Ibn al-Nadīm (*Kitāb al-fihrist*, I, 262.23–27) and Ibn al-Qifṭī (*Tāʾrīḫ al-ḥukamāʾ*, 77.1–5), Ibrāhīm al-Quwayrī was a logician and the teacher of Abū Bišr Mattā ibn Yūnus (d. 328/940). He wrote commentaries on Aristotle's *Categories* (*Categoriae*), *On Interpretation* (*De interpretatione*), and *Prior Analytics* (*Analytica priora*).

Theodore the Syncellus.[170]
Dārīʿ, the monk.[171]
Hibā.[172]
Biṯyūn [Pethion].[173]
Ṣalībā.[174]
Ayyūb al-Ruhāwī.[175]
Ṯābit ibn Qumaʿ.[176]
Ayyūb and Simʿān [Simeon]:[177] they translated Ptolemy's astronomical tables[178] and other ancient books for Muḥammad ibn Ḫālid ibn Yaḥyā ibn Barmak.
Bāsīl, who served (Ṭāhir ibn al-Ḥusayn) Ḏū l-Yamīnayn.[179]
Ibn Šahdā al-Karḫī: he translated badly from Syriac into Arabic. Among his translations was Hippocrates's *Book on Embryos* [*Kitāb al-aǧinnah*].[180]

---

170 The Arabic term *al-sunqal* transliterates the Syriac rendition ܣܘܢܩܠܐ of the Greek σύγκελλος (*synkellos*). Nothing is known about this translator. He may be the Theodore who, together with his brother Theophanes and Michael the Syncellus, was active at Mar Saba but left for Constantinople after the sack of the monastery in 198/813. See Brock, 'Syriac into Greek', 201.

171 Unidentified.

172 Hibā of Edessa (d. AD 457) was the head of the School of Edessa and the author of the first Syriac translation of Porphyry's *Isagoge*. See Lucas van Rompay, 'Hiba', *GEDSH*.

173 Unidentified, perhaps a doublet of the Pethion mentioned above. Ibn Abī Uṣaybiʿah only lists a single translator of this name; see below, II.2.

174 Unidentified. This is an Arabic transliteration of the Eastern Syriac name Saliba, Slibo or Sliba (ܨܠܝܒܐ).

175 Ayyūb al-Ruhāwī or Job of Edessa was a Greek–Syriac translator of the first half of the third/ninth century and a Christian physician in the service of the ʿAbbāsid caliphs and the governor of Ḫurāsān, ʿAbd Allāh ibn Ṭāhir. His translations come in for frequent criticism in Ḥunayn ibn Isḥāq's *Epistle*; see See Barbara H. Roggema, 'Iyob of Edessa', *GEDSH*; Gutas, *Greek Thought*, 72; and Meyerhof, 'New Light', 703–704.

176 Unidentified, possibly a garbled repetition of the mathematician and translator Ṯābit ibn Qurrah (see above, I.6 with n. 66), who is listed below.

177 Unidentified.

178 Presumably his *Handy Tables* (*Πρόχειροι κανόνες*).

179 Unidentified. For chronological reasons, it is unlikely that this Bāsīl corresponds to Basīl al-Muṭrān, listed below. Ṭāhir ibn al-Ḥusayn (d. 207/822) was an important ally of the caliph al-Maʾmūn during the war with his brother and predecessor al-Amīn. He became the first in a line of important governors of Ḫurāsān. See Charles Bosworth, 'Ṭāhir b. al-Ḥusayn', *EI2*. On the role of the Ṭāhirids as sponsors of scholars and translators, see Gutas, *Greek Thought*, 129–130.

180 Ḥunayn ibn Isḥāq credited Ibn Šahdā (written Sahdā in the *Epistle*) with a Syriac translation of Galen's *On Sects* (*De sectis*) (*al-Risālah*, 4.19–5.1/Ar. and 4/Ger.); *On the Medical Art* (*Ars medica*) (*al-Risālah*, 6.1–2/Ar. and 5/Ger.); and *On the Pulse for Beginners* (*De pulsibus ad tirones*) (*al-Risālah*, 6.13–14/Ar. and 5/Ger.).

Abū ʿAmr Yūḥannā ibn Yūsuf al-Kātib: he was one of the translators and translated Plato's book on the education of boys.[181]

Ayyūb ibn al-Qāsim al-Raqqī: he translated from Syriac into Arabic; among his translations was (Porphyry's) *Isagoge*.[182]

Midlāǧī/Marlāḥī[?][183] in our own time: he knows Syriac well, but stammers when speaking Arabic. He translates under the supervision of ʿAlī ibn Ibrāhīm al-Dahakī[184] from Syriac into Arabic, and Ibn al-Dahakī corrects his translations.

Dādīšūʿ: he translated for Isḥāq ibn Sulaymān ibn ʿAlī al-Hāšimī from Syriac into Arabic.[185]

Qusṭā ibn Lūqā al-Baʿlabakkī: he was a good translator and had a good command of Greek, Syriac, and Arabic. He translated some things and corrected many (other) translations.[186] […]

Ḥunayn (ibn Isḥāq).[187]

Isḥāq (ibn Ḥunayn ibn Isḥāq).[188]

Ṯābit (ibn Qurrah).[189]

Ḥubayš (ibn al-Ḥasan al-Aʿsam).[190]

ʿĪsā ibn Yaḥyā.[191]

---

181 Unidentified. In spite of their similar names, this translator is unlikely to be the same person as Yūḥannā ibn Yūsuf ibn Ḥāriṯ ibn al-Biṭrīq al-Qass (see above, I.13 with n. 143), a geometer who also translated from Greek.

182 Almost nothing is known about Ayyūb ibn al-Qāsim al-Raqqī (fl. c. 820). Raqqah was a Ṣābian (see above, n. 64) centre, and he might have belonged to this religious group. His translation of the *Isagoge* has not survived. See Rescher, *The Development*, 96.

183 Unidentified.

184 Possibly Abū l-Qāsim ʿAlī ibn Ibrāhīm ibn Muḥammad al-Dahakī, a contemporary of Ibn al-Nadīm. According to Yāqūt al-Ḥamawī (*Muʿǧam al-udabāʾ*, IV, 1641–1642, n. 711), al-Dahakī was a man of letters associated with the vizier Abū l-Faḍl al-ʿAbbās al-Šīrāzī who read the *Book of Songs* (*Kitāb al-aġānī*) with its author, Abū l-Faraǧ al-Iṣbahānī.

185 Unidentified.

186 Qusṭā ibn Lūqā (d. c. 300/912–913) was a Christian translator from Baʿlabakk who also wrote medical and mathematical works. See Donald R. Hill, 'Ḳusṭā b. Lūḳā', *EI2*.

187 See above, 255–256, and I.5 with n. 52.

188 See above, I.6 with n. 63.

189 See above, I.6 with n. 64.

190 The nephew of Ḥunayn ibn Isḥāq, Ḥubayš ibn al-Ḥasan al-Dimašqī (fl. second half of third/ninth century), nicknamed al-Aʿsam, was a prominent translator and physician who also wrote original works on medicine and pharmacology. See Gotthard Strohmaier, 'Ḥubaysh b. al-Ḥasan al-Dimashḳī', *EI Three*.

191 Nothing is known about the life of ʿĪsā ibn Yaḥyā, one of Ḥunayn's students and collaborators, apart from two brief entries in Ibn Abī Uṣaybiʿah (*ʿUyūn al-anbāʾ*, II/1, 505.1–2 and II/1, 508.4–5), and Ḥunayn's references to his translations in his *Epistle* (Ḥunayn ibn Isḥāq, *al-*

Al-Dimašqī.[192]
Ibrāhīm ibn al-Ṣalt (Abū Nūḥ).[193]
Ibrāhīm ibn ʿAbd Allāh.[194]
Yaḥyā ibn ʿAdī.[195]
Al-Nafīsī.[196]

The Names of the Translators from Persian into Arabic

Ibn al-Muqaffaʿ[197] [...]
Most of the Nawbaḫt family[198] [...]
Mūsā and Yūsuf, the sons of Ḫālid:[199] they served Dāʾūd ibn ʿAbd Allāh ibn Ḥumayd ibn Qaḥṭabah[200] and translated for him from Persian into Arabic.

Risālah, index, s.v. ʿĪsā ibn Jaḥjā). See Sezgin, GAS, III, 257; and Ullmann, *Die Medizin*, 30, 57.

192 Most likely the translator Abū ʿUṯmān Saʿīd ibn Yaʿqūb al-Dimašqī; see above, I.6 with n. 65.

193 Ibrāhīm ibn al-Ṣalt was a physician and translator of a number of medical and astronomical works who belonged to the Church of the East. Almost nothing is known about his life.

194 Ibrāhīm ibn ʿAbd Allāh al-Kātib or al-Nāqil (fl. second half of third/ninth century) was a Christian scholar and contemporary of the next name on Ibn al-Nadīm's list, Abū Zakariyāʾ Yaḥyā ibn ʿAdī (d. 363/974). See Genequand, *Alexander of Aphrodisias*, 31.

195 Abū Zakariyāʾ Yaḥyā ibn ʿAdī, a Syrian Orthodox philosopher and theologian, who also translated a number of works by Aristotle. Ibn al-Nadīm used Yaḥyā ibn ʿAdī's personal library when he compiled the information for his *Catalogue*. See Gerhard Endress, 'Yaḥyā b. ʿAdī', EI2, and Endress, *The Works*.

196 This is probably ʿĪsā ibn Ibrāhīm al-Nafīsī (fl. fourth/tenth century), a physician at the court of the Ḥamdanid ruler of Aleppo Sayf al-Dawlah (r. 333/944–356/967) who translated several philosophical works from Syriac into Arabic, among them the Syriac version of the pseudo-Aristotelian *On the Universe* (*De mundo*). See Peters, *Aristoteles Arabus*, 62; and Ibn al-Qifṭī, *Taʾrīḫ al-ḥukamāʾ*, 250.15–16 (who refers to him as ʿĪsā al-Nafīsī).

197 Ibn al-Muqaffaʿ (d. c. 139/757), a prose writer and secretary, and one of the first translators of belletristic works from Persian into Arabic. He is mainly known for his translation of the *Kalīlah wa-Dimnah*. See Derek Latham, 'Ebn Al-Moqaffaʿ, Abū Moḥammad ʿAbd Allāh Rōzbeh', *Encyclopaedia Iranica*.

198 A prominent Shiʿite family of Persian origin who achieved great influence at the court of several ʿAbbāsid caliphs. Their eponymous forefather, al-Nawbaḫt (fl. mid-second/eighth century), joined the court of the caliph al-Manṣūr as court astrologer, an office subsequently held by several of his descendants. The most important member of the family in the field of translation was Abū Sahl ibn Nawbaḫt, mentioned below. See Sean Anthony, 'Nawbakti Family', *Encyclopaedia Iranica*.

199 Unidentified.

200 The Qaḥāṭibah, descendants of Qaḥṭabah ibn Šabīb (d. 132/749), were a prominent Arab family settled in Ḫurāsān who played an important role in the ʿAbbāsid revolution. Dāʾūd ibn ʿAbd Allāh ibn Ḥumayd ibn Qaḥṭabah (d. 159/776) was governor of Egypt during the reign of al-Manṣūr.

Al-Tamīmī, whose name was ʿAlī ibn Ziyād, with the patronym Abū l-Ḥasan: he translated from Persian into Arabic, and among his translations were the *Astronomical Tables* of al-Šahriyār.[201]

Al-Ḥasan ibn Sahl (ibn Nawbaḫt)[202] [...]

Al-Balāḏurī, Aḥmad ibn Yaḥyā ibn Ǧābir[203] [...]: he translated from Persian into Arabic.

Ǧabalah ibn Sālim, the secretary of Hišam,[204] mentioned above: he translated into Arabic from Persian.[205]

Isḥāq ibn Yazīd[206] translated from Persian into Arabic and among his translations was a book about the history of Persia known as the *Book of Baḫtiyār* [*Baḫtiyār Nāmah*].

Among the Persian translators are: Muḥammad ibn al-Ǧahm al-Barmakī;[207] Hišām ibn al-Qāsim;[208] Mūsā ibn ʿĪsā al-Kisrawī;[209] Zādawayh ibn Šāh-

---

201 This Abū l-Ḥasan ʿAlī ibn Ziyād al-Tamīmī remains unidentified. The *Astronomical Tables* mentioned in this entry, now lost, were a royal astrological canon written in Pahlavi, compiled AD c. 450 and revised during the reign of Ḫusraw I Anūširwān (r. AD 531–579). On this work and its influence, see Jamil Ragep, 'Astronomy', *EI Three*.

202 Abū Sahl ibn Nawbaḫt (fl. c. 153/770–193/809) worked as astrologer under the caliphs al-Manṣūr, al-Mahdī (r. 158/775–169/785), and al-Hādī (169/785–170/786). Under Hārūn al-Rašīd, he took a position as translator in the caliphal library (*ḫizānat al-ḥikmah*), where he produced at least seven translations of Persian works (Ibn al-Nadīm, *Kitāb al-fihrist*, I, 274.7–12). See Pingree, 'Abu Sahl Nawbakt', *Encyclopaedia Iranica*.

203 This is Aḥmad ibn Yaḥyā ibn Ǧābir ibn Dāʾūd al-Balāḏurī (d. c. 279/892–893), the Muslim historian of the third/ninth century. See above, I.3.

204 Theodor Nöldeke rejected the identification of this Hišām with the Umayyad caliph Hišām ibn ʿAbd al-Malik on the grounds that this would result in an implausibly early date for the translation. Instead, he suggested the historian Hišām ibn Muḥammad ibn al-Kalbī (d. 206/821). See Nöldeke, *Geschichte*, 475.

205 Ǧabalah ibn Sālim translated two Pahlavi works, the *Romance of Bahrām Čōbīn* (*Kitāb Bahram Šūš*) and the *Book of Rustam and Isfandiyār* (*Kitāb Rustam wa-Isfandiyār*) (Ibn al-Nadīm, *Kitāb al-fihrist*, I, 305.9–10). See Nöldeke, *Geschichte*, 474–475.

206 Unidentified.

207 Muḥammad ibn al-Ǧahm al-Barmakī was an expert in logic and astronomy who composed several books on astrological elections (*iḫtiyārāt*) for the caliph al-Maʾmūn (see Ibn al-Qifṭī, *Taʾrīḫ al-ḥukamāʾ*, 284.1–4). According to Ḥamzah al-Iṣfahānī (*Taʾrīḫ sinī mulūk al-arḍ*, 10.2), he translated a book on Persian kings.

208 Hišām ibn al-Qāsim was the translator or compiler of a history of the Sasanian kings; see Ḥamzah al-Iṣfahānī, *Taʾrīḫ sinī mulūk al-arḍ*, 10.6–7.

209 Mūsā ibn ʿĪsā al-Kisrawī (d. after 350/961) was an Iranian historian. Ibn al-Nadīm does not give the titles of his translations, but lists two of his own books (*Kitāb al-fihrist*, I, 128.9–11). Ḥamzah al-Iṣfahānī (*Taʾrīḫ sinī mulūk al-arḍ*, 16.3–8) quotes from al-Kisrawī's translation of the Pahlavi *Book of Lords* (*Ḫwadāy-nāmag*). See Gutas, *Greek Thought*, 40; and Rubin, 'Ḥamzah al-Iṣfahānī's Sources', 43–44.

awayh al-Iṣfahānī;[210] Muḥammad ibn Bahrām ibn Miṭyār al-Iṣfahānī;[211] Bahrām ibn Mardān Šāh,[212] the priest of Šāpūr, a Persian city; ʿUmar ibn al-Farruḫān[213] [...]

Indian and Nabataean Translators

Mankah al-Hindī:[214] he belonged to the group of Isḥāq ibn Sulaymān ibn ʿAlī al-Hāšimī[215] and translated from the Indian language into Arabic.

Ibn Dahn al-Hindī:[216] he administered the hospital of the Barmak family.[217] He translated into Arabic from the Indian language.

---

210 Zādawayh or Dādawayh ibn Šāhawayh (Zādūyah or Dādūyah ibn Šahūyah al-Iṣfahānī) was an Iranian scholar and the author of a history of Persian notables; see al-Bīrūnī, *Chronologie*, 44.2. According to Ḥamzah al-Iṣfahānī (*Taʾrīḫ sinī mulūk al-arḍ*, 10.3–4), he translated a Pahlavi history of the Persian kings. See also Justi, *Iranisches Namenbuch*, 378.

211 Muḥammad ibn Bahrām ibn Miṭyār al-Iṣfahānī was an Iranian historian; according to Ḥamzah al-Iṣfahānī (*Taʾrīḫ sinī mulūk al-arḍ*, 10.4–5), he translated or compiled another history of the Persian rulers.

212 Bahrām ibn Mardān Šāh, a Zoroastrian priest (*mobed*), translated the *Book of Lords* (*Ḫwadāy-nāmag*). According to Ḥamzah al-Iṣfahānī (*Taʾrīḫ sinī mulūk al-arḍ*, 10.7–8), he also revised a history of the Sasanian kings (*Kitāb taʾrīḫ mulūk Banī Sāsān*). See Djalal Khaleghi Motlagh, 'Bahrām b. Mardānšāh', *Encyclopaedia Iranica*.

213 Abū Ḥafs ʿUmar ibn Farruḫān al-Ṭabarī (d. c. 200/815), known in Latin as Omar Tiberiades, was a Persian astrologer and translator. He translated into Arabic the Pahlavi version of Dorotheus's *Pentateuch*, composed a commentary on Ptolemy's *Tetrabiblos*, and wrote several astrological treatises; see King, 'Astronomy', 294, 296.

214 Mankah (Sanskrit: Māṇikya) was one of the physicians who translated Indian works for the caliph Hārūn al-Rašīd and others, among them the *Suśruta-Saṃhitā* (*Kitāb Susrud* in Arabic), i.e., Suśruta's medical handbook (see Kahl, *The Sanskrit, Syriac and Persian Sources*, 14–16), and a book entitled *Names of Indian Medicines* (*Asmāʾ ʿaqāqīr al-Hind*, see Ibn al-Nadīm, *Kitāb al-fihrist*, I, 303.6–7, 11–12; and Ibn Abī Uṣaybiʿah, *ʿUyūn al-anbāʾ*, II/2, 890.1–891.5). See also Sezgin, GAS III, 197–198, 200–201; and Ullmann, *Die Medizin*, 106.

215 Isḥāq ibn Sulaymān ibn ʿAlī al-Hāšimī was a high-ranking court official during the reign of Hārūn al-Rašīd and al-Amīn.

216 According to Ibn al-Nadīm, the director of the hospital of Ǧondēšāpūr, Ibn Dahn the Indian translated several books from Sanskrit into Arabic, among them Vāgbhata's *Aṣṭāṅgahṛdaya-Saṃhitā* (*Compendium of the Eightfold Essence* [*of Medicine*]; see Kahl, *The Sanskrit, Syriac and Persian Sources*, 21–22), translated into Arabic as *Istankār al-jāmiʿ*; and Ravigupta's *Siddhasāra* (*The Perfect Quintessence* [*of Medicine*]; see Kahl, *The Sanskrit, Syriac and Persian Sources*, 22–24), which in Arabic is called *Kitāb Sindistāq* or *Sindhašar*. See Ibn al-Nadīm, *Kitāb al-fihrist*, I, 303.7, 8–9; Sezgin, GAS, III, 198–199; and Ullmann, *Die Medizin*, 105.

217 On the Barmak family, see above, II.1 with n. 153.

Ibn Waḥšīyah:[218] he translated from Nabataean into Arabic and translated many books, as people said […][219]

II.2. Ibn Abī Uṣaybiʿah, *The Best Accounts of the Classes of Physicians* (*ʿUyūn al-anbāʾ fī ṭabaqāt al-aṭibbāʾ*), II/1, 507.1–517.10 (ed. Savage-Smith et al.); tr. Savage-Smith et al., III/1, 541–555, adapted by Vagelpohl and Sánchez.

The following text is part of the ninth chapter of the book, which focuses on the translators of medical works and evinces obvious similarities with the previous text (II.1) from Ibn al-Nadīm's *Catalogue*. Both Ibn Abī Uṣaybiʿah and Ibn al-Nadīm seem to have relied on the same unidentified source, which, in view of the transliteration system used in both texts and the religious affiliation of the translators listed, was probably originally written by a Christian Syriac author and later translated into Arabic. The second part of Ibn Abī Uṣaybiʿah's account lists sponsors of the Greek–Arabic translation movement and has no parallel in Ibn al-Nadīm; the main source for this list of patrons is Ḥunayn ibn Isḥāq's *Epistle*.[220]

Chapter Nine: Physicians who Translated Works on Medicine and Other Subjects from Greek into Arabic, and Their Patrons

Ǧūrǧis was the first of those who began the work of translating medical and other works from Greek into Arabic. He undertook this task at the invitation of the caliph al-Manṣūr, who treated hin with the utmost generosity.[221] […]

Ḥunayn ibn Isḥāq knew four languages—Arabic, Syriac, Greek, and Persian—including rare words as well as common vocabulary. His works are superb examples of the translator's art.[222]

---

218 Ibn Waḥšīyah (d. 318/930–931) was the alleged translator of several Nabatean works. The most famous is the so-called *Nabatean Agriculture* (*Kitāb al-filāḥah al-nabāṭīyah*), which is now considered a compilation rather than a translation. See Toufic Fahd, 'Ibn Waḥshiyya', *EI2*; Hämeen-Anttila, *The Last Pagans*, 3–33; and the chapter by Isabel Toral in this volume.

219 Another translation: Ibn al-Nadīm, *The Fihrist*, 586–590 (English).

220 On these sponsors and the phenomenon of patronage in general, see Meyerhof, 'New Light'; Gutas, *Greek Thought*, 121–150; and Micheau, 'Mécènes et médecins'.

221 Ǧūrǧīs ibn Buḫtīšūʿ was a renowned Christian physician from Ǧondēšāpūr and director of the hospital of this town. In 148/765–766, he was summoned to Baghdad to attend on the caliph al-Manṣūr. See Dominique Sourdel, 'Bukhtīshūʿ', *EI2*.

222 See above, 255–256, and I.5 with n. 52.

Isḥāq ibn Ḥunayn was conversant with all the languages that his father knew, and was comparable to him as a translator. His style was mellifluous and his diction fluent. However, Ḥunayn has more works to his credit, both original compositions and translations, than Isḥāq.[223] [...]

Ḥubayš al-Aʿsam, the son of a sister of Ḥunayn ibn Isḥāq, and his pupil. An admirable translator, worthy to be ranked with Ḥunayn and Isḥāq.[224] [...]

ʿĪsā ibn Yaḥyā ibn Ibrāhīm was another pupil of Ḥunayn ibn Isḥāq. An excellent translator: Ḥunayn praised him and held his translation work in high regard. ʿĪsā's translations are closely modelled on those of Ḥunayn. He also produced original compositions.[225]

Qusṭā ibn Lūqā al-Baʿlabakkī was a translator experienced in languages and accomplished in the philosophical and other sciences.[226] [...]

Ayyūb, known as al-Abraš ['the Speckled'], was an unproductive and mediocre translator, until late in life, when he produced a number of translations worthy of Ḥunayn.[227]

Māsarǧīs was a translator from Syriac into Arabic, and renowned for his knowledge of the art of medicine. He is the author of works entitled *On the Effectiveness of Foods, their Beneficial and Harmful Effects* and *On the Effectiveness of Drugs, their Beneficial and Harmful Effects*.[228]

ʿĪsā ibn Māsarǧīs followed in his father's footsteps. He is the author of *On Colours* and *On Scents and Flavours*.[229]

Šahdā [or Sahdā] al-Karḫī, a native of Karḫ [a district of Baghdad], was much like the aforementioned as a translator.[230]

Ibn Šahdā [or Sahdā] al-Karḫī: as a translator, he produced work similar to that of his father. The translations he produced later in life, however, are superior to those of his father, although still undistinguished. He translated

223 See above, I.6 with n. 63.

224 See above, II.1 with n. 190.

225 See above, II.1 with n. 191.

226 See above, II.1 with n. 186.

227 Unidentified. In Ḥunayn ibn Isḥāq's *Epistle* (*al-Risālah*, 3.21–22), the nickname al-Abraš ('the freckled') is associated with Ayyūb al-Ruhāwī (see above, II.1 with n. 175), but a little further down (296), Ibn Abī Uṣaybiʿah explicitly pointed out that this Ayyūb al-Abraš and Ayyūb al-Ruhāwī are not the same person.

228 Māsarǧīs, also called Māsarǧawayh (Persian: Māsargōye), was a physician of Judeo-Persian origin. He is one of the few physicians known from the Umayyad period, but the exact dates of his life are unknown. See Albert Dietrich, 'Māsardjawayh', *EI2*.

229 ʿĪsā was the son of Māsarǧīs and also a physician. Almost nothing is known about his life. See Albert Dietrich, 'Māsardjawayh', *EI2*.

230 Presumably the father of the next translator on the list, who is mentioned several times in Ḥunayn's *Epistle*.

from Syriac into Arabic. Among other things, he translated Hippocrates's work *On Embryos*.[231]

Al-Ḥaǧǧāǧ ibn Maṭar translated for the caliph al-Ma'mūn.[232] Among other works, he produced a translation of the book of Euclid [i.e., the *Elements*], which was subsequently revised and corrected by Ṯābit ibn Qurrah al-Ḥarrānī.[233]

Ibn Nāʿimah—that is, ʿAbd al-Masīḥ ibn ʿAbd Allāh al-Ḥimṣī al-Nāʿimī—was a generally undistinguished translator, although he produced some respectable work on occasion.[234]

Zarūyā ibn Mānaḥūh al-Naʿimī al-Ḥimṣī produced facile translations, which, however, did not measure up to those of his predecessors.[235]

Hilāl ibn Abī Hilāl al-Ḥimṣī: his translations are sound, but are lacking in literary style and elegance.[236]

Fiṯyūn [Pethion] the translator: I have found his translations barbarously ungrammatical; his grasp of the Arabic language was most imperfect.[237]

Abū Naṣr ibn Nārī ibn Ayyūb produced few translations, and those he did produce are considered less reliable than those of other translators.[238]

Basīl al-Muṭrān was a prolific and passably good translator.[239]

Iṣṭifan ibn Basīl was not far short of the standard of Ḥunayn ibn Isḥāq himself as far as his ability as a translator goes; however, Ḥunayn's style is smoother and more graceful.[240]

Mūsā ibn Ḫālid the Translator:[241] I have found many works translated by this scholar, including Galen's *Sixteen Books*[242] and other works, but in terms of skill he fell far short of the standard of Ḥunayn.

---

231 See above, II.1 with n. 180.

232 See above, I.6 with n. 61.

233 See above, I.6 with n. 64.

234 See above, II.1 with n. 151.

235 Unidentified. See above, II.1 with n. 156.

236 See above, II.1 with n. 157.

237 See above, II.1 with n. 159.

238 Unidentified. See above, II.1 with n. 160.

239 See above, II.1 with n. 161.

240 See above, II.1 with n. 165.

241 An otherwise little-known earlier translator whose works were revised by Ḥunayn ibn Isḥāq; see below, II.7. Ibn al-Nadīm (in II.1 above) lists two brothers, Mūsā and Yūsuf ibn Ḫālid, who translated from Persian into Arabic. Since the translation mentioned in this entry presumably took place from Greek or Syriac into Arabic, it seems unlikely that the two Mūsās are the same person.

242 The so-called *Sixteen Books* were sixteen treatises by Galen that the Alexandrian scholars established as the core corpus for the study of medicine. See Pormann and Savage-Smith, *Medieval Islamic Medicine*, 13, 85.

Usṭāṯ was a mediocre translator.[243]

Ḫayrūn ibn Rābiṭah was not renowned for the quality of his translations.[244]

Theodore the Syncellus: I am familiar with his translations of various philosophical works. They are quite acceptable in terms of quality.[245]

Sergius of Rēšʿaynā was a native of the city of Raʾs al-ʿAyn. He translated many works, but was undistinguished as a translator. Some of his works were revised by Ḥunayn, and those are excellent, whereas those works that Ḥunayn did not revise are of indifferent quality.[246]

Ayyūb al-Ruhāwī[247] is not the Ayyūb al-Abraš[248] who was mentioned earlier. This Ayyūb was an excellent translator with a good knowledge of languages, although his Syriac was better than his Arabic.

Yūsuf the translator: Abū Yaʿqūb Yūsuf ibn ʿĪsā, known as 'the doctor' and 'the translator', and also dubbed 'the sleepless',[249] was a pupil of ʿĪsā ibn Ṣaharbuḫt.[250] Yūsuf was a native of Ḫūzistān, and suffered from a speech impediment. His translations are unimpressive.

Ibrāhīm ibn al-Ṣalt was a mediocre translator. His work is similar to that of Sergius of Rēšʿaynā.[251]

243 See above, II.1 with n. 163.

244 Unidentified. See above, II.1.

245 See above, II.1 with n. 170.

246 Sergius of Rēšʿaynā (d. 536), a distinguished scholar of philosophy and medicine, was one of the most prolific Syriac translators, with numerous theological, medical, and philosophical works to his name. Ḥunayn ibn Isḥāq (*al-Risālah*, index, s.v. Sergios ar-Raʾsʿainī) lists and frequently criticises his many medical translations. See Griffith, 'Sergios of Rešʿaina'.

247 See above, II.1 with n. 175.

248 See above, 294 with n. 227.

249 According to Ibn al-Qifṭī (*Taʾrīḫ al-ḥukamāʾ*, 392.1–11) and Ibn Abī Uṣaybiʿah (*ʿUyūn al-anbāʾ*, II/1, 506.1–5), Yūsuf al-Sāhir ('the insomniac'), also known as Yūsuf al-Qass ('the priest'), worked as a physician during the caliphate of al-Muktafī (r. 298/902–295/908) and received the nickname 'the insomniac' either because of his habit of studying during the night or due to a tumour on his forehead that prevented him from sleeping; see Ullmann, *Die Medizin*, 124, 319. This person may be identical with the Yūsuf al-Ḫūrī ('the priest') criticised by Ḥunayn ibn Isḥāq (*al-Risālah*, 30.1–2) for his deficient Syriac translation of Galen's *Simple Medicines* (*De simplicium medicamentorum temperamentis et facultatibus*).

250 ʿĪsā ibn Ṣaharbuḫt was a Christian physician from Ǧondēšāpūr, a student of Ǧurǧis ibn Ǧibraʾīl ibn Buḫtīšūʿ and the author of a treatise entitled *On the Power of Simple Drugs* (*Kitāb quwā l-adwiyah al-mufradah*). He is also credited with at least one translation into Arabic, that of Ḥunayn ibn Isḥāq's Syriac epitome on Galen's *Commentary on Hippocrates's Aphorisms* (*In Hippocratis aphorismos commentarii*) (Ibn Abī Uṣaybiʿah, *ʿUyūn al-anbāʾ*, II/1, 492.7–8, 505.6–7). See Gerrit J. Reinink, 'Ṣharbokht bar Msargis', *GEDSH*.

251 See above, II.1 with n. 162.

Ṯābit the Translator was also mediocre, albeit better than Ibrāhīm ibn al-Ṣalt. He was not a prolific translator. Galen's *On Juices* is among the works he translated.[252]

Abū Yūsuf the Secretary also was an undistinguished translator. He translated a number of works of Hippocrates.[253]

Yūḥannā ibn Buḫtīšūʿ translated numerous works into Syriac, but is not known to have done any translation into Arabic.[254]

Al-Biṭrīq lived in the time of al-Manṣūr. The caliph ordered him to undertake the task of translating a number of ancient works. He did so, translating numerous books, but as a translator he is not up to the standard of Ḥunayn ibn Isḥāq. I have seen Arabic versions of many of the medical works of Hippocrates and Galen that are by al-Biṭrīq.[255]

Yaḥyā ibn al-Biṭrīq belonged to the circle of al-Ḥasan ibn Sahl. He did not know much of either Arabic or ancient Greek, as he was a Byzantine Roman who was fluent in the contemporary vernacular form of Greek and knew how to write it in its characteristic cursive script instead of the discrete letters of ancient Greek writing.[256]

Tūmā al-Ruhāwī [Thomas of Edessa]: whenever Ḥunayn had a great deal of translation work in hand and was pressed for time, he would ask Tūmā al-Ruhāwī to do some of the translation, subsequently revising what he had done.[257]

Manṣūr ibn Bānās was approximately as good a translator as Thomas of Edessa. His Syriac was stronger than his Arabic.[258]

---

252 This is Ṯābit ibn Qurrah (see above, I.6 with n. 64). His translation of this work is mentioned by Ḥunayn ibn Isḥāq (*al-Risālah*, 36.8).

253 Unidentified.

254 Yūḥannā ibn Buḫtīšūʿ (d. c. 300/912), a Christian physician of the Church of the East, was the son of Buḫtīšūʿ ibn Ǧibraʾīl. Ḥunayn ibn Isḥāq noted that Yūḥannā sought his help when translating Galen's *On Antidotes* (*De antidotis*) (*al-Risālah*, 38.11–12). See Sezgin, GAS, III, 258; and Ullmann, *Die Medizin*, 111.

255 See above, I.11 with n. 105.

256 See above, II.11 with n. 191.

257 Of Tūmā al-Ruhāwī (Thomas of Edessa) we only know that he was a contemporary of Ḥunayn ibn Isḥāq, for whom he translated the last part of Galen's *A Man's Ability to Know the Imperfections of his Soul* (*De cuiuslibet animi peccatorum dignotione*) into Arabic (*al-Risālah*, 48.19–49.4; see also Ibn al-Nadīm, *Kitāb al-fihrist*, I, 291.6–7, and Ibn al-Qifṭī, *Taʾrīḫ al-ḥukamāʾ*, 131.15).

258 Almost nothing is known about this man. According to Ḥunayn ibn Isḥāq (*al-Risālah*, 49.6–9, where the name is spelled Ibn Aṯānās), Ibn Bānās was the Ṣābian author of a deficient Syriac translation of Galen's *Character Traits* (*De moribus*).

ʿAbdīšūʿ ibn Bahrīz, the metropolitan of Mosul.[259] He was a friend of Ǧibrāʾīl ibn Buḫtīšūʿ[260] and one of his translators.

Abū ʿUṯmān Saʿīd ibn Yaʿqūb al-Dimašqī was a translator of the first order who served ʿAlī ibn ʿĪsā exclusively.[261]

Abū Isḥāq Ibrāhīm ibn Baks [or Bakkuš] was a renowned physician who translated many works into the Arabic language. His translations, too, are highly esteemed.[262]

Abū l-Ḥasan ʿAlī ibn Ibrāhīm ibn Baks [or Bakkuš] was also a well-known physician, and a reliable translator as well.[263]

The patrons of these translators, other than caliphs, are listed below:[264]

Šabrīšūʿ ibn Quṭrub was from Ǧondēšāpūr and was always generous to translators, constantly bestowing presents upon them; he sought to attract them and obtain books from them by paying them as much as he could manage. It was Syriac translations that he prized, rather than Arabic ones. He was of Ḫūzistānī origin.[265]

Muḥammad ibn Mūsā al-Munaǧǧim ['the Astrologer'] was one of the sons of Mūsā ibn Šākir, who were noted arithmeticians and celebrated for their skill, learning, and writings on the mathematical sciences.[266] This one,

---

259 See above, II.1 with n. 149.

260 Ǧibrāʾīl or Ǧibrīl ibn Buḫtīšūʿ (d. 212/827), the son of Buḫtīšūʿ ibn Ǧurǧīs, was court physician under Hārūn al-Rašīd and al-Maʾmūn. See Dominique Sourdel, 'Bukhtīshūʿ', *EI2*; Ullmann, *Die Medizin*, 109; Baum, 'Gabriel ibn Bukhtishu'; Meyerhof, 'New Light', 717–718; Micheau, 'Mécènes et médecins', 155–157; and Gutas, *Greek Thought*, 118.

261 See above, I.6 with n. 65.

262 See above, I.13 with n. 142.

263 Like his father, ʿAlī ibn Ibrāhīm ibn Baks (d. 394/1003–1004) was a physician and worked at the ʿAḍudī hospital in Baghdad (Ibn al-Qifṭī, *Taʾrīḫ al-ḥukamāʾ*, 235.17–236.4).

264 The information in the following list of patrons is essentially based on Ḥunayn ibn Isḥāq's *Epistle*.

265 Šīršūʿ (or Šīrīšūʿ) ibn Quṭrub, a contemporary of Ibn Isḥāq, was a physician from Ǧondēšāpūr. Ḥunayn ibn Isḥāq mentioned him as the sponsor of a Syriac translation of Galen's *On Sects*; see below, III.3. See Meyerhof, 'New Light', 719; and Micheau, 'Mécènes et médecins', 162–163.

266 Mūsā ibn Šākir was an astronomer and companion of the caliph al-Maʾmūn since before his accession. Mūsā's three sons, Muḥammad, Aḥmad, and al-Ḥasan, played a central role in the intellectual life of third-/ninth-century Baghdad, especially as sponsors of translations and as scholars of mathematics and mechanics. Muḥammad ibn Mūsā (d. 259/873) is mentioned by Ḥunayn ibn Isḥāq as the commissioner of numerous translations (see *al-Risālah*, index, s.v. Muḥammad ibn Mūsā). On the Banū Mūsā family, see Donald R. Hill, 'Mūsā, Banū', *EI2*; and Gutas, *Greek Thought*, 133–134.

Muḥammad, was a most generous patron to Ḥunayn ibn Isḥāq, who translated numerous medical works for him.

ʿAlī ibn Yaḥyā, known as Ibn al-Munağğim, was a secretary and companion of the caliph al-Ma'mūn, who held him in high regard. He was interested in medicine and commissioned the translation of numerous works on that subject.[267]

Tādurā *al-usquf* [Theodore the Bishop] was a bishop in al-Karḫ in Baghdad. He was an avid book-collector and accordingly sought to win the hearts of translators. In this way he acquired many books. A number of Christian physicians composed important works for him, which they published under his patronage.[268]

Muḥammad ibn Mūsā ibn ʿAbd al-Malik: a number of medical books were translated for him. He himself was a man of considerable learning and prepared summaries of various books, retaining their valuable parts and discarding those that were of no interest.[269]

ʿĪsā ibn Yūnus, the Secretary and Accountant, was a distinguished scholar of Iraq. He was a keen collector of ancient books and had a deep interest in Greek learning.[270]

ʿAlī, known as al-Fayyūmī, so dubbed because of his tenure as governor of the Fayyum. He was generous to translators, who grew fat on his bounty.[271]

---

267 The Banū Munağğim were a Zoroastrian family of Persian origin who joined the court of al-Manṣūr as astrologers, whence their name. They enjoyed high status for several generations, standing out as one of the most influential families in the intellectual life of the third/ninth–fourth/tenth centuries, and were important sponsors of the translation movement. Among other works, ʿAlī ibn Yaḥyā commissioned Ḥunayn ibn Isḥāq's famous *Epistle* (see above, 255–256). See Meyerhof, 'New Light', 714; and Micheau, 'Mécènes et médecins', 164–167. On the Munağğim family, see Manfred Fleischhammer, 'Munadjdjim, Banū', *EI2*; and Gutas, *Greek Thought*, 128.

268 See above, II.1 with n. 158.

269 The identification of this person is conjectural, but he may be Abū ʿImrān Mūsā ibn ʿAbd al-Malik ibn Hišām al-Iṣbahānī (Muḥammad is not part of his name). Mūsā ibn ʿAbd al-Malik (d. 246/860–861) was a court official in charge of land tax and the caliphal chancery under al-Mutawakkil. See Ibn Ḫallikān, *Wafayāt al-aʿyān*, v, 337–341, n. 750.

270 Unidentified, but possibly the result of a misreading or scribal error. According to recension B of Ḥunayn's *Epistle*, his son Isḥāq translated Galen's *The Usefulness of the Pulse* (*De usu pulsuum*) for a sponsor named ʿĪsā ibn Yūnus, who may be the person meant here. See Bergsträßer, *Neue Materialien*, 34.

271 This translator, who also appears in Ḥunayn's *Epistle* as the sponsor of his Syriac translation of Galen's *On the Parts of the Art of Medicine* (*De partibus artis medicativae*) (*al-Risālah*, 31.20–32.1), is otherwise unknown. Since al-Fayyūmī requested a Syriac translation, he must have been a Christian. See Meyerhof, 'New Light', 720; and Micheau, 'Mécènes et médecins', 163.

Aḥmad ibn Muḥammad, known as Ibn al-Mudabbir, the Secretary: an unfailing source of ample money and benefits for translators.[272]

Ibrāhīm ibn Muḥammad ibn Mūsā, the Secretary: he took great interest in the translation of Greek works into the Arabic language. He was extremely generous toward men of learning and scholarship, including translators in particular.[273]

ʿAbd Allāh ibn Isḥāq: he also was deeply interested in having works translated and collecting them.[274]

Muḥammad ibn ʿAbd al-Malik (ibn) al-Zayyāt: he spent nearly 2,000 dinars a month on translators, scholars, and copyists. Many books were translated under his patronage,[275] including Greek works, and a number of distinguished physicians translations for him, including Yūḥannā ibn Māsawayh,[276] Ǧibrāʾīl ibn Buḫtīšūʿ,[277] Buḫtīšūʿ ibn Ǧibrāʾīl ibn Buḫtīšūʿ,[278] Dāʾūd ibn Sarābiyūn,[279] Salmawayh ibn Bunān,[280] Elīšaʿ,[281] Isrāʾīl ibn Zakariyāʾ ibn al-Ṭayfūrī,[282] and Ḥubayš ibn al-Ḥasan.[283]

---

272 Aḥmad ibn Muḥammad Ibn al-Mudabbir (d. 270/883 or 271/884) was a man of letters and high-ranking official at the courts of al-Wāṯiq (r. 227/842–232/847) and al-Mutawakkil. Ḥunayn ibn Isḥāq translated a sample of Galen's *Commentary on the Hippocratic Aphorisms* for him (*al-Risālah*, 40.9–13). See H.L. Gottschalk, 'Ibn al-Mudabbir', *EI2*; Meyerhof, 'New Light', 714; and Micheau, 'Mécènes et médecins', 171–172.

273 Ibrāhīm ibn Muḥammad ibn Mūsā might be the son of the aforementioned Muḥammad ibn Mūsā al-Munaǧǧim; see Meyerhof, 'New Light', 715. Ḥunayn ibn Isḥāq translated the Hippocratic lemmata from Galen's *Commentary on Hippocrates's Prognosticon* (*In Hippocratis Prognosticum commentarii*) into Arabic for him (*al-Risālah*, 40.23–41.1).

274 Almost nothing is known about ʿAbd Allāh ibn Isḥāq. Ḥunayn ibn Isḥāq wrote that he either translated anew or revised for him a translation of Galen's *That the Best Physician is also a Philosopher* (*Quod optimus medicus sit quoque philosophus*) (*al-Risālah*, 46.23–47.1, with Bergsträßer, *Neue Materialen*, 37).

275 Ibn al-Zayyāt (d. 233/847) was vizier under the ʿAbbāsid caliphs al-Muʿtasim and al-Wāṯiq, but fell into disgrace under al-Mutawakkil, who had him imprisoned and tortured. A poet himself, Ibn al-Zayyāt was an important patron of the arts and sciences who supported authors such as al-Ǧāḥiẓ (d. 255/869) and sponsored translations, among them Galen's *On the Voice* (*De voce*); see Ḥunayn ibn Isḥāq, *al-Risālah*, 24.17–20. See also Dominique Sourdel, 'Ibn al-Zayyāt', *EI2*; Meyerhof, 'New Light', 715; and Micheau, 'Mécènes et médecins', 172–174.

276 See above, 1.5 with n. 53.

277 See above, n. 260.

278 Buḫtīšūʿ, the son of Ǧibrīl ibn Buḫtīšūʿ, succeeded his father as court physician. See Dominique Sourdel, 'Bukhtīshūʿ', *EI2*; and Ullmann, *Die Medizin*, 109.

279 Max Meyerhof ('New Light', 719) suggested that Dāʾūd ibn Sarābiyūn was the Dāʾūd al-Mutaṭabbib who commissioned five Syriac Galen translations from Ḥunayn ibn Isḥāq (*al-Risālah*, 3.23, 6.2–6, 48.19–49.2, 49.18, 49.21). The *Epistle* states that Ḥunayn was in his thirties when he translated Galen's *On the Medical Art* for Dāʾūd (*al-Risālah*, 6.3–4). According to Ibn al-Qifṭī (*Taʾrīḫ al-ḥukamāʾ*, 431.5–8), Dāʾūd, a native of Beth Garmai, was

II.3. Ibn al-Nadīm, *The Catalogue* (*Kitāb al-fihrist*), I, 242.7–11 (ed. Flügel); tr. Rosenthal, *The Classical Heritage*, 47, adapted by Vagelpohl and Sánchez.

Although the Greek–Arabic translation movement is traditionally associated with the ʿAbbāsid caliphate, there are indications that there were already isolated efforts to translate individual Greek and Syriac philosophical and scientific works under their predecessors, the Umayyad dynasty. One example, the early eighth-century Arabic translation of the medical *Handbook* of Ahrun, was discussed above.[284] Also credited with sponsoring translations, particularly of alchemical writings, and with authoring original works on the subject, is the son of the Umayyad caliph Yazīd I (r. 60/680–64/683), Ḫālid ibn Yazīd ibn Muʿāwiyah.[285] While the association between Ḫālid ibn Yazīd and alchemy or translations may not stand up to historical scrutiny,[286] the requirements of a multi-lingual state made the translation of political, administrative, and commercial documents a daily necessity.[287]

---

a court physician during the reign of al-Hādī and the brother of Yūḥannā ibn Sarābiyūn, the author of the *Pandectae* (*al-Kunnāš*). Manfred Ullmann has questioned the dating of this work, which contains quotations from authors who lived in the second half of the third/ninth century. He argued that Ibn Sarābiyūn's *Kunnāš* probably appeared at the end of that century, and that Dāʾūd and Yūḥannā ibn Sarābiyūn were separated by a century; see Ullmann, 'Yūḥannā ibn Sarābiyūn', 280. Ibn Abī Uṣaybiʿah mentioned Dāʾūd ibn Sarābiyūn several times, often in connection with the Buḫtīšūʿ family (*ʿUyūn al-anbāʾ*, II/1, 343, 353, 439, 444, 517). See also Micheau, 'Mécènes et médecins', 159–161.

280 Salmawayh ibn Bunān (d. 226/840–841) was a prominent Christian physician in the service of the caliph al-Muʿtaṣim. He commissioned Ḥunayn ibn Isḥāq to produce a number of Syriac and Arabic translations (see for example below, III.5). Ḥunayn wrote a short epistle addressed to him as an introduction to his Syriac translation of Galen's *Habits* (*De consuetudinibus*). See Ibn Abī Uṣaybiʿah, *ʿUyūn al-anbāʾ*, II/1, 423.15–433.10; and Ullmann, *Die Medizin*, 112.

281 Unidentified. A certain Elīsaʿ commissioned the Syriac translation of Galen's *Classification of the Diseases of the Internal Parts* (*De locis affectis*) from Sergius of Rēšʿaynā; see Ḥunayn ibn Isḥāq, *al-Risālah*, 12.21–22.

282 Physician to a high-ranking official at the court of the caliph al-Mutawakkil, al-Fatḥ ibn Ḫāqān (d. 247/861–862), and son of the physician Zakariyāʾ ibn ʿAbd Allāh al-Ṭayfūrī (see below, III.4 with n. 355). Ḥunayn ibn Isḥāq reported that he asked him to finish his Syriac translation of Galen's *Classification of the Diseases of the Internal Parts* (*al-Risālah*, 13.5–6). On this physician (and his father), see Ibn Abī Uṣaybiʿah, *ʿUyūn al-anbāʾ*, II/1, 406.4–408.3; Meyerhof, 'New Light', 719; and Micheau, 'Mécènes et médecins', 161–162.

283 See above, II.1 with n. 190.

284 See above, I.2 with n. 30.

285 See also al-Ǧāḥiẓ, *al-Bayān wa-l-tabyīn*, I, 328.1–2, who maintained that he was the first to translate works on astronomy, medicine, and chemistry.

286 See Ullmann, 'Ḫālid ibn al-Yazīd', 214–217, who traces the origins of this apocryphal story in the first half of the third/ninth century and its subsequent transmission.

287 See Gutas, *Greek Thought*, 23–24.

Another report: Ḫālid ibn Yazīd ibn Muʿāwiyah used to be called the sage of the Marwānids [i.e., the Umayyads]. He was a distinguished man who had great interest in and love for the sciences. He was interested in alchemy. By his command, a group of Greek philosophers who resided in the capital of Egypt and mastered the Arabic language well were brought to him. He commanded them to translate books about alchemy from Greek and Coptic into Arabic. This was the first translation from a foreign language in Islam.[288]

II.4. al-Masʿūdī, *The Meadows of Gold and Mines of Gems* (*Murūǧ al-ḏahab wa-maʿādin al-ǧawhar*), V, 211.16–23, par. 3446 (ed. Pellat); tr. Gutas, *Greek Thought*, 30–31, adapted by Vagelpohl and Sánchez.

Al-Masʿūdī, the author of the *Book of Admonition and Revision* quoted above (I.1), finished his main work, a world history entitled *The Meadows of Gold and Mines of Gems* (*Murūǧ al-ḏahab wa-maʿādin al-ǧawhar*), in 332/943, revising it in 336/947 and again in 345/956. In the following sample, al-Masʿūdī credits the second ʿAbbāsid caliph, al-Manṣūr, as the first caliphal patron of translations. This marks the beginning of the systematic and state-sponsored translation efforts that have come to be known as the translation movement.[289]

(Al-Manṣūr) was the first caliph to favour astrologers and to act on the basis of astrological prognostications. He had in his retinue the astrologer Nawbaḫt the Zoroastrian, who converted to Islam upon (al-Manṣūr's) instigation and who is the progenitor of this family of the Nawbaḫts.[290] Also in his retinue were the astrologer Ibrāhīm al-Fazārī, the author of an ode to the stars and other astrological and astronomical works,[291] and the astrologer ʿAlī ibn ʿĪsā the Astrolabist.[292]

288 Another translation: Ibn al-Nadīm, *The Fihrist*, 581 (English).

289 On the role of astrology and what Dimitri Gutas termed 'early ʿAbbāsid imperial ideology' during this formational period of the translation movement, see Gutas, *Greek Thought*, 28–60.

290 See above, II.1 with n. 202.

291 Al-Fazārī (fl. second half of the second/eighth century), a pioneering Arab astronomer and astrologer in the service of al-Manṣūr, who helped cast the horoscope that determined the best time to found the city of Baghdad. See Julio Samsó, 'al-Fazārī', *EI Three*; and Sezgin, *GAS*, VI, 122–124.

292 ʿAlī ibn ʿĪsā al-Asṭurlābī (the astrolabist), a contemporary and colleague of al-Fazārī; we know precious little about his life. See Sezgin, *GAS*, VI, 143–144.

He was the first caliph to have books translated from foreign languages into Arabic, among them *Kalīlah wa-Dimnah*[293] and *Sindhind*.[294] There were also translated for him books by Aristotle on logic and other subjects, the *Almagest* by Ptolemy,[295] the *Arithmetic* (by Nicomachus of Gerasa),[296] the book by Euclid (on geometry),[297] and other ancient books from classical Greek, Byzantine Greek, Pahlavi [*al-fahlawīyah*, Middle Persian], Neopersian [*al-fārisīyah*], and Syriac. These (translated books) were published among the people, who examined them and devoted themselves to knowing them.[298]

II.5. Ṣāʿid al-Andalusī,[299] *The Classes of Nations* (*Ṭabaqāt al-umam*), 48.10–17, 48.20–49.2 (ed. Cheikho); tr. Gutas, *Greek Thought*, 94, adapted by Vagelpohl and Sánchez.

In his only surviving work, *The Classes of Nations* (*Ṭabaqāt al-umam*), the Andalusian Muslim author Ṣāʿid al-Andalusī sketches a rudimentary history of the progress of the philosophical sciences from people to people and from India to Muslim Spain. The present text describes the role of the ʿAbbāsid caliph al-Maʾmūn, traditionally credited as a crucial supporter of the sciences and of translations from Greek and other languages into Arabic, in amplifying the efforts of his predecessor al-Manṣūr.

Then the caliphate devolved upon the seventh ʿAbbāsid caliph, ʿAbd Allāh al-Maʾmūn, the son of Hārūn al-Rašīd, the son of al-Mahdī, the son of al-Manṣūr,

293 *Kalīlah wa-Dimnah* is the title of the Arabic translation of the *Pañcatantra*, a Sanskrit mirror for princes in the form of animal fables, which was originally composed around AD 300. Around the middle of the sixth century AD, it was translated into Middle Persian for the Sasanian king Ḫusraw I Anūširwān by his physician, Burzōe, who added an autobiographical introduction. The second/eighth-century Arabic translator of this Persian version, Ibn al-Muqaffaʿ, added his own preface and inserted within Burzōe's introduction a section on the uncertainty of religions. See Carl Brockelmann, 'Kalīla Wa-Dimna', *EI2*.

294 The *Sindhind* was the Arabic title of a Sanskrit astronomical work, probably entitled *Mahāsiddhānta*, which was brought to Baghdad in 156/773 by a delegation from Sind, the easternmost province of the ʿAbbāsid state. The Arabic translation may have been produced by the abovementioned al-Fazārī in collaboration with an Indian scholar. See David Pingree, 'Sindhind', *EI2*.

295 See above, I.7.

296 The handbook on number theory, the *Introduction to Arithmetic* (*Introductio arithmeticae*) by Nicomachus of Gerasa (d. AD c. 120).

297 Euclid's *Elements*; see above, I.6.

298 Other translations: Kennedy, *The Caliphate*, 138 (English); and al-Masʿūdī, *Les Prairies d'Or*, V, 1378–1379, par. 3446 (French).

299 See G. Martinez-Gros, 'Ṣāʿid al-Andalusī', *EI2*.

who completed what his forefather al-Manṣūr had started. He applied himself to acquiring knowledge in the right places and mined them at their sources, thanks to his high-mindedness and the capacity of his excellent spirit.

He approached the Byzantine kings, gave them great gifts, and asked them to grant him the philosophical books they had. They sent him the books they had by Plato, Aristotle, Hippocrates, Galen, Euclid, Ptolemy, and other philosophers. He appealed to the most skilled translators and commissioned them to produce the most precise translations, and they translated for him as well as they could. He then encouraged people to read them and motivated them to teach them.

The intellectual marketplace flourished during his reign and wisdom ruled during his age. [...] He did the same with all other scholars, jurists, *ḥadīṯ* compilers, theologians, linguists, historians, literary scholars, and genealogists. People from all fields of practice and learning brought many branches of philosophy to perfection during his reign, established for their successors the medical curriculum, and laid the foundation of education. As a result, the ʿAbbāsid state almost rivalled the Byzantine empire in its heyday and period of greatest unity.

II.6. Ibn al-Nadīm, *The Catalogue* (*Kitāb al-fihrist*), I, 271.9–14 (ed. Flügel); tr. Vagelpohl and Sánchez.[300]

In addition to caliphs and administrators, scholars and other private individuals were instrumental in sustaining the Greek–Arabic translation movement, even after state support dried up. Most famous among them were Mūsā ibn Šākir, an astronomer of obscure origin who already belonged to the inner circle of the caliph al-Maʾmūn before his accession in 198/813, and his three sons Muḥammad, Aḥmad, and al-Ḥasan, all of them renowned scientists in their own right.[301]

The Banū Mūsā: Muḥammad, Aḥmad, and al-Ḥasan were the sons of Mūsā ibn Šākir. [...] These people belonged to those who distinguished themselves in the study of the ancient sciences, made them the object of their desire and themselves their followers. They dispatched people to the Byzantine territory who sent (the ancient sciences) [i.e., scholarly books] to them, attracted translators from all lands and places with generous payments, and uncovered wonders of learning. The sciences with which they were the most concerned were geome-

300 Parallel: Ibn al-Qifṭī, *Taʾrīḫ al-ḥukamāʾ*, 315.22–316.4.

301 See above, II.2 with n. 266. See also Meyerhof, 'New Light', 714–715; Micheau, 'Mécènes et médecins', 167–170; and Gutas, *Greek Thought*, 133–134.

try, mechanics, dynamics, and music, and they were the least (concerned) with astronomy.[302]

II.7. Ibn Ǧulǧul, *The Classes of the Physicians and Sages* (*Ṭabaqāt al-aṭibbā' wa-l-ḥukamā'*), 68.2–69.11 (ed. Fu'ād Sayyid); tr. Vagelpohl and Sánchez.[303]

In an age rich in accomplished translators, Ḥunayn ibn Isḥāq[304] stood out both because of his linguistic abilities in at least three languages (Greek, Syriac, and Arabic), and his astounding productivity: over the course of his career as a physician and translator, which spanned at least half of the third/ninth century, he almost single-handedly translated the corpus of medical and philosophical writings of the physician and philosopher Galen into Syriac and Arabic.[305] The present text, a short biography from Ibn Ǧulǧul's *Classes of Physicians*, gives a brief if not always reliable overview of his career and accomplishments, emphasizing his relationship to the 'Abbāsid caliph al-Mutawakkil.

(Ḥunayn ibn Isḥāq was) a disciple of Yūḥannā ibn Māsawayh,[306] an expert in the Arabic language and very proficient in Greek, excellent in both languages to such a degree that he was able to detect rhetorical infelicities in both languages.[307] He went from Baghdad to Persia while the grammarian Ḫalīl ibn Aḥmad,[308] may God have mercy upon him, was there. Ḥunayn became his student until he excelled in the Arabic language. He brought the *Book of* (*the letter*) *'Ayn* to Baghdad.[309]

---

302 Another translation: Ibn al-Nadīm, *The Fihrist*, 645 (English).

303 Parallels: Ibn al-Qifṭī, *Ta'rīḫ al-ḥukamā'*, 171.2–16; and Ibn Abī Uṣaybi'ah, *'Uyūn al-anbā'*, II/1, 473.6–14. Cf. also the partial and abbreviated version in Ṣā'id al-Andalusī, *Ṭabaqāt al-umam*, 36.18–20, which does not mention the caliph al-Mutawakkil or the team of translators.

304 See above, 255–256, and I.5 with n. 52.

305 Ḥunayn documented his translations in the *Epistle* (*al-Risālah*); see above, 255–256.

306 See above, I.5 with n. 53.

307 Ar. *'ilal al-lisānayn*, literally 'the illnesses of both tongues'. The term *'illah* ('illness') was used to refer to mistakes and rhetorical blunders that had a negative impact on the eloquence of speech, e.g., repetition, mispronunciation, verbosity, and lack of clarity (see al-Ǧāḥiẓ, *al-Bayān wa-l-tabyīn*, I, 13). In prosody, it denotes deviations from the ideal, 'healthy' meter; see Willem Stoetzer, 'Prosody (*'arūḍ*)', *Encyclopaedia of Arabic Literature*, II, 622.

308 The Arab philologist al-Ḫālīl ibn Aḥmad al-Farāhīdī (d. betw. 160/776 and 175/791) was a pioneering lexicographer whose work laid the foundations of the first Arabic dictionary, the *Book of* (*the letter*) *'Ayn* (*Kitāb al-'ayn*), which was compiled after al-Ḫalīl's death. See Rudolf Sellheim, 'al-Khalīl b. Aḥmad', *EI2*.

309 Ḥunayn's alleged language studies under al-Ḫalīl and his involvement in bringing the *Book of* (*the letter*) *'Ayn* are without a doubt apocryphal: al-Ḫalīl was long dead by the time

He was then appointed as a translator and entrusted with (translating). The person appointing him was (the ʿAbbāsid caliph) al-Mutawakkil, and he assigned to him experienced scribes who were experts in translation—they translated and Ḥunayn revised what they had translated—such as Iṣṭifan ibn Basīl,[310] Ḥubayš,[311] Mūsā ibn Abī Ḫālid the translator,[312] and Yaḥyā ibn Hārūn.[313]

Ḥunayn served (the caliph) al-Mutawakkil as a physician and remained in his favour during his reign. He always wore the sash (worn by Christians). He learned Greek in Alexandria and was an excellent translator. He was the one who clarified the meanings of Hippocrates's and Galen's works, explained them in the best possible way, and clarified those that were obscure and explained their problems. He wrote useful, masterly, and erudite books. He attended (in particular) to the works of Galen and followed the model of the Alexandrians by composing (his books) in the form of questions and answers, which he did very well.

II.8. Ibn Abī Uṣaybiʿah, *The Best Accounts of the Classes of Physicians* (*ʿUyūn al-anbāʾ fī ṭabaqāt al-aṭibbāʾ*), II/1, 465.2–467.14 (ed. Savage-Smith); tr. Savage-Smith et al. III/1, 492–495, adapted by Vagelpohl and Sánchez.

The following episode throws some light on the relationship between Ḥunayn ibn Isḥāq and his teacher, the physician and translator Yūḥannā ibn Māsawayh.[314] Rudely dismissed from Yūḥannā ibn Māsawayh's lecture circle because he had annoyed his teacher with his constant questions, Ḥunayn left Baghdad and returned some two years later, far surpassing his former teacher's abilities, after reportedly having studied Greek in Alexandria or somewhere in the Byzantine realm.[315]

Yūsuf ibn Ibrāhīm[316] said: Ḥunayn ibn Isḥāq was diligent and keenly interested in learning the art of medicine, but his initial experience in that con-

---

Ḥunayn took up his studies. The association of the two, which may have been spread by his students, was probably meant to burnish Ḥunayn's linguistic credentials.

310 See above, II.1 with n. 165.

311 Ḥunayn's nephew and collaborator; see above, II.1 with n. 190.

312 This must be Mūsā ibn Ḫālid; see above, II.2 with n. 241.

313 Unidentified.

314 See above, I.5 with n. 53.

315 See Strohmaier, 'Homer in Bagdad', 196; Gutas, *Greek Thought*, 138; and Gutas, 'Scholars as Transmitters', 683.

316 Yūsuf ibn Ibrāhīm ibn al-Dāyah (fl. first half third/ninth century), the father of the Ṭūlūnid historian commonly known by the same name, was a foster brother of the caliph

nection was unfortunate. Yūḥannā ibn Māsawayh gave courses in the subject that were attended by educated persons of every kind. I used to see Ḥunayn ibn Isḥāq, the translator, reading, under Yūḥannā's guidance, *On Sects*, the title of which in both Greek and Syriac is 'harāsīs' [*Peri haireseōn*].[317] At that time, Ḥunayn was constantly asking questions, much to Yūḥannā's annoyance. Another factor that did not endear him to his teacher was the fact that Ḥunayn came of a family of money-changers in al-Ḥīrah, for the people of Ǧondē-šāpūr, and in particular its physicians, looked down on the people of al-Ḥīrah and did not encourage tradesmen's sons to enter their profession. It so happened one day that Ḥunayn asked Yūḥannā a question about a passage that he had read, which he was having some trouble understanding. Yūḥannā lost his temper. 'The people of al-Ḥīrah are not fit to learn the art of medicine!' he said. 'You would do better to go to so-and-so, who is a relative of yours, and borrow fifty dirhams from him. For one dirham you can buy some little baskets; for three more, you can buy some orpiment. Spend the rest on copper coins of Kufa and al-Qādisīyah, coat them with orpiment, put them into the baskets, and then sit by the roadside crying "Fine coins for alms and gratuities!" You will make a better living selling such coins than you ever will by practising this profession. Leave my house!' So Ḥunayn departed from home in tears, humiliated and downcast. We saw no more of him for two years. [...]

When (Isḥāq ibn al-Ḫaṣī[318] [a friend of the narrator]) fell ill, I went to visit him. There at his home, what should I see but a man with such a head of hair that it partially covered his face, so that I could not tell who he was. He was walking up and down, reciting some Greek poetry by Homer, the greatest of all the Greek poets, and the sound of his voice put me in mind of Ḥunayn, whom I had not seen for two years and more. 'That's Ḥunayn!' I said to Isḥāq ibn al-Ḫaṣī. He said no it wasn't, but did not sound as though he meant it. I addressed the hirsute man tentatively: 'Ḥunayn?', and he acknowleged that it was indeed

al-Mu'taṣim and secretary of Ibrāhīm ibn al-Mahdī. Ibn al-Dāyah is the author of a lost collection of anecdotes about his patron from which authors such as Ibn Abī Uṣaybi'ah and Abū l-Faraǧ al-Iṣbahānī often quote; see Franz Rosenthal, 'Ibn al-Dāyah', *EI2*; and Bruning, 'Yūsuf b. Ibrāhīm b. al-Dāyā'.

317 Galen's *On Sects*. The name *harāsīs* is a transliteration of its Greek title, (*Περί*) *αἱρέσεων*.

318 Isḥāq ibn al-Ḫaṣī ('the son of the eunuch') was the son of Ḫaršā, a Byzantine slave of the caliph Hārūn al-Rašīd; Isḥāq's father was castrated, hence the name. Isḥāq was taught Greek by his mother, who also instructed him in Greek and Byzantine culture; his story is told in Ibn Abī Uṣaybi'ah, *'Uyūn al-anbā'*, II/1, 465.13–466.5 (the passage is left out in this text).

he. 'The son of that whore Risālah,'[319] he went on, 'said that no ʿAbādī[320] was capable of learning the art of medicine. May I renounce the Christian religion if I undertake the study of medicine before I have achieved a more comprehensive mastery of the Greek language than anyone else in this age! No one knows about this apart from my brother here, and if it had occurred to me that you might realize who I was, I should have kept out of your way. But now that my disguise no longer deceives you, I must ask you to keep my identity to yourself.'

After that, it was a good three years, perhaps closer to four, before I saw him again. I had gone to call on Ǧibrīl ibn Buḫtīšūʿ,[321] who had just come down from al-Maʾmūn's camp shortly before the caliph's death,[322] and there in his house was none other than Ḥunayn. He had been translating parts of one of Galen's works on anatomy: a Roman editor had taken the work in question and divided it up into sections, and Ḥunayn had translated a number of these sections for Ǧibrīl. Ǧibrīl was addressing him with the utmost respect and said to him 'Rabban Ḥunayn'; 'Rabban' means 'teacher'. Seeing my astonishment at this spectacle, Ǧibrīl said to me: 'You need not consider that I am showing this young fellow undue respect. I tell you, if God preserves him, he will outshine not only Sergius but other translators as well.' I should explain here that the Sergius to whom Ǧibrīl was alluding was Sergius of Rēšʿaynā,[323] who was the first to translate a number of Greek scientific works into the Syriac language. I remained at Ǧibrīl's house for quite a long time after Ḥunayn had left, but when I did leave, I found him outside the front door, where he had been waiting for me to appear. After greeting me, he said: 'I once asked you not to divulge what I had been up to, but now I should like you to make the matter known, including what you have just heard Abū ʿĪsā say about me.' 'I shall be making Yūḥannā feel like a fool when I tell him how I heard Abū ʿĪsā praise you to the skies,' I said. 'To make him feel even more of a fool, show him this,' said Ḥunayn, tak-

---

319 Ibn Abī Uṣaybiʿah (*ʿUyūn al-anbāʾ*, II/1, 444.14–18) reported that Yūḥannā ibn Māsawayh was the son of a Slav slave girl by the name of Risālah, whom his father Māsawayh had received as a gift from his patron, Ǧibrīl ibn Buḫtīšūʿ.

320 According to Ibn Abī Uṣaybiʿah, Ḥunayn ibn Isḥāq belonged to a group of Christian Arabs from Ḥīrah named ʿAbādīyah (*ʿUyūn al-anbāʾ*, II/1, 464.8–9, where Ibn Abī Uṣaybiʿah noted that the term is vocalized with *fatḥah*; modern scholars usually transcribe the name as ʿIbādīyah). This term is related to the verb *ʿabada*, 'to serve, worship'. Bar Hebraeus (*Taʾrīḫ muḫtaṣar al-duwal*, 250.3–6) explained that this Christian faction called themselves ʿAbādiyah because the *ʿabād* serve only God, whilst the *ʿabīd* ('servants, slaves') serve both the Creator and the created (*al-ḫāliq wa-l-maḫlūq*).

321 See above, II.2 with n. 260.

322 Al-Maʾmūn died in 218/833.

323 See above, II.2 with n. 246.

ing from his sleeve a copy of the translation he had done for Ǧibrīl, 'without telling him that it was I who translated it, and then, once you see that he thinks it really excellent, tell him.' I hastened to show the manuscript to Yūḥannā that very day, without so much as going home first.

Having read those selections, which covered that part of the work that the Greeks know as the 'factors',[324] he expressed the utmost astonishment. 'Is it possible for someone in our present age to be inspired by Christ, do you think?' he said. 'No,' I replied, 'neither in our present age nor in any other. Christ has never inspired anyone; it is he himself who was inspired.' 'Spare me the sermon,' said Yūḥannā, 'I am sure this could have been produced only with the help of the Holy Ghost.' 'As a matter of fact,' I said, 'it is the work of Ḥunayn ibn Isḥāq, whom you ordered out of your house and advised to sell doctored copper coins.' 'Impossible!' he gasped. At length, however, he realized that I must be speaking the truth, and he begged me to use my good offices to effect a reconciliation between himself and Ḥunayn. I was able to bring this about, and from then on Yūḥannā treated Ḥunayn with the utmost respect and generosity, showering him with benefits. I observed this to be the case without interruption until I left Iraq in the year 225/840.

11.9. Ibn Abī Uṣaybiʿah, *The Best Accounts of the Classes of Physicians* (*ʿUyūn al-anbāʾ fī ṭabaqāt al-aṭibbāʾ*), II/1, 472.2–16 (ed. Savage-Smith et al.); tr. Savage-Smith et al., III/1, 501–502, adapted by Vagelpohl and Sánchez.

Besides some information about Ḥunayn's sons, who took up similar careers more or less successfully, the following passage remarks further on Ḥunayn's accomplishments as a translator. His translations were stylistically superior to those of his predecessors and other contemporaries and enjoyed a correspondingly higher success with their audience.

I say: Ḥunayn had two sons, Dāwūd and Isḥāq. Both of them studied the art of medicine, and their father composed manuals of instruction for them, designed to teach them its basic principles. He also translated many of the works of Galen for their benefit. I have not found any evidence that Dāwūd was noteworthy as a physician, nor does he seem to have left any written works as monuments to his skill and proficiency, apart from a single *Compendium*. Isḥāq, in contrast, became well known as a distinguished practitioner of the

324 It is not clear what the reference is; see Ibn Abī Uṣaybiʿah, *The Best Accounts* (ed. Savage-Smith et al.), III/1, 495, n. 17.

art of medicine and the author of numerous works on the subject. He also translated many Greek works into Arabic, but as a translator he dedicated himself primarily to the works of the philosophers. such as Aristotle and others.[325]

Their father, Ḥunayn, on the other hand, was interested mainly in translating medical works, and more particularly the works of Galen. We may say that, in general, all Galen's works that we have at our disposal are translated versions by Ḥunayn, or else corrected by him in cases where the Arabic version is the work of another translator. Works that were translated by some other translator, such as Usṭāṯ,[326] Ibn Bakkuš,[327] al-Biṭrīq,[328] Abū Saʿīd ʿUṯmān al-Dimašqī,[329] or others, are less highly prized and are deemed less desirable than those that were translated or revised by Ḥunayn. This is owing to the fluency and attractiveness of his style, and also because he possessed extensive knowledge of the opinions expressed by Galen in his works and and an effortless mastery of all this material.

I have seen some of Galen's *Sixteen Books*[330] in translation from Greek into Syriac by Sergius (of Rēšʿaynā) the physician,[331] and from Syriac into Arabic by Mūsā ibn Ḫālid the Translator.[332] When I read them and studied their phraseology closely, it was quite apparent to me that they were a far cry indeed from the *Sixteen Books* in the translations of Ḥunayn. There is no comparison; the one version is stammering, the other, eloquence, the one is the earth, the other, the Pleiades.

II.10. Ibn al-Nadīm, *The Catalogue* (*Kitāb al-fihrist*), I, 289.15–18 (ed. Flügel); tr. Vagelpohl and Sánchez.[333]

As Ibn al-Nadīm's note here shows, Ḥunayn ibn Isḥāq's success meant that his contemporaries almost automatically assumed Arabic Galen translations to be by Ḥunayn, even though Ḥunayn himself translated mostly from Greek into Syriac. Many Arabic Galen translations were in fact authored by his collaborators and then corrected and revised by Ḥunayn.

---

325 See above, I.6 with n. 63.
326 See above, II.1 with n. 163.
327 For Ibn Bakkuš or Bakkūš or Baks, see above, I.13 with n. 142.
328 See above, I.11 with n. 106, and II.1 with n. 149.
329 See above, I.6 with n. 65.
330 See above, II.2 with n. 242.
331 See above, II.2 with n. 246.
332 See above, II.2 with n. 241.
333 Parallels: Ibn al-Qifṭī, *Taʾrīḫ al-ḥukamāʾ*, 128.18–129.2; cf. also 177.11–14.

Muḥammad ibn Isḥāq said: Ḥunayn had the good fortune that what Ḥubayš ibn al-Ḥasan al-Aʿsam,[334] ʿĪsā ibn Yaḥyā,[335] and others translated into Arabic was attributed to Ḥunayn. When we consult the catalogue of Galen's books that Ḥunayn produced for ʿAlī ibn Yaḥyā [i.e., his *Epistle*],[336] we learn that most of what Ḥunayn translated was into Syriac, but he may have corrected and examined the Arabic of other people's translations.[337]

II.11. Ibn Abī Uṣaybiʿah, *The Best Accounts of the Classes of Physicians* (*ʿUyūn al-anbāʾ fī ṭabaqāt al-aṭibbāʾ*), II/1, 478.4–479.7 (ed. Savage-Smith et al.); tr. Savage-Smith et al. III/1, 509–510.

Ḥunayn ibn Isḥāq's success as a translator and a physician in the service of the ʿAbbāsid caliphs came at a price: the enmity of his competitors. The following text forms part of what purports to be an autobiographical text by Ḥunayn, quoted at length in Ibn Abī Uṣaybiʿah's *Best Accounts*.[338] The narrator, supposedly Ḥunayn, rails against his detractors, who have attempted to blacken his reputation and even inflict physical harm. Besides the colourful detail, it is interesting because of the defence Ḥunayn mounts of his achievements as a translator, which rest on two key accomplishments: his full mastery of the target language, including the niceties of terminology and style, and the accessibility of his translations to all readers, regardless of their medical and philosophical background knowledge.

How should I not be hated, with the numbers of those who envy me so great and slanderous remarks about me at gatherings of important persons so prevalent? Large sums of money were expended on attempts on my life; those who reviled me grew strong, and those who honoured me were despised. I, for my part, was guilty of no offence; I had not wronged any of them in any way. But they saw that I had outdone them, that I knew more than they did and had achieved greater things than they, that I had translated great scientific works for them from languages that they were far from mastering, or of which they knew nothing at all. They perceived that my translations were admirable models of elegance and clarity, free of errors or defects, containing no bias in favour of any religion,

334 See above, II.1 with n. 190.
335 See above, II.1 with n. 191.
336 See above, n. 5.
337 Another translation: Ibn al-Nadīm, *The Fihrist*, 682 (English).
338 See Cooperson, 'The Purported Autobiography', 237; Salama-Carr, 'Translation', 392; Strohmaier, 'Byzantinisch-arabische Wissenschaftsbeziehungen', 182; and Olsson, 'The Reputation of Ḥunayn b. Isḥāq', 29–31.

and nothing ambiguous or ungrammatical. In saying this, I rely on the authority of acknowledged Arab experts on eloquence, scholars who are learned in all aspects of grammar and are familiar with uncommon expressions. None of them has ever detected an error of usage, an incorrect vowelling, or an injudicious shade of meaning, but only the most mellifluous and readily intelligible style. My works can be understood, admired and appreciated even by someone who is not himself a physician, knows nothing of philosophical method, and does not profess the Christian religion or any other religion; indeed, just such lay individuals regurlarly pay me substantial sums for works that I have translated, preferring them to the versions produced by others. I may also say unhesitatingly that other educated men of various religious persuasions have appreciated me, have shown me respect, have gratefully accepted what I gave them, and have repaid me handsomely.

But these Christian physicians, whom I have known all their lives and who learned their art under my tutelage, most of them, are the very ones who want to see my blood shed, even though I am indispensable to them. 'Who is Ḥunayn?' they have been known to say. 'He only translates these works because he is paid to translate them, after the fashion of any artisan; we fail to see any difference. A blacksmith may make a sword for a mounted warrior for a dinar or so, earning perhaps a hundred dinars a month by his trade. Now, Ḥunayn is a servant who makes the instrument that we use, but he does not use it himself, just as the blacksmith may be skilful at the craft of making a sword, but not at using one. A blacksmith will never be a mounted warrior. Similarly, a translator has nothing to say about the art of medicine. Ḥunayn has no skill in matters of illness and has never diagnosed a disease. He wants to be like us, that's all, and to be known as Ḥunayn the Physician rather than Ḥunayn the Translator. He would do better to stick to his craft and desist from talking about our art; that would be advantageous for him, as he would get his hands on more of our money and we should show him more respect if he would stop taking pulses, examining phials of urine, and prescribing medicines.' They say, 'Whenever Ḥunayn calls on a patient, regardless of whether the people of the house belong to the cream of society or the common folk, they snigger when he comes in, and they laugh behind his back when he leaves.'[339]

339 Another translation (partial): Cooperson, 'The Purported Autobiography', 237–239.

III. *Methods: Procedures and Approaches*

III.1. Ḥunayn ibn Isḥāq, *Epistle* (*al-Risālah*), 2.12–3.2/Ar., preface (ed. Bergsträßer); tr. Vagelpohl and Sánchez.

Ḥunayn ibn Isḥāq, the master translator and physician, has already figured in several previous texts (e.g., II.7–11). In the following passage, taken from the introduction to his *Epistle*, Ḥunayn explains what kind of information he collected in the *Epistle* and also sets out some important criteria for assessing the quality of a translation.[340]

To that[341] you [i.e., Ibn al-Munaǧǧim] replied that even if this were the case, we and all other people interested in (this field) who read (Galen's) books in Syriac and Arabic need to know which of these books were translated into Syriac and Arabic and which ones were not; which ones I have been charged with translating and which ones other people; which ones have been translated before me by others, which I then retranslated or revised; who the translators were who translated each book for which someone else was responsible, the level of competence of each of these individual translators, and for whom the (works) were translated; who the people were for whom I translated each of the books I was charged with translating, and at which age I translated them—for these are two things one needs to know, since (the quality of) a translation depends on the competence of the book's translator and the person for whom it was translated; for which of the books that have not yet been translated a Greek manuscript is available and for which there is none or a partial one—for this is something one needs to know to take care of the translation of texts that are available and to seek out those that are not.[342]

III.2. Ḥunayn ibn Isḥāq, *Epistle* (*al-Risālah*), 47.12–48.6/Ar., no. 115 (ed. Bergsträßer);[343] tr. Gutas, *Greek Thought*, 179, adapted by Vagelpohl and Sánchez.

In the following text, Ḥunayn recounts the complex translation history of Galen's major logical work, *On Demonstration* (*De demonstratione*), of which only fragments survive

340 See Salama-Carr, 'Translation', 392; and Gutas, *Greek Thought*, 140.

341 I.e., to Ḥunayn's suggestion that the information about Galen's writings his addressee was requesting could be found in Galen's own bibliographical writings.

342 Other translations: Montgomery, *In Praise of Books*, 186–187 (English); Salama-Carr, *La traduction*, 58–59 (French, partial); and Ḥunayn ibn Isḥāq, *al-Risālah*, 2/Ger. (German).

343 Parallel: Ibn Abī Uṣaybiʿah, *ʿUyūn al-anbāʾ*, II/1, 275.3–276.1.

in the original Greek. It illustrates the importance Ḥunayn and his contemporaries attached to this text and the lengths to which they would go in search of manuscripts.

None of our contemporaries has up to this point come across a complete Greek manuscript of (Galen's) *On Demonstration*, despite the fact that Ǧibrīl[344] spared no effort looking for it, just as I myself mounted a thorough search for it. While searching for it, I travelled in northern Mesopotamia, all of Syria, Palestine, and Egypt until I reached Alexandria. I only found about half of it in Damascus, but the books were not consecutive and incomplete.[345]

Ǧibrīl had also found some books that did not completely overlap with the ones I found. Ayyūb[346] translated for him what he found. I for one did not like (the idea of) translating any of them unless I had read them all, because (the work) was incomplete and defective, and because the soul desires and longs to see this work in its entirety.

I then translated what I had found into Syriac, namely a small part of Book 2, most of Book 3, about half of Book 4 from its beginning, and Book 9, except a bit at the beginning, which is lost. Of the remaining books I found (all) until the end of the work, except Book 15; there is (also) a gap at the end (of the work).

III.3. Ḥunayn ibn Isḥāq, *Epistle* (*al-Risālah*), 4.19–5.9/Ar., no. 3 (ed. Bergsträßer); tr. Rosenthal, *The Classical Heritage*, 20, adapted by Vagelpohl and Sánchez.

Galen's important introductory treatise on the medical schools of his day and their doctrinal differences, *On Sects*, was also translated several times into Syriac before Ḥunayn produced an Arabic version for his sponsor Muḥammad ibn Mūsā, an eminent mathematical and mechanical scholar and court official.[347] In this passage, Ḥunayn describes the elaborate process of manuscript collection and collation he employed to establish a reliable textual basis for his translations.

(Galen's *On Sects*) was translated into Syriac before me by a certain Ibn Šahdā from al-Karḫ,[348] who was a weak translator. When I was a young man of twenty or a little older, I translated it for a physician from Ǧondēšāpūr named Šīrīšūʿ

344 Ǧibrīl ibn Buḫtīšūʿ; see above, II.2 with n. 260.

345 Gutas, *Greek Thought*, 179.

346 See above, II.1 with n. 175.

347 See above, II.2 with n. 266. See also Meyerhof, 'New Light', 714–715; and Gutas, *Greek Thought*, 133–134.

348 An early ninth-century translator from Greek into Syriac who is mentioned twice more in the *Epistle*; see above, II.2 with n. 180, and Meyerhof, 'New Light', 704.

ibn Quṭrub,[349] from a very faulty Greek manuscript. Later, when I was about forty, my pupil Ḥubayš[350] asked me to correct the translation. Meanwhile, a number of Greek manuscripts had accumulated in my possession. I collated these manuscripts and thereby produced a single correct copy. Next, I collated the Syriac text with it and corrected it. I am in the habit of doing this with everything I translate. A few years later, I translated the Syriac text into Arabic for Abū Ǧaʿfar Muḥammad ibn Mūsā.[351]

III.4. Ḥunayn ibn Isḥāq, *Epistle* (*al-Risālah*), 17.18–18.17/Ar., no. 20 (ed. Bergsträßer); tr. Rosenthal, *The Classical Heritage*, 20–21, adapted by Vagelpohl and Sánchez.

Another of Ḥunayn's sponsors, the Christian translator and physician Salmawayh ibn Bunān,[352] requested a revised Syriac version of Galen's comprehensive therapeutic work, *The Method of Healing* (*Methodus medendi*), of which an older Syriac translation by the distinguished philosopher, physician, and translator Sergius of Rēšʿaynā[353] existed. The attempt to save some money by revising the existing translation ended in failure, and Ḥunayn translated it from scratch. In addition to the colourful details of this translation's destruction, the report again illustrates the efforts Ḥunayn made to establish a reliable Greek source text for a work that was already hard to find in his day.

Salmawayh urged me to correct the second part (of Sergius's Syriac translation of Galen's *Method of Healing*) for him, in the hope that this would be easier and better than translating it (anew). So he sat down opposite me with part of the seventh book, he with the Syriac text and I with the Greek. He read the Syriac text aloud and whenever I noticed something that conflicted with the Greek text, I told him about it and he then set out to correct it, until this became too much for him and he realized that an entirely new translation would be simpler and better stylistically, as well as producing a tidier result. So he asked me to translate these books, and I translated them all. (This happened when) we were in al-Raqqah[354] at the time of al-Ma'mūn's campaigns. He handed (the

349 See above, II.2 with n. 265.
350 Ḥunayn's student and nephew; see above, II.1 with n. 190.
351 Another translation: Ḥunayn ibn Isḥāq, *al-Risālah*, 4/Ger. (German).
352 See above, II.2 with n. 280.
353 See above, II.2 with n. 246.
354 An important urban centre in north-eastern Syria with a caliphal palace built by al-Ma'mūn's father, the caliph Hārūn al-Rašīd. It was the site of al-Ma'mūn's headquarters during the campaign he waged against the Byzantines in the year 215/830.

translation) to Zakariyāʾ ibn ʿAbd Allāh, known as al-Ṭayfūrī,[355] when he was about to return to Baghdad, in order to have it copied there. However, fire broke out on board the ship on which Zakariyāʾ was travelling, the book was burned, and no copy of the translation survives.

A few years later I then translated the work from the beginning for Buḫtīšūʿ ibn Ǧibrīl.[356] For the last eight books, a number of Greek manuscripts were at my disposal. I produced a (single) correct copy from them and translated it with the utmost accuracy and in the best possible style. For the first six books only a single manuscript, and a very faulty one at that, was at my disposal. I was therefore unable to restore these books in the manner required. Later, I came across another manuscript and collated the text with it and corrected it as much as possible. It would be better if I could collate a third manuscript with it, if only I were fortunate enough to find one. For Greek manuscripts of this work are rare, since it does not belong to the works that were read in the school of Alexandria.[357]

From the Syriac manuscripts of my translation Ḥubayš ibn al-Ḥasan translated this work (into Arabic) for Muḥammad ibn Mūsā. Then, after he had translated the work, he asked me to review the last eight books for him and fix the problems I found. I agreed to do this for him and did so successfully.[358]

III.5. Ibn Abī Uṣaybiʿah, *The Best Accounts of the Classes of Physicians* (*ʿUyūn al-anbāʾ fī ṭabaqāt al-aṭibbāʾ*), II/1, 488.8–16 (ed. Savage-Smith); tr. Savage-Smith et al., III/1, 519–520, adapted by Vagelpohl and Sánchez.

A brief comment in Ḥunayn ibn Isḥāq's autobiographical narrative reported by Ibn Abī Uṣaybiʿah, parts of which have been quoted above (II.11), gives an insight into the commercial value of translations. Since he was paid for his work by weight, Ḥunayn apparently used unusually thick paper and wrote in a large and widely spaced hand,

---

355 A contemporary of Ḥunayn and fellow physician who went on to serve at the court of the caliph al-Muʿtaṣim. He was father of the physician Isrāʾīl ibn Zakariyāʾ al-Ṭayfūrī (see above, II.2 with n. 282); see Meyerhof, 'New Light', 719; and Micheau, 'Mécènes et médecins', 161–162.

356 Buḫtīšūʿ ibn Ǧibrīl (d. 256/870), court physician to a succession of caliphs, was the son of the physician Ǧibrīl ibn Buḫtīšūʿ. Jealous of Ḥunayn's success, he may have been the instigator of the court intrigue alluded to in II.11 above and III.5 immediately below; see Micheau, 'Mécènes et médecins', 157–159.

357 I.e., it was not part of the official medical curriculum at the late antique school of Alexandria. On the School of Alexandria and medical teaching there, see Overwien, 'Medizinische Lehrwerke'.

358 Other translations: Ḥunayn ibn Isḥāq, *al-Risālah*, 14–15/Ger. (German); Garofalo, 'La traduzione araba', 73–74 (Italian); and Micheau, 'Mécènes et médecins', 151–152 (French, partial).

not just to ensure their longevity, but also to maximize his profits. Ibn Abī Uṣaybiʿah's note also reveals that Ḥunayn relied on the services of a secretary to produce clean copies of translated texts for his patrons.

(Ḥunayn reports how he tried to placate his enemies after he had been exonerated.) Moreover, I would translate books for them, as meticulously as always, without charging them anything, so anxious was I to show my good will toward them, whereas formerly it had been usual, whenever I translated a work for one of them for me to be paid a purse of silver dirhams equal in weight to the translated work.

I [i.e., Ibn Abī Uṣaybiʿah] say: I have found large numbers of these works, and have purchased a good many of them. They are written in *muwallad* Kūfic script in the handwriting of al-Azraq, Ḥunayn's secretary. The letters are written very large, with broad strokes, and the lines are widely spaced. The paper is very heavy, being three or four times as thick as the paper manufactured nowadays, while the sheets are trimmed to about a third the size of a sheet of Baghdādī paper. Ḥunayn had his work published in this fashion to make them bulkier and increase their weight, inasmuch as he was paid weight for weight in silver dirhams. It is thus clear that he used that particular kind of paper deliberately. Small wonder, then, that the manuscripts have lasted so well for so long.

III.6. Ibn al-Nadīm, *The Catalogue* (*Kitāb al-fihrist*), I, 22.7–15 (ed. Flügel); tr. Rosenthal, 'R. Walzer', review, 253, with additions, adapted by Vagelpohl and Sánchez.

Ibn al-Nadīm quotes the following passage from the preface of a work on scriptures and revealed books. The author of the preface, an otherwise unknown translator who seems to have worked during the reign of al-Maʾmūn, claims to have translated not just the book itself, but also Jewish, Christian, and Ṣābian scriptures. Irrespective of its authenticity and historicity, the text makes some interesting observations about translation methods, especially of religious texts. The author is at pains to emphasize that he translated as literally as possible without violating the norms of Arabic grammar and syntax.[359]

Aḥmad ibn ʿAbd Allāh ibn Sallām said: I translated into Arabic the beginning of this book, along with the scriptures and the Torah and the Gospels and the

359 The sensitivity of religious translations continued to be a concern; see al-Ǧāḥiẓ's remarks below, IV.5d.

books of the prophets and the disciples, from Hebrew and Greek and Ṣābian,[360] the languages of the people of each book, letter by letter, and did not seek to embellish or adorn any expression because I was afraid that I might change the text. I did not add to, or take away from what I found in the book I translated. There are, however, in some of these texts things that come first in that language but are not properly expressed when translated into Arabic unless they are postponed, while others come later but are not proper unless they are put first to straighten out the (text) in Arabic.[361] For example, when someone says *āb mē-ār* [i.e., 'bring water' in Persian], which translates into Arabic as *mā' hāti*, I postponed *mā'* and brought forward *hāti* [i.e., to produce the idiomatic *hāti l-mā'*], and I did the same with other languages where appropriate when translating into Arabic. Heaven forbid that I add to or take away from (the text) in this book anything except in the manner I have mentioned and explained.[362]

III.7. Al-Ṣafadī, *The Pouring Rain* (*Kitāb al-ġayṯ al-musaǧǧam*), 46.12–25 (Cairo ed. 1305 AH); tr. Rosenthal, *The Classical Heritage*, 17–18, adapted by Vagelpohl and Sánchez.

Arabic authors were well aware of the distinction between, and the respective advantages and disadvantages of, literal or word-for-word translation on the one hand and free or sense-for-sense translation on the other.[363] The following late source is quoted not so much because it offers any genuine insight into the methods of the earlier and later translators it purports to discuss—their translation methods were in fact much more complex than the author claims[364]—but because it offered a conveniently simple and clear-cut narrative of the development of translation methods and became authoritative 'by virtue of sheer frequency of repetition'.[365]

The translators used two methods of translation. One of them is that of Yūḥannā ibn al-Biṭrīq,[366] Ibn al-Nā'imah al-Ḥimṣī,[367] and others, namely (the translator) studies each Greek word and its meaning, chooses an Arabic word of corresponding meaning and uses it. Then he turns to the next word and pro-

360 The language of the Ṣābian religious community; see above, n. 67.
361 I.e., the word order of these languages differed.
362 Another translation: Ibn al-Nadīm, *The Fihrist*, 42 (English).
363 The historical context of this methodological distinction is set out by Brock, 'Aspects'.
364 See Endress, *Proclus Arabus*, 154, and Mattock, 'The early translations', 74–75. See also Gutas, *Greek Thought*, 142.
365 Mattock, 'The early translations', 74.
366 See above, I.11 with n. 105.
367 See above, II.2 with n. 151.

ceeds in the same manner, until in the end he has gone over the entire text he wishes to translate. This method is bad for two reasons. First, it is impossible to find Arabic expressions corresponding to all Greek words, and through this method many Greek words therefore remain untranslated. Second, certain (syntactical) constructions and predications in the one language do not always necessarily correspond to similar ones in the other; besides, the use of metaphors, which are frequent in every language, causes additional mistakes.[368]

The second method is that of Ḥunayn ibn Isḥāq, al-Ǧawharī,[369] and others. Here the translator considers a (whole) sentence, ascertains its (full) meaning, and then expresses it in the other language with a sentence identical in meaning, irrespective of whether the (individual) words match or differ. This method is superior, and hence there is no need to improve the works of Ḥunayn ibn Isḥāq, except those in the mathematical sciences, which he had not mastered, in contrast to those on medicine, logic, natural science, and metaphysics, whose Arabic translations require no corrections at all. On the other hand, Euclid has been improved by Ṯābit ibn Qurrah al-Ḥarrānī,[370] as have the *Almagest*[371] and the intermediate works.[372] [373]

IV. *Problems: Difficulties and Criticisms*

IV.1. Al-Bīrūnī, *India* (*Taḥqīq mā li-l-Hind*), I, 76.6–10 (ed. Sachau); tr. Sachau, *India*, I, 159, adapted by Vagelpohl and Sánchez.

The Persian polymath Abū Rayḥān Muḥammad ibn Aḥmad al-Bīrūnī (d. c. 440/1048) left his imprint on almost all branches of knowledge, among them historiography, pharmacology, physics, philosophy, and theology. In addition, he stands out as almost the only Muslim scholar who developed a deep and lasting interest in Indian culture. His contacts with Indian scholars enabled him to gather extensive information about Indian religion, culture, philosophy, and science. He compiled this material in a comprehensive account entitled *The Verification of What is Said about India* (*Taḥqīq mā li-l-Hind min maqūlah*), which he completed in 420/1030.[374]

368 The same arguments were put forward by the grammarian al-Sīrāfī; see below, IV.8c.
369 An otherwise unknown translator, presumably a contemporary of Ḥunayn ibn Isḥāq.
370 On the translation history of Euclid's *Elements*, see I.6 above.
371 On the translation history of Ptolemy's *Almagest*, see I.7 above.
372 I.e., the works students were required to read between Euclid's *Elements* and Ptolemy's *Almagest*.
373 Another translation: Rosenthal, 'R. Walzer', 253–254 (English).
374 See Michio Yano, 'al-Bīrūnī', *EI Three*.

In the following section, al-Bīrūnī identifies a potential major problem of translation: wilful manipulation by the translator. The context is as follows: when he produced his Middle Persian version of the *Kalīlah wa-Dimnah*, the translator Burzōe added an autobiographical introduction; Ibn al-Muqaffaʿ then added his own preface to his Arabic translation, which was based on Burzōe's Persian text, and also inserted a section about the uncertainty of religion into Burzōe's introduction, which al-Bīrūnī here dismisses as Manichaean propaganda.

They [i.e., the Indians] have numerous other branches of knowledge and have a nearly boundless literature, but I could not attain full knowledge of it. I wish I could translate the book *Pañcatantra*, known among us as the book of *Kalīlah and Dimnah*.[375] It went back and forth repeatedly between Persian and Hindi, then Arabic and Persian in translations of people who are not free from the suspicion of having altered the text. For instance, ʿAbd Allāh ibn al-Muqaffaʿ[376] has added the chapter about Burzōe with the intention of raising doubts in the minds of people of feeble religious belief, and of suborning them to spread the doctrines of the Manichaeans. If he is suspected of adding something, he is hardly free from (suspicion) as a translator.

IV.2. Al-Ǧāḥiẓ, *Treatise on Clarity and Clarification* (*Kitāb al-bayān wa-l-tabyīn*), III, 29.9–12 (ed. Hārūn); tr. Vagelpohl and Sánchez.

Al-Ǧāḥiẓ is considered one of the fathers of Arabic prose. The *Treatise on Clarity and Clarification* (*Kitāb al-bayān wa-l-tabyīn*), one of his major works, is a long reflection on speech, nature, and the ways in which God communicates with humans. Translation is a topic that al-Ǧāḥiẓ frequently addresses in this and several other works, usually to cast doubt on the possibility of faithfully conveying meaning from language to language without alteration.[377]

In a passage about the relative merits of the literary style and eloquence of Persians and Arabs, al-Ǧāḥiẓ attempts to discredit the writings cited as models of eloquence by the apologists of Persian style by insinuating that they could just as easily have been manufactured by their translators into Arabic, among them again Ibn al-Muqaffaʿ.

We are not in a position to know whether the treatises circulating among the people are (actually) Persian, whether they are authentic and not manufac-

375 See above, II.4 with n. 293.
376 See above, II.1 with n. 197.
377 See Montgomery, 'al-Jāḥiẓ'.

tured, ancient and not recent, since people such as Ibn al-Muqaffaʿ,[378] Sahl ibn Hārūn,[379] Abū ʿUbayd Allāh,[380] ʿAbd al-Ḥamīd,[381] and Ġaylān (al-Dimašqī)[382] could have produced such treatises and made up such stories.[383]

IV.3. Ibn al-Qifṭī, *History of Learned Men* (*Taʾrīḫ al-ḥukamāʾ*), 29.17–30.19 (ed. Lippert); tr. Vagelpohl and Sánchez.[384]

Two frequent polemical themes in narrations about the provenance of translated books were that previous Christian rulers, aware of the dangers that such books might pose to religious beliefs, had locked away ancient Greek texts, and that handing such texts over to the Muslim rulers who demanded them would fatally undermine the Islamic state (see also IV.4 below). Ibn al-Qifṭī combines these two motifs in the following story.

Someone other than Ibn Isḥāq[385] said: (the caliph) al-Maʾmūn wrote a letter to the Byzantine king (in which he) was arrogant, insulted the heathen religion, and demanded from him philosophical books from the works of Aristotle. The Roman king tried to find them, but could not find any trace of them in his territory. This distressed him and he said: 'The Muslim ruler demands from me the

378 See above, II.1 with n. 197. In IV.1 above, al-Bīrūnī accused Ibn al-Muqaffaʿ of adding a chapter to the book *Kalīlah wa-Dimnah* in order to support the Manichaean cause.

379 Sahl ibn Hārūn (d. 215/830) was an Iranian man of letters who worked as secretary for Yaḥyā ibn Ḫālid al-Barmakī and the caliph Hārūn al-Rašīd. He translated several works from Pahlavi, among them the romance *Wāmiq and ʿAḏra*ʾ. See Mohsen Zakeri, 'Sahl b. Hārūn b. Rāhawayh', *EI2*.

380 This may be a scribal error for Abū ʿUbaydah, i.e., the philologist Abū ʿUbaydah Maʿmar ibn al-Muṯannā (d. c. 210/824–825), the son of a Persian Jew who converted to Islam. Abū ʿUbaydah is one of the most important scholars in the field of Arabic philology; see Reinhard Weipert, 'Abū ʿUbayda', *EI Three*.

381 See above, I.4 with n. 46.

382 Also of Iranian descent, the scholar Ġaylān al-Dimašqī (fl. second half second/eighth century) was mainly known for his advocacy of the doctrine of free will (*qadarīyah*), which led to his execution by the caliph Hišām ibn ʿAbd al-Malik. See Charles Pellat, 'Ġaylān b. Muslim', *EI2*.

383 See Webb, '"Foreign Books"', 41, and Enderwitz, *Gesellschaftlicher Rang*, 99–100. All of these authors were accused of being *šuʿūbī*, i.e., non-Arab Muslims who extolled the sophistication of Persian culture and mocked the Arabs, whose traditions and lore they considered uncivilized. On the role of al-Ǧāḥiz in these polemics, see Enderwitz, *Gesellschaftlicher Rang*, 162–163.

384 Parallel: cf. the somewhat different and shorter version in Ibn Nubātah, *Sarḥ al-ʿuyūn*, 242.9–19.

385 I.e., someone other than Ibn al-Nadīm, the author of the *Catalogue*, from which Ibn al-Qifṭī had quoted immediately before this passage; the phrase signals a quotation from a different source.

knowledge of my ancient Greek ancestors, but I do not find (any of) it. What excuse do I have, and what value will this Byzantine nation still have in the eyes of the Muslims?' He started to ask around and investigate.

One of the monks assigned to a monastery far away from Constantinople visited him and said: 'I have the knowledge you seek.' (The emperor) replied: 'Tell me (about it)!' He said that there was some building in such-and-such a location, the contents of which each king keeps locked when he is made king. He said that people say that it contains the treasures of the previous kings. Each king keeps it locked up so that people do not say that he needs what is in it because of his bad administration and opens it. The monk told him: 'This is not the case. All there is in this place is a temple in which the ancient Greeks worshipped (their gods) before the religion of the Messiah was established. When it was established in these areas in the days of Constantine, the son of Helen, philosophical books were collected from the people, stored in that building, and its door was closed, and the kings have kept it closed, as you have heard.'[386]

The king assembled the administrators of his state, informed them about this, and asked their advice on whether to open the building. They advised him to do so, and he asked the monk about how to retrieve (the books) if they are in Muslim territory, and whether (these books) are dangerous in this world or sinful in the next. The monk replied: 'Retrieve them, you will be rewarded for it! When they are introduced to a religious community, they cannot but rock its foundations.'

(The emperor) went to the building, opened it, and found everything as the monk had said. In it they found many books and without knowing or inspecting them, they took from among them five (camel) loads and sent them to al-Ma'mūn. Al-Ma'mūn handed them over to the translators, who rendered them from Greek into Arabic. Later on, people found out after al-Ma'mūn's reign that he had requisitioned them and did everything to acquire a large number of them.

When the books were sent to al-Ma'mūn, some of them arrived intact and some in fragmentary form. The fragmentary books are to this day still fragmentary, and no one has found an intact version.

386 A variant of this legend that presents Constantine the Great (r. 306–337) and his mother as enemies of philosophy who ordered medical and philosophical books to be burned is reported by 'Abd al-Ǧabbār, *Taṯbīt*, 161.16–162.2. See Stern, "Abd al-Jabbār's Account'.

IV.4. Al-Maqdisī, *Brief Argument for Those Who Lost the Way* (*Muḫtaṣar al-ḥuǧǧah 'alā tārik al-maḥaǧǧah*), 661.6–663.13 (ed. Hārūn); tr. Vagelpohl and Sánchez.

Abū l-Fatḥ Naṣr ibn Ibrāhīm al-Maqdisī (d. 490/1096–1097) was a Šāfi'ite scholar and transmitter of *ḥadīṯ* from Nablus who made his career in Jerusalem and Damascus. His *Brief Argument for Those Who Lost the Way* (*Muḫtaṣar al-ḥuǧǧah 'alā tārik al-maḥaǧǧah*) is a work on religious doctrine and ethics that, among other things, warns against the dangers that logic and philosophy in general pose for Islam.[387]

The following polemical story again emphasizes the corrosive potential of ancient Greek learning. According to al-Maqdisī, the Byzantine authorities were all too happy to get rid of these dangerous writings and help the vizier Yaḥyā ibn Ḫālid ibn Barmak[388] use them to corrupt Islam.[389]

One of the innovations (the 'Abbāsid caliphs) introduced was to disseminate Greek books in Islamic lands, which were translated into Arabic and circulated among the Muslims. The reason why these books were brought from Byzantium into Islamic lands was Yaḥyā ibn Ḫālid ibn Barmak. For the Greek books were in Byzantine territory, and the king of Byzantium had feared that if his people were to study them, they would abandon the Christian religion and return to that of the Greeks; their authority would splinter and their community would break apart. Hence, he collected the books in a place and erected a building above them to conceal them with stone and plaster so that they were inaccessible.

After Yaḥyā ibn Ḫālid, who was a heretic [*zindīq*],[390] assumed a position of power in the 'Abbāsid state, he heard about the books inside the building in Byzantium. He then ingratiated himself with the Byzantine king of his time with gifts without asking for anything in return. When he kept doing this, the Byzantine king assembled the nobles and told them: 'This Arab official has sent me numerous presents without demanding anything from me, but I do not think that he would do so without seeking something, and I fear that it is something that would be difficult and troublesome for me.' When Yaḥyā's emissary arrived (again), (the king) said to him: 'Tell your master to let me know if there

387 On al-Maqdisī, see al-Ḏahabī, *Sīrat a'lām al-nubalā'*, XIX, 136, n. 72.

388 See above, I.7 with n. 69.

389 See Balty-Guesdon, 'Le Bayt al-ḥikma', 136.

390 The term *zindīq* (pl. *zanādiqah*), is difficult to translate here. Borrowed from Persian, the term originally means Manichaean, but was later used to refer to freethinkers, heretics, and unbelievers in general; see François de Blois, 'Zindīḳ', *EI2*.

is anything he wants from me.' After the emissary informed Yaḥyā (about it), he sent him back with this message: 'My request is that the books underneath that building be sent to me. I will copy from them what I need and then we will return them.'

When the Byzantine (king) read this message, he was overjoyed. He assembled the nobles, bishops, and monks and said to them: 'I have already told you that this Arab official must have a request. He has finally revealed it to me, and it is one that is very easy for me (to fulfil). I have already formed my opinion about it. Hear me out, and if you agree I will proceed with it, and if you disagree we will discuss it until we reach an agreement.' They asked: 'What is (his request)?' He replied: 'He wants the Greek books to copy from them what he wants and return them.' They asked: 'And what is your opinion?' He replied: 'I know that our ancestors would not have erected a building above these books unless they had feared that if these books were to fall into the hands of the Christians and they read them, it would cause their faith to perish and their community to be destroyed. I think that we should send him (the books) and ask him not to return them, so that they [i.e., the Muslims] suffer the tribulations these books will bring them, while we rid ourselves from the evil (of these books), for I am not certain that one of my successors will not venture to disseminate them to the people, as a result of which they will fall into the state we fear.' They said: 'An excellent proposal you have made, O King, carry it out!'

He then sent the books to Yaḥyā ibn Ḫālid, and when they arrived, he gathered all the heretics [*zindīq*] and philosophers (to read) them. One of those books that was published was *The Definition of Logic*.[391]

IV.5a. Al-Ǧāḥiẓ, *The Book of Animals* (*Kitāb al-ḥayawān*), I, 74.14–75.4 (ed. Hārūn); tr. Rosenthal, *The Classical Heritage*, 18.

The *Book of Animals* (*Kitāb al-ḥayawān*), a phenomenal literary and scholarly tour de force to explain God's creation, is a compilation of zoological knowledge, lore, and anecdotes in which al-Ǧāḥiẓ often relied on Ibn al-Biṭrīq's translation of two of Aristotle's zoological works.[392]

391 Ibn al-Nadīm's *Catalogue* does not list a book by the title *The Definition of Logic* (*Ḥadd al-manṭiq*), and the work remains a mystery. The sunnī theologian and belletrist Ibn Qutaybah (d. 276/889), a sharp critic of his contemporaries' fascination with Greek philosophy, also refers to this book; see *Adab al-kātib*, 2.9 and 5.9. Soravia, 'Ibn Qutayba', 551, n. 34, maintains that the title refers to Aristotle's *Categories*.

392 *The History of Animals* (*Historia animalium*) and *The Generation of Animals* (*De generatione animalium*), which were combined in the Arabic tradition into the *Book of Animals*

The following passages represent al-Ǧāḥiẓ's most sustained discussion of translation and its problems.[393] Without clearly indicating where his own sympathies lay, the author presented arguments both in opposition to and in defence of translation through the figures of an unnamed 'Apologist of Books' and an equally unnamed 'Apologist of the Poetry of the Arabs.'[394]

Starting off with his thoughts on the history of Arabic poetry and the poetic accomplishments of the Arabs, al-Ǧāḥiẓ first reports the claim of the Apologist of Books that translating poetry from one language into another is impossible (IV.5a–b), which he considered false.[395] In opposition to this, the Apologist of the Poetry of the Arabs then launches into a wide-ranging discussion of translation in general and lists numerous arguments to show that faithful translation requires abilities and knowledge that no one possesses, least of all the translators he dealt with, who are a far cry from the original authors they translate, such as Aristotle and Plato (IV.5c).[396] These problems, he then points out, are bad enough when it comes to translations of scholarly texts, but they can lead to catastrophic results in the case of translations of religious texts: they can endanger the very salvation of believers who put their trust in them (IV.5d).[397] The Apologist of the Poetry of the Arabs then returns to the problem of translators' qualifications and their correct understanding of texts that have been transmitted from language to language (IV.5e). In addition, translations suffer from errors introduced by copyists that distort a text's meaning (IV.5f). He finally summed up these and other reasons why readers should not trust translations, including intentional alterations and 'lies' introduced by translators:[398] books turn out to be as untranslatable as poetry.

(The Apologist of Books said:) Only the Arabs and people who speak Arabic have a correct understanding of poetry. Poems do not lend themselves to translation and ought not to be translated. When they are translated, their poetic

(*Kitāb al-ḥayawān*). See Peters, *Aristoteles Arabus*, 47–48; and Kruk, 'La zoologie aristotélicienne'.

393 For an analytical outline and interpretation of the following passages from al-Ǧāḥiẓ's *Book of Animals*, see Montgomery, *In Praise of Books*, 451–459, esp. 455–457.

394 See Montgomery, 'Al-Ǧāḥiẓ, Falsafa and Hippocrates Arabicus', 77, and Montgomery, *In Praise of Books*, 455–457.

395 See Salama-Carr, 'Translation', 387; and Montgomery, 'Al-Ǧāḥiẓ, Falsafa and Hippocrates Arabicus', 77.

396 See Kraus, 'Zu Ibn al-Muqaffa'', 2–4; and Salama-Carr, 'Translation', 387–390.

397 See Salama-Carr, 'Translation', 388–389.

398 See Endress, 'Die wissenschaftliche Literatur', 4; Endress, 'The Circle of al-Kindī', 44; and Gutas, *Greek Thought*, 137–138.

structure is rent; the meter is no longer correct; poetic beauty disappears; and nothing worthy of admiration remains in the poems. It is different with prose. Accordingly, original prose is more beautiful and appropriate than prose renderings of metric poetry.

IV.5b. Al-Ǧāḥiẓ, *Book of Animals* (*Kitāb al-ḥayawān*), I, 75.10–16 (ed. Hārūn); tr. Vagelpohl and Sánchez.

(The Apologist of the Poetry of the Arabs said:) The books of the Indians have been translated, as have the wisdom of the Greeks and Persian belles-lettres. Some (of these works) have been improved (in this process), some have not been diminished. But if the wisdom of the Arabs were to be translated, the marvel that is poetic meter would render it worthless. In addition, if people were to translate it, they would not find anything there that non-Arabs have not already said in the books they have composed about their way of living, their intellectual exploits, and wise sayings. These books have been carried from nation to nation, from century to century, from language to language, until they reached us, their final heirs and readers. It is true that books register the legacy of the past more comprehensively than buildings or poetry.

IV.5c. Al-Ǧāḥiẓ, *Book of Animals* (*Kitāb al-ḥayawān*), I, 75.18–77.4 (ed. Hārūn); tr. Vagelpohl and Sánchez.

(The Apologist of the Poetry of the Arabs) then said: Some of those who come to the aid of, guard, and vindicate poetry claim that the translator never conveys what a sage says with the particular nuances of its meanings, the true essence of its teachings, the intricacies of its succinct expressions, and the undertones of its terms. (They argue) that he cannot render it accurately and faithfully, and that the translator is subject to the same constraints as an intermediary or a go-between. How could he convey it and deliver and impart its true meaning with precision and veracity, unless he has full knowledge of its semantic nuances, of how to use the inflections of its words, and how to interpret its (rhetorical) devices in the same way as the author and composer of the book? By God Almighty, how could Ibn al-Biṭrīq,[399] Ibn Nāʿimah,[400] Ibn Qurrah,[401] Ibn

399 See above, I.11 with n. 105, and II.1.

400 See above, II.1 with n. 151.

401 Ṯābit ibn Qurrah; see above, I.6 with n. 64.

Bahrīz,[402] Theophilus (ibn Tūmā),[403] Ibn Wahīlā[404] or Ibn al-Muqaffaʿ[405] be considered equal to Aristotle? And how (could) Ḫālid[406] compare with Plato?

The clarity with which the translator translates has to match his knowledge of the discipline he translates. He needs to be the most skilful person in both the original and the target languages and master both to the same extent. If he happens to be a fluent speaker of both languages, we know that he will surely do injury to both of them because each of the two languages pulls on, borrows from, and interferes with the other. How could someone's command of two languages at the same time be the same as that of someone who only speaks one? He only possesses one (verbal) faculty, and if he speaks one language only he will lose the faculty of (speaking) both of them; and the same happens if he speaks more than two languages. This same reasoning applies to translation in all languages.

The more difficult and arcane a field of knowledge and the fewer the scholars who cultivate it, the more arduous (the task of) the translator, and the likelier that he will commit mistakes. You will never find a translator who is the equal of any of these scholars.

IV.5d. Al-Ǧāḥiẓ, *Book of Animals* (*Kitāb al-ḥayawān*), I, 77.6–11, 78.2–5 (ed. Hārūn); tr. Vagelpohl and Sánchez.

(The Apologist of the Poetry of the Arabs said:) What we have said so far applies to (the translation of) books on geometry, astrology, arithmetic, and music. What if these were religious books speaking about God Almighty—how do we (distinguish between) that which is permissible and that which is not when you want to speak about the true meanings of the natures [i.e., of created beings], an issue bound up with God's oneness; or (when you want to) speak about the forms the (divine) message (can take) and its implications for these forms, an issue that belongs to (distinguishing between) what is permissible with regard to God and what is not, and what is permissible with regard to the people and what is not?

What we have just mentioned is but a small number (of examples) from a wide range (of problems). When the translator does not know these (issues), his

402 See above, II.1 with n. 155.

403 Theophilus ibn Tūmā al-Ruhāwī (d. 169/785), otherwise known as Theophilus of Edessa, a Maronite scholar and translator; see Lucas van Rompay, 'Theophilos of Edessa', *GEDSH*.

404 Unidentified, possibly the result of a scribal error.

405 See above, II.1 with n. 197.

406 See above, II.3 with n. 286.

interpretation of religious discourse will be erroneous, and errors in religious matters are more harmful than errors in mathematics, medicine, philosophy, chemistry [or alchemy], or other things on which humans spend their lives.

IV.5e. Al-Ǧāḥiẓ, *Book of Animals* (*Kitāb al-ḥayawān*), I, 78.6–14 (ed. Hārūn); tr. Vagelpohl and Sánchez.

(The Apologist of the Poetry of the Arabs said:) If the author of a translation does not attain perfection in this, his errors will be proportional to his shortcomings. And what about the translator's knowledge of the difference between a conclusive proof and one that only seems so, what about his knowledge of the celestial signs? The invisible boundaries? The correction of linguistic oversights and scribal errors? The ravings that sometimes fill the prologues (of books)? We know that prologues are mandatory, and they should be properly organized and resemble the extended string (of a necklace). Ibn al-Biṭrīq[407] and Ibn Qurrah[408] do not understand this when it is described, arranged, organized, and explained in detail (for them) by a fellow scholar or a skilful medical expert. What chances then does a book have (of being understood) that has been passed from language to language and from pen to pen, and has been written in the scripts of different religious communities and nations?

IV.5f. Al-Ǧāḥiẓ, *Book of Animals* (*Kitāb al-ḥayawān*), I, 78.18–79.3 (ed. Hārūn); tr. Vagelpohl and Sánchez.

(The Apologist of the Poetry of the Arabs said:) And finally, there are the defects introduced by different kinds of copyists. For their original is not free from errors, and then someone makes a copy of it and adds his own errors to those already in his copy—he will not decrease them in any case—, and then he will (simply) blame the person who has left this mass of errors, since he is incapable of correcting problems he has not identified in his copy (in the first place).

IV.5g. Al-Ǧāḥiẓ, *Book of Animals* (*Kitāb al-ḥayawān*), VI, 280.3–11 (ed. Hārūn); tr. Endress, 'The Circle of al-Kindī', 43–44.

(The Apologist of the Poetry of the Arabs said:) You have maintained that you found mention of the shooting-stars in the books of the ancient philosophers,

407 See above, I.11 with n. 105, and II.1.

408 Ṯābit ibn Qurrah; see above, I.6 with n. 64.

as in Aristotle's *Meteorology*,[409] where he holds forth 'On the shooting-stars', 'On the comets', as also 'On the rainbow' and the 'halo' surrounding the moon at night. Indeed, if you seek support from sources like this, you should be made aware of the lies of the translators and their inventions, also of the corrupt text of the book with regard to (the uncertainties of) the interpretation, to the ignorance of the translators as to how to render words of one language in another, and to the miserable copies of the book, old as it is and not to be relied upon because of the many alterations and corruptions. This is known and ascertained![410]

IV.6. Al-Bīrūnī, *On Pharmacology* (*Kitāb al-ṣaydanah*), 14.8–15.2/Ar. (ed. Meyerhof); tr. Vagelpohl and Sánchez.

Al-Bīrūnī's[411] *On Pharmacology* (*Kitāb al-ṣaydanah*) is one of the most important contributions to this discipline in medieval Islam. It consists of an alphabetically ordered list of 1,116 entries on medical substances, including Greek and Latin terms. In the introduction to the treatise, the author comments on two problems that beset translation: the difficulty of translating specific terms in different languages, and the problematic nature of the Arabic script.

I was fortunate that since my childhood, I have been extremely keen on acquiring knowledge in accordance with my age and situation. As evidence, let it suffice that (once) a Byzantine man came to our area and I used to visit him with seeds, pips, fruit, plants, and other (such things), ask him for their names in his language, and write them down. But the Arabic script has a major defect: the similarity of the shapes of letters that have the same form and the necessity of marking them with diacritical dots and vowel signs, which, if left out, would obscure their meaning. When this combines with carelessness in comparing (a copy of a text with its original) and negligence in correcting (a text) by collation—this practice is common among our people—then it does not make a difference (any more) whether a book is available or not, whether one knows what is in it or not.

If it were not for this defect, it would suffice (just) to transcribe the Greek names that occur in the books of Dioscurides, Galen, Paul (of Aegina), and

409 The Arabic translation of this work was produced by Ibn al-Biṭrīq; see Peters, *Aristoteles Arabus*, 39–40. The Arabic version was edited by Pieter L. Schoonheim; see Aristotle, *Kitāb al-āṯār al-ʿulwīyah*.

410 Other translations: Endress, 'Die griechisch-arabischen Übersetzungen', 104–105; and Endress, 'Die wissenschaftliche Literatur', 4 (German).

411 See above, IV.1.

Oribasius that have been translated into Arabic.[412] But we cannot rely on them and are not safe from alterations in their readings. There is another problem with the translations: leaving in Greek (the names of) some of the drugs that are available in our area and for which there are Arabic terms, which then need to be explained once the books has been translated; for example (the Greek equivalents of) mountain celery, wild carrots, barberry, goatsbeard, and such. They did not translate them into Arabic, just as they did not translate the titles of the books on logic, like, for example the *Isagoge, Categories, On Interpretation, Prior Analytics, Posterior Analytics.*[413] (This) only doubled the hate and coldness (towards the ancient sciences) on the part of its opponents.[414]

IV.7. Al-Fārābī, *The Book of Particles* (*Kitāb al-ḥurūf*), 112.1–9, 112.20, 114.13–115.1 (ed. Mahdi); tr. Mahdi, *The Book of Particles*, adapted by Vagelpohl and Sánchez.

Al-Fārābī[415] was one of the most important philosophers of the classical period of Islam. His writings covered not only philosophical topics but also other scientific disciplines, such as mathematics and physics. Al-Fārābī's family hailed from either Ḫurāsān or Transoxania, his ethnic background was either Persian or Turkish, and he studied with Syriac-speaking scholars in Baghdad. He was thus acutely aware of the subtleties of language and translation, as the following excerpt from his *Book of Particles* (*Kitāb al-ḥurūf*) shows. In it, he discusses one of the most vexing problems that translators of philosophical, especially metaphysical and logical texts, into Arabic had to deal with: the absence of the copula in Arabic.[416]

From its origin, Arabic did not have a word that fulfils the role of *hast*[417] in Persian, *estin*[418] in Greek, or the counterparts of these two in the other languages. This (word), however, is indispensable in the theoretical sciences and in the

412 On the translations of Dioscorides and Oribasius, see above, I.12.

413 I.e., these titles were usually transliterated in Arabic rather than translated.

414 Another translation: Meyerhof, 'Das Vorwort', 198–199/Ger. (German).

415 See above, I.13 with n. 138.

416 Semitic languages, such as Arabic, use personal pronouns to link subjects and predicates, whereas Indo-European languages such as Persian and Greek use copular verbs: where the latter languages use *hast* or *estin* ('it is'), Arabic simple uses the pronoun *huwa*, 'he/it'. As al-Fārābī observed, this pronoun can be nominalised by adding the definitive article *al-* to form a neologism that denotes existence, namely *al-huwa*, 'that which is'. According to al-Fārābī, this notion can be better expressed with the passive participle *al-mawǧūd*, 'to be found, to exist', which is derived from the verb *waǧada*, 'to find'.

417 *Hast* (or *ast*) is the 3rd person singular present of the Persian verb *būdan*, 'to be'.

418 *Estin* (ἐστίν) is the 3rd person singular present active indicative of the verb *einai* (εἶναι), 'to be'.

art of logic. So when philosophy was transmitted to the Arabs, philosophers who spoke Arabic and expressed the concepts that are part of philosophy and logic felt a need for it. But as they did not find in the Arabic language from its origin a term with which to translate passages in which *estin* is used in Greek and *hast* in Persian and make it take on the role of these words in the passages in which other nations use them, some thought that they should use the (pronoun) *huwa* in place of the Persian *hast* and the Greek *estin* […], whereas others thought that in place of those (foreign) words they should use the word *al-mawǧūd* [i.e., 'that which exists'] instead of (the word) *al-huwa* [i.e., 'that which is']. […]

Because this word [i.e., *al-mawǧūd*] became extremely misleading due to its current use in Arabic and its particular (morphological) pattern, some people thought that they should avoid using it and used in its place our word *huwa* [i.e., 'he (is)'] and in place of *al-wuǧūd* [i.e., 'existence'] *al-huwīyah* [i.e., 'being'].[419] But since in Arabic the word *huwa* is neither a noun nor a verb and we therefore cannot derive a verbal noun at all from it; since one needs a term with which to signify the meanings one seeks to signify in the theoretical sciences; and since one needs to derive from it such (concrete and abstract nouns) as 'man' and 'manliness', and 'human' and 'humanness', some people thought that they should avoid (these words) and use *al-mawǧūd* instead of *huwa*, and *al-wuǧūd* instead of *al-huwīyah*. I for one think that a person is free to use whichever of these two (pairs of terms) he wishes.[420]

IV.8a. Al-Tawḥīdī, *The Book of Enjoyment and Conviviality* (*Kitāb al-imtāʿ wa-l-muʾānasah*), I, 111.11–112.6 (ed. Amīn and al-Zayn); tr. Vagelpohl and Sánchez.[421]

Abū Ḥayyān al-Tawḥīdī (d. 414/1023), a state official and scholar who held various offices until he became the vizier of the Buyid sultan Ṣamṣām al-Dawlah in 373/983, was renowned for his refined style and breadth of knowledge. Both are on display in his *adab* works, in which he addressed a wide variety of topics.[422]

In *The Book of Enjoyment and Conviviality* (*Kitāb al-imtāʿ wa-l-muʾānasah*), al-Tawḥīdī revived the genre of philosophical seances to discuss major philosophical issues, including the relationship between language and logic, and the role of trans-

---

419 Again, the common verbal noun *al-wuǧūd*, derived from the verb *waǧada*, is contrasted with the neologism *al-huwīyah*, derived from the pronoun *huwa*.

420 Other translations: Endress, 'Die wissenschaftliche Literatur', 21–22 (German, partial); and Endress, 'Die griechisch-arabischen Übersetzungen', 137–138 (German, partial).

421 Parallel: al-Tawḥīdī, *al-Muqābasāt*, 71.21–72.11.

422 See Samuel M. Stern, 'Abū Ḥayyān al-Tawḥīdī', *EI2*.

lation. The following samples are taken from the eighth *soirée*, which starts out with a critique of philosophers who exploit their knowledge for commercial ends. The translator and philosopher Abū Bišr Mattā ibn Yūnus[423] in particular comes in for biting criticism.[424] He also happens to be one of the participants in the famous debate on the primacy of logic or the grammar of individual languages, which al-Tawḥīdī then reported on the authority of the philologist ʿAlī ibn ʿĪsā al-Rummānī (d. 384/994).[425]

The key question of the debate was whether logic, which purports to be universal, or the grammatical rules of a particular language, in this case Arabic, are sufficient to determine whether a given proposition is true or false. Abū Bišr's interlocutor, the philologist and grammarian Abū Saʿīd al-Sīrāfī (d. 368/979),[426] challenged the universality of (Greek) logic. Abū Bišr maintained that individual languages have different means to establish the truth value of statements. In the following excerpts, al-Sīrāfī first argues that the process of translation inevitably distorts the meaning it seeks to convey (IV.8a). He then attacks Abū Bišr's naive belief that his command of Arabic does not need to be perfect to translate his sources into Arabic (IV.8b), lectures him on the complexities of language and the various ways different languages express meanings, and questions his ability to explain Aristotle in a language he does not fully understand (IV.8c). He then ridicules logicians for their universalist pretensions and the obscurity of their translations, which are the result of their imperfect grasp of the languages involved (IV.8d).

(Abū Saʿīd al-Sīrāfī) said: So you are not inviting us (to study) the science of logic, but to learn the Greek language while you do not know Greek (yourself)? How can you invite us (to study) a language of which you are not in perfect command, (a language) that has been wiped out a long time ago, whose speakers have passed away, and when the people who used to converse in it and communicate their intentions by its inflexions have perished? Although you translate from Syriac, what do you say about concepts that have changed by being translated from Greek to another language, Syriac, and then from that language to another, Arabic?

(Abū Bišr) Mattā said: Even though the Greeks have passed away with their language, translation has preserved the intentions (of the authors), conveyed the concepts and faithfully transmitted the true meanings.

---

423 See above, I.13 with n. 144.

424 Al-Tawḥīdī, *Kitāb al-imtāʿ*, I, 104.13–14.

425 The debate has been analysed by a number of scholars. See for example Mahdi, 'Language and Logic'; Endress, 'Grammatik und Logik', 163–299; Kühn, 'Die Rehabilitierung der Sprache', 343–344; Abed, *Aristotelian Logic*, xiv–xvi; Vagelpohl, 'The ʿAbbasid Translation Movement', 256–258; and Salama-Carr, 'Translation', 390–391.

426 See Geneviève Humbert, 'al-Sīrāfī', *EI2*.

Abū Saʿīd (al-Sīrāfī) said: If we admit that translation is truthful and not misleading, puts straight and does not distort, is neither obscure nor garbled, has not omitted or added (anything), has not put first or put last (anything),[427] has neither misrepresented the sense of the general and the particular, nor of the most particular and the most general—even though this does not happen and is not part of the natural characteristics of languages nor the scope of concepts—then you seem to say that the only competent authority is the minds of the Greeks, the only demonstration that which they have produced, and the only truth that which they have brought to light.[428]

IV.8b. Al-Tawḥīdī, *Book of Enjoyment and Conviviality* (*Kitāb al-imtāʿ wa-l-muʾānasah*), I, 115.7–15 (ed. Amīn and al-Zayn); tr. Vagelpohl and Sánchez.[429]

(Abū Saʿīd al-Sīrāfī said:) But if you inevitably need a little of the (Arabic) language for translation, you also inevitably need a great deal of it to carry out the translation, to inspire confidence and to protect against the attendant imperfection.

(Abū Bišr) Mattā said: It is sufficient for me to know of your (Arabic) language the noun, the verb, and the particle: this is enough to enable me (to convey) the ideas that the Greeks have elaborated for me.

(Abū Saʿīd al-Sīrāfī) said: You are wrong, because you need (to know) the characteristics and (the way) to assemble this noun, verb, and particle according to the order inherent in the natural disposition of (Arabic) speakers. You also need (to know) in addition to this the vocalization of these nouns, verbs, and particles: making mistakes and altering the vowels is the same as (introducing) errors and flaws in consonants.[430] [431]

427 I.e., the translation has not changed the word order.

428 Other translations: Margoliouth, 'The Discussion', 114 (English); and Endress, 'Grammatik und Logik', 244–245 (German).

429 Parallel: al-Tawḥīdī, *al-Muqābasāt*, 75.8–16.

430 Like other Semitic languages, Arabic has a consonantal script. Vowel marks, which are optional, may be added to aid understanding. The lack of proper vocalization sometimes results in ambiguities or misunderstandings. See above, IV.6.

431 Other translations: Margoliouth, 'The Discussion', 117–118 (English); and Endress, 'Grammatik und Logik', 249–250 (German).

IV.8c. Al-Tawḥīdī, *Book of Enjoyment and Conviviality* (*Kitāb al-imtāʿ wa-l-muʾānasah*), I, 115.15–116.9 (ed. Amīn and al-Zayn); tr. Vagelpohl and Sánchez.[432]

(Abū Saʿīd al-Sīrāfī said:) This is a subject to which you, your friends, and your party do not pay attention. But there is a secret that you fail to grasp and which has not been revealed to your intellect: you know that no one language matches another one in all respects with the definitions of its characteristics in nouns, verbs, and particles; in its composition; putting (elements) first and putting (them) last;[433] its use of metaphor and actual meaning; its duplication and simplification (of nouns); extending and narrowing (of meanings); its verse, prose, and rhymed prose; its meter and the deviation (from it); and other things it would take too long to mention. I do not think that anyone who possesses a smidgen of intelligence or a tiny share of fair judgment objects to this assessment or doubts its correctness. How then can you trust anything that was translated for you, given this description? Rather, you are more in need of learning Arabic than the Greek concepts, given that the concepts are not Greek or Indian, as the languages are Persian, Arabic, or Turkish. On top of this, you claim that the concepts are accessible through intelligence, study, and reflection, and then nothing remains but the rules of language. Why then do you disparage Arabic while explaining the books of Aristotle in it, even though you are ignorant of its essence?[434]

IV.8d. Al-Tawḥīdī, *Book of Enjoyment and Conviviality* (*Kitāb al-imtāʿ wa-l-muʾānasah*), I, 121.6–10 (ed. Amīn and al-Zayn); tr. Vagelpohl and Sánchez.[435]

(Abū Saʿīd al-Sīrāfī said:) All these (rules of language) are regulated by the succession (of scholars), the transmission (of knowledge), audition (of authoritative texts), accepted usage, and analogy that proceeds without distortion from a known starting point. The logicians were seized by pride, because they assumed that one can learn and investigate the concepts only with their method, their insight, and their effort. They translated a language in which they are weak and imperfect [i.e., Greek] into another in which they are (also) weak and imperfect [i.e., Arabic]. They raised this (sort of) translation to an

---

432 Parallel: al-Tawḥīdī, *al-Muqābasāt*, 75.16–76.5.

433 I.e., the word order.

434 Other translations: Margoliouth, 'The Discussion', 118 (English); and Endress, 'Grammatik und Logik', 250–251 (German).

435 Parallel: al-Tawḥīdī, *al-Muqābasāt*, 80.11–15.

art and claimed that the grammarians have to do (only) with words, not with concepts.[436]

IV.9. Al-Bīrūnī, *On Pharmacology* (*Kitāb al-ṣaydanah*), 13.2–18/Ar. (ed. Meyerhof); tr. Vagelpohl and Sánchez.

In this second passage from the introduction to al-Bīrūnī's *On Pharmacology*,[437] the author praises Arabic as a scientific language that has been enriched by the texts translated into it and compares it with his native language, the Ḫwārizmian dialect of Persian, and to Persian, both of which he considers inferior.

Sciences from (all) corners of the world have been translated into Arabic, and thus became adorned and delighted (people's) hearts. The charms of their linguistic expression (in Arabic) now flow through (people's) arteries and veins (as if it was their blood), even though each people delights in the language to which they are accustomed and habituated and which they use in their dealings with their intimates and peers. I judge this from my own self, which has been moulded by a language that, had a science been immortalized in it, would have seemed as preposterous as a camel on a roof gutter or a giraffe in a herd of pure-bred horses. I then switched to Arabic and Persian, and I am a stranger and amateur in each, but prefer to be mocked in Arabic than praised in Persian. People who carefully read a scientific book that has been translated into Persian will know that what I say is true: how its beauty has disappeared, its meaning obscured, its intent darkened, and its usefulness vanished, since this language is only good for tales about Persian kings and for evening parties.[438]

V. *Why Do We Translate?*

V.1. Al-Kindī, *On First Philosophy* (*Fī l-falsafah al-ūlā*), I, 103.1–7 (ed. Abū Rīdah); tr. Ivry, *First Philosophy*, 58.

Abū Yūsuf Yaʿqūb ibn Isḥāq al-Kindī, known as the 'philosopher of the Arabs', was a companion of the caliphs al-Maʾmūn and al-Muʿtaṣim and one of the most important

436 Other translations: Margoliouth, 'The Discussion', 122–123 (English); and Endress, 'Grammatik und Logik', 258 (German).

437 See above, IV.6.

438 Another translation: Meyerhof, 'Das Vorwort', 196–197/Ger. (German).

thinkers of the third/ninth century.[439] He wrote on a wide range of topics, especially philosophy, and played a central role in the translation movement.[440]

The first section of *On First Philosophy* (*Fī l-falsafah al-ūlā*), one of his most famous philosophical works, which he dedicated to the caliph al-Muʿtaṣim, is essentially a paean to Greek philosophy and an exhortation to profit from this rich tradition. It includes the following appeal, which encapsulates the spirit of openness with which he approached Greek philosophy and science.[441]

Aristotle, the most distinguished of the Greeks in philosophy, said: 'We ought to be grateful to the fathers of those who have contributed any truth, since they were the cause of their existence, let alone (being grateful) to the sons; for the fathers are their cause, while they are the cause of our attaining the truth.'[442] How beautiful is that which he said in this matter!

We ought not to be ashamed of appreciating the truth and of acquiring it wherever it comes from, even if it comes from races distant and nations different from us. For the seeker of truth, nothing takes precedence over the truth, and there is no disparagement of the truth, nor belittling either of him who speaks it or of him who conveys it. (The status of) no one is diminished by the truth; rather does the truth ennoble all.

439 On al-Kindī and his attitude towards Greek philosophy, see Adamson, 'Al-Kindī'.
440 See Jean Jolivet and Roshdi Rashed, 'al-Kindī', *EI2*; and Endress, 'The Circle of al-Kindī'.
441 See Gutas, *Greek Thought*, 158–159.
442 According to Alfred Ivry, this is an adaptation of the (rather free) Arabic translation of Aristotle's *Metaphysics*, 993a30–b4; see Ivry, 'Al-Kindī's First Philosophy', 20.

Part Three: Original Texts

1. *History: Origins, Genres, Works*

1.1. Al-Masʿūdī, *Kitāb al-tanbīh*, 282.12; 283.2–5 (ed. de Goeje).

كتب من حضر من الكتّاب [...] وزيد بن ثابت الأنصاريّ ثمّ الخزرجيّ من بني غنم بن مالك ابن النجّار يكتب إلى الملوك ويجيب بحضرة النبيّ صلعم وكان يترجم للنبيّ صلعم بالفارسيّة والروميّة والقبطيّة والحبشيّة تعلّم ذلك بالمدينة من أهل هذه الألسن.

1.2. Ibn Ǧulǧul, *Ṭabaqāt al-aṭibbā'*, 61.2–5 (ed. Fu'ād Sayyid).

ماسرجويه كان يهوديّ المذهب سريانيّاً وهو تولّى في الدولة المروانيّة تفسير كتاب أهرن بن أعين القسّ إلى العربيّة ووجده عمر بن عبد العزيز في خزائن الكتب فأمر بإخراجه ووضعه في مصلّاه فاستخار الله في إخراجه إلى المسلمين للانتفاع به فلمّا تمّ له في ذلك أربعين صباحاً أخرجه إلى الناس وبثّه في أيديهم.

1.3. al-Balāḏurī, *Futūḥ al-buldān*, 193.1–9 (ed. de Goeje).

نقل ديوان الروميّة

قالوا ولم يزل ديوان الشام بالروميّة حتّى ولي عبد الملك بن مروان فلمّا كانت سنة ٨١ أمر بنقله وذلك أنّ رجلاً من كتّاب الروم احتاج أن يكتب شيئاً فلم يجد ماء فبال في الدواة فبلغ ذلك عبد الملك فأدّبه وأمر سليمان بن سعد بنقل الديوان فسأله أن يعينّه بخراج الأردنّ سنة ففعل ذلك وولّاه الأردنّ فلم تنقض السنة حتّى فرغ من نقله وأتى به عبد الملك فدعا بسرجون كاتبه فعرّض ذلك عليه فغمّه وخرج من عنده كئيباً فلقيه قوم من كتّاب الروم فقال اطلبوا المعيشة من غير هذه الصناعة فقد قطعها الله عنكم.

1.4. Al-Balāḏurī, *Futūḥ al-buldān*, 300.8–301.9 (ed. de Goeje).

نقل ديوان الفارسيّة

وحدّثني المدائنيّ عليّ بن محمّد بن أبي سيف عن أشياخه قالوا لم يزل ديوان خراج السواد وسائر العراق بالفارسيّة فلمّا ولي الحجّاج العراق استكتب زادان فرّوخ بن بيري وكان معه صالح بن عبد الرحمن مولى بني تميم يخطّ بين يديه بالعربيّة والفارسيّة وكان أبو صالح من سبيّ سجستان فوصل زادان فرّوخ صالحاً بالحجّاج وخفّ على قلبه فقال له ذات يوم إنّك سببي إلى الأمير وأراه قد استخفّني ولا آمن أن يقدّمني عليك وأن تسقط فقال لا تظنّ ذلك هو أحوج إليّ منه إليك لأنّه لا يجد من يكفيه حسابه غيري فقال والله لو شئت أن أحوّل الحساب إلى العربيّة لحوّلته قال فحوّل منه شطراً حتّى أرى ففعل فقال له تمارض فتمارض فبعث إليه الحجّاج طبيبه فلم ير به علّة وبلغ زادان فرّوخ ذلك فأمره أن يظهر.

ثمّ إنّ زادان فرّوخ قتل أيّام عبد الرحمن بن محمّد بن الأشعث الكنديّ وهو خارج من منزل كان فيه إلى منزله أو منزل غيره فاستكتب الحجّاج صالحاً مكانه فأعلمه الذي كان جرى بينه وبين زادان فرّوخ في نقل الديوان فعزم الحجّاج على أن يجعل الديوان بالعربيّة وقلّد ذلك صالحاً فقال له مردانشاه بن زادان فرّوخ كيف تصنع بدهويه وبيستويه قال أكتب عشر ونصف عشر قال فكيف تصنع بويد قال أكتبه أيضاً والويد النيّف والزيادة تزاد فقال قطع الله أصلك من الدنيا كما قطعت أصل الفارسيّة وبذلت له مائة ألف درهم على أن يظهر العجز عن نقل الديوان ويمسك عن ذلك فأبى ونقله فكان عبد الحميد بن يحيى كاتب مروان ابن محمّد يقول لله درّ صالح ما أعظم منّته على الكتّاب.

وحدّثني عمر ابن شبّة قال حدّثني أبو عاصم النبيل قال أنبأ سهل بن أبي الصلت قال أجّل الحجّاج صالح بن عبد الرحمن أجلاً حتّى قلب الديوان.

1.5. Ibn Ǧulǧul, *Ṭabaqāt al-aṭibbāʾ*, 65.2–5 (ed. Fuʾād Sayyid).

يوحنّا ابن ماسويه: مسيحيّ المذهب سريانيّ قلّده الرشيد ترجمة الكتب القديمة الطبّيّة ممّا وُجد بأنقرة وعمورية وبلاد الروم حين سباها المسلمون ووضعه أميناً على الترجمة ووضع له كتّاباً حذّاقاً يكتبون وخدم هارون والأمين والمأمون وبقي على ذلك إلى أيّام المتوكّل.

I.6. Ibn al-Nadīm, *Kitāb al-fihrist*, I, 265.20–24 (ed. Flügel).

الكلام على كتابه فى أصول الهندسة واسمه الأسطروشيا ومعناه أصول الهندسة: نقله الحجّاج بن يوسف بن مطر نقلين أحدهما يعرف بالهارونيّ وهو الأوّل ونقلاً ثانياً وهو المأمونيّ ويعرف بالمأمونيّ وعليه يعوّل.

ونقله إسحق بن حنين وأصلحه ثابت بن قرّة الحرّانيّ ونقل أبو عثمان الدمشقيّ منه مقالات رأيت منها العاشرة بالموصل في خزانة عليّ بن أحمد العمرانيّ وأحد غلمانه أبو الصقر القبيصيّ ويقرأ عليه المجسطيّ في زماننا.

I.7. Ibn al-Nadīm, *Kitāb al-fihrist*, I, 267.29–268.4 (ed. Flügel).

الكلام على كتاب المجسطيّ: هذا الكتاب ثلث عشرة مقالة وأوّل من عنى بتفسيره وإخراجه إلى العربيّة يحيى بن خالد بن برمك ففسّره له جماعة فلم يتقنوه ولم يرض ذلك فندب لتفسيره أبا حسّان وسلم صاحب بيت الحكمة فأتقناه واجتهدا في تصحيحه بعد أن أحضرا النقلة المجوّدين فاختبرا نقلهم وأخذا بأفصحه وأصحّه.

وقد قيل إنّ الحجّاج بن مطر نقله أيضاً فأمّا الذي عمله النيريزيّ وأصلح ثابت الكتاب كلّه بالنقل القديم ونقل إسحق هذا الكتاب وأصلحه ثابت نقلاً غير مرضى لأنّ إصلاحه الأوّل أجود.

I.8. Ibn Abī Uṣaybiʿah, *ʿUyūn al-anbāʾ*, II/2, 886.3–888.2 (ed. Savage-Smith et al.).

وكان من بعد صنجهل الهنديّ جماعة في بلاد الهند ولهم تصانيف معروفة في صناعة الطبّ وغيرها من العلوم مثل باكهر[443] راحه[444] صكه[445] داهر أنكر زنكل[446] جبهر[447] أندي[448] جاري[449] كلّ

443 باكهر Ibn al-Nadīm, *Kitāb al-fihrist* (ed. Flügel): راكهر (ed. Tağaddud): راكهّر (ed. Sayyid)
444 راحه Ibn al-Nadīm, *Kitāb al-fihrist* (ed. Flügel): راجه (ed. Tağaddud, Sayyid)
445 أنكر Ibn al-Nadīm, *Kitāb al-fihrist* (ed. Tağaddud, Sayyid): أنكو (ed. Flügel)
446 زنكل Ibn al-Nadīm, *Kitāb al-fihrist* (ed. Flügel): رنكل (ed. Tağaddud, Sayyid)
447 جنهر Ibn al-Nadīm, *Kitāb al-fihrist* (all eds)
448 أندي Ibn al-Nadīm, *Kitāb al-fihrist* (ed. Tağaddud, Sayyid): اندو (ed. Flügel)
449 جبارى Ibn al-Nadīm, *Kitāb al-fihrist* (all eds)

هؤلاء أصحاب تصانيف وهم من حكماء الهند وأطبّائهم ولهم الأحكام الموضوعة في علم النجوم والهند تشتغل بمؤلّفات هؤلاء فيما بينهم ويقتدون بها ويتناقلونها وقد نقل كثير منها إلى اللغة العربيّة.

ووجدت الرازيّ قد نقل في كتابه الحاوي وفي غيره عن كتب جماعة من الهند مثل كتاب سيرك[450] الهنديّ وهذا الكتاب فسّره عبد الله بن عليّ من الفارسيّ إلى العربيّ لأنّه نقل أوّلاً من الهنديّ إلى الفارسيّ وعن كتاب سسرد وفيه علامات الأدواء ومعرفة علاجها وأدويتها وهو عشر مقالات أمر يحيى بن خالد بتفسيره وكتاب ندان في علامات أربعمائة وأربعة أدواء ومعرفتها بغير علاج وكتاب سندهشار[451] وتفسيره كتاب صفوة النجح وكتاب فيما اختلفت فيه الهند والروم في الحارّ والبارد وقوى الأدوية وتفصيل السنة وكتاب تفسير أسماء العقّار بأسماء عشرة وكتاب أستانكر[452] الجامع وكتاب علاجات الحبالى للهند وكتاب مختصر في العقاقير للهند وكتاب نوفشل[453] فيه مائة داء ومائة دواء وكتاب روسى الهنديّة في علاجات النساء وكتاب السكّر للهند وكتاب رأي[454] الهنديّ في أجناس الحيّات وسمومها وكتاب التوهّم في الأمراض والعلل لأبي قبيل[455] الهنديّ.

1.9. Ibn al-Nadīm, *Kitāb al-fihrist*, I, 243.3–14 (ed. Flügel).

ذكر السبب الذي من أجله كثرت كتب الفلسفة وغيرها من العلوم القديمة في هذه البلاد

أحد الأسباب في ذلك أنّ المأمون رأى في منامه كأنّ رجلًا أبيض اللون مشرباً حمرة واسع الجبهة مقرون الحاجب أجلح الرأس أشهل العينين حسن الشمائل جالس على سريره قال المأمون وكأنّي بين يديه قد ملئت له هيبة فقلت من أنت قال أنا أرسطاليس فسررت به وقلت أيّها الحكيم أسئلك قال سَلْ قلت ما الحسن قال ما حسن في العقل قلت ثمّ ماذا قال ما حسن في الشرع قلت ثمّ ماذا قال ما حسن عند الجمهور قلت ثمّ ماذا قال ثمّ لا ثمّ.

وفي رواية أخرى قلت زدني قال من نصحك في الذهب فليكن عندك كالذهب وعليك بالتوحيد.

فكان هذا المنام من أوكد الأسباب في إخراج الكتب فإنّ المأمون كان بينه وبين ملك الروم مراسلات وقد استظهر عليه المأمون فكتب إلى ملك الروم يسئله الإذن في إنفاذ ما من مختار من العلوم القديمة المخزونة المدّخرة ببلد الروم.

450 شيرك Ibn Abī Uṣaybiʿah, *ʿUyūn al-anbāʾ* II/2, 886, app. crit. n. 15
451 سندستاق Ibn al-Nadīm, *Kitāb al-fihrist* (ed. Flügel): سستاق (ed. Tağaddud, Sayyid)
452 اسانكر Ibn Abī Uṣaybiʿah, *ʿUyūn al-anbāʾ* II/2, 886, app. crit. n. 22
453 توقشتل Ibn al-Nadīm, *Kitāb al-fihrist* (all eds)
454 راى Ibn al-Nadīm, *Kitāb al-fihrist* (ed. Flügel): باى (ed. Tağaddud): بابي (ed. Sayyid)
455 لتوقشتل Ibn al-Nadīm, *Kitāb al-fihrist* (all eds)

فأجاب إلى ذلك بعد امتناع فأخرج المأمون لذلك جماعة منهم الحجّاج بن مطر وابن البطريق وسلم صاحب بيت الحكمة وغيرهم فأخذوا ممّا وجدوا ما اختاروا فلمّا حملوه إليه أمرهم بنقله فنقل وقد قيل إنّ يوحنّا بن ماسويه ممّن نفذ إلى بلد الروم.

I.10. Ibn al-Qifṭī, *Ta'rīḫ al-ḥukamā'*, 61.4–13 (ed. Lippert).

ولمّا أخرجت الكتب من بلاد الروم إلى المأمون أخرج من هذا الكتاب الجزء الأوّل لا غير يشتمل على سبع مقالات ولمّا ترجم الكتاب دلّت مقدّمته على أنّه ثماني مقالات وأنّ المقالة الثامنة تشتمل على معاني المقالات السبع وزيادة واشترط فيها شروطاً مفيدة وفوائد يرغب فيها ومن ذلك الزمان وإلى يومنا يبحث أهل هذا الشأن عن هذه المقالة فلا يطّلعون لها على خبر ولا شكّ أنّها كانت من ذخائر الملوك لعزّة هذه العلوم عند ملوك يونان وكنت قد ذاكرت بعض من يعاني شيئاً من هذا العلم في زماننا أو يدّعيه بأمر هذه المقالة فقال لي قد وجدت وأخذ في وصفها فذكر ما لم يطابق كلام مؤلّفها في وصفها فعلمت أنّه يجهل الأصل والفرع فأضربت عنه وتركته بجهله.

I.11. Ibn Ǧulǧul, *Ṭabaqāt al-aṭibbā'*, 67.6–11 (ed. Fu'ād Sayyid).

يوحنّا ابن البطريق الترجمان مولى المأمون أمير المؤمنين كان أميناً على الترجمة حسن التأدية للمعاني بكيء اللسان في العربيّة وترجم كثيراً من كتب الأوائل وهو ترجم كتاب أرسطاطاليس إلى الإسكندر المعروف بسرّ الأسرار وهو كتاب السياسة في تدبير الرئاسة.

ذكر يوحنّا أنّه مشى في طلبه وقصد الهياكل في البحث عنه حتّى وصل إلى هيكل عبد الشمس الذي كان بناه هرمس الأكبر لنفسه يمجّد الله تعالى فيه قال فظفرت فيه براهب متناسك ذي علم بارع وفهم ثاقب فتلطّفت به وأعملت الحيلة عليه حتّى أباح لي مصاحف الهيكل المودعة فيه فوجدت في جملتها المطلوب الذي أمرني أمير المؤمنين بطلبه مكتوباً بالذهب فرجعت إلى الحضرة المنصورة ظافراً بالمراد.

I.12. Ibn Abī Uṣaybi'ah, *'Uyūn al-anbā'*, II/2, 928.4–930.9 (ed. Savage-Smith et al.).

وهو يقول في أوّل كتابه هذا إنّ كتاب ديسقوريدس ترجم بمدينة السلام في الدولة العبّاسيّة في أيّام جعفر المتوكّل وكان المترجم له إصطفن بن بسيل الترجمان من اللسان اليونانيّ إلى اللسان العربيّ

وتصفّح ذلك حنين بن إسحاق المترجم فصحّح الترجمة وأجازها فما علم إصطفن من تلك الأسماء اليونانيّة في وقته له اسماً في اللسان العربيّ فسّره بالعربيّة وما لم يعلم له في اللسان العربيّ اسماً تركه في الكتاب على اسمه اليونانيّ اتّكالاً منه على أن يبعث الله تعالى بعده من يعرف ذلك ويفسّره باللسان العربيّ إذ التسمية لا تكون إلّا بالتواطئ من أهل كلّ بلد على أعيان الأدوية بما رأوا أن يسمّوا ذلك إمّا باشتقاق وإمّا بغير ذلك من تواطئهم على التسمية فاتّكل إصطفن على شخوص يأتون بعده ممّن قد عرف أعيان الأدوية التي لم يعرف هو لها اسماً في وقته فيسمّيها على قدر ما سمع في ذلك الوقت فيخرج إلى المعرفة.

قال ابن جلجل: وورد هذا الكتاب إلى الأندلس وهو على ترجمة إصطفن منه ما عرف له اسماً بالعربيّة ومنه ما لم يعرف له اسماً فانتفع الناس بالمعروف منه بالمشرق وبالأندلس إلى أيّام الناصر عبد الرحمن بن محمّد وهو يومئذ صاحب الأندلس فكاتبه أرمانيوس الملك مالك قسطنطينيّة أحسبُ في سنة سبع وثلاثين وثلثمائة وهاداه بهدايا لها قدر عظيم فكان في جملة هديّته كتاب دسقوريدس مصوّر الحشائش بالتصوير الروميّ العجيب وكان الكتاب مكتوباً بالإغريقيّ الذي هو اليونانيّ وبعث معه كتاب هروسيس صاحب القصص وهو تأريخ للروم عجيب فيه أخبار الدهور وقصص الملوك الأول وفوائد عظيمة.

وكتب أرمانيوس في كتابه إلى الناصر أنّ كتاب دسقوريدس لا تجتني فائدته إلّا برجل يحسن العبارة باللسان اليونانيّ ويعرف أشخاص تلك الأدوية فإن كان في بلدك من يحسن ذلك فزتَ أيّها الملك بفائدة الكتاب وأمّا كتاب هروسيس فعندك في بلدك من اللطينيّين من يقرأه باللسان اللطينيّ وإن كشفتهم عنه نقلوه لك من اللطينيّ إلى اللسان العربيّ.

قال ابن جلجل: ولم يكن يومئذ بقرطبة من نصارى الأندلس من يقرأ اللسان الإغريقيّ الذي هو اليونانيّ القديم فبقي كتاب ديسقوريدس في خزانة عبد الرحمن الناصر باللسان الإغريقيّ ولم يترجم إلى اللسان العربيّ وبقي الكتاب بالأندلس والذي بين أيدي الناس بترجمة إصطفن الواردة من مدينة السلام بغداد فلمّا جاوب الناصر أرمانيوس الملك سأله أن يبعث إليه برجل يتكلّم بالإغريقيّ واللطينيّ ليعلّم له عبيداً يكونون مترجمين فبعث أرمانيوس الملك الناصر براهب كان يسمّى نقولا فوصل إلى قرطبة سنة أربعين وثلثمائة.

وكان يومئذ بقرطبة من الأطبّاء قوم لهم بحث وتفتيش وحرص على استخراج ما جهل من أسماء عقاقير كتاب ديسقوريدس إلى العربيّة وكان أبحثهم وأحرصهم على ذلك من جهة التقرّب إلى الملك عبد الرحمن الناصر حسداي بن شبروط الإسرائيليّ وكان نقولا الراهب عنده أحظى الناس

وأخصّهم به وفسّر من أسماء عقاقير كتّاب ديسقوريدس ما كان مجهولاً وهو أوّل من عمل بقرطبة ترياق الفاروق على تصحيح الشجاريّة التي فيه وكان في ذلك الوقت من الأطبّاء الباحثين عن تصحيح أسماء عقاقير الكتّاب وتعيين أشخاصه محمّد المعروف بالشجّار ورجل كان يعرف بالشبانسيّ وأبو عثمان الجزّار الملقّب باليابسة ومحمّد بن سعيد الطبيب وعبد الرحمن بن إسحاق بن هيثم وأبو عبد الله الصقلّيّ وكان يتكلّم باليونانيّة ويعرف أشخاص الأدوية.

قال ابن جلجل: وكان هؤلاء النفر كلّهم في زمان واحد مع نقولا الراهب أدركتُه وأدركتُ نقولا الراهب في أيّام المستنصر وصحبتُهم في أيّام المستنصر الحكم وفي صدر دولته مات نقولا الراهب فصحّ ببحث هؤلاء النفر الباحثين عن أسماء عقاقير كتّاب ديسقوريدس تصحيح الوقوف على أشخاصها بمدينة قرطبة خاصّة بناحية الأندلس ما أزال الشكّ فيها عن القلوب وأوجب المعرفة بها بالوقوف على أشخاصها وتصحيح النطق بأسمائها بلا تصحيف إلّا القليل منها الذي لا بال به ولا خطر له وذلك يكون في مثل عشرة أدوية.

1.13. Ibn Suwār, annotation to the translation of Aristotle's *Sophistical Refutations* in Aristotle, *Manṭiq Arisṭū*, III, 1017.5–1018.13 (ed. Badawī), and Aristotle, *Les Catégories*, 198–199 (ed. Georr).

نسخت هذا النقل من نسخة بخطّ الشيخ أبي الخير الحسن ابن سوار رضي الله عنه وفي آخرها ما هذه حكايته:

نسخت هذا النقل من نسخة خيّل إليّ أنّها بخطّ أبي نصر الفارابيّ كان النصف الأوّل منها مصحّحاً جيّداً والنصف الثاني مسقاماً.

قال الشيخ أبو الخير الحسن بن سوار رضي الله عنه: لمّا كان الناقل يحتاج في تأدية المعنى إلى فهمه باللغة التي منها ينقل إلى أن يكون متصوّراً له كتصوّر قائله وإلى أن يكون عارفاً باستعمال اللغة التي منها ينقل والتي إليها ينقل وكان أثانس الراهب غير فهم بمعاني أرسطوطالس فيه داخل نقله الخلل لا محالة ولمّا كان من نقل هذا الكتّاب من السريانيّة بنقل أثانس إلى العربيّة ممّن قد ذكر اسمه لم يقع إليهم تفسير له عوّلوا على أفهامهم في إدراك معانيه فكلّ اجتهد في إصابته الحقّ وإدراك الغرض الذي إيّاه قصد الفيلسوف فغيّروا ما نقلوه من نقل أثانس إلى العربيّة فلأنّنا أحببنا الوقوف على ما وقع لكلّ واحد منهم كتبنا جميع النقول التي وقعت إلينا ليقع التأمّل لكلّ واحد منها ويستعان ببعضها على بعض في إدراك المعنى.

وقد كان الفضل يحيى بن عديّ فسّر هذا الكتاب تفسيراً رأيت منه الكثير وقدّرته نحو من ثلثيه بالسريانيّة والعربيّة وأظنّ تمّمه ولم يوجد في كتبه بعد وفاته وتصرّفت بي الظنون في أمره فتارة أظنّ أنّه أبطله لأنّه لم يرتضيه وتارة أظنّ أنّه سرق وهذا أقوى في نفسي ونقل هذا الكتاب النقل المذكور قبل تفسيره إيّاه فلذلك لحق نقله اعتياص ما لأنّه لم يشارف المعنى وأتبع السريانيّ في النقل.

وقد وجد في وقتنا هذا تفسير الإسكندر الأفروديسيّ له باليونانيّة يعجز من أوّله كرّاسة ولم يخرج منه إلّا اليسير.

واتّصل بي أنّ أبا إسحق إبراهيم بن بكّوش نقل هذا الكتاب من السريانيّ إلى العربيّ وأنّه كان يجتمع مع يوحنّا القسّ اليونانيّ المهندس المعروف بابن فتيلة على إصلاح مواضع منه من اليونانيّ ولم يقع إليّ.

وقيل إنّ أبا بشر رحمه الله أصلح النقل الأوّل أو نقله نقلاً آخر ولم يقع إليّ.

وكتبت هذه الجملة ليعلم من يقع إليه هذا الكتاب صورة أمره والسبب في إثباتي جميع النقول على السبيل المسطور.

11. *People: Translators and Sponsors*

11.1. Ibn al-Nadīm, *Kitāb al-fihrist*, I, 244.1–245.10 (ed. Flügel).

أسماء النقلة من اللغات إلى اللسان العربيّ

اصطفن القديم ونقل لخالد بن يزيد بن معوية كتب الصنعة وغيرها.

البطريق وكان في أيّام المنصور وأمره بنقل أشياء من الكتب القديمة.

ابنه أبو زكريّاء يحيى بن البطريق وكان في جملة الحسن بن سهل.

الحجّاج بن مطر فسّر للمأمون وهو الذي نقل المجسطيّ وأقليدس.

ابن ناعمة واسمه عبد المسيح بن عبد الله الحمصيّ الناعميّ.

سلّام الأبرش من النقلة القدماء في أيّام البرامكة ويوجد بنقله السماع الطبيعيّ كذا حكى سيّدنا أبو القاسم عيسى بن عليّ بن عيسى أيّده الله.

حبيب بن بهريز مطران الموصل فسّر للمأمون عدّة كتب.

زرويا (؟) بن ما حوه (؟) الناعميّ الحمصيّ.

هلال بن أبي هلال الحمصيّ.

تذاري.

فثيون.

أبو نصر بن آوى[456] بن أيّوب.

بسيل المطران.

أبو نوح بن الصلت.

أسطاث.

حيرون.

اصطفن بن باسيل.

ابن رابطة.

تيوفيلي.

شملي.

عيسى بن نوح.

قويري واسمه إبراهيم ويكنى أبا اسحق.

تذرس السنقل.

داريع الراهب.

هبا.[457]

بثيون.

صليبا.

أيّوب الرهاويّ.

ثابت بن قمع.

أيّوب وسمعان فسّرا زيج بطلميوس لمحمّد بن خالد بن يحيى بن برمك وغير ذلك من الكتب القديمة.

باسيل وكان يخدم ذا اليمنيين.

ابن شهدا الكرخيّ نقل من السريانيّ إلى العربيّ نقلاً رديئاً فممّا نقل كتاب الأجنّة لبقراط.

أبو عمرو يوحنّا بن يوسف الكاتب أحد النقلة ونقل كتاب فلاطون في آداب الصبيان.

---

456 ناري Ibn Abī Uṣaybiʿah, *ʿUyūn al-anbāʾ*, II/1, 511.1

457 هبا corrected: هيّا Ibn al-Nadīm, *Kitāb al-fihrist* (ed. Flügel, Taǧaddud, Sayyid)

أيّوب ابن القاسم الرقّيّ نقل من السريانيّ إلى العربيّ ومن نقله كتاب إيساغوجي.

مرلاحي (؟) في زماننا جيّد المعرفة بالسريانيّة عفطيّ الألفاظ بالعربيّة ينقل بين يدي عليّ بن إبراهيم الدهكيّ من السريانيّ إلى العربيّ ويصلح نقله ابن الدهكيّ.

داديشوع كان يفسّر لإسحق ابن سليمان بن عليّ الهاشميّ من السريانيّة إلى العربيّة.

قسطا بن لوقا البعلبكّيّ جيّد النقل فصيح باللسان اليونانيّ والسريانيّ والعربيّ وقد نقل أشياء وأصلح نقولاً كثيرة [...]

حنين.

إسحق.

ثابت.

حبيش.

عيسى بن يحيى.

الدمشقيّ.

إبراهيم بن الصلت.

إبراهيم بن عبد الله.

يحيى بن عديّ.

النفيسيّ.[458] [...]

أسماء النقلة من الفارسيّ إلى العربيّ

ابن المقفّع [...]

آل نوبخت أكثرهم [...]

موسى ويوسف أبناء خالد وكانا يخدمان داود بن عبد الله بن حميد بن قحطبة وينقلان له من الفارسيّة إلى العربيّة.

التميميّ واسمه عليّ بن زياد ويكنى أبا الحسن نقل من الفارسيّ إلى العربيّ فممّا نقل زيج الشهريار.

---

458 النفيسيّ Ibn al-Nadīm, *Kitāb al-fihrist* (ed. Tağaddud, Sayyid): التفليسيّ (ed. Flügel; in the critical apparatus Flügel notes the reading البلقيسي)

الحسن بن سهل [...]
البلاذريّ أحمد بن يحيى بن جابر [...] وكان ناقلاً من اللسان الفارسيّ إلى العربيّ.
جبلة بن سالم كاتب هشام [...] وكان ناقلاً إلى العربيّ من الفارسيّ.
إسحق بن يزيد نقل من الفارسيّ إلى العربيّ فمّا نقل كتاب سيرة الفرس المعروف بباختيار[459] نامه.
ومن نقلة الفرس محمّد بن الجهم البرمكيّ، هشام بن القاسم، موسى بن عيسى الكسرويّ،[460]
زادويه بن شاهويه الإصفهانيّ، محمّد بن بهرام بن مطيار الإصفهانيّ، بهرام بن مردان شاه موبد مدينة سابور[461] من بلد فارس، عمر بن الفرّخان [...]

نقلة الهند والنبط

منكه الهنديّ وكان في جملة إسحق بن سليمان بن عليّ الهاشميّ ينقل من اللغة الهنديّة إلى العربيّة.
ابن دهن الهنديّ وكان إليه بيمارستان البرامكة نقل إلى العربيّ من اللسان الهنديّ.
ابن وحشيّة ينقل من النبطيّة إلى العربيّة وقد نقل كتباً كثيرة على ما ذكر [...]

II.2. Ibn Abī Uṣaybiʿah, *ʿUyūn al-anbāʾ*, II/1, 507.1–517.10 (ed. Savage Smith et al.).

الباب التاسع في طبقات الأطبّاء النقلة الذين نقلوا كتب الطبّ وغيره من اللسان اليونانيّ إلى اللسان العربيّ وذكر الذين نقلوا لهم

جورجس وهو أوّل من ابتدأ في نقل الكتب الطبّيّة إلى اللسان العربيّ عندما استدعاه المنصور وكان كثير الإحسان إليه ...
حنين بن إسحاق كان عالماً باللغات الأربع غريبها ومستعملها العربيّة والسريانيّة واليونانيّة والفارسيّة ونقله في الغاية من الجودة.
إسحاق بن حنين كان أيضاً عالماً باللغات التي يعرفها أبوه وهو يلحق به في النقل وكان إسحاق عذب

459 بباختيار corrected: باختيار Ibn al-Nadīm, *Kitāb al-fihrist* (ed. Flügel): بجداد (ed. Tağaddud)
460 الكسروي corrected: الكرديّ Ibn al-Nadīm, *Kitāb al-fihrist* (ed. Flügel): الكرويّ (ed. Tağaddud)
461 سابور corrected: نيسابور Ibn al-Nadīm, *Kitāb al-fihrist* (ed. Flügel)

العبارة فصيح الكلام وكان حنين مع ذلك أكثر تصنيفاً ونقلاً [...]

حبيش الأعسم وهو ابن أخت حنين بن إسحاق وتلميذه ناقل مجوّد يلحق بحنين وإسحاق [...]

عيسى بن يحيى بن إبراهيم كان أيضاً تلميذاً لحنين بن إسحاق وكان فاضلاً أثنى عليه حنين ورضي نقله وقلّده فيه وله مصنّفات.

قسطا بن لوقا البعلبكّيّ كان ناقلاً خبيراً باللغات فاضلاً في العلوم الحكميّة وغيرها [...]

أيّوب المعروف بالأبرش كان قليل النقل متوسّطه وما نقله في آخر عمره يضاهي نقل حنين.

ماسرجيس كان ناقلاً من السريانيّ إلى العربيّ ومشهوراً بالطبّ وله من الكتب كتاب قوى الأطعمة ومنافعها ومضارّها كتاب قوى العقاقير ومنافعها ومضارّها.

عيسى بن ماسرجيس كان يلحق بأبيه وله من الكتب كتاب الألوان كتاب الروائح والطعوم.

شهدى[462] الكرخيّ من أهل الكرخ وكان قريب الحال في الترجمة.

ابن شهدى[463] الكرخيّ كان مثل أبيه في النقل ثمّ إنّه في آخر عمره فاق أباه ولم يزل متوسّطاً وكان ينقل من السريانيّ إلى العربيّ ومن نقله كتاب الأجنّة لأبقراط.

الحجّاج بن مطر نقل للمأمون ومن نقله كتاب أقليدس ثمّ أصلح نقله فيما بعد ثابت بن قرّة الحرّانيّ.

ابن ناعمة واسمه عبد المسيح بن عبد الله الحمصيّ الناعميّ كان متوسّط النقل وهو إلى الجودة أميل.

زرويا[464] بن مانحوه[465] الناعميّ الحمصيّ كان قريب النقل وما هو في درجة من قبله.

هلال بن أبي هلال الحمصيّ كان صحيح النقل ولم يكن عنده فصاحة ولا بلاغة في اللفظ.

فثيون[466] الترجمان وجدت نقله كثير اللحن ولم يكن يعرف شيئاً من علم العربيّة أصلاً.

أبو نصر بن ناريّ بن أيّوب كان قليل النقل ولم يعتدّ بنقله كغيره من النقلة.

بسيل المطران نقل كتباً كثيرة وكان نقله أميل إلى الجودة.

إصطفن بن بسيل كان يقارب حنين بن إسحاق في النقل إلّا أنّ عبارة حنين أفصح وأحلى.

موسى بن خالد الترجمان وجدت من نقله كتباً كثيرة من الستّة عشر لجالينوس وغيرها وكان لا يصل إلى درجة حنين أو يقرّب منها.

---

462 شهدى corrected according to Ibn al-Nadīm, *Kitāb al-fihrist*, I, 244.14: سَهدا Ḥunayn ibn Isḥāq, *al-Risālah*, 4.20 and 6.2

463 See previous note.

464 زرويا conjectured: زروبا Ibn al-Nadīm, *Kitāb al-fihrist*, I, 244.8

465 مانحوه: ماجوه Ibn al-Nadīm, *Kitāb al-fihrist*, I, 244.8

466 فثيون corrected according to Ibn al-Nadīm, *Kitāb al-fihrist*, I, 244.9

أسطاث كان من النقلة المتوسّطين.

خيرون بن رابطة ليس له شهرة بجودة النقل.

تذرس السنقل وجدت له نقلاً في الكتب الحكمية لا بأس به.

سرجس الرأسيّ من أهل مدينة رأس العين نقل كتباً كثيرة وكان متوسّطاً في النقل وكان حنين يصلح نقله فما وجد بإصلاح حنين فهو الجيّد وما وجد غير مصلح فهو وسط.

أيّوب الرهاويّ ليس أيّوب الأبرش المذكور أوّلاً ناقل جيّد عالم باللغات إلّا أنّه بالسريانيّة خير منه بالعربيّة.

يوسف الناقل هو أبو يعقوب بن يوسف بن عيسى المتطبّب الناقل ويلقّب بالناعس[467] وهو تلميذ عيسى بن صهر بخت وكان يوسف الناقل من خوزستان وكانت في عبارته لكنة وليس نقله بكثير الجودة.

إبراهيم بن الصلت كان متوسّطاً في النقل يلحق بسرجس الرأسيّ.

ثابت الناقل كان متوسّطاً في النقل إلّا أنّه يفضل إبراهيم بن الصلت وكان مقلّاً من النقل ومن نقله كتاب الكيموسين لجالينوس.

أبو يوسف الكاتب كان متوسّطاً في النقل ونقل عدّة كتب من كتب بقراط.

يوحنّا بن بختيشوع نقل كتباً كثيرة إلى السريانيّ فأمّا إلى العربيّ فما عرف بنقله شيء منها.

البطريق كان في أيّام المنصور وأمره بنقل أشياء من الكتب القديمة وله نقل كثير جيّد إلّا أنّه دون نقل حنين بن إسحاق وقد وجدت بنقله كتباً كثيرة في الطبّ من كتب أبقراط وجالينوس.

يحيى بن البطريق كان في جملة الحسن بن سهل وكان لا يعرف العربيّة حقّ معرفتها ولا اليونانيّة وإنّما كان لطينيّاً يعرف لغة الروم اليوم وكتابتها وهي الحروف المتّصلة لا المنفصلة اليونانيّة القديمة.

توما[468] الرهاويّ كان إذا كثرت على حنين الكتب وضاق عليه الوقت استعان به في نقلها ثمّ يصلحها بعد ذلك.

منصور بن باناس[469] طبقته في النقل مثل توما[470] الرهاويّ وكان بالسريانيّة أقوى منه بالعربيّة.

---

467 الساهر Ibn al-Qifṭī, *Taʾrīḫ al-ḥukamāʾ*, 392.1

468 توما corrected according to Ḥunayn ibn Isḥāq, *al-Risālah*, 49.3; Ibn al-Qifṭī, *Taʾrīḫ al-ḥukamāʾ*, 131.15; Ibn al-Nadīm, *Kitāb al-fihrist*, I, 291.6

469 اثاناس Ḥunayn ibn Isḥāq, *al-Risālah*, 49.7–8

470 See the note on Tūmā al-Ruhāwī above (n. 468).

عبد يسوع بن بهريز مطران الموصل كان صديقاً لجبريل بن بختيشوع وناقلاً له.

أبو عثمان سعيد بن يعقوب الدمشقيّ أحد النقلة المجيّدين وكان منقطعاً إلى عليّ بن عيسى.

أبو إسحاق إبراهيم بن بكس[471] كان من الأطبّاء المشهورين وترجم كتباً كثيرة إلى لغة العرب ونقله أيضاً مرغوب فيه.

أبو الحسن عليّ بن إبراهيم بن بكس كان أيضاً طبيباً مشهوراً وكان مثل أبيه في النقل.

فأمّا الذين كان هؤلاء النقلة ينقلون لهم خارجاً عن الخلفاء

فنهم شبرشوع[472] بن قطرب من أهل جنديسابور وكان لا يزال يبرّ النقلة ويهدي إليهم ويتقرّب إلى تحصيل الكتب منهم بما يمكنه من المال وكان يريد السريانيّ أكثر من العربيّ وهو أحد الخوز.

ومنهم محمّد بن موسى المنجّم وهو أحد بني موسى بن شاكر الحسّاب المشهورين بالفضل والعلم والتصنيف في العلوم الرياضيّة وكان محمّد هذا أبرّ الناس بحنين بن إسحاق وقد نقل له حنين كثيراً من الكتب الطبّيّة.

ومنهم عليّ بن يحيى المعروف بابن المنجّم أحد كتّاب المأمون وكان نديماً له وعنده فضل ومال إلى الطبّ فنقلوا له منه كتباً كثيرة.

ومنهم تادُرى[473] الأسقف كان أسقفاً في الكرخ ببغداد وكان حريصاً على طلب الكتب متقرّباً إلى قلوب نقلها لحصل منها شيئاً كثيراً وصنّف له قوم من الأطبّاء النصارى كتباً لها قدر وجعلوها باسمه.

ومنهم محمّد بن موسى بن عبد الملك نقلت له كتب طبّيّة وكان من جملة العلماء الفضلاء يلخّص الكتب ويعتبر جيّد الكلام فيها من رديئه.

ومنهم عيسى بن يونس الكاتب الحاسب من جملة الفضلاء بالعراق وكان كثير العناية بتحصيل الكتب القديمة والعلوم اليونانيّة.

ومنهم عليّ المعروف بالفيّوم اشتهر باسم المدينة التي كان عاملها وكانت النقلة يحصلون من جانبه ويمتارون من فضله.

471 بكوش Ibn al-Qifṭī, *Taʾrīḫ al-ḥukamāʾ*, 37.15–16, 40.18, 107.5

472 شيريشوع Ḥunayn ibn Isḥāq, *al-Risālah*, 5.2

473 ثياذورى Ḥunayn ibn Isḥāq, *al-Risālah*, 12.21

ومنهم أحمد بن محمّد المعروف بابن المدبّر الكاتب وكان يصل إلى النقلة من ماله وأفضاله شيء كثير جدّاً.

ومنهم إبراهيم بن محمّد بن موسى الكاتب وكان حريصاً على نقل كتب اليونانيّين إلى لغة العرب ومشتملاً على أهل العلم والفضل وعلى النقلة خاصّة.

ومنهم عبد الله بن إسحاق وكان أيضاً حريصاً على نقل الكتب وتحصيلها كثيرة.

ومنهم محمّد بن عبد الملك ابن الزيّات وكان يقارب عطاؤه للنقلة والعلماء والنسّاخ في كلّ شهر ألفي دينار ونقل باسمه كتب عدّة. وكان أيضاً ممّن نقلت له الكتب اليونانيّة وترجمت باسمه جماعة من أكابر الأطبّاء مثل يوحنّا بن ماسويه وجبريل بن بختيشوع وبختيشوع بن جبريل بن بختيشوع وداوّد بن سرابيون وسلمويه بن بنان واليسع وإسرائيل بن زكريّاء بن الطيفوريّ وحبيش بن الحسن.

II.3. Ibn al-Nadīm, *Kitāb al-fihrist*, I, 242.7–11 (ed. Flügel).

حكاية أخرى: كان خالد بن يزيد بن معاوية يسمّى حكيم آل مروان وكان فاضلاً في نفسه وله همّة ومحبّة للعلوم خطر بباله الصنعة فأمر بإحضار جماعة من فلاسفة اليونانيّين ممّن كان ينزل مدينة مصر وقد تفصّح بالعربيّة وأمرهم بنقل الكتب في الصنعة من اللسان اليونانيّ والقبطيّ إلى العربيّ وهذا أوّل نقل كان في الإسلام من لغة إلى لغة.

II.4. Al-Masʿūdī, *Murūǧ al-ḏahab*, V, 211.16–23, par. 3446 (ed. Pellat).

وكان أوّل خليفة قرّب المنجّمين وعمل بأحكام النجوم وكان معه نوبخت المجوسيّ المنجّم وأسلم على يديه وهو أبو هاؤلاء النوبختيّة وإبراهيم الفزاريّ المنجّم صاحب القصيدة في النجوم وغير ذلك من علوم النجوم وهيآت الفلك وعليّ بن عيسى الأسطرلابيّ.

وكان أوّل خليفة ترجمة له الكتب من اللغة العجميّة إلى العربيّة منها كتاب كليلة ودمنة وكتاب السندهند وترجمت له كتب أرسطاطاليس من المنطقيّات وغيرها وترجم له كتاب المجسطيّ لبطلميوس وكتاب أقليدس وكتاب الأرثماطيقي وسائر الكتب القديمة من اليونانيّة والروميّة والفهلويّة والفارسيّة والسريانيّة وأخرجت إلى الناس فنظروا فيها وتعلّقوا إلى علمها.

11.5. Ṣāʿid al-Andalusī, *Ṭabaqāt al-umam*, 48.10–17, 48.20–49.2 (ed. Cheikho).

ثمّ لمّا أفضت الخلافة إلى الخليفة السابع منهم عبد الله المأمون بن هارون الرشيد ابن محمّد المهدي بن أبي جعفر المنصور تمّم ما بدأ به جدّه المنصور فأقبل على طلب العلم في مواضعه واستخرجه من معادنه بفضل همّته الشريفة وقوّة نفسه الفاضلة.

فداخل ملوك الروم وأتحفهم بالهدايا الخطيرة وسألهم صلته بما لديهم من كتب الفلاسفة فبعثوا إليه بما حضرهم من كتب أفلاطون وأريسطاطاليس وأبقراط وجالينوس وأوقليدس وبطلميوس وغيرهم من الفلاسفة فاستجاد لها مهرة التراجمة وكلّفهم إحكام ترجمتها فترجمت له على غاية ما أمكن ثمّ خصّ الناس على قراءتها ورغّبهم في تعليمها.

فنفقت سوق العلم في زمانه وقامت دولة الحكمة في عصره [...] وكذلك كانت سيرته مع سائر العلماء والفقهاء والمحدّثين والمتكلّمين وأهل اللغة والأخبار والمعرفة بالشعر والنسب فأتقن جماعة من ذوي الفنون والتعلّم في أيّامه كثيراً من أجزاء الفلسفة وسنّوا لمن بعدهم منهاج الطبّ ومهّدوا أصول الأدب حتّى كادت الدولة العبّاسيّة تضاهى الدولة الروميّة أيّام اكتمالها وزمان اجتماع شملها.

11.6. Ibn al-Nadīm, *Kitāb al-fihrist*, I, 271.9–14 (ed. Flügel).

بنو موسى: محمّد وأحمد والحسن بنو موسى بن شاكر [...] وهؤلاء القوم ممّن تناهى في طلب العلوم القديمة وبذل فيها الرغائب وأتعبوا فيها نفوسهم وأنفذوا إلى بلد الروم من أخرجها إليهم فأحضروا النقلة من الأصقاع والأماكن بالبذل السنيّ فأظهروا عجائب الحكمة وكان الغالب عليهم من العلوم الهندسة والحيل والحركات والموسيقى والنجوم وهو الأقلّ.

11.7. Ibn Ǧulǧul, *Ṭabaqāt al-aṭibbāʾ*, 68.2–69.11 (ed. Fuʾād Sayyid).

حنين بن إسحاق تلميذ يوحنّا بن ماسويه عالماً بلسان العرب فصيحاً باللسان اليونانيّ جدّاً بارعاً في اللسانين بلاغة بلغ بها تمييز علل اللسانين ونهض من بغداد إلى أرض فارس وكان خليل بن أحمد النحويّ رحمه الله بأرض فارس فلزمه حنين حتّى برع في لسان العرب وأدخل كتاب العين بغداد.

ثمّ اختير للترجمة وائتمن عليها وكان المتخيّر لها جعفر المتوكّل على الله ووضع له كتّاباً نحارير عالمين بالترجمة كانوا يترجمون ويتصفّح حنين ما ترجموا كأسطفن بن بسيل وحبيش وموسى ابن أبي خالد الترجمان ويحيى بن هارون.

وخدم حنين بالطبّ المتوكّل على الله وحظي في أيّامه وكان يلبس زنّاراً وتعلّم ويعلم اللسان اليونانيّة بإسكندريّة وكان جليلاً في ترجمة وهو الذي أوضح معاني كتب بقراط وجالينوس ولخّصها أحسن تلخيص وكشف ما استغلق منها وأوضح مشكلها وله تواليف نافعة متقنة بارعة وعمد إلى كتب جالينوس فاحتذى فيها حذو الإسكندرانيّين فصنعها على سبيل المسألة والجواب فأحسن في ذلك.

11.8. Ibn Abī Uṣaybiʿah, *ʿUyūn al-anbāʾ*, 11/1, 465.2–467.14 (ed. Savage-Smith et al.).

قال يوسف بن إبراهيم: أوّل ما حصل لحنين بن إسحاق من الاجتهاد والعناية في صناعة الطبّ هو أنّ مجلس يوحنّا بن ماسويه كان من أعمر مجلس يكون في التصدّي لتعليم صناعة الطبّ وكان يجتمع فيه أصناف أهل الأدب. قال يوسف: وذلك أنّي كنت أعهد حنين بن إسحاق الترجمان يقرأ على يوحنّا بن ماسويه كتاب فرق الطبّ المرسوم باللسان الروميّ والسريانيّ بهراسيو وكان حنين إذ ذاك صاحب سؤال وذلك يصعب على يوحنّا وكان يباعده أيضاً من قبله أنّ حنيناً كان من أبناء الصيارفة من أهل الحيرة وأهل جنديسابور خاصّة ومتطبّبوها يحرفون عن أهل الحيرة ويكرهون أن يدخل في صناعتهم أبناء التجّار فسأله حنين في بعض الأيّام عن بعض ما كان يقرأ عليه مسألة مستفهم لما يقرأ فحرد يوحنّا وقال: ما لأهل الحيرة وتعلّم صناعة الطبّ! صر إلى فلان قرابتك حتّى يهب لك خمسين درهماً تشتري منها قفافاً صغاراً بدرهم وزرنيخاً بثلاثة دراهم واشتر بالباقي فلوساً كوفيّة وقادسيّة وزرنخ القادسيّة في تلك القفاف واقعد على الطريق وصحّ الفلوس الجياد للصدقة والنفقة وبع الفلوس فإنّه أعود عليك من هذه الصناعة! ثمّ أمر به فأخرج من داره فخرج حنين باكياً مكروباً. وغاب عنّا حنين فلم نره سنتين.

[...]

واعتلّ إسحاق بن الخصيّ علّة فأتيته عائداً فإنّي لفي منزله إذ بصرت بإنسان له شعرة قد جلّلته وقد ستر وجهه عنّي ببعضها وهو يتردّد وينشد شعراً بالروميّة لأوميرس رئيس شعراء الروم فشبهت نغمته بنغمة حنين وكان العهد بحنين قبل ذلك الوقت بأكثر من سنتين فقلت لإسحاق بن الخصيّ: هذا حنين! فأنكر ذلك إنكاراً يشبّه الإقرار فهتفت بحنين فاستجاب لي وقال: ذكر ابن رسالة الفاعلة أنّه من المحال أن يتعلّم الطبّ عباديّ وهو بريء من دين النصرانيّة إنّه رضي أن يتعلّم الطبّ حتّى يحكم اللسان اليونانيّ إحكاماً لا يكون في دهره من يحكمه إحكامه وما اطّلع عليّ أحد غير أخي هذا ولو علمت أنّك تفهمني لاستترت عنك لكنّي عملت على أنّ حليتي قد تغيّرت في عينك وأنا أسألك أن تستر أمري.

فبقيت أكثر من ثلاث سنين وإنّي لأظنّها أربعاً لم أره ثمّ إنّي دخلت يوماً على جبريل بن بختيشوع وقد انحدر من معسكر المأمون قبل وفاته بمدّة يسيرة فوجدت عنده حنيناً وقد ترجم له أقساماً قسمها بعض الروم في كتاب من كتب جالينوس في التشريح وهو يخاطبه بالتبجيل ويقول له يا ربّن حنين وتفسير ربّن المعلّم فأعظمت ما رأيت وتبيّن ذلك جبريل فيّ فقال لي: لا تستكثرن ما ترى من تبجيلي هذا الفتى فوالله لئن مدّ له في العمر ليفضحن سرجس وسرجس هذا الذي ذكره جبريل هو الرأس عينيّ وهو أوّل من نقل شيئاً من علوم الروم إلى اللسان السريانيّ وليفضحن غيره من المترجمين وخرج من عنده حنين وأقمت طويلاً ثمّ خرجت فوجدت حنيناً ببابه ينتظر خروجي فسلّم عليّ وقال لي: قد كنت سألتك ستر خبري والآن فأنا أسألك إظهاره وإظهار ما سمعت من أبي عيسى وقوله فيّ فقلت له: أنا مسودّ وجه يوحنّا بما سمعت من مدح أبي عيسى لك فأخرج من كمّه نسخة ما كان دفعه إلى جبريل وقال لي: تمام سواد وجه يوحنّا يكون بدفعك إليه هذه النسخة وسترك عنه علم من نقلها فإذا رأيته قد اشتدّ عجبه بها أعلمه أنّها إخراجي. ففعلت ذلك من يومي وقبل انتهائي إلى منزلي.

فلمّا قرأ يوحنّا تلك الفصول وهي التي يسمّيها اليونانيّون الفاعلات كثر تعجّبه وقال: أترى المسيح أوحى في دهرنا هذا إلى أحد فقلت له في جواب قوله: ما أوحى المسيح في هذا الدهر ولا في غيره إلى أحد ولا كان المسيح إلّا أحد من يوحى إليه فقال لي: دعني من هذا القول ليس هذا الإخراج إلّا إخراج مؤيد بروح القدس فقلت له: هذا إخراج حنين بن إسحاق الذي طردته من منزلك وأمرته أن يشتري فلوساً فحلف بأنّ ما قلت له محال ثمّ صدق القول بعد ذلك وسألني التلطّف لإصلاح ما بينهما ففعلت ذلك وأفضل عليه إفضالاً كثيراً وأحسن إليه ولم يزل مبجّلاً له حتّى فارقت العراق في سنة خمس وعشرين ومائتين.

11.9. Ibn Abī Uṣaybiʿah, *ʿUyūn al-anbāʾ*, II/1, 472.2–16 (ed. Savage-Smith et al.).

أقول: وكان لحنين ولدان داؤد وإسحاق واشتغلا بصناعة الطبّ وصنّف لهما كتباً طبّيّة للمبادئ للتعليم ونقل لهما كتباً كثيرة من كتب جالينوس فأمّا داؤد فإنّي لم أجد له شهرة بنفسه بين الأطبّاء ولا يوجد له من الكتب ما يدلّ على براعته وعلمه وإن كان الذي يوجد له إنّما هو كنّاش واحد وأمّا إسحاق فإنّه اشتهر وتميّز في صناعة الطبّ وله تصانيف كثيرة ونقل إسحاق من الكتب اليونانيّة إلى اللغة العربيّة كتباً كثيرة إلّا أنّ جلّ عنايته كانت مصروفة إلى نقل الكتب الحكميّة مثل كتب أرسطوطاليس وغيره من الحكماء.

وأمّا حنين أبوه فكان ملتهماً بنقل الكتب الطبّيّة وخصوصاً كتب جالينوس حتّى أنّه في غالب الأمر لا يوجد شيء من كتب جالينوس إلّا وهي بنقل حنين أو بإصلاحه لما نقل غيره فإن رؤي شيء منها وقد تفرّد بنقله غيره من النقلة مثل أسطاث وابن بكس والبطريق وأبي سعيد عثمان الدمشقيّ وغيرهم فإنّه لا يعتنى به ولا يرغب فيه كما يكون بنقل حنين وإصلاحه وإنّما ذلك لفصاحته وبلاغته ولمعرفته أيضاً بآراء جالينوس في كتبه ولتمهّره فيها.

ووجدت بعض الكتب الستّة عشر لجالينوس وقد نقلها من الروميّة إلى السريانيّة سرجس المتطبّب ونقلها من السريانيّة إلى العربيّة موسى بن خالد الترجمان فلمّا طالعتها وتأمّلت ألفاظها تبيّن لي بين نقلها وبين الستّة عشر التي هي نقل حنين تباين كثير وتفاوت بيّن وأين الألكن من البليغ والثرى من الثريّا.

II.10. Ibn al-Nadīm, *Kitāb al-fihrist*, I, 289.15–18 (ed. Flügel).

قال محمّد بن إسحق: من سعادات حنين أنّ ما نقله حبيش بن الحسن الأعسم وعيسى ابن يحيى وغيرهما إلى العربيّ يُنحل إلى حنين وإذا رجعنا إلى فهرست كتب جالينوس الذي عمله حنين إلى عليّ بن يحيى علمنا أنّ الذي نقل حنين أكثره إلى السريانيّ وربّما أصلح العربيّ من نقل غيره أو تصفّحه.

II.11. Ibn Abī Uṣaybiʿah, *ʿUyūn al-anbāʾ*, II/1, 478.4–479.7 (ed. Savage-Smith et al.).

فأقول كيف لا أبغض ويكثر حاسدي ويكثر ثلبي في مجالس ذي المراتب وتبذل في قتلي الأموال ويعزّ من شتمني ويهان من أكرمني كلّ ذلك بغير جرم لي إلى واحد منهم ولا جناية لكنّهم لمّا رأوني فوقهم وعالياً عليهم بالعلم والعمل ونقلي إليهم العلوم الفاخرة من اللغات التي لا يحسنوها ولا يهتدون إليها ولا يعرفون شيئاً منها في نهاية ما يكون من حسن العبارة والفصاحة ولا نقص فيها ولا زلل ولا ميل لأحد من الملل ولا استغلاق ولا لحن باعتبار أصحاب البلاغة من العرب الذين يقومون بمعرفة وجوه النحو والغريب ولا يعثرون على سيّئة ولا شكلة ولا معنى لكنّ بأعذب ما يكون من اللفظ وأقربه إلى الفهم يسمعه من ليس صناعته الطبّ ولا يعرف شيئاً من طرقات الفلسفة ولا من ينتحل ديانة النصرانيّة وكلّ الملل يستحسنه ويعرف قدره حتّى إنّهم قد يغرمون عليّ ما كان من الذي أنقل الأموال الكثيرة إذ كانوا يفضلون هذا النقل على نقل كلّ من قبلي وأيضاً فأقول ولا أخطئ إنّ سائر

أهل الأدب وإن اختلفت مللهم محبّون لي مائلون إليّ مكرمون لي يأخذون ما أفيدهم بشكر ويجازوني بكلّ ما يصلون إليه من الجميل.

فأمّا هؤلاء الأطبّاء النصارى الذين أكثرهم تعلّموا بين يدي ونشأوا قدامي هم الذين يرومون سفك دمي على أنّهم لا بدّ لهم منّي فمرّة يقولون: من هو حنين؟ إنّما حنين ناقل لهذه الكتب ليأخذ على نقله الأجرة كما يأخذ الصنّاع الأجرة على صناعتهم ولا فرق عندنا بينه وبينهم لأنّ الفارس قد يعمل له الحدّاد السيف في المثل بدينار ويأخذ هو من أجله كلّ شهر مائة دينار فهو خادم لأداتنا وليس هو عامل بها كما أنّ الحدّاد وإن كان يحسن صنعة السيف إلّا أنّه ليس يحسن يعمل به فما للحدّاد وطلب الفروسيّة كذلك هذا الناقل ما له وللكلام في صناعة الطبّ ولم يحكم في عللها وأمراضها وإنّما قصده في ذلك تشبيه بنا ليقال حنين الطبيب ولا يقال حنين الناقل والأجود له لو أنّه لزم صناعته وأمسك عن ذكر صناعتنا لقد كان يكون أجدى عليه فيما كنّا سنوصله إليه من أموالنا ونحسن إليه ما أمكننا وذلك يتمّ له بترك أخذ المجسّ والنظر في قوارير الماء ووصف الأدوية ويقولون: إنّ حنيناً ما يدخل إلى موضع من دور الخاصّة والعامّة إلّا يهزؤون به ويتضاحكون منه عند خروجه.

III. *Methods: Procedures and Approaches*

III.1. Ḥunayn ibn Isḥāq, *al-Risālah*, 2.12–3.2/Ar., preface (ed. Bergsträßer).

فكان من جوابك في ذلك أن قلت إنّه وإن كان الأمر على ما وصفت فإنّ بنا وسائر أهل هذا الغرض ممّن يقرأ الكتب بالسريانيّة والعربيّة حاجة إلى أن نعلم ما ترجم من هذه الكتب إلى اللسان السريانيّ والعربيّ وما لم يترجم وما كنت أنا المتولّي لترجمته دون غيري وما تولّى ترجمته غيري وما سبقني إلى ترجمته غيري ثمّ عدت فيه فترجمته أو أصلحته ومن تولّى ترجمة كتاب كتاب من الكتب التي تولّى ترجمتها غيري ومبلغ قوّة كلّ واحد من أولائك المترجمين في الترجمة ولمن ترجمت ومن الذين ترجمت أنا لهم كلّ واحد من تلك الكتب التي تولّيت ترجمتها وفي أيّ حدّ من سنّي ترجمته لأنّ هذين أمرين قد يحتاج إلى معرفتهما إذ كانت الترجمة إنّما تكون بحسب قوّة المترجم للكتاب والذي ترجم له وأيّ تلك الكتب ممّا لم يترجم إلى هذه الغاية وجدت نسخته باليونانيّة وأيّها لم توجد له نسخة أو وجد البعض منه فإنّ هذا أمر يحتاج إليه ليعنى بترجمة ما قد وجد منها وبطلب ما لم يوجد.

III.2. Ḥunayn ibn Isḥāq, *al-Risālah*, 47.12–48.6/Ar., no. 115 (ed. Bergsträßer).

ولم يقع إلى هذه الغاية إلى أحد من أهل دهرنا لكتاب البرهان نسخة تامّة باليونانيّة على أنّ جبريل قد كان عني بطلبه عناية شديدة وطلبته أنا غاية الطلب وجلت في طلبه بلاد الجزيرة والشام كلّها وفلسطين ومصر إلى أن بلغت الإسكندريّة فلم أجد منه شيئاً إلّا بدمشق نحواً من نصفه إلّا أنّها مقالات غير متوالية ولا تامّة.

وقد كان جبريل أيضاً وجد منه مقالات ليست كلّها المقالات التي وجدت بأعيانها وترجم له أيّوب ما وجد وأمّا أنا فلم تطب نفسي بترجمة شيء منها إلّا باستكمال قراءتها لما هي عليه من النقصان والاختلال وللطمع وتشوّق النفس إلى وجود تمام هذا الكتاب.

ثمّ إنّي ترجمت ما وجدت إلى السريانيّة وهو جزء يسير من المقالة الثانية وأكثر المقالة الثالثة ونحو نصف المقالة الرابعة من أوّلها والمقالة التاسعة خلا شيئاً من أوّلها فإنّه سقط وأمّا سائر المقالات الأخر فوجدت إلى آخر الكتاب خلا المقالة الخامسة عشرة فإنّ في آخرها نقصاناً.

III.3. Ḥunayn ibn Isḥāq, *al-Risālah*, 4.19–5.9/Ar., no. 3 (ed. Bergsträßer).

وقد كان ترجمه قبلي إلى السريانيّ رجل يقال له ابن شهدا من أهل الكرخ وكان ضعيفاً في | الترجمة ثمّ إنّي ترجمته وأنا حدث من أبناء عشرين سنة أو أكثر قليلاً لمتطبّب من أهل جندي سابور يقال له شيريشوع بن قطرب من نسخة يونانيّة كثيرة الأسقاط ثمّ سألني بعد ذلك وأنا من أبناء أربعين سنة أو نحوها حبيش تلميذي إصلاحه بعد أن كانت قد اجتمعت له عندي عدّة نسخ يونانيّة فقابلت تلك بعضها ببعض حتّى صحّت منها نسخة واحدة ثمّ قابلت بتلك النسخة السريانيّ وصحّحته وكذلك من عادتي أن أفعل في جميع ما أترجمه ثمّ ترجمته من بعد سنيّات إلى العربيّة لأبي جعفر محمّد بن موسى.

III.4. Ḥunayn ibn Isḥāq, *al-Risālah*, 17.18–18.17/Ar., no. 20 (ed. Bergsträßer).

وقد كان سلمويه أذأرني على أن أصلح له هذا الجزء الثاني وطمع أن يكون ذلك أسهل من الترجمة وأجود فقابلني ببعض المقالة السابعة ومعه السريانيّ ومعي اليونانيّ وهو يقرأ على السريانيّة وكنت كلّما مرّ بي شيء مخالف لليونانيّ خبّرته به فجعل يصلح حتّى | كبر عليه الأمر وتبيّن له أنّ الترجمة من الرأس أرخى وأبلغ وأنّ الأمر يكون فيها أشدّ انتظاماً فسألني ترجمة تلك المقالات فترجمتها عن آخرها وكنّا بالرقّة في أيّام غزوات المأمون ودفعها إلى زكريّا بن عبد الله المعروف بالطيفوريّ لمّا أراد الانحدار

إلى مدينة السلام لتنسخ له هناك فوقع حريق في السفينة التي كان فيها زكريّا فاحترق الكتاب ولم يبق له نسخة.

ثمّ إنّي بعد سنين ترجمت الكتاب من أوّله لبختيشوع بن جبريل وكانت عندي للثماني المقالات الأخيرة منه عدّة نسخ باليونانيّة فقابلت بها وصحّحت منها نسخة وترجمتها بغاية ما أمكنني من الاستقصاء والبلاغة فأمّا الستّ المقالات الأول فلم أكن وقعت لها إلّا على نسخة واحدة وكانت مع ذلك نسخة كثيرة الخطأ فلم يمكنّي لذلك تخلّص تلك المقالات على غاية ما ينبغي ثمّ إنّي وقعت على نسخة أخرى فقابلت بها وأصلحت ما أمكنني إصلاحه وأخلو إلى أنّى أقابل به ثالثة إن اتّفقت لي نسخة ثالثة فإنّ نسخ هذا الكتاب باليونانيّة قليلة وذلك أنّه لم يكن ممّا يقرأ في كتّاب الإسكندريّة وترجم هذا الكتاب من النسخ السريانيّة التي ترجمتها حبيش بن الحسن لمحمّد بن موسى ثمّ إنّه سألني بعد ترجمته لها أن أتّصفح له المقالات الثماني الأخيرة وأصلح ما وجدت من الأسقاط فأجبته إلى ذلك وأجدت فيه.

III.5. Ibn Abī Uṣaybiʿah, *ʿUyūn al-anbāʾ*, II/1, 488.8–16 (ed. Savage-Smith et al.).

وصرت أنقل لهم الكتب على الرسم بغير عوض ولا جزاء وأسارع إلى جميع محابّهم بعد أن كنت إذا نقلت لأحدهم كتاباً أخذت منه وزنه دراهم.

أقول: وجدت من هذه الكتب كتباً كثيرة وكثيراً منها اقتنيته وهي مكتوبة مولّد الكوفيّ بخطّ الأزرق كاتب حنين وهي حروف كبار بخطّ غليظ في أسطر متفرّقة وورقها كلّ ورقة منها بغلظ ما يكون من هذه الأوراق المصنوعة يومئذ نحو ثلاث ورقات أو أربع وذلك في تقطيع مثل ثلث البغداديّ وكان قصد حنين بذلك تعظيم حجم الكتاب وتكثير وزنه لأجل ما يقابل به من وزنه دراهم وكان ذلك الورق يستعمله بالقصد ولا جرم أنّ لغلظه بقي هذه السنين المتطاولة من الزمان.

III.6. Ibn al-Nadīm, *Kitāb al-fihrist*, I, 22.7–15 (ed. Flügel).

قال أحمد بن عبد الله بن سلّام: ترجمت صدر هذا الكتاب والصحف والتوراة والإنجيل وكتب الأنبياء والتلامذة من لغة العبرانيّة واليونانيّة والصابئة وهي لغة أهل كلّ كتاب إلى لغة العربيّة حرفاً حرفاً ولم أبتغ في ذلك تحسين لفظ ولا تزيينه مخافة التحريف ولم أزد على ما وجدته في الكتاب الذي نقلته ولم أنقص إلّا أن يكون في بعض ذلك من الكلام ما هو متقدّم بلغة أهل ذلك الكتاب فلا يستقيم لفظه في النقل إلى العربيّة إلّا أن يؤخّر ومنه ما هو مؤخّر لا يستقيم إلّا أن يقدّم ليستقيم ذلك بالعربيّة

وهو مثل قول من يقول آب ميار فإنّ ترجمته بالعربيّة ماء هات فأخّرت الماء وقدّمت هات وكذلك اللغات فيما يستقيم إذا نقل إلى العربيّة وأعوذ بالله أن أزيد في ذلك أو أنقص منه إلّا على هذا الوجه الذي ذكرته وبيّنته في هذا الكتاب.

III.7. Al-Ṣafadī, *Kitāb al-ġayṯ al-musaǧǧam*, 46.12–25 (Cairo ed. 1305 AH).

وللترجمة في النقل طريقان أحدهما طريق يوحنّا بن البطريق وابن الناعمة الحمصيّ وغيرهما وهو أن ينظر إلى كلّ كلمة مفردة من الكلمات اليونانيّة وما تدلّ عليه من المعنى فيأتي بلفظة مفردة من الكلمات العربيّة ترادفها في الدلالة على ذلك المعنى فيثبتها وينتقل إلى الأخرى كذلك حتّى يأتي على جملة ما يريد تعريبه وهذه الطريقة رديئة لوجهين أحدهما أنّه لا يوجد في الكلمات العربيّة كلمات تقابل جميع الكلمات اليونانيّة ولهذا وقع في خلال هذا التعريب كثير من الألفاظ اليونانيّة على حالها الثاني أنّ خواصّ التركيب والنسب الإسناديّة لا تطابق نظيرها من لغة أخرى دائماً وأيضاً يقع الخلل من جهة استعمال المجازات وهي كثيرة في جميع اللغات.

الطريق الثاني في التعريب طريق حنين ابن إسحاق والجوهريّ وغيرهما وهو أن يأتي إلى الجملة فيحصل معناها في ذهنه ويعبّر عنها من اللغة الأخرى بجملة تطابقها سواء ساوت الألفاظ أم خالفتها وهذه الطريق أجود ولهذا لم تحتج كتب حنين بن إسحاق إلى تهذيب إلّا في العلوم الرياضيّة لأنّه لم يكن قيّماً بها بخلاف كتب الطبّ والمنطق والطبيعيّ والإلاهيّ فإنّ الذي عرّبه منها لم يحتج إلى إصلاح فأمّا أوقليدس فقد هذّبه ثابت بن قرّة الحرّانيّ وكذلك المجسطيّ والمتوسّطات بينهما.

IV. *Problems: Difficulties and Criticisms*

IV.1. Al-Bīrūnī, *Taḥqīq mā li-l-hind*, I, 76.6–10 (ed. Sachau).

ولهم فنون من العلم أخر كثيرة وكتب لا تكاد تحصى ولكنّي لم أحط بها علماً وبودّي أن كنت أتمكّن من ترجمة كتاب پنچ تنتر وهو المعروف عندنا بكتاب كليلة ودمنة فإنّه تردّد بين الفارسيّة والهنديّة ثمّ العربيّة والفارسيّة على ألسنة قوم لا يؤمن تغييرهم إيّاه كعبد الله بن المقفّع في زيادته باب برزويه فيه قاصداً تشكيك ضعفي القوائد في الدين وكسرهم للدعوة إلى مذهب المنانيّة وإذا كان متّهماً فيما زاد لم يخل عن مثله فيما نقل.

IV.2. Al-Ǧāḥiẓ, *Kitāb al-bayān wa-l-tabyīn*, III, 29.9–12 (ed. Hārūn).

ونحن لا نستطيع أن نعلم أنّ الرسائل التي بأيدي الناس للفرس أنّها صحيحة غير مصنوعة وقديمة غير مولّدة إذ كان مثل ابن المقفّع وسهل بن هارون وأبي عبيد الله وعبد الحميد وغَيلان يستطيعون أن يولّدوا مثل تلك الرسائل ويصنعوا مثل تلك السير.

IV.3. Ibn al-Qifṭī, *Taʾrīḫ al-ḥukamāʾ*, 29.17–30.19 (ed. Lippert).

قال غير ابن إسحق فراسل المأمون ملك الروم وكان قد استطال عليه وأذلّ دين الكفر وطلب منه كتب الحكمة من كلام أرسطوطاليس فطلبها ملك الروم فلم يجد لها ببلاده أثراً فاغتمّ بذلك وقال يطلب منّي ملك المسلمين علم سلفي من يونان فلا أجده أيّ عذر يكون لي أم أيّ قيمة تبقى لهذه الفرقة الروميّة عند المسلمين وأخذ في السؤال والبحث.

فحضر إليه أحد الرهبان المنقطعين في بعض الأديرة النازحة عن القسطنطينيّة وقال له عندي علم ما تريد فقال له أدركني فقال إنّ البيت الفلانيّ في موضع كذا الذي يقفل كلّ ملك عليه قفلاً إذا ملّك ما فيه قال فيه على ما يقال مال الملوك المتقدّمين وكلّ ملك يجيء يقفل عليه حتّى لا يقال قد احتاج إلى ما فيه لسوء تدبيره ففتحه فقال له الراهب ليس الأمر كذلك وإنّما في ذلك الموضع هيكل كانت يونان تتعبّد به قبل استقرار ملّة المسيح فلمّا تقرّرت ملّته بهذه الجهات في أيّام قسطنطين بن ألانة جمعت كتب الحكمة من أيدي الناس وجعلت في ذلك البيت وأغلق بابه وقفّل الملوك عليه إقفالاً كما سمعت.

فجمع الملك مقدّمي دولته وعرّفهم الأمر واستشارهم في فتح البيت فأشاروا بذلك فاستشار الراهب في تسييرها إذا وجدت إلى بلد الإسلام وهل عليه في ذلك خطر في الدنيا أم إثم في الأخرى فقال له الراهب سيّرها فإنّك تثاب عليه فإنّها ما دخلت في ملّة إلّا وزلزلت قواعدها.

فسار إلى البيت وفتحه ووجد الأمر فيه كما ذكر الراهب ووجدوا فيه كتباً كثيرة فأخذوا من جانبها بغير علم ولا فحص خمسة أحمال وسيّرت إلى المأمون فأحضر لها المأمون المترجمين فاستخرجوها من الروميّة إلى العربيّة ثمّ تنبّه الناس بعد ذلك على تطلّبها بعد المأمون وتحيّلوا إلى أن حصّلوا منها الجملة الكثيرة.

ولمّا سيّرت الكتب إلى المأمون جاء بعضها تامّاً وبعضها ناقصاً فالناقص منها ناقص إلى اليوم لم يجد أحد تمامه.

IV.4. Al-Maqdisī, *Muḫtaṣar al-ḥuǧǧah*, 661.6–663.13 (ed. Hārūn).

فأوّل الحوادث التي أحدثوها إخراج كتب اليونانيّة إلى أرض الإسلام وترجمت بالعربيّة وشاعت في أيدي المسلمين. وسبب خروجها من أرض الروم إلى بلد الإسلام يحيى بن خالد بن برمك وذلك أنّ كتب اليونانيّة كانت ببلد الروم وكان ملك الروم خاف ⟨على الروم⟩ إن نظروا في كتب اليونانية تركوا دين النصرانيّة ويرجعون إلى اليونانيّة فتتشتّت كلمتهم وتتفرّق جماعتهم فجمع الكتب في موضع وبنى عليها بناء مطمساً بالحجارة والجصّ حتّى لا يوصل إليها.

فلمّا أفضت رئاسة دولة بني العبّاس إلى يحيى بن خالد وكان زنديقاً بلغه خبر الكتب التي في المبنى ببلد الروم فصانع ملك الروم الذي كان في وقته بالهدايا ولا يلتمس منه حاجة فلمّا أكثر عليه جمع الملك بطارقته وقال لهم: إنّ هذا الرجل خادم العربيّ قد أكثر عليّ من هداياه ولا يطلبني حاجة وما أراه يلتمس إلّا حاجة وأخاف أن تكون حاجة تشقّ عليّ وقد شغل بالي. فلمّا جاءه رسول يحيى قال له: قل لصاحبك إن كانت له ⟨حاجة⟩ فاليذكرها. فلمّا أخبر الرسول يحيى ردّه إليه وقال له: حاجتي الكتب التي تحت البناء يرسلها إليّ أخرج فيها بعض ما أحتاج وأردّها إليه.

فلمّا قرأ الروميّ كتابه استطار فرحاً وجمع البطارقة والأساقفة والرهبان وقال لهم: قد كنت ذكرت لكم عن خادم العربيّ أنّه لا يخلو من حاجة وقد أفضح بحاجته وهي أخفّ الحوائج عليّ وقد رأيت رأياً فاسمعوه فإن رضيتموه أمضيه وإن رأيتم خلافه تشاورنا في ذلك حتّى تتّفق كلمتنا. فقالوا: وما هو؟ قال: حاجته الكتب اليونانيّة يستخرج منها ما أحبّ ويردّها. قالوا: فما رأيك؟ قال: قد علمت أنّه ما بنى عليها من كان قبلنا إلّا أنّه خاف إن وقعت في أيدي النصارى وقرؤوها كان سبباً لهلاك دينهم وتبديل جماعتهم وأنا أرى أن أبعث بها إليها وأسأله أن لا يردّها يبتلون بها ونسلم من شرّها فإنّي لا آمن أن يكون من بعدي من يجترئ على إخراجها إلى الناس فيقعوا فيما خيف عليهم. فقالوا: نعم الرأي ما رأيت أيّها الملك فأمضه!

فبعث بالكتب إلى يحيى بن خالد فلمّا وصلت إليه جمع عليها كلّ زنديق وفيلسوف فممّا أخرج منها كتاب حدّ المنطق.

IV.5a. Al-Ǧāḥiẓ, *Kitāb al-ḥayawān*, I, 74.14–75.4 (ed. Hārūn).

وفضيلة الشعر مقصورة على العرب وعلى من تكلّم بلسان العرب والشعر لا يستطاع أن يترجم ولا يجوز عليه النقل ومتى حوّل تقطّع نظمه وبطل وزنه وذهب حسنه وسقط موضع التعجّب لا كالكلام المنثور والكلام المنثور المبتدأ على ذلك أحسن وأوقع من المنثور الذي تحوّل من موزون الشعر.

IV.5b. Al-Ǧāḥiẓ, *Kitāb al-ḥayawān*, I, 75.10–16 (ed. Hārūn)

وقد نقلت كتب الهند وترجمت حكم اليونانيّة وحوّلت آداب الفرس فبعضها ازداد حسناً وبعضها ما انتقص شيئاً ولو حوّلت حكمة العرب لبطل ذلك المعجز الذي هو الوزن مع أنّهم لو حوّلوها لم يجدوا في معانيها شيئاً لم تذكره العجم في كتبهم التي وضعت لمعاشهم وفطنهم وحكمهم. وقد نقلت هذه الكتب من أمّة إلى أمّة ومن قرن إلى قرن ومن لسان إلى لسان حتّى انتهت إلينا وكنّا آخر من ورثها ونظر فيها. فقد صحّ أنّ الكتب أبلغ في تقييد المآثر من البنيان والشعر.

IV.5c. Al-Ǧāḥiẓ, *Kitāb al-ḥayawān*, I, 75.18–77.4 (ed. Hārūn).

ثمّ قال بعض من ينصر الشعر ويحوطه ويحتجّ له إنّ الترجمان لا يؤدّي أبداً ما قال الحكيم على خصائص معانيه وحقائق مذاهبه ودقائق اختصاراته وخفيّات حدوده ولا يقدر أن يوفيها حقوقها ويؤدّي الأمانة فيها ويقوم بما يلزم الوكيل ويجب على الجريّ وكيف يقدر على أدائها وتسليم معانيها والإخبار عنها على حقّها وصدقها إلّا أن يكون في العلم بمعانيها واستعمال تصاريف ألفاظها وتأويلات مخارجها ومثل مؤلّف الكتاب وواضعه فمتى كان رحمه الله تعالى ابن البطريق وابن ناعمة وابن قرّة وابن فهريز[474] وثيفيل وابن وهيلى وابن المقفّع مثل أرسطاطاليس ومتى كان خالد مثل أفلاطون؟

ولا بدّ للترجمان من أن يكون بيانه في نفس الترجمة في وزن علمه في نفس المعرفة وينبغي أن يكون أعلم الناس باللغة المنقولة والمنقول إليها حتّى يكون فيهما سواء وغاية ومتى وجدناه أيضاً قد تكلّم بلسانين علمنا أنّه قد أدخل الضيم عليهما لأنّ كلّ واحدة من اللغتين تجذب الأخرى وتأخذ منها وتعترض عليها وكيف يكون تمكّن اللسان منهما مجتمعين فيه كتمكّنه إذا انفرد بالواحدة وإنّما له قوّة واحدة

474 بهريز: Ibn al-Nadīm, *Kitāb al-fihrist*; Ibn Abī Uṣaybiʿah, *ʿUyūn al-anbāʾ*.

فإن تكلّم بلغة واحدة استفرغت تلك القوّة عليهما وكذلك إن تكلّم بأكثر من لغتين وعلى حساب ذلك تكون الترجمة لجميع اللغات.

وكلّما كان الباب من العلم أعسر وأضيق والعلماء به أقلّ كان أشدّ على المترجم وأجدر أن يخطئ فيه ولن تجد البتّة مترجماً يفي بواحد من هؤلاء العلماء.

IV.5d. Al-Ǧāḥiẓ, *Kitāb al-ḥayawān*, I, 77.6–11, 78.2–5 (ed. Hārūn).

هذا قولنا في كتب الهندسة والتنجيم والحساب واللحون فكيف لو كانت هذه الكتب كتب دين وإخبار عن الله عزّ وجلّ بما يجوز عليه ممّا لا يجوز عليه حتّى يريد أن يتكلّم على تصحيح المعاني في الطبائع ويكون ذلك معقوداً بالتوحيد ويتكلّم في وجوه الإخبار واحتمالاته للوجوه ويكون ذلك متضمّناً بما يجوز على الله تعالى ممّا لا يجوز وبما يجوز على الناس ممّا لا يجوز [...] والذي ذكرنا قليل من كثير ومتى لم يعرف ذلك المترجم أخطأ في تأويل كلام الدين والخطأ في الدين أضرّ من الخطأ في الرياضة والصناعة والفلسفة والكيمياء وفي بعض المعيشة التي يعيش بها بنو آدم.

IV.5e. Al-Ǧāḥiẓ, *Kitāb al-ḥayawān*, I, 78.6–14 (ed. Hārūn).

وإذا كان المترجم الذي قد ترجم لا يكمل لذلك أخطأ على قدر نقصانه من الكمال وما علم المترجم بالدليل عن شبه الدليل وما علمه بالأخبار النجوميّة وما علمه بالحدود الخفيّة وما علمه بإصلاح سقطات الكلام وأسقاط الناسخين للكتاب وما علمه ببعض الخطرفة لبعض المقدّمات؟ وقد علمنا أنّ المقدّمات لا بدّ أنّ تكون اضطراريّة ولا أن تكون مرتّبة وكالخيط الممدود. وابن البطريق وابن قرّة لا يفهمان هذا موصوفاً منزّلاً ومرتّباً مفصّلاً من معلّم رفيق ومن حاذق طبّ فكيف بكتاب قد تداولته اللغات واختلاف الأقلام وأجناس خطوط الملل والأمم؟

IV.5f. Al-Ǧāḥiẓ, *Kitāb al-ḥayawān*, I, 78.18–79.3 (ed. Hārūn).

ثمّ يصير إلى ما يعرض من الآفات لأصناف الناسخين وذلك أنّ نسخته لا يعدمها الخطأ ثمّ ينسخ له من تلك النسخة من يزيده من الخطأ الذي يجده في النسخة ثمّ لا ينقص منه ثمّ يعارض بذلك من يترك ذلك المقدار من الخطأ على حاله إذا كان ليس من طاقته إصلاح السقط الذي لا يجده في نسخته.

IV.5g. Al-Ǧāḥiẓ, *Kitāb al-ḥayawān*, VI, 280.3–11 (ed. Hārūn).

وزعمتم أنّكم وجدتم ذكر الشهب في كتب القدماء من الفلاسفة وأنّه في الآثار العلويّة لأرسطاطاليس حين ذكر القول في الشهب مع القول في الكواكب ذوات الذوائب ومع القول في القوس والطوق الذي يكون حول القمر بالليل فإن كنتم بمثل هذا تستعينون وإليه تفزعون فإنّا نوجدكم من كذب التراجمة وزيادتهم ومن فساد الكتّاب من جهة تأويل الكلام ومن جهة جهل المترجم بنقل لغة إلى لغة ومن جهة فساد النسخ ومن أنّه قد تقادم فاعترضت دونه الدهور والأحقاب فصار لا يؤمن عليه ضروب التبديل والفساد وهذا الكلام معروف صحيح.

IV.6. Al-Bīrūnī, *Kitāb al-ṣaydanah*, 14.8–15.2/Ar. (ed. Meyerhof).

قد حظيت في غريزتي منذ حداثتي بفرط الحرص على اقتناء المعارف بحسب السنّ والحال ويكفي شاهداً عليه أنّ رومياً حلّ أرضنا فكنت أجيء بالحبوب والبزور والثمار والنبات وغيرها وأسأله عن أسمائها بلغته وأحرّرها ولكنّ للكتابة العربيّة آفة عظيمة هي تشابه صور الحروف المزدوجة فيها واضطرارها في التمايز إلى نقط العجم وعلامات الإعراب التي إذا تركت استبهم المفهوم منها فإذا انضاف إليه إغفال المعارضة وإهمال التصحيح بالمقابلة وذلك من الفعل عامّ قومنا يساوي به وجود الكتّاب وعدمه بل علم ما فيه وجهله.

لولا هذه الآفة لكفى نقل ما في كتب ديسقوريذس وجالينوس وبولس وأوريباسيوس المنقولة إلى العربيّ من الأسامي اليونانيّة إلّا أنّا لا نثق بها ولا نأمن التغايير في نسخها وللتراجمة فيها خيانة أخرى هو ترك بعض ما يوجد في أرضنا من العقاقير وفي لغة العرب اسم لها على حاله باليونانيّة حتّى يُحوِج بعد الترجمة إلى تفسير كالكرفس الجبليّ والجزر البرّيّ والزرشك ولحية التيس وأمثالها فإنّهم لم ينقلوها إلى العربيّة كما لم ينقلوا أسماء كتب المنطق من المدخل والمقولات والعبارة والقياس والبرهان فتضاعف البغض والبرودة فيها من جانب الخصوم.

IV.7. Al-Fārābī, *Kitāb al-ḥurūf*, 112.1–9, 112.20, 114.13–115.1 (ed. Mahdi).

وليس في العربيّة منذ أوّل وضعها لفظة تقوم مقام هست في الفارسيّة ولا مقام استين في اليونانيّة ولا مقام نظائر هاتين اللفظتين في سائر الألسنة وهذه يحتاج إليها ضرورة في العلوم النظريّة وفي صناعة المنطق. فلمّا انتقلت الفلسفة إلى العرب واحتاجت الفلاسفة الذين يتكلّمون بالعربيّة ويجعلون عبارتهم

عن المعاني التي في الفلسفة وفي المنطق بلسان العرب ولم يجدوا في لغة العرب منذ أوّل ما وضعت لفظة ينقلوا بها الأمكنة التي تستعمل فيها استين في اليونانيّة وهست بالفارسيّة فيجعلوها تقوم مقام هذه الألفاظ في الأمكنة التي يستعملها فيها سائر الأمم فبعضهم رأى أن يستعمل لفظة هو مكان هست بالفارسيّة واستين باليونانيّة. [...] ورأى آخرون أن يستعملوا مكان تلك الألفاظ بدل الهو لفظة الموجود. [...]

ولأنّ هذه اللفظة بحيث ما هي عربيّة وبنيتها عندهم هذه البنية صارت مغلطة جدّاً رأى قوم أن يتجنّبوا استعمالها واستعملوا مكانها قولنا هو ومكان الوجود الهويّة ولأنّ لفظة هو ليست باسم ولا كلمة في العربيّة ولذلك لا يمكن فيها أن نعمل منها مصدراً أصلاً وكان يحتاج في الدلالة على هذه المعاني التي يلتمس أن يدلّ عليها في العلوم النظريّة إلى اسم وكان يحتاج إلى أن يعمل منه مثل الرجل والرجوليّة والإنسان والإنسانيّة رأى قوم أن يتجنّبوها ويستعملوا الموجود مكان هو والوجود مكان الهويّة. وأمّا أنا فإنّي أرى أنّ الإنسان له أن يستعمل أيّهما شاء.

IV.8a. Al-Tawḥīdī, *Kitāb al-imtāʿ wa-l-muʾānasah*, I, 111.11–112.6 (ed. Amīn and al-Zayn).

قال: أنت إذاً لست تدعونا إلى علم المنطق إنّما تدعو إلى تعلّم اللغة اليونانيّة وأنت لا تعرف لغة يونان؟ فكيف صرت تدعونا إلى لغة لا تفي بها وقد عفت منذ زمان طويل وباد أهلها وانقرض القوم الذين كانوا يتفاوضون بها ويتفاهمون أغراضهم بتصاريفها؟ على أنّك تنقل من السريانيّة فما تقول في معانٍ متحوّلة بالنقل من لغة يونان إلى لغة أخرى سريانيّة ثمّ من هذه إلى أخرى عربيّة؟

قال متّى: يونان وإن بادت مع لغتها فإنّ الترجمة حفظت الأغراض وأدّت المعاني وأخلصت الحقائق.

قال أبو سعيد: إذا سلّمنا لك أنّ الترجمة صدقت وما كذبت وقوّمت وما حرّفت ووزنت وما جزفت وأنّها ⟨ما⟩ التاثت ولا حافت ولا نقصت ولا زادت ولا قدّمت ولا أخّرت ولا أخلّت بمعنى الخاصّ والعامّ ولا ⟨بأخصّ الخاصّ ولا⟩ بأعمّ العامّ—وإن كان هذا لا يكون وليس هو في طبائع اللغات ولا في مقادير المعاني—فكأنّك تقول: لا حجّة إلّا عقول يونان ولا برهان إلّا ما وضعوه ولا حقيقة إلّا ما أبرزوه.

IV.8b. Al-Tawḥīdī, *Kitāb al-imtāʿ wa-l-muʾānasah*, I, 115.7–15 (ed. Amīn and al-Zayn).

(قال أبو سعيد:) وإذا لم يكن لك بدّ من قليل هذه اللغة من أجل الترجمة فلا بدّ لك أيضاً من كثيرها من أجل تحقيق الترجمة واجتلاب الثقة والتوقّي من الخلّة اللاحقة.

فقال متّى: يكفيني من لغتكم هذه الاسم والفعل والحرف فإنّي أتبلّغ بهذا القدر إلى أغراض قد هذّبتها لي يونان.

قال: أخطأت لأنّك في هذا الاسم والفعل والحرف فقير إلى وصفها وبنائها على الترتيب الواقع في غرائز أهلها وكذلك أنت محتاج بعد هذا إلى حركات هذه الأسماء والأفعال والحروف فإنّ الخطأ والتحريف في الحركات كالخطأ والفساد في المتحرّكات.

IV.8c. Al-Tawḥīdī, *Kitāb al-imtāʿ wa-l-muʾānasah*, I, 115.15–116.9 (ed. Amīn and al-Zayn).

(قال أبو سعيد:) وهذا باب ⟨أنت وأصحابك ورهطك عنه في غفلة على أنّ هاهنا سرّاً ما علق⟩ بك ولا أسفر لعقلك وهو أن تعلم أنّ لغة من اللغات لا تطابق لغة أخرى من جميع جهاتها بحدود صفاتها في أسمائها وأفعالها وحروفها وتأليفها وتقديمها وتأخيرها واستعارتها وتحقيقها وتشديدها وتخفيفها وسعتها وضيقها ونظمها ونثرها وسجعها ووزنها وميلها وغير ذلك ممّا يطول ذكره وما أظنّ أحداً يدفع هذا الحكم أو يشكّ في صوابه ممّن يرجع إلى مسكة من عقل أو نُصَيب من إنصاف فمن أين يجب أن تثق بشيء ترجم لك على هذا الوصف؟ بل أنت إلى تعرّف اللغة العربيّة أحوج منك إلى تعرّف المعاني اليونانيّة على أنّ المعاني لا تكون يونانيّة ولا هنديّة كما أنّ اللغات تكون فارسيّة وعربيّة وتركيّة ومع هذا فإنّك تزعم أنّ المعاني حاصلة بالعقل والفحص والفكر فلم يبق إلّا أحكام اللغة فلِمَ تزري على العربيّة وأنت تشرح كتب أرسطوطاليس بها مع جهلك بحقيقتها؟

IV.8d. Al-Tawḥīdī, *Kitāb al-imtāʿ wa-l-muʾānasah*, I, 121.6–10 (ed. Amīn and al-Zayn).

(قال أبو سعيد:) وكلّ ذلك محصور بالتتبّع والرواية والسماع والقياس المطّرد على الأصل المعروف من غير تحريف وإنّما دخل العجب على المنطقيّين لظنّهم أنّ المعاني لا تعرف ولا تستوضح إلّا بطريقهم ونظرهم وتكلّفهم فترجموا لغة هم فيها ضعفاء ناقصون ⟨بترجمة أخرى هم فيها ضعفاء ناقصون⟩ وجعلوا تلك الترجمة صناعة وادّعوا على النحويين أنّهم مع اللفظ لا مع المعنى.

IV.9. Al-Bīrūnī, *Kitāb al-ṣaydanah*, 13.2–18/Ar. (ed. Meyerhof).

وإلى لسان العرب نقلت العلوم من أقطار العالم فازدانت وحلت في الأفئدة وسرت محاسن اللغة منها في الشرايين والأوردة وإن كانت كلّ أمّة تستحلي لغتها التي ألفتها واعتادتها واستعملتها في مآربها مع آلّافها وأشكالها وأقيس هذا بنفسي وهي مطبوعة على لغة لو خُلِد به علم لاستغرب استغراب البعير على الميزاب والزرافة في العراب ثمّ منتقلة إلى العربيّة والفارسيّة فأنا في كلّ واحدة دخيل ولها متكلّف والهجو بالعربيّة أحبّ إليّ من المدح بالفارسيّة وسيعرف مصداق قولي من تأمّل كتاب علم قد نقل إلى الفارسيّ كيف ذهب رونقه وكسف باله واسودّ وجهه وزال الانتفاع به إذ لا تصلح هذه اللغة إلّا للأخبار الكسرويّة والأسمار الليليّة.

V. ***Why Do We Translate?***

V.1. Al-Kindī, *Fī l-falsafah al-ūlā*, I, 103.1–7 (ed. Abū Rīdah).

فأمّا أرسطوطالس مبرّز اليونانيّين في الفلسفة فقال: ينبغي لنا أن نشكر آباء الذين أتوا بشيء من الحقّ إذ كانوا سبب كونهم فضلاً عنهم إذ هم سبب لهم وإذ هم سبب لنا إلى نيل الحقّ فما أحسن ما قال في ذلك.

وينبغي لنا أن لا نستحي من استحسان الحقّ واقتناء الحقّ من أين أتى وإن أتى من الأجناس القاصية عنّا والأمم المباينة ⟨لنا⟩ فإنّه لا شيء أولى بطالب الحقّ من الحقّ وليس بخس الحقّ ولا تصغير بقائله ولا بالآتي به ولا أحد بخس بالحقّ بل كلّ يشرفه الحقّ.

Bibliography

Primary Sources

ʿAbd al-Ǧabbār, *Taṯbīt dalāʾil al-nubuwwah*, ed. ʿAbd al-Karīm ʿUṯmān, Beirut: Dār al-ʿarabīyah li-l-ṭibāʿah wa-l-našr, 1966

ʿAbd al-Laṭīf al-Baġdādī, *Kitāb al-ifādah wa-l-iʿtibār fī l-umūr al-mušāhadah wa-l-ḥawādiṯ al-muʿāyanah bi-arḍ miṣr*, Paris: Imprimerie impériale, 1810

Abū Hilāl al-ʿAskarī, *Dīwān al-maʿānī*, ed. Muḥammad Abduh and Muḥammad Maḥmūd al-Tarkazī al-Šinqīṭī, 2 vols, Cairo: Maktabat al-qudsī, 1933

Apollonius, *Conics Books V to VII: The Arabic Translation of the Lost Greek Original in the Version of the Banū Mūsā*, ed. and tr. Gerald J. Toomer, New York: Springer, 1990

Aristotle, *Kitāb al-āṯār al-ʿulwīyah = Aristotle's Meteorology in the Arabico-Latin Tradition*, ed. Pieter L. Schoonheim, Leiden: Brill, 2000

Aristotle, *Manṭiq Arisṭū*, ed. ʿAbd al-Raḥmān Badawī, 3 vols, Cairo: Maṭbaʿat Dār al-kutub al-miṣrīyah, 1948–1952

Aristotle, *Les Catégories d'Aristote dans leurs versions syro-arabes*, ed. Khalil Georr, Beirut: Institut français de Damas, 1948

Badawī, ʿAbd al-Raḥmān, *al-Uṣūl al-yūnānīyah li-l-naẓarīyāt al-sīyāsīyah fī l-islām*, Cairo: Maṭbaʿat Dār al-kutub al-miṣrīyah, 1954

al-Balāḏurī, *Kitāb futūḥ al-buldān*, ed. Michael J. de Goeje, Leiden: Brill, 1866

al-Balāḏurī, *The Origins of the Islamic State, Being a Translation from the Arabic, Accompanied with Annotations, Geographic and Historic Notes of the Kitâb futûh al-buldân of al-Imâm abu-l Abbâs Ahmad ibn-Jâbir al-Balâdhuri*, tr. Philip K. Hitti, 2 vols, New York: Columbia University, 1916

Bar Hebraeus, *Taʾrīḫ muḫtaṣar al-duwal*, ed. Anṭūn Ṣāliḥānī, 2nd edn, Al-Ḥāzimīyah: Dār al-rāʾid al-lubnānī, 1994

al-Bīrūnī, *Taḥqīq mā li-l-Hind*, ed. Eduard Sachau, 2 vols, London: Trübner & Co., 1887

al-Bīrūnī, *India = Alberuni's India: An Account of the Religion, Philosophy, Literature, Geography, Chronology, Astronomy, Customs, Laws and Astrology of India about A.D. 1030*, tr. Eduard Sachau, 2 vols, London: Kegan Paul, Trench, Trübner & Co., 1910

al-Bīrūnī, *Chronologie orientalischer Völker* [*al-Āṯār al-bāqiyah ʿan al-qurūn al-ḫāliyah*], ed. Eduard Sachau, Leipzig: F.A. Brockhaus, repr. 1923

al-Bīrūnī, *Kitāb al-ṣaydanah* = Meyerhof, Max, 'Das Vorwort zur Drogenkunde des Bērūni', *Quellen und Studien zur Geschichte der Naturwissenschaften und der Medizin*, 3 (1932), 157–208

al-Ḏahabī, *Sīrat aʿlām al-nubalāʾ*, ed. Šuʿayb Arnāʾūṭ et al., 25 vols, Beirut: Muʾassasat al-risālah, 1985

al-Fārābī, *Kitāb al-ḥurūf = Alfarabi's Book of Letters (Kitāb al-Ḥurūf): Commentary on Aristotle's Metaphysics*, ed. Muhsin Mahdi, Beirut: Dār al-mašriq, 1969

al-Fārābī, *The Book of Particles*, tr. Muhsin Mahdī, unpublished

Forster, Regula, *Das Geheimnis der Geheimnisse: Die arabischen und deutschen Fassungen des pseudo-aristotelischen Sirr al-asrār, Secretum secretorum*, Wiesbaden: Reichert, 2006

al-Ǧāḥiẓ, 'Ḥuǧaǧ al-nubuwwah', in *Rasāʾil al-Ǧāḥiẓ*, ed. ʿAbd al-Salām Muḥammad Hārūn, 4 vols, Cairo: Maktabat al-ḫānǧī, 1964–1979, III, 223–281

al-Ǧāḥiẓ, 'al-Radd ʿalā l-naṣārā', in *Rasāʾil al-Ǧāḥiẓ*, ed. ʿAbd al-Salām Muḥammad Hārūn, 4 vols, Cairo: Maktabat al-ḫānǧī, 1964–1979, III, 301–351

al-Ǧāḥiẓ, *Kitāb al-ḥayawān*, ed. ʿAbd al-Salām Muḥammad Hārūn, 8 vols, Cairo: Maktabat al-ḫānǧī, 1965–1969

al-Ǧāḥiẓ, *al-Bayān wa-l-tabyīn*, ed. ʿAbd al-Salām Muḥammad Hārūn, 4 vols, Cairo: Maktabat al-ḫānǧī, 1998

Ḥamzah al-Iṣfahānī, *Taʾrīḫ sinī mulūk al-arḍ wa-l-anbiyāʾ*, ed. Yūsuf Yaʿqūb Miskūnī, Beirut: Dār maktabat al-ḥayāt, 1961

Ḥunayn ibn Isḥāq, *al-Risālah* = Gotthelf Bergsträßer, *Ḥunain ibn Isḥāq über die syrischen und arabischen Galen-Übersetzungen*, Abhandlungen für die Kunde des Morgenlandes 17/2, Leipzig: F.A. Brockhaus, 1925

Ḥunayn ibn Isḥāq, *Ḥunayn ibn Isḥāq on His Galen Translations*, ed. and tr. John C. Lamoreaux, Provo UT: Brigham Young University Press, 2016

Ibn al-Abbār, *al-Takmilah li-kitāb al-ṣilah*, ed. ʿAbd al-Salām al-Harās, 4 vols, Beirut: Dār al-fikr, 1995

Ibn Abī Uṣaybiʿah, *ʿUyūn al-anbāʾ fī ṭabaqāt al-aṭibbāʾ* = *A Literary History of Medicine: The* ʿUyūn al-anbāʾ fī ṭabaqāt al-aṭibbāʾ *of Ibn Abī Uṣaybiʿah*, ed. and tr. Emilie Savage-Smith et al., 5 vols, Leiden/Boston: Brill, 2019

Ibn al-ʿAdīm, *Buġyat al-ṭalab*, ed. Suhayl Zakkār, 12 vols, Beirut: Dār al-fikr, 1988

Ibn Ǧulǧul, *Ṭabaqāt al-aṭibbāʾ wa-l-ḥukamāʾ*, ed. Ayman Fuʾād Sayyid, Beirut: Muʾassasat al-risālah, 1985

Ibn Ǧulǧul, *Libro de la explicación* = *Ibn Ŷulŷul, Libro de la explicación de los nombres de los medicamentos simples tomados del libro de Dioscórides*, ed. and tr. Ildefonso Garijo, Córdoba: Universidad de Córdoba, 1992

Ibn Ḥaǧar al-ʿAsqalānī, *Tahḏīb al-tahḏīb*, ed. Ibrāhīm Zaybaq and ʿĀdil Muršid, 4 vols, Beirut: Muʾassasat al-risālah, 1996

Ibn Ḫallikān, *Wafayāt al-aʿyān*, ed. Iḥsān ʿAbbās, 8 vols, Beirut: Dār ṣādir, 1972

Ibn al-Nadīm, *Kitāb al-fihrist*, ed. Gustav Flügel, 2 vols, Leipzig: Verlag von F.C.W. Vogel, 1870

Ibn al-Nadīm, *The Fihrist of al-Nadīm*, tr. Bayard Dodge, 2 vols, New York/London: Columbia University Press, 1970

Ibn al-Nadīm, *Kitāb al-fihrist li-l-Nadīm*, ed. Riḍā Taǧaddud, Tehran: Maṭbaʿat-e dānešgāh-e Tehrān, 1971

Ibn al-Nadīm, *al-Fihrist li-l-Nadīm*, ed. Ayman Fuʾād Sayyid, 2 vols in 4, London: Muʾassasat al-furqān li-l-turāṯ al-islāmī, 2009

Ibn Nubātah, *Sarḥ al-ʿuyūn fī šarḥ risālat Ibn Zaydūn*, ed. M. Abū l-Faḍl Ibrāhīm, Cairo: Dār al-fikr al-ʿarabī, 1964

Ibn al-Qifṭī, *Taʾrīḫ al-ḥukamāʾ*, ed. Julius Lippert, Leipzig: Dieterich'sche Verlagsbuchhandlung, 1903

Ibn Qutaybah, *Adab al-kātib* = *Ibn Kutaiba's Adab-al-kâtib*, ed. Max Grünert, Leiden: Brill, 1900

Ibn Qutaybah, *Muškil* = *Taʾwīl muškil al-Qurʾān*, ed. Sayyid Aḥmad Ṣaqr, Cairo: Dār al-turāṯ, 1973

al-Kindī, ʻFī l-falsafah al-ūlāʼ, in *Rasāʾil al-Kindī al-falsafiyah*, ed. Muḥammad ʿAbd al-Hādī Abū Rīdah, 2 vols, Cairo: Dār al-fikr al-ʿarabī, 1950–1954, I, 81–162

al-Kindī, *First Philosophy* = *Al-Kindi's Metaphysics: A Translation of Yaʿqūb ibn Isḥāq al-*

Kindī's Treatise "On First Philosophy" (fī al-Falsafah al-Ūlā), tr. Alfred L. Ivry, Albany NY: State University of New York Press, 1974
al-Maqdisī, *Muḫtaṣar al-ḥuǧǧah ʿalā tārik al-maḥaǧǧah*, ed. Muḥammad Ibrāhīm Muḥammad Hārūn, Riyadh: Dār aḍwāʾ al-salaf, 2005
al-Maqrīzī, *Kitāb al-mawāʿiẓ wa-l-iʿtibār fī ḏikr al-ḫiṭaṭ wa-l-āṯār*, ed. Ayman Fouad Sayyid, London: Furqān Foundation, 2013
al-Masʿūdī, *Kitāb al-tanbīh wa-l-išrāf*, ed. Michael J. de Goeje, Leiden: Brill, 1894
al-Masʿūdī, *Le Livre de l'avertissement et de la revision*, tr. Bernard Carra de Vaux, Paris: Imprimerie nationale, 1896
al-Masʿūdī, *Les Prairies d'Or*, tr. Charles Pellat, 5 vols, Paris: Geuthner, 1962–1997
al-Masʿūdī, *Murūǧ al-ḏahab wa-maʿādin al-ǧawhar*, ed. Charles Pellat, 6 vols, Beirut: al-Ǧāmiʿah al-lubnānīyah, 1966–1979
Orosius, *Kitāb Hurūsiyūs* (*Traducción árabe de las Historiae adversus paganos de Orosio*), ed. Mayte Penelas, Madrid: Universidad Autónoma, 1998
al-Rāzī, *al-Kitāb al-ḥāwī fī l-ṭibb*, 25 vols, Hyderabad: Maṭbaʿat maǧlis dāʾirat al-maʿārif al-ʿuṯmānīyah, 1955–1971
al-Ṣafadī, *Kitāb al-ġayṯ al-musaǧǧam fī šarḥ Lāmīyat al-ʿaǧam*, Cairo: al-Maṭbaʿah al-azharīyah al-miṣrīyah, 1305 AH
Ṣāʿid al-Andalusī, *Kitāb ṭabaqāt al-umam*, ed. Louis Cheikho, Beirut: Imprimerie catholique, 1912
al-Ṭabarī, ʿAlī ibn Rabban, *Firdaws al-ḥikmah fī l-ṭibb li-Abī l-Ḥasan ʿAlī ibn Sahl Rabban al-Ṭabarī*, ed. Muḥammad Zubayr al-Ṣiddīqī, Berlin: Aftāb, 1928
al-Ṭabarī, Muḥammad ibn Ǧarīr, *Annales quos scripsit Abu Djafar Mohammed Ibn Djarir At-Tabari*, ed. Michael J. de Goeje, 15 vols, Leiden: E.J. Brill, 1879–1901
al-Ṭabarī, Muḥammad ibn Ǧarīr, *Taʾrīḫ al-rusul wa-l-mulūk*, ed. Muḥammad Abū l-Faḍl Ibrāhīm, 10 vols, 2nd edn, Cairo: Dār al-maʿārif, 1969
al-Tawḥīdī, *Kitāb al-imtāʿ wa-l-muʾānasah*, ed. Aḥmad Amīn and Aḥmad al-Zayn, 3 vols, Cairo: Laǧnat al-taʾlīf wa-l-tarǧamah wa-l-našr, 1939–1942
al-Tawḥīdī, *al-Muqābasāt*, ed. Ḥasan al-Sandūbī, Kuwait: Dār siʿād al-ṣabāḥ, 1992
Yāqūt al-Ḥamawī, *Muʿǧam al-udabāʾ*, ed. Iḥsān ʿAbbās, 7 vols, Beirut: Dār al-ġarb al-islāmī, 1993

Secondary Literature

Abed, Shukri B., *Aristotelian Logic and the Arabic Language in Alfārābī*, Albany NY: State University of New York Press, 1991
Adamson, Peter, 'Al-Kindī and the Reception of Greek Philosophy', in *The Cambridge Companion to Arabic Philosophy*, ed. Peter Adamson and Richard C. Taylor, Cambridge: Cambridge University Press, 2005, 32–51
Aydin, Sami, *Sergius of Reshaina: Introduction to Aristotle and His Categories Addressed to Philotheos*, Leiden: Brill, 2016

Balty-Guesdon, Marie-Geneviève, 'Le Bayt al-ḥikma de Baghdad', *Arabica*, 39 (1992), 131–150

Baum, Wilhelm, 'Gabriel ibn Bukhtishu', in *Biographisch-bibliographisches Kirchenlexikon XXII*, ed. Friedrich Wilhelm and Traugott Bautz, Nordhausen: Bautz, 2003, cols 125–128

Bergsträßer, Gotthelf, *Neue Materialien zu Ḥunain ibn Isḥāq's Galen-Bibliographie*, Abhandlungen für die Kunde des Morgenlandes 19/2, Leipzig: F.A. Brockhaus, 1932

Berti, Vittorio, *L'Au-delà de l'âme et l'en-deça du corps: Approches d'anthropologie chrétiennes de la mort dans l'église syro-orientale*, Fribourg: Academic Press, 2015

Brentjes, Sonja, 'Textzeugen und Hypothesen zum arabischen Euklid in der Überlieferung von al-Ḥaǧǧāǧ b. Yūsuf b. Maṭar (zwischen 786 und 833)', *Archive for History of Exact Sciences*, 47 (1994), 53–92

Brock, Sebastian, 'Aspects of Translation Technique in Antiquity', *Greek, Roman and Byzantine Studies*, 20 (1979), 69–87

Brock, Sebastian, 'Syriac into Greek at Mar Saba', in *The Sabaite Heritage in the Orthodox Church*, ed. Joseph Patrich, Leuven: Peeters, 2001, 201–208

Bruning, Jelle, 'Yūsuf b. Ibrāhīm b. al-Dāyā (3rd/9th c.): A Study into his Life and Works', *Journal of Semitic Studies*, 57 (2012), 97–120

Cabo González, Ana María, 'Ibn al-Hayṯam, 'Abd al-Raḥmān', in *Biblioteca de al-Andalus*, III: *De Ibn al-Dabbaq a Ibn Kurz*, Almería: Fundación Ibn Tufayl, 2004, 327

Cooperson, Michael, 'The Purported Autobiography of Ḥunayn ibn Isḥāq', *Edebiyât*, 7 (1997), 235–249

Daiber, Hans, 'The Meteorology of Theophrastus in Syriac and Arabic Translation', in *Theophrastus: His Psychological, Doxographical, and Scientific Writings*, ed. William W. Fortenbaugh, New Brunswick NY/Oxford: Transaction Publishers, 1992, 166–293

Dunlop, Douglas M., 'The Translations of al-Biṭrīq and Yaḥyā (Yuḥannā) b. al-Biṭrīq', *Journal of the Royal Asiatic Society*, 3–4 (1959), 140–150

EI2 = *The Encyclopaedia of Islam: New Edition*, edited by H.A.R. Gibb et al., 13 vols, Leiden: E.J. Brill, 1954–2009

EI Three = Encyclopaedia of Islam, Three, ed. Kate Fleet et al., Leiden: Brill, 2007–

Emmerick, Ronald E., 'Rapigupta's Siddhasāra in Arabic', in *Studien zur Geschichte und Kultur des vorderen Orients*, ed. Hans Robert Roemer and Albrecht Noth, Leiden: Brill, 1981, 28–32

Encyclopaedia Iranica, ed. Ehsan Yarshater, London: Routledge & Paul, 1982–

Encyclopaedia Judaica, ed. Michael Berenbaum et al., 2nd edn, Detroit: Thomson Gale, 2007–

Encyclopedia of Arabic Literature, ed. Julie Scott Meisami and Paul Starkey, 2 vols, London/New York: Routledge, 1998

Enderwitz, Susanne, *Gesellschaftlicher Rang und ethnische Legitimation: Der arabische*

Schriftsteller Abū ʿUṯmān al-Ǧāḥiẓ (gest. 868) über die Afrikaner, Perser und Araber in der islamischen Gesellschaft, Freiburg: Klaus Schwarz Verlag, 1979

Endress, Gerhard, *Proclus Arabus: Zwanzig Abschnitte aus der Institutio Theologica in arabischer Übersetzung*, Beirut: Orient-Institut der Deutschen Morgenländischen Gesellschaft, 1973

Endress, Gerhard, *The Works of Yaḥyā ibn ʿAdī: An Analytical Inventory*, Wiesbaden: L. Reichert, 1977

Endress, Gerhard, 'Grammatik und Logik: Arabische Philologie und griechische Philosophie im Widerstreit', in *Sprachphilosophie in Antike und Mittelalter*, ed. Burkhard Mojsisch, Amsterdam: B.R. Grüner, 1986, 163–299

Endress, Gerhard, 'Die griechisch-arabischen Übersetzungen und die Sprache der arabischen Wissenschaften', in *Symposium Graeco-Arabicum II*, ed. Gerhard Endress and Marika Schmeink, Amsterdam: B.R. Grüner, 1989, 103–145

Endress, Gerhard, 'Die wissenschaftliche Literatur', in *Grundriß der arabischen Philologie. Band III: Supplement*, ed. Wolfdietrich Fischer, Wiesbaden: Dr. Ludwig Reichert Verlag, 1992, 3–152

Endress, Gerhard, 'The Circle of al-Kindī: Early Arabic Translations and the Rise of Islamic Philosophy', in *The Ancient Tradition in Christian and Islamic Hellenism: Studies in the Transmission of Greek Philosophy and Sciences*, ed. Gerhard Endress and Remke Kruk, Leiden: Research School CNWS, 1997, 43–76

Frank, Richard M., 'The Jeremias of Pethion ibn Ayyûb al-Sahhâr', *The Catholic Biblical Quarterly*, 21 (1959), 136–170

Garofalo, Ivan, 'La traduzione araba della Methodus medendi, libri I–VI', *Galenos*, 10 (2016), 73–106

GEDSH = *Gorgias Encyclopedic Dictionary of the Syriac Heritage*, ed. Sebastian P. Brock et al., Piscataway NJ: Gorgias Press, 2011

Genequand, Charles, *Alexander of Aphrodisias on the Cosmos*, Leiden: Brill, 2001

Goulet, Richard and Aouad, Maroun, 'Alexandre d'Aphrodisias', in *Dictionnaire des philosophes antiques*, I: *Abam(m)on à Axiothéa*, ed. Richard Goulet, Paris: Éditions du Centre National de la Recherche Scientifique, 1989, 125–139; id., 'Alexandre d'Aphrodisias', in *Dictionnaire des philosophes antiques: Supplément*, 2003, 61–70

Graf, Georg, *Geschichte der christlichen arabischen Literatur*, 5 vols, Città del Vaticano: Biblioteca Apostolica Vaticana, 1944–1953

Griffith, Sidney H., 'Sergios of Rešʿaina', in *The Oxford Dictionary of Byzantium*, ed. Aleksandr P. Kazhdan, 3 vols, Oxford: Oxford University Press: 2011, III, 1879–1888

Gutas, Dimitri, *Greek Thought, Arabic Culture: The Graeco-Arabic Translation Movement in Baghdad and Early ʿAbbāsid Society (2nd–4th/8th–10th centuries)*, London/New York: Routledge, 1998

Gutas, Dimitri, 'The "Alexandria to Baghdad" Complex of Narratives. A Contribution to the Study of Philosophical and Medical Historiography among the Arabs', *Documenti e studi sulla tradizione filosofica medievale*, 10 (1999), 155–193

Gutas, Dimitri, 'Scholars as Transmitters of Philosophical Thought', in *Philosophy in the Islamic World. Volume 1: 8th–10th Centuries*, ed. Ulrich Rudolph, Rotraud Hansberger, and Peter Adamson, Leiden/Boston MA: Brill, 2007, 680–726

Gutas, Dimitri, 'A New "Edition" of Ḥunayn's *Risāla*', *Arabic Sciences and Philosophy*, 28 (2018), 279–284

Hämeen-Anttila, Jaakko, *The Last Pagans of Iraq: Ibn Waḥshiyya and his Nabatean Agriculture*, Leiden: Brill, 2006

Hugonnard-Roche, Henri, 'Un manuscrit savant, mémoire de quatre siècles de philologie: le Parisinus ar. 2346', *Revue des mondes musulmans et de la Méditerranée*, 99–100 (2002), 147–155

Ivry, Alfred L., 'Al-Kindī's First Philosophy and Aristotle's Metaphysics', in *Essays on Arabic Philosophy and Science*, ed. George F. Hourani, Albany NY: State University of New York Press, 1975, 15–24

Justi, Ferdinand, *Iranisches Namenbuch*, Marburg: N.G. Elwert'sche Verlagsbuchhandlung, 1895

Kahl, Oliver, *The Sanskrit, Syriac and Persian Sources in the Comprehensive Book of Rhazes*, Leiden: Brill, 2015

Kapp, Albert Georg, 'Arabische Übersetzer und Kommentatoren Euklids, sowie deren math.-naturwiss. Werke auf Grund des Ta'rīkh al-Ḥukamā' des Ibn al-Qifṭī [part 1]', *Isis*, 22 (1934–1935), 150–172

Käs, Fabian, 'Eine neue Handschrift von Ḥunain ibn Isḥāqs Galenbibliographie', *Zeitschrift für Geschichte der arabisch-islamischen Wissenschaften*, 19 (2010–2011), 135–193

Kennedy, Hugh, *The Caliphate*, London: Pelican, 2016

King, David A., 'Astronomy', in *Religion, Learning and Science in the 'Abbāsid Period*, ed. M.J.L. Young, John D. Latham, and Robert B. Serjeant, Cambridge: Cambridge University Press, 1990, 274–289

Kraus, Paul, 'Zu Ibn al-Muqaffaʿ', *Rivista degli studi orientali*, 14 (1934), 1–20

Kruk, Remke, *The Arabic Version of Aristotle's Parts of Animals: Book XI–XIV of the Kitāb al-Ḥayawān*, Verhandelingen der Koninklijke Nederlandse Akademie van Wetenschappen, Afd. Letterkunde, n.s., 97, Amsterdam/Oxford: North-Holland Publishing Company, 1979

Kruk, Remke, 'La zoologie aristotélicienne. Tradition arabe', in *Dictionnaire des philosophes antiques: Supplément*, ed. Richard Goulet, Paris: Éditions du Centre National de la Recherche Scientifique, 2003, 329–334

Kühn, Wilfried, 'Die Rehabilitierung der Sprache durch den arabischen Philologen as-Sirafi', in *Sprachphilosophie in Antike und Mittelalter*, ed. Burkhard Mojsisch, Amsterdam: B.R. Grüner, 1986, 301–402

Levi Della Vida, Giorgio, 'La traduzione araba delle storie di Orosio', *Al-Andalus*, 19 (1954), 257–293

Mahdi, Muhsin, 'Language and Logic in Classical Islam', in *Logic in Classical Islamic Culture*, ed. Gustave E. von Grunebaum, Wiesbaden: Otto Harrassowitz, 1970, 51–83

Margoliouth, David S., 'The Discussion between Abū Bishr Mattā and Abu Saʿīd al Sīrāfī on the Merits of Logic and Grammar', *Journal of the Royal Asiatic Society*, 1 (1905), 79–129

Mattock, John N., 'The early translations from Greek into Arabic: an experiment in comparative assessment', in *Symposium Graeco-Arabicum II*, ed. Gerhard Endress and Marita Schmeink, Amsterdam: B.R. Grüner, 1989, 73–102

Meyerhof, Max, 'New Light on Ḥunain Ibn Isḥâq and his Period', *Isis*, 8 (1926), 685–724

Meyerhof, Max, 'Das Vorwort zur Drogenkunde des Bērūni', *Quellen und Studien zur Geschichte der Naturwissenschaften und der Medizin*, 3 (1932), 157–208

Meyerhof, Max, 'Die Materia Medica des Dioskurides bei den Arabern', *Quellen und Studien zur Geschichte der Naturwissenschaften und Medizin*, 3 (1933), 72–84

Meyerhof, Max, 'Mediaeval Jewish Physicians in the Near East, from Arabic Sources', *Isis*, 28 (1938), 432–460

Micheau, Françoise, 'Mécènes et médecins à Baghdad au IIIe/IXe siècle. Les commanditaires des traductions de Galien par Ḥunayn ibn Isḥāq', in *Les Voies de la science grecque: Études sur la transmission des textes de l'antiquité au dix-neuvième siècle*, ed. Danielle Jacquart, Geneva: Librairie Droz, 1997, 147–179

Montgomery, James E., 'al-Jāḥiẓ', in *Arabic Literary Heritage*, ed. Shawkat Toorawa and Michael Cooperson, Detroit MI: Layman, Brucoli & Clark, 2005, 231–242

Montgomery, James E., *Al-Jāḥiẓ: In Praise of Books*, Edinburgh: Edinburgh University Press, 2013

Montgomery, James E., 'Al-Ǧāḥiẓ, Falsafa and Hippocrates Arabicus', in *Philosophy and Medicine in the Formative Period of Islam*, ed. Peter Adamson and Peter E. Pormann, London: Warburg Institute, 2017, 75–103

New Pauly = *Brill's New Pauly*, ed. Hubert Cancik et al., Leiden: Brill, 2005–

Nöldeke, Theodor, *Geschichte der Perser und Araber zur Zeit der Sasaniden aus der arabischen Chronik des Tabari*, Leiden: Brill, 1973 (reprint of the 1st edn, 1879)

Olsson, Joshua T., 'The Reputation of Ḥunayn b. Isḥāq in Contemporaneous and Later Sources', *Journal of Abbasid Studies*, 3 (2016), 29–55

Overwien, Oliver, 'Medizinische Lehrwerke aus dem spätantiken Alexandria', *Les études classiques*, 80 (2012), 157–186

Pellat, Charles, 'Nouvel essai d'inventaire de l'œuvre ǧāḥiẓienne', *Arabica*, 31 (1984), 117–164

Penelas, Mayte, 'A Possible Author of the Arabic Translation of Orosius' Historiae', *Al-Masāq*, 13 (2001), 113–135

Peters, Francis E., *Aristoteles Arabus: The Oriental Translations and Commentaries on the Aristotelian Corpus*, Leiden: E.J. Brill, 1968

Pormann, Peter E. and Savage-Smith, Emilie, *Medieval Islamic Medicine*, Edinburgh: Edinburgh University Press, 2007

Rescher, Nicholas, *The Development of Arabic Logic*, Pittsburgh PA: University of Pittsburgh Press, 1964

Richter-Bernburg, Lutz, 'Islamischer Hellenismus zwischen Nil und Indus. Perioden des Hellenisierungsprozesses in der arabisch-islamischen Kultur', in *Symposium Graeco-Arabicum II*, ed. Gerhard Endress and Marita Schmeink, Amsterdam: B.R. Grüner, 1989, 44–64

Richter-Bernburg, Lutz, 'Abū Bakr Muḥammad al-Rāzī's (Rhazes) Medical Works', *Medicina nei secoli. Arte e scienza*, 6 (1994), 377–392

Rosenthal, Franz, 'R. Walzer, Galen: On Medical Experience', review, *Isis*, 36 (1946), 251–255

Rosenthal, Franz, *The Classical Heritage in Islam*, London/New York: Routledge, 1975

Rubin, Zeev, 'Ḥamzah al-Iṣfahānī's Sources for Sasanian History', *Jerusalem Studies in Arabic and Islam*, 35 (2008), 27–58

Sahner, Christian C., 'From Augustine to Islam: Translation and History in the Arabic Orosius', *Speculum*, 88 (2013), 905–931

Salama-Carr, Myriam, *La Traduction à l'époque abbasside: L'école de Ḥunayn Ibn Isḥāq et son importance pour la traduction*, Paris: Didier Érudition, 1990

Salama-Carr, Myriam, 'Translation as Seen by Al-Jāḥiẓ and Ḥunayn Ibn Isḥāq: Observer versus Practitioner', in *Across the Mediterranean Frontiers: Trade, Politics and Religion, 650–1450*, ed. Dionisius A. Agius and Ian R. Netton, Turnhout: Brepols, 1997, 385–393

Sezgin, GAS = Sezgin, Fuat, *Geschichte des arabischen Schrifttums*, 17 vols, Leiden: Brill, 1967–2015

Signes Codoñer, Juan, 'La diplomacia del libro en Bizancio: algunas reflexiones en torno a la posible entrega de libros griegos a los árabes en los siglos VIII–X', *Scrittura e civiltà*, 20 (1996), 153–187

Soravia, Bruna, 'Ibn Qutayba en al-Andalus. La préface à l'Adab al-Kātib dans le commentaire d'Ibn al-Sīd al-Baṭalyawsī', *Al-Qanṭara*, 25 (2004), 539–565

Sprengling, Martin, 'From Persian to Arabic', *The American Journal of Semitic Languages and Literatures*, 56 (1939), 175–224

Stern, Samuel M., ''Abd al-Jabbār's Account of How Christ's Religion was Falsified by the Adoption of Roman Customs', *The Journal of Theological Studies*, n.s., 19 (1968), 128–185

Strauß, Bettina, 'Giftbuch des Šānāq. Eine literaturgeschichtliche Untersuchung', in *Quellen und Studien zur Geschichte der Naturwissenschaften*, 4/2 (1934), 1–64 (study), and 1–64 (Arabic edition)

Strohmaier, Gotthard, 'Homer in Bagdad', *Byzantinoslavica*, 41 (1980), 196–200

Strohmaier, Gotthard, 'Byzantinisch-arabische Wissenschaftsbeziehungen in der Zeit des Ikonoklasmus', in *Studien zum 8. und 9. Jahrhundert in Byzanz*, ed. Helga Köpstein and Friedhelm Winkelmann, Berlin: Akademie-Verlag, 1983, 179–183

Strohmaier, Gotthard, 'Ḥarrān—die Stadt des Sin in islamischer Zeit', in *Babylon: Wissenskultur in Orient und Okzident*, ed. Eva Cancik-Kirschbaum et al., Berlin: De Gruyter, 2011, 305–315

Touwaide, Alain, 'Translation and Transliteration of Plant Names in Ḥunayn ibn Isḥāq's and Iṣṭifān b. Bāsil's Arabic Version of Dioscorides, De Materia Medica', *al-Qanṭara*, 30 (2009), 557–580

Ullmann, Manfred, *Die Medizin im Islam*, Leiden: Brill, 1970

Ullmann, Manfred, 'Yūḥannā ibn Sarābiyūn. Untersuchungen zur Überlieferungsgeschichte seiner Werke', *Medizinhistorisches Journal*, 6 (1971), 278–296

Ullmann, Manfred, *Die Natur- und Geheimwissenschaften im Islam*, Leiden: Brill, 1972

Ullmann, Manfred, 'Ḫālid ibn al-Yazīd und die Alchemie: Eine Legende', *Der Islam*, 55 (1978), 181–218

Ullmann, Manfred, *Wörterbuch zu den griechisch-arabischen Übersetzungen des 9. Jahrhunderts*, Wiesbaden: Harrassowitz, 2002

Ullmann, Manfred, *Untersuchungen zur arabischen Überlieferung der Materia medica des Dioskurides*, Wiesbaden: Harrassowitz, 2009

Ullmann, Manfred, *Die Nikomachische Ethik des Aristoteles in arabischer Übersetzung*, 2 vols, Wiesbaden: Harrassowitz, 2011–2012

Vaccari, Alberto, 'Le versioni arabe dei Profeti', *Biblica*, 3 (1922), 401–423

Vagelpohl, Uwe, 'The ʿAbbasid Translation Movement in Context. Contemporary Voices on Translation', in *ʿAbbasid Studies II: Occasional Papers of the School of ʿAbbasid Studies*, ed. John Nawas, Leuven: Peeters, 2010, 245–267

Walker, Alicia, *The Emperor and the World: Exotic Elements and the Imaging of Middle Byzantine Imperial Power, Ninth to Thirteenth Centuries C.E.*, Cambridge: Cambridge University Press, 2012

Walzer, Richard, *Greek into Arabic: Essays on Islamic Philosophy*, Oxford: Bruno Cassirer, 1962

Webb, Peter, '"Foreign Books" in Arabic Literature: Discourses on Books, Knowledge and Ethnicity in the Writings of al-Ǧāḥiẓ', *Journal of Arabic and Islamic Studies*, 12 (2012), 16–55

Zadeh, Travis, *The Vernacular Qur'an: Translation and the Rise of Persian Exegesis*, Oxford: Oxford University Press, 2012

Zimmermann, Fritz W., 'The Origins of the so-called Theology of Aristotle', in *Pseudo-Aristotle in the Middle Ages: The Theology and Other Texts*, ed. Jill Kraye, W.F. Ryan, and Charles B. Schmitt, London: Warburg Institute, 1986, 110–240

CHAPTER 6

The *Nabatean Agriculture* by Ibn Waḥšiyya, a Pseudo-Translation by a Pseudo-Translator: The Topos of Translation in Occult Sciences

Isabel Toral

Part One: Essay

The text introduced in this chapter, the *Kitāb al-Filāḥa al-Nabaṭiyya* (*The Book of Nabatean Agriculture*) by Abū Bakr Aḥmad ibn Waḥšiyya (early tenth century AD), constitutes an exception in the structure of this volume. On the one hand, it shares with the other textual testimonies herein the quality of being a text that, in its original context, was regarded and presented as a translation; on the other, it has proven to be impossible to identify a single genuine source text for the *Filāḥa al-Nabaṭiyya*. This means that the preface, which claims that the text is a translation of an approximately 20,000-year-old original Babylonian source, is not reliable; we must therefore assume that there has been no actual linguistic-transfer operation that could be reconstructed and evaluated. From this perspective, the *Filāḥa al-Nabaṭiyya* fits into the category of pseudo- or fictitious translations, and one might wonder whether such a testimony can be of any value in a volume studying the social and cultural history of the translation of *real* scientific texts.[1] However, I will argue in the following that the *Filāḥa al-Nabaṭiyya* is very relevant to a better understanding of the function of translation and translated texts in the cultural system of ʿAbbāsid and Buyid Baghdad. To this end, I shall first introduce the concept of pseudo-translation as developed in translation studies; I shall then present the text by Ibn Waḥšiyya and sketch the current state of research into it and the complexities of its textual history. I shall further discuss the extent to which the concept of 'pseudo-translation' serves as an analytical tool in evaluating the relevance of *Filāḥa al-Nabaṭiyya* in the context of translation history. Finally, I shall translate and comment on the preface, which is one of the most significant peritextual passages in the *Filāḥa al-Nabaṭiyya*.

1 It does not come as a surprise therefore that the *Filāḥa al-Nabaṭiyya* is not discussed in Gutas, *Greek Thought, Arabic Culture*.

 | DOI:10.1163/9789004472648_007

Pseudo-translation and Translation Studies

The concept of pseudo-translation is closely connected with the translational theories developed by Gideon Toury[2] in the field of *descriptive translation studies*. Based on Even-Zohar's polysystem theory,[3] this approach focuses on the socio-cultural aspects of translations and studies their function and connectedness within a given cultural system. Toury's notion of translation is target-oriented, so he assumes that 'translations are facts of target cultures; on occasion facts of a special status, sometimes even constituting identifiable (sub)systems of their own, but of the target culture in any event'.[4] He also introduced the notion of 'assumed translations', and stated that any text regarded or presented as a translation in its original cultural context must be treated as such for translation studies.[5] Because they are oriented to a specific culture, pseudo-translations are the object of research in (descriptive) translation studies as much as genuine translations, the only difference being that, though regarded as translations, no genuine source text exists for them.[6]

Toury's focus on the presentation and framing of the translated text brings out the agency and intentionality of a pseudo-translation's author, and this dimension makes pseudo-translation particularly relevant for translation studies. What then are the textual strategies found in a text of this type intended to enhance its status as translation and so to persuade the reader of its authenticity? Toury mentions peritextual elements (e.g., title, the name of a foreign author, a preface claiming translation status) and internal textual strategies, since many of these pseudo-translated texts have been already produced as translations.[7] The latter are particularly significant for translation studies, since they reflect those elements commonly associated with translations in the target culture (e.g., foreign and exotic names, settings, situations, cultural terms and customs, text genres, even grammatical mistakes and stylistic peculiarities); pseudo-translations even tend to overdo and exaggerate these features to enhance their resemblance to genuine translations.[8] Therefore, by accommodating the text to pre-existing notions and expectations of translated texts, pseudo-translations reveal a great deal about actual translations and even

2 Toury, *Descriptive Translation Studies*; and id., 'Enhancing Cultural Changes'.

3 Cf. Shuttleworth, 'Polysystem Theory'.

4 Toury, *Descriptive Translation Studies*, 29.

5 Ibid., 45.

6 Ibid., 5; cf. also Robinson, 'Pseudotranslation', and Gürçağlar, 'Pseudotranslation on the Margin of Fact and Fiction'.

7 Toury, *Descriptive Translation Studies*, 40–53.

8 Ibid., 48.

throw their key features into greater relief. Furthermore, they bear out the status and role of genuine translations, which pseudo-translations only put to use.

The motives that lead to the composition of pseudo-translations are partly the same as those for translations: foremost are the introduction of innovations (cultures are commonly more permissive and open to novelties when these are imported from abroad, since non-domestic origin makes them look less menacing)[9] and representational objectives (both translations and pseudo-translations have always been crucial in the construction of cultural identities, for the self and for the other alike).[10] Other motives are specific to pseudo-translation: the urge to participate in the existing prestige of translations and the evasion of censorship by the transfer of authorial responsibility.

The theories of Gideon Toury are still our main reference in the study of pseudo-translation. They have the disadvantages however that they have been developed on the basis of modern literary texts, and that they imply that pseudo-translation and translations are mutually exclusive categories. Andrea Rizzi has argued, in contrast, that it is particularly difficult to reach a clear demarcation between cases when the pseudo-translation is actually drawn from a group of texts and when the notion of translation in the target culture differs from our idea of fixed, one-to-one correspondence of texts.[11] As a case in point, she presents the *Historia Imperiale*, a text that claims to be a translation from Latin into vernacular Italian, authored by Mateo Maria Boiardo (1441–1494). In 1723, it was unveiled as being a patchwork conflation of diverse Latin historical texts and thus a 'forgery'; it had nevertheless been regarded, used, and valued as an authentic 'translation' for several centuries. Rita Copeland has also studied many examples of medieval texts that lie in a grey zone between pseudo-translation and translation, sometimes mixing real and fictitious translations from diverse sources, and sometimes just imitating, importing, and translating a foreign literary model and genre.[12] In these cases, the transitions between literal translation (word for word), linguistic translation (between two languages), and cultural translation (between two cultures) are fluid. As we will see in what follows, the peculiar complexities of the multi-layered *Filāḥa al-Nabaṭiyya* suggest that this text fits rather into this hybrid model between pseudo-translation and translation.

9 Ibid., 43; cf. also Toury, 'Enhancing Cultural Changes'.
10 Venuti, *The Scandals of Translation*, 69–75.
11 Rizzi, 'When a Text Is Both a Pseudotranslation and a Translation'.
12 Copeland, *Rhetoric, Hermeneutics, and Translation in the Middle Ages*.

The al-Filāḥa al-Nabaṭiyya—*a Pseudo-translation?*

According to the work's preface, Abū Bakr Aḥmad ibn 'Alī ibn Qays al-Kasdānī al-Qussīnī, alias Ibn al-Waḥšiyya, claims to have rescued several books from oblivion and then translated from the 'Chaldean language' (supposedly an old variant of Syriac) into Arabic in 291/903–904, and to have dictated this translation to his pupil Abū Ṭālib Aḥmad ibn Ḥusayn al-Zayyāt in 318/930.[13] These books allegedly belong to the heritage, neglected and dating back millennia, of the 'Nabateans' or *Nabaṭ*, a term then used in Arabic for the Aramean peasant population in Iraq.[14] The *Kitāb al-Filāḥa al-Nabaṭiyya*, a monumental book of 1,494 pages in the printed edition,[15] is presented as the translation of only one of these books, whose original author is given as a certain, otherwise unknown, Quṯāmā, who based his work on that of a certain Ṣaġrīṯ, in the edition of Yanbūšād.[16] The period during which these authors are claimed to have lived is fantastically ancient: approximately 21,000 years before Ibn Waḥšiyya. The spurious attribution to these authors, and the very complex textual history of this multilayered work gave rise to a scholarly debate that is also of interest, because it neatly demonstrates the ambivalent status of this text between translation and pseudo-translation.[17]

After it had been brought to attention of the scholarly world by Etienne Quatremère in 1835, the book enjoyed a short period of enthusiastic reception as genuine testimony of Babylonian agriculture,[18] but the *Filāḥa al-Nabaṭiyya* was eventually completely rejected as a scandalous and useless 'forgery' of Islamic times by Alfred von Gutschmid in 1861 and Theodor Nöldeke in 1875.[19] They both pointed to its sources of more recent date, which included much Graeco-

13 Preface, section 1.

14 *Nabaṭī* in Arabic meant at that time 'ancient Syriac', i.e., 'Aramaic' or 'native Babylonian', and had no connection to the Nabateans of Roman times.

15 The only critical edition to date, that used here, is by Taufīq Fahd (Toufic Fahd). Almost forty-two manuscripts and many abridgements are extant. For a survey of the manuscripts, see the introduction in the edition, 9–19, and the corresponding section of the 'Filāḥa Project' website (http://www.filaha.org/author_Ibn_wahshiyah.html, accessed 13 July 2021).

16 Fahd edition, preface, section 14, and index. These names also appear in other texts attributed to Ibn Waḥšiyya, e.g., in the Alphabet booklet; see Toral-Niehoff and Sundermeyer, 'Going Egyptian in Medieval Arabic Culture', 251. According to Hämeen-Anttila, *The Last Pagans of Iraq*, 19–21, they are encoded names. I am using here the versions of the names proposed by Hämeen-Anttila.

17 Hämeen-Anttila, *The Last Pagans of Iraq*, 3–9.

18 Particularly celebrated as such by Daniel A. Chwolson in *Über die Überreste altbabylonischer Literatur* (1859).

19 Gutschmid, 'Die nabatäische Landwirtschaft'; and Nöldeke, 'Noch Einiges über die „nabatäische Landwirtschaft"'.

Roman material[20] conflated with what they believed to be 'pagan' magic and superstitions. This verdict overshadowed study of the text for a long time. Manfred Ullmann still discredited the work as a disgusting mixture of genuine ancient material and *purer Schwindel* ('pure fake') in 1972, although he admitted that it might contain some valuable late antique material.[21] This state of affairs changed significantly with the publication in 2010 of *The Last Pagans of Iraq* by Jaakko Hämeen-Anttila, who studied the *Filāḥa al-Nabaṭiyya* in detail and took the book seriously, namely for its evidence of late antique agricultural knowledge in Mesopotamia and of the survival of pagan beliefs and practices. Summarizing, he concluded that the work is a genuine translation by Ibn Waḥšiyya, a real person, of a late Syriac compendium put together by a certain Quṯāmā in late antique Mesopotamia. Quṯāmā had collected diverse oral and written sources, including several previous Syriac translations from Greek, current in late antiquity in the pagan community of northern Iraq.[22]

From the perspective of translation studies, this brief outline of the reception history of the *Filāḥa al-Nabaṭiyya* in European research shows how it was first accepted as a highly valued genuine translation, then demoted to the status of a useless forgery and pseudo-translation, and then again elevated to the rank of a genuine translation of a pseudo-translation. This ambivalence neatly reflects the problems arising when we try to draw a sharp distinction between translation and pseudo-translation—as already indicated by Andrea Rizzi—particularly when we do so with texts that draw upon various sources and mix genuine and invented translations, a phenomenon that seems to be rather frequent in premodern times.[23]

On the other hand, the status of the *Filāḥa al-Nabaṭiyya* in its original context, i.e., in the system of scientific texts of its period, is a completely different issue. As mentioned, the book contains not only agricultural and botanical knowledge, but also numerous magical recipes, and frequent references to astral theology, astrology, and late Babylonian divinities and rituals; it even quotes pagan and pseudo-pagan invocations and hymns extensively, which makes the text a curious hybrid between scientific handbook and magical treatise.[24] This explains why the author Ibn Waḥšiyya was not only credited as a specialist in agronomy and Nabatean culture and religion, but also as a magi-

20 The diverse textual layers are reconstructed in Hämeen-Anttila, *The Last Pagans of Iraq*, 11–33.

21 Ullmann, *Die Natur- und Geheimwissenschaften im Islam*, 440–443.

22 Hämeen-Anttila, *The Last Pagans of Iraq*, 22.

23 Rizzi, 'When a Text Is Both a Pseudotranslation and a Translation'.

24 For these references, see Hämeen-Anttila, *The Last Pagans of Iraq*, 109–310.

cian and alchemist:[25] books on astrology, magical alphabets,[26] and poisons[27] were attributed to him, and his work was quoted extensively in magical handbooks. The entry in the *Fihrist* by Ibn al-Nadīm suggests that Ibn Waḥšiyya was not considered an esteemed author overall, which is hardly surprising, since this is a status that he shared with all authors in occult sciences, an area of knowledge of ambiguous cultural prestige. I have not however come across any commentary that indicates that he was perceived as a veritable forger, or that his works were considered as mere valueless fabrications or pseudo-translations. On the contrary, his *Filāḥa al-Nabaṭiyya* had a remarkable career as one of the most quoted books in the Arabic tradition, particularly in agronomic literature. The classic work in this genre, the renowned twelfth-century *Kitāb al-Filāḥa* by Ibn al-ʿAwwām, quotes it extensively, indeed 540 times.[28] This indicates that the *Filāḥa al-Nabaṭiyya* was regarded as a real translation, or at least as a valuable part of the corpus of Hermetic and pseudo-Hermetic texts and magic, where framing as translation was quite common.[29]

For this volume, it is important to emphasize that the preface of the *Filāḥa al-Nabaṭiyya* is a unique and very valuable testimony for scientific translation in ʿAbbāsid times in one of two ways: either as a uniquely extensive peritext to a genuine translation, and then as an authentic egodocument of a translator; or as a fictive and exaggerated preface to a pseudo-translation or an introduction to a hybrid text. In either case, the preface evinces a textual strategy that aims both to convince the reader of the authenticity of the translation, by drawing on existing notions of translation, and to enhance the status of Ibn Waḥšiyya as a competent translator, by meeting the reader's expectations.[30]

25 The author appears in two sections of the *Fihrist* by Ibn al-Nadīm (ed. Flügel; 372, 423), once in the eighth *maqāla*, on magic, where we also find his book on agriculture, and then in the tenth *maqāla*, on alchemy.

26 Toral-Niehoff and Sundermeyer, 'Going Egyptian in Medieval Arabic Culture'.

27 Levey, *Medieval Arabic Toxicology*.

28 Butzer, 'The Islamic Traditions of Agroecology'.

29 For the many translations and pseudepigrapha to be found in Arabic magic, see Ullmann, *Die Natur- und Geheimwissenschaften im Islam*, 359–426.

30 Toral-Niehoff, 'Warum geheimes Wissen'.

Part Two: Text in Translation

Ibn-Waḥšīyya al-Nabaṭī, *al-Filāḥa al-nabaṭīya*, 5–10 (ed. Fahd); tr. Hämeen-Anttila, *The Last Pagans of Iraq*, 93–99, slightly adapted.[31]

Preface

1. This is the Book of Nabatean Agriculture, which was translated from the language of the Kasdānians[32] into Arabic by Abū Bakr Aḥmad ibn ʿAlī ibn Qays al-Kasdānī al-Qussīnī,[33] who is known by the name of Ibn Waḥšiyya. (He translated it) in the year 291 according to the counting of the Arabs from the *hijra*. He dictated it to Abū Ṭālib Aḥmad ibn Ḥusayn ibn ʿAlī ibn Aḥmad ibn Muḥammad ibn ʿAbdalmalik al-Zayyāt in the year 318 according to the counting of the Arabs from the *hijra*. (Ibn Waḥšiyya) said to him:[34]
2. Know, my son, that I found this book among the other books of the Kasdānians that I have found, and its title in Arabic would be 'The Book of Making the Soil Prosper and Keeping in Order the Crops, the Trees, and the Fruits and Repelling Calamities from Them'. I found the book too massive and too long, so it occurred to me to abbreviate it. Yet on second consideration, this was wrong, not right, as my original aim and objective was to bring the sciences of this nation—I mean the Nabateans and the Kasdānians from among them[35]—to the people, and to promulgate them so that they [my contemporaries] would know the measure of their [the Kasdānians'] understanding and the favours that God (He is Blessed and Exalted) has shown them, in that they could comprehend the useful and recondite sciences and discover what other nations were unable to.

31 The translator frequently placed in parentheses the Arabic word(s) that correspond to his version; these have been excised here, as the complete Arabic text is given in Part Three. The division of the preface into numbered sections is by the present author.

32 The Arabic *Kasdāniyyīn* (cf. Hebrew *Kasdīm*) and *Kaldāniyyīn* (cf. Aramaic *Kaldanāyye*) are both variants of the ethnic term that has been translated in the Latin Bible as *Chaldaioi* (Chaldeans).

33 Added in MSS Beyazit 4064 and Leiden, vocalized thus in Beyazit, and read *al-Qasīṭī* or *al-Qusayṭī* according to M. Plessner; cf. however *al-Fihrist*: *min ahl Qussīn*, whence Hämeen-Anttila's reading.

34 The first part could be characterized as a 'peritext to the peritext', since it introduces the main part of the preface, which consists of Ibn Waḥšiyya's allocution to his pupil and is kept in the first person singular.

35 The connection between Kasdānians and Nabateans remains unclear. In general, the Kasdānians appear to be the forefathers of the Nabateans.

3. In this way I have come to their book and (this) time, when memory of them has faded away, what they have told has been deleted, and what they taught has vanished, so that no more than a mention of them and of some of their sciences remains, just like fairy tales, and that even those who mention them have no (actual) knowledge of them.
4. When I realized this, I started looking for their books and I did find (some of them) among people who were the remnants of Kasdānians and (still) retained their religion, habits, and language. So, I found some books which they had in their possession, although they are extremely careful to hide them, to keep them concealed, and to deny them [to others], because they are afraid to divulge them.
5. Before that, God (He is Exalted, Majestic, and Lofty) had provided me with a knowledge of their language—which is ancient Syriac—to an extent possessed by very few. That is because I am one of them, that is to say, of the progeny of them, and God (He is Exalted) has given me property and money (praise be to Him for that!) so that I was able to gain access to those of their books that I wanted, thanks to what I have just explained: that I am one of them, that I know their language, and that I have plenty of money. So, I made use of affability, generosity, and delicate tricks until I had access to all the books that I could.
6. The one who had these books in his possession thought that he needed me to understand the contents of these books. This is because all these people who are their descendants are like cows and donkeys, and they are unable to understand anything of the sciences of their forebears, except that the man with whom I found the collection of these books is distinguished from the others and is apart from the donkeyness of all the others. I rebuked him for his excessive eagerness to keep these books hidden and these sciences in secrecy and said: 'You are exceedingly careful, in fact, to wipe out the name of your people into oblivion and to bury their merits. In your action you follow the example of those who have lived before you, but in doing so, both you and they are unjust to those earlier scholars of yours, who are also my scholars and forebears as well. By this very action, you have obscured mention of our forebears and concealed their sciences and their merits from people. What if I translate these books, or some of them, into Arabic so that people can read them? They would then know the measure of our sciences and could make use of what our forebears have invented and that would become a sort of pride for us and an indication of our excellence.'
7. The man to whom I said all this found my words very repugnant and answered: 'Abū Bakr, do you want to argue against the way of our elders

and forebears and their admonitions to us to keep our religion and habits hidden?' I replied to him: 'Nay, it is you who are being unjust to our elders and your forebears. No indeed! They charged us to keep secret our religion and the use of *šarīʿa* [religious law], because they knew the opposition it would raise if it became known to others, and this is why they were on their guard with their religion. By my life, it is right to keep our religion secret, but it is different with the sciences, which are useful to people and which are now being forgotten! If others were to know them and know who invented them, these (forebears of ours) would regain prestige and honour in their minds! The sciences are one case and religion and *šarīʿa* another! These sciences do not come under the commission of secrecy!'

8. He answered me: 'But what use would there be (from our point of view) to promulgate the rarities of (almost) forgotten sciences, even if they were of use to people? Go ahead and disseminate them among them; they will benefit from them, but you know how they think of our religion! Nay! It was well that our forebears concealed both *šarīʿa* and the sciences from them, as they [the others] deserve neither of them!' I said to him: 'Even if they really ordered the sciences to be concealed, I do disagree with both our forebears and yours. Nevertheless, I do share their opinion when it comes to concealing *šarīʿa*. If they did not order the concealing of the sciences, I agree with them, without any disagreement. Now listen to me! Can't you see how exceedingly ignorant all people are in our present time and into what weakness and negligence these religious habits and *šarīʿa*s, which are now preponderant among them, have made them enter? They have become like dumb animals, and even worse than that, inferior to them in some matters! By God, it is my wish to defend them that brings me to promulgate some of our sciences to them, so that they might stop defaming the Nabateans and awaken from their sleep and be resurrected for a while from their death [of ignorance]: after all, everyone has been fashioned and made to be disposed to understand everything. There used to be among them [the ancient Nabateans], one after another, men of extreme acumen and lavish talent. Since they are like this, it is he who conceals the sciences from them [their descendants] and deviates their course from them, who treats them unjustly: they have this understanding and talent, but it is left unused, and they have become what they are because they have not been taught the sciences and their ways and manners and inventions. Now obey me, my dear man, and let me translate into Arabic my selection of these books! You are not a whit more eager to obey our forebears than I am, or more persistent in keeping

concealed what has to be concealed. You should also listen to these sciences yourself, as you don't understand them because of your eagerness to obey your forebears, claiming that they have ordered them to be kept concealed. If you yourself had had the opportunity to peruse some of these books, that would have profited you considerably and been most useful to you. Think of what I am saying to you and you will realize that it is as I say, and your intellect will find it right.'

9. So, he obeyed me and let me see these books. I started reading them to him, and he repeated what I had read to him and tried to comprehend it, until one day he said to me: 'By God, Abū Bakr, you have revived me, may God reward you for my sake!' I answered him: 'What use does a man have of books that are hidden and unattainable to him, so that he cannot read them or learn from them? They are no more valuable to him than stones and mud bricks!' He approved of my words and followed my opinion. Then I started translating these books of the Nabateans one after another and I read them to him in Arabic, so that he gained more and more understanding, and he was fascinated by them. Finally, he thanked me with all his heart and accepted my opinion concerning the matter (of promulgating the books). Yet this did not happen without me showering dirhams and dinars on him, so that he followed the joint lead of craving money and accepting the argument, deeming worthy what he heard and finding the benefit within himself.

10. The first book that I translated into Arabic was *The Book of Dāwānāy the Babylonian Concerning the Secrets of the Sphere and the Decrees over Events, Proceeding from the Movements of the Stars*. This is an important and valuable book, very precious, but I could not translate it completely, so I translated only its beginning: I found out that it was about two thousand double-pages of material called *raqq* [parchment] of the same size as the largest sheets of paper that is nowadays used. It was written in a most beautiful hand, very correctly, clearly, and flawlessly. So, by God, my son, I was unable to make a complete translation of it, only because of its length, for no other reason.

11. Together with it, I translated their book on the cycles, which is *The Great Book of Cycles*.

12. Then I translated this book and others after (I had translated) some other books. With 'this book' I mean *The Book of Nabatean Agriculture*. I gave a complete and unabridged translation of it, because I liked it and I saw the great benefits in it and its usefulness in making the earth prosper, caring for trees, and making orchards and fields thrive, and also because of the discussions in it on the special properties of things, countries, and times,

as well as on the proper times of labours during the seasons, on the differences of the natures of (different) climates, on their wondrous effects, the grafting of trees, on their planting and care, on repelling calamities from them, on making use of plants and herbs, on curing with them and keeping maladies away from the bodies of animals and repelling calamities from trees and plants with the help of each of the plants, and on some uncommon qualities that can be caused by combining different things that alone do not have the same effect, either similar or different (to the effects of the components). When I realized this, I made a complete translation of the book.

13. Now I have dictated this to my son [i.e., disciple] Abū Ṭālib Aḥmad ibn al-Ḥusayn ibn ʿAlī ibn Aḥmad ibn Muḥammad ibn ʿAbdalmalik al-Zayyāt, and I have charged him not to withhold it from anyone who asks to see it, wishing to make use of it. It indeed is useful to everyone, of momentous benefit to them in their lives, but I have also charged him to keep other things concealed.

14. I found out that this *Book of Agriculture* is attributed to three ancient Kasdānian sages. They say that one of them began it, the second added other things to it, and the third made it complete. (The book) was written in ancient Syriac and it comprised some 1,500 double-pages. Concerning the one who began this, it is said that he was a man who appeared in the seventh millennium of the 7,000 years of (the rule of) Saturn. That is the millennium in which Saturn was in partnership with the Moon. His name was Ṣaġrīṯ. The one who added other things to that was a man who appeared in the end of these millennia, and his name was Yanbūšād; and the third who made it complete was a man who appeared after 4,000 years had elapsed from the cycle of the Sun in this cycle, I mean the cycle that belonged to Saturn, i.e., the 1,000 years in which the two earlier men had appeared. I counted the interval between the two times and it came out to be 21,000 years. The name of this third man was Quṯāmā. He said that he appeared after 4,000 years had elapsed from the cycle of the Sun, which lasts for 7,000 years, so between them there was the period I have mentioned. Both of the two who added to what the first, Ṣaġrīṯ, had composed, added in their books something to every chapter that Ṣaġrīṯ had written, but they changed nothing from what he had said and written and spoken about the things that he mentioned, nor did they alter the order in which he presented his material. They merely added to everything that he had put down, according to what they found out and invented after him. So, the beginning and preface of the book are by Ṣaġrīṯ.

Commentary

Except for the first, introductory passage (section 1), which is written in the third person singular and is probably a later addition by Ibn al-Zayyāt, the structure of the preface has the form of a personal allocution by Ibn Waḥšiyya to his disciple, in which he mainly reports a dialogue he had with an unnamed person who had the books in his custody (6, 7, 8). The dialogue dramatizes the translator's difficulties in acquiring his material and simultaneously functions as legitimation and framing for the translation. The insertion of dialogues was a very frequent strategy in the Arabic textual tradition and often served didactic purposes.[36] The preface is followed by the authentic or fabricated preface of the supposed original, a laudatory pagan hymn attributed to the wise Ṣaġrīṯ.[37]

The information about technical details is frequently very precise. The text is particularly accurate when it comes to names (except that of the mysterious custodian of the legacy) and the exact dates of translation and dictation (1, 13). It also reports the details of the textual history of the original text and even gives precise names and dates of the lifetime of these foreign authors (14). However, since in order to express pre-Islamic chronology Ibn Waḥšiyya refers to complex astrological data that presuppose knowledge of astronomy, it remains unclear whether he expects the average reader to understand these technical minutiae, or he is just suggesting or intending to say that they lived 'very long ago'.[38] Ibn Waḥšiyya also enumerates meticulously the other works he translated and gives the reasons why he either translated them completely or abbreviated them (10, 11, 12, 13, 14). He even describes the physical shape of the books (how many folios, material) (10, 14). All this indicates a manuscript and bookish culture where these details were regarded as important and indicated authenticity.

On the other hand, the text remains vague when it comes to linguistic details: he relates that he 'transported' (*naqalahu*)[39] the text from the language of the Kasdāniyyīn into Arabic, but later he refers continuously to the *Nabaṭ* as the people whose legacy he wants to rescue, and explains that their language is *suryāniyya al-qadīma* ('ancient Syriac') (1, 5); he is clearly referring to some

---

36 Cf. the exhaustive recent study on dialogues in Arabic literature by Regula Forster, *Wissensvermittlung im Gespräch*.

37 In *al-Filāḥa al-nabaṭīya*, 10–12 (ed. Fahd); Hämeen-Anttila, *The Last Pagans of Iraq*, 151–155.

38 Hämeen-Anttila, *The Last Pagans of Iraq*, 99, n. 34, calculates a total of 18,000 years before the days of Ibn Waḥšiyya.

39 *nql* is the commonly used root in the text denoting 'to translate'.

variant of Aramaic (the Aramaic and pseudo-Aramaic onomastics in the main text confirm this impression).[40] This was probably not the kind of information in which the reader was supposed to be interested; alternatively, it may have been deemed sufficient to evoke an idea of 'exotic people from the past related to the Nabaṭ'. A very important point is the equation of 'translating into Arabic' with 'making public and accessible', which is highly telling for the status of unquestioned scientific *lingua franca* that Arabic had attained in the tenth century: if one wanted to be read by 'the people', one had to write in Arabic, or have one's work translated into it (6, 8).

The text further reflects the difficulties encountered when translating religious references from non-Islamic and/or heretical contexts, and indicates a milieu where people could not overtly manifest their beliefs but practised some sort of dissimulation, *taqiyya* (7). It also indicates a society obsessed with occult wisdom and esoterica, as was the case in tenth-century Baghdad.[41] Concealment, secrecy, and censorship are constant themes in the text. Given that the main text is replete with pagan reminiscences, it appears that enhancing its status as 'translated' was a good strategy for avoiding scandal and disclaiming responsibility. Furthermore, it reflects a common textual strategy in occult literature, the 'legend' of 'discovering' hidden books, which has its roots in Hellenistic models and aims to emphasize the antiquity of the mysterious wisdom unearthed.

Another recurrent theme is the usefulness of the knowledge contained in the book, which counterbalances the dangers of translating foreign and morally questionable knowledge (8 and *passim*). There is also an interesting passage in which Ibn Waḥšiyya argues in favour of translating science because of its utility, as opposed to translating religion, which might reflect contemporaneous debates. Agricultural knowledge and the introduction of new techniques were useful for society, and advantage ought to be taken of these resources independently of their origins (7). Religion, by contrast, is a matter of ethics and faith, and much more dangerous. This point fits also into the function of translation and pseudo-translation as a factor in and means of anchoring innovation, as emphasized by Toury.[42]

Finally, the preface underlines the representational function of the text: it serves to 'save the honour' of the Nabateans, apparently a denigrated people, a motivation that has conferred on Ibn Waḥšiyya the reputation of being

---

40 Hämeen-Anttila, *The Last Pagans of Iraq*, 19–21, regards the personal names as encoded.

41 See Kraemer, *Humanism in the Renaissance of Islam*, 20–25.

42 Toury, 'Enhancing Cultural Changes'.

a Nabatean *šuʿūbī*[43] or 'nationalist'.[44] The issue of representation and the construction of a 'Nabatean' identity are indeed central to the agenda of the *Filāḥa al-Nabaṭiyya* and demonstrate the importance of translations in the development of identity discourses. The issue is also indicative however of how multicultural ʿAbbāsid society in the tenth century was, with diverse groups competing to establish their places in the ranking of cultural authority; they did so by having recourse to real or pretended ancient wisdom, knowledge, and history, which would be put in circulation principally by means of translations and/or pseudo-translations.

43 Goldziher, *Muhammedanische Studien*, 158.

44 Hämeen-Anttila, *The Last Pagans of Iraq*, 33–52. 'Nationalism' seems a somewhat unfortunate, anachronistic term; 'ethnic pride' or 'cultural identity discourse' might be more fitting.

Part Three: Original Text

The following text is based on the standard edition published by Toufic Fahd in 1993, 5–10,1. For these passages, Fahd used four central manuscripts: Istanbul, Esad Efendi 2490, dated 1064/1654; Istanbul, Fatih 3613, dated 803/1401; Leiden, composite manuscript, dated 872/1467; and Istanbul Beyazit, Umumi 19052, before the seventeenth century; in addition, he used various partial manuscripts, which he lists in the introduction to his edition.[45] In total, he used for the whole edition twenty-one manuscripts of the roughly forty manuscripts that are preserved.[46] The manuscript situation is complicated, and a new edition that includes a broader material basis is much needed.

In what follows I do not reproduce Fahd's critical apparatus. I have also removed significantly his punctuation and added helpful vocalization, some hamzas, and section numbers to facilitate reference to the translation.

1. هذا الكتاب الفِلاحة النبَطيّة نقله من لسان الكَسْدانيّين إلى العربيّة أبو بكر أحمد بن علي بن قيس الكَسْداني القُسِّيني[47] المعروف بابن وحْشيّة في سنة إحدى وتسعين ومائتين من تأريخ العرب من الهجرة، وأملاه على أبي طالب أحمد بن الحُسين بن علي بن أحمد بن محمّد بن عبد الملك الزيّات في سنة ثماني عشرة وثلثمائة من تأريخ العرب من الهجرة. فقال له

2. "اعلَم يا بُنيّ أنّني وجدتُ هذا الكتاب في جُمْلة ما وجدتُ من كُتب الكَسْدانيين مُتَرْجماً بِترجمة معناها بالعربيّة "كتاب إفلاح الأرض وإصلاح الزرع والشجر والثِمار ودفْع الآفات عنها" فاستكبرتُه واستطلتُه وخطَرَ ببالي اختصارُه. ثمّ فكّرتُ فإذا ذلك خطأٌ غيرُ صوابٍ مِن أجل أن قصدي الأوّل وغرضي إنّما هو إيصالُ علوم هؤلاء القوم أعني النبَط الكَسدانيين منهم إلى الناس وبثُّها فيهم ليعلموا مقدارَ عقولهم ونِعَم الله تبارك وتعالى عندهم في إدراك العلوم النافعة الغامضة واستنباطِ ما عجز عنه غيرهم من الأُمَم.

3. وذلك أنّني وصلتُ الى كُتبهم في زمان قد دَرَس فيه ذِكرُهم ونُسخت فيه أخبارُهم وعدم إعلامُهم حتّى لم يبقَ إلا ذكرُهم فقطْ وذكرُ بعض علومهم ذكراً كالخُرافات بلا معرفة ممن يذكرُها بها.

45 Cf. Fahd, *al-Filāḥa al-nabaṭīya*, 9–30, where he describes the difficult manuscript situation and his governing principles in constituting the text.

46 For a recent update, see the webpage on Ibn Waḥšīyya of the Filāḥa Text project: http://www.filaha.org/author_Ibn_wahshiyah.html (accessed 13 July 2021).

47 The edition by Fahd has قسيتي, corrected by Hämeen-Anttila (93). This is a reference to the town of Qussīn; see n. 32.

4. فلمّا رأيتُ ذلك اجتهدتُ في طلب كتبهم فوجدتُها عند قوم هم بقَايا الكَسْدانيين وعلى دينهم وسُنّتهم ولغتهم ووجدتُ ما وجدتُ عندهم من الكتب وهم في نهاية الكِتمان والإخفاء والجحود لها والجزَع من إظهارها.

5. وكان الله تعالى عزّ وجلّ قد يُرزّقني قبل ذلك من المعرفة بلُغتهم التي هي السُريانيّة القديمة ما لم أره مع كثيرٍ أحد. وذلك أنّني منهم أعني من نسل بعضهم ومكّنني ألله تعالى من المال والدنانير فله الحمد فوصلتُ إلى ما أحببتُ من كتبهم بهذه الوجوه التي عددتُها من أنّني منهم وأنّني عارفٌ بلغتهم وأنّني متمكّن المال. فاستعملتُ المدارة والبذل ولطيف الحيلة إلى أن وصلتُ إلى ما أمكن من كتبهم.

6. ونظر الذي هو في يده أنّه محتاجُ إليّ في فهْم ما فيها إذ كانت الكافةُ من هؤلاء القوم الذين هم بقاياهم كالبقَر والحَمِير والعاجزين عن فهْم شيء من علوم أسلافهم إلاّ أنّ الإنسان الذي وجدتُ هذه الكتب مجموعة عنده يتميّز عن هذه الجُملة وينفصل عن حِماريّة هذه الكافة. فلُمْته على الإفراط في كِتمان هذه الكتب وخَبْءِ هذه العلوم وقلتُ له "إنّك تَزيد من الإحتياط بفِعل شيء هو دَرْس ذكر قومك وطمر محاسنهم. وهذا الذي تَعملُه اقتديتَ فيه بمن مضى قبلك. وأنت والماضون مُخطئون في ذلك على من تقدّم من علمائكم الذين هم عُلمائي وأسلافي معكم. وهذا الفعل كان الطريق الى اندراسِ ذكر أسلافنا وغيبوبة علومهم عن الناس واختفاء محاسنهم عنهم. ولو نقلتُ هذه الكتب أو بعضها إلى العربيّة حتّى ينظر الناس فيها عرفوا مقدار علومنا وانتفعوا بما وضع أسلافنا وصار في ذلك ضرب من الفخْر لنا والتنْبيهُ على فضلنا."

7. فاستبشع الرجلُ الذي خاطبتهُ بهذا ما كلّمته به جدًّا وقال لي "يابا بكر أتُريد أن تخالف رسْمَ شيوخنا وأسلافنا ووصاياهم إيّانا بكتمان ديننا وسُنّتنا؟" قلتُ له "إنّك هو ذا تخطيء على شيوخنا أسلافك! لا، فهم وصَّوا بكِتمان الدين واستعمال الشريعة لما عملوا من مضادّته لما ظهر في الناس واحتاطوا لدينهم بذلك. ولعَمري إنّ كتمانه صوابٌ. فأمّا العلوم النافعة للناس الدارسة عنهم التي لو علموها وعرفوا من هم واضعوها لكبروا في نفوسهم وعظموا عندهم. فإنّ هذه العلوم غير جارية مجرى الدين والشريعة ولا داخلة في الوصيّة بالكِتمان."

8. قال: "وأيّ فائدةٌ في إظهار غريب علوم دارسة وإن كانت نافعة للناس فبثُّها فيهم فينتفعون بها ودِيننا عندهم بالصورة التي تعلم بل نِعْمَ ما رأى أسلافُنا من كتمان الدين والعلوم عنهم جميعاً إذ كانوا لا يستحقّونها جميعاً." قلتُ له "فإنّي أخالف أسلافنا وأسلافك في كتمان العلوم وأُوافقهم في كتمان الشريعة إن كانوا أمروا بكتمان العلوم وإن كانوا لم يأمروا بذلك فأنّي

مُوافق لهم غير مخالفٍ. يا هذا ألا ترى وتعلم في زمانك هذا ما عليه كافة الناس من فرط الجهل وما أدخلت هذه الأديان والشرائع الظاهرة فيهم عليهم من العياء والغفلة حتّى صاروا كالبهائم أو شرّ منها أو دونها في بعض الأحوال؟ فوالله إنّ الغيْرة على الناس تحملني على إظهار بعض علومنا لهم لعلّهم أن ينتهوا عن ثلْب النبط وينتبهوا من رقدتهم ويعيشوا قليلاً من موتهم إذ كان كلّ الناس مهيئين مشكلين لفهم كلّ شيء وكان فيهم الواحد بعد الآخر في نهاية الذكاء وجودة القريحة. فمثل هؤلاء من ظلمهم كتمانهم العلم وأزواه عنهم إذ كانوا مع تلك الأفهام التي فيهم والقرائح التي لهم مغفلين قد صاروا كذلك لعدم سماع العلوم ووجوه طرقها واستنباطها. فاطْعني يا أيّها الرجل ودعْني أنقل إلى العربيّة بعض ما أرى نقله من هذه الكتب فلستَ بأحرص منّي على طاعة أسلافنا ولا أشدّ مواظبة على كتمان ما ينبغي أن يُكتَم. وأيضاً فإنّ لك في سماع هذه العلوم التي لم تفهمها من شدّة حرصك على طاعة أسلافك فيما تزعم عنهم أنّهم أمروا بكتمانها. ولو قد نظرتَ في بعضها لكان لك في ذلك أعظم الفوائد وأجزل المنافع تنتفع بها. فأفكِر فيما أقولُه لك فإنّك تجِده كما أقول ويراه عقلُك صواباً."

9. فأطاعني وأمكنني من الكتب فجعلتُ أقرأها عليه فيستعيد ما أقرأه عليه ويتفهّمه إلى أن قال لي في بعض الأيّام "أحْييْتَني والله يابا بكر فجزاك الله عني خيراً." قلتُ له "فما يصنعُ الإنسان بكتب مخبوءة مرفوعة عنده لا يقرأها ولا يتفهّمها فهي كائنة عنده بمنزلة الحِجارة والمدر." فصدّقني فيما قلتُه له واتّبع قبول رأْيي فيما رأيتُ. وابتدأتُ أنقل كتاباً بعد كتابٍ من كُتب النبَط وأقرأه عليه بالعربيّة فيزداد فهماً إلى فهمه ويُعجبه ذلك. فلم أزل به حتّى شكرني أتمّ شكر وعرف صواب رأْيي وصِحّته في ذلك. لكن لم يستول ذلك معه إلّا ببِذْل الدراهم والدنانير له حتّى انقاد لاجتماع الرغْبة بالمال مع إلزام الحجّة له واستحسانٍ بما يسمع وموقع الفائدة له من نفسه.

10. فكان أوّل كتاب نقلتُه إلى العربيّة "كتاب دواناي البابلي في أسرار الفَلَك والأحكام على الحوادث من حركات النجوم". وهو كتابٌ عظيمُ المحلّ والقدر نفيس ولم يستول نقلُه كلّه بل نقلتُ منه صدراً لأنّني وجدتُه في نحو ألفيْ ورقة من الورق المسمّى الرقّ في مقادير أتمّ ما يكون من الكاغد الطلحيّ الموجود في زماننا هذا بأحسن خَطّ وأصحّه و أقومه وأبينه. فعجزت والله يا بُنيّ عن استتمام نقلِه لطُوله فقط لا لغير ذلك.

11. ونقلتُ معه كتابهم في الأدوار وهو "الأدوار الكبير".

12. ونقلت هذا الكتاب مع غيره بعد عدّة كتب أعني بهاذا الكتاب "كتاب الفلاحة". ونقلتُه كلّه علي تمامِه وكماله لاستحساني له وعُظْم ما رأيتُ من فائدته وجميل موقعه في إفلاح الأرض وعلاج

الشجر وزكاء الثِمار وتجويدها وزَكاء الزروع والكلام على خواصّ الأشياء وخواصّ البُلدان والأزمنة ومَواقع أفعال فُصول الأزمنة واختلاف طِباع الأهوية وعجيب أفعالها وتراكيب الشجر وغروسها وإفلاحها ودفع الآفات عنها واستخراج منافع المنابت والحشائش والمداواة بها ودفع العاهات عن أبدان الحيوانات ودفع آفات الشجر والمنابت بعضها ببعض وطرائف ما ركّبوا من الأشياء حتّى حدث عنها أشياء هي غيرها إما قريبةٌ منها أو بعيدةٌ. فلما رأيتُ ذلك فيه أكملتُ نقله.

13. وهآنذا الآن قد أمليتُه على ابني أبي طالب أحمد بن الحسين بن علي بن أحمد بن محمّد بن عبد الملك الزَيّات ووصّيتهُ بأن لا يمنعه أحداً يلتمسه طالباً للانتفاع به فإنّه نافعٌ لجميع الناس عظيم المنفعة لهم في معايشهم مع وصيّتي له بكِتمان أشياء أخر غيره.

14. ووجدتُ كتّاب الفلاحة هذا منسوباً الى ثلاثة من الحُكماء الكسْدانيين القُدماء ذكروا أنّ أحدَهم ابتدأه وأنّ الثاني أضاف إلى ذلك المبتدأ شيئاً آخر وأنّ الثالثَ تمّمه. وكان مكتوباً بالسُريانيّة القديمة في نحو ألف وخمس مائة ورقة. فأمّا الأوّل الذي ابتدأه فذكروا أنّه رجلٌ ظهر في الألف السابع من سبعة آلاف من سني زُحَل وهي الألف التي يشارك فيها زُحَل القَمر كان اسمه صغريث وأنّ الذي أضاف اليه شيئاً آخر رجُل ظهر في آخر هذه آلاف كان اسمه ينبوشاد وأنّ الثالث الذي تمّمه رجل ظهر بعد مُضيّ أربعة آلاف سنة من دور الشمس في هذه الدوْرة أعني في الدورة التي كان لزُحَل فيها تلك الألف التي ظهر فيها الرجلان. فنظرتُ إلى ما بين الزمانيين فإذا هو إحدى وعشرون ألف سنة. وذلك إنّ اسم هذا الثالث كان قوثامى. وقال إنّه ظهر بعد مُضيّ أربعة آلاف من دور الشمس التي هي سبعة آلاف سنة فكان بينهما ما ذكرت لك من المدّة. وكانت زيادة كلّ واحد من الاثنيْن على ما ألّفه الأوّل الذي كان اسمه صغريث زيادة في كلّ باب من الأبواب التي رسمها صغريث في كتبهم لم يغيّروا شيئاً من قوله ورسمه الذي رسمه وتكلّم به على المعاني التي ذكرها وترتيبهُ الذي رتّبه. وإنّما زادوا على كلّ شيء دوّنه بحسب استخراجهم واستنباطهم بعده. فصار صدر الكتّاب وابتداؤه لصغريث.

Bibliography

Baker, Mona, and Gabriela Saldanha (eds), *Routledge Encyclopedia of Translation Studies*, 2nd edn, London: Routledge, 2009

Bermann, Sandra, and Catherine Porter (eds), *A Companion to Translation Studies*, Blackwell Companions to Literature and Culture, New York: John Wiley & Sons, 2014

Butzer, Karl W., 'The Islamic Traditions of Agroecology: Crosscultural Experience, Ideas and Innovations', *Ecumene*, 1/1 (2016), 7–50

Chwolson, Daniel A., *Über die Überreste altbabylonischer Literatur in arabischen Übersetzungen*, Petersburg, 1859

Copeland, Rita, *Rhetoric, Hermeneutics, and Translation in the Middle Ages: Academic Traditions and Vernacular Texts*, new edn, Cambridge Studies in Medieval Literature 11, Cambridge: Cambridge University Press, 1995

Fahd, Taufīq *See* Ibn Waḥšiyya al-Nabaṭī

Forster, Regula, *Wissensvermittlung im Gespräch: Eine Studie zu klassisch-arabischen Dialogen*, Islamic History and Civilization 149, Leiden/Boston: Brill, 2017

Goldziher, Ignaz, *Muhammedanische Studien*, 2 vols, Halle: Niemeyer, 1888

Gürçağlar, Şehnaz Tahir, 'Pseudotranlastion on the Margin of Fact and Fiction', in Bermann and Porter, *A Companion to Translation Studies*, 516–527

Gutas, Dimitri, *Greek Thought, Arabic Culture: The Graeco-Arabic Translation Movement in Baghdad and Early 'Abbāsid society (2nd–4th/8th–10th centuries)*, repr. Abingdon: Routledge, 2005

Gutschmid, Alfred von, 'Die nabatäische Landwirtschaft und ihre Geschwister', *Zeitschrift der Deutschen Morgenländischen Gesellschaft*, 15 (1861), 1–110

Hämeen-Anttila, Jaakko, *The Last Pagans of Iraq: Ibn Waḥshiyya and His Nabatean Agriculture*, Islamic History and Civilization 63, Leiden/Boston: Brill, 2010

Ibn al-Nadīm, *Kitāb al-Fihrist: Erster Band, den Text enthaltend. Zweiter Band, die Anmerkungen und Indices enthaltend* (repr. of the 1. Aufl., Leipzig, 1871–1872), ed. Gustav Flügel, Frankfurt am Main: Universität Frankfurt, Institut fur Geschichte der Arabisch-Islamischen Wissenschaften, 2005

Ibn-Waḥšīyya al-Nabaṭī, Abū-Bakr Aḥmad, *Al-Filāḥa al-nabaṭīya*, Ṭab'a 1, ed. Toufic Fahd, Damascus: al-Ma'had al-'Ilmī al-Faransī li-l-Dirāsāt al-'Arabīya, 1998

Kraemer, Joel L., *Humanism in the Renaissance of Islam: The Cultural Revival during the Buyid Age*, 2nd rev. edn, Brill paperbacks, Leiden/New York/Cologne: Brill, 1992

Levey, Martin, *Medieval Arabic Toxicology: The Book on Poisons [As-sumūm, engl.] of Ibn Wahshiya [Aḥmad Ibn-'Alī Ibn-Waḥšijja] and its Relation to Early Indian and Greek Texts*, Transactions of the American Philosophical Society, N.S. 56,7, Philadelphia PA, 1966

Nöldeke, Theodor, 'Noch Einiges über die „nabatäische Landwirtschaft"', *Zeitschrift der Deutschen Morgenländischen Gesellschaft*, 29 (1875), 445–455

Rizzi, Andrea, 'When a Text Is Both a Pseudotranslation and a Translation', in Bermann and Porter, *A Companion to Translation Studies*, 154–161

Robinson, Douglas, 'Pseudotranslation', in Baker and Saldanha, *Routledge Encyclopedia of Translation Studies*, 183–184

Shuttleworth, Mark, 'Polysystem Theory', in Baker and Saldanha, *Routledge Encyclopedia of Translation Studies*, 176–179

Toral-Niehoff, Isabel, 'Warum geheimes Wissen nicht vermittelt werden soll—oder doch? Der Einführungsdialog der ‚Nabatäischen Landwirtschaft'', in *Performanz von Wissen: Strategien der Wissensvermittlung in der Vormoderne*, ed. Therese Fuhrer and Almut-Barbara Renger, Bibliothek der klassischen Altertumswissenschaften, Reihe 2, N.F. 134, Heidelberg: Winter, 2012, 195–207

Toral-Niehoff, Isabel, and Annette Sundermeyer, 'Going Egyptian in Medieval Arabic Culture: The Long-Desired Fulfilled Knowledge of Occult Alphabets by Pseudo-Ibn Waḥshiyya', in *The Occult Sciences in Pre-Modern Islamic Culture*, ed. Nader El-Bizri and Eva Orthmann, Baden-Baden: Ergon, 2018, 249–264

Toury, Gideon, *Descriptive Translation Studies and Beyond*, Benjamins Translation Library 4, Amsterdam/Philadelphia PA: J. Benjamins, 1995

Toury, Gideon, 'Enhancing Cultural Changes by Means of Fictitious Translations', in *Translation and Cultural Change: Studies in History, Norms and Image-projection*, ed. Eva Hung, Benjamins Translation Library 61, Amsterdam/Philadelphia PA: J. Benjamins, 2005, 3–17

Ullmann, Manfred, *Die Natur- und Geheimwissenschaften im Islam*, Handbuch der Orientalistik: Der Nahe und der Mittlere Osten, Ergänzungsband 6, Zweiter Abschnitt, Leiden: Brill, 1972

Venuti, Lawrence, *The Scandals of Translation: Towards an Ethics of Difference*, London/New York: Routledge, 1998

CHAPTER 7

Translations into Greek in the Byzantine Period

Anthony Kaldellis

Part One: Essay

What we call 'Byzantium' was nothing other than the direct continuation of the ancient Roman empire in the east that survived until 1453, when the Ottoman Turks conquered Constantinople. Its main language was Greek, and its official religion was Orthodox Christianity. Ancient paganism was suppressed by law and abandoned by the population. Until the sixth century, the highest level of the state operated mostly in Latin, but after that it too switched to Greek. Until the seventh century, the empire was home to many speakers of Syriac and Coptic, and some speakers of Armenian, Gothic, Thracian, Arabic, and possibly Isaurian. By then, most of these linguistic communities had begun to translate the Bible, along with other Christian texts, from Greek. But the Arab conquests of the seventh century stripped the empire of most of its linguistic diversity; henceforth, it would be predominantly Greek-speaking, even though the identity of the majority of the population was Roman (so they called their state *Romanía*, or 'Romanland'). Our concern here will be exclusively with translations into Byzantine Greek, of which I have separately prepared a catalogue.[1] A few of those texts contain self-reflexive comments on the goals, methods, contexts, and implications of translation as an act of intellectual and cultural transference; they are quoted in the documents presented below.

Compared with other post-classical cultures, which had to access ancient thought for the most part through translations, Byzantium was in a 'uniquely unique' position, insofar as ancient philosophical, scientific, and technical knowledge was recorded and perfected in Greek texts, and Byzantium, a Greek-speaking society whose elite education was based on ancient *paideia*,[2] was the only post-classical culture that had direct and unmediated access to the surviving original sources. It did not need translations to access ancient thought. In addition to being the only medieval culture that preserved and transmitted

1 This 'Catalogue of Translations into Byzantine Greek from Texts in Any Other Language, 300–1453 AD' is posted on my academia.edu page. I do not intend to publish it formally, but I will periodically update it with new findings.

2 For the Byzantine history of Greek and the coexistence of the spoken vernacular and high-register elite literacy, see Kaldellis, 'The Byzantine Phase'.

 | DOI:10.1163/9789004472648_008

ancient Greek sources in their original language, the decisions made by Byzantine scholars about what to keep and study—that is, the way in which the culture formed its own canon of 'classical' thought and literature—fundamentally shaped the transmission of that corpus and determined the survival and loss of texts and authors. To be sure, many ancient texts had already been lost before the end of antiquity, whereafter Byzantine filters of transmission kicked in. But *our* canon of classical texts consists largely of what the Byzantines decided to keep after that: it is reconstructed from Byzantine manuscripts because it was *their* classical canon (barring accidents of survival and loss).[3] The degree to which the Byzantines capitalized on this advantage and cultivated the scientific patrimony that they preserved, or failed to do so, is a related but separate question.

Byzantium therefore needed translations only in order to access contemporary intellectual developments taking place in neighboring cultures, primarily in the Near East (in Syriac, Arabic, and Persian) and in the Latin West. But for various reasons, the Byzantines did not show much interest in what was happening there, and, by the time they did, they had been overtaken intellectually, probably before they even knew it. What were those reasons?

As the intellectual heirs of the ancient Greeks, henceforth referred to as Hellenes, the Byzantines inherited a body of knowledge and inquiry that had never placed much value on the acquisition and investigation of foreign languages, not even Latin.[4] For example, in the early period (fourth to seventh centuries), when the empire's eastern provinces were home to millions of speakers of Aramaic and Syriac, not a single native speaker of Greek is known who learned how to speak Syriac.[5] This aloofness continued into later times. The Byzantines inherited a strong view of the superiority of their language. While they saw Latin as their own 'ancestral' language, they also regarded it as a 'narrow' tongue, unable to express fine distinctions of thought. The idea that even more 'barbarous' languages could compete would have seemed ludicrous. Thus, the Byzantines saw no reason to praise their own intellectuals or emperors for knowing, learning, or speaking foreign languages; the only exception to this after AD 600 was for having a smattering of legal Latin, which lay at the heart of Byzantine law, and it too was rare. To be sure, the Byzantines were Chris-

3 For a case-study focusing on historiography, see Kaldellis, 'The Byzantine Role'. For the meagre survival of ancient works in science and philosophy, see the statistics and tables given by Goulet and Rudolph, 'Introduction'.

4 Momigliano, *Alien Wisdom*; and Moatti, *The Birth of Critical Thinking*, 80–81.

5 Butts, *Language Change*, 38.

tians, which means that a translated text, the Septuagint, lay at the heart of their religious and intellectual life.[6] But that (ancient) text was regarded as canonical and in need of no improvement. No Byzantine thought to learn Hebrew in order to access a more authentic original.

As Romans, the Byzantines generally disdained non-Romans, whom they regarded as 'barbarians'. The only civilization that some of them were willing to regard as a peer, even if not an equal, was the Persian. Information about Persia was processed primarily through historiography and ethnography, genres that, in a single instance, led to the translation of Persian lore for use by the historian Agathias (late sixth century, Text 1). But ethnography went out of fashion in Byzantium around the same time that Sasanian Persia became extinct. After the seventh century, the Roman empire was surrounded by barbarians who had dismembered and occupied its lands, including the Germanic tribes who took over the West and the 'Saracens' (i.e., Muslim Arabs) who conquered Egypt and Syria. Nothing about those groups at the time of the conquests indicated to the Romans, who believed themselves to be more civilized and boasted exclusive possession of classical *paideia*, that they posed any intellectual challenge. In fact, ancient literature acted as a marker of social stratification *among* Romans, and therefore all the more so between elite Romans and mere barbarians. For centuries after the Germanic and Arab conquests, there was no awareness among the Romans of any potential *intellectual* threat arising from a foreign direction, no idea that the Franks would produce an Abelard or the Muslims an al-Fārābī.

The linguistic conventions of elite literature reinforced this exclusivist attitude. Precisely those authors who might have been the most interested in scientific and philosophical matters adhered to a linguistic code that valorized a high register and Attic purity of diction; failing that, they aimed at a respectable ecclesiastical koine. Getting the language right for the Byzantines was just as important as the content of a text, and for some purposes it was even more important. This entailed suppressing all post-classical idioms, grammar, syntax, and vocabulary—for example, the thousands of Latin words that had entered spoken Greek in late antiquity. Those were to be avoided, along with foreign words, as intrusive blemishes on the Hellenic purity of the text. If a foreign, Latinate, or demotic word had to be mentioned, it was self-consciously glossed and attributed to a different linguistic tradition or register, a reflex that contained and almost sterilized the damage that its mention caused. This attitude was not conducive to engaging with a body of knowledge in a foreign language.

6 Magdalino and Nelson (eds), *The Old Testament in Byzantium.*

The existence of multiple, concurrent registers of Greek meant that texts could be translated from one register to another within the same language, a process that the Byzantines called *metaphrasis*. Thus, texts written in elevated and inaccessible Greek, such as the histories of Anna Komnene and Niketas Choniates, could be rewritten in a more accessible ecclesiastical koine, whereas lower-register texts could be elevated to satisfy more demanding tastes, as in the corpus of saints' *Lives* assembled by Symeon Metaphrastes.[7] While intra-Greek translation is a fascinating topic in its own right, we will here focus on translations from different languages.

Beyond linguistic considerations, Byzantium was blocked for religious reasons from engaging with the learning generated by its first intellectual peer, the Arabic-Persian world of Islam. The Roman Orthodox viewed the Muslim Arabs, who had suddenly burst upon the scene, initially as weird heretics of some sort and later as fully fledged infidels. It also took the eastern Romans a long time to recognize the intellectual advances that were taking place in the medieval West. Initially, the lands of the former western Roman empire were ruled by uncouth barbarians, from whom nothing serious could be expected. Then, over time, these western 'Latins', as the Byzantines regarded them, drifted further and further away from the original Patristic definitions of the faith to the point where they could be regarded as heretics, or at least verging upon it—a view reciprocated by many Latins and especially by the Church of Rome. This erected further barriers of religious suspicion around Latin and Arabic traditions of learning; even their secular thought could be perceived as tainted. The insistence on Orthodoxy in the empire was such that intellectuals who dabbled in suspicious fields, including aspects of ancient Hellenic philosophy, had to defend their studies with loud confessions of faith and protestations of contempt for the foolish and vain learning of the Hellenes. For example, the most famous polymath of the eleventh century, Michael Psellos, did this, and was nevertheless rightly regarded as heterodox in his thinking.[8] In the later period, Demetrios Kydones complained that his study of Latin and translation of Latin texts were taken by his countrymen as propaganda on behalf of the Latin Creed (Text 19), and they were entirely correct in that suspicion.

The Byzantines' dominant religious ideology also adversely impacted their engagement with the ancient scientific tradition, not only with contemporary foreign developments. Hellenic thought and Orthodoxy did not sit comfortably next to each other in the complex patrimony that the Byzantines had inherited from antiquity. While the study of ancient literature, philosophy, and science

7 Efthymiadis, 'Rewriting'.
8 Kaldellis, 'Byzantine Philosophy'.

was prized among the educated, the dominant paradigm explicitly and resolutely rejected any notion that potentially conflicted with Orthodox doctrine, and the punitive powers of both state and Church could be brought to bear on those who ventured too far into 'Hellenic' territory. Thus, Hellenic *paideia* in Byzantium may have been prized in the abstract, but was policed on all sides by repudiations of the folly and short-sightedness of the ancients and declarations of firm (if not always sincere) commitment to the official truth. Although some looked back with admiration on the scientific accomplishment of the ancients, almost none suggested that it was a foundation on which additional discoveries could or should be built, at least insofar as they did not contribute to the elaboration of Orthodox doctrine (a dangerous business in its own right). Christianity was seen at best as the capstone of ancient thought, a culmination that also imposed limits on scientific progress by refuting its allegedly materialist errors and excesses. Byzantine thinkers were not looking to extend or improve upon ancient thought in scientific or philosophical directions, if the latter were unconstrained in advance by Orthodoxy.[9]

Despite these factors, of course, the Byzantines did translate foreign works, as not every person conformed to the prevailing pattern, not all genres were subject to the same constraints, and some circumstances emerged during Byzantium's long history that called for more open engagement. However, the Byzantines translated less than we might otherwise expect from an intellectually advanced post-classical culture, and I have tried to explain why. It is time now to look at what they did translate, when, and why. Most translations, however, had little to do with *scientific* texts.

I note first the ongoing need for translations in various legal, administrative, bureaucratic, and diplomatic spheres. Before the early sixth century, the highest levels of the bureaucracy and court, even in Constantinople, continued to operate in Latin, so interaction between the state and its Greek-speaking subjects often required (routine) translations.[10] But gradually during the fifth and especially the early sixth century, the operations of the state switched to Greek. In the sixth century, a sizable body of Roman law was translated into Greek for the purposes of instruction, and these translations later became official law (especially in the form of the *Basilika*, issued c. 900).[11] Church politics was another arena that required translations, especially in the communications between the Church of Rome and the four patriarchates of the east (Con-

9 Cf. Gutas and Siniossoglou, 'Philosophy and "Byzantine Philosophy"'.

10 Millar, *A Greek Roman Empire*.

11 Stolte, 'Legal Thought'.

stantinople, Antioch, Alexandria, and Jerusalem).[12] This was the case both for their correspondence and for the 'live' translations that took place at Church Councils. Translation was required both at the Councils and subsequently for the written versions of their creeds, canons, and *Acta* (minutes). As the Councils debated doctrinal matters, translation here impinged on theological-philosophical issues and could cause misunderstandings (for example, how was one to render into Latin the distinctions between *hypostasis*, *ousia*, and *prosopon* in the Trinity?). In the middle and late Byzantine periods, the imperial court maintained translators 'on staff' to facilitate diplomacy and write or translate treaties.[13] After the mid-eighth century, the Church of Rome was also treated as a foreign power from the standpoint of translation.

When it came to the translation of philosophical, theological, and scientific works in the (still) unified empire of late antiquity, the main flow was from Greek into Latin, but as we have seen there were vectors of translation from Latin into Greek too. The eastern bishops occasionally needed to know what Rome thought about a particular matter. More importantly, the Romanization of the East, i.e., its adoption of a Roman identity, entailed the wholesale importation and translation of Roman law, as noted above, as well as translations of Eutropius's brief history of Rome, works on horse medicine (possibly for army use), and probably military manuals as well (e.g., likely a translation of Vegetius).[14]

The genre that produced the greatest number of translations in all linguistic directions was undoubtedly hagiography (the lives of saints), which was fuelled by an insatiable desire within the broader Christian community for sensational accounts of heroic martyrdoms and miraculous feats. Greek was the main source language for this genre, but it also received some translations, primarily from Latin and Syriac, and also possibly a handful from Coptic and Armenian.[15] Some of the translations from Latin were made in southern Italy and Sicily, so along the periphery of what would be the Byzantine empire, and those from Syriac were made in the eastern provinces that the empire would lose after the Arab conquests.[16] Christian communities in the east that switched to the use

12 Koltsiou-Nikita, 'Ο ρόλος της μετάφρασης'.

13 See a number of papers in Balard et al. (eds), *Byzance et le monde*, esp. the section 'Le lettre diplomatique'.

14 Roberto, 'Il *Breviarium* di Eutropio'; McCabe, *A Byzantine Encyclopaedia*, 13; and Allmand, *The* De Re Militari *of Vegetius*, 3.

15 For all aspects of Byzantine hagiography, see Efthymiades (ed.), *The Ashgate Research Companion.*

16 Lequeux, 'Latin Hagiographical Literature'; and Peeters, *Orient et Byzance*, 165–218, esp. 179–180 for Syriac (a preliminary list).

of Arabic in later centuries translated some of their own saints' lives into Greek too, after which these texts could circulate in Byzantium. Saints' lives could be written in a high literary style, but most were not: their goal was edification, not verbal purity, so the linguistic anomalies introduced by translation were less of a problem. Moreover, these texts aimed to convey the same set of messages regarding Christian virtue and not channel some kind of 'alien wisdom', so it was unnecessary for the translator to reflect on the fact of translation (though some did).[17] It is not clear whether the exotic setting of some of the translated lives was an attraction. We may surmise that this was the case with *Barlaam and Ioasaph*, a hagiographical tale whose ultimate ancestor was the story of the Buddha, which was translated from Georgian in the late tenth century by the abbot of the monastery of the Georgians (Iviron) on Mt Athos.[18]

Because Greek was spoken outside the Byzantine state, some translations into Greek were made by individuals or within communities that were not part of *Romanía*, for example the post-Roman Melkites of the caliphate. Some of these translations were historically significant, for example of the Qur'ān, which was imported to Byzantium and enabled its theologians to write refutations of Islam, as well as the translation of the Syriac chronicle that lies behind much of the reporting on Near Eastern history found in Theophanes the Confessor (c. 815). Others were less consequential, such as the saints' lives mentioned above.[19] The Byzantines benefited from these texts but were not responsible for their translation. It would therefore be inaccurate to credit them to Byzantium, for they were made under linguistic and religious conditions significantly different from those that characterized the civilization that we study under that name. If we want to argue for an 'expansive' Byzantium in order to claim them, then we have to put our money where our mouth is and begin to include the history of those post-Roman Christians in our conception of Byzantium. So far, there has been no move to do so, probably rightly. These texts, then, were translations into Greek, but not *Byzantine* translations. Let us turn now to the latter.

The first sign that Constantinopolitan intellectuals knew that the Arabs had made intellectual advances is indirect and comes from the tenth century. A story was circulating by then according to which Leon the Philosopher and his students (in the early ninth century) were revealed to be so superior to their Arab counterparts that the caliph offered Leon a job at his court, prompting

---

17 See especially the *Martyrdom of the Sixty New Martyrs*.

18 See the edition by Volk for a thorough study.

19 Ulbricht, *Coranus Graecus*; and Hoyland, *Theophilus of Edessa's Chronicle*.

the emperor Theophilos to make him a generous counter-offer.[20] Historically, the story is impossible nonsense. Yet while it continues to assert Roman intellectual superiority, it also reveals a hint of insecurity about Arab advances. The first explicit statement by a Byzantine that the Arabs had in fact pulled ahead is a rant on that very topic attributed to the philosopher Ioannes Italos by his teacher Michael Psellos, possibly in the 1060s—over two centuries too late.[21]

It is perhaps no coincidence that the next generation witnessed the scientific and literary output of Symeon Seth, a bilingual Greek-Arabic scholar, sponsored by the emperor Alexios I Komnenos (1081–1118). He translated the *Kalīlah wa-Dimnah* under the title *Stephanites and Ichnilates* (Text 9); relied on Arabic medical treatises (probably al-Rāzī) in his *Refutation of Galen* and his treatise on the properties of various foods (Text 3); and was possibly the Greek translator of al-Rāzī's work on smallpox, of which there had been an outbreak in Byzantium a few decades previously (Text 2).[22] In sum, one person, writing at one particular historical moment, was responsible for a great deal of intellectual imports. In his prefaces, Symeon insists on the need to be inclusive when it comes to scientific knowledge, and he also notes that foreign works—even those in the 'Saracen' language—contain crucial supplements and corrections to Galen. For his part, Alexios I was ruling an empire that had just lost Asia Minor to the Seljuk Turks, over whom he was trying to establish some kind of nominal suzerainty while forming an alliance with their former overlord, the Seljuk sultan Malik Shāh. In his case, an appropriation of Muslim learning by his court might have been intended to boost his credibility among these new constituents. The translation of the *Sinbad Tales* for the benefit of Gabriel, the Roman governor of post-Byzantine Melitene who presented himself as an imperial officer surrounded by Syrians, Armenians, Arabs, and Turks, was made in a similar context at the same time (c. 1100)[23] (Text 11).

Even so, Seth's œuvre was but one moment in a longer history of translations from Arabic into Greek that was largely restricted to technical disciplines, including astronomy (Texts 4–6), astrology, medicine, pharmacology, dream interpretation (Text 7), geomancy, and alchemy.[24] There was a quiet hum of translation (or absorption and adaptation) of such works taking place in the

20 Gutas, *Greek Thought, Arabic Culture*, 180; and Sypiański, 'Arabo-Byzantine Relations'.

21 Kaldellis, *Hellenism in Byzantium*, 220.

22 Long in Gutas, Kaldellis, and Long, 'Intellectual Exchanges', 92–98.

23 Toth, 'Authorship and Authority'.

24 For a preliminary survey, see Mavroudi, *A Byzantine Book*, 392–430; and for a preliminary list, with a discussion of the social and ideological context of the translations, see Gutas, 'Arabic into Byzantine Greek'.

background, mostly between the ninth and eleventh centuries, though in some fields (especially astronomy) it would continue until the fourteenth. I call this a 'quiet hum' because, taken together, these translations do not constitute a huge corpus; they often do not even admit that they are translations (this is usually a modern inference), or, if they do admit it, they do not reflect on the epistemological, cultural, or linguistic implications of that fact; some leave it unclear whether they were made from Arabic, Persian, Syriac, or Arabic via Latin; many are not translations as such, but treatises written on the basis of foreign materials or after only a consultation of them; and, finally, they do not seem to have much impacted the mainstream of Byzantine intellectual or literary life, their impact being confined to experts in the disciplines in question. We can talk about the late eleventh-century moment in greater detail because we can associate it with specific texts and individuals, and also because Symeon Seth self-consciously reflected on what he was doing as a translator and importer of Arabic wisdom.

How was that background hum of translation possible, given the strictures of Byzantine intellectual life noted at the beginning of this essay? First, we cannot be sure that all of these Greek translations were in fact made in Byzantium; it is possible that many were made in bilingual environments in the early caliphate and entered the Byzantine manuscript tradition later. Second, some of these technical disciplines were not expected to adhere to high standards of prose in the first place—they were not really 'literature'—and so they could throw foreign words around freely. Third, to a certain degree these disciplines constituted ancient traditions of expertise in which the subject matter—and getting it 'right'—overrode the cultural idiom of their transmission in any one language. Thus, contributions to knowledge could be welcome in them from all quarters, as there was nothing 'Arabic' or 'Greek' about the movement of the stars or the properties of plants. Finally, in the eyes of the Church, astrology in particular was questionable to begin with;[25] adding Arabic sources to it would not be the straw that broke the camel's back.

Only a few of these translations have prefaces that contain even a modicum of reflection on why the work was carried out at all (Texts 2–11). The translations that do have such prefaces likely have them because they were dedicated to an emperor or other Byzantine lord (e.g., Text 11) and protocol required that he be praised as either the sponsor, recipient, or beneficiary of the project; this, in turn, called for a brief justification of the project. We cannot know in these cases—as with many works of Byzantine literature presented to emperors—

25 Magdalino, *L'Orthodoxie des astrologues*.

whether the translation was originally requested by the emperor, or whether it was only presented to him by a scholar seeking some favour: each scenario suggests a significantly different explanation for the decision to translate. The justifications given are, moreover, so broad that they are not helpful in situating the work in a specific moment of intellectual history. Medical translations have practical uses and foreign advances fill the gaps in ancient science (Text 2); or the justification is eclecticism, to sample the best learning that the foreign nations have to offer, possibly in a gesture toward more comprehensive coverage (Texts 3 and 7); or else, in the case of moral literature, because the work is by a foreign philosopher and not yet available among the Romans (Texts 9–11). However, the astronomical tables prepared by Chioniades and Chrysokokkes (Texts 4, 6) seem to be addressed to experts in the field or students, not to rulers.

We turn now to the most concentrated burst of translation in Byzantium, namely from Latin, which flourished between the late thirteenth and the early fifteenth centuries, involving about half a dozen translators, especially Manuel Holobolos, Maximos Planoudes, Demetrios Kydones (who famously converted to Catholicism), and Manuel Kalekas. This can be dubbed the late Byzantine Latin Translation Movement. It took place at a particular moment in Byzantine history and channelled the tensions in the empire's fraught relationship with the Catholic West.[26]

It took a long time for the Byzantines to realize that the 'Celts', 'Franks', or 'Italians' were their intellectual peers, and by the time they did, the latter had advanced beyond them in decisive areas, including logic, science, and metaphysics. The original points of contention were the doctrinal disputes between what we call the Catholic and Orthodox churches (which at the time were usually called the Latin or Roman and Greek churches). It turned out that the Latins could hold their own theologically, just as they could defeat the Byzantines in battle, which they did decisively when they captured Constantinople in 1204. This created a deep sense of insecurity in Byzantine imperial ideology and its intellectual and theological correlates. The emperor-in-exile Theodoros II Laskaris (1254–1258), who presided over a number of doctrinal disputes, boasted that the Hellenes—meaning the ancient Greeks, whom Laskaris took to be ancestors of his own people, and not all pagan everywhere—were naturally superior in all branches of philosophy, whereas the Latins had contributed nothing of their own. He was judging both sides

26 The authors, texts, and methods of the Translation Movement have been studied in a series of brilliant and indispensable articles by Fisher. I cite only two here: 'Planoudes, Holobolos', and 'Ovid's Metempsychosis'.

by their ancient accomplishments. But elsewhere Laskaris revealed a fear that 'philosophy' might be slipping away from the modern-day descendants of the Hellenes and going over to the side of the barbarians.[27]

The Byzantines retook Constantinople in 1261, but the theological debates did not abate. Especially in the aftermath of Aquinas (d. 1274), Byzantine thinkers began to realize that their side had to understand the sources of Latin strength. To be sure, no one planned to do this systematically, at least not initially. The first known translators, Holobolos and Planoudes, took up ancient logical, philosophical, and scientific works (for example, by Boethius and Macrobius), theological treatises (by Augustine), but also literature (by Cicero, Ovid, and, of all things, Juvenal). The literary works appealed to the scholars of that time, who were, incidentally, pioneer philologists of the ancient Greek classics. These translations were initially not dedicated to emperors, but were rather addressed to fellow scholars, or 'friends', who are represented as eager to know more. For example, Holobolos justified his translation of Boethius's work *Concerning Dialectical Topics* as a minor addition to philosophy, coming from the inferior Latin language, to the ancient corpus that was already available to Greek-speakers—'not because we need it, but in order to have a superabundance' (Text 12). Yet his translation contains a stunning coda in which Holobolos, an ardent opponent of Union with the Catholic Church, advises the reader to use Boethius's logic to 'strangle the arrogant Latins', a clear reference to polemical theological intent. His translation of the pseudo-Aristotelian *On Plants* had no such polemical intent. In the preface, Holobolos traces the history of the text, lost from Greek, translated into Arabic, then into Latin, and finally now back into Greek by him (Text 13). Holobolos expresses no derogatory comments toward the Latins or the Arabs, but still he makes it clear that the text was, through his translation, coming back to its original intellectual 'home'.

Maximos Planoudes remained aloof from the theological controversies, but his translation of Augustine's *De trinitate* was of obvious and urgent doctrinal relevance, and its accuracy became itself a topic of debate among later controversialists (Texts 15–18). Especially interesting is the suggestion by Demetrios Kydones that Planoudes was not free to speak candidly because he feared the anti-Union emperor Andronikos II Palaiologos (Text 15). The Latin Translation Movement kicked into high gear in the fourteenth century, when the two Kydones brothers, Demetrios and Prochoros, embarked on a more systematic translation of key works of Catholic philosophy, including more Boethius,

27 Kaldellis, *Hellenism in Byzantium*, 374–375.

Augustine, and plenty of Aquinas. Unlike Holobolos and Planoudes, the Kydones brothers were open champions of Union, and Demetrios in particular wrote a detailed account of how he came to learn Latin, why he began to translate Latin works into Greek, and how this led him to convert to Catholicism (Text 19). This *Apologia* is by far the longest and most detailed account that we have from any Byzantine writer giving an autobiographical account of how he came to learn a foreign language and translate its canonical texts into Greek. Kydones's personal narrative was absolutely wrapped up in his agitation on behalf of Catholicism and Union. Along the way, Kydones also exposed what he took to be the complacency and lazy superiority that kept his fellow countrymen mired in outdated ideas. Many of them assumed that the Latins were still ignorant barbarians and that they, who could read Plato in the original, were still masters of the intellectual domains.

Kydones's remonstrations, interested and partisan as they were, fell upon deaf ears. Some late Byzantine intellectuals did convert and went the way of Rome, but the mainstream of Byzantine culture armoured itself behind a new definition of Orthodoxy (Hesychasm) that largely opposed Western intellectual currents, and, thereby insulated, prepared itself to ride out the centuries of Turkish rule that lay ahead.

The Latin Translation Movement had different contours and a different context from previous Byzantine translations from Arabic (or Persian). The latter were for the most part understood as foreign contributions to fields and genres whose history stretched back to classical antiquity: they slotted easily into existing traditions of learning. A star-chart or alchemical formula here, or a new dream manual or book of botany there, even if 'Saracen' in origin, did not cause major intellectual disruption. Their consumption was usually restricted to experts in these obscure fields, or to the emperors who sponsored this appropriation of foreign knowledge. There was no overt polemic with Islam in all this and zero likelihood or expectation that Byzantine culture had to change in order to accommodate Islamic learning more broadly. Nor did it ever.

The Latin Translation Movement, by contrast, took place at a time when the empire was fighting for its survival and desperately required the assistance of the Latin West in fighting back the Turks. The question of Union lay at the heart of this political dilemma, but it also required the Byzantines to retrace their intellectual steps back to the original formulation of Christian doctrine in late antiquity and then move forward to the present along alternative Latin paths, from Augustine and Boethius to Abelard and Aquinas. This cut to the heart of Byzantine Orthodoxy and identity. Translations were written by and for the many intellectuals, theologians, scholars, and churchmen who were engaged in this fierce contest for the Byzantine soul. It is also possible that some of the translations were made not by Byzantines, but by Western missionaries in Constantinople, who were zealously trying to save Greek heretics or schismatics from their errors.[28] This should give a sense of how tense the context was. The only equivalent on the Arabic/Persian side was (yet another) treatise on the astrolabe that was both written and translated by Šams al-Dīn al-Buḫārī, during a visit to the court of Andronikos II Palaiologos (1282–1328) (Text 5). This is good to have, but was but a minor footnote to the intellectual history of the period. By contrast, the translations of Aquinas that were made during the fourteenth century would prove fundamental to Byzantine thinking about Union, both for and against.[29]

In the end, the case for Union collapsed, as the Byzantines remained staunchly and overwhelmingly anti-Latin; Western armies proved ineffectual against the Turks, even when they were fighting to save Byzantium, which was not often; and, under Ottoman rule, the post-Byzantine Orthodox Romans could, for a very long time, effectively ignore developments in the West.

---

28 Fisher, '*Homo Byzantinus*'.

29 Plested, 'Aquinas in Byzantium'.

Part Two: Texts in Translation

Overview

I. Translation of Persian History and Lore (Sixth Century)
II. Scientific Translations from Arabic and Persian
  (a) *Medical*
  (b) *Dietary*
  (c) *Astronomical*
  (d) *Dream interpretation*
  (e) *Apocalyptic*
III. Moral and Didactic Literature Translated from Arabic and Syriac
IV. The Late Byzantine Latin Translation Movement
  (a) *Translations by Manuel Holobolos* (*c. 1245–c. 1310*)
  (b) *Translations by Maximos Planoudes* (*c. 1255–c. 1305*)
  (c) *Translations by Demetrios Kydones* (*c. 1324–c. 1398*)

I. *Translation of Persian History and Lore* (*Sixth Century*)

Text 1. Agathias, *Histories*, 4.30.2–5 (ed. Keydell, 162.1–6); tr. Frendo, *Agathias: The Histories*, 133–134.

Agathias was a poet, lawyer, and historian from Asia Minor who studied in Alexandria and then practised law in Constantinople, where, in about 580, he wrote a continuation of the *Wars* of Prokopios. The latter had ended his coverage in 551, so Agathias recounted the events of the years 552–559. His narrative focuses, among other military fronts, on war between Rome and Persia and digresses in two passages (*Histories*, 2.23–32, 4.24–30) on the broad sweep of Persian history and Zoroastrian religion. Agathias claims that he found this material in a translation of works kept at the Sasanian court that was made by Sergios (a diplomatic interpreter who is otherwise unknown). Agathias's interests here were not strictly speaking scientific, but this is the most significant passage from early Byzantium that expresses an interest, mediated explicitly through translation, in a foreign culture and its lore.

I have kept my promise and given a complete chronological record of the reigns of the kings of Persia. It is, I think, a true and an accurate one since it is based on Persian sources. Sergios the interpreter managed in fact during a stay in Persia to prevail upon the keepers of the royal archives to grant him access to the relevant literature. He did so, as it happens, in response to frequent requests from me. Fortunately, when he stated that his sole purpose was to preserve even among our nation the memory of what they, the Persians, knew and cherished, they immediately obliged, thinking that it would enhance the prestige of their kings if the Romans too were to learn what kind of men they were together with their numbers and the order and manner in which the succession has been maintained. What Sergios did then was to take the names and dates and principal events and put them into good Greek, a task for which he was peculiarly well-fitted, being far and away the best translator of his day, so much so that his talents had won him the admiration of Chosroes himself and made him the acknowledged master of his subject in both empires.[30] After having made what must have been an extremely accurate translation, he was as good as his word and most obligingly brought me all his material, urging me to fulfil the purpose for which it had been entrusted to him. And that is exactly what I have done. Consequently even if there are some discrepancies between my account of the reign of Kavad[31] and Procopius's version of it, we must follow the authority of the Persian documents and credit their contents with greater veracity.

II. *Scientific Translations from Arabic and Persian*

(a) *Medical*

Text 2. Al-Rāzī, *Discourse on Smallpox* [...] *Translated into our Greek Language*, preface, 3.1–27 (ed. Kouzis); tr. Kaldellis.

We have an as-yet undated Byzantine translation of *Kitāb al-ǧudarī wa-l-ḥaṣbah* by al-Rāzī (d. 925). The Greek title is *Discourse on Smallpox by the Admirable Doctor Razes Translated into our Greek Language from the Tongue of the Syrians* (by 'Syrians' the Byzantines often meant Arabs or Muslims). If the emperor mentioned in the preface is Alexios I Komnenos (1081–1118), then the translator may be Symeon Seth (about whom see below, under *Dietary*), but this a guess. The preface is unusual in offering a long justification, for the need to supplement Galenic medicine (which was canonical in Byzantium) with the advances made by 'Saracens'. The first sentences contain a blatant contradiction between an acknowledgment of Galen's universal coverage of medical matters and his omissions. The overt rationale behind making the translation is to increase medical knowledge and give doctors the means to treat smallpox.

Everyone, O God-fearing and eternally revered emperor, who has meticulously read Galen's books and treatises would grant that he overlooked no component

30 I.e., the Sasanian emperor Khosrow I Anushirwan (531–579).
31 I.e., the Sasanian emperor Kavad (488–531).

of the medical art, nor did anything that contributes to the care and healing of bodies escape his notice. His treatises have been written with no ordinary care, and he has likewise expounded his teaching regarding pestilential diseases. But he notes it merely in passing in some of his works, only as the matter happened to arise, as it was not his purpose to expound a full therapy for it. Thus it is not surprising that some of today's doctors have nothing significant to say about how to respond to this disease, seeing as they have received no adequate teaching about it that is consistent with the art of their leader. What is surprising is that he who put the medical art into order, finding it previously confused, and who brought precision to matters that were previously vague, should have overlooked a matter that is so much more pressing than all the others. There is almost no person who has not been afflicted by this disease—its causes will be explained in the proper place. Indeed, in the years after Galen, men who were learned in Greek, who excelled in both speech and deeds, and who elevated their thoughts to dwell on the loftiest of subjects introduced many (people) to other matters that are also worthy of study, but none of them focused on this issue. I do not know why that was the case: either they lacked the will, or had the will but lacked the ability. We, however, even though we are not able to contribute anything from our own knowledge, will obey your command and set forth the most beautiful and perfect treatise written in the Saracen language by the wise Razes regarding the pestilential disease. We have translated it into the Greek language and bring it before your Majesty, so that your godlike soul may not lack this either, seeing as it is a teaching beneficial to the soul and providential for all in common. And if we are judged by you, our virtuous lord and emperor, to be well-meaning and acceptable, may the good repute for it fall upon you, who instigated and moved us to it and who has led us in every way toward prosperity of all kinds. So let us begin to translate the wise man's teaching into Greek, calling upon the Subsistent Wisdom of God to become our guide. We will traverse the theory laid before us without making any mistakes, deviating to neither side as if we were walking upon a rope.

(b) *Dietary*

Text 3. Symeon Seth, *Alphabetical Treatise on the Powers of Various Foods*, preface, 1.1–10 (ed. Langkavel); tr. Kaldellis.

Symeon Seth was a Byzantine scholar of the second half of the eleventh century who knew Arabic and brought its lore into his scientific writings.[32] He also translated works

32 On Symeon in general, see Gutas, Kaldellis, and Long, 'Intellectual Exchanges', 92–98.

from Arabic into Greek, including the *Kalīlah wa-Dimnah.* He dedicated some of these works to emperors, including Alexios I Komnenos (1081–1118). His *Alphabetical Treatise on the Powers of Various Foods* (*Σύνταγμα κατὰ στοιχεῖον περὶ τροφῶν δυνάμεων*) is not a translation as such, but relied on materials in Arabic. In the preface, Symeon offers a justification for including foreign materials.

Many men, O greatest of emperors, including learned men, have written about the properties of foods, not only Greeks but Persians, Arabs, and Indians too. About some they have gone on at an excessive length, whereas others they have either overlooked entirely or left only a hazy account of them. I therefore thought that it was necessary to collect the best and truest from among all of them, seeing as a treatise of this kind is absolutely necessary, for it contributes maximally to the preservation of health.

(c) *Astronomical*

Text 4. Gregorios Chioniades, *Persian System of Astronomy*, preface, 36.3–20 (ed. Pingree); tr. Pingree, modified.

Gregorios Chioniades (fl. c. 1300, previous name Georgios), a native of Constantinople, translated a number of *zīj* books (tables to be used in astronomical calculations) in his *Persian System of Astronomy* (*Περσικὴ σύνταξις ἀστρονομίας*).[33] The story behind Chioniades's studies, including his travels to and studies in Persian lands east of Trebizond, probably at Tabriz, was told later by Chrysokokkes (see Text 6 below). In his preface, Chioniades does not call his work a translation, but a recording of the oral instructions imparted to him by his teacher Šams al-Dīn al-Buḫārī (b. 1254; see also the next entry). The justification that he gives for this labour is to preserve knowledge from oblivion and to make it more generally intelligible to future generations.

In Greek this book is called an astronomical system [*syntaxis*], but it is called a *zêz* in the language of the Persians. Based on the oral instruction of Samps Bukharis, a man of the Persian race who has studied the entire rational curriculum to its limit, I listened to this teaching of the science (that is) better than any others, which I reduced to writing in order to remember it so that even a long time hence this wonderful science may not be hidden in the depths of forgetfulness; I also did this so that what we have been so enthusiastic about may not, to future generations, appear unintelligible, as do so many things. Because

33 On Chioniades in general, see Pingree, 'Gregory Chioniades', 141–142.

we have laboured on this book as much as was necessary to make it understandable, it will be completely clear even to those who look at it just a little. Indeed, if what has been read clear-sightedly without any instruction, since it is well taught, is still not understood, (it is) not because of us or because of the subject, which, as if spontaneously, has acquired some difficulty, but rather the zealous student will master this book more easily than others that deal with these matters. The teaching runs thus. The scientists of the Arabs and Persians call this book a *zêz*, and that deals with the arrangement of the stars, wandering and fixed, etc.

Text 5. Šams al-Dīn al-Buḫārī, *On the Astrolabe*, preface, 170.16–171.26 (ed. Fisher, 'Arabs, Latins and Persians'); tr. by Fisher, modified.

On the Astrolabe, by Šams al-Dīn al-Buḫārī, was translated into Greek by the author himself from the original Arabic or Persian, which is not extant, apparently after he travelled to the Byzantine court. We do not have a full published edition of the Greek text. Šams al-Dīn al-Buḫārī was probably the teacher of Gregorios Chioniades in Tabriz (see both the previous and next entries). The treatise used to be attributed to Chioniades, but Tihon ('Les tables astronomiques' and 'Traités byzantins') showed that the preface was by Šams al-Dīn, and Fisher ('Arabs, Latins and Persians') argued that the rest of the work was his too, a gift to the Byzantine emperor Andronikos II Palaiologos (1282–1328).[34] In this case, neither the original work (obviously) nor the translation was written by a Byzantine, but the latter was at least done in Greek and possibly in Constantinople. The language and imagery of the translation is not Byzantine.

Moreover, after God created all things, and his prophets appeared like stars in heaven, by whom mankind in the world is led upon its way as sailors at sea are also led upon their way by ⟨the stars⟩, this desire came to me, whose name is Siamps, since I have a name with the same meaning as 'sun' in the Arabic language, though ⟨I do⟩ not by any means ⟨have⟩ that form. Therefore, I took up my own search to find something with a partial resemblance to the sun. Just as the moon does not have its own light but takes ⟨its light⟩ from the sun, because I indeed knew this, I strove ⟨to take light⟩ so to speak from the ruler of the Byzantines ⟨or⟩ sun whose name is 'Victory of Men' [Andro-nikos] and whose surname is 'Words of the Ancients' [Palaio-logos], just as the moon does from

34 Tihon, 'Les tables astronomiques', 'Traités byzantins'; and Fisher, 'Arabs, Latins and Persians'.

the sun—⟨I did this⟩ in consideration of my name. I wished therefore to find something, your Greatness, to be equivalent either to my name or to your disposition. I found no better midpoint between these two things than the astrolabe.

Text 6. Georgios Chrysokokkes, *Commentary on the System of the Persians*, preface, 334–336 (ed. Lambros); tr. Kaldellis (first paragraph), and Mercier, 'The Greek "Persian Syntaxis"', 35–36 (second paragraph, modified).

Georgios Chrysokokkes (fl. c. 1347), a scholar active in Constantinople and Trebizond, wrote an account of the studies of Chioniades (c. 1300) among the Arabs and Persians and his subsequent importation and translation of 'Persian' astronomical knowledge.[35] But those translations lacked commentaries, which Chrysokokkes aimed to provide in his own *Commentary on the System of the Persians* (*Ἐξήγησις εἰς τὴν σύνταξιν τῶν Περσῶν*). It is unlikely that Chrysokokkes himself had direct independent access to Arabic or Persian texts.[36]

The most wise doctor, sir Georgios Chrysokokkes set forth [the following] for his own brother, sir Ioannes Charsanites, on the *Syntaxis* ['System'] of the Persians. As you well know, my dearest Ioannes, some time ago I was studying this Persian manual and had as my teacher a certain priest by the name of Manuel, whose city was Trebizond. You yourself were often in our company and took pleasure in the discussion [...] [Manuel explained to Chrysokokkes] how the *Syntaxis* was originally conveyed out of Persia and was translated by someone into the Greek language. He told us, therefore, that there was a certain Chioniades who was raised in Constantinople. He acquired mastery over the whole of mathematics [*mathemata*] and also fell in love with a different teaching [*mathesis*] through which one might acquire wisdom, and he was also a precise practitioner of medicine. So, when he learned from someone that unless he travelled to Persia he would not be able to obtain that which he desired, he dropped everything as quickly as he could and was off. He passed through Trebizond on his way and spent quite a bit of time in the company of the Grand Komnenos.[37] When he explained his project to him, he received much succour from him in turn. He received many supplies and went off to Persia.

35 For Chioniades, see the two entries above.

36 In general, see Kunitzsch, 'Das Fixsternverzeichnis'; and Mercier, 'The Greek "Persian Syntaxis"'.

37 The rulers of the empire of Trebizond called themselves 'Grand Komnenoi', as they were descendants of that Byzantine imperial family. Their state was a splinter Byzantine state

In a short while, he was taught by the Persians, having both consorted with the king and met with consideration from him. Then he desired to study astronomical matters, but found that they were not taught. For it was the rule with the Persians that all subjects were available to those who wished to study except astronomy, which was for Persians only. He searched for the cause, which was that a certain ancient opinion prevailed among them concerning the mathematical sciences, namely, that their king will be overthrown by the Romans after consulting the practice of astronomy, whose foundation would first be taken from the Persians. He was at a loss as to how he might come to share this wonderful thing. In spite of being wearied, and having much served the Persian king, he had scarcely achieved his objective when, by royal command, the teachers were gathered. Soon Chioniades shone in Persia and was thought worthy of the king's honour. Having gathered many treasures and organized many subordinates, he again reached Trebizond with his many books on the subject of astronomy. He translated these by his own lights, making a noteworthy effort. There are in fact other books of the Persian *Syntaxis* that he translated, those having certain examples with the years systematically at the beginning. However, he handed on the *Syntaxis* alone, the best and most accurate of all, as our teacher said, who appeared to be telling the truth. He translated separately the commentary, which was taken from the Persians by word of mouth alone. In this way, the *Syntaxis*, called the *Handy*, was produced.

(d) *Dream Interpretation*

Text 7. *Oneirokrites of Achmet*, preface, 1.3–14 (ed. Drexl); tr. Oberhelman, *Oneirocriticon of Achmet*, 85.

The *Oneirokrites of Achmet*, a book about dream interpretation written between the ninth and eleventh centuries, is based on a variety of Arabic sources, through which it has indirect access to ancient Greek material.[38] The preface does not evince self-consciousness about the foreign sources used by the author.

During my many and hard toils in examining for my master[39] the accurate interpretation of dreams (for he used to be very zealous about such matters), I have

in north-eastern Asia Minor, formed after the break-up of the Byzantine empire during the Fourth Crusade in 1204.

38 In general, see Mavroudi, *A Byzantine Book*.

39 The end of the work reveals that this patron was an emperor. Mavroudi, *A Byzantine Book*, 61, proposes that it was Leon VI (886–912), though the grounds for this are conjectural.

discovered that great and true precision has been achieved by some, namely the Indians, Persians, and Egyptians: for, having weighed accurately and put in fine detail the truth, they have set in records they kept the present explanation. I have extracted summarily from each of these three sources, and have arranged in particular chapters selected dream-symbols and their interpretations, so that my master, too, having learned the interpretation and result in that very chapter, may make inferences about, and perceive, what is true; thus, he may so experience the sweetness, profundity, satisfaction, and power of the wisdom contained therein: for through this wisdom the foreknowledge of the fulfilment of future events is obtained.

(e) *Apocalyptic*

Text 8. *The Apocalypse of the Prophet Daniel*, preface, 171.4–172.16 (ed. Boudreaux); tr. Fisher, 'Alexios of Byzantium', 179–182 (modified).

The Apocalypse of the Prophet Daniel was translated in about 1245 from Arabic by a certain Alexios, who is otherwise unknown. The translation has an elaborate preface in which Alexios situates his project in the long-term history of the text, starting with the production of the Septuagint.[40]

Ptolemy Philadelphos, king of Egypt, began ruling in the [Seleucid] year 38 [= 275 BC]. During his reign, after enslaving the Hebrews, he compels the translation of all their holy texts from the Hebrew language into Greek, and seventy men distinguished in wisdom among the Hebrews make the translation. Among all these (texts) was also this book of the prophet Daniel, a marvellous and well-accepted book that contains a vision and proclaims future (events) until the end of the world according to fate (as revealed in) the celestial portents that occur. In the days of Constans, emperor of Constantinople (and) descendant of Heraclius [i.e., Konstas II, AD 641–668], Moabias, the leader of the Arabs [i.e., caliph Muʿāwiya, r. AD 661–680], advancing with a huge military force, went as far as Rhodes [AD 673], plundering the territory of the Byzantines, with the result that he subjugated it and laid waste its entire seacoast. After hearing these things and mustering his troops, the emperor went to Phoenicia in order to deploy his army against this Moabias and join battle. The Byzantines are defeated, and Constans himself, rescued with difficulty, returned to the City [i.e., Constantinople] in disgrace. Elated by his victory, Moabias ravaged and despoiled the surrounding territory, even up to Constantinople itself

40 The basic study now (with translation of the preface) is by Fisher, 'Alexios of Byzantium'.

[AD 674–678],[41] with the result that he chanced upon this book and marvelled at its contents, gave (it) to the Arabs, and had (it) translated into their own language, as it has remained up to this day. Chancing upon this (book) in the year 6753 [= AD 1245], a certain Alexios of Byzantium, being enslaved as a prisoner to the kings of the Arabs, knowing the writing system and the language of both [i.e., the Arabs and Romans], and having read the narrative, longed to translate this (narrative) for the Romans exactly as (it) is; (it) bears witness in support of (the translator) and agrees with (him), while he chokes off the vulgar language of the Arabs. He prays that (his) desire to bring this (project) to conclusion be granted him by God, Amen.

III. *Moral and Didactic Literature Translated from Arabic and Syriac*

Text 9. Symeon Seth, *Stephanites and Ichnilates*, preface, 151.1–5 (ed. Sjöberg); tr. Kaldellis.

Stephanites and Ichnilates is a translation by Symeon Seth (c. 1080s) of the popular moralizing animal fable *Kalīlah wa-Dimnah*.[42] The text was ultimately of Indian origin, but was translated by Symeon Seth from the Arabic version made from a Middle Persian intermediary. The beginning of the text, containing the dedication, explains something of the circumstances of the translation, but does not reveal the proximate Arabic context.

A book about the affairs of life. It is laid out through mythical archetypes that match each situation. It is by one of the philosophers in India. It was written at the instigation of the king there, in the local ancestral language and its script. It was translated into Greek in Constantinople at the command of the famous emperor, lord Alexios Komnenos [AD 1081–1118].

Text 10. Anomymous poem on *Stephanites and Ichnilates* (ed. Haskins, *Studies*, 175); tr. Kaldellis.

There is an allegorical iambic poem, in 36 verses, possibly dating to the twelfth century, based on the Greek text produced by Symeon Seth.[43] The following briefer poem in Greek appears to (mis)attribute Seth's translation to Eugenius, an admiral and scholar of twelfth-century Norman Sicily. So while this (Greek) poem is not Byzantine as such,

41 Muʿāwiya did not personally accompany this expedition.

42 For a study, see Condylis-Bassoukos, *Stéphanitès kai Ichnélatès*.

43 See Haskins, *Studies*, 176–177. See also the discussion by Roilos, *Amphoteroglossia*, 135.

but belongs rather to the post-Byzantine Greek-speaking population of Norman Sicily, it comments explicitly on the translated aspect of the text, noting its Indian wisdom, Persian *paideia*, and the barbarous Arabic sound of the original text.

A book of fables drawn from Indian wisdom,
added to (the store of) Persian culture,
focuses on our actions in a riddling way,
making them conducive to the art of living;
it is here translated into the language of the Greeks
from the Arabic and barbarous nonsense
for [by?] the wise, glorious, and great
emir and the king of Sicily and
Calabria, and the Prince of Italy.
[One verse here is obscure, possibly corrupt.]
The noble Eugenios, he of Palermo,
gave this book to us,
as a gift, full of wise teachings.

Text 11. Michael Andreopoulos, *The Book of the Philosopher Syntipas* (*Sinbad Tales*), dedicatory poem and prologue, 1.1–2.9 (ed. Jernstedt and Nikitin); tr. Toth, 'Authorship and Authority', 88, 92, modified.

Michael Andreopoulos translated *The Book of the Philosopher Syntipas*, the *Sinbad Tales*, from Syriac for the benefit of the post-Byzantine *doux* (military governor) of Melitene, Gabriel, c. 1100. Even though Melitene was not controled by the empire after the Seljuk conquest of Asia Minor (1071), Gabriel was a former imperial official and identified with the Roman empire. The work was a Sinbad tale that was translated from Persian into Arabic, then into Syriac, and finally into Greek. The text comes with a dedicatory epigram and a prologue.

Dedicatory poem

The beginning of the book reads as follows. The volume that you behold, my friend, is that of the fable-writer Syntipas [Sinbad], according to [the traditions of] the Syrians, or, rather, the learned Persian authors. I, Michael Andreopoulos, a Christian, and the most humble grammarian, have translated the book, originally written in Syriac, into the Greek that you see before you, and have also written it in my own hand. This book is my own endeavour, and a commission of the illustrious patrician Gabriel, the revered *doux* of the city [Melitene] that derives its name from the word 'honey' [*meli* in Greek], and a truly ardent servant of Christ. He ordered it to be written because the Romans did not possess such a volume. Above all, this story derides evildoers and, towards its end, praises righteous deeds.

Prologue

The prologue of the prototype or collation [*antibolaion*] of the Syriac *Book of Syntipas the Philosopher*, which reads as follows. 'A philosophical tale of the Persian king Cyrus, and his legitimate son, and the son's teacher Syntipas as well as of the king's seven philosophers and one of his wives, a slanderous and cunning woman, who, as the boy's stepmother, launched accusations and lied against him, appealing to the authority of the king. Mousos the Persian composed this story for the benefit of his readers.'

IV. *The Late Byzantine Latin Translation Movement*

(a) *Translations by Manuel Holobolos* (*c. 1245–c. 1310*)

Text 12. Manuel Holobolos, *Translation of Boethius' De topicis differentiis*; preface, 392–393 (ed. Megas), tr. Fisher, 'Planoudes, Holobolos', 86; coda, 145.62–65 (ed. Nikitas), tr. Kaldellis.

Holobolos was a leading orator and scholar of the generation after the Byzantine reconquest of Constantinople from the Latins (1261), and an opponent of the emperor Michael VIII Palaiologos's attempt to force Union with the Catholic Church upon the Orthodox Church. He was also the first in a series of Byzantine scholars to learn Latin and translate Latin philosophical and theological texts into Greek. Among these was Boethius's *De topicis differentiis*. In the preface, he explains that he made a translation because some friends asked him to do so; he also defensively insists that Latin is inferior to Greek in its high register and that the Greek tradition needs no help from Latin texts, but that some of the latter may be translated for the purpose of having an 'overabundant' supply of good things. In the coda to the translation, Holobolos added an odd polemical comment about using the Latins' own learning to strangle them with.

Preface

For these reasons, O Friendship, I have fulfilled your assignment, gladly accepted the toil associated with it, and extracted from the language of Italy [Latin] the treatise by the Roman author Boethius *Concerning Dialectical Topics*, which is eagerly studied by the Italians. I have conveyed the treatise into Greek and grafted it from the wild olive of the (Italians') language—as one might say because it is both unproductive in its literary texts and also earthy in its speech—onto the cultivated olive—I mean, onto our rhetorical language—which is evergreen, luxuriant, and, as the Psalmist says, fruitful [Psalm 51.10]. (I did this) not because our own dialectic now needs even the minimal support from that (Latin) source—no more than the great light-bringing (sun) needs, so to speak, the light of a lamp—but so that what we already possess in sufficiency we might also have to a superabundant degree and so that the nature of the rhetorically sophisticated works produced by the Italians might not entirely escape the notice of those who are also children of Ausonia [i.e., the Byzantines] and who of all people are especially zealous to acquire literary works. But let these remarks suffice for my prologue. I have a few cursory observations about the philosopher Boethius [d. ca. 524 AD], namely that he was a Christian as well as a philosopher, a contemporary of Gregory the Great, the Dialogos [i.e., pope Gregory I, 590–604 AD],[44] who was at the helm of the Church in Rome when Maurice [582–602 AD] completed his imperial reign.

Coda

These, then, are the *Dialectical Topics*, of which the preceding work has given a sufficient account, and we have gone through it with the help of God. Now take these syllogisms, my friend, wrap them around the arrogant Italians, and strangle them.

Text 13. Manuel Holobolos, *Translation of Pseudo-Aristotle's De plantis*, preface, 814a10–815b9 (ed. Drossaart Lulofs and Poortman); tr. Fisher, 'Manuel Holobolos, Alfred of Sareshal', *passim*, modified, with gaps filled in by Kaldellis.

Holobolos also translated, from Latin, the pseudo-Aristotelian *De plantis*. This text, originally written in ancient Greek and combined with one by Theophrastos, had later been translated into Syriac, then Arabic, and then Latin, until finally Holobolos translated it 'back' into Greek. His preface comments on the importance of this translation and on the background of his quest to find the work and translate it, if need be; he also discusses his method of translation, facing a text that was so many steps removed from its original. The Latin translation from which he worked was that of Alfred of Sareshal (twelfth–thirteenth centuries), but Holobolos mentions Alfred as the translator from the Arabic only in a marginal note.[45]

---

44 This is obviously an error, as Boethius and Gregory the Great were not contemporaries.

45 'Βαλφρέδος ὁ ἐκ τῆς Ἀραβικῆς βίβλου μεταβαλὼν τὸ παρὸν βιβλίον.'; Fisher, 'Manuel Holobolos, Alfred of Sareshal', 196. See also Fisher, 'Manuel Holobolos and the Role of Bilinguals', 216–220.

In the ever-blooming field of Aristotle's philosophy, many varied writings have grown up in the manner of the best sorts of plants, and they have brought forth fruit that is very productive for those of us who study them and very prolific for those of us who apply them. In addition to the other writings, there existed this one, *On Plants*, itself like a graceful plant, small, but burgeoning with great delights for those who love learning, and I do not only mean our Ausonians [i.e., the Byzantines], who are superbly and incomparably renowed in literary studies, in wisdom, and in quick intelligence [tr. Fisher], but also the Italians; and not only the latter but any people of the same kind who appear among the blessed Arabians [i.e., Arabia Felix] and the Indians. But in some way—I do not know exactly how—this plant not only became desiccated among us, which would be a change for the worse, as tends to happen over time, but it was even pulled up from the roots and veered into non-existence, as if it had chanced upon no gardener and never seen the light [tr. Kaldellis].

Like us, the Latin-speaking world for a long time suffered the loss of this material. It was known only to the Arabs, and they alone enjoyed its fruit in abundance. However, a certain man, a Celt by birth [i.e., Alfred of Sareshal], advanced to such a degree in the learning of Latin-speakers that he yielded primacy of place to no one, and then he deemed this accomplishment a slight thing, although it was so great, if he did not also acquire an acquaintance with the Arabs second only to his acquaintance with Westerners' Latin and develop his skill to the highest degree. He devotes himself entirely to the task and determines to advance his intention to the limit and to gain his objective completely. He then undertook a foreign journey, employed the Arabs as teachers, applied diligence equal to his natural aptitude, and reputedly came into happy possession of so much education that he was able to translate both the Arabs' literature into Latin and Latin literature into Arabic. This scholar then translates the work *On Plants* in addition to some others and transplants, so to speak, a noble plant from Arabic literary meadows into Latin gardens. The Latin world rejoiced in this and boasted about it exceedingly, for beyond all others they pride themselves on their earnest attention to the writings of Aristotle [...] A long time after this, I too came to desire this plant. For in my opinion, literature is of all things in the world the most worthy of honour. I longed constantly to be sated with its fruits and to possess its contents. Hope urged on my desire and eventually prayer replaced hope. But for a long time, my longings were unfulfilled and my hope remained only hope [...] [But his prayers were answered.] For a certain Latin from someplace or another suddenly appeared to us, a man of considerable literary education and learning; he enters the room, greets us pleasantly, chats in a friendly fashion, then places in our hands the book we longed for! I was delighted at this event, like a gift from God! I took the book with a burning heart; I became wholly engrossed in it. I read its contents with care [...] As if in a glass darkly [1Corinthians 13.12], right out of the starting box I recognized Aristotle and his style of examining and arguing against the philosophers of old. In the name of literature itself and of those who love it, I thought a grave injustice had been done, and I worried that I would be accused of having an evil heart if I did not translate this work into the Greek language and show it numbered among Aristotle's writings when I had the opportunity to do it, by God's grace. I encountered great difficulty and obscurity of meaning because of the frequent retranslations from our language into that of the West, from that into Arabic, from Arabic back again into Latin, and finally from Latin into our language [...] Since every literary work comes to us either in a literal or in a free version when it is translated from one language into another, and since a free translation maintains clarity but sacrifices accuracy, while a literal translation, although accurate, descends into obscurity, we have kept to accuracy and settled for the method that is quite trustworthy and beyond criticism, just as we have done in our other translations. But to avoid the appearance of tolerating obscurity, which we acknowledge is an absolute vice in literature, we have guided the literary wayfarer by means of explanations and marginal annotations if there is any infelicity or awkwardness resulting from obscurity, and we have given him a helping hand out of difficult and rough sections into a smooth and level path while offering grateful acknowledgment to a power beyond our own [tr. Fisher].

(b) *Translations by Maximos Planoudes* (*c. 1255–c. 1305*)

Text 14. Gregorios, epigram on Planoudes's Latin translations; ed. and tr. Fisher, 'Ovid's *Metamorphoses*', 54.

Planoudes was a leading teacher, classical scholar, and translator of numerous texts from Latin. 'After Planoudes' death sometime around 1305, his student Gregorios composed a brief epigram that celebrates Planoudes' translations from Latin and accompanied a visual representation of one of them.'

Stranger, in the fullness of his intellectual capacity, Maximus, servant of the Muses, attained wisdom and set its utmost limits by translating me from the speech of Ausonia [here this means Latin, i.e., ancient Italy] into the Greek language, as a proof of his wisdom and as a joy to future generations.

The prefaces to Planoudes's translations do not reflect on the purposes, context, or goals of translation. However, as some of them were translations of Latin texts with important doctrinal content, which was a topic of debate between the Orthodox and Catholic churches, later readers did comment on the choices he made as a translator and his biases. Key passages by later readers discussing his translation of Augustine's *De trinitate* have been identified and translated by Fisher. The following passages reveal the context of theological polemic in which many of these translations were made.

Text 15. Demetrios Kydones, on Maximos Planoudes's translation of Augustine's *De trinitate*, as reported by Bessarion, *Regarding the Procession of the Holy Spirit*, 312B–C (ed. Migne); tr. Fisher, 'Planoudes' *De trinitate*', 56.

'In a refutation of Planoudes's essay on the procession of the Holy Spirit, the uniate [pro-Union and Catholic] Cardinal Bessarion [in the fifteenth century] quotes Demetrios Kydones [a fourteenth-century Byzantine advocate of Union with the Latin Church, on whom see below].'

What Kydones wrote with his own hand against Planoudes: One would wonder at this man who attempts through these paltry, weak arguments to undermine the book of Saint Augustine over which he labored greatly as he translated it into Greek. For if wishing to harm the readers he submitted to the toil of translation, he is truly unnatural and deserves great blame from the pious, in that he toiled to harm the souls of the faithful. But if knowing that the book would be beneficial to those who would read it he chose to translate it and to do his fellow Greeks a kindness, why in turn as if out of regret does he warn his readers to guard against as harmful what he translated for their benefit? The cause of this unnatural behavior was his fear of the one who was then emperor,[46] whom he knew to be without mercy toward those who said that the Holy Spirit proceeded from the Father and the Son; he desired through these arguments to escape prison and chains.

Text 16. Bessarion, on Maximos Planoudes's translation of Augustine's *De trinitate*, *Regarding the Procession of the Holy Spirit*, 317C–D (ed. Migne); tr. Fisher, 'Planoudes' *De trinitate*', 57.

'After quoting Kydones, Cardinal Bessarion includes his own brief discussion of Planoudes's puzzling behavior in first translating Augustine's *De trinitate* and then attacking western trinitarian views.'

And so you were going to advocate for the Latins rather than to oppose them, O best and wisest of men. For I would not accuse you of the opposite, who

46 I.e., the emperor Andronikos II Palaiologos (1282–1328).

are so well versed in every branch of knowledge, so outstanding in philosophy and insight, so well traveled in all learning, so esteemed in everything, training in the highest degree in the Latin language in addition to Greek, and with the abundance of your ability having translated into our language both many other marvelous things and the great work *De Trinitate* of the great and wondrous Augustine, in which in addition to many other things also the divine procession of the Holy Spirit from the Father and the Son is presented in theological terms in the best and wisest and most inspired manner. How could you yourself oppose him and words as great as his, when you were his foster child and admirer? And to do that, with these arguments (or rather puny protests) that even a child could refute? But wishing to indulge the wrath of some people, I think, or to avoid it, you took this course, both to escape the wrath and more than that to protect the truth by the weakness of your own arguments, making obvious the superiority of the truth even to those who are unable to comprehend it.

Text 17. Anonymous Dominican, on Maximos Planoudes's translation of Augustine's *De trinitate, Contra Graecos*, 421–422 (ed. Dondaine); tr. Fisher, 'Planoudes' *De trinitate*', 58.

'An anonymous Dominican at Constantinople expressed a very different assessment of Planoudes's translation at the time of the translator's death [...] With marked sarcasm the anonymous author complains [...]'.

And in order to be able to defend the errors in which (how sad!) they obstinately had persisted, some of them misrepresented texts, as did the 'venerable' (ha!) monk named Maximus, who died recently in Constantinople. For he translated from Latin into Greek the book *On the Trinity* by the great doctor St Augustine, whom the Fifth Ecumenical Council included among the Doctors of the Church, and yet the aforementioned 'venerable' (ha!) monk passed over in silence and suppressed the procession of the Holy Spirit from the Son as transmitted in that work, thus operating as a forger and not as a translator, and telling a lie not only to mankind but even to God.

Text 18. Gennadios II Scholarios, on Maximos Planoudes's translation of Augustine's *De trinitate, Treatise I on the Procession of the Holy Spirit*, 2, 228.34–229.14 (ed. Jugie et al.); tr. Fisher, 'Planoudes' *De trinitate*', 59–60.

'Gennadios II Scholarios, patriarch of Constantinople immediately after the fall of the city to the Turks [...] directly addresses the theological problem presented by

Planoudes's translation of the Latin verb *procedere*, first surveying the vocabulary used in an acceptable Orthodox statement of the procession of the Holy Spirit and then continuing [...]'.

For the assemblage of our teachers proclaims aloud that the Spirit both is of the Son and goes forth [*proïenai*] from the Son, being made manifest and being given to the Apostles through him [i.e., the Son], and in this way the Spirit is of the Son just as of the Father, and in addition his [i.e., the Spirit's] first cause [*aitias*] is from there [i.e., from the Father], in his eternal existence not set apart from the Son, but remaining in him by nature although distinguished from him in hypostasis and this is said to 'go forth' [*proïenai*] through him [i.e., the Son]. Therefore Augustine speaks very well in agreeing with them in respect to this understanding. He agrees well in everything, except if *procedere* is always translated with 'proceed' [*ekporeuesthai*]; but when it [i.e., *procedere*] is applied to the Father alone, it should be translated with 'proceed' [*ekporeuesthai*] and when applied to the Son, or to the Father and the Son, with 'go forth' [*proïenai*]. And thus that wise Planoudes should have translated his [i.e., Augustine's] words in *On the Trinity*. But he, who always translated *procedere* with 'proceed' [*ekporeuesthai*], either was inept in our theology and for this reason held to the teacher's [i.e., Augustine's] single expression, or he had the Latins' understanding [i.e., of the question] and corrupted his translation, thinking that everyone would attribute it to Augustine whenever they heard him say that he [i.e., the Spirit] proceeds [*ekporeuesthai*] from the Son and from both [i.e., the Father and the Son].

(c) *Translations by Demetrios Kydones (c. 1324–c. 1398)*

Text 19. Demetrios Kydones, *Apologia 1*; paraphrase by Kaldellis.

Demetrios Kydones was the most vocal proponent of Union with the Catholic Church in the fourteenth century, and he personally converted to Catholicism. In his *Apologia*, he explained how he came to learn Latin, convert, and translate a number of Latin texts into Greek. This is the key text for the late Byzantine Latin Translation Movement. It is also the most detailed and personal account of 'vocational' translation that we have from the whole of Byzantine history. It cannot all be quoted here, so I offer only a paraphrase and some highlights. Kydones also talks about his translations in some of his letters (e.g., *Letter* L25).[47]

47 Wright, 'Towards an Edition', 18.

Because of my education, I secured a position at the court of Ioannes VI Kantakouzenos [AD 1347–1354]. I screened petitioners who wanted to see him, including Westerners, but I had trouble understanding them without an interpreter; also, I wanted to converse with the philosophers among them, so I found a tutor to teach me Latin [a Spanish Dominican]. He gave me Aquinas's *Summa contra gentiles* as a reading exercise, but I thought that it needed to be made known to my friends, so I translated some parts of it. I brought these to the emperor's attention, and he said that I should translate the whole thing. I did, and the emperor arranged for it to be copied. My house became a centre of Latin intellectual activity, and the Westerners wanted more to be translated so that the Byzantines did not think that the West was lagging behind in learning. The Byzantines tended to regard the Latins as merchants, sailors, and fighters, not as intellectuals. I therefore translated more and more of their works. The scribes now prospered, because the emperor paid for copies to be made, adding these works to his personal library. I would have done more had my duties at the court not kept me busy. But some criticized me for making the Byzantines look bad by comparison; they didn't like foreign knowledge; and they objected to the idea that one had to know Latin to be truly educated. I was also accused of agitating on behalf of the Latin Creed. My translations had made it clear that the Latins were indeed philosophical and now my countrymen had to deal with their works. It became apparent that the Latins understood the subtleties of Plato and Aristotle better than my countrymen. So I was attacked, even though I sought only to increase our store of knowledge and learning. Moreover, I was primarily interested in the thoughts contained in these texts, not so much in their rhetorical and stylistic ornament. I sought the truth. [This leads to the tale of his conversion to Catholicism.]

Text 20. Demetrios Kydones, *Translation of Passages from Augustine*, concluding remark (section 389), 138 (ed. Kalamakis); tr. Kaldellis.

Kydones concluded a translation of a selection of passages from Augustine with the following comment.

This is all that I found. If anyone searches for the rest and finds it, and then adds what he finds to the present materials, he would confer no small boon on lovers of learning; at the same time, he will please us without making a great effort and he may correct what we have translated. For in many places the Latin text was corrupt, wherefore it was necessary to infer what was being said based on the thought of this blessed man.

Part Three: Original Texts

I. *Translation of Persian History and Lore (Sixth Century)*

Text 1. Agathias, *Histories* 4.30.2–5, 162.1–26 (ed. Keydell).

ἥ τε γὰρ τῶν Περσικῶν βασιλέων διαδοχὴ καὶ ὁ τῶν ἐνιαυτῶν κατάλογος καὶ συλλήβδην φάναι ἅπαν μοι τὸ ἐπηγγελμένον ξυνετελέσθη. οἶμαι δὲ ἄγαν ἀληθῆ ταῦτα καθεστάναι καὶ ἐς τὸ ἀκριβὲς ἐκπεπονημένα, ὡς δὴ ἐκ τῶν Περσικῶν βίβλων μεταληφθέντα. Σεργίου γὰρ τοῦ ἑρμηνέως ἐκεῖσε γενομένου καὶ τοὺς τῶν βασιλικῶν ἀπομνημονευμάτων φρουρούς τε καὶ ἐπιστάτας μεταδοῦναί οἱ τῆς περὶ ταῦτα γραφῆς ἐκλιπαρήσαντος (πολλάκις γὰρ αὐτὸν ἐς τοῦτο προτρέψας ἐτύγχανον), προσθέντος δὲ καὶ τὴν αἰτίαν, ὡς οὐκ ἄλλου του χάριν ζητοίη τῶνδε τυχεῖν ἢ ὥστε καὶ παρ᾽ ἡμῖν ἀνάγραπτα ἔσεσθαι τὰ σφίσιν ἐγνωσμένα καὶ τίμια, παρέσχον εὐθὺς εὖ γε ποιοῦντες ἐκεῖνοι, οὐκ ἄχαρι τὸ χρῆμα εἶναι ἡγούμενοι, ἀλλὰ καὶ πρὸς εὐκλείας τοῖς σφῶν βασιλεῦσιν ἐσόμενον, εἰ μέλλοιεν καὶ παρὰ Ῥωμαίοις γιγνώσκεσθαι, ὁποῖοί τε γεγόνασι καὶ ὁπόσοι, καὶ ὅπως ἡ τοῦ γένους αὐτοῖς ἀποσέσωσται τάξις. λαβὼν οὖν ὁ Σέργιος τά τε ὀνόματα καὶ τοὺς χρόνους καὶ τῶν ἐπ᾽ αὐτοῖς γεγενημένων τὰ καιριώτερα καὶ μεταβαλὼν εὐκόσμως εἰς τὴν Ἑλλάδα φωνήν (ἦν γὰρ δὴ ἑρμηνέων ἄριστος ἁπάντων καὶ οἷος ὑπ᾽ αὐτοῦ Χοσρόου θαυμάζεσθαι, ὡς ἐν ἑκατέρᾳ πολιτείᾳ τὰ πρωτεῖα λαχὼν τῆς ἐπιστήμης), εἰκότως οὖν ἀκριβεστάτην ποιησάμενος τὴν μετάφρασιν ἀπεκόμισέ τέ μοι ἅπαντα μάλα πιστῶς καὶ φιλίως καὶ προὔτρεψε διανύειν τὴν αἰτίαν, ἐφ᾽ ἧπερ αὐτὰ καὶ παρείληφε. καὶ τοίνυν διήνυσται. ὥστε εἰ καὶ Προκοπίῳ τῷ ῥήτορι ἔνια τῶν ἐπὶ Καβάδῃ ἀφηγηθέντων ἑτέρως ἀπήγγελται, ἀλλ᾽ ἡμῖν ἀκολουθητέον τοῖς Περσικοῖς χειρογράφοις καὶ τῶν ἐν αὐτοῖς φερομένων ὡς ἀληθεστέρων ἀντιληπτέον. τούτου δὴ οὖν ἡμῖν τοῦ πόνου τελεώτατα ἐξειργασμένου, φέρε πρὸς αὐτὴν αὖθις ἐπανίωμεν τὴν τῆς ἱστορίας ξυνέχειαν.

II. *Scientific Translations from Arabic and Persian*

Text 2. Al-Rāzī, *Discourse on Smallpox* […] *Translated into our Greek Language*, preface, 3.1–27 (ed. Kouzis).

Ὅτι μὲν οὐδέν τι τῶν συνιστώντων τὴν ἰατρικὴν τέχνην παραλέλοιπε Γαληνός, θεόσεπτε καὶ ἀεισέβαστε αὐτοκράτωρ, οὐδέ τι τῆς τῶν σωμάτων θεραπείας τε καὶ ἐπιμέλειας τὴν αὐτοῦ διάνοιαν ἐκπέφευγε, πᾶσιν οἶμαι διωμολόγηται, ὅσοι τοῖς ἐκείνου συγγράμμασι τε καὶ πονήμασι ἐμμελῶς ἐντετύχηκασιν. Οὐ μὴν ὥσπερ γε τοῖς ἄλλοις πεπόνηται αὐτῷ αἱ πραγματεῖαι, οὕτω δὴ καὶ ἡ περὶ τῆς λοιμικῆς νόσου ἐξήπλωται διδασκαλία. Ἀλλὰ μέμνητο μὲν καὶ ταύτης ἓν τισι λόγοις κατὰ πάροδον, καὶ ἐκ τοῦ προήκοντος, ἐπεὶ μηδὲ σκοπὸν ἔθετο θεραπείαν τινὰ διδάξαι. Ὥστε οὐ θαυμαστὸν εἰ μηδὲν τι ἀξιόλογον ἔνιοι τῶν νῦν ἰατρῶν περὶ τὴν ταύτης ἐπιμέλειαν ἐνδεδείξατο, ἅτε μηδένα ἱκανὴν διδασκαλίαν καὶ ἀπηρτισμένην πρὸς τοῦ τῆς τέχνης καθηγητοῦ παραλαβόντες. Ἐκεῖνο μὲν οὖν θαύματος ἄξιον, ὅπως ὁ τὴν ἰατρικὴν τέχνην συγκεχυμένην οὖσαν διοργανώσας, καὶ περιορίσας τὰ ἀόριστα, κατέλειπε πρᾶγμα τῶν ἄλλων ἁπάντων ἀναγκαιότατον. Οὐδεὶς γοῦν σχεδὸν ἀνθρώπων, ὃς οὐχ ἁλωτὸς τῷ τοιούτῳ νοσήματι καθέστηκεν· ἡ δὲ αἰτία λελέξεται ἐν τῷ προσήκοντι. Καὶ μέντοι καὶ μετὰ τοὺς χρόνους τοῦ Γαληνοῦ ἄνδρες ἐλλόγιμοι τὴν ἑλληνικὴν φωνὴν ἔργῳ καὶ λόγῳ διαπρέψαντες, καὶ τὸν νοῦν πρὸς τὰ ὑψηλότερα τῶν μαθημάτων μετεωρίσαντες, εἰς ἄλλα μὲν τινα σπουδῆς κἀκεῖνα ἄξια ὅτι πολλοὺς ἐξενηνόχασι, ταύτης δὲ οὐδεὶς τούτων ἐμνήσθη, οὐκ οἶδα καθ' ὁπότερον, εἴτε οὐκ ἦν αὐτοῖς βούλησις, εἴτε οὐ προσῆν τῇ βουλήσει καὶ δύναμις. Ἡμεῖς δὲ εἰ καὶ μηδὲν νῦν οἴκοθεν συνεισφέρομεν, ἀλλ' οὖν πειθόμενοι τῷ προστάγματί σου, τὴν παρὰ τοῦ σοφοῦ Ραζῆ καλλίστην καὶ τελειοτάτην περὶ τῆς λοιμικῆς νόσου πραγματείαν σαρακηνικαῖς λέξεσιν ἐκτεθειμένην, εἰς τὴν ἑλληνίδα μεταμειψάμενοι φωνήν, ἠνέγκαμεν τῷ κράτει σου, ὡς ἂν μηδὲ ταύτης ἄμοιρος ἡ θεόμορφός σου εἴη ψυχή, ψυχοφελοῦς οὔσης τῆς σπουδῆς, καὶ κοινοφελοῦς τῆς πρόνοιας. Εἰ δὲ καὶ εὐγνώμονες καὶ δεκτέοι κριθείημέν σοι τῷ ἀγαθῷ δεσπότῃ καὶ βασιλεῖ, σοὶ χάρις ἂν εἴη τῷ προτρέψαντι και κινήσαντι, καὶ καθ' ἑκάστην πρὸς πᾶν ἀγαθὸν ὁδηγοῦντι. Ἀρξώμεθα οὖν τὴν τοῦ σοφοῦ διδασκαλίαν ἐξελληνίσαι, τὴν ἐνυπόστατον Θεοῦ σοφίαν ὁδηγὸν γενέσθαι ἐπικαλούμενοι, ἀπταίστως διελθεῖν τὴν προκειμένην θεωρίαν, καὶ μὴ ἐπὶ θάτερα μετακλίνειν, ἀλλ' ὥσπερ ἐπὶ κάλων βαίνοντες ἀτέρμα διοδεύοντες ἔνθεν κἀκεῖθεν, ἀρρεπεῖς τε καὶ ἄτρεπτοι, μήτε λέξιν ἐπαμφοτερίζουσαν διαλεθεῖν ἡμᾶς, μηδ' αὖ νόημα ἀμφιλαγὲς καὶ ὑψιπετὲς διαδιδράσκειν. Οὗτος τοίνυν ὁ θαυμαστὸς ἀνὴρ εἰς δύο ἤδη τέμνων τὴν νόσον, εἰς ιε΄ τὴν ὅλην πραγματείαν ἐμερίσατο.

Text 3. Symeon Seth, *Alphabetical Treatise on the Powers of Various Foods*, preface, 1.1–10 (ed. Langkavel).

Πολλῶν καὶ λογίων, ὦ μέγιστε βασιλεῦ, οὐχ Ἑλλήνων μόνον, ἀλλὰ καὶ Περσῶν καὶ Ἀγαρηνῶν καὶ Ἰνδῶν περὶ τροφῶν δυνάμεων συγγραψαμένων, καὶ ἐπί τινων μὲν πόρρω τοῦ δέοντος τὸν λόγον προσαγόντων, ἔνια δὲ ἢ πάντως ἀμνημόνευτα καταλιπόντων ἢ μνήμης ἀμυδρᾶς ἀξιωσάντων, δεῖν ᾠήθην ἐξ ἁπάντων τὰ κάλλιστα καὶ τὰ τῆς ἀληθείας ἐχόμενα ἐρανίσασθαι, ἀναγκαίας οὔσης τῆς τοιαύτης πραγματείας, ὡς τὰ μέγιστα πρὸς τὴν τῆς ὑγείας σωτήρησιν ἐπιτελούσης.

Text 4. Gregorios Chioniades, *Persian System of Astronomy*, preface, 36.3–20 (ed. Pingree).

Τὸ βιβλίον τόδε καθ' Ἕλληνας μὲν σύνταξις λέγεται, κατὰ δὲ τὴν τῶν Περσῶν διάλεκτον ζῆζι. ἀπὸ φωνῆς τοίνυν τοῦ Σάμψ Πουχάρης, ἀνδρὸς τὸ γένος Πέρσου πᾶσαν λογικὴν παιδείαν εἰς ἄκρον ἐξησκημένου, ταύτην περὶ τῆς ἐπιστήμης τῆς κρείττονος τῶν ἄλλων [ταύτης] τὴν διδασκαλίαν ἀκήκοα, ἣν καὶ εἰς μνήμην γραφῇ παραδέδωκα ὡς ἂν μὴ τῷ κρόνῳ καὶ αὖθις ἡ θαυμασία ἐπιστήμη τοῖς τῆς λήθης βυθοῖς ἐναποκρυβῇ καὶ τὸ σπουδασθὲν ἐπὶ τοσοῦτον ἡμῖν ὡς τὰ τῶν πολλῶν ἀνόνητον τοῖς ἐσομένοις ἀναφανῇ. ὅτι δὲ πρὸς σαφήνειαν καθ᾽ ὅσον ἔδει τὸ βιβλίον τοῦτο ἐξείργασται παρ᾽ ἡμῶν, ἐντεῦθεν πάντως καὶ τοῖς μικρὰ συνορῶσι δῆλον γενήσεται. εἰ δέ γε καὶ τοῦτο οὕτω πρὸς εὐκρίνειαν λεχθὲν δίχα διδαχῆς, ἐπεὶ καὶ τὰ καλὰ διδακτά, οὐκ ἔτι καταληπτόν, οὐκ ἀφ᾽ ἡμῶν ἢ ἀπ᾽ αὐτοῦ τοῦ πράγματος ἅτε κεκτημένου καθ᾽ αὑτὸ τὴν δυσχέρειαν, πλὴν ἀλλὰ ῥᾷον τούτου ἢ τῶν ἄλλων ὅσα περὶ τῶν τοιούτων διαλαμβάνει ὁ σπουδαῖος περιγενήσεται. Ἔχει δὲ ἡ διδασκαλία ὧδε· οἱ τῶν Ἀράβων καὶ Περσῶν ἐπιστήμονες τὸ βιβλίον τόδε ζῆζι καλοῦσιν, αὐτὸ δὲ τοῦτο διαλαμβάνει τήν τε τάξιν τῶν ἀστέρων τῶν πλανωμένων καὶ ἀπλανῶν.

Text 5. Šams al-Dīn al-Buḫārī, *On the Astrolabe*, preface, 170.16–171.26 (ed. Fisher, 'Arabs, Latins and Persians').

ὕστερον τοίνυν τοῦ δημιουργήσαντος τὰ πάντα θεοῦ· καὶ τῶν αὐτοῦ προφητῶν τῶν ἀναφανέντων δίκην ἀστέρων τῶν ἐν τῷ οὐρανῷ, δι᾽ ὧν ὁδηγοῦνται οἱ ἐν τῷ κόσμῳ ἄνθρωποι ὡς οἱ ἐν θαλάσσῃ πλέοντες καὶ δι᾽ αὐτῶν ὁδηγούμενοι, ἐπῆλθέ μοι αὕτη ἡ προθυμία, ᾧ ὄνομα Σιάμψ. ὁ τῷ ἡλίῳ ὁμωνυμῶν κατὰ τὴν Ἀρράβων φωνήν· οὐ μὴν καὶ κατ᾽ εἶδος· ἐφ᾽ ᾧ καὶ ἐζήτησα αὐτὸς τῷ ἡλίῳ εὑρεῖν ἰδέαν παρεμφερῆ· οἷα δὴ καὶ ἡ σελήνη μὴ ἔχουσα φῶς οἰκεῖον, ἀλλ᾽ ἐξ ἡλίου τοῦτο λαμβάνουσα· διὰ τῆς ἐπιστήμης τοίνυν ταύτης, ἐσπούδασα ὡς ἐξ ἡλίου τοῦ βασιλέως Αὐσόνων, οὗ καὶ τὸ ὄνομα νίκη ἀνδρῶν· καὶ ἡ ἐπωνυμία ἐκ λόγων τῶν παλαιῶν, λαβεῖν ὡς ἐξ ἡλίου διὰ τοῦ ὀνόματος φωτισμὸν ὥσπερ ἡ σελήνη· ἐβουλήθην τοίνυν εὑρεῖν ἐξισούμενόν τι μέγιστε· ἢ ὀνόματι τῷ ἐμῷ, ἢ τρόπῳ τῷ σῷ· κρεῖττον δ᾽ οὐχ εὗρον μέσον ἀμφοῖν, ἢ τὸν ἀστρόλαβον.

Text 6. Georgios Chrysokokkes, *Commentary on the System of the Persians*, preface, 334–336 (ed. Lambros).

τοῦ σοφωτάτου ἰατροῦ κυροῦ Γεωργίου τοῦ Χρυσοκόκκη εἰς τὴν σύνταξιν τῶν Περσῶν ἐκτεθεῖσα πρὸς τὸν αὐτοῦ ἀδελφὸν κὺρ Ἰωάννην τὸν Χαρσανίτην. Πάλαι μου καὶ αὐτός, ὡς οἶσθα, ὦ φίλτατε Ἰωάννη, τοῦ περσικοῦ τοῦδε προχείρου σπουδάζοντος μάθησιν διδασκάλῳ χρώμενου τινὶ ἱερεῖ ᾧ ὄνομα Μανουήλ, πόλεως ὄντι τῆς Τραπεζοῦντος καὶ αὐτός δέ μοι πολλάκις συνὴς καὶ τοῖς λεγομένοις ἡδόμενος καὶ πόρρω τῆς τούτων καταλήψεως εἶναι μὴ ἀνεχόμενος, ἥψω μὲν τοῦ μαθήματος, ἐμοὶ δ᾽ ἐπέτρεψας μεθοδικώτερον ὑποδειγματίσασθαι τὰ λεγόμενα ἐμοῦ τὰς τούτων ἐφόδοις ψιλὰς παρατιθέντος καὶ τῶν διδασκαλικῶν ἐκείνων φωνῶν ὡς οἷόν τε ὑπομιμνήσκοντος ἄκουε. πρῶτον τοίνυν ἐπιμνησθῆναι, τῶν ἐκείνου τινός, ὅπως ἐκ Περσίδος ἐκομίσθη αὕτη ἡ σύνταξις καὶ παρὰ τίνος εἰς τὴν Ἑλλάδα μετηνέχθη γλῶτταν. ἔλεγε τοίνυν ἐκεῖνος ὅτι Χιονιάδης τις ἐν Κωνσταντινουπόλει τραφεὶς καὶ πάντων ἐν καταλήψει τῶν μαθημάτων γενόμενος, εἰς ἔρωτα πεσὼν καὶ ἑτέρας μαθήσεως διαλέκτου δι᾽ ἧς σοφίαν πορίσαιτο καὶ ἰατρικὴν ἀκριβῶς ἀσκήσειεν, ἐπειδὴ παρά τινων ἤκουσεν, ὡς, εἰ μὴ εἰς Περσίδα ἀφίκοιτο, τοῦ ποθουμένου οὐ τεύξεται. πάντων καταφρονήσας ᾗ τάχους εἶχε τῆς ὁδοῦ εἴχετο. εἰς Τραπεζοῦντα ὡς ἐν παρόδῳ ἐλθὼν καὶ τῷ μεγάλῳ Κομνηνῷ ὁμιλήσας χρόνον συχνόν, εἶτα καὶ τὰ τοῦ πράγματος κοινωνησάμενος, μεγίστης ἠξιώθη κηδεμονίας. πολλὰ γὰρ παρ᾽ ἐκείνου λαβὼν ἀναλώματα, εἰς Περσίδα ἀφίκετο. ἐν ὀλίγῳ δὲ τὰ τῶν Περσῶν παιδευθεὶς καὶ τῷ τούτων βασιλεῖ ὁμιλήσας, προμηθείας τε παρ᾽ ἐκείνου τετυχηκώς, ἐπειδὴ τὸ τῆς ἀστρονομίας [μάθημα] μανθάνειν ἐβούλετο μέν, οὐκ εἶχε δὲ τὸν διδάξαντα· νόμος γὰρ ἐν Περσίδι πάντα μὲν τὰ μαθήματα τοῖς βουλομένοις ἐξεῖναι μανθάνειν, ἀστρονομίαν δὲ μόνοις τοῖς Πέρσαις. ὁ δὲ τὴν αἰτίαν ἐξετάσας καὶ μαθὼν δόξαν εἶναι τινα ἐπικρατήσασαν παρ᾽ αὐτοῖς, ὡς φθαρήσεσθαι τὴν ἐκείνων βασιλείαν ὑπὸ ρωμαίων τῇ τέχνῃ τῆς ἀστρονομίας χρώμενων παρ᾽ ἐκείνων πρότερον ταύτης λαβόντες τὰς ἀφορμὰς διηπορεῖτο πῶς ἂν τοῦ τοιούτου μέτασχοι καλοῦ. ὅμως πολλὰ μοχθήσας καὶ πολλὰ δουλεύσας τῷ βασιλεῖ Περσῶν, μόλις τοῦ ποθουμένου τετύχηκε. προστάγματι γὰρ βασιλικῷ τοὺς διδασκάλους συναγαγὼν ἐν ὀλίγῳ μέγας ἐν Περσίδι Χιονιάδης ἐγένετο καὶ τιμῆς ἠξιοῦτο βασιλικῆς. χρήματα δὲ πολλὰ συλλεξάμενος καὶ πολλοὺς ὑπηκόους κτησάμενος εἰς τὴν Τραπεζοῦντα πάλιν ἀφίκετο, πολλὰ βιβλία τοῦ τῆς ἀστρονομίας μαθήματος ἔχων μεθ᾽ ἑαυτοῦ. οἰκείᾳ δὲ γνώμῃ ταῦτα ἐξελληνίσας, μνήμης ἄξιον ἔργον ἐποίησεν. εἰσὶ μὲν οὖν καὶ ἕτερα βιβλία τῆς Συντάξεως τῶν Περσῶν, ἅπερ αὐτὸς ἐξελλήνισεν, ὑποδείγματα τινα μεθοδικὰ ἐν ἀρχῇ ἔχοντα ἐποχῶν. ταύτην δὲ μόνην τὴν Σύνταξιν, ἣν καὶ ὡς κρείττονα πασῶν καὶ ἀκριβεστέραν παρέδωκεν, ὡς ὁ ἡμῶν διδάσκαλος ἔλεγεν ἀληθεύων ἐφαίνετο, χωρὶς ἑρμηνείας ἐξελλήνισεν, οὕτω ταύτην δεξάμενος ἐκ Περσῶν διὰ ζώσης μόνης ἑρμηνευομένην φωνῆς. οὕτως ἐκομίσθη αὕτη ἡ Σύνταξις, ἣ καὶ πρόχειρος λέγεται. ἄξιον τοίνυν ἐκθέσθαι πρότερον τὰ κεφάλαια, ὧν τὰς ἑρμηνείας ἔμπροσθεν λέξομεν.

Text 7. *Oneirokrites of Achmet*, preface, 1.3–14 (ed. Drexl).

Πολλὰ κοπιάσας πρὸς τὸ ἀνερευνᾶν τῷ δεσπότῃ μου τὴν ἀκριβῆ ἑρμηνείαν τῶν ὀνειράτων, καθὼς αὐτὸς δι᾽ ἐπιθυμίας εἶχε πολλῆς, εὗρον ἐκ τῶν ποιησάντων τὴν τοιαύτην ἀκρίβειαν κατὰ ἀλήθειαν, ἤτοι Ἰνδῶν καὶ Περσῶν καὶ Αἰγυπτίων, οἳ τὴν ἀλήθειαν ἀκριβολογησάμενοι καὶ λεπτολογήσαντες ἐξέθεντο καὶ ἐλογογράφησαν τὴν παροῦσαν ἑρμηνείαν. καὶ ἐξ ἑκάστου τούτων ἐκλεξάμενος κεφαλαιωδῶς ἐξεθέμην τῶν τριῶν τὰς κρίσεις καὶ λύσεις ἐν ἑκάστῳ κεφαλαίῳ, ὡσὰν καὶ ὁ ἐμὸς δεσπότης γνοὺς τὴν περὶ τὸ αὐτὸ κεφάλαιον ἑκάστου κρίσιν καὶ λύσιν συλλογίσηται καὶ μάθῃ τὸ ἀληθὲς καὶ πειραθῇ τοῦ γλυκέος καὶ βαθέος καὶ πεποθημένου καὶ δυνατοῦ τῆς παρούσης σοφίας, δι᾽ ἧς ἡ τοῦ μέλλοντος ἔκβασις προγινώσκεται.

Text 8. *The Apocalypse of the Prophet Daniel*, preface, 171.4–172.16 (ed. Boudreaux).

Πτολεμαῖος ὁ Φιλάδελφος ὁ βασιλεὺς Αἰγύπτου ὑπῆρχε βασιλεύσας ἔτη λη'· ἐν τῇ αὐτοῦ βασιλείᾳ οὗτος Ἑβραίους καταδουλωσάμενος εἰς τὴν ἑλληνίδα μεταβαλεῖν ἀναγκάζει φωνὴν ἐκ τῆς ἑβραΐτιδος πάσας τὰς θείας γραφὰς καὶ ποιοῦνται τὴν ἑρμηνείαν ἄνδρες ο' ἐν σοφίᾳ παρὰ τοῖς Ἑβραίοις ἐπαιρόμενοι. ἐν τούτοις ἅπασιν ἦν καὶ ἡ βίβλος ἥδε Δανιὴλ τοῦ προφήτου, ὀπτασίαν ὑπάρχουσα βίβλος θαυμαστὴ καὶ ἔγκριτος, τὰ ἐπερχόμενα ἀναγγέλλουσα μέχρι τῆς συντελείας τοῦ κόσμου κατὰ τύχην τῶν συμβαινόντων αἰθερίων τεράτων. Ἐν δὲ ταῖς ἡμέραις Κῶνστα βασιλέως Κωνσταντινουπόλεως, ἐγγόνου δὲ Ἡρακλείου, Μοαβίας ὁ τῶν Ἀράβων ἀρχηγὸς μετὰ δυνάμεως πλείστης ἐξελθὼν κατῆλθε μέχρι καὶ τῆς Ῥόδου τὴν γῆν τῶν Ῥωμαίων ληΐζων, ὡς καὶ ταύτην κατέστρεψε καὶ πᾶσαν τὴν παραλίαν ταύτης ἐλεηλάτησε. ὁ δὲ βασιλεὺς ταῦτα ἀκούσας καὶ λαὸν συναθροίσας ἦλθεν εἰς Φοίνικα, ὡς τούτῳ τῷ Μοαβίᾳ ἀντιπαρατάξασθαι καὶ πόλεμον συνάψαι. ἡττῶνται Ῥωμαῖοι, ὁ δὲ αὐτὸς Κώνστας μόλις διασωθεὶς ὑπέστρεψεν ἐν τῇ πόλει μετ' αἰσχύνης. ὁ δὲ Μοαβίας ἐπαρθεὶς τῇ νίκῃ ἐπόρθησεν κατεδαφήσας τὴν περίχωρον μέχρι καὶ αὐτῆς τῆς Κωνσταντινουπόλεως, ὡς καὶ τὴν βίβλον ταύτην ἐντυχὼν καὶ τὰ γεγραμμένα θαυμάσας, δοὺς τοῖς Ἄραψιν μετέφρασεν τῇ ἰδίᾳ διαλέκτῳ, μείνασαν ἐν τούτοις μέχρι καὶ τὴν σήμερον. Ἐν δὲ τῷ ‚ϛψνγ' ἔτει ἐντυχὼν τήνδε Ἀλέξιός τις ἀπὸ Βυζάντιον δοῦλος αἰχμάλωτος ὑπάρχων βασιλεῦσι Ἀράβων τὰς γραφὰς καὶ τὴν διάλεκτον ἄμφων γινώσκων καὶ τὴν διήγησιν ἀναγνοὺς ἐπόθησε τοῦ μεταφράσαι ταύτην Ῥωμαίοις καθὼς ὑπόκειται συμμαρτυροῦσαν καὶ συμφωνοῦσαν ἄγχων τὴν τῶν Ἀράβων πεζικὴν γλῶτταν, ὃς καὶ εὔχεται τοῦ δωρηθῆναι αὐτῷ παρὰ θεοῦ προθυμίαν τοῦ εἰς τέλος ταύτην ἐξάξειν· ἀμήν.

III. *Moral and Didactic Literature Translated from Arabic and Syriac*

Text 9. Symeon Seth, *Stephanites and Ichnilates*, preface, 151.1–5 (ed. Sjöberg).

Συγγραφὴ περὶ τῶν κατὰ τὸν βίον πραγμάτων· ἐκτεθεῖσα διὰ μυθικῶν παραδειγμάτων ἁρμοζόντων ἑκάστῃ ὑποθέσει· ὑπό τινος τῶν ἐν Ἰνδίᾳ φιλοσόφων· προτροπῇ τοῦ ἐν αὐτῇ βασιλεύοντος· γραφεῖσα μὲν τῇ πατρικῇ διαλέκτῳ καὶ τοῖς γράμμασιν· ἐξελληνισθεῖσα δὲ ἐν Κωνσταντινουπόλει προστάξει τοῦ ἀοιδίμου βασιλέως κυροῦ Ἀλεξίου τοῦ Κομνηνοῦ.

Text 10. Anomymous poem on *Stephanites and Ichnilates* (ed. Haskins, *Studies*, 175).

μυθικὴ βίβλος ἐξ Ἰνδικῆς σοφίας,
προσενεχθεῖσα πρὸς Περσικὴν παιδείαν,

αἰνιγματωδῶς συντείνουσα τὰς πράξεις,
πρὸς βιωτικὴν συντείνουσα τὰς πράξεις·
ἡ μεταβληθεῖσα πρὸς γλῶσσαν τῶν Ἑλλήνων
ἐξ Ἀραβικοῦ καὶ βαρβαρώδους ὕθλου
παρὰ τοῦ σοφοῦ, ἐνδόξου καὶ μεγάλου
τοῦ καὶ Ἀμηρᾶ, καὶ ῥιγὸς Σικελίας
Καλαβρίας τε πρίνκιπος Ἰταλίας·
οὗσπερ εὐρικῶς, ὡς γνωστικοὺς τοῖς πᾶσιν
τοῦτο δέδωκε πρὸς ἡμᾶς τὸ βιβλίον,
ὥσπερ δώρημα, διδασκαλίας πλέον,
Εὐγενὴς Εὐγένιος, ὁ τῆς Πανόρμου.

Text 11. Michael Andreopoulos, *The Book of the Philosopher Syntipas* (*Sinbad Tales*), dedicatory poem and prologue, 1.1–2.9 (ed. Jernstedt and Nikitin).

Dedicatory poem

Ἡ ἀρχὴ τῆς βίβλου ἔνε οὕτως·
Τοῦ μυθογράφου Συντίπα κατὰ Σύρους,
μᾶλλον δὲ Περσῶν τοὺς σοφοὺς λογογράφους,
αὕτη πέφυκεν ἣν βλέπεις δέλτος, φίλε.
Ἣν καὶ συρικοῖς τοῖς λόγοις γεγραμμένην
εἰς τὴν παροῦσαν αὐτὸς ἑλλάδα φράσιν
μετήγαγόν τε καὶ γέγραφα τὴν βίβλον,
τῶν γραμματικῶν ἔσχατός γε τυγχάνων,
Ἀνδρεόπωλος Μιχαήλ, Χριστοῦ λάτρις,
ἔργον τεθεικὼς προστεταγμένον τόδε
παρὰ Γαβριὴλ τοῦ μεγιστάνων κλέους,
δουκὸς σεβαστοῦ πόλεως μελωνύμου,
ὅς ἐστι Χριστοῦ θερμὸς ὄντως ἱκέτης.
Ὃς καὶ διωρίσατο γραφῆναι τάδε,
ὅτι γε μὴ πρόσεστι Ῥωμαίων βίβλοις.
Ἡ συγγραφὴ γὰρ ἥδε τοὺς κακεργάτας
διασύρει μάλιστα καὶ πρὸς τῷ τέλει
πράξεις ἐπαινεῖ τὰς καλῶς εἰργασμένας.

Prologue

Πρόλογος τοῦ πρωτοτύπου ἤτοι τοῦ ἀντιβολαίου, τῆς συρικῆς βίβλου τῆς λεγομένης Συντίπα τοῦ φιλοσόφου, ἔχων αὐταῖς λέξεσιν οὕτως· Διήγησις ἐμφιλόσοφος, συγγραφεῖσα παρ' ἡμῶν περὶ τοῦ τῶν Περσῶν βασιλέως Κύρου καὶ τοῦ γνησίου τούτου παιδὸς καὶ τοῦ αὐτοῦ διδασκάλου Συντίπα, ἔτι δὲ καὶ περὶ τῶν τοῦ βασιλέως ἑπτὰ φιλοσόφων καὶ τῆς μιᾶς αὐτοῦ τῶν ἄλλων πονηρᾶς καὶ ἀναιδοῦς γυναικὸς καὶ ἧς τῷ βασιλεῖ κατὰ τοῦ υἱοῦ προέθετο διαβολῆς καὶ σκεωρίας οἷα τούτου μητρυιὰ τυγχάνουσα. Ἥντινα διήγησιν προϊστόρησε Μοῦσος ὁ Πέρσης πρὸς τὴν τῶν ἐντυγχανόντων ὠφέλειαν.

IV. *The Late Byzantine Latin Translation Movement*

Text 12. Manuel Holobolos, *Translation of Boethius' De topicis differentiis*, preface, 392–393 (ed. Megas); coda, 145.62–65 (ed. Nikitas).

Preface

Διά τοι ταῦτα τὴν ὑμετέραν, ὦ φιλότης, ἐπιταγὴν ἐκπληρῶν καὶ χαριέντως τὸν ὑπὲρ ταύτης μόχθον ἀναδεχόμενος, τὴν τοῦ Λατίνου Βοετίου *Περὶ τόπων διαλεκτικῶν* πραγματείαν σπουδαζομένην παρ' Ἰταλοῖς, τῆς τούτων φωνῆς ἀποθλίψας πρὸς τὴν Ἑλληνίδα μεταπεπόρθμευκα, ἐξ ἀγριελαίου ἄν τις εἴποι διά τε τὸ μὴ τῶν λέξεων γόνιμον καὶ τὸ τῆς φράσεως χθαμαλόν, πρὸς τὴν καλλιέλαιον μετεγκεντρίσας, τὴν ἡμετέραν φημὶ λεκτικήν, τὴν ἀειθαλῆ τε καὶ τηλεθόωσαν καὶ τὴν ψαλτῳδουμένην κατάκαρπον, οὐχ ὡς δεομένης τάχα τῆς παρ' ἡμῖν διαλεκτικῆς καὶ τῆς ἐντεῦθεν κἂν τὸ βραχὺ συγκροτήσεως, εἴπερ οὐδ' ὁ μέγας φωσφόρος, ἵν' οὕτως εἴποιμι, λυχνιαίου χρῄζει φωτός, ἀλλ' ἵν' ὧνπερ ἔχοιμεν ἱκανῶς τούτων καὶ ὑπερεκπερισσοῦ εὐποροίημεν καὶ τὰ παρ' Ἰταλοῖς σεμνολογούμενα, τίνα πεφύκασι, μὴ πάνυ καὶ τοὺς τῶν Αὐσόνων παῖδας διαλανθάνοιεν, οἷς εἴπερ τισὶν ἄλλοις ἡ τῶν λόγων κτῆσις ἐξαιρέτως πέφυκε περισπούδαστον. Ἀλλὰ ταῦτα μὲν προοιμίου χάριν ἀρκείσθω μοι· περὶ δὲ τοῦ φιλοσόφου Βοετίου βραχέως καὶ ἐπιτροχάδην λέγομεν, ὡς χριστιανὸς ἦν ὁ ἀνὴρ καὶ φιλόσοφος, σύγχρονος τῷ μεγάλῳ Γρηγορίῳ τῷ Διαλόγῳ, τῷ τότε τοὺς τῆς ἐν Ῥώμῃ ἐκκλησίας οἴακας ἰθύνοντι, ὅτε καὶ Μαυρίκιος ἐν τῇ Κωνσταντινουπόλει τὰ τῆς βασιλείας σκῆπτρα διήνυε.

Coda

Οὗτοί εἰσιν οἱ διαλεκτικοὶ τόποι, περὶ ὧν ὅσον ὁ τοῦ προκειμένου ἔργου λόγος ἀπῄτει σὺν Θεῷ διεξεληλύθαμεν. Τῶν συλλογισμῶν ταῖσδε πλεκτάναις, φίλε, τραχηλιῶντας Ἰταλοὺς κρατῶν πνίγε.

Text 13. Manuel Holobolos, *Translation of Pseudo-Aristotle's De plantis*, preface, 814a10–815b9 (ed. Drossaart Lulofs and Poortman).

τῷ τῆς Ἀριστοτελικῆς ἀειθαλεῖ φιλοσοφίας λειμῶνι πολλαὶ μὲν καὶ ἄλλαι δέλτοι κατὰ τὰ εὐγενῆ παραπεφύκασι τῶν φυτῶν, παμφόρους μὲν οἷς θεωροῦμεν, πολύχοας δὲ καὶ οἷς διαπραττόμεθα προϊέμεναι τοὺς καρπούς. ἦν δὲ πρὸς ταῖς ἄλλαις καὶ ἡ περὶ φυτῶν ὅσα καὶ φυτὸν εὐφυὲς βραχὺ μέν, ἀλλὰ συχναῖς ταῖς χάρισι βρύον καὶ κάλλιστοις ἐκφορίοις τοὺς τῶν λόγων ἐραστὰς δεξιούμενον, οὐ μόνον λέγω τοὺς ἡμετέρους τοὺς πάνυ, τοὺς περιφανεῖς τὰ ἐς λόγους καὶ τὴν σοφίαν ἀπαραμίλλους καὶ τὴν σύνεσιν Αὔσονας ἀλλὰ καὶ ὅσους τοιούτους οἶδεν ἡ τῶν εὐδαιμόνων Ἀράβων τρέφειν καὶ τῶν Ἰνδῶν. συχνὸν γὰρ ἐνέβαλεν ἵμερον τῶν ἑαυτοῦ συγγραμάτων τοῖς ὁπουδήποτε φιλοσόφοις ὁ ἐκ Σταγείρων ἄκρος φιλόσοφος. ἀλλὰ τὸ τοιοῦτον φυτὸν οὐκ οἶδ᾽ ὅπως οὐ μόνον αὔανσιν πέπονθε παρ᾽ ἡμῖν, ἀλλοίωσιν τινα τάχα περὶ το χεῖρον, οἷα φιλεῖ χρόνος ποιεῖν ἀλλά καὶ προθέλυμνον ἐκκοπὴν καὶ εἰς τὸ μὴ ὂν ἀποστροφήν, ὡς οὐδ᾽ ἂν ποτε φυτηκόμον εἶδεν, οὐδ᾽ εἰ προῆλθεν εἰς φῶς. ὑπέστη τὰ ἶσα καὶ Ἰταλὸς μακρῶν χρόνων σὺν ἡμῖν χηρεύων τοῦ χρήματος. τοῖς Ἄραψιν ἐγνωρίζετο μόνοις, καὶ μόνοι τῶν τούτου καρπῶν ἀφθόνως ἀπήλαυον. ἀλλὰ τις Κελτὸς μὲν τὸ γένος, ἐς τοσοῦτον δὲ τῆς τῶν Ἰταλῶν σοφίας ἐλάσας ὡς τῶν πρωτείων παραχωρεῖν μηδενί, εἶτα βραχὺ τοῦτο, καίπερ ὂν [μὴ] τοσοῦτον, καθ᾽ ἑαυτὸν οἰηθείς, ἂν μὴ καὶ τὴν τῶν Ἀράβων μετὰ τὴν τῶν Ἰταλῶν μεταλάβοι γνῶσιν καὶ τὴν καλλίστην ἕξιν ἀσκήσειεν, ὅλος γίνεται τοῦ ἔργου καὶ εἰς πέρας τὴν γνώμην ἀγαγεῖν θέλει καὶ τύχειν παντοίως τοῦ ἐφετοῦ. στέλλεται λοιπὸν τὴν ἀλλοδαπήν, χρῆται τοῖς Ἄραψι παιδευταῖς καὶ τῇ εὐφυίᾳ συμμετρήσας σπουδὴν τοσοῦτον εὐμοίρησε παιδείας, φασίν, ὡς δύνασθαι καὶ τὰ τῶν Ἀράβων εἰς Ἰταλοὺς καὶ τὰ τῶν Ἰταλῶν μεταπορθμεύειν εἰς Ἄραβας. οὗτος οὖν πρὸς τισιν ἄλλαις καὶ τὴν περὶ φυτῶν τήνδε βίβλον μεθερμηνεύει, κἀκ τῶν λογικῶν Ἀραβικῶν λειμώνων ὅσα καὶ φυτὸν εὐγενὲς εἰς τὰς Ἰταλικὰς μετακηπεύει φυταλιάς. ἔχαιρε γοῦν Ἰταλὸς ἐπὶ τοῦτῳ καὶ τὰ μεγάλα ἐκόμπαζε· ταῖς Ἀριστοτελικαῖς γὰρ φιλοτιμεῖσθαι βίβλοις ὑπὲρ πάντας βρενθύεται· καὶ τοσοῦτον ἐνεκαυχᾶτο τῷ χρήματι, Ξέρξης οὐχ ἧττον ἐκεῖνος εἰς ἣν ὑπεσκιάσθη καὶ ἣν χρυσέαν ἔδειξε πλάτανον. ἦλθον εἰς ἐπιθυμίαν (μακρὸς ἐξότου χρόνος) κἀγὼ τοῦ φυτοῦ· λόγοι μοι τῶν τοὺ κόσμου πάντων χρῆμα τὸ τιμιώτατον, ἐπόθουν ἐμφορηθῆναι τῶν τούτου καρπῶν, τῶν νοημάτων εἰς ἕξιν ἐλθεῖν. ἐψυχαγώγει μοι τὴν ἐπιθυμίαν ἐλπίς, καὶ τὴν ἐλπίδα πάλιν εὐχὴ διεδέχετο. ἀλλὰ μέχρι πολλοῦ τὰ τῆς ἐπιθυμίας ἀτέλεστα. ἡ ἐλπὶς καὶ μόνον ἐλπίς· τὰ δὲ τῆς εὐχῆς ἐπετείνετο. τί δ᾽ ἡ τῶν ἀγαθῶν ἁπάντων παντελὴς αἰτία καὶ ὑπερτελὴς καὶ προτέλειος; ἠρεμίαν τῇ ἐπιθυμίᾳ χορηγεῖ, τῇ ἐλπίδι πέρας, καὶ τὰ τῆς αἰτήσεως τῇ εὐχῇ. Ἰταλὸς γάρ τις ἔκποθεν ἡμῖν αἰφνιδίως ἀναφανείς, ἱκανῶς ἔχων λογικῆς παιδείας καὶ γνώσεως, πρόσεισιν ἐς ἡμᾶς, ἀσπάζεται χαριέντως, τὰ φιλικὰ προσλαλεῖ, εἶτα τὴν ποθουμένην βίβλον ἡμῖν ἐγχειρίζεται. ηὐφράνθη αὐτὸς ὡς ἐφ᾽ ἑρμαίῳ τῷ χρήματι. ἐδεξάμην τὴν βίβλον ἐκθύμως. ὅλος αὐτῆς ἐγενόμην, τὸν ἐγκείμενον νοῦν ἀνελεξάμην ἐπιμελῶς, καὶ οἷον ὡς ἐν ἐσόπτρῳ τῇδε πρώτης αὐτῆς βαλβῖδος τὸν Ἀριστοτέλη ἀνέγνων καὶ τοὺς Ἀριστοτελικοὺς κατὰ τῶν πάλαι φιλοσόφων ἀντιλογικοὺς ἐτασμούς. ἀδικεῖν δ᾽ ᾠήθην, μὰ τοὺς λόγους, τοὺς λόγους αὐτοὺς καὶ τοὺς αὐτῶν ἐραστάς, καὶ γραφῆναι βασκάνου ψυχῆς ηὐλαβήθην, ἂν μὴ ταύτην ἱκανῶς ἔχων ὑπὸ τῆς ἄνωθεν χάριτος πρὸς τὴν Ἑλλάδα μεθερμηνεύσαιμι γλῶτταν καὶ ταῖς λοιπαῖς Ἀριστοτελικαῖς δέλτοις δείξαιμι συνάριθμον. τῷ τοι καὶ σὺν θεῷ τὸ ἔργον διήνυσα, καὶ εἰς κοινὴν προὐθέμην τὴν βίβλον τοῖς αἱρουμένοις ἀπόλαυσιν, πολλὴν εὑρὼν τὴν δυσχέρειαν καὶ τὴν τῶν νοημάτων σύγχυσιν διὰ τὰς συχνὰς τῶν μεθερμηνεύσεων μεταβολὰς ἐκ τῆς ἡμετέρας εἰς τὴν τῶν Ἰταλῶν, εἰς τοὺς Ἄραβας ἐκεῖθεν, καὶ πάλιν ἐξ Ἀράβων εἰς Ἰταλούς, καὶ εἰς ἡμᾶς ἐκ τούτων ἐπ᾽ ἔσχατον. οὐ γὰρ διπρόσωπος, ὅσον ἐπὶ ταύτῃ τῇ βίβλῳ, ὁ Σταγειρίτης ὤφθη κατὰ τὸν θρυλλούμενον Ἴανον ἐκεῖνον, ὅν καὶ εἰς χρόνον ἄγει λόγος ἰσχνός, ἀλλὰ πολύμορφος κατὰ τὸν Πρωτέα τὸν Φάριον. ἐπεὶ δὲ πᾶσα δέλτος ἐκ μίας πρὸς ἑτέραν μεταφερομένη διάλεκτον ἢ κατὰ λέξιν ἢ κατ᾽ ἔννοιαν προχωρεῖ, καὶ ἡ μὲν ὑστέρα τὸ σαφὲς ἔχουσα τοῦ ἀκριβοῦς ἀπολείπεται, ἡ δ᾽ ἑτέρα καὶ πρότερα προσόντος τοῦ ἀκριβοῦς εἰς ἀσάφειαν καταστρέφεται, ἡμεῖς ὡς ἀσφαλεστέρῳ καὶ ἀδιαβλήτῳ τούτῳ τῷ μέρει στοιχοῦντες, καθὼς καὶ ἐν ἑτέραις πεποιήκαμεν δέλτοις, τῆς ἀκριβείας ἐχόμεθα. ἵνα δὲ μὴ φανοίμεθα καὶ χώραν τῇ ἀσαφείᾳ διδόντες, ἣν κακίαν παντελῆ τοῖς λόγοις ἐγνώκαμεν, ἐπανερμηνείαις τισὶ καὶ σχολικαῖς παρασημειώσεσιν, εἴ τί που ἐκ τῆς ἀσαφείας ἄχαρι καὶ ἀτερπὲς ἐψυχαγωγήσαμεν, κἀκ τῆς σκληρᾶς καὶ τραχείας τοὺς λογικοὺς ὁδίτας εἰς τὴν λείαν τε καὶ πλατεῖαν ἐχειραγωγήσαμεν, καὶ τῷ ὑπὲρ ἡμᾶς τὴν εὐχαριστίαν καὶ τήν, ὡς εἰκός, ἀνθομολόγησιν ἀπενείμαμεν. τοῦτο γὰρ ἀφ᾽ ἡμῶν ἀπαιτεῖ τῶν ὑπὸ χεῖρα ὁ μόνος σοφὸς καὶ δυνατὸς θεὸς κατὰ τὸν θεῖον ἀπόστολον. τίνος γὰρ ἄλλου δεῖται παρ᾽ ἡμῶν, ὃς διά τινος τῶν προφητῶν περὶ ἑαυτοῦ λέγει τὸν οὐρανὸν πληροῦν καὶ τὴν γῆν, καὶ οὗ ἡ οἰκουμένη καὶ τὸ πλήρωμα αὐτῆς, κατὰ τὸν Δαυίδ, καὶ μεθ᾽ οὗ πᾶς ἀγρὸς παντοίοις ὡραϊσμένος φυτοῖς;

Text 14. Gregorios, epigram on Planoudes's Latin translations (ed. Fisher, 'Ovid's *Metamorphoses*', 54).

Αὐσονίης διαλέκτου μ᾽, ὦ ξεῖν᾽, ἐκμετενεγκὼν
Μάξιμος ἀρτίφρων μουσοπόλος σοφίης
ψαύσας ἄκρα γε πείρατα θῆκεν ἐς Ἑλλάδα φωνήν,
δεῖγμα τε ἧς σοφίης ἐσ⟨σ⟩ομένοις δὲ χάριν.

Text 15. Demetrios Kydones, on Maximos Planoudes's translation of Augustine's *De trinitate*, as reported by Bessarion, *Regarding the Procession of the Holy Spirit* 312B–C (ed. Migne).

Δημητρίου τοῦ Κυδώνη κατὰ Πλανούδη ἐξ οἰκειοχείρων γραφέντα. Θαυμάσειέ τις τὸν ἄνδρα τοῦτον, τὸν διὰ τῶν ὀλίγων τούτων καὶ ἀσθενῶν ἐπιχειρημάτων πειρώμενον τὸ τοῦ ἁγίου Αὐγουστίνου βιβλίον ἀνατρέπειν, ἐφ᾽ ᾧ πόλλ᾽ ἐμόγησεν ἐπὶ τὴν Ἑλλάδα μετενεγκών. εἰ μὲν γὰρ βλάψαι βουλόμενος τοὺς ἀναγινώσκοντας τὸν ἐν τῇ ἑρμηνείᾳ πόνον ὑπέστη, ἄτοπος ὄντως καὶ πολλῆς τοῖς εὐσεβέσιν ἄξιος μέμψεως, πονεῖν ἐφ' ᾧ ταῖς ψυχαῖς τῶν ἄλλων λυμήνασθαι· εἰ δ' ἐπιστάμενος, ὠφέλιμον τοῖς ἐντευξομένοις ἔσεσθαι τὸ βιβλίον, μετενεγκεῖν τοῦτο εἵλετο, καὶ τοῖς ὁμοφύλοις χαρίσασθαι, τί πάλιν ὥσπερ ἐκ μεταμέλειας, ὡς βλαβερὰ φυλάττεσθαι παραινεῖ, ἃ δι' ὠφελειαν τῶν ἐντυγχανόντων ἡρμήνευσεν; ἀλλὰ τῆς ἀτοπίας ταύτης αἴτιον ὁ τοῦ τότε βασιλεύοντος φόβος· ὃν εἰδὼς ἀπαραίτητον ὄντα τοῖς τὸ Πνεῦμα τὸ ἅγιον ἐκ Πατρὸς καὶ Υἱοῦ λέγουσιν ἐκπορεύεσθαι, ἠθέλησε διὰ τῶν ἐπιχειρημάτων τούτῳ εἱρκτῆς καὶ δεσμῶν ἑαυτὸν ἀπαλλάξαι.

Text 16. Bessarion, on Maximos Planoudes's translation of Augustine's *De trinitate, Regarding the Procession of the Holy Spirit* 317C–D (ed. Migne).

καὶ οὕτω συνηγορήσων Λατίνοις, ἢ ἐλέγξων ἐλήλυθας, ἀνδρῶν ἄριστε καὶ σοφώτατε. οὐ γὰρ ἂν ἐγὼ σου καταγνοίην τοὐναντίον, οὕτω δὲ διαβεβηκότος ἐν φιλοσοφίᾳ καὶ θεωρίᾳ, καὶ διὰ πάντων μὲν μαθημάτων ὁδεύσαντος, εὐδοκιμήσαντος δ' ἐν ἅπασι, πρὸς δὲ τῇ Ἑλλάδι καὶ τὴν Λατίνων γλῶτταν ἐς ἄκρον ἐξησκημένου, καὶ περιουσίᾳ δυνάμεως ἄλλα τε πολλὰ θαυμαστὰ καὶ τὸ μέγα *Περὶ Τριάδος* τοῦ μεγάλου καὶ θαυμαστοῦ Αὐγουστίνου ἔργον εἰς τὴν ἡμετέραν μετενεγκότος φωνὴν ἐν ᾧ πρὸς ἄλλοις πολλοῖς καὶ ἡ ἐκ Πατρὸς καὶ Υἱοῦ τοῦ Ἁγίου Πνεύματος θεία ἐκπόρευσις ἄριστα τε καὶ σοφώτατα θεολογεῖται καὶ ἐνθεώτατα· ᾧ καὶ τοῖς τοσούτοις αὐτοῦ λόγοις πῶς ἂν αὐτὸς ἐναντιοῦσθαι δύναιο, τρόφιμος ὢν ἐκείνῳ καὶ ἐραστής; καὶ ταῦτα τούτοις τοῖς λόγοις, ἢ ῥηματίοις, ἃ κἂν παῖς ἐξελέγξειν; ἀλλ' οἶμαι τῷ τινων θυμῷ χαρίσασθαι, ἢ ἐκκλῖναι βουλόμενος, ταύτην εὗρες ὁδόν, ἐκεῖνον τε φυγεῖν, καὶ τῇ ἀληθείᾳ μᾶλλον ἀμῦναι τῇ ἀσθενείᾳ τῶν λόγων, τὴν ἐκείνης περιουσίαν καὶ τοῖς μὴ συνορᾶν δυναμένοις ἀνακαλύπτων.

Text 17. Anonymous Dominican, on Maximos Planoudes's translation of Augustine's *De trinitate, Contra Graecos*, 421–422 (ed. Dondaine).

Et ut defendere possint errores in quibus pertinaciter, proch dolor [!] perseverant, quidam eorum scripturas corrumpunt, sicut fecit kalogerus nomine Maximus, qui nuper obiit in Constantinopoli. Nam de latino in grecum transtulit librum De trinitate magni doctoris sancti Augustini, quem quinta ycumenica synodus connumerat doctoribus orthodoxis; processione vero Spiritus Sancti a Filio in eodem libro traditam, dictus kalogerus subticuit et suppressit, profecto falsarii et non translatoris functus officio, et non solum hominibus mentitus sed Deo.

Text 18. Gennadios II Scholarios, on Maximos Planoudes's translation of Augustine's *De trinitate*, *Treatise I on the Procession of the Holy Spirit*, 2, 228.34–229.14 (ed. Jugie et al.).

Εἶτα ἐκ τούτων ἁπάντων συμπεραίνει καὶ τοῦ Πατρὸς καὶ τοῦ Υἱοῦ πνεῦμα εἶναι τὸ ἐν Τριάδι λεγόμενον Πνεῦμα ἅγιον καὶ ἐκ Πατρὸς τοῦτο καὶ Υἱοῦ προϊέναι. Ταῦτα τοίνυν καὶ πάνυ καλῶς λέγει. Τὸ Πνεῦμα γὰρ καὶ τοῦ Υἱοῦ εἶναι καὶ ἐκ τοῦ Υἱοῦ προϊέναι τοῖς ἀποστόλοις δι' αὐτοῦ φανερούμενον καὶ διδόμενον, καὶ οὕτως εἶναι τὸ Πνεῦμα τοῦ Υἱοῦ, καθάπερ καὶ τοῦ Πατρός, πλὴν τῆς ἐκεῖθεν αἰτίας, οὐδ' ἐν τῇ ἀϊδίῳ ὑπάρξει τοῦ Υἱοῦ διϊστάμενον, ἀλλὰ μένον μὲν ἐν αὐτῷ κατὰ φύσιν, διακρινόμενον δὲ αὐτοῦ καθ' ὑπόστασιν, καὶ οὕτω δι' αὐτοῦ λεγόμενον, προϊέναι, ὁ τῶν ἡμετέρων διδασκάλων κύκλος βοᾷ. Οὐκοῦν καὶ Αὐγουστῖνος ἄριστα λέγει, πρὸς τὴν αὐτὴν ἐκείνοις συντρέχων διάνοιαν· συνδραμεῖται δὲ καὶ ἐν πᾶσι καλῶς, εἰ μὴ ἀεὶ τὸ προκέδερε πρὸς τὸ ἐκπορεύεσθαι μεταβάλλοιτο, ἀλλ' ὅτε μὲν πρὸς τὸν Πατέρα ἀναφέρεται μόνον πρὸς τὸ ἐκπορεύεσθαι ἑρμηνεύοιτο, ὅτε δὲ πρὸς τὸν Υἱόν, ἢ τὸν Πατέρα καὶ τὸν Υἱόν, εἰς τὸ προϊέναι. Οὕτω δὲ καὶ τὸν σοφὸν ἐκεῖνον Πλανούδην τοὺς περὶ Τριάδος λόγους ἑρμηνεύειν ἐχρῆν· ὃ δέ, τὸ προκέδερε ἀεὶ πρὸς τὸ ἐκπορεύεσθαι μεταβεβληκώς, ἢ τῆς καθ' ἡμᾶς θεολογίας ἄπειρος ἦν, καὶ διὰ τοῦτο τῇ λέξει μόνῃ τοῦ διδασκάλου προσεῖχεν, ἢ τὰ Λατίνων φρονῶν ἐκακούργει τὴν ἑρμηνείαν, οἰόμενος πάντας τοῖς Αὐγουστίνου λόγοις ἐνδώσειν, ἐπειδὰν ἀκούωσιν αὐτὸν ἐκ τοῦ Υἱοῦ καὶ ἐξ ἀμφοτέρων ἐκπορεύεσθαι λέγοντα.

Text 19. Demetrios Kydones, *Apologia I*.

[The translation, as explained above, is a paraphrase of a longer text, which cannot be given here.]

Text 20. Demetrios Kydones, *Translation of Passages from Augustine*, concluding remark (section 389), 138 (ed. Kalamakis).

Ταῦτα εὕρηται μόνα· εἰ δέ τις καὶ τὰ λοιπὰ ζητήσας εὕροι καὶ τοῖς παροῦσι τὰ εὑρημένα προσθείη, οὐκ ὀλίγον ὠφελείας συνεισοίσει τοῖς φιλολόγοις, ἅμα δὲ καὶ ἡμῖν χαριεῖται οὐκ ὀκνῶν καὶ τὰ παρ' ἡμῶν ἑρμηνευθέντα διορθώσασθαι· ἐν πολλοῖς γὰρ ἦσαν ἐφθαρμένα τὰ ῥωμαϊκὰ γράμματα· ὅθεν τὴν τοῦ μακαρίου τούτου διάνοιαν ἀναλαμβάνοντας ἐχρῆν ἑρμηνεύειν.

Bibliography

Primary Sources

Agathias, *Histories* (*Ἱστορίαι*), ed. R. Keydell, *Agathiae Myrinaei Historiarum Libri Quinque*, Berlin: de Gruyter, 1967; tr. J.D. Frendo, *Agathias: The Histories*, Berlin/New York: de Gruyter, 1975

al-Rāzī, *Discourse on Smallpox by the Admirable Doctor Razes Translated into our Greek Language from the Tongue of the Syrians* (*Ῥαζῆ θαυμαστοῦ ἰατροῦ λόγος περὶ λοιμικῆς ἐξελληνισθεὶς ἀπὸ τῆς Σύρων διαλέκτου πρὸς τὴν ἡμετέραν*), ed. A.P. Kouzis, *Ραζῆ λόγος περὶ λοιμικῆς: Ἐξελληνισθεὶς ἐκ τῆς Σύρων διαλέκτου πρὸς τὴν ἡμετέραν*, Athens: Royal Press of Chiotis and Rouseas, 1909; online at http://soranos.lib.uoc.gr/metadata/c/e/9/metadata-003-0000229.tkl# (accessed 15 July 2012)

Anonymous Dominican, *Contra Graecos*, ed. A. Dondaine, '"Contra Graecos": Premiers écrits polémiques des dominicains d'Orient', *Archivum fratrum praedicatorum*, 21 (1951), 320–446

The Apocalypse of the Prophet Daniel (no Greek title), ed. P. Boudreaux in *Catalogus Codicum Astrologorum Graecorum*, 8.3: *Codices Parisini*, Brussels: Lamertin, 1912, 171–179

Barlaam and Ioasaph (variously titled in the Greek, from *Διήγησις Βαρλαὰμ καὶ Ἰωάσαφ* to *Ἱστορία ψυχωφελὴς ἐκ τῆς ἐνδοτέρας τῶν Αἰθιόπων χώρας, τῆς Ἰνδῶν λεγομένης, πρὸς τὴν ἁγίαν πόλιν μετενεχθεῖσα διὰ Ἰωάννου μοναχοῦ μονῆς τοῦ ἁγίου Σάβα*), ed. R. Volk, *Die Schriften des Johannes von Damascus*, 6.1–2: *Historia animae utilis de Barlaam et Ioasaph (spuria)*, Berlin: de Gruyter, 2006–2009

Bessarion, *Regarding the Procession of the Holy Spirit* (*Περὶ τῆς ἐκπορεύσεως τοῦ ἁγίου Πνεύματος*), in J.-P. Migne (ed.), *Patrologia Graeca* 161, cols 321–407

Chioniades, Gregorios, *Persian System of Astronomy* (*Περσικὴ σύνταξις ἀστρονομίας*), ed. and tr. D. Pingree, *The Astronomical Works of Gregory Chioniades*, 1.1: *The Zîj al-ʿAlâʾî: Text, Translation, Commentary*; 1.2: *The Zîj al-ʿAlâʾî: Tables*, Corpus des astronomes byzantines 2, Amsterdam: Gieben, 1985–1986

Chrysokokkes, Georgios, *Commentary on the System of the Persians* (*Ἐξήγησις εἰς τὴν σύνταξιν τῶν Περσῶν*), partial edn by H. Usener, *Ad historiam astronomiae symbola*, Bonn: C. Georgi, 1876, 27–37 (= id., *Kleine Schriften*, Leipzig: Teubner, 1912–1914), 3, 350–371; improved edition of the preface by S. Lambros, 'Τὰ ὑπ' ἀριθμὸν 91 καὶ 92 κατάλοιπα', *Νέος Ἑλληνομνήμων*, 15 (1921), 332–336

Gennadios II Scholarios, *Treatise I on the Procession of the Holy Spirit* (*Περὶ τῆς ἐκπορεύσεως τοῦ ἁγίου Πνεύματος*), in *Oeuvres complètes de Georges* [Gennadios] *Scholarios*, ed. M. Jugie et al., Paris: Maison de la bonne presse, 1928–1933, here 2 (1929), 1–268

Holobolos, Manuel, *Translation of Boethius's De topicis differentiis*, ed. D.Z. Nikitas, *Boethius, De topicis differentiis Καὶ οἱ βυζαντινὲς μεταφράσεις τῶν Μανουὴλ Ὁλοβώλου καὶ Προχόρου Κυδώνη*, Corpus Philosophorum Medii Aevi: Philosophi Byzantini 5, Athens/Paris/Brussels, 1990, 95–145

Holobolos, Manuel, *Preface to the Translation of Boethius's De topicis differentiis* (*Βοετίου φιλοσόφου λατίνου περὶ τόπων διαλεκτικῶν διαίρεσις ἀρίστη μεταγλωττισθεῖσα παρὰ τοῦ ἀξιολογωτάτου ῥήτορος κυροῦ Μαξίμου τοῦ Ὀλοβώλου*), ed. A. Megas, *Maximos Planudes: Boethii de philosophiae consolatione in linguam graecam translata*, Λατινο-ελληνική Βιβλιοθήκη 9, Thessaloniki, 1996, 391–394

Holobolos, Manuel, *Translation of Pseudo-Aristotle's De plantis* (*Περὶ φυτῶν*), ed. H.J. Drossaart Lulofs and E.L.J. Poortman, *Nicolaus Damascenus, De plantis: Five Translations*, Aristoteles Semitico-Latinus 4, Amsterdam/New York: North-Holland, 1989, 563–590; partial tr. E. Fisher, 'Manuel Holobolos, Alfred of Sareshal, and the "Anonymous" Greek Translator of ps.-Aristotle's *De Plantis*', *Classica et Mediaevalia*, 57 (2006), 189–211

Kydones, Demetrios, *Apologia*, ed. G. Mercati, 'Apologie della propria fede. 1. Ai Greci Ortodossi', in *Notizie di Procoro e Demetrio Cidone, Manuele Caleca e Teodoro Meliteniota ed altri appunti per la storia della teologia e della letteratura Bizantina del secolo xiv*, Studi e Testi 56, Vatican City, 1931, 359–403; English tr. (by Ruper Ederer, based, it seems, on a German translation) in J. Likoudis, *Ending the Byzantine Greek Schism, Containing the 14th c. Apologia of Demetrios Kydones for Unity with Rome and St. Thomas Aquinas' Contra errores graecorum*, New Rochelle NY: Catholics United for the Faith, 1992

Kydones, Demetrios, *Translation of Passages from Augustine* (*Τοῦ μακαρίου Αὐγουστίνου ἐπισκόπου Ἱππῶνος κεφάλαια ἐκ τῶν αὐτοῦ λόγων παρεκβληθέντα, ἑρμηνευθέντα δὲ ἐκ τοῦ λατινικοῦ παρ' ἐμοῦ Δημητρίου*), ed. D. Kalamakis, *Ἀνθολόγιον ἐκ τῶν ἔργων Αὐγουστίνου Ἵππωνος ἐξελληνισθὲν ὑπὸ Δημητρίου τοῦ Κυδώνη*, Athens: Parnassos Literary Society, 1996, 55–138

Martyrdom of the Sixty New Martyrs Who Were Martyred under the Tyranny of the Arabs in the Holy City of Our God Christ, ed. A. Papadopoulos-Kerameus, Pravoslavnij Palestinskij Sbornik 12, St Petersburg, 1892, 1–7

Oneirokrites of Achmet (*Ὀνειροκρίτης*), ed. F. Drexl, *Achmetis Oneirocriticon*, Leipzig: Teubner, 1925; tr. S. Oberhelman, *The Oneirocriticon of Achmet: A Medieval Greek and Arabic Treatise on the Interpretation of Dreams*, Lubbock TX: Texas Tech University Press, 1991

Planoudes, Maximos, *Translation of Augustine's De trinitate* (*Τοῦ μακαριωτάτου Αὐγουστίνου ἐπισκόπου Ἱππῶνος περὶ Τριάδος βιβλία πέντε πρὸς τοῖς δέκα, ἅπερ ἐκ τῆς Λατίνων διαλέκτου εἰς τὴν Ἑλλάδα μετήνεγκεν ὁ σοφώτατος καὶ τιμιώτατος μοναχῶν κύριος Μάξιμος ὁ Πλανούδης*), ed. M. Papathomopoulos, G. Rigotti, and I. Tsavare, *Αὐγουστίνου: Περὶ Τριάδος βιβλία πεντεκαίδεκα ἅπερ ἐκ τῆς Λατίνων διαλέκτου εἰς τὴν Ἑλλάδα μετήνεγκε Μάξιμος ὁ Πλανούδης*, Βιβλιοθήκη Α. Μανούση 3, Athens, 1995

Šams al-Dīn al-Buḫārī, *On the Astrolabe* (*Ποίημα τοῦ Σιάμψ τοῦ Πέρσου περὶ τῆς διδασκαλίας τοῦ ἀστρολάβου*), excerpt edn by A. Delatte, Anecdota atheniensia 2, Liège: Vaillant-Carmanne, 1939, 263–271; preface ed. and tr. E. Fisher, 'Arabs, Latins and Persians Bearing Gifts: Greek Translations of Astronomical Texts, ca. 1300', *Byzantine and Modern Greek Studies*, 36 (2012), 161–177

Sinbad Tales, translated by Michael Andreopoulos as *The Book of the Philosopher Syntipas* (*Βίβλος Συντίπα τοῦ Φιλοσόφου*), ed. V. Jernstedt and P. Nikitin, *Michaelis Andreopuli liber Syntipae*, Mémories de l'Académie impériale des sciences de St. Pétersbourg, VIII série, Classe historico-philologique 11.1, St Petersburg, 1912

Symeon, Seth, *Alphabetical Treatise on the Powers of Various Foods* (*Σύνταγμα κατὰ στοιχεῖον περὶ τροφῶν δυνάμεων*), ed. B. Langkavel, *Simeonis Sethi Syntagma de alimentorum facultatibus*, Leipzig: Teubner, 1868, 18–140

Symeon, Seth, *Stephanites and Ichnilates* (*Τὰ κατὰ Στεφανίτην καὶ Ἰχνηλάτην*), ed. L.-O. Sjöberg, *Stephanites und Ichnelates: Überlieferungsgeschichte und Text*, Studia Graeca Upsaliensia 2, Stockholm: Almqvist and Wiksell, 1962

Secondary Literature

Allmand, C., *The* De Re Militari *of Vegetius: The Reception, Transmission and Legacy of of a Roman Text in the Middle Ages*, Cambridge: Cambridge University Press, 2011

Balard, M., et al. (eds), *Byzance et le monde éxterieur: Contacts, relations, échanges*, Paris: Sorbonne, 2005

Butts, A.M., *Language Change in the Wake of Empire: Syriac in Its Greco-Roman Context*, Winona Lake IN: Eisenbrauns, 2016

Condylis-Bassoukos, H., *Stéphanitès kai Ichnélatès, traduction grecque* (XIe *siècle*) *du livre Kalīla wa-Dimna d'Ibn al-Muqaffaʿ* (VIIIe *siècle*)*: Étude lexicologique et littéraire*, Leuven: Peeters, 1997

Efthymiades, S. (ed.), *The Ashgate Research Companion to Byzantine Hagiography*, 1: *Periods and Places*, Farnham: Ashgate, 2011; 2: *Genres and Contexts*, Farnham: Ashgate, 2014

Efthymiades, S., 'Rewriting', in *The Oxford Handbook of Byzantine Literature*, ed. S. Papaioannou, Oxford: Oxford University Press, forthcoming

Fisher, E., 'Planoudes, Holobolos, and the Motivation for Translation', *Greek, Roman, and Byzantine Studies*, 43 (2002), 77–104

Fisher, E., 'Ovid's *Metamorphoses*, Sailing to Byzantium', in *Remusings: Essays on the Translation of Classical Poetry*, special issue of *Classical and Modern Literature*, 27.1 (2008, for 2007), 45–67

Fisher, E., 'Planoudes' *De trinitate*, the Art of Translation, and the Beholder's Share', in *Orthodox Readings of Augustine*, ed. G. Demacopoulos and A. Papanikolaou, Crestwood NY: St. Vladimir's Press, 2008, 41–61

Fisher, E., 'Alexios of Byzantium and the Apocalypse of Daniel: A Tale of Kings, Wars and Translators', in *Bizans ve Çevre Kültürler / Byzantium and the Surrounding Cul-*

tures: Festschrift in honor of S. Yıldız Ötüken, ed. S. Doğan and M. Kadiroğlu, Istanbul: Arkeoloji ve Sanat Yayınları, 2010, 177–185

Fisher, E., 'Manuel Holobolos and the Role of Bilinguals in Relations Between the West and Byzantium', in *Knotenpunkt Byzanz: Wissensformen und kulturelle Wechselbeziehungen*, ed. A. Speer and P. Steinkrüger, New York/Berlin: de Gruyter, 2012, 210–222

Fisher, E., 'Ovid's Metempsychosis: The Greek East', in *Ovid in the Middle Ages*, ed. J.G. Clark et al., Cambridge: Cambridge University Press, 2011, 26–47

Fisher, E., 'Arabs, Latins and Persians Bearing Gifts: Greek Translations of Astronomical Texts, ca. 1300', *Byzantine and Modern Greek Studies*, 36 (2012), 161–177

Fisher, E., '*Homo Byzantinus* and *Homo Italicus* in Late Thirteenth-Century Byzantium', in *Dante and the Greeks*, ed. J.M. Ziolkowski, Washington DC: Dumbarton Oaks, 2014, 63–82

Goulet, R., and U. Rudolph, 'Introduction', in *Entre Orient et Occident: La philosophie et la science gréco-romaines dans le monde arabe*, ed. P. Derron, Geneva: Fondation Hardt, 2011, xi–xxxvii

Gutas, D., *Greek Thought, Arabic Culture: The Graeco-Arabic Translation Movement in Baghdad and Early 'Abbāsid Society (2nd–4th/8th–10th centuries)*, London: Routledge, 1998

Gutas, D., 'Arabic into Byzantine Greek: Introducing a Survey of the Translations', in *Knotenpunkt Byzanz: Wissensformen und kulturelle Wechselbeziehungen*, ed. A. Speer and P. Steinkrüger, Berlin: de Gruyter, 2012, 246–262

Gutas, D., A. Kaldellis, and B. Long, 'Intellectual Exchanges with the Arab World', in *The Cambridge Intellectual History of Byzantium*, ed. A. Kaldellis and N. Siniossoglou, Cambridge: Cambridge University Press, 2017, 79–98

Gutas, D., and N. Siniossoglou, 'Philosophy and "Byzantine Philosophy"', in *The Cambridge Intellectual History of Byzantium*, ed. A. Kaldellis and N. Siniossoglou, Cambridge: Cambridge University Press, 2017, 271–295

Haskins, H., *Studies in the History of Mediaeval Science*, Cambridge MA: Harvard University Press, 1927

Hoyland, R.G., *Theophilus of Edessa's Chronicle and the Circulation of Historical Knowledge in Late Antiquity and Early Islam*, Liverpool: Liverpool University Press, 2011

Kaldellis, A., *Hellenism in Byzantium: The Transformations of Greek Identity and the Reception of the Classical Tradition*, Cambridge: Cambridge University Press, 2007

Kaldellis, A., 'The Byzantine Role in the Making of the Corpus of Classical Greek Historiography: A Preliminary Investigation', *Journal of Hellenic Studies*, 132 (2012), 71–85

Kaldellis, A., 'Byzantine Philosophy Inside and Out: Orthodoxy and Dissidence in Counterpoint', in *The Many Faces of Byzantine Philosophy*, ed. K. Ierodiakonou and B. Bydén, Athens: The Norwegian Institute at Athens, 2012, 129–151

Kaldellis, A., 'The Byzantine Phase and Reception of Ancient Greek', in *Encyclopedia of Ancient Greek Language and Linguistics*, ed. G. Giannakis, 3 vols, Leiden: Brill, 2014, I, 248–252

Koltsiou-Nikita, A., 'Ο ρόλος της μετάφρασης και των ερμηνέων στο θεολογικό διάλογο ελληνόφωνης Ανατολής και λατινόφωνης Δύσης', in *Φιλία και Κοινωνία: Τιμητικός τόμος στον καθηγητή Γρηγόριο Δ. Ζιάκα*, Thessaloniki: Banias, 2008, 301–323

Kunitzsch, P., 'Das Fixsternverzeichnis in der "Persischen Syntaxis" des Georgios Chrysokokkes', *Byzantinische Zeitschrift*, 57 (1964), 382–411

Lequeux, X., 'Latin Hagiographical Literature Translated into Greek', in *The Ashgate Research Companion to Byzantine Hagiography*, 1: *Periods and Places*, ed. S. Efthymiades, Farnham: Ashgate, 2011, 386–399

Magdalino, P., *L'Orthodoxie des astrologues: La science entre le dogme et la divination à Byzance (VIIe–XIVe siècle)*, Paris: Lethielleux, 2006

Magdalino, P., and R. Nelson (eds), *The Old Testament in Byzantium*, Washington DC: Dumbarton Oaks, 2010

Mavroudi, M., *A Byzantine Book on Dream Interpretation: The Oneirocriticon of Achmet and Its Arabic Sources*, Leiden: Brill, 2002

McCabe, A., *A Byzantine Encyclopaedia of Horse Medicine: The Sources, Compilation, and Transmission of the Hippiatrica*, Oxford: Oxford University Press, 2007

Mercier, R., 'The Greek "Persian Syntaxis" and the Zīj-i Ilkhānī', *Archives internationales d'histoire des sciences*, 112 (1984), 35–60

Millar, F., *A Greek Roman Empire: Power and Belief under Theodosius II, 408–450*, Berkeley CF: University of California Press, 2006

Moatti, C., *The Birth of Critical Thinking in Republican Rome*, tr. J. Lloyd, Cambridge: Cambridge University Press, 2015

Momigliano, A., *Alien Wisdom: The Limits of Hellenization*, Cambridge: Cambridge University Press, 1971

Peeters, P., *Orient et Byzance: Le tréfonds oriental de l'hagiographie byzantine*, Brussels: Société des Bollandistes, 1950

Pingree, D., 'Gregory Chioniades and Palaeologan Astronomy', *Dumbarton Oaks Papers*, 18 (1964), 133–160

Plested, M., 'Aquinas in Byzantium', in *The Cambridge Intellectual History of Byzantium*, ed. A. Kaldellis and N. Siniossoglou, Cambridge: Cambridge University Press, 2017, 542–556

Roberto, U., 'Il *Breviarium* di Eutropio nella cultura greca tardoantica e bizantina: la versione attribuita a Capitone Licio', *Medioevo Greco*, 3 (2003), 241–271

Roilos, P., *Amphoteroglossia: A Poetics of the Twelfth-Century Medieval Greek Novel*, Washington DC: Center for Hellenic Studies, 2005

Stolte, B., 'Legal Thought', in *The Cambridge Intellectual History of Byzantium*, ed. A. Kaldellis and N. Siniossoglou, Cambridge: Cambridge University Press, 2017, 141–166

Sypiański, J., 'Arabo-Byzantine Relations in the 9th and 10th Centuries as an Area of Cultural Rivalry', in *Byzantium and the Arab World: Encounter of Civilizations*, Thessaloniki: Aristotle University of Thessaloniki, 2013, 465–478

Tihon, A. (= Duhoux-Tihon), 'Les tables astronomiques Persanes à Constantinople dans la première moitié du XVIe siècle', *Byzantion*, 57 (1987), 471–487

Tihon, A. (= Duhoux-Tihon), 'Traités byzantins sur l'astrolabe', *Physis*, 32 (1995), 323–357

Toth, I., 'Authorship and Authority in the *Book of the Philosopher Syntipas*', in *The Author in Middle Byzantine Literature: Modes, Functions, Identities*, ed. A. Pizzone, Berlin: de Gruyter, 2014, 87–103

Ulbricht, M., *Coranus Graecus: Die älteste überlieferte Koranübersetzung in der «Ἀνατροπὴ τοῦ Κορανίου» des Niketas von Byzanz (9. Jh.): Einleitung—Text—Übersetzung—Kommentar*, PhD dissertation: Freie Universität Berlin, 2015

Volk, R., *Die Schriften des Johannes von Damascus*, 6.1–2: *Historia animae utilis de Barlaam et Ioasaph (spuria)*, Berlin: de Gruyter, 2006–2009

Wright, C., 'Towards an Edition of Demetrios Kydones' Autograph Translation of Thomas Aquinas' *Summa theologiae, Prima Pars*', *Nicolaus*, 40 (2013), 15–30

CHAPTER 8

The Statements of Medieval Latin Translators on Why and How They Translate Works on Science and Philosophy from Arabic

Charles Burnett

Part One: Essay

In this chapter I shall be concentrating on texts translated from Arabic into Latin from the mid-tenth to the late thirteenth century. The emphasis will be on the statements of the translators themselves, or on comments of others regarding the translators, concerning the reasons for translating texts from Arabic. Thus the cultural and political situation that provided the context for the translations, and (for the most part) translation theories and methods, will only be referred to in passing.[1]

The earliest Latin texts based on Arabic sources form a corpus of texts on astrology, composed in Catalonia and dating to the second half of the tenth century, that includes literal translations and Latin redactions and elaborations of these translations, and in which the Arabic names of the twenty-eight lunar mansions figure prominently. This corpus is known as the *Alchandreana* from the alleged source of the doctrine, Alchandreus (possibly Alexander the Great).[2] The only reference to translation is made in one of the Latin redactions—*Proportiones*, ch. 48 (Text 1). Here, the redactor (Miró Bonfill?) states that he is revealing material taken not from Christian or Hebrew sources, but from 'Chaldean' ones. The Chaldei in the *Alchandreana*, as elsewhere,[3] are the Mesopotamian experts in astrology. The term does not specify a language,

1 The pioneer in discovering and transcribing Arabic–Latin translators' prefaces was Haskins, *Studies in the History of Mediaeval Science*, which remains valuable. For a larger and differing selection of Latin texts from that given here, with English summaries rather than translations, see Gázquez, *The Attitude of the Medieval Latin Translators*. For more specific studies, see Ricklin, '*Arabes contigit imitari*', and the articles collected in Burnett, *Arabic into Latin in the Middle Ages*. For prefaces referring to translation theory and method, see Burnett, 'Translating from Arabic into Latin in the Middle Ages: Theory, Practice, and Criticism'.

2 Juste, *Les* Alchandreana *primitifs*.

3 E.g., Abū Maʿšar, *The Great Introduction to Astrology*, V, ch. 11, in Abū Maʿšar, *Liber Introductorii Maioris*, ed. Lemay, I, 326–327; V, 199; and VIII, 85.

 | DOI:10.1163/9789004472648_009

but the language must be Arabic, and the claim is that it is the first time that the doctrine is presented in Latin. The *Alchandreana* include elements of Hebrew, and fragments of Christian works (Martianus Capella and the *Excerpta Plinii*), but it is the recent discovery of Arabic writings that is emphasized here. The reference to sanctuaries ('adyta', from the Greek ἄδυτος, meaning 'not to be entered [sc. by the uninitiated]') is a theme that will recur later on, evoking both the hiddenness and the sacredness of the doctrine that is being translated.

Slightly later (after 978), but in the same context (Catalonia), a corpus of texts on the astrolabe was assembled.[4] The translator of one of these texts—sometimes identified, on no solid grounds, with Lupitus of Barcelona[5]—uses a similar high-flown language in a long introduction accompanying this corpus (Text 2). Having explained how important the science of astronomy is (and, incidentally, having poured scorn on the art of the Chaldeans), he turns to the authorities on astronomy, and singles out Ptolemy as the greatest of them, both because his doctrine ('studium') is clearer, and because he uses a more lucid ('lucidius') style of writing. He invented an instrument useful for investigating the heavens and, in general, for pursuing every aspect of astronomy and geometry. This instrument, he says, is the astrolabe (which he calls by an Arabic name, *basṭ al-kura*), and it represents the configurations of the universe according to 'architectonic reasoning'.[6] The translator is trying to convince his reader of the importance of the text that he is translating. He emphasizes that the material is not new, but is rather a re-introduction of the wisdom of the Ancients, which has been neglected and forgotten. It is so important that he feels he must not add anything of his own, but rather should translate the Arabic faithfully. Exactly the same order of material (or even of words) should be preserved in Latin as in the Arabic. As we shall see, the conceit that translation is the rediscovery of something lost is a leitmotif.

Not until well into the next century do we have further translations of works from Arabic into Latin. This time the field is natural science and medicine, and the place of translating is Southern Italy. We have several anonymous translations from Arabic, and translations from Greek.[7] The most prolific of the translators is Constantine the African. His large number of versions of Arabic medical writings, many written by doctors from his own region of Ifriqiya (roughly, modern Tunisia), were made between c. 1057 and before 1099, in the environment of the medical school at Salerno and later at the Benedictine abbey of

4 These texts are discussed and edited in Millás Vallicrosa, *Assaig d'història*, 271–302.

5 Ibid., 195.

6 For this new approach, see Borrelli, *Aspects of the Astrolabe*.

7 See Burnett, 'Physics before the *Physics*'.

Montecassino. He is notorious for having disguised the Arabic original of his texts, giving the impression that they were based on works by Greek authors. In his preface to the *Pantegni* (his version of *The Complete Book of Medicine* of ʿAlī ibn al-ʿAbbās al-Maǧūsī), he claims that he is 'the author because he made a unity out of the books of many' ('auctor, quia ex multorum libris coadunator').[8] However, in three works he explicitly states that he has translated the texts from Arabic, albeit reluctantly. In the preface to his translation of Isḥāq ibn Sulaymān al-Isrāʾīlī's *On Urines* (Text 3), Constantine laments the absence of a work on the subject matter in Latin, which is his reason for turning to an Arabic text, for which he gives the name of the author (Isḥāq al-Isrāʾīlī). Even so, he lays stress on the fact that the true authors were the Ancients, and that Isḥāq was merely a collector and excerptor, and hence performing the same role as he claimed for himself in the *Pantegni*. In another confessed translation (Galen's *Commentary on Hippocrates's Aphorisms*), his avowed reluctance to translate is based not so much on the fact that he has to turn to Arabic, as that his ancient author, Galen, is too important a philosopher for him to dare to translate him (Text 4). Here the language from which Galen is translated is irrelevant.

Stephen of Antioch, who produced a new translation of the *The Complete Book of Medicine* (or *Royal Book*) of ʿAlī ibn al-ʿAbbās al-Maǧūsī, is very critical of Constantine as a translator.[9] His criticism, in the preface to his own translation, reads like a blue-print for what constitutes a good (one might even say 'honest') translation (Text 5a). The bad translator puts his own name as author in place of that of the original author, and tries to justify this by missing out things, rearranging the order of material, and altering the exposition. He thereby does a great injustice to the real author. Stephen himself will adopt a different policy. He will distinguish clearly between what belongs to the original author and what belongs to the translator, and he will follow the Arabic closely. It is necessary to do this, Stephen adds, because the book on medicine that he is translating is so important for preserving human life. He is translating a text on medicine first for the health of the body, with a view to proceeding to what is necessary for the soul later, i.e., philosophy.

Stephen belongs to a small group of translators who had access to Arabic books in Antioch in the second quarter of the twelfth century, a generation after the Crusaders first entered the Middle East. He advertises the riches of philosophical works in Arabic, which he intends to devote himself to translating. Stephen makes further comments about his translation method, especially

---

8 Jacquart, 'Le sens donné par Constantin l'Africain à son œuvre', 79.

9 This criticism is explored in detail in Burnett, 'The Legend of Constantine the African'. For an edition and translation of all the prefaces of Stephen, see Burnett, 'Antioch as a Link', 22–40.

in regard to Arabic terms of medicaments that have no equivalent in Latin, which he decides to transcribe as they are, but to add Latin endings to them. He also adds a substantial glossary, in three columns, respectively of the Greek, Latin, and Arabic equivalents of *materia medica*. In this, however, we see him using yet other texts: a Greek manuscript of Dioscorides, *De re medica*, and, apparently, an Arabic manuscript of the same work (Text 5b). From his translation, with inserted commentary, of Ibn al-Hayṯam's *On the Configuration of the World*, and the translation of Ptolemy's *Almagest* by his colleague Ebdelmessie ('Abd al-Masīḥ), it is clear that several Arabic manuscripts of mathematical works (including astronomy) were also available. Stephen contrasts the poverty of the Latins with the wealth that the Orient can provide ('quod [...] Oriens habebat'),[10] and exhorts his readers to consult the 'truth of the Arabs' ('Arabum veritas').[11]

It must have been the attraction of finding such learning that drew Adelard of Bath to the Middle East at approximately this time. For he tells us at the beginning of his popular work on natural science, the *Quaestiones naturales*, that he spent 'seven years pursuing the studies of the Arabs' ('Arabum studia') and relates events that occurred in Tarsus and Mamistra, both of which were part of the principality of Antioch.[12] He credits his 'Arabic teachers' ('magistri Arabici')[13] with inculcating the importance of using reason (contrasting this with 'authority', which, he claims, ruled supreme in the French schools). He also refers to what he has learnt from the Arabs as 'something new' ('aliquid Arabicorum studiorum novum'). Unlike Pseudo-Lupitus and Constantine, he is not saying that he is rediscovering the wisdom of the Ancients, but, like them (and, indeed, like Stephen of Antioch), he is afraid that novelties will be scorned by his readers and adds rather cheekily: 'Don't blame me; blame my teachers' (Text 6). In his original work on the astrolabe (written c. 1150), addressed to the future King Henry II (reigned 1154–1189), he reiterates that he will 'write in Latin what I have learnt in Arabic' ('quod Arabice didici Latine subscribam').[14]

None of Adelard's own translations from Arabic, in the fields of geometry, astronomy, astrology, and magic, has a preface, so we do not have his own account of his choice of particular authors and texts. We do not know the sources of the Arabic manuscripts or the identities of the Arabic teach-

10 Ibid., 37 (Text 5b).
11 Ibid., 29 (Text 5a).
12 See Adelard of Bath, *Conversations with His Nephew*.
13 Ibid., 103.
14 From the preface to Adelard's *De opere astrolapsus*, edited in Burnett, *The Introduction of Arabic Learning into England*, 92.

ers and scholars (beyond the mysterious ʿAbd al-Masīḥ) to which and whom Stephen and Adelard had access. There was clearly a rich potential in Antioch for exchange of learning between at least three languages, Arabic, Greek and Latin, but it does not seem to have been fully exploited. Apart from Stephen's translation of al-Maǧūsī Stephen and ʿAbd al-Masīḥ's work had virtually no impact in the West. Adelard fared better, thanks, perhaps, to his association with the School of Chartres. For the rest of the twelfth century and the thirteenth, it is to the West that one must turn: first, to Northeast Spain, and its connections to the Muslim kingdom of Saragossa; then to Toledo, with its connections to the rest of Muslim Spain and North Africa. In both areas one can observe sustained and successful programs for translating texts from Arabic into Latin.

∴

The story of the translation movement in Spain and the West has been told several times.[15] The main translators in Toledo, Gerard of Cremona and Dominicus Gundissalinus, did not write prefaces to their translations, and we only know of their motives indirectly. The situation in north-east Spain is different, and here a related group of translators appear to have followed a common policy: paying attention to the Latinity of their translations, adding prefaces, and situating their translations within a larger programme of research. These translators are Hermann of Carinthia; his colleague, Robert of Ketton; Hermann's student, Rudolph of Bruges; and two other translators who can be associated either by common interests and terminology (Hugo Sanctelliensis, with Hermann) or a common patron (Plato of Tivoli, with Rudolph). It is probable that they all benefitted from the use of Arabic manuscripts available in the Saragossan kingdom of the Banū Hūd. This is stated explicitly in the preface by Hugo Sanctelliensis to Ibn al-Muṯannā's commentary on al-Ḫwārizmī's *Astronomical Tables*, in which he mentions that his patron, Michael, bishop of Tarazona (r. 1119–1151) found the Arabic manuscript in the 'Rotense armarium' (Text 7);[16] this must be the library of Rueda de Jalón, the last stronghold of the Banū Hūd (until its capture in 1140), some 55 kilometres from Tarazona. The manuscript was discovered in 'the more secret depths', implying the hiddenness of the knowl-

15 Among recent accounts one may include Santoyo, *La traducción medieval en la Península Ibérica* and Gutas, 'What was there in Arabic for the Latins to Receive?'.

16 The preface is edited in Haskins, *Studies*, 73, within a chapter on 'Translations of Hugo Sanctallensis' (67–81); see most recently, Santoyo, 'El Normando Hugo de Cintheaux (*Hugo Sanctelliensis*)'.

edge in a similar way to the translator of the *Alchandreana*. Hugo suggests here (and elsewhere) that his patron has asked him to write on a subject, but, feeling unqualified, he has turned to the Arabs. In other prefaces Hugo refers to works that he has already translated or works that he intends to translate, as if he has on his bookshelf Arabic manuscripts of at least twenty-four texts on astronomy, astrology, and various forms of divination (not counting the 'ten books on astrological judgements' to which he refers in the preface to his translation of Pseudo-Ptolemy's *Centiloquium*, and at least one work—a commentary on al-Ḫwārizmī by al-Farġānī—that he finds too defective to be used).

While Hugo addresses his prefaces (where there is an addressee) to his bishop, Michael, Hermann and Robert had a more complete programme and a wider audience in mind. In his preface to an astrological work, *The Forty Chapters* of al-Kindī, addressed to Hermann, Robert mentions, as their common task, translations of Euclid (the *Elements*), Theodosius's 'measurement of the world', and a book on proportions (probably Aḥmad ibn Yūsuf's *On Proportions and Proportionality*) as a preparation for translating the *Almagest*, which is their ultimate aim (Text 8). In the preface to his translation of Ptolemy's *Planisphere*, addressed to Thierry, the chancellor of Chartres Cathedral and a leading educator of the time, Hermann mentions two classical works by Ptolemy, on astronomy and astrology, and Arabic adaptations of these, which he implies are more convenient to use than the ancient texts (Text 9). In referring to al-Battānī as 'very conveniently' ('commodissime') abbreviating Ptolemy's *Almagest*, he might have had in mind Plato of Tivoli's translation of al-Battānī's *De scientia astrorum* (translated in 1138). For Plato, in his preface to this work, says that 'I have discovered that al-Battānī is his [Ptolemy's] perfect imitator among the Arabs, in that he compendiously abbreviates the prolixity of Ptolemy.'[17]

Hermann himself had translated the Arabic astrological text that he mentions: Abū Maʿšar's *Great Introduction to Astrology*. Robert, in turn, said that Hermann had suggested that he [Robert] should translate the *Forty Chapters* of al-Kindī as the most convenient (again 'commodissimus') and truthful ('veracissimus') astrologer. In his *De essentiis*, an original work on the cosmos and on man, Hermann refers in his preface, also addressed to Robert, to the learning that they had both acquired from the Arabs (Text 10). Like Miró Bonfill, he refers to the sanctuaries ('adyta') in which Robert and he had been studying Arabic texts, thus endowing knowledge of the universe with an air of sanctity.

Robert and Hermann also share the impression that Arabic is prolix and cannot be translated literally. Robert has reduced the Arabic of the *Forty Chapters* to the 'compact style of the Latins' (Text 8). Hermann writes at more length,

17 See Text 14 below.

in his preface to Abū Maʿšar's *Great Introduction to Astrology*, addressed to Robert, saying that he wishes to be pardoned for not being the 'faithful interpreter' ('fidus interpres') advocated by Boethius, since Arabic is such a prolix language (Text 11). But sometimes the effusiveness of the Arabic language can be an advantage: in the preface to a commentary on Isaac's *On Fevers* (*De febribus*), probably written by Richard the Englishman (early thirteenth century) the author claimed that Greek, because of its terseness, was difficult to understand, whereas Arabic, because of its tendency to expand the text, was more suitable for students (Text 13).[18]

In his preface to the *De scientia astrorum* of al-Battānī, Plato of Tivoli points out the convenience of the Arabic author (in this case, al-Battānī) for presenting ancient wisdom on astronomy, in the context of a complete lack of scientific writing in the Latin tradition: one does not find a single name of a scientific authority among the Latins ('in Rome'), in contrast to Egypt and Greece, and especially Arabia, where one can point to 'al-Ḫwārizmī, Māšā'allāh, and al-Battānī together with very many others' (Text 14).

⁂

The most comprehensive programme of translation is that which took place in Toledo in the second half of the twelfth century. Here Dominicus Gundissalinus, an archdeacon resident at the cathedral (d. after 1190), and Gerard of Cremona, nicknamed 'the Master' (1114–1187), a canon of the cathedral, undertook the translation of texts on psychology and metaphysics (Gundissalinus), and on logic, arithmetic, geometry, astronomy, natural science, medicine, and alchemy (Gerard). Both translators were involved—as translator (Gerard) and reviser (Gundissalinus)—in the very significant Arabic text concerning the range of human knowledge: al-Fārābī's *On the Classification of the Sciences*. Here al-Fārābī, who lived in Baghdad in the early tenth century, described the nature and the divisions of logic, mathematics, natural science, metaphysics, and the social sciences (politics, law, and rhetoric). Al-Fārābī's book provided the main source for Gundissalinus's own work *On the Division of Philosophy* (*De divisione philosophiae*); it also reminded Latin readers of which sciences were lacking textbooks in Latin. The programme that Gerard of Cremona followed in translating works on logic and (Aristotelian) philosophy corresponds with the programme outlined by al-Fārābī.[19]

---

18 Piero Morpurgo refers to this passage in *L'idea di natura nell'Italia Normannosveva*, 106.

19 See Burnett, 'The Coherence'. While Gerard himself does not mention following al-Fārābī,

Unfortunately, neither Gundissalinus nor Gerard of Cremona left any first-person testimonies on the reasons why they translated what they did, and we rely on a handful of indirect statements. For Gundissalinus, we have the preface that Avendauth added to the translation of Avicenna's *On the Soul* (*De anima*), in which he explains how he himself translated the Arabic text into the vernacular language (probably the local Romance, rather than spoken Arabic), and Dominicus translated this into good Latin (Text 15). This is a precious witness to what appears to have been quite a common phenomenon, the 'translation by four hands'.[20] The dedication also tells us that the translation is of matters previously unknown, and that the Arabic author has collected in one place material from three books of Aristotle.

For Gerard of Cremona, we have the testimony of Daniel of Morley (who purports to have attended a lecture of his), and a kind of obituary written by his students (*socii*). In the preface to his one surviving work—*Philosophia* or *Liber de naturis inferiorum et superiorum*—Daniel tells us that he was attracted to Toledo by the rumour that Arabic learning, which was almost entirely based on the quadrivium (i.e., the mathematical sciences), thrived there (Text 16). He was not disappointed. He promises to write down in Latin an account of the universe that he learnt from a Mozarab, Galippus (the Latinized form of the Arabic name Ġālib) 'in the Toledan tongue' ('quod a Galippo mixtarabe in lingua Tholetana didici, Latine subscribitur').[21] He begs his readers to accept the simple and clear opinions of the Arabs rather than the teachings that the Latins deliberately make obscure in order to hide their own ignorance. Finally, he justifies taking wisdom from pagans, from the Biblical precedent of the Jews taking the jewels of the Egyptians when they escaped from captivity. At the end of the *Philosophia* he reports on a lecture on astrology that he heard Gerard giving (though none of Gerard's extant translations are in fact on astrology) and mentions the same Ġālib as collaborating with Gerard on his translation of the *Almagest*: 'Gerard of Toledo, who Latinized the *Almagest*, while Ġālib the Mozarab interpreted' ('Girardus Tholetanus, qui Galippo mixtarabe interpretante, Almagesti latinavit').[22]

For a more circumstantial account of Gerard's aims and motifs we may turn to the *Vita* written by his students (Text 17).[23] Here again we find a reference to

---

Alfred of Shareshill, who continued to translate Aristotle's works on natural philosophy from the point where Gerard left off, does, in the gloss accompanying his translation of Aristotle's *Meteora*; see Alfred of Sareshel, *Commentary on the* Metheora *of Aristotle*, 37, and Burnett, 'The Coherence', 261–262.

20 D'Alverny, 'Les traductions à deux interprètes'; and Zonta, 'The Jewish Mediation'.

21 Daniel of Morley, *Philosophia*, 213.

22 Ibid., 245.

23 For an analysis of the *Vita*, see Burnett, 'The Translator as an Authority'.

the 'poverty of the Latins' ('Latinorum penuria', comparable to the 'blindness of the ignorance of Latin culture' in Plato of Tivoli's preface to al-Battānī); the desire to study Ptolemy's *Almagest*; and the discovery of books (manuscripts) written in Arabic. Gerard's seventy-odd translations mark the high point in the translation movement in Toledo.

∵

The thirteenth century was largely a period for filling in gaps left by the translators of the twelfth century, for translating commentaries on Aristotle's works, and for supplementing or replacing Arabic–Latin translations of works by Greek authors with translations directly from the Greek. Where translators' prefaces exist, they tend to be substantial; the translator shows knowledge of a wide range of Arabic texts, and of the transmission of the subject to which he is contributing.

Toledo continued to be a centre of translation into the thirteenth century. Mark, a canon of Toledo Cathedral (attested there between 1194 and 1216), while best known for his translation of the Qur'ān and the religious writings of Ibn Tūmart, also translated some medical texts and, like Gerard, found Arabic manuscripts for these texts in Toledo. In the preface to his translation of the first of these texts (Galen's *De tactu pulsus*, Text 18), Mark draws attention both to the originality of the Arabic texts, and the absence of any previous translation into Latin.

Hermann the German, also in Toledo, clearly saw his task as completing the interpretation of Aristotle's works on logic (which included the *Rhetoric* and *Poetics* in the Arabic tradition), but found that he could not do this without the aid of Arabic commentaries and summaries. Thus, in his preface to his translation of Aristotle's *Rhetoric* (written between 1243 and 1256, Text 19), he expresses his own experience of the difficulty of recovering Aristotle's works on rhetoric and poetry and the helpfulness of the Arabic commentaries, as well as of the value of preliminary translations of commentaries on or summaries of these texts from Arabic into Latin, which might eventually be complemented by full translations of the original Greek texts made by other scholars.

Michael Scot (d. before 1236) moved from Toledo to Italy (Bologna) and eventually became the astrologer of Emperor Frederick II. He is the scholar who was most involved in the translations of Averroes's Large Commentaries on Aristotle's works on natural science. In general, he did not add prefaces to these commentaries or to any of his translations, but, exceptionally, he wrote a short note at the beginning of his translation of Averroes's *Large Commentary on Aristotle's De caelo*, in which he addresses Stephen of Provins, probably because

Stephen had been assigned by Pope Gregory IX the task of examining the orthodoxy of Aristotle's works on natural philosophy (Text 20).

Michael Scot also translated Avicenna's *Abbreviatio de animalibus* for Frederick II, and it is probably under the emperor's aegis that translations of the commentaries of Averroes were made. Frederick's son, King Manfred of Sicily, continued his father's interest in natural philosophy, but concentrated on translations from Greek. But in a letter addressed to the scholars of Paris (written in 1263), Manfred also mentions translations from Arabic (Text 21). In this letter, he considers translation as providing 'new clothes' for ancient texts, and uses a sexual metaphor, in referring to the preservation of the virginity of the originals.

As a final example of the statements of a translator, one could refer to one of the longest and latest prefaces to a translation from Arabic into Latin. This is the translation of Abū Bakr ibn al-Zakarīyā' al-Rāzī's *Kitāb al-Ḥāwī*, made in 1282 by Faraǧ ibn Sālim (Faragut). The preface includes many of the conceits that we have met with in other translators' prefaces (Text 22): here we have praise of an Arabic author's skill in compiling a large amount of information into one book, the lack of such a book in Latin, the hiddenness of the Arabic original, the acquiring of the Arabic manuscript by a Christian patron, and the preservation of the virginity of the original text.

∵

Common to several of these statements of translators from Arabic between the tenth and thirteenth centuries are the following (I give the number(s) of the text(s) in which each of these statements can be found).

1. The Arabs have texts that are lost or inexistent among the Latins (the poverty of the Latins): 3, 17, 18, 22.
2. Translating from Arabic introduces new ideas to the Latins: 5a, 6.
3. Reluctance of 'the Moderns' to accept new ideas: 6.
4. The Arabic text sought after has originated in the depths of a library, which can be described as a sanctuary, or *adytum*: 1, 7, 10, 18, 21.
5. The Latin scholar has benefitted from Arabic teachers: 6, 16.
6. A patron has procured the manuscript for the translator and/or ordered the translation to be made: 7, 15, 22.
7. While translation involves an exchange of words, what is important is the transfer of the subject matter or *res* that the words present: 8, 11, 12.
8. The subject matter is hidden in the obscurity of the Arabic: 21, 22.
9. Translating will 'open doors' to learning or new methodology: 2, 3, 8.
10. The 'virginity' of the original text should not be 'corrupted': 21, 22.

11. The translator is following an academic programme: 8, 18.
12. The Arabs are expert at compiling several ancient Greek texts into one, easy to use, text, or at summarizing succinctly a single but complicated Greek book: 3, 8, 14, 15, 21, 22.
13. The Arabs are good at explaining the ancient Greek texts: 13, 19.
14. Arabic learning is more comprehensible than the Western Latin tradition of learning: 6, 16.
15. But if the Arabs streamline ancient Greek texts, often the Arabic text itself has to be shortened, since the Arabic language is by nature more prolix than Latin: 8, 11.
16. The translation from Arabic is provisional, with the expectation that comparisons with Greek will be made: 5b, 19.

And finally:

17. Translating from Arabic is good for one's soul: 3.

Part Two: Texts in Translation

The following texts illustrate why and how medieval Latin translators translated works on science and philosophy from Arabic.[24] All translations are the author's own unless otherwise indicated.

Text 1. New material on astrology, translated from the language of the Chaldeans. From an anonymous (Miró Bonfill?) adaptation of a Latin translation of an Arabic astrological text, written in Catalonia before AD 984.

Alchandreana, Proportiones, ch. 48 (ed. Juste, *Les* Alchandreana *primitifs*, 554).

Now, since I have written many things concerning the duties of the Moon, I shall attempt to explain something briefly: not what I have found up to now in the pages of the Christians, nor what has been transmitted to the wisdom of the Hebrews from their ancestors, but what has been taken in modern times from the very sanctuaries of Chaldean wisdom by ever-wakeful labour, and translated from the Chaldean language[25] into Latin by a very wise master of the Chaldeans who is my instructor.

Text 2. The importance of ancient wisdom on the astrolabe. From the anonymous preface (beginning 'Ad intimas summe philosophie') to an astrolabe text within the corpus of the earliest Latin astrolabe texts, written in Catalonia between c. 980 and c. 1000.

Ed. Millás Vallicrosa, *Assaig*, 274–276.

This is the Wazzalcora,[26] obtained by a divine mind, which in Latin means 'flat sphere', which by another name is also called 'the astrolabe of Ptolemy'. In this Wazzalcora the whole form of the celestial sphere is described by a natural reasoning following the roundness of the sphere, and all things are noted correctly by the 'architectonic' reasoning of celestial figurations, which I shall describe more lucidly [*lucidius*] a little later on. How useful this is, both what follows will demonstrate and the truth of the matter will prove. It has pleased me above all

24 Prefaces and dedications are complete unless otherwise indicated.

25 I.e., the language of the ancient astrologers, considered to come from Mesopotamia, but here referring to Arabic. Cf. Horace, *Carmina*, Book 1, Ode XI.

26 I.e., *basṭ al-kura* ('the levelling out of the sphere'). The transliteration shows the typical Spanish ambiguity between *b* and *v* (*vv*).

to sketch a few of the effects of its usefulness, so that the wise soul will know what to seek from this. And I will keep to the fixed order of the Arabic source from which it is derived, by the help of God, not as if showing anything new, but as if recommending something neglected or forgotten. Whoever reads this, re-reads the inventions of the Ancients. In present circumstances I am not able to invent new doctrines. Hence, I have refused to avoid the duty of a faithful interpreter,[27] but rather ⟨I have decided⟩ simply to interpret how it is in the Arabic, and having said this, to speak in order, about each thing that is written on it [the astrolabe], and how we are trained by it. But at the end of the completed work ⟨I shall speak⟩ about its construction, for the reason that it needs more effort, and the rationale of the construction is more open after the theory has been grasped.

Text 3. Turning to the Arabs as a last resort. Constantine the African's preface to his translation of *On Urines* by Isḥāq ibn Sulaymān al-Isrā'īlī (c. 855–955), written probably in the abbey of Montecassino before 1099 and dedicated to a fellow monk and convert from Islam, Johannes Afflacius.

Omnia opera Ysaac, fol. clvi r (ed. Bloch, *Montecassino*, I, 103).

Among Latin books I was able to find no author who provided a reliable and authoritative knowledge concerning urines. Hence, I turned to Arabic, in which I found a wonderful book with information of this kind. This book I, Constantine the African, a monk of Montecassino, decided to translate into Latin, so that I might obtain a reward for my soul from my efforts and might widen the path for those intending to learn about urines. This book has been collected and excerpted from ancient authors. From it one can easily approach the knowledge of urine, and also its different types and indications. It was composed in the Arabic language by Isaac, the adoptive son of Solomon, and he divided it into ten parts.

Text 4. Hesitation in translating a great authority. From Constantine the African's preface to his translation of Galen's commentary on Hippocrates's *Aphorisms*.

Ed. Bloch, *Montecassino*, I, 104.

27 'Fidus interpres' is the term used by Boethius in defending literal translation in his second commentary to the *Isagoge* of Porphyry; Boethius, *In Isagogen Porphyrii Commentorum Editio secunda*, 135.

Although you have persistently entreated me, my son Azo,[28] to translate from Arabic into Latin some of Galen's smaller works, I refused to do this for a long time, hesitating to translate the works of such a great philosopher.

Text 5a. A blue-print for translators. From Stephen of Antioch's first preface to his translation of 'Alī ibn al-'Abbās al-Maǧūsī's *The Royal Book* (*of Medicine*), completed in 1127. Stephen criticizes an earlier translation of the same work, which must be that of Constantine the African, and advocates his own method of translating.

Ed. Burnett, 'Antioch as a Link', 26–30.

Therefore, following the command of Solomon, I strove for the sake of wisdom to investigate thoroughly not only the Latin language, but also Arabic, so that, the more I should have the knowledge of different languages, the more expressly I should understand the substance, the dimensions, and the quality of what I had once learnt as a mere beginner in the cradle of philosophy. So, I stumbled upon a certain book that is called among the Arabs the *Completion of Medicine* and the *Royal Book*. On investigating whether Latin had any of this book, I found that it lacked the latter and greater part of it, but that the other part had been vitiated through the crafty deceit of the translator. For he took away the name of the author and the title, and put himself, who was the translator, as also the creator [*inventor*] of the book, and he entitled the book with his own name. In order to do this more easily, he missed out many necessary things, and, changing the orders of many things, he put forward many things in a different way, observing this alone: that he added nothing at all of his own invention. In this he showed himself clearly to us to have been the translator rather than the author. Therefore, he should rather be condemned, who, when he translates the book of another author from one language into another, by some over-confidence or impudence has not blushed either to take away from the author what he had laboured over, or to usurp that material for himself.

Since this procedure does not please us at all, and the book is very necessary for the life of man, and outstanding for its learning, we also approach the task of translating this book, but following another method: ascribing to the author what is his, and to ourselves what belongs to the interpreter, so that what is

28 Azo (Atto), a fellow monk at Montecassino, and former chaplain to the German empress, Agnes, who is credited with improving the Latin of Constantine's translations.

lacking in Latin is supplied, what was placed in the wrong position is restored to its proper position, and what was put forward wrongly is translated as it is in Arabic.

We have, then, proposed to devote the effort of our labour first to these books, although the Arabic language has, hidden within it, other things more noble than these: namely, all the secrets of philosophy, to the translating of which, afterwards, if divine kindness permits, we shall devote our skill, once it has become practised. For we put these easier subjects first, so that there is a path for us to the difficult subjects, and we provide first what is necessary for bodies, so that, when healing has been provided for these by the art of medicine, what belongs to the excellence of the mind, being much more lofty, should follow. Thus, in the face of the wicked and well-prepared tenacity of the Moderns to defend what they have once learnt, we beg that, if we shall seem to have made mistakes in these things, they should prove us wrong, if they can, by consulting the truth of the Arabs, and [...] we demand that they should at least cease from criticizing us.

Text 5b. Any deficiencies in the translation from Arabic should be made up by local knowledge and comparison with Greek. From Stephen of Antioch's second preface to his translation of 'Alī ibn al-'Abbās al-Maǧūsī's *The Royal Book* (*of Medicine*).

Ed. Burnett, 'Antioch as a Link', 35–38.

But since we are translating this work from the Arabic language, and almost all the names of the medicaments placed here are put forward in the language of the Arabs, and we hardly have common Latin words for them, we put them forward as they are in Arabic—sometimes even those which are known to us, in all cases where they are unknown—but we have declined them according to the form of Latin declensions. For if we did not use this method, there would be no way for us to translate at present, since we have no one who knows both languages. I have preferred, then, to seem a little infirm than not to transmit knowledge, since it is less of a task for someone to ask about a few things than about everything, when Latin has the first part incomplete, and the second part not at all, or at least some disordered piece of work which rather gets in the way of knowledge.[29] But we have not altogether consigned the reader to error and

29 Stephen is referring to the same incomplete translation of Constantine the African as in 5a above.

worry, but, as was possible for us and as the Orient could provide, at the end of the whole work we have added a compendium of all the medicaments that are in Dioscorides,[30] having on this side the names in Greek, on that side, in Arabic, so that the man into whose hands this work comes can ask what each thing is, if he finds a Greek, or indeed an Arab. But as for the rest, which are only in Arabic, we have either put them forward with their proper definitions, when we were able to do so, or have been completely silent about them, intending to make an effort, if in any way divine grace gives us the possibility, to insert the Latin names among the Greek and Arabic ones.

Text 6. 'Something new from the studies of the Arabs.' Adelard of Bath's justification of Arabic studies at the beginning of his *Quaestiones naturales* (between 1107 and 1133), which purports to give the method of approaching natural science that Adelard had learnt on a seven-year quest for the 'studies of the Arabs'.

Adelard of Bath, *Conversations with His Nephew*, 83 (ed. Burnett).

A certain nephew of mine, investigating the causes of things, was tying them in knots rather than unravelling them. He urged me to put forward something new from the studies of the Arabs. As the others agreed with him, I undertook the following treatise, which I know will be useful to its auditors, but whether it gives pleasure, I know not. For the present generation suffers from this ingrained fault: that it thinks that nothing should be accepted that is discovered by the 'Moderns'. Hence it happens that whenever I wish to publish my own discovery, I attribute it to another person, saying: 'Someone else said it, not I.' Thus, lest I have no audience at all, ⟨I say:⟩ some teacher came with all my opinions, not I.

Text 7. Discovering a manuscript in the hidden depths of an Arabic library. From Hugo Sanctelliensis's preface to Ibn al-Muṯannā's commentary on al-Ḫwārizmī's *Astronomical Tables* (before 1151), addressed to his patron, Michael, bishop of Tarazona 1119–1151.

Ed. Haskins, *Studies*, 73.

Because, then, my lord bishop of Tarazona, I, the Sanctellian, am not able to fulfil your request by myself, I dare to offer to Your Worthiness a translation of

30 Stephen is referring to the comprehensive pharmacological handbook, Dioscorides, *De re medica* (written between AD 50 and 70).

this commentary [...] that your insatiable desire for philosophizing deserved to find in the library of Rueda,[31] among the more secret depths of the bookshelves.

Text 8. A programme of learning and an Arabic astrological work reduced to Latin brevity. From Robert of Ketton's preface to his translation of al-Kindī's *Iudicia*, addressed to Hermann of Carinthia, whilst they were collaborating 'on the banks of the Ebro', c. 1138–1143.

Oxford, Bodleian Library, MS Digby 51, fol. 55r (collated by Charles Burnett with other manuscripts).

Although after Euclid I would have more willingly sweated over the *Measurement of the World* of Theodosius[32] and a book on proportions, so that a more convenient way should be opened up to the *Almagest*,[33] to which above all our study aspires, nevertheless, lest our friendship should grow weak because of my inactivity, when your wishes are demanding nothing unreasonable, dear Hermann (second to no Latin astronomer of our time), being prepared to obey completely, I have translated him whom your care has often noted to be the most convenient and truthful judge among astrologers, following your wish, relying not less on the faculties of our friendship than on ⟨our⟩ expertise. And, striving greatly in this to please both you and other students of this science, having untangled the surface meaning of the words, I have included in the compact style of the Latins the seed and effect of things, the sum total of the effects of the stars, and certain results of prognostication.

Text 9. Arabic works replace those of the Greeks. From Hermann of Carinthia's preface to his translation of Ptolemy's *Planisphaerium*, dedicated to Thierry, chancellor of Chartres Cathedral (d. c. 1150).

Hermann of Carinthia, *De essentiis*, 348 (ed. Burnett).

---

31 I.e., Rueda de Jalón, the last stronghold of the Banū Hūd (until its capture by the Christians in 1140).

32 I.e., Theodosius of Bithynia (second century BC), whose *Sphaerica* was transmitted from Greek into Arabic and into Latin as a kind of appendix to Euclid's *Elements*; but the 'cosmometria' could refer to his *De habitationibus*, which considered astronomical phenomena as seen from different parts of the globe.

33 Ptolemy's astronomical masterpiece, the *Almagest* (AD c. 160), was the goal of several translators, and at least three translations into Latin were made in the twelfth century.

(Ptolemy) brought together two volumes in the Greek language: for the first [celestial motion] the *Sintasis*, for the second [celestial effect] the *Tetrastis* [*Tetrabiblos*],[34] called in Arabic the *Almagesti* and *Alarba*,[35] of which al-Battānī[36] most conveniently [*commodissime*] abbreviated the former, while Abū Maʿšar[37] no less conveniently amplified the latter.

Text 10. Hidden studies. From Hermann of Carinthia's preface to his original work, *De essentiis* (*On the Essences*), written in Béziers in 1143, which includes the fruit of his studies of Arabic natural science and mathematics, and was addressed to his colleague, Robert of Ketton.

Hermann of Carinthia, *De essentiis*, 70 (ed. Burnett).

You remember, I think, that when we went forth from our inner sanctuaries into the public display of Minerva [wisdom],[38] the multitude of people milling around were gaping at us with open mouths, not valuing us so much as individuals as admiring the trappings and decorations that long vigils and our most earnest labour had acquired for us from the depths of the treasuries of the Arabs.

Text 11. The prolixity of Arabic. From Hermann of Carinthia's preface to his translation of Abū Maʿšar's *Great Introduction to Astrology* (1140), addressed to his colleague, Robert of Ketton.

Abū Maʿšar, *Liber Introductorii Maioris*, VIII, 3 (ed. Lemay).

You [Robert] have indeed experienced how difficult it is to translate from so overflowing [*fluxus*] a kind of speaking as Arabic anything befitting Latin speech, especially in those cases that demand such a strict correspondence to the subject matter [*res*].

---

34 While Hermann refers to the original Greek title of Ptolemy's *Almagest* (*Μαθηματικὴ Σύνταξις* or *Ἡ Μεγάλη Σύνταξις*), his Greek title for Ptolemy's astrological work, the *Τετράβιβλος* (also known as the *Ἀποτελεσματικά*) seems to have been invented by him.

35 These are the usual names for the Arabic titles of Ptolemy's major works: *al-majisṭi*, derived from *Ἡ Μεγίστη* (*Σύνταξις*), and *al-arbaʿa*, 'the four (books)'.

36 Al-Battānī (c. 858–929), active in Raqqa, whose astronomical tables (with their canons) were regarded as an alternative to those of al-Ḫwārizmī (c. 780–850) in the Islamic West.

37 Abū Maʿšar (787–886), the foremost Arabic astrologer, whose *Great Introduction to Astrology* was translated twice into Latin in the mid-twelfth century.

38 Minerva, as Roman goddess of wisdom, symbolizes some unspecified kind of school. Hermann is evidently speaking metaphorically here.

Text 12. The difficulty of translating from Arabic into Latin. Emir Eugenius's preface to his translation of Ptolemy's *Optics*.

L'Optique de Claude Ptolémée (ed. Lejeune), 5; tr. Michael Angold; for another translation, see Smith, *Ptolemy's Theory of Visual Perception*, 62.

Since I consider Ptolemy's *Optics* as essential both for those interested in the sciences and for those investigating natural phenomena, I have not hesitated to take on the burden of translating it into Latin in the present volume. Because each language has its own idiom, it is not easy to translate from one language to another. This is especially so for the faithful translator [*fidus interpres*], but still more when translating from Arabic into either Greek or Latin, since there are such differences between them, in terms of both vocabulary and syntax. Hence, since there are certain ⟨propositions⟩ in this work that are not immediately clear, I have thought it proper to provide a short explanation of the author's train of thought, which is more easily intelligible in the Arabic version, so that the way is made easier for the reader.

Text 13. The advantage of the expansiveness of Arabic. From the commentary of Richard the Englishman(?) on Isḥāq al-Isrā'īlī's *De febribus* (*On Fevers*).

Würzburg, Universitätsbibliothek, MS M. p. med. Q. 1, fol. 138rb.

Likewise, the books of Isḥāq bring about clarity for another reason: because they are translated from the Arabic language, which is very prolix—more so than the Greek language—and because brevity creates obscurity, ⟨while⟩ expansiveness ⟨leads to⟩ explanation and clarity, and this and other books of Isḥāq are translated from the Arabic language, they bring about clarity and explanation. But others, since they are translated from Greek, which is characterized by brevity, are difficult and obscure, as is manifest in the books of natural science of Aristotle, such as in the book *De anima*.

Text 14. The blindness of Latin ignorance and the richness of Arabic scientific literature. From Plato of Tivoli's preface to his translation of the *De scientia astrorum* of al-Battānī (c. 855–c. 955), translated in Barcelona in 1138.

Paris, BNF, MS lat. 7266, fol. 48r.

Among all the studies of the liberal arts that the Greeks, and even before them, the Egyptians, are known to have discovered, that teaching which professes knowledge of the stars both is, and is considered rightly, the chief. We would not hesitate to prove this with the irrefutable arguments of demonstrations, except that one would depart far from one's plan, and it is ⟨anyway⟩ most confidently established among those who profess philosophy. For where ⟨else is⟩ there such subtlety in discovery, such firmness in demonstrations, such pleasure in practical performance, such benefit in results? All the more so should the blindness of the ignorance of Latin culture be deplored, all the more the negligence of its laziness be criticized. Occupied with easier and certainly less worthy studies, ⟨Latin culture⟩ has either feared to approach the subtle elegance of this science, or has avoided it out of contempt. Granted that in the happy outcome of wars, and the expansion of the empire, Rome has surpassed not only Egypt and Greece, but also all nations, however many there are. But in the training in the arts, and in the speculations of the disciplines, although some insolently compare her to Greece, and others more insolently prefer her, she is far inferior not only to Egypt and Greece, but also to Arabia. This can both 1) be easily understood in the other arts, which, when the Latins have them, they have borrowed not from themselves but from elsewhere, and 2) especially is made evident in the aforementioned science of the stars, of which Latin does not dare to present, I do not say an author, but even a translator of whom she can boast. The Egyptians have, among many others, Hermes as the leader in this art; the Greeks Aristotle, Abracaz,[39] Ptolemy, and innumerable others; the Arabs have al-Ḫwārizmī, Māšā'allāh, and al-Battānī, together with very many others. Our Latins have not a single author—but in place of books, mad ravings, dreams, and old wives' tales.

Moved by this, I, Plato of Tivoli, have decided to enrich the narrowness of our language, where it is especially deficient, from the riches of a foreign language as far as I can. But since after long and diligent thought I found nothing in Greek or Arabic that related to this science more perfect than the work of Ptolemy that is called the *Almagest*, since in that work everything depends on the causes of events, numerical ratios, the depiction of geometrical figures, and the firmness of demonstrations, and since I discovered that al-Battānī is his perfect imitator among the Arabs, in that he compendiously abbreviates the prolixity of Ptolemy, emending the blemishes, albeit very uncommon, that he imputes not to him [Ptolemy] but to their origin in Hipparchus, asserting that a stable mechanism cannot be built on a weak foundation, however exceptional.

---

39 I.e., Hipparchus (d. after 127 BC), the predecessor of Ptolemy in the calculation of the positions of the stars and planets.

This al-Battānī I decided to translate and offer to the ears of the Latins by my own labour and with God's favour. If, by chance, the reader encounters any difficulty in this work, he should not think this is caused by the fault of the translator, but rather by the weightiness of the subject matter. For this book is most weighty also in Arabic, both because the science ⟨of the stars⟩ is very subtle and the arguments are convoluted, and because in many places the geometrical demonstrations are deliberately left out, as an exercise meant not for beginners but for advanced students. Therefore, I call upon God, the giver of the science, to be my helper.

Text 15. Translation with four hands. Avendauth's preface to the translation by Dominicus Gundissalinus (d. after 1190) and himself of Avicenna's *De anima.*

Avicenna Latinus, Liber de anima, 4.

To John, the most reverend archbishop of Toledo and primate of the Spains,[40] Avendauth, the Israelite philosopher,[41] gives homage, recognizing the debt that is due to him [...] Therefore, Sir, I have attempted to put into effect your order to translate the book of Avicenna the philosopher concerning the soul, so that by your support and our labour the Latins become certain about what hitherto has been unknown, namely, whether the soul exists, and what and how it is, according to essence and effect, proved by the truest arguments. Thus you have the book, translated from Arabic, with me taking the lead and rendering each word in the vernacular language, and archdeacon Dominicus turning the words into Latin. In this book, you should know that whatever Aristotle said in his books *On the Soul*, *On Sense and What is Sensed*, and *On Intellect and What is Intellected*[42] has been collected by the author.

Text 16. Preference of the learning of the Arabs to that of the French. From Daniel of Morley's preface to his *Philosophia.*

Daniel of Morley, *Philosophia*, 212–213 (ed. Maurach).

40 I.e., John of Castelmoron-sur-Lot, archbishop of Toledo 1152–1166. The plural 'Spains' was a common designation of the various Christian provinces in the Iberian peninsula.

41 I.e., Abraham ibn Daud, the Jewish philosopher and theologian, who fled from the Almohades to Toledo and died c. 1180.

42 *On the Intellect and What is Intellected* is not a work by Aristotle, but may refer to books of this title in the Peripatetic tradition, such as those by Alexander of Aphrodisias or al-Kindī.

When, some time ago, I went away from England to study, I stopped a while in Paris. There I saw some asses occupying the chairs in the schools, pretending to be very important. They had two or three desks in front of them, bearing immovable tomes, and were painting Roman Law in golden letters. With leaden styluses in their hands, they inserted asterisks and obeluses in their books with a grave and reverent air. But because they did not know anything, they were no better than marble statues: by their silence alone they wished to seem wise, and as soon as they tried to say anything, I found them completely unable to express a word. When I discovered things were like this, I did not want to get infected by a similar petrifaction, and I was seriously worried that the liberal arts, which illuminate the Bible, were being skipped over, or passed over in summaries. But when I heard that the doctrine of the Arabs, which is devoted almost entirely to the quadrivium, was all the fashion in Toledo in those days, I hurried there as quickly as I could, so that I could hear the wisest of the philosophers in the world. Eventually my friends begged me to come back from Spain; so, on their invitation, I arrived in England, bringing a precious multitude of books with me […]

As happens when friends first meet again, I was repeatedly questioned by my lord the bishop about the wonderful things in Toledo and the teaching there. His last question, about the movements of the celestial bodies, led the talk round to astrology [*astronomia*[43]]. He mentioned, for instance, that some things in our sublunar region seem to be subservient to their superiors, as if under a bond of fealty. Because the shortness of time did not allow me to answer these questions sufficiently, I decided to present the following treatise for his scrutiny. Its first book is about the lower part of the universe, its second about the higher. I beg the reader earnestly, that, just because everything is set out in black and white, he should not despise the simple and clear opinions of the Arabs, but should note that Latin philosophers make heavy weather of these subjects quite unnecessarily, and, through their ignorance, have put forward figments of their imagination veiled in obscure language, so that their unsteady floundering in this subject might be covered by a blanket of unintelligibility. For no one should be upset over what is said here, that when dealing with the creation of the world, I do not call to witness the Catholic Fathers but the pagan philosophers, because, although such people are not numbered amongst the faithful, nevertheless some of their words, since they can be fully trusted, should be drawn into our teaching. Hence, we too, who are freed from

43 Gundissalinus and the writers of Gerard of Cremona's *Vita* also use *astronomia* and *astrologia* in the sense of astrology and astronomy respectively.

Egypt in a mystical way, are ordered by the Lord to take from the Egyptians bronze and silver vessels, to enrich the Hebrews. Therefore, with our Lord's instruction and help, let us take from the philosophers of the pagans, wisdom and eloquence, and in this way let us despoil them in their faithlessness, so that we might be enriched faithfully by their spoils.

Text 17. The life of a translator. Gerard of Cremona's students' short *Vita* of their master, written after his death in 1187; it accompanies a list of the books he translated and an eight-line verse eulogy.

Ed. Burnett, 'The Coherence', 255–256.

Although from his very cradle he (Gerard) had been educated in the lap of philosophy and had arrived at the knowledge of each part of it according to the study of the Latins, nevertheless, because of his love for the *Almagest*, which he did not find at all amongst the Latins, he made his way to Toledo, where, seeing an abundance of books in Arabic on every subject and, pitying the poverty he had experienced among the Latins concerning these subjects, out of his desire to translate, he thoroughly learnt the Arabic language and in this way, trustworthy in each—i.e., the subject-matter and the language (as Aḥmad ⟨ibn Yūsuf al-Dāya⟩ in his letter *On Ratio and Proportion* says: 'It is necessary that the interpreter, in addition to the excellence that he has acquired from the knowledge of the languages from which and into which he translates, should also have knowledge of the subject which he translates')—in the manner of a prudent man who, walking through green meadows, weaves a crown from flowers—not from all of them, but from the more beautiful—he read through the writings of the Arabs, from which he did not cease until the end of his life to transmit to Latin, as if to a beloved heir, in as plain and intelligible way as was possible for him, books on many subjects—whatever he esteemed as the most choice. He went the way of all flesh in the seventy-third year of his life, in the year of Our Lord Jesus Christ 1187.

Text 18. The search for Arabic books in the libraries of Toledo. The preface of Mark, a canon of Toledo Cathedral between c. 1194 and 1216, to his translation of Galen's *De tactu pulsus*.

Ed. D'Alverny, 'Marc de Tolède, traducteur d'Ibn Tūmart', 259–260.

When I was already studying the art of medicine and the books on this subject written in Latin, the teachers and scholars there, who already were aware that

I knew Arabic and had thoroughly read the works of Galen and other teachers on the same subject, urged me insistently to choose, under the inspiration [*intuitus*] of God, some books, among the large number of texts that the Arabs, drawing from the fonts of the Greeks, had translated into Arabic, and to translate them into Latin for the benefit of scholars. Since it was not right that I should oppose them, when their request was so favourable, I returned to Toledo as quickly as possible, and made efforts to put into practice what they sought. When I was thinking about this and trying to work out a solution, it occurred to me that the most obvious text to translate was the book of Iohannicius,[44] which I found to be the most perfect and useful one amongst them [the books], being the one that is first read, belonging to 'introductions' [*ysagoge*]. With the help of God, I translated this book. Then, after this, according to the Gospel precept 'Seek and you shall find, knock and it will opened to you' (Luke, XI, 9), searching diligently in the libraries [*armaria*] of the Arabs for another book to translate, I found Galen's *About the Pulse*, *About the Usefulness of the Pulse*, and *On Problematic Movements*, all in one volume, and my mind urged me to bring these to the notice of the Latins.

Having called upon the name of God, I took up my pen, and translated these books from Arabic into Latin, since they are very useful for scholars and teachers of this profession, as I believe. As for *On the Pulse*, although the book on this subject written by Philaretus is available, this one (of Galen) is thought to be much more useful and precious to readers as gold is more precious than lead. But as for *About the Usefulness of the Pulse* and *On Problematic Movements*, there has never been any indication that they had ever been translated into Latin either from Greek or from Arabic.

Text 19. The translation of Arabic texts as a preliminary to translating the original Greek. From Hermann the German's preface to his translation of Aristotle's *Rhetoric* (between between 1243 and 1256).

Aristoteles Latinus, I, 211 (ed. Lacombe).[45]

---

44 I.e., Ḥunayn ibn Isḥāq (808–873), the prolific translator of medical and philosophical works from Greek (and Syriac) into Arabic in Baghdad. The work mentioned is known as the *Ysagoge Iohannitii* and is the first text of the medical teaching corpus known later known as the *Articella*.

45 See also Woerther, 'Citer/traduire', where a partial edition and a detailed summary of the text are provided. I am grateful to Frédérique Woerther for her advice on the text and the translation.

I undertook the task of the present translation of the *Rhetoric* and *Poetics* of Aristotle from Arabic into Latin a long time ago, on the instigation of the venerable father John, bishop of Burgos and chancellor of the king of Castile. But I was not able to finish it up to now because of interruptions. In my opinion, the Latins, being eager above all other nations in their pursuit of the rational part of philosophy, should welcome this in the present state of affairs, so that they hereby have the complement of the curriculum of logic according to the intention of Aristotle. That these two books belong to logic no one doubts who has investigated the books of the famous Arabs, al-Fārābī, Avicenna, Averroes, and some others. [...] The writings of these aforementioned men are most useful for understanding the present works.

No one should be surprised or annoyed by the difficulty and almost crudity of the translation. For it had been translated in an even more difficult and crude way from Greek into Arabic—such that al-Fārābī, who first tried to draw some sense out from the *Rhetoric* by writing a gloss, passed over and left out many Greek examples because of their obscurity. For the same reason he produced many uncertain explanations, and for this reason too, as Avicenna and Averroes think, he did not continue his glossing up to the end of the work. These two men also say this at the end of the treatises that they composed on the model of Aristotle. This is what we have been able to understand and take from the translation that has come down to us of these volumes of Aristotle. And this is the reason why even until today these two books have been almost completely neglected amongst the Arabs, and I was scarcely able to find anyone who wished to devote enough effort to studying these books with me.

Therefore, let those pardon ⟨me⟩ who have accused this work of mine of being imperfect (perhaps rightly); and if they do not wish to seek any benefit from this, they can refute and abandon it. Nevertheless, I would advise them to prefer to have these codices translated in this way than to leave them completely neglected; for nothing is less civilized than having nothing at all. But it is possible that, once one has these ⟨texts⟩, through gradual increases in some way one can finally and more easily share in the perfection of the desired aim. This happened in the case of the book of Nicomachus that the Latins call the *Ethics* of Aristotle. For this work too I translated from Arabic into Latin, as far as I could. But afterwards, the reverend father master Robert Grosseteste ('thickhead', but of subtle intelligence), bishop of Lincoln, translated it more completely,[46] from the first font from which it had flowed, i.e., Greek, and com-

46 Robert Grosseteste (c. 1168–1253), bishop of Lincoln 1235–1253, was well known for his Greek studies, which included the first full translation of Aristotle's *Nicomachean Ethics*.

mented on it, weaving his own notes into the commentaries of the Greeks. Thus, if it pleases the Giver of all knowledge, it could happen in these little efforts worked on by us in a preliminary, though weak, fashion, that He who lives and reigns eternally in the perfect Trinity, might think it worthy to bring to pass the same ⟨perfection⟩. Amen.

Text 20. A supplement to Aristotle's *De caelo*. The dedicatory preface of Michael Scot (d. before 1236) to his translation of Averroes's *Large Commentary on Aristotle's De caelo et mundo.*

Averrois Cordubensis commentum magnum super libro De celo et mundo Aristotelis, I, 1 (ed. F.J. Carmody et al.).

I especially commend to you, Stephen of Provins,[47] this work, which I, Michael Scot, have given to Latin from the words of Aristotle, and if Aristotle left out anything in this book concerning the construction of the world, you can receive its supplement from the book of al-Biṭrūǧī,[48] which I similarly have given to Latin, and you will be skilled in it.

Text 21. New clothes for old texts. From Manfred, king of Sicily's letter to the scholars of Paris in 1263, advertising translations that he had commissioned from Greek and Arabic.

Ed. Gauthier, 'Notes sur les débuts (1225–1240) de premier Averroïsme', 323–324.

While I was continuously turning over in my mind and examining with close scrutiny the volumes of books of which differing and variously distinctive manuscripts fill the bookcases of our treasuries, various compilations from ⟨the works of⟩ Aristotle and other philosophers, produced a long time ago with Greek and Arabic vocabulary, in the disciplines of eloquence and mathematics, came to our notice. They were still woven in the web of their original words and covered with the clothing of those ancient garments that the first age had woven for them, and by the fault or deed of man had not been brought to the notice of the Latin language. Wishing, then, that the reverend ancient authority of such great works should be rejuvenated among us with the transmitting

47 Stephen of Provins was one of the three scholars commissioned by Pope Gregory IX to examine the works of Aristotle on natural philosophy in 1231.

48 Al-Biṭrūǧī's *Kitāb fī l-hay'a*, written between c. 1185 and 1190 in Córdoba, was translated soon afterwards (in 1217) by Michael Scot in Toledo.

organ of the voice, with many benefits, I thought that they should be translated immediately by chosen men, skilled in speaking both languages, whilst the virginity of the words should be faithfully kept intact.

Text 22. A single medical compendium that substitutes for a library of texts. From Farağ ibn Sālim (Faragut)'s preface to his translation of Abū Bakr ibn al-Zakarīyā' al-Rāzī's *Kitāb al-Ḥāwī*, made in 1282.

Ed. Fischer and Weisser, 'Das Vorwort zur lateinischen Übersetzung von Rhazes' Liber continens (1282)'. See also Borghese and Grévin, 'Aspects linguistiques', 88–89.

Abū Bakr, when he noticed that this redounded in some way not only on the famous excellence of Hippocrates and Galen, but even more on his own insufficiency and that of other authors of medicine, like an exceptional student of medical philosophy, not giving way to laziness, but taking up tasks willingly, and not even complaining about the troubles of old age, he planned this book in his old age and completed what he had planned [...]

He diligently inserted the authoritative statements of them all into this book, to which he gave the name 'Elhauy' in the Arabic language, i.e., 'containing the whole' or 'bringing together all things', because all things that are dispersed in the books of other authors under some prolix obscurity of words are inserted succinctly and clearly in this book, so that scholars—having considered the prolixity of the art, the fallacy of experience, and above all the brevity of life,[49] finding in it the whole of Hippocrates and Galen speaking in their own words, with ⟨the words of⟩ all other expert doctors succeeding them, and also the opinions of the author himself [al-Rāzī]—content with this compilation and otherwise extending their study concerning it, are not forced to turn repeatedly the pages of many volumes of medicine.

Frequent and assiduous reading of this book has made very many of the Arabs skilled in the science of medicine. However, outside ⟨the Arabic world⟩ it was for a long time unknown. But after long cycles of years, the most Christian lord Charles, king of Jerusalem and Sicily,[50] moved by the fame of the same book, and led on by his passion for such handy usefulness, not desiring to benefit himself alone, but all worshippers of Christ, chose to mix with the cares of war his liberal studies' concern for the translation of this book, so that what had been acquired from heaven for him in the triumphs of battles on the side of

49 Here Faragut echoes the opening words of Hippocrates's *Aphorisms*.

50 Charles I of Anjou, king of Sicily 1266–1285 and king of Jerusalem 1277–1285.

the recognition of kingdoms and great titles, he might partly return to the peoples, in honour of Him who is the creator and preserver of the peoples. Having obtained the aforesaid book from the king of Tunisia through official messengers, he employed a faithful servant, skilled as much in Arabic as in Latin, and in this very book in which so much usefulness was hidden under the darkness of the Arabic tongue, he commanded and caused the light of a Latin translation to be lit.

Nevertheless, he subjected this ⟨translation⟩ to the examination, the overseeing, and correction of his own medical staff, professors of the art of medicine and others teaching in Naples and Salerno, and he granted them not summary but ordinary time to do this. All these, one by one making a public presentation concerning such a translation on a day chosen for this, with one voice and a single agreeing opinion, praised the interpreter and the author as being most worthy of the pronouncements of praise. For if the proper meaning of the Arabic words may seem anywhere perhaps to have been corrupted by the Latin translator, nevertheless the true and complete meaning of this compilation has not lost the grace of virginity.

Part Three: Original Texts

In many cases I have silently altered the orthography and punctuation of the editions I have used, to make them easier to read.

Text 1. *Alchandreana, Proportiones*, ch. 48. Ed. Juste, *Les* Alchandreana *primitifs*, 554.

Nunc quoniam perscripsi plura ex officiis Lunae, temptabo quoddam explicare compendiose, non Christianorum hactenus paginis a me repertum, nec Hebraeorum sagacitati a patribus eorum traditum, set moderno sub tempore vigilanti industria Caldaeorum ab ipsis adytis sapientiae discerptum et a sapientissimo Caldaeorum magistro me imbuente in Latinum ex Caldaeo eloquio translatum.

Text 2. From the anonymous preface ('Ad intimas summe philosophie') to an astrolabe text. Ed. Millás Vallicrosa, *Assaig*, 274–276.

Est autem Wazzalcora divina mente comparata, quod Latine sonat 'plana spera', que etiam alio nomine astrolapsus Ptolomei, in qua Wazzalcora secundum celi rotunditatem formata naturali ratione tota celestis sphere describitur forma et omnia ritu celestium figurationum architectonica ratione notantur, quam paulo post lucidius describam. Que quam utilis sit, et subsequens series demonstrabit et rei veritas probabit. Placuit autem mihi in primis pauca de eius utilitatis effectibus perstringere, ut prudens animus quod ex hac petat, sciat. Et Arabici fontis, unde Dei amminiculatione derivata est, stabilem observem seriem, non quasi novum aliquid ostendendo, sed quasi neglectum aut oblitum commonendo. Qui hec legit, inventa Priscorum relegit. Nequeo modo nova invenire studia; dehinc nolui fidi interpretis officium precavere, sed potius, ut in Arabico habetur, simpliciter interpretari; quo dicto, singillatim de eius inscriptionibus et quomodo ipsa exercitemur seriatim dicere. In calce autem completi opusculi de eius constructione, eo quod maioris sit studii, et quod apertior sit constructionis ratio post perceptam scientiam.

Text 3. Constantine the African's preface to his translation of Isḥāq al-Isrāʾīlī's *On Urines*. *Omnia opera Ysaac*, fol. clvi r (ed. Bloch, *Montecassino*, I, 103).

In Latinis quidem libris nullum auctorem invenire potui qui de urinis certam et autenticam cognitionem dederit. Unde ad linguam Arabicam me diverti, in qua quendam librum in huiusmodi notitia admirandum repperi. Quem ego,

Constantinus Africanus, Montis Cassianensis monachus, Latinae linguae ad transferendum destinavi dare, ut de labore premium animae adipiscerer, et introducendis ampliarem ad cognoscendam urinam iter. Liber iste de antiquis collectus et excerptus est auctoribus, per quem cognoscendae urinae patent aditus et etiam divisiones ipsius et significationes. Fecit autem eum filius adoptivus Salomonis Ysaac in Arabica lingua, et eum divisit in decem particulas.

Text 4. From Constantine the African's preface to his translation of Galen's commentary on Hippocrates's *Aphorisms*. Ed. Bloch, *Montecassino*, I, 104.

[…] Licet petitionibus tuis continuis, fili mi Azo, mihi saepius diceres, ut ex opusculis Galieni aliqua Latinae linguae traducerem ex Arabica lingua, diu tamen multum denegavi, hesitans tanti transferre opera philosophi.

Text 5a. From Stephen of Antioch's prefaces to his translation of 'Alī ibn al-'Abbās al-Mağūsī's *The Complete Book of Medicine*. Ed. Burnett, 'Antioch as a Link', 26–30.

Ipse igitur Salomonicum sequens mandatum, non solum Latinam, verum Arabicam quoque linguam sapientie gratia rimari studui, ut quanto diversarum scientiam linguarum nossem, tanto id quod in cunabulis philosophie rudis aliquando didiceram, quid esset, quantum qualeque expressius intelligem. Impegi itaque in quendam qui apud Arabes 'medicine complementum' dicitur et 'regalis' librum. De quo si quid haberet Latinitas requirens, eius ultimam et maiorem deesse sensi partem, alteram vero interpretis callida depravatam fraude. Nomen etenim auctoris titulumque subtraxerat, seque qui interpres extiterat et inventorem libri posuit, et suo nomine titulavit. Que ut facilius posset, et in libri prologo et in aliis multa pretermisit pluribus necessaria locis, multorumque ordines commutans, nonnulla aliter protulit, hoc uno tantum observato: nichil prorsus ex suis addidit. In quo manifeste nobis innuit ipsum interpretem potius quam scriptorem fuisse. Quare magis arguendus est qui, cum alterius librum de lingua in linguam transferret, confidentia seu impudentia nescio qua et illi quod elaboraverat abrogare et sibi usurpare non erubuit.

Que quoniam nobis minime placent, liberque iste pernecessarius hominis est vite scientiaque conspicuus, aggredimur et ipsi, sed alia via librum hunc transferre: auctori quod suum est, nobis quod interpretis, ascribentes, ut et que desunt in Latino suppleantur, que transposita, suis ordinibus reddantur, que aliter prolata, prout sunt in Arabico transferantur.

His igitur in libris nostri primum consumere laboris proposuimus operam, tametsi alia his preclariora lingua habeat apud se Arabica recondita, omnis scilicet philosophie arcana, quibus deinceps, si divina dederit benignitas, exercitatum dabimus transferendis ingenium. Leviora enim hec preferimus, ut ad difficilia via nobis sit, et que corporibus necessaria sunt tempore preponimus, ut his sanitate proposita arte medicine, que ad animi attinent excellentiam longe altiora subsequentur. Rogamus itaque modernorum improbam et ad ea que semel didicerunt defendendum paratam pertinaciam, ut, si quid in his errase illis videbimur, consulta Arabum veritate, si poterint, nos arguant, et [...] nobis detrahere desinant poscimus.

Text 5b. From Stephen of Antioch's second preface to his translation of 'Alī ibn al-'Abbās al-Maǧūsī's *The Complete Book of Medicine*. Ed. Burnett, 'Antioch as a Link', 35–38.

Sed quoniam opus hoc ex Arabica transferimus lingua, omniaque hic fere posita medicaminum nomina Arabum proferuntur lingua, et nos Latina parum habebamus assueta, prout sunt in Arabico ea proferimus etiam que cognita nobis sunt nonnumquam, que incognita ubique, sed ad Latine formam declinationis inclinata. Nisi etenim sic, nulla esset nobis ad presens transferendi via, cum qui utrasque nosset nullum haberemus linguas. Malui igitur paulo infirmus videri quam scientiam non transferre, quoniam minus est pauca quam omnia aliquem interrogare, cum primam quidem partem sed non plenam, secundam non omnino aut certe turbidum quid, quod magis obesset scientie haberet Latinitas. Nec vero omnino lectorem errori et sollicitudini remisimus, sed quod posse nobis fuit et Oriens habebat, in totius operis fine omnium que apud Diascoridem sunt medicaminum breviarium subdidimus, hinc eorum nomina Grece, illinc Arabice habens, ut in cuius venerit hoc opus manus, quid queque res sit aut Grecum si invenit aut certe Arabem, sit illi posse consulere. Reliqua vero que Arabica sunt tantum, aut propriis protulimus diffinitionibus, *quando* quidem potuimus, aut omnino siluimus, studium daturi ut si quovis modo nobis posse gratia dederit divina, inter Greca et Arabica nomina inseramus Latina.

Text 6. From Adelard of Bath's preface to his *Quaestiones naturales*. Adelard of Bath, *Conversations with His Nephew*, 83 (ed. Burnett).

Nepos quidam meus in rerum causis magis implicans quam explicans, aliquid Arabicorum studiorum novum me proponere exhortatus est. Cui cum assentirent ceteri, ego tractatum subscriptum recepi, quem quidem auditoribus suis utilem fore scio, iocundum nescio. Habet enim hec generatio ingenitum vitium, ut nichil quod a modernis reperiatur putet esse recipiendum. Unde fit ut si quando inventum proprium publicare voluerim, persone id aliene imponens inquam: 'Quidam dixit, non ego'. Itaque, ne omnino non audiar, omnes meas sententias dominus quidam invenit, non ego.

Text 7. From Hugo Sanctelliensis's preface to his translation of Ibn al-Muṯannā's commentary on al-Ḫwārizmī's *Astronomical Tables*. Ed. Haskins, *Studies*, 73.

Quia, ergo, mi domine Tirassonensis antistes, ego Sanctelliensis tue petitioni ex me ipso satisfacere non possum, huius commenti translationem, [...] in Rotensi armario et inter secretiora bibliotece penetralia tua insatiabilis filosophandi aviditas meruit repperiri, tue dignitati offerre presumo.

Text 8. From Robert of Ketton's preface to his translation of al-Kindī, *Iudicia*. Oxford, Bodleian Library, MS Digby 51, fol. 55r (collated by Charles Burnett with other manuscripts).

Quemadmodum post Euclidem Theodosii *Cosmometrie* libroque proportionum libentius insudarem, unde commodior ad Almaiesti, quo precipuum nostrum aspirat studium, pateret accessus, tamen, ne per meam segnitiem nostra surdesceret amicitia, vestribus nutibus nichil preter equum postulantibus, mi Hermanne, nulli Latinorum huius nostri temporis astronomico secunde, penitus parere paratus, eum quem commodissimum et veracissimum inter astrologos iudicem vestra quam sepe notavit diligentia, voto vestro serviens transtuli, non minus amicitie quam peritie facultatibus innisus. In quo tum vobis tum ceteris huius scientie studiosis placere plurimum studens, enodato verborum vultu rerum semen et effectum atque summam stellarium effectuum pronosticationisque quorumlibet eventuum Latine brevitati diligenter inclusi.

Text 9. From Hermann of Carinthia's preface to his translation of Ptolemy's *Planisphaerium*. Hermann of Carinthia, *De essentiis*, 348 (ed. Burnett).

Duo Ionica lingua collegit volumina, in primam ⟨scientiam⟩ Sintasim, in secundam Tetrastim—Arabice dicta Almagesti et Alarba, quorum Almagesti quidem Albeteni commodissime restringit, Tetrastim vero Albumasar non minus commode exampliat.

Text 10. From Hermann of Carinthia's preface to his *De essentiis*. Hermann of Carinthia, *De essentiis*, 70 (ed. Burnett).

Meministi, opinor, dum nos ex adytis nostris in publicam Minerve pompam prodeuntes, circumflua multitudo inhianter miraretur, non tanti personas pensans quantum cultus et ornatus spectans quos ex intimis Arabum thesauris diutine nobis vigilie laborque gravissimus acquisierit.

Text 11. From Hermann of Carinthia's preface to his translation of Abū Maʿšar's *Great Introduction to Astrology*. Abū Maʿšar, *Liber Introductorii Maioris*, VIII, 3 (ed. Lemay).

Expertus quippe tu nichilominus quam grave sit ex tam fluxo loquendi genere quod apud Arabes est Latine orationi congruum aliquid commutari atque in his maxime que tam artam rerum imitationem postulant.

Text 12. Emir Eugenius's preface to his translation of Ptolemy's *Optics*. *L'*Optique *de Claude Ptolémée*, 5 (ed. Albert Lejeune).

Cum considerarem Optica Tholomei necessaria utique fore scientiam diligentibus et rerum percrutantibus naturas, laboris onus subire et illa in presenti libro Latine interpretari non recusavi. Verumptamen, quia universa linguarum genera proprium habent ydioma, et alterius in alterum translatio, fideli maxime interpreti, non est facilis, et presertim Arabicam in Grecam aut Latinam transferre volenti tanto difficilius est quanto maior diversitas inter illas, tam in verbis et nominibus quam in litterali compositione, reperitur. Unde, quia in hoc opere quedam forte non manifesta apparent, dignum duxi intentionem auctoris, ab Arabico libro evidentius intellectam, breviter exponere, ut lectoribus via levior efficiatur.

Text 13. From the commentary of Richard the Englishman(?) on Isḥāq al-Isrāʾīlī's *De febribus*. Würzburg, Universitätsbibliothek, MS M. p. med. Q. 1, fol. 138rb.

Item alia de causa habent libri Ysaac declarationem quia translati sunt a lingua Arabica que multam habet prolixitatem plus quam lingua Greca, et quia brevitas creat obscuritatem, longitudo explanationem et claritatem, et iste et alii libri Ysaac a lingua Arabica sunt translati, ideo habent claritatem et explanationem. Alia autem, cum sint a Greco translata, scilicet cuius modus est brevis, habent difficultatem et obscuritatem, ut patet in libris naturalibus Aristotelis, ut in libro *De anima*.

Text 14. From Plato of Tivoli's preface to his translation of al-Battānī's *De scientia astrorum*. Paris, BNF, MS lat. 7266, fol. 48r (with corrections).[51]

Inter universa liberalium artium studia que Grecos, queque etiam prius invenisse constat Egiptios, que stellarum scientiam profitetur disciplina, et est et habetur iure princeps. Quod inconcussis demonstrationum rationibus asserere non gravaremur, nisi et a proposito longe recederet et apud philosophie professores indubitata fide constaret. Ubi enim tanta in inventione subtilitas, tanta in demonstrationibus firmitas, tanta in exercitiis iocunditas, tanta in proventione utilitas? Quo magis Latinitatis ignorantie cecitas deploranda, magisque desidie negligentia redarguenda est que levioribus et certe indignioribus studiis occupata, huius scientie subtilem elegantiam, aut ex desperatione attemptare timuerit, aut ex contemptu fastidierit. In bellorum quidem felicitate, in imperii dilatatione, Roma non solum Egiptum et Greciam, verum etiam omnes quotquot sunt nationes antecessit, in artium vero gymnasiis, in disciplinarum speculationibus, licet quidam insolenter eam Grecie conferant, quidam insolentius preferant, non tamen Egipto vel Grecia, set etiam Arabia longe inferior extitit. Hoc cum in ceteris artibus facile deprehendi possit, quas, si habent Latini, non a se, set aliunde mutuati sunt, tum vel maxime in prememorata astrorum scientia declaratur, cuius non dico auctorem, set ne interpretem quidem quo se iactet, audet ostentare Latinitas. Habent inter multos in hac arte precipuum Hermetem Egiptii, Aristotilem, Abracaz, Ptolomeum ceterosque innumerabiles Greci, Arabes cum compluribus Alchorismi, Messehella, Albatheni, nostri Latini auctorem quidem nullum, pro libris, deliramenta, sompnia, fabulas aniles.

Hac causa permotus, ego, Plato Tiburtinus, nostre lingue angustias qua maxime deficiebat, ex aliene lingue thesauris pro ingenii facultate ditare constitui. Verum cum post longam et diligentem deliberationem, nichil in Greco aut Arabico quod quidem ad hanc spectaret scientiam opere Ptolomei quod Almagesti dicitur perfectius invenirem, quippe ubi singula eventuum causis, numerorum proportione, figurarum descriptione geometricarum, demonstrationum firmitate subnixa sunt, cumque eius imitatorem perfectum inter Arabes Albateni deprehenderem, quippe Ptolomei prolixitatem compendiose coartat, eiusque nevos emendans, qui quidem rarissimi sunt, non ipsi set Abracaz radici imputat, asserens super debile fundamentum, quamvis egregium, mechanicum stabile edificium architectari non posse.

Hunc inquam Albateni meo labore, deo propitiante, transferendum et Latinis auribus offerendum, censui. In quo opere si qua forte difficultas lectorem

51 I am grateful to Rosa Comes for her collation of the manuscripts of the preface.

offenderit, ne hanc estimet interpretis vitio accidisse set materie gravitate. Est enim liber ille etiam in Arabico gravissimus tum quia scientia subtilissima est, rationes perplexe, tum quia plerisque locis demonstrationes geometrice ex industria subtrahuntur tamquam negotio non rudibus set peritis instituto. Deum igitur scientie datorem adiutorem invoco.

Text 15. Avendauth's preface to the translation by Dominicus Gundissalinus and himself of Avicenna's *De anima*. *Avicenna Latinus. Liber de anima*, 4 (ed. van Riet).

Reverentissimo Toletanae sedis archiepiscopo et Yspaniarum primati Iohanni, Avendehut Israelita philosophus, gratum debitae servitutis obsequium. [...] Quapropter iussum vestrum, Domine, de transferendo librum Avicennae philosophi de Anima effectui mancipare curavi, quatinus vestro munere et nostro labore Latinis fieret certum quod hactenus extitit incognitum, scilicet an sit anima, et quid et qualis sit, secundum essentiam et effectum, rationibus verissimis comprobatum. Habetis ergo librum vobis praecipiente et me singula verba vulgariter proferente, et Dominico archidiacono singula in Latinum convertente, ex Arabico translatum, in quo quicquid Aristoteles dixit in libro suo *De anima*, et *De sensu et sensato*, et *De intellectu et intellecto*, ab auctore libri sciatis esse collectum.

Text 16. From Daniel of Morley's preface to his *Philosophia*. Daniel von Morley, *Philosophia*, 212–213 (ed. Maurach).

Cum dudum ab Anglia me causa studii excepissem et Parisiis aliquamdiu moram fecissem, videbam quosdam bestiales in scolis gravi auctoritate sedes occupare, habentes coram se scamna duo vel tria et desuper codices importabiles, aureis litteris Ulpiani traditiones representantes, necnon et tenentes stilos plumbeos in manibus, cum quibus asteriscos et obelos in libris suis quadam reverentia depingebant. Qui, dum propter inscitiam suam locum statue tenerent, tamen volebant sola taciturnitate videri sapientes, sed tales, cum aliquid dicere conabantur, infantissimos reperiebam. Cum hoc, inquam, in hunc modum se habere deprehenderem, ne et ego simile damnum incurrerem, artes, que scripturas illuminant, non in transitu salutandas vel sub compendio pretereundas mecum sollicita deliberatione tractabam. Sed quoniam doctrina Arabum, que in quadruvio fere tota existit, maxime his diebus apud Tholetum celebratur, illuc, ut sapientiores mundi philosophos audirem, festinanter properavi. Vocatus vero tandem ab amicis et invitatus ut ab Hyspania redirem, cum pretiosa multitudine librorum in Angliam veni [...]

Cum itaque, ut fit in primo amicorum conventu, a domino episcopo de mirabilibus et disciplinis Tholetanis satis quesitum esset, ad ultimum, de motibus supercelestium corporum scrutabundus inquirens, ad astronomiam sermonem direxit. Inter cetera vero quedam de sublunaribus istis adiecit, que suis superioribus quadam necessitatis obedientia videntur servire; sed quia me brevitas temporis ad presens questionibus eius non satisfacere sinebat, ideo illius discretionis examini hunc presentum tractatum presentandum decrevi, cuius prima pars continent de inferiori parte mundi, secunda de superiori.

Exorandus igitur atque multipliciter exhortandus est, ut, quamvis hic nichil contineatur obscurum, non idcirco planas atque dilucidas Arabum sententias contempnere festinet, sed attendat quod Latini philosophi circa talia inutiliter laborantes obscura per ignorantiam figmenta quibusdam ambagibus obvoluta protulerunt, ut ita sub umbra ambiguitatis error incertus tegeretur. Neminem etiam moveat quod de creatione mundi tractans super his que dicuntur, non patres catholicos sed gentiles philosophos in testimonium voco, quia, licet tales inter fideles non connumerentur, quedam tamen eorum verba, cum sint fide plena, ad nostram doctrinam trahenda sunt. Unde et nos, qui mystice liberati sumus ab Egypto, a Domino iubemur mutuari ab Egyptiis vasa aerea et argentea, ut ditemus Hibreos. Mutuemur ergo, Domino iubente et auxiliante, a philosophis gentium sapientiam et eloquentiam, et sic eos in infidelitate sua spoliemus, ut eorum spoliis fideliter ditemur.

Text 17. From Gerard of Cremona, *Vita*. Ed. Burnett, 'The Coherence', 275–276.

Et cum ab istis infantie cunabulis in gremiis phylosophie educatus esset et ad cuiuslibet partis ipsius notitiam secundum Latinorum studium pervenisset, amore tamen Almagesti, quem apud Latinos minime reperit, Toletum perrexit, ubi librorum cuiuslibet facultatis habundantiam in Arabico cernens et Latinorum penurie de ipsis quam noverat miserans, amore transferendi linguam edidicit Arabicam, et sic de utroque—de scientia videlicet et ydiomate—confisus, quemadmodum Hametus in epistula sua *De proportione et proportionalitate* refert: 'Oportet ut interpres preter excellentiam quam adeptus est ex notitia lingue de qua et in quam transfert, artis quam transfert scientiam habeat,' more prudentis qui virida prata perlustrans, coronam de floribus—non de omnibus sed de pulcrioribus—connectit, scripturam revolvit Arabicam, de qua plurium facultatum libros quoscunque valuit elegantiores Latinitati tamquam dilecte heredi, planius ac intelligibilius quo ei possibile fuit, usque ad finem vite sue transmittere non cessavit. Viam autem universe carnis ingressus est anno vite sue .lxxiii., in anno domini nostri Ihesu Christi .mclxxxvii.

Text 18. Mark of Toledo's preface to his translation of Galen's *De tactu pulsus*. Ed. d'Alverny, 'Marc de Tolède', 259–260.

Cum iam in arte medicine studerem, librosque facultatis huius in eloquio Latino, magistri illic regentes et scolares, quibus iam notum erat me linguam novisse Arabicam et litteras Galieni aliorumque doctorum in eadem perlegisse doctrina, diligenter instabant, exorantes ut inter numerosam codicum multitudinem eligerem quos de Grecorum fonte Arabes derivantes in Arabicam transtulere linguam, nonnullos intuitu Dei et commodo scolarium in Latinum transferem sermonem, quibus in tam favorabili petitione, cum nec deberem nec fas erat obviare, quam citius Toletum rediens, eam exquisitionem moderare curavi. Michi itaque super hoc excogitanti atque deliberanti, Iohannicii liber quem penes eos perfectiorem et utiliorem repperi, qui primus utpote Ysagogarum legitur, prima fronte transferendus occurrit, quem Domino adiuvante transtuli. Deinde post hunc, iuxta illud Evangelicum: Querite et invenietis, pulsate et aperietur vobis, in armariis Arabum studiose querens alium quem transferrem librum, inveni Galieni *De pulsu* ac *De pulsus utilitate* atque *Motibus membrorum liquidis* uno volumine contentos, pulsavitque animus ut hos in Latinorum deducerem notitiam.

Invocato igitur Dei nomine stilum accipiens, hos libros de Arabico in linguam transtuli Latinam, scolaribus ac magistris huius professionis perutiles, ut credo. *De pulsu* vero, licet habeatur liber quem Philaretus edidit, tanto tamen hic legentibus utilior ac pretiosior habetur, quanto aurem plumbo pretiosius existit. *De pulsus* autem *utilitate* et *De motibus membrorum liquidis* a seculo non est auditum quod in Latinam linguam de Greco sive de Arabico a quoquam fuisset translatum.

Text 19. From Hermann the German's preface to his translation of Aristotle's *Rhetoric*. Ed. Lacombe, *Aristoteles Latinus*, I, 211; and Woerther, 'Citer/traduire', 183–185.

Opus presentis translationis *Rethorice* Aristotilis et eius *Poetrie* ex Arabico eloquio in Latinum iam dudum intuitu venerabilis patris Johannis Burgensis episcopi et regis Castelle Cancellarii inceperam, sed propter occurrentia impedimenta usque nunc non potui consummare. Suscipiant ergo ipsum Latini, precipui inter ceteras nationes secundum statum presentis temporis zelatores et cultores partis philosophie rationalis, ut estimo, ut sic habeant complementum logici negotii secundum Aristotelis intentionem. Quod autem hi duo libri logicales sint, nemo dubitat qui libros perspexerit Arabum famosorum, Alfarabii videlicet et Avicenne et Avenrosdi et quorundam aliorum [...] Verumptamen dictorum virorum scripta non minimum utilia sunt ad opera presentia intelligendum.

Nec miretur quisquam vel indignetur de difficultate vel quasi ruditate translationis, nam multo difficilius et rudius ex Greco in Arabicum est translata, ita quod Alfarabius, qui primus conatus est ex *Rethorica* aliquem intellectum glosando elicere, multa exempla Greca propter ipsorum obscuritatem pertransiens derelinguit, et propter eandem causam multa dubie exposuit et, ut Avicenna et Avenrosd estimant, propter hanc etiam causam glosam usque ad finem negotii non perduxit. Et isti quoque duo viri in finibus tractatuum suorum, quos imitantes Aristotilem composuerunt, sic inquiunt; hoc est quod intelligere et excipere potuimus de translatione que pervenit ad nos horum voluminum Aristotilis, ideoque usque hodie etiam apud Arabes hi duo libri quasi neglecti sunt, et vix unum invenire potui qui mecum studendo in ipsis vellet diligentius laborare.

Veniam igitur concedant qui forsitan non immerito poterunt hunc meum laborem de imperfectione redarguere; et si eis non placuerit quicquam fructus ex eo querere, possunt ipsum deserere redargutum. Sane tamen ipsis consulo ut malint hos codices habere sic translatos, quam penitus derelictos. Nichil enim pura privatione incultius, sed potest quoquomodo hiis habitis per paulatina incrementa finis tandem desiderate perfectionis facilius impertiri, quemadmodum contingit in libro Nichomachie quem Latini Ethicam Aristotelis appellant. Nam et hunc prout potui in Latinum verti eloquium ex Arabico. Et postmodum reverendus pater magister Robertus Grossicapitis sed subtilis intellectus Linkoniensis episcopus ex primo fonte unde emanaverat, Greco videlicet, ipsum est completius interpretatus et Grecorum commentis proprias annectens notulas commentatus. Sic, si totius scientie largitori placuerit, contingere poterit in his opusculis primordialiter a nobis, etsi debiliter, elaboratis, quod ipse patrare dignetur, qui vivit et regnat eternaliter in perfecta trinitate. Amen.

Text 20. Michael Scot's dedicatory preface to his translation of Averroes's *Large Commentary on Aristotle's De caelo et mundo*. *Averrois Cordubensis commentum magnum*, I, 1 (ed. Carmody and Arnzen).

Tibi, Stephane de Provino, hoc opus, quod ego Michael Scotus dedi Latinitati ex dictis Aristotelis, specialiter commendo, et si aliquid Aristoteles dimisit de constitutione mundana in hoc libro, recipies eius supplementum ex libro Alpetraugii, quem similiter dedi Latinitati, et es in eo exercitatus.

Text 21. From King Manfred's letter to the scholars of Paris. Ed. Gauthier, 'Notes sur les débuts (1225–1240) de premier Averroïsme', 323–324.

Dum librorum ergo volumina quorum multifarie multisque modis distincta cyrographa divitiarum nostrarum armaria locupletant, sedula meditatione revolvimus et accurata contemplatione pensamus, compilationes varie, ab Aristotile aliisque philosophis sub Grecis Arabisque vocabulis antiquitus edite, in sermocinalibus et mathematicis disciplinis, nostris aliquando sensibus occurrerunt, quas adhuc originalium dictionum ordinatione consertas et vetustarum vestium quas eis etas prima contexerat, operimento contectas vel hominis defectus aut operis ad Latine lingue notitiam non perduxit. Volentes igitur ut reverenda tantorum operum senilis auctoritas apud nos non absque multorum commodis vocis organo traduce iuvenescat, ea per viros electos et utriusque lingue prolatione peritos instanter duximus verborum fideliter servata virginitate transferri.

Text 22. Excerpts from Farağ ibn Sālim (Faragut)'s preface to Abū Bakr ibn al-Zakarīyā' al-Rāzī's *Kitāb al-Ḥāwī*. Ed. Fischer and Weisser, 'Das Vorwort zur lateinischen Übersetzung von Rhazes' Liber continens (1282)', 224–226.

[...] attendens Bubikyr istud, non solum ad famosam Ypocratis et Galieni excellentiam, sed magis ad suam et aliorum medicine auctorum insufficientiam quodammodo redundare, tamquam egregius philosophie medicinalis alumpnus, otia non affectans, sed labores assumens voluntarios et etiam incommoda senectutis non causans, librum hunc in senectute sua excogitavit componere et excogitatum complevit [...]

omnium auctoritates [...] diligenter inseruit in hoc libro, cui Arabico ydiomate nomen inposuit Elhauy, id est totum continens vel omnia coadunans, eo quod omnia, que in aliorum libris auctorum sub quadam prolixa verborum obscuritate dispersa sunt, succincte et manifeste in ipso reperiuntur inserta, ut scolares, artis prolixitate, experimenti fallacia, vite insuper brevitate pensatis, totum Ypocratis et Galieni in propria ipsorum auctoritate necnon aliorum omnium peritorum eis succedentium medicorum et etiam ipsius auctoris propositum invenientes in eo, hac compilatione contenti et studium circa ipsam de cetero ampliantes, multifaria medicine volumina volvere et revolvere non cogantur.

Frequens equidem et assidua lectio huius libri quamplures de Arabibus in scientia medicine peritos reddidit et famosos. Latius tamen diu fuit incognitus. Sed post longa annorum curricula christianissimus Dominus Karolus, Iherusalem et Sicilie Rex, eiusdem libri fama conmotus et ardore tam captabilis utilitatis illectus, non sibi solum sed christicolis omnibus prodesse desiderans, bellicis curis liberalium studiorum sollicitudinem ad ipsius libri translationem miscere delegit, ut in partem recognitionis regnorum et titulorum ingentium, que divinitus sibi bellorum fuerant acquisita triumphis, referret aliquid populis in eius honorem qui populorum est conditor et servator. Predicto itaque libro a rege Tunisii per sollempnes nuntios conquisito virum fidelem adhibuit lingue tam Arabice quam Latine peritum, et in libro ipso, in quo sub Arabice lingue tenebris tanta nobis occultabatur utilitas, mandavit et fecit lucernam Latine translationis accendi.

Hanc nichilominus artis medicine professorum familiarium medicorum suorum ac aliorum Neapoli et Salerni regentium examinationi, provisioni et correctioni supposuit, et eis ad id non summarium, set ordinarium tempus indulsit. Qui omnes et singuli de translatione huiusmodi statuto ad hoc die relationem in publico facientes, una voce et una omnium concordi sententia dignissimos laudum preconiis extulerunt interpretem cum auctore. Nam si Arabica verborum proprietas a Latino interprete forsan alicubi videatur corrupta, vera tamen et integra compilationis huius sententia virginitatis gratiam non amisit.

Bibliography

Primary Sources

Abū Maʿšar, *Liber Introductorii Maioris ad Scientiam Judiciorum Astrorum*, ed. Richard Lemay, 9 vols, Naples: Istituto Universitario Orientale, 1996

Adelard of Bath, *Conversations with His Nephew:* On the Same and the Different, Questions on Natural Science, *and* On Birds, ed. and tr. Charles Burnett with the collaboration of Italo Ronca, Pedro Mantas España, and Baudouin van den Abeele, Cambridge: Cambridge University Press, 1998

Alfred of Shareshill (Sareshel), *Commentary on the* Metheora *of Aristotle*, ed. James K. Otte. Leiden: E.J. Brill, 1988

Aristoteles Latinus, ed. George Lacombe, 2 vols, Rome: La Libreria dello Stato, 1939

Averrois Cordubensis commentum magnum super libro De celo et mundo Aristotelis, ed. Francis James Carmody and Rüdiger Arnzen, 2 vols, Leuven: Peeters, 2003

Avicenna Latinus. Liber de anima seu sextus de naturalibus I-II-III, ed. Simone van Riet, Leuven/Leiden: E. Peeters and E.J. Brill, 1972

Boethius, *In Isagogen Porphyrii Commentorum Editio secunda*, ed. Georg Schepss and Samuel Brandt, Vienna/Leipzig: F. Tempsky, 1906

Daniel of Morley, *Philosophia*, ed. Gregor Maurach, in Gregor Maurach, 'Daniel von Morley, „Philosophia"', *Mittellateinisches Jahrbuch*, 14 (1979), 204–255

Hermann of Carinthia, *De essentiis*, ed. and tr. Charles Burnett, Leiden: E.J. Brill, 1982

Isḥāq al-Isrāʾīlī, *Omnia opera Ysaac*, Lyon: Barthélemi Trot, 1515

Ptolemy, L'Optique *de Claude Ptolémée dans la version latine d'après l'arabe de l'émir Eugène de Sicile*, ed. Albert Lejeune, Louvain: Bibliothèque de l'Université, Bureaux du recueil, 1956; reissued Leiden/New York: E.J. Brill, 1989; A. Mark Smith, *Ptolemy's Theory of Visual Perception: an English Translation of the* Optics *with Translation and Commentary*, Transactions of the American Philosophical Society 86, part 2, Philadelphia PA: University of Pennsylvania Press, 1996

Secondary Literature

d'Alverny, Marie-Thérèse, *La Connaissance de l'Islam dans l'Occident medieval*, Aldershot: Ashgate Variorum, 1994

d'Alverny, Marie-Thérèse, 'Marc de Tolède, traducteur d'Ibn Tūmart', *Al-Andalus*, 16 (1951), 259–307; repr. in ead., *La Connaissance de l'Islam*, article II

d'Alverny, Marie-Thérèse, 'Les traductions à deux interprètes: d'arabe en langue vernaculaire et de langue vernaculaire en latin', in *Traductions et traducteurs au Moyen Âge*, ed. Geneviève Contamine, Paris: CNRS, 1989, 193–206

Bloch, Herbert, *Montecassino in the Middle Ages*, 3 vols, Rome: Edizioni di Storia e Letteratura, 1986

Borghese, Gian Luca and Benoît Grévin, 'Aspects linguistiques de la diplomatie sicilienne au XIIIe siècle (1220–1290)', in *Les langues de la négociation: Approches historiennes*, ed. Dejanirah Couto and Stéphane Péquignot, Rennes: Presses Universitaires de Rennes, 2017, 77–96

Borrelli, Arianna, *Aspects of the Astrolabe: 'architectonica ratio' in Tenth- and Eleventh-century Europe*, Stuttgart: Franz Steiner Verlag, 2008

Burnett, Charles, 'Antioch as a Link between Arabic and Latin Culture in the Twelfth and Thirteenth Centuries', in *Occident et Proche-Orient: contacts scientifiques au temps des croisades*, ed. Anne Tihon, Isabelle Draelants, and Baudouin van den Abeele, Turnhout: Brepols, 2000, 1–78; repr. with corrections in id., *Arabic into Latin in the Middle Ages*, article IV

Burnett, Charles, *Arabic into Latin in the Middle Ages: The Translators and their Intellectual and Social Context*, Farnham/Burlington: Ashgate Variorum, 2009

Burnett, Charles, 'The Coherence of the Arabic-Latin Translation Program in Toledo in the Twelfth Century', *Science in Context*, 14 (2001), 249–288; repr. with corrections in id., *Arabic into Latin in the Middle Ages*, article VII

Burnett, Charles, *The Introduction of Arabic Learning into England*, London: The British Library, 1997

Burnett, Charles, 'The Legend of Constantine the African', in *The Medieval Legends of Philosophers and Scholars*, Florence: SISMEL, 2013, 77–94

Burnett, Charles, 'Physics before the *Physics*: Early Translations from Arabic of Texts concerning Nature in MSS British Library, Additional 22719 and Cotton Galba E IV', *Medioevo*, 27 (2002), 53–109; repr. with corrections in id., *Arabic into Latin*, article II

Burnett, Charles, 'The Translator as an Authority', in *Translation and Authority—Authorities in Translation*, ed. Pieter De Leemans and Michèle Goyens, Turnhout: Brepols, 2016, 53–67

Burnett, Charles, 'Translating from Arabic into Latin in the Middle Ages: Theory, Practice, and Criticism', in *Éditer, traduire, interpreter: essais de methodologie philosophique*, ed. Steve G. Lofts and Philipp W. Rosemann, Louvain-la-Neuve: Peeters, 1997, 55–78

Fischer, Klaus-Dietrich and Ursula Weisser, 'Das Vorwort zur lateinischen Übersetzung von Rhazes' Liber continens (1282)', *Medizinhistorisches Journal*, 21 (1986), 211–241

Gauthier, René, 'Notes sur les débuts (1225–1240) de premier Averroïsme', *Revue des sciences philosophiques et théologiques*, 66 (1982), 321–374

Gázquez, José Martínez, *The Attitude of the Medieval Latin Translators towards the Arabic Sciences*, Florence: SISMEL, 2016

Gutas, Dimitri, 'What was there in Arabic for the Latins to Receive? Remarks on the Modalities of the Twelfth-Century Translation Movement in Spain', in *Wissen über Grenzen*, 3–21

Haskins, Charles Homer, *Studies in the History of Mediaeval Science*, 2nd edn, Cambridge MA: Harvard University Press, 1927

Jacquart, Danielle, 'Le sens donné par Constantin l'Africain à son œuvre: les chapitres introductifs en Arabe et en Latine', in *Constantine the African and 'Alī ibn al-'Abbās al-Mağūsī: the* Pantegni *and Related Texts*, ed. Charles Burnett and Danielle Jacquart, Leiden: E.J. Brill, 1995, 71–90

Juste, David, *Les* Alchandreana *primitifs: étude sur les plus anciens traités astrologiques latins d'origine arabe* (*x*[e] *siècle*), Leiden: Brill, 2007

Millás Vallicrosa, José María, *Assaig d'història de les idees físiques i matemàtiques a la Catalunya medieval*, Barcelona: la Institucío Patxot, 1931

Morpurgo, Piero, *L'idea di natura nell'Italia Normannosveva*, Bologna: CLUEB, 1993

Ricklin, Thomas, '*Arabes contigit imitari.* Beobachtungen zum kulturellen Selbstverständnis der iberischen Übersetzer der ersten Hälfte des 12. Jahrhunderts', in *Wissen und Grenzen*, 47–67

Santoyo, Julio-César, 'El Normando Hugo de Cintheaux (*Hugo Sanctelliensis*), traductor en Tarazona (ca. 1145)', in *Antes se agotan la mano y la pluma que su historia; Magis deficit manus et calamus quam eius hystoria*, ed. Constance Carta, Sarah Finci, and Dora Mancheva, I: *Edad media*, San Millán de la Cogolla: Cilengua, 2016, 340–357

Santoyo, Julio-César, *La traducción medieval en la Península Ibérica* (*siglos III–XV*), León: Universidad de León, 2009

Translating at the Court: Bartholomew of Messina and Cultural Life at the Court of Manfred, King of Sicily, ed. Pieter De Leemans, Leuven: Leuven University Press, 2014

Wissen über Grenzen: Arabisches Wissen und lateinisches Mittelalter, ed. Andreas Speer and Lydia Wegener, Miscellanea mediaevalia 33, Berlin/New York: Walter de Gruyter, 2006

Woerther, Frédérique, 'Citer/traduire. La traduction arabo-latine de la *Rhétorique* d'Aristote par Hermann l'Allemand et les citations d'al-Fārābī, Ibn Sīnā et Averroès', *Documenti e studi sulla tradizione filosofica medievale*, 28 (2017), 177–218

Zonta, Mauro, 'The Jewish Mediation in the Transmission of Arabo-Islamic Science and Philosophy to the Latin Middle Ages. Historical Overview and Perspectives of Research', in *Wissen über Grenzen*, 89–105

CHAPTER 9

Latin Translators from Greek in the Twelfth Century on Why and How They Translate

Michael Angold and Charles Burnett

Part One: Essay

This chapter will be limited to the twelfth century for three reasons. Firstly, this was the period in which the transmission of science and philosophy from both Greek and Arabic into Latin went hand in hand, if not in competition with each other—as the thirteenth century progressed Arabic texts became less and less a source for new knowledge, and Greek sources became progressively more sought after. Secondly, it was also a period of close contact between Byzantium and the West, thanks partly to the intense discussions between the Western and Eastern Churches, in which some Greek–Latin translators were involved. Thirdly, it was a period in which Greek, Arabic, and Latin culture still flourished side by side in Sicily and Southern Italy—an area that retained the Classical designation of Magna Graecia—as well as in the Crusader States.

The twelfth century marked the beginning of a substantial transmission of knowledge from both Arabic and Greek into Latin.[1] This was made possible by the opening of new territories to Latin scholars through expansion of the Christian realm in the Iberian peninsula, in Sicily, and in the Middle East. It was encouraged by the rise of secular education, in cathedral schools and, progressively, in universities. It was promoted by patrons in high positions and put into practice by scholars of a high calibre. It is one of these scholars, Burgundio of Pisa (d. 1193), a lawyer, papal adviser, and ambassador, who gave the most detailed account of the translating enterprise, in the prologue to his translation of the commentary of John Chrysostom on the Gospel of St John. Although this account occurs in the preface to a theological work, it is relevant to the entire translating enterprise of the twelfth century, and so is worth quoting

1 For Greek–Latin translations of scientific texts, Haskins in his *Studies in the History of Mediaeval Science* remains valuable, especially the section on 'Translations from the Greek' (141–241). See also Minio-Paluello, *Opuscula*; Berschin, *Griechisch-lateinisches Mittelalter*; and the table of Greek–Latin translations in Pasnau, *The Cambridge History of Medieval Philosophy*, II, 793–801.

 | DOI:10.1163/9789004472648_010

at length here (Text 1). In it, we find described (1) the purpose of the translation: immediately to seek absolution for his own sins and those of his son, but ultimately to fill in a gap in Latin literature; (2) the circumstances of the translation: failing to buy a copy of the text, he had to borrow manuscripts from which transcriptions had to be made quickly, while the translating itself could be carried out over a longer period of time; (3) the method of translating: here a strict *verbum de verbo* translation method is advocated, complemented by notes and explanations, and supported by a long list of precedents, which becomes a virtual history of translations made from Greek into Latin from the *Twelve Tables* of Roman law to a text on urines by Theophilus Protospatharius. Burgundio gives Boethius's method as one of these precedents, and refers to Horace's description of the *fidus interpres* ('faithful interpreter').[2] His advocacy of a literal method of translating reflected what was becoming more and more the norm among translators of the twelfth century—most notably that of Gerard of Cremona (d. 1187), who advanced it to an extreme in Arabic–Latin translations—and was followed scrupulously by Bartholomew of Messina (fl. 1258–1266) and William of Moerbeke (d. c. 1286) when translating from Greek in the later thirteenth century.[3]

Burgundio of Pisa was one of the 'three wise men who knew both languages [Greek and Latin]', the others being James of Venice and Moses of Bergamo; they participated in the dispute between the Eastern and Western Churches that took place in Constantinople in 1136.[4] While Moses of Bergamo's writings included theology, local history, and the grammar of the Greek language, James of Venice and Burgundio were both pioneers in the translation of Aristotle's works, and must have been aware of each other's achievements. Burgundio was responsible for *On Generation and Corruption* and parts of the *Nicomachean Ethics*; James of Venice translated the *Posterior Analytics*, *Sophistici Elenchi*, *Physics*, *De anima*, several sections of the *Parva naturalia* and *Metaphysics*, as well as glosses to some of these works. We do not have any of James's own prefaces, but we do have a preface written by a certain 'Ioannes' to his own translation of the *Posterior Analytics* (Text 2). 'Ioannes' was writing, probably in Southern Italy, shortly after James, but before 1159, when John of Salisbury refers to his

2 Horace, *Ars poetica*, vv. 133–134. See Mundt and Cohen in this volume, and Schwarz, 'The Meaning of *Fidus interpres*'.

3 For Bartholomew, see *Translating at the Court*; for William, see Gudrun Vuillemin-Diem's introductions to her editions of William's translations of Aristotle's *Metaphysica, Lib. I–XIV* and *Meteorologica*. Burgundio gave a shorter account of his translation method, and the circumstances of his translation, in his prologue to the same Chrysostom's *Ninety Homilies on Matthew*; see Classen, *Burgundio von Pisa*, 71–72.

4 Anselm of Havelberg, quoted in Haskins, *Studies*, 197.

translation; 'Ioannes' is highly critical of the two translations that precede his own, for different reasons.[5] John of Salisbury similarly comments on the problems of translating the *Posterior Analytics* (Text 3). He emphasizes the work's importance for scientific method (demonstrative argument), and the lack of a single reliable text in Latin—hence the proliferation of translations.

While James and Burgundio may have been visiting Constantinople in 1136, other Latin translators from Northern Italy had a more permanent position there. Their priority was texts relevant to healing the breach between the Eastern and Western Churches, but at least two of them were also translating scientific texts. In his preface to his translation of Achmet's *Oneirocriticon*, Leo Tuscus cannot avoid referring to the contemporary church controversies. The preface is full of metaphors and similes—starting with that of the Emperor Manuel sitting on a bronze horse—and its main message is that 'Greek philosophers should speak Latin' in order to make their doctrines clear, whether in the debate over the relation of Father and Son, or in dream-interpretation (Text 4).[6] Another member of the group is Pascalis Romanus, who wrote the *Liber thesauri occulti* based on several Greek sources (Text 5). The leitmotif of his preface is that one has a duty to reveal hidden knowledge. In his translation of the magical text *Kirannides*, his concern for clarity leads him to prefer the *sensus de sensu* to the *verbum de verbo* method of translating (Text 6).

Another visitor to Constantinople was Henricus Aristippus, who was probably part of an embassy sent by William I in 1154.[7] The embassy was to the same emperor to whom Leo Tuscus referred, Manuel I Komnenus, whose reign from 1143 to 1180 encompassed all the personalities mentioned here, and who was evidently a learned man. Aristippus received some manuscripts from him. But he spent most of his career in Sicily, where he was archdeacon of Catania from 1156 and a *familiaris* of King William I from 1160, until he fell out of favour and died in 1162. Aristippus contributed to the rediscovery of Aristotle's works on natural science by translating the fourth book of the *Meteorologica*.[8] According to the testimony of Gerard of Cremona's pupils, Gerard already knew that this book had been translated when he came to translate the first three books.[9] It is likely that the reference is to Aristippus's translation, and it is also possible

5 Bloch, 'John of Salisbury', 269.
6 See Mavroudi, *A Byzantine Book on Dream Interpretation*, 111–115.
7 For the date, see Angold, 'The Norman Sicilian Court', 151–152.
8 Aristotle, *Meteorologica*, ed. Rubino.
9 See Burnett, 'The Coherence of the Arabic-Latin Translation Program in Toledo in the Twelfth Century', 279: 'Liber Aristotilis methaurorum tractatus .iii.; quartum autem non transtulit eo quod sane invenit eum translatum.'

that the fourth book was translated separately from the rest because its subject is different from that of the first three, dealing with the mixing of qualities and the elements rather than events in the atmosphere. There is, however, no preface indicating Aristippus's motives for translating the work.

The contrary is the case for Aristippus's translation of two of Plato's dialogues, the *Meno* and the *Phaedo*. Since these were the first works of Plato to be translated into Latin since Calcidius translated the first two thirds of the *Timaeus* at some point in the fourth century, one would expect Aristippus to justify the translation in one or both of his prefaces, and indeed, he does so in his preface to the *Meno* (Text 7). Aristippus emphasizes that Plato is the source of the philosophizing of Aristotle and his followers, and apologizes for the literalness of the translation, which he imputes to his own lack of skill; following Plato's model, he extols the value of friendship. In the preface to the *Phaedo* (Text 8), friendship is again a leitmotif, but Aristippus's emphasis is on education: on teaching and learning, on having a good library at hand. Here, unlike in other prefaces, Aristippus is not drawing attention to the lack of educational resources among the Latins—he rather paints a picture of plenitude. And the translation of the *Phaedo* has been made not to fill a gap in learning, but as a token of friendship—its choice perhaps due to the fact that Aristippus's namesake, the 'happy philosopher', is mentioned in the dialogue (as he is also in the *Meno*).[10]

Belonging to the same context is the translation of Ptolemy's *Almagest* made directly from Greek (Text 9). The translator used one of the manuscripts that had been brought from Constantinople by Henricus Aristippus in 1154. But in this case the anonymous translator and Aristippus had evidently fallen out with each other. Instead, the Anonymus acknowledges the help he has received from the emir Eugenius, whose interests in the mathematical sciences were closer to his own.[11] Nevertheless, a common tone prevails between Aristippus's and the Anonymus's prefaces: friendship and mutual obligations, and a proper education dominate both. The preface to the *Almagest* criticizes both those (identified with churchmen) who deny or condemn the practical mathematical sciences, and those who constantly have their heads in the clouds and do not value earthly matters. It lauds mathematics and points to some texts that contributed to its learning: Euclid's *Data*, *Optics*, and *Catoptrics*, and Proclus's *Elementatio physica*.[12]

10 See Carlini, 'Vigilia greca normanna', 50–55.

11 For his translation of Ptolemy's *Optics* from Arabic, see Burnett, 'The Statements of Medieval Latin Translators on Why and How they Translate Works on Science and Philosophy from Arabic', in this volume.

12 The Anonymus may also have been responsible for a translation of Euclid's *Elements*; Murdoch, 'Euclides Graeco-Latinus', 260–263.

In the case of these Latin translations from Greek we thus have a close-knit network. Some of the translators knew one another. The texts that have prefaces are addressed to friends rather than potentates. The translators worked in the same academic environment, albeit in different cities. Constantinople was the centre where they met, or from which they obtained the manuscripts of the texts that they translated. They shared a predilection for extravagant language, rich in Classical allusions, no doubt because, as scholars of Greek, they were educated in the Classical authors in both Greek and Latin.

Part Two: Texts in Translation

The translation of Texts 1–6 is the responsibility of Charles Burnett, that of Texts 7–9 of Michael Angold. We are grateful to Carlotta Dionisotti and Daniel Hadas for reading, correcting, and improving the texts and their translations.

Text 1. An essay on translation. Burgundio of Pisa's translation of the commentary of John Chrysostom on the Gospel of St John (AD 1174).

Classen, *Burgundio von Pisa*, 84–102.

Burgundio, judge and citizen of Pisa, gives greetings to all his faithful brothers in Christ.

When I was sent by my fellow citizens to the Emperor Manuel in Constantinople, as an ambassador for the public business of my city, I lost a son of mine called Hugolinus, whom I had brought with me on the journey, snatched away by illness. For the redemption of his soul, I decided to turn from Greek into Latin the explanation of the Gospel of St John the Evangelist, wonderfully composed by the blessed John Chrysostomus, the patriarch of Constantinople, both because I had already produced the commentary of the same father John Chrysostom on the Gospel of St Matthew the Evangelist, translated in its entirety for Pope Eugenius the Third of blessed memory, and because there was a because there was a great lack among the Latins of commentary on this John, the Evangelist. For I found no other author expounding it continuously except St Augustine. Since I was not able to do this on the spot, owing to the public duties occupying me, and since I could not find the book (thought to be most eagerly desired by very many people and especially by monks, both because of the sharpness of its ideas and because of the thoughtful instructions of its moral principles) for sale, so that I could take it back with me to Pisa to translate it, having received two exemplars that I could borrow from two monasteries, I handed the book to be transcribed by two copyists, one starting from the beginning, the other from the middle, and having it thus soon in my hands I faithfully emended it, as I listened to it being read, day and night, whenever the opportunity arose.

When I had finished the business of my city, receiving permission to return from the emperor, and coming to Messina, I stopped there for a while, and, writing it in my own hand, I began to translate the book there, and thus for the whole journey via Naples and Gaeta, wherever I stopped, grasping free time for myself, I continuously translated the book, and then, against hope, over a period of two years, thanks to God, I completed the translation of the whole book *de verbo ad verbum* from Greek into Latin.

[description of the contents of the book and Chrysostom's contribution]

I was afraid that if I, Burgundio, were to take the sense of the commentary of this Holy Father and write it [literally: dictate it] in my own style, I would in some way change the deep meaning in what these two most wise men [i.e., John the Evangelist and John Chrysostom] were saying, and would run the risk of falling into some deviation in such an important matter (since we are dealing with the words of the Christian Faith). So I chose the more difficult path and decided to keep in my translation both (a) words with the same meaning, and (b) the same style and order that is among the Greeks. But the ancient interpreters among both the Greeks and the Latins are said to have done the same thing all the time. For the seventy-two Jewish interpreters, chosen by the chief priest as the six men from each tribe who knew the Hebrew and Greek languages best, and sent from Jerusalem to Alexandria, translated the whole Old Testament *de verbo ad verbum* from Hebrew into Greek for Ptolemy, king of Egypt. But Aquila, Theodosion, Symmachus, and many other Christian Greeks, suspected that the seventy-two interpreters, writing for the aforesaid king who was an idol-worshipper, could have substituted some things for others; so then they learnt the Hebrew language in order thus to correct their translation. But after a long and most careful investigation of the translation, they finally found that the seventy-two interpreters had hardly erred in anything. They claimed that they had merely added some superfluous things that the correctors, since they did not dare to alter the text of the seventy-two interpreters in any way, signified should be cancelled by putting an obelus outside the text in the margin of the book. Similarly, of certain Hebrew words with two meanings, for which the seventy-two interpreters adopted one meaning in their translation, the more recent correctors, not presuming to alter anything of the hallowed text of the interpreters, noted the other meaning by using an asterisk in a similar way, which they themselves translated *de verbo ad verbum* from Hebrew into Greek.

Our ancient fathers translated this entire production of the seventy-two interpreters, which is more correct than the other modern ⟨versions⟩, *de verbo ad verbum* from Greek into Latin. But St Jerome made a new translation (which we now use) of the sacred prophet Isaiah (not following the Septuagint), which he comments on, but mentioning there in a critical way the aforesaid old translation of Isaiah translated into Latin from the aforesaid translation of the seventy-two interpreters. But St Basil, expounding the aforesaid Prophet Isaiah, writes a literal commentary in a wonderful way on the version of the seventy-two interpreters, and I, Burgundio the judge, translating this commentary word for word for Eugenius III, of blessed memory, have followed almost in every detail our old translation made from the aforesaid version of the seventy-two interpreters, since I found the new translation of St Jerome was not explained there [in St Basil], nor was I able to follow it in any way in that commentary.

The Psalter also has been translated *de verbo ad verbum* from Greek into Latin, and I reckon that those different versions of it that are current among the Latins—the Roman, and that of the Franks—have resulted from the equivocation of the Greek words, for which the interpreters now assume this meaning, now that. But the whole New Testament—all its books, including the letter of St Paul to the Romans and the Gospel of St Mark—has without doubt been translated word for word from Greek into Latin.

The Roman decemviri translated the laws of Solon, the most famous philosopher and greatest friend of Plato, word for word, and bringing them to Rome made from them the ancient law of the *12 Tables*. Also the Emperor Justinian, a most famous man and most wise in every way, rendering the book of the *Digest* and the *Codex* word for word, translated them from Latin into Greek, so that it might be clear not only to the Latins but also to the Greeks how the eloquence of the Emperor Diocletian, Papinianus the Jurist, and other emperors and jurists differed from one another, both in ideas and in elevation of language. The *Book of Institutes* also, both in Greek and in Latin, neither has one word too many nor one word too few. The aforesaid Justinian also ordered that the *New Constitutions* [the *Novels*], which he composed afterwards assiduously in the Greek language for new cases emerging every day, be translated word for word from Greek into Latin, so that they should be obeyed throughout the whole world.

A certain patriarch from Constantinople translated the dialogue of St Gregory the pope word for word from Latin into Greek and from that book the Greeks call him 'Gregory the Dialogist'. But also Plato in the *Timaeus* was translated word for word from Greek into Latin and furnished with a commentary by Calcidius. But the very wise Priscian too, not at all following Donatus, in the accidents of the eight parts of speech, the species both of proper and common nouns, in Atticisms and in accents, transfers word for word everything from Greek into Latin in whatever cases he can, and takes all things from their source. Also Boethius, the most famous philosopher, translating Porphyry, Aristotle in his *Categories*, *De interpretatione*, *Topics*, and *Sophistici Elenchi*, and Nicomachus in his *Arithmetic* word for word, brought them from the Greek into the Latin language.

But also the wisest of the doctors, Hippocrates in his *Aphorisms* and Galen in his *Art*, were translated in a wonderful way word for word from Greek into Latin. John ⟨Scotus Eriugena⟩, on the command of King Charlemagne, in respect to Dionysius the Areopagite, who was both most profound because of the acumen of his ideas, and most eloquent in the loftiness of his style and the thoughtful fashioning of new words—although he knew that Dionysius presented to the Latins in this way would be rather 'heavy', for the same word structure could not be kept among the Latins in the translation—nevertheless presumed to render him, and the marginal glosses that Maximus the Greek had made over him, in a *de verbo ad verbum* Latin voice. Also the book on urines of Theophilus reached the knowledge of the Latins faithfully rendered word for word.

It is right that the author of a book should shine out no less in the ornament of his style and the felicity of his words than in what he has to say. For a wall both from its stones, and from the excellent way that they have been put together, makes those who look at the artefact amazed by the talent of the architect. But a painter too is elevated with paeans of praise not only when his figuration of an image is beautiful but also when it is ugly, as long as he has followed in every way the figuration or appearance of the prototype. Thus, also the high and sublime style of the letters of Paul, by comparison with the lowly and rough speech of Aristotle and others, is brought forward by everybody for admiration. If (equivalent) words [*dictiones*] can be found and the particular nature [*ydioma*] of each language does not prevent it, and if one does not wish to establish one's own glory and to pretend that what belongs to others is one's own, the translation *de verbo ad verbum* should not be rejected by a diligent and completely faithful translator. For if you wish foreign material to be thought your own and under your control, 'you will not care to render word for word like a *fidus interpres*', as Horace says; rather, taking the sense [*sententia*] of that matter, you will explain it by the combination of your own words, and so you will not be a translator, but you will seem to have composed your own words from yourself, which is what both Cicero and Terence attest that they did. They were imbued with the highest wisdom, but, disdaining to preserve the cases and configurations of the Greeks in their writing, they used not to stick to the words of the Greeks, but by their eloquence kept the beauty and elegance of the contents in their translations. Otherwise, if they had rendered word by word in a style that was not native, they would easily have been caught stealing the contents from someone else. Even St Jerome, who attacks the word for word translation excessively, says that he, 'in the translation of Greek texts, will not render them *verbum de verbo*, but *sensus de sensu*, except in the case of Holy Scriptures, where even the order of words is a mystery'. But Horace also says: '⟨Accept the Latin words⟩ if they descend from a Greek source, but distorted sparingly'.[13]

---

13 Horace, *Ars poetica*, vv. 52–53.

Since, then, this translation of mine is Holy Script, and in this work of mine I seek not glory but the pardon of the Lord for my sins and those of my son,[14] I have decided with good reason that, preserving among the Latins too the glory of his own work for our father, John Chrysostom, the *verbum ex verbo* method of translation should be followed. ⟨The only concession I have made is⟩ to make up for the occasional deficiency of (Latin) words by adding two or three words, and to bend any form of words that results in a barbarism, a metaplasm, schema, or trope in Latin into the correct and proper form of speech […]

[the divisions and structure of the book]

When this book was complete in all its parts, compared faithfully with the Greek exemplar from which it was taken, and studiously emended as well as I could, when many friends asked me to give it to them to be copied, because they were words of faith, I decided first to show it to be examined by the lord pope, so that, relying on his authority, I might after that confidently hand it over to be copied for everyone […]

[Burgundio describes the arrangements he has made to have the translation checked and approved before being released to the general public.]

Text 2. Criticism of earlier translations. Aristotle's *Posterior Analytics*, translated by 'Ioannes'.

Haskins, *Studies*, 229; Aristoteles Latinus, IV/1–4, xliv.

Though encumbered by many occupations, my love for you has compelled me to translate the *Posterior Analytics* of Aristotle from Greek into Latin. I have tackled this all the more keenly as I knew that this book contains in itself many fruits of learning, and it was clear that knowledge of it was not evident among the Latins in our times. For the translation of Boethius is not available complete among us, and that which is found of it is darkened by the fault of corruption. But as for the translation of James ⟨of Venice⟩, the French masters show by their silence that it is covered with the darkness of obscurity; although they have that translation and commentaries translated by the same James, they do not dare to profess knowledge of that book. Therefore, if Latin culture comes to know anything of use from my translation, it ought to attribute that to your demand. For I have approached this not out of the hope for gain or empty glory, but so that, by giving something to Latin culture, I might satisfy your wish. But

14 The reference is to his son Hugolinus, whose death in Constantinople he mentions at the beginning of the prologue.

if in anything I should seem to have departed from the path of reason, I will not blush to emend it by the admonition of you or of other masters.

Text 3. The inaccessibility of Aristotle's *Posterior Analytics*. John of Salisbury, *Metalogicon*, 4, 6.

Johannes Saresberiensis, *Metalogicon*, ed. Hall, 226–227; tr. Hall, 293–294.

[...] but the subtle science of the *Posterior Analytics* is accessible to few talents, which clearly arises from many causes. For it contains the art of demonstrating, which is more difficult than the other means of argument. Then, because of being seldom used, it has now fallen almost into disuse, because the practice of demonstration barely survives ⟨and then⟩ only among mathematicians, and among these almost only amongst geometers. Also, use of this discipline [geometry] is not frequent among us, except perhaps in the region of Iberia and the nearest parts of Africa. For those peoples practise geometry above all ⟨disciplines⟩ for the sake of astronomy—likewise Egypt and some peoples of Arabia. Additionally, this book by which the demonstrative discipline is transmitted is much more disturbed than the others, by the transposition of words, the jumbling up of letters and the unfamiliarity of the examples, which are borrowed from different disciplines. Finally (something which is not a criticism of the author), it has been so corrupted by the fault of the copyists that there are almost as many stumbling blocks as chapters, and one is lucky if there there are no more stumbling blocks than chapters. Hence the blame for its difficulty redounds onto the translator from many people, asserting that the book has come down to us not correctly translated.

Text 4. The difficulties of persuading Greek philosophers to speak Latin. The preface to Leo Tuscus's translation of Ahmed's *Oneirocriticon* (1176).

Haskins, *Studies*, 217–218.[15]

To Hugo Etherianus his teacher and blood brother, Leo Tuscus, translator of imperial letters, about dreams and oracles.

15 For the text, see Mavroudi, *A Byzantine Book on Dream Interpretation*, which includes a summary of the preface on p. 115. The Latin text is discussed in Camozzi Pistoja, 'The *Oneirocriticon of Achmet*'.

Although, best of teachers, I follow the unconquered Emperor Manuel as he puts the Persian to flight in the territories of Bithynia and Lycaonia as they trip over the twining ivy,[16] nevertheless I do not forget the memorable dream seen by you—that you saw the undefeatable aforementioned man nobly sitting on a bronze horse perched on top of a column that the Thracians call the Augustiana situated in Byzantium. In the same place, when some most learned Latins were present, when he was reading in Latin in a certain book, and he seemed to favour you alone when you were addressing him. At that time, what the dream portended was hidden from both of us, but a little book published by you about the lesser position of the Son of Man in respect to God the Father[17] afterwards revealed an oracle of this kind under its wraps. Indeed, that bronze soulless noisy-footed ⟨horse⟩ was an airing of that most profound and resonating issue that was on the lips of the Greeks, namely the claim that the Word of God in as much as it was incarnate remained equal to the Father, that claim was completely without reason or truth like the above-mentioned four-footed animal. But what solved the controversy was the decree of imperial mercy, at the urgent request of the little book, leaving a few embers of scandal against his will. From that time then, I beat my breast in worry, I churned fiery pieces of coal in my heart, thinking that it would be useful if I persuaded the Dream-book of the Greeks when speaking to philosophers, to speak clearly in Latin, and succinctly unlocked the unimpeded clarity of the dreaming fiction, by your favour, to the wishes of our Tuscans, whom I ardently desire to be sprinkled with a more abundant flow so that they might be irrigated sufficiently, and made more fruitful. For the Chinese, as they say, sprinkle their trees with water, when they strive to obtain the richest silk that China creates. But it is not easy to travel on this kind of sea, whose breadth is so wide that it spreads from ear to ear, not spontaneously welcoming a weak oarsman. For not only does the subject of subtle investigations repel even those who creep along the ground because of the weakness of their feet, like the inhabitants of the antipodes, and those who do not move their tongues, like most aquatic creatures, but it does not even allow navigation in a boat made of a single plank.[18] Therefore, I assigned to this translation a ⟨form of⟩ speech matching the translation of imperial let-

16 A quotation from Ovid, *Metamorphoses*, 10.99.

17 Hugo's book *De filii hominis minoritate ad patrem deum* ('About the lesser position of the Son of Man in respect to God the Father').

18 Leo Tuscus could have picked up his unusual vocabulary from Pliny, *Natural History*, book 6, 23.26 (105): 'Regio, ex qua piper monoxylis lintribus convehunt' ('The region from which they convey pepper on boats made of one plank'), but this hardly elucidates his obscure phrase.

ters, sometimes staying up all night, but not without hope of success. I will subject the whole work to the judgment of your wisdom, thus considering of equal weight myself as author and you as judge. For I have no doubt your examination will recognize whatever a dry and bloodless argumentation requires. For my most beloved nephew, Fabricius, being himself also knowledgeable of Greek literature and with a keen nose for the rose-garlands of dream-tales, by writing to me frequently, raises me from the surrounding turmoil, and provides no little help. So, the title will not be without a subject nor the publication without strength, and it will, I think, show itself impenetrable to the arrows of the peoples of Tyre and Sidon. Therefore, whoever is worn out by the intricacies of knotty dreams, if he demands that they are solved by anything written here, he should beware of trying to know more than is sufficient, lest he corrupts the inscriptions of the Apolline city with cracks of deviations and bramble-weeds. But I put forward this book to be examined by you alone, being most skilled in both languages, so that whatever things in it are dry and submerged in darkness, by being lit up by you may be made known to the world, finally undergirded with perfection.

Text 5. The importance of revealing secret treasures. From the preface of Pascalis Romanus, *Liber thesauri occulti* (AD 1165).

Oxford, Bodleian, MS Digby 103, and London, British Library, MS Harley 4025; Haskins, *Studies*, 218. See also Collin-Roset, 'Le "Liber thesauri occulti"', 141–147.

A hidden treasure lies in the heart of the wise man and is therefore desirable, but in a hidden treasure and in hidden wisdom there is almost no usefulness. Thus, hidden things should be revealed and those that are occult should be laid open. Therefore, from the very many unknown and hidden things I have chosen only to describe the covering and an appropriate revelation of one thing: namely of a dream according to its genus and species, so that a profound and almost inscrutable thing should be distinguished ultimately in a clear order. Its teaching is very necessary for philosophers and learned men, lest by chance, after they are sought out, they are found to be silent or making errors […] This book is collected from divine and human writings, based both on experience and on reasoning about the matter, from Latins, Greeks, Chaldeans, Persians, and the annals of Pharaoh and Nebuchadnezzar, in which their dreams are explained in a manifold way […] Therefore, the length of the preface should not delay us further, nor should any reply get in our way, but, putting aside every care, let us swiftly hasten to reveal the desired treasure.

Text 6. A defence of translating the sense not the words. The preface to Pa⟨scalis Romanus⟩'s translation of the *Kirannides* (AD 1169)

Oxford, Bodleian, MS Ashmole 1471, f.143v; Haskins, *Studies*, 219–220.

In the name of Christ, Amen. Here begins the book of Kirannis, the daughter of Ypocration.

Pa., the lowest cleric, gives greetings to his most erudite lord and master Ka.

I admire and commend the sharpness of your wisdom, which, while it is learned and expert in matters that are above the nature of our sphere and has already, as it were, penetrated beyond the seven heavens by contemplation, now also does not disdain to consider the lowest earthly experiences. For you asked me to translate this medical book from Greek eloquence into the Latin language. It was an easy thing to say, but difficult to fulfil. But, moved by your charitable love and your kindness, I did not hesitate to obey. Since the translations from the Arabic language into Greek are various, as you know, I have imitated studiously and faithfully in everything the Greek book that you have given me. I did not wish to miss out those two prologues, however rough, as if taken from the most ancient inscriptions [*tituli*], by bringing together not the words, which are of a barbaric sterility, but their sense, ⟨which is⟩ useful. So, if you find anything alien from common sense, it should not be imputed to unfaithfulness or malice, but to common error. For no one is so wise that he can be found to have no trace of ignorance. I want you to know, however, that among the Greeks there is a book of Alexander the Great concerning the seven plants of the seven planets, and another one called the mystery of Thessalus, addressed to Hermes, i.e., Mercury, about the twelve plants attributed to the twelve signs, and about another seven plants (ordered) among the seven other stars, which, if by chance they should come into my hands or yours, because they imitate the celestial worthiness, should rightly be preferred to this work. Thus, this book is translated in Constantinople under the command of Manuel the Emperor in the year of the world 6,677, AD 1169, second indiction.

Text 7. The importance of translating Plato. Henricus Aristippus's preface to his translation of Plato's *Meno*.

Plato Latinus, *Meno*, ed. Kordeuter, 5–6.

> Poems are like pictures: one will captivate you
> The closer you stand, while another, the further away you are.[19]

It follows that you should attend closely to the writings of Plato, so that you are more and more captivated and delighted by their style. Who of those who are like you is unaware, who of those well-versed in the Scriptures denies, or which investigator of prophetic dogma doubts how lofty, how inscrutable, and how well-founded are the perceptions and opinions of Plato, who illuminated the moral character and the wisdom of the Athenians, who founded the Academy and its philosophy, who tamed the tyranny of the Sicilian court, from whom Aristotle derived his subtlety, and from whom both the Peripatus and the Lyceum acquired the beginning of their philosophizing? Consequently, the closer you approach his works, so much the more will you admire them. Each word raises questions. Should you solve one of them, you will take fright from the Hydra's heads that rise up. This man, unique in ethics, and without par in theological matters, grounds all his discourses in a sweetness of morality, fortifies them with the solidity of divinity, and, though starting from different first principles, brings all his treatises to a single conclusion. In fact, he refers all things to the powers of the gods, as if he believed that all things proceeded from them. Therefore, may you approach, beloved and revered friend, and sense the smell of this Platonic perfume, and be revived by it, and may you be filled with the sweetness of such a fragrance, and may you accept Plato's *Meno* in that spirit of devotion in which it is known to have been requested by you and given by me. In translating this, I have concentrated more on the literal meaning, lest, trusting too heavily on my weak intelligence, I impose the wrong meanings. Consequently, it will perhaps seem to belong to a clumsier and less elegant oration in the Latin language, seeing that drinks acquire a more bitter taste when decanted into a variety of containers.

In answer to your request for a translation of one of the Platonic dialogues I have chosen to translate the *Meno*. To be sure, a discourse on virtue was believed to be both the more fitting and the more acceptable to a student of virtue. What would seem more pleasing than virtue to the one whose perceptions and strength are acts of virtue, whose works bloom with virtue, and whose speeches are redolent of virtue? Once you have read and reread the present work and have confirmed it by the virtue of memory, you cannot by virtue of your knowledge but be aware of how much virtue Plato himself brings

19 Horace, *Ars poetica*, vv. 361–362.

to his discourse on virtue. I would not like it to escape your notice to what extent I have prefered you to the great number of entreaties ⟨I have received⟩. For I was about to translate at the orders of my Lord William, the most glorious king of the Sicilians, the works of Gregory of Nazianzus, who composed as many discourses as the Athenian Plato. I was also preparing to translate into Latin Diogenes Laertius's *On the Lives of the Philosophers*, at the request of Maio, the Great Admiral of Sicily,[20] and of Hugh, archbishop of the see of Palermo.[21] I postponed these ⟨projects⟩, preferring to acquiesce to your proposal. Is this not living proof of how attentively one hearkens to the pleas of one's friends, for whom one will sometimes even overlook the instructions of one's superiors? For, whatever is inspired by affection is the more quickly accomplished.

Text 8. Translation in the context of cultural richness. Henricus Aristippus's preface to his translation of Plato's *Phaedo*.

Plato Latinus, *Phaedo*, ed. Minio-Paluello and Drossaart Lulofs, 89–90. For another translation, see Mallette, *The Kingdom of Sicily*, 154–156.

Henry Aristippus, archdeacon of Catania, to Roboratus Fortuna,[22] greetings!

Teaching and learning are the two linchpins that preserve human knowledge, while refutation and being refuted are the two that stimulate it when it falters. Both (pairs) are equally praiseworthy, delightful, and useful, but the second is all the more desirable seeing that it is more agreeable to be recalled from error than it is to recall. You have always embraced each pair with the double affect of love [i.e., loving and being loved] diligently and gently: the first ⟨of the two pairs⟩ you apply gently to your ⟨friends⟩, the second, diligently to yourself. It is for this reason, my Roboratus, that—speaking rather boldly to your face—I would accuse your friendship of being too hasty. For all haste is unwise, because too much haste leads to more trouble. Why are you in such a rush? Why are you getting ready to return home? To a man of education 'every land is his homeland, just as all the empty spaces of the world lie open for a bird'.[23]

20 William I of Sicily's chief minister 1154–1160.

21 Archbishop of Palermo 1150–1161.

22 'The man strengthened by fortune.' This is a play on the name Robert. Various candidates have been suggested, including Robert of Cricklade and Robert of Selby.

23 Ovid, *Fasti*, I, vv. 493–494.

You have in Sicily the library of Syracuse and the Greek library; there is no lack of Latin philosophy; Theoridus of Brindisi,[24] who is skilled in Greek literature, is ready to assist; your Aristippus is here, whom you can use as a whetstone, if not as a cutting edge. You have at your disposal the philosopher Hero's *Mechanics*,[25] which provides such a subtle treatment of the vacuum: how great is its power and how swift is the movement of something passing through it. You have Euclid's *Optics*, in which he discourses so truthfully and admirably about the nature of seeing that he substantiates conjectures by means of demonstrative reasoning. On the principles of the sciences, you have Aristotle's *Apodictice* [*Posterior Analytics*], in which, using axioms drawn from nature and from the senses, he treats of what is above and beyond nature and the senses. You have in your hands works of philosophy by Anaxagoras, Aristotle, Themistius, Plutarch, and the other philosophers of renown. You will perhaps have come across the gist of these, at least while you effectively applied yourself to the study of medicine. But I can, in any case, offer you theological, mathematical, and meteorological theorems.

But why do I waste time enumerating the marvellous ⟨works⟩ of great names? In England you will be able to provide what you can compare to, and match with, these and everything of this kind. But, when you have provided all those things, will you be able to provide a King William?[26] For he holds in his hands a likeness of the round world.[27] His court is a school; his retinue a gymnasium; his every word a philosophical gem; his questions impossible to unravel; his solutions leave no possibility unaccounted for; his study has left nothing untried. To his dominion bow down Sicily, Calabria, Lucania, Campania, Apulia, Libya, and Africa, while Dalmatia, Thessaly, Greece, Rhodes, Crete, Cyprus, Cyrene, and Egypt have felt his victorious right hand. The noble deeds that he has thus far achieved are rendered the more illustrious and much more radiant in their splendour by ⟨the memory of⟩ his father, the Great Roger.[28] You have been able to obtain his goodwill all the quicker for being on good terms with his household officers and for your virtue and scholarship, being considered worthier of appreciation thanks to our good offices.

24 Given Aristippus's penchant for variants of names, this may be the mysterious Theorianos, who was later in the service of the Byzantine Emperor Manuel I Komnenos, but seems to have had a South Italian background; see *Oxford Dictionary of Byzantium*, III, s.v. Theorianos.

25 Hero's *Pneumatics* are most likely intended.

26 William I of Sicily (1154–1166).

27 I.e., his realm is a microcosm of the whole world. The Latin text is not clear however.

28 Roger II of Sicily (1101–1154).

If it should so happen that you are not swayed by your Aristippus's words of warning and you do not desist from the journey on which you are embarked, go, farewell, and take care that you remain true to yourself. Receive as a comfort for such a difficult journey such provisions I can give you: Plato's *Phaedo* on the immortality of the soul, which has been translated from Greek into Latin syllables. I started it on campaign, when the aforesaid king was besieging the Samnite city of Benevento, and I finished it at Palermo. Whenever you take it into your hands—which I know you will often do—think of me, whom you will leave in tears when you depart, bereft of the presence of so good a friend.

You will find in the present dialogue the most subtle arguments about the meditation on death, on which it will delight you greatly to meditate—arguments which even escaped the philosopher Cleombrotus of Ambracia[29]—, and the mutual generation of contraries, the rarity of extremes and the prevalence of means, an enquiry into causes, the immortality of souls, the position of the earth, the true earth, the pleasure of paradise, the infernal places of punishment, the death of Socrates, and other aspects of the highest philosophy, which are as worthy of admiration as they are of study.

Text 9. The search for Ptolemy's *Almagest* to complete a curriculum in mathematics. The preface to the anonymous translation of the *Almagest* made from Greek into Latin in c. 1160.

Haskins, *Studies*, 191–193. For another, partial, translation, see Mallette, *The Kingdom of Sicily*, 156–158. For a French translation and commentary, see Lemay, 'De la scolastique à l'histoire', 433–439.

The Ancients are said to have had the custom of depicting two of the Graces gazing upon the third, who, with face averted, stretches out her hand towards them. Pondering upon the hidden meaning of this ⟨composition⟩ for a long time, I have held in my memory that a single gracious act should return a twofold profit. Therefore, ⟨because I am⟩ mindful of your generous gift, through which you have imparted to a thirsty mind, as if it were a draught of living water, knowledge of what Aristotle calls the most exact of the arts, I have sought within myself in a state of some anxiety how I could repay your kindness. For 'a pure and overflowing fountain of intelligence'[30] and a heart that was home to

29 Cleombrotus was notoriously inspired by Plato's *Phaedo* to commit suicide; see Williams, 'Cleombrotus of Ambracia'. Aristippus seems to be suggesting that he had got hold of the wrong end of the stick.

30 Juvenal, *Satires*, 10.119.

the arts really did not need those things that I had acquired—as a late-comer to philosophy—by a painstaking exercise of a slight talent. I also understood perfectly well that there was little chance of help from anybody who held these things in contempt. I therefore became more and more vexed in my mind, finding it increasingly difficult to keep my vow the more any opportunity of fulfilling it was denied me. But, at last, God in his mercy heard the never-ending clamour of my desire, providing in my opinion a fitting means of repayment, which surpasses your gift—to presume upon your good nature—in so far as the end is more praiseworthy than the means. For what is apparent to any educated man cannot escape your sharp mind: that the science of numbers and measurements constructs a bridge, which serves as a virtual introduction to the science of the stars. On the part of this branch of knowledge, which explores the movement of the stars, Claudius Ptolemy wrote in thirteen books, being a man deeply versed in the study of the stars, who was an ornament[31] of the Ancients and is a mirror for the Moderns. These are called the *Mathematics* or the *Megiste syntaxis* by the Greeks, by the Saracens, however, by the corrupt name of *Almagest*.

I was studying medicine at Salerno, when I heard that one of the ambassadors sent by the king of Sicily to Constantinople, Aristippus by name, had received these books thanks to the generosity of the emperor and had conveyed them to Palermo. Fired by the hope of ⟨obtaining⟩ something so long and ardently desired, I did not shudder at the thought of howling Scylla, I passed through Charybdis, I negotiated Etna flowing with lava, as I sought out the man, who, I hoped, would furnish me with the object of my desires. At last I found him near the Pergusan spring, where he was examining the wonderful phenomena of Etna, at no small danger to himself. Leaving aside other undisclosed considerations, the obvious reason preventing him from translating for me the aforementioned work was a lack of knowledge of astronomy. Having already made a diligent study of Greek, I first got to grips with Euclid's *Data*, *Optics*, and *Catoptrics*, and Proclus's *Elementatio physica*. I then tackled Ptolemy's aforementioned work, God's grace providing me with a generous guide in the shape of Eugenius, a man most expert in the Greek and Arabic tongues, and by no means ignorant of the Latin. I managed to translate it into Latin despite the opposition of the ill-tempered man.[32] I was often tempted to give up the work that I had begun, because my mind was overburdened by the terrible labour

31 Literally: 'file' (*lima*). The thought behind this is that a file was used to put the finishing touch to a sculpture.

32 Evidently, Aristippus himself.

involved, but spurring me on were, on the one hand, the most powerful feelings of the love I have for you, which serve to overcome any difficulty, and, on the other, my constant awareness of the need to repay your gift. For I would not have been able to carry out this task, had I only been motivated by hope of gain or glory, since it is agreed that where art becomes the object of ridicule and shame an artist is left without hope. For it is not possible for a man who does not admire the art to admire the artist.

You also have heard about some presumptuous people in these times, who, when judging a case about which they are ignorant, declare, lest they appear uneducated, that whatever they do not know is of no practical or moral value. As the Arabs put it, 'There is no greater enemy of art than the man who has no experience of it.' They persevere in decrying the arts the more pertinaciously, the more certainly they see praise ⟨of the arts⟩ threatening them with the opprobrium of ignorance. I do not need to mention those who, although spurred on by a love of honour, persecute honourable studies. I would prefer to call these pious sinners, for, while they strive to eradicate the putrefaction of unhealthy pursuits, they do not keep their undiscriminating hands from philosophy, the nursemaid of healthy intellectual pursuits, but even bring serious charges against that part of it which, sought out and cultivated by the most brilliant minds, the more refined and pure it is, the worthier it is of the name of wisdom and the more pleasingly it restricts human intelligence by shackling it to the most agreeable speculation. Accordingly, their error—albeit disguised as honourable—but dependent on their malicious testimony, inevitably leads the uneducated, who form the majority, to a neglect of the liberal arts, so that the science of numbers and measurements is considered superfluous and without use, while the study of astronomy is deemed idolatry.

This is evidently the way that the pious and prudent man [Boethius] felt, when he said: 'This is therefore the fourfold path [*quadrivium*] that must be travelled by those who wish to be led from our carnal senses to the greater certainties of rational understanding.'[33] Remigius also bears witness to this when he says: 'When all the arts were despised by the Egyptians, Abraham taught them astrology.'[34] So no doubt the Lord reproved the saints Moses and Daniel because ⟨of what they said⟩ of the science of the stars: namely that it is stupid to gaze upon the works of the Creator and through observing admire more delightfully His ineffable power and wisdom; also that it will clearly be a crime, even a sin, to elevate the mind to an understanding of the Creator by trying to

33 Boethius, *De institutione arithmetica*, I, 1.

34 Remigius of Auxerre, *Commentum in Martianum Capellam*, ed. Lutz, 153.2.

understand Creation and to make the Creator, who is beyond all understanding, in some way comprehensible through the observation of the sensible world.[35] O minds that are so blind that they are entirely ignorant of the philosopher's path! For the invisible things of God are understood by the creatures of the world [Romans 1, 20]. Nor can the human mind properly perceive the truth of things that cannot be sensed, unless it has first forged for itself a method and competence through the prelude of sensible things. It is for this reason that this programme of studies has been established and carefully followed by educated men, so that not only could they first brush aside the clouds of inborn ignorance by painstaking investigation of created things, by studying all disciplines, but especially those which, when contemplated without any deviance of error or stumbling block of doubt, illuminate again the mind's eye, as Boethius witnesses. Then, imbued with this strength, they have the confidence to apply the cutting edge of their intelligence to the discipline of theology, ⟨which is why⟩ from its order in the curriculum they designated theology 'metaphysics'.

But, ⟨men of⟩ our ⟨time⟩—obviously eagles[36]—hardly need this kind of gigantic effort, being most able, without any help from created things, to fix their eyes on the rays of the highest light and penetrate efficiently the secrets of the Almighty and the highest truth. Hardly out of their babbling infancy, they soar to the heavens, and inhabiting this world is beneath them. They pass their lives above the clouds and they vaunt their repose in the lap of the highest wisdom. They make a mockery of mundane wisdom and deride it as the ramblings of the empty-headed. But you, who are a most clear-thinking man, have a quite different attitude of mind. You perfectly correctly consider the arts to be sacred and worth cultivating for their own sake, sweet to the knowledgeable, and to be adored by the uneducated. Your acute intelligence knows well that the perfection of happiness consists in a fulness of knowledge: that access to happiness is gained by making progress through knowledge, given that the mind is above all corrupted by laziness, while study is a friend to virtue. I believe that you consider it wonderful that a man should outshine another in that ⟨faculty⟩ in which he is superior to the other animate beings. Hence you judge that one should rise and apply oneself to the greatest of one's abilities so that the mind, illuminated by every kind of knowledge, prepares itself for happiness, and is adorned with the good that is proper to it.

It is because of you and those like you that I have undertaken this present work, and if I please you and them, my endeavours have found their consum-

35 The author is being ironic.

36 More irony.

mation. Enemies of the arts may mock and sneer; they may judge what they do not know and may declare the study of the stars to be a form of madness. I have to confess that it is for me a sweet madness. It is a sweet thing for me to proclaim with Ovid:

> O happy souls, who first sought to understand these things
> and climb to their heavenly home.[37]

So I pray that you will regard my work with favour and accept it kindly, since it was with the greatest care fashioned for you. Because I applaud that discrimination that you employed in the instruction of geometry, there is, however, one thing above all that I urge and request: that you allow only those who are worthy to consult this work, and that you deny access to those who are unworthy. Surely communicating things of worth to the unworthy deprives them of their worth!

37 Ovid, *Fasti*, I, vv. 297–298.

Part Three: Original Texts[38]

Text 1. Burgundio of Pisa's translation of the commentary of John Chrysostom on the Gospel of St John (AD 1174). Classen, *Burgundio von Pisa*, 84–102; Haskins, *Studies*, 151, n. 36 (partial).

Omnibus in Christo fidelibus fratribus Burgundio iudex civis Pisanus in Domino salutem.

Cum Constantinopoli pro negotiis publicis patrie mee a concivibus meis ad imperatorem Manuelem missus legati munere fungerer et quendam filium meum Hugolinum nomine quem mecum duxeram in itinere morbo arreptum amiserim, pro redemptione anime eius explanationem evangelii sancti Iohannis evangeliste a beato Iohanne Crisostomo Constantinopoleos patriarcha mirabiliter editam de Greco in Latinum vertere statui sermonem, tum quia eiusdem sancti patris Iohannis Crisostomi commentationem supra evangelium sancti Matthei evangeliste iam pridem beate memorie tertio Eugenio pape integre translatam tradideram, tum quia huius Iohannis evangeliste expositionis penuria apud Latinos maxima erat. Nullum enim alium nisi sanctum Augustinum eum continue exponentem inveni. Cumque ibidem id negotiis communibus imminentibus facere minime possem et eum librum, a quampluribus et maxime a monachis tam sententiarum acumine quam etiam morum perpensa instructione avidissime haberi desideratum, venalem, ut mecum Pisas transferendum referrem, nullatenus invenirem, duobus exemplariis a duobus monasteriis in commodatum acceptis, duobus scriptoribus, uno a capite altero a medietate incipiente, librum tradidi transcribendum, et eum brevi ita adeptus nocte ac die cum vacabat diligenter auscultans fideliter emendavi. Negotiis vero mee civitatis peractis, licentiam redeundi ab imperatore accipiens, Messanam veniens, ibique moram faciens, manibus meis scribens inibi librum transferre incepi, et sic per totam viam Neapoli et Gaete et ubicumque moram faciebam vacationem michi extorquens iugiter transferebam, et deinde contra spem per duos continuos annos, Deo auctore, totum librum de verbo ad verbum de Greco in Latinum transferens integre consummavi [...]

Verens igitur ego Burgundio ne, si sententiam huius sancti patris commentationis assumens meo eam more dictarem, in aliquo alterutrorum horum duorum sapientissimorum virorum sententiis profundam mentem mutarem

38 In many cases I have silently altered the orthography and punctuation of the editions I have used, to make them easier to read.

et in tam magna re, cum sint verba fidei, periculum lapsus alicuius alteritatis incurrerem, difficilius iter arripiens et verba significatione eadem et stilum et ordinem eundem qui apud Grecos est in hac mea translatione servare disposui. Sed et veteres tam Grecorum quam et Latinorum interpretes hec eadem continue egisse perhibentur. Nam septuaginta duo interpretes Iudei, ut sapientiores tam Hebrea quam Greca lingua sex per unamquamque tribum a principe sacerdotum electis et a Ierusalem Alexandriam directi, de verbo ad verbum de Hebreo in Grecum Ptolomeo regi Egypti totum vetus testamentum transtulerunt. Sed et Aquila et Theodosion et Symmachus et ceteri quamplures Christiani Greci, qui linguam Hebraicam suspitione, ne predicti septuaginta duo interpretes predicto regi ut ydolatre transcribentes alia pro aliis transtulerint, ut ita eorum translationem corrigerent, postea didicerunt, post longam et diligentissimam translationis eorum investigationem vix eos tandem in aliquo peccasse invenerunt, superflua quedam eos apposuisse autumantes, que cum predictorum septuaginta duorum interpretum textum in aliquo mutare non auderent, sub obelo de foris in margine libri posito cancellanda esse significaverunt. Similiter et dictionum quarundam Hebraicarum equivocarum, quarum unam significationem predicti septuaginta duo interpretes in sua translatione apposuerant, hii moderni correctores nichil de solempni predictorum interpretum serie alterare presumentes, alterum sensum sub asterisco similiter notaverunt, quem[39] et ipsi de verbo ad verbum de Hebreo in Grecum transtulerunt.

Hanc igitur septuaginta duorum interpretum editionem ceterorum modernorum correctiorem antiqui patres nostri de verbo ad verbum ex Greco in Latinum totam transtulerunt. Verumptamen beatus Ieronimus novam quandam qua nunc nos utimur sancti Ysaie prophete, non septuaginta duo sequens, translationem fecit quam ipse commentatur, predicte vero antique Ysaie translationis ex septuaginta duorum interpretum editione in Latinum translate inibi ceu detrahens, memoriam faciens. Sanctus vero Basilius, predictum Ysaiam prophetam exponens, septuaginta duorum interpretum editionem mirabiliter ad litteram commentatur, eiusque hanc commentationem ego Burgundio iudex domino tertio Eugenio beate memorie pape de verbo ad verbum transferens, ex predicta septuaginta duorum interpretum editione factam antiquam nostram translationem in omnibus fere sum prosecutus, cum sancti Ieronimi novam suam editionem nullatenus ibi expositam invenirem nec eam sequi ullo modo in ea[40] commentacione possem.

39 *qui* Classen.

40 *in ea* Haskins, *mea* Classen.

Psalterium quoque de verbo ad verbum ex Greco in Latinum translatum est sermonem, et diverse ille que eius proferuntur apud Latinos edi⟨c⟩tiones Romana et Francorum, ex equivocatione Grecarum dictionum ortas esse perpendo, interpretibus modo hanc, modo illam in eis assumentibus significationem. Novum vero testamentum totum non solum in aliis libris omnibus sed et in epistola sancti Pauli ad Romanos et in evangelio sancti Marci procul dubio de verbo ad verbum ex Graeco in Latinum transversum est sermonem.

Decem quoque viri Romani leges Solonis, clarissimi philosophi et Platonis amicissimi, de verbo ad verbum Athenis transtulerunt et eas Romam adducentes inde legem antiquam xii tabularum confecerunt. Sed et Iustinianus imperator, clarissimus vir per omnia prudentissimus, librum Digestorum et Codicis verbum verbo reddens de Latino in Grecum transtulit sermonem, ut eloquencia Dioclitiani imperatoris et Papiniani iurisconsulti et ceterorum imperatorum et iurisconsultorum qualiter in legibus condendis ab invicem discrepant, non solum ex sentenciarum sed ex stili altitudine non modo Latinis sed etiam Grecis liquido pateret. Liber quoque institutionum tam in Greco quam in Latino ne una quidem superhabundat vel deficit dictione. Novellas etiam autentice constitutiones quas novis cotidie emergentibus casibus postea assidue Greca lingua predictus Iustinianus composuit, de verbo ad verbum de Greco in Latinum translatas toti orbi terrarum obediendas direxit.

Quidam etiam patriarcha Constantinopolitanus dyalogum sancti Gregorii pape e verbo ad verbum de Latino in Grecum transtulit et ex eo libro illum 'Gregorium dialogum' Greci vocant. Sed et Plato in Thymeo de verbo ad verbum de Greco in Latinum a Calcidio translatus est et commentatione explanatus. Sed et Priscianus, vir sapientissimus, Donatum minime sequens, et in accidentibus octo parcium orationum et speciebus tam propriorum quam appellativorum nominum et in attacismis et in accentibus de verbo ad verbum omnia in quibuscumque potest ex Greco in Latinum transfert et omnia ex illorum fonte trahit. Sed et Boetius, clarissimus philosophus, Porphirium et Aristotilem in cathegoriis et periermeniis et topicis et elenchis, et Nichomacum in arismeticis transferens de verbo ad verbum, eos ex Greca Latine reddidit lingue. Sed et sapientissimi medicorum, Ypocras in afforismis et Galienus in libro techni, mirabiliter de verbo ad verbum de Greco in Latinum translati sunt. Sed et Iohannes rogatu Karoli regis magnum Dyonisium Ariopagitam tam sententiarum acumine profundissimum quam stili altitudine et novarum dictionum perpensa compositionis compagine facundissimum, etsi eum ita Latinis redditum noverat effici onerosiorem—non enim eadem apud Latinos poterat compositionis complexio in translationem servari—ipsum tamen et quas supra eum beatus Maximus Grecus marginales fecerat clausulas, non aliter quam de verbo ad verbum Latina presumpsit edere voce. Liber quoque de urinis Theophili verbo ad verbum fideliter reddito in Latinorum pervenit cognitionem.

Congruum enim est auctorem libri non minus quam ex sententiis ex ornatu etiam stili verborumque pariter pollere lepore. Nam et paries sicut ex lapidibus ita et ex optima compaginis constructione architectoris ingenii stuporem artem examinantibus infert. Sed et pictor non ex pulchra solum sed ex turpi ymaginis figuratione a prudentibus laudum preconiis elevatur, dummodo protypi, id est exemplaris, figurationem vel effigiem omnifariam fuerit prosecutus. Sic quoque altus et sublimis epistolarum Pauli stilus Aristotilis et aliorum humilium et hyspidorum dictaminum comparatione ab omnibus in admirationem deducitur. Non igitur de verbo ad verbum translatio, si et dictiones inveniantur et ydioma alterutrius lingue minime impediat, et non quis suam velit statuere gloriam et ea que aliorum sua esse simulare, a diligenti et fideli omnino est interprete respuenda. Si enim alienam materiam tuam tuique iuris vis esse putari, 'non verbo verbum', ut ait Horatius, 'curabis reddere ut fidus interpres', immo eius materiei sententiam sumens tui eam dictaminis compagine explicabis, et ita non interpres eris sed ex te tua propria composuisse videberis, quod et Tullius et Terentius se fecisse testantur. Summa enim sapientia imbuti, casibus et figuris servire Grecorum in sua oratione dedignantes, non verbis Grecorum adherebant, sed sua eloquentia sententie decorum atque elegantiam in suis translationibus conservabant. Alioquin, si verbum verbo reddidissent, ex stilo non nostrate eam alii surripuisse facile deprehenderentur. Sed et ipse sanctus Ieronimus, qui nimium in hanc verbi ad verbum translationem invehitur, ait se 'interpretatione Grecorum absque scripturis sanctis, ubi et verborum ordo mysterium est, non verbum e verbo sed sensum exprimere de sensu'. Sed et ipse Horatius ait: 'Si de Greco fonte cadant, parce detorta'.

Cum igitur hec mea translatio scriptura sancta sit et in hoc meo labore non gloriam sed peccatorum meorum et filii mei veniam Domini expetam, merito huic sancto patri nostro Iohanni Crisostomo sui operis gloriam et apud Latinos conservans, verbum ex verbo statui transferendum, deficientiam quidem dictionum intervenientem duabus vel etiam tribus dictionibus adiectis replens, ydioma vero quod barbarismo vel metaplasmo vel scemate vel tropo fit, recta et propria sermocinatione retorquens [...]

Hunc igitur librum omnibus partibus completum et cum Greco exemplari a quo sumptus est fideliter collatum et melius quam potui studiose emendatum, cum plures amicorum me peterent dari transcribendum, quia verba fidei erant domino pape prius statui ostendere examinandum, ut eiusdem auctoritate subnixum, omnibus confidenter possem extunc eum tradere exemplandum.

Text 2. Aristotle's *Posterior Analytics*, translated by 'Ioannes'; ed. Haskins, *Studies*, 229, and ed. Minio-Paluello, xliv.

Vallatum multis occupationibus me dilectio vestra compulit ut Posteriores Analeticos Aristotelis de Greco in Latinum transferrem. Quod eo affectuosius agressus sum quod cognoscebam librum illum multos in se scientie fructus continere et certum erat notitiam eius nostris temporibus Latinis non patere. Nam translatio Boetii apud nos integra non invenitur, et id ipsum quod de ea reperitur vitio corruptionis obfuscatur. Translationem vero Iacobi obscuritatis tenebris involvi silentio suo peribent Francie magistri, qui quamvis illam translationem et commentarios ab eodem Iacobo translatos habeant, tamen notitiam illius libri non audent profiteri. Eapropter, siquid utilitatis ex mea translatione sibi noverit Latinitas provenire, postulationi vestre debebit imputare. Non enim spe lucri aut inanis glorie ad transferendum accessi, sed ut aliquid conferens Latinitati vestre morem gererem voluntati. Ceterum, si in aliquo visus fuero rationis tramitem excessisse, vestra vel aliorum doctorum ammonitione non erubescam emendare.

Text 3. John of Salisbury, *Metalogicon*, 4, 6; ed. Hall, 226–227.

[...] posteriorum vero analeticorum subtilis quidem scientia est et paucis ingeniis pervia. Quod quidem ex causis pluribus evenire perspicuum est. Continet enim artem demonstrandi, que pre ceteris rationibus disserendi ardua est. Deinde hec utentium raritate iam fere in desuetudinem abiit eo quod demonstrationis usus vix apud solos mathematicos est et in his fere apud geometras dumtaxat. Sed huius quoque discipline non est celebris usus apud nos, nisi forte in tractu Hibero vel confinio Africe. Etenim gentes iste astronomie causa geometriam exercent pre ceteris—similiter Aegyptus et nonnulle gentes Arabie. Ad haec liber quo demonstrativa traditur disciplina ceteris longe turbatior est et transpositione sermonum, traiectione litterarum, desuetudine exemplorum, que a diversis disciplinis mutuata sunt, et postremo, quod non contingit auctorem, adeo scriptorum depravatus est vitio ut fere quot capita tot obstacula habeat. Et bene quidem, ubi non sunt obstacula capitibus plura. Unde a plerisque in interpretem difficultatis culpa refunditur, asserentibus librum ad nos non recte translatum.

Text 4. The preface to Leo Tuscus's translation of Ahmed's *Oneirocriticon* (ad 1176); ed. Haskins, *Studies*, 217–218.

Ad Hugonem Eterianum doctorem suum et utraque origine fratrem Leo Tuscus imperatoriarum epistolarum interpres de sompniis et oraculis.

Quamquam, optime preceptor, invictum imperatorem Manuel per fines sequar Bithinie Licaonieque fugantem Persas flexipedum hederarum complectentes vestigia, tamen memorandi non sum oblitus sompnii a te visi, quod dictum inexpugnabilem virum eneo in equo supra columpnam quam Traces dicunt Augustiana Bizantii sito nobiliter sedere conspicabaris. Eodem autem in loco, doctissimis quibusdam astantibus Latinis, Romana oratione cum in quodam legeret libello, interpellanti tibi soli favorem prestitisse visus est. Latuit tunc utrumque nostrum ea quidem quid portenderet visio, at vero eiusmodi oraculum editus per te de Filii hominis minoritate ad Patrem Deum libellus tempore post revelavit sub tegumentis. Profecto eneus ille sonipes anima carens altissime sonantissimeque questionis erat que inter Grecos versabatur ventilatio, verbum scilicet Dei secundum quod incarnatum Patri equale perstans, rationis veritatisque radicitus expers, ut quadrupes nominatus. Solvit autem illam controversiam clamitante illo libello augustalis clementie decretum pauco scandali fomento contra voluntatem illius relicto. Ex eo igitur tempore pectus sollicitudine percussi, sub corde ignitos versavi carbones, cogitando utile esse si onirocriti Grecorum philosophis ariolanti loqui Latine persuaderem enucleatim atque inoffensam perspicuitatem figmenti sompnialis, tuo favore, nostrorum Tuscorum desiderio breviter reserarem, quos quidem fluctu percupio aspergi undiosiore, ut irrigentur affatim, efficianturque fecundiores. Nam Seres, ut fertur, arbores suas undis aspergunt quando uberiorem lanuginem quam Sericum creat admittere nituntur. Ceterum haut facile est in huiusmodi versari pelago cuius latitudo ad aures usque dehiscit, non sponte remigem asciscens invalidum. Non solum enim subtilibus expositum investigationibus et illos repellit qui debilitate pedum serpunt, ut antipodes, et eos qui non movent linguas, ut pleraque aquatilium, set neque monoxilo se navigari lintre patitur. Quamobrem loquelam imperatoriorum interpretationibus apicum obsequentem per excubias interdum huic translationi non irrita spe addixi, totum opus sapientie tue dicaturus iudicio, mei quidem auctoris, tui vero probatoris equilibre pensans meritum. Nam tuum examen cognoscere non sum ambiguus quicquid arida exsanguisque poscit ratiocinatio. Desiderantissimus enim nepos Fabricius, Grecarum sciolus et ipse litterarum sompnialium figmentorum odoratus rosaria, scribendi assiduitate me a confluentibus elevat prestatque non mediocre adiumentum, atque idcirco neque nomen sine subiecto neque sine viribus erit editio, Sidoneis Tirrenisque sagittis parum penetrabilis apparitura, ut arbitror. Ergo quisquis nodosorum sompniorum fatigatur involucris, si per aliquod hic scriptorum absolvi postulet, caveat pretemptare plus nosse quam sat est, ne titulos depravet Apollinee Urbis ambagum rimis herbidisque sentibus. Ego autem tui solius utrarumque linguarum peritissimo examine volumen hoc subpono, ut in eo que arescunt ac caligant per te illustrata orbi demum succincta perfectione vulgentur.

Text 5. From the preface of Pascalis Romanus, *Liber thesauri occulti* (AD 1165); ed. Haskins, *Studies*, 18, and ed. Collins-Roset, 'Le "Liber thesauri occulti"', 141–142.

Thesaurus occultus requiescit in corde sapientis et ideo desiderabilis, et in thesauro occulto et in sapientia abscondita nulla pene utilitas. Ergo revelanda sunt abscondita et patefacienda que sunt occulta. Quare de plurimis ignotis et occultis unius tantummodo elegi tegumentum aptamque revelationem describere, videlicet sompnii secundum genus et species eius, quo res profunda et fere inscrutabilis ad summum patenti ordine distinguatur. Eius namque doctrina philosophis et doctis viris valde necessaria est, ne forte cum exquisiti fuerint muti vel fallaces inveniantur [...] Collectus autem est liber iste ex divina et humana scriptura, tam ex usu experimenti quam ex ratione rei de Latinis et Grecis et Caldaeicis et Persis et Pharaonis et Nabugodonosor annalibus, in quibus multipharie sompnia eorum sunt exposita [...] Non itaque longitudo prohemii nos amplius protrahat nec responsio aliqua impediat, sed omni cura seposita succincte ad thesaurum desiderabilem aperiendum properemus.

Text 6. The preface to Pa⟨scalis Romanus⟩'s translation of the *Kirannides* (AD 1169); ed. Haskins, *Studies*, 219–220, and ed. Delatte, 11–12.

In Christi nomine amen incipit liber Kirannis, Ypocrationis filie.

Eruditissimo domino magistro Ka. Pa. infimus clericus.

Admiror et commendo sagacitatem tue prudentie, que cum docta et experta sit in his que super naturam nostri circuli sunt et que iam quasi ultra .vii. celos contemplando penetravit, modo etiam infima experimenta terrena conspicere non dedignatur. Rogasti enim me ut hunc librum medicinalem de Greco eloquio in Latinum sermonem transferrem. Res quidem facilis fuit ad dicendum, sed difficilis ad perficiendum. Verum, caritativo amore tuoque beneficio permotus obedire non renui. Et quoniam diverse sunt translationes de Agarenica lingua in Grecam, ut nosti, librum Grecum quem mihi dedisti studiose et fideliter per omnia emulatus sum, ipsos etiam duos prologos, quamvis asperos, velud de antiquissimis titulis abstractos preterire nolui, non verba, que de sterilitate barbarica sunt, sed sensum utilitatis recolligendo. Si quid ergo reperieris alienatum ab communi sensu, non infidelitati vel malitie sed communi errori deputatur. Nullus enim tam sapiens qui absque titulo inscientie reperiatur. Volo tamen te scire quod est apud Grecos quidam liber Alexandri Magni de .vii. herbis .vii. planetarum, et alter qui dicitur Thessali mysterium ad Hermem, id est Mercurium, de .xii. herbis .xii. signis attributis et de .vii. aliis herbis per .vii. alias stellas, qui si forte pervenerint ad manus meas vel tuas, quia celestem dignitatem imitantur, recte huic operi preponentur. Transfertur itaque liber iste Constantinopoli Manuele imperante anno mundi .VI^{m}.DCLXXVII., anno vero Christi .MCLXIX., indictione secunda.

Text 7. Henricus Aristippus's preface to his translation of Plato's *Meno*; ed. Kordeuter, 5–6.

'Ut pictura poesis: erit que si propius stes
Te capiat magis, et quedam si longius abstes.'

Platonicis ergo propius assistas scriptis, ut te magis ac magis ritu suo capiant et oblectent. Quis enim tui similium ignorat, quis scripturarum frequenti lectione fretus hesitat, quis prophetici dogmatis ambigit perscrutator, quam alta, quam perinscrutabilia, quanta subnixa firmitudine Platonis sensa et sententie consistant, qui Atticorum illustravit mores et sapientiam, qui Academicam instituit philosophiam, qui Sicule mansuefecit aule tyrannidem, a quo Aristotelica subtilitas est derivata, a quo tam Peripatus quam Licium exordium sumpsit philosophandi? Hec ergo quanto magis appropinquaveris, tanto magis admiraberis. Singula quippe verba suas ostendent questiones. Quodsi unam solveris, Ydre capita surgentia formidabis. Unicus namque hic in ethicis, singularis in theologicis, universos sermones morum condit dulcedine, divinitatis vallat et soliditate, diversis principiis inchoatos uno fine cunctos concludit tractatus. Omnia nimirum refert ad superum numina, quasi ab eis cuncta processisse crediderit. Accedas itaque, dilecte ac venerande, et Platonice nardi gustes odorem reficereque ac tante replearis flagrantie suavitate atque Menonem Platonis ea recipias devotione qua et a te postulatus et a me datus esse cognoscitur. In quo transferendo magis dictionum idiomati deservivi, ne de propria confisus parvitate ingenii alienos intruderem sensus. Unde et agrestioris forsan et minus compte videbitur orationis in Latiari, transfusis siquidem in varia vasa poculis gustus acerbior.

Aliquem tibi sane transferri Platonicorum dialogorum petenti Menonem potius interpretatus sum. Virtutis nimirum alumno sermo de virtute et magis competere et magis gratus esse credebatur. Cuius intuitus, cuius actus virtutis vigor, cuius opera virtute virent, cuius sermones virtutem redolent, huic quid virtute gratius videretur? Quanta igitur virtute hic de virtute disputet Plato, tue virtus scientie non ignorabit, ubi lectum et relectum presens opusculum virtute firmaveris memorie. In quo te quantis pretulerim precibus, te latere nolim. Iussu namque domini mei, glorisissimi Siculorum regis Guilelmi, Gregorii Nazanzeni opuscula translaturus eram, qui eodem numero quo et Atheniensis Plato dictavit sermones. Rogatus item a Maione magno Sicilie admirabili atque ab Hugone Epanormitane sedis archipontifice Laertium Diogenem De vita et conversatione dogmateque philosophorum in Ytalicas transvertere sillabas parabam. Quibus ad tempus postpositis tuo potius acquievi consilio. In quo manifestius ostenditur quam diligenter amicorum votis obeditur, propter quos nonnunquam dominorum iussum preteritur. Promptius enim agitur quidquid ex animo venit.

Text 8. Henricus Aristippus's preface to his translation of Plato's *Phaedo*; ed. Minio-Paluello and Drossaart Lulofs, 89–90.

Enricus Aristippus Cathinensis Archidiaconus Roborato fortune salutem dicit.

Humanam scientiam duo articuli iugem servant, docere et doceri, labentem duo erigunt, redarguere et redargui; quorum utraque tam laudabilia quam iocunda et utilia, sequentia vero illorum eo appetibiliora quo commodius ex errore revocari quam revocare. Bina hec bino semper caritatis affectu diligenter et dulciter amplexatus es: precedencia erga tuos dulciter, subsequentia ergo te diligenter. Hinc est, mi Roborate, ut fronte tua—confidentius loquor—tuam deprehendam nimie festinationis amicitiam. Omnis quippe festinatio improvida; quod si nimia fuerit, in maius vergit vitium. Quo enim properas? Quo repatriare paras? Sapienti 'omne solum patria est, ut volucri vacuo quicquid in orbe patet'.

Habes in Sicilia Siracusanam et Argolicam bibliothecam; Latina non deest philosophia; Theoridus assistit Brundusinus, Graiarum peritissimus litterarum; Aristippus tuus presens est, cuius si non acie verum cote fungi poteris. Habes Eronis philosophi Mechanica pre manibus, qui tam subtiliter de inani disputant, quanta eius virtus quantaque per ipsum delationis celeritas! Habes Euclidis Optica, qui tam vere et mirabiliter de opinione videndi disserit, ut opinabilia raciocinatione probet demonstrativa. Habes de scientiarum principiis Aristotilis Apodicticen, in qua supra naturam et sensum de anxiomatis a natura et sensu sumptis disceptat. Philosophica Anaxagore, Aristotilis, Themistii, Plutarchi, ceterorumque magni nominis philosophorum in manibus tuis sunt. At fortassis horum summam nactus es, dum saltem medicine studio efficacem adhibuisti operam. Atqui theologica, mathematica, metheorologica tibi propono theoreumata.

Ceterum quid in enumerandis magnorum nominum cunctor miraculis? Hiis et huiusmodi universis in Anglia quod conferas et recompenses dare potes. Verum cum omnia dederis, regemne dabis Gulielmum? Orbis enim teretis manibus instar habet, cuius curia scola, comitatus cuius gignasium, cuius singula verba philosophica apofthegmata, cuius questiones inextricabiles, cuius solutiones nichil indiscussum, cuius studium nil relinquit intemptatum, cuius dominatui applaudunt Sicania, Calabria, Lucania, Campania, Apulia, Libia, Affrica, cuius victricem dextram sentiunt Dalmatia, Tessalia, Grecia ⟨R⟩odus, Creta, Ciprus, Cirene, et Egiptus, cuius itatenus preclara facinora magnus ille Rogerius genitor illustriora et multo fulgore radiantiora reddidit. Huius familiaritatem tanto citius adipisci vales quanto domesticis suis existis familiarior, quantoque dignius pre complicibus nostris tua meretur virtus et scientia.

Quod si talibus Aristippi tui minus moveris monitis et a cepto non desistis itinere, vade, vale, cave ne teipsum fallas; et, qualia largiri queo efodia, tanti solatium itineris recipe Fedonem Platonis de immortalitate anime ab Argivis in Ytalicas translatum sillabas, quem in castris, dum rex pretaxatus Samnitum urbem Beneventum obsedisset, ceptum Eponarmi consummavi. Quotiensque in manus eum arripueris—scio quippe quod id frequentius ages—mei recordare, quem tanti presentia orbatum amici in lacrimis cum abscesseris relinques.

Invenies quippe in presenti dialogo, in quibus te meditari plurimum oblectabit, subtilissima de mortis meditatione—que etiam Cleobrotum Ambraciotem philosophum fefellerunt—argumenta, de contrariorum mutua generatione, de extremorum raritate et mediorum numerositate, de causarum inquisitione, de animarum immortalitate, de situ terre, de vera terra, de paradisi amenitate, de locis inferni penalibus, de morte Socratis, de reliquis altissime philosophie articulis, tam admiratione quam studio dignis.

Text 9. The preface to the anonymous translation of the *Almagest* made from Greek into Latin in c. 1160; ed. Haskins, *Studies*, 191–193.

Eam pingendi Gratias Antiqui feruntur habuisse consuetudinem, ut unam quidem vultum aversam, due quibus illa manum porrigeret aspectarent. Cuius mysterii non ignarus dudum memoriter teneo gratiam simplam profectam duplam reverti oportere. Tui ergo boni muneris memor, quo earum quas Aristotiles acrivestatas vocat artium doctrina quasi haustu aque vive animum sitientem liberaliter imbuisti, olim quidem anxie queritabam quid tue dignum benivolentie referre possem. Nec enim eis que philosophie tardus assecla longo pauperis exercitio vene conquisieram 'purus ingenii torrens fons' et domus artium pectus penitus indigebat. Opes quoque apud earum contemptorem minimum promereri non dubius intelligebam. Angebatur ergo in dies magis ac magis animus meus, eo molestius sustinens votum quo complendi voti absolutius facultas negabatur. Verum diutini clamorem desiderii superna tandem pietas exaudivit, dignum, ut arbitror, plene tribuens remunerationis instrumentum, quod tuum tanto, ut tua pace loquar, precedit munus, quanto finis eo quod ad finem laudabilius. Nec enim tuum latet acumen quod omni sapienti liquet, numerorum mensurarumque scientiam ad eam que astrorum quasi quandam introductionis prestruere pontem. Huius vero eam partem que siderum motus speculatur, veterum lima speculum modernorum Claudius Ptolomeus astrorum scientie peritissimus .xiii. perscripsit libris. Qui a Grecis quidem mathematica seu megisti sintaxis, a Sarracenis vero almegesti corrupto nomine appellantur.

Hos autem cum Salerni medicine insudassem audiens quendam ex nuntiis regis Sicilie quos ipse Constantinopolim miserat, agnomine Aristipum, largitione susceptos imperatoria Panormum transvexisse, rei diu multumque desiderate spe succensu, Scilleos latratus non exhorrui, Caribdim permeavi, ignea Ethne fluenta circuivi, eum queritans a quo mei finem sperabam desiderii. Quem tandem inventum prope Pergusam fontem Ethnea miracula satis cum periculo perscrutantem, cum occulte quidem alia, manifeste vero mens scientie siderum expers prefatum mihi opus transferre prohiberent, Grecis ego litteris diligentissime preinstructus, primo quidem in Euclidis Dedomenis, Opticis, et Catoptricis, Phisicaque Procli Elementatione prelusi. Dehinc vero prefatum Ptolomei aggressus opus, expositorem propitium divina mihi gratia providente Eugenium, virum tam Grece quam Arabice lingue peritissimum, Latine quoque non ignarum, illud contra viri discoli voluntatem Latine dedi orationi. In quo nimirum mea mens infando pressa labore inceptum sepe destituisset opus, nisi superande difficultatis auctor potentissimus amor tui tuumque munus animum crebra mutui repetitione pulsarent. Neque enim questus spe motus aut gloria istum potui laborem substinere, cum liquido constet spei locum artifici non relinqui, ubi ars ludibrio et dedecori est. Neque enim artificem mirari potest qui artem non miratur.

Sensisti vero et tu nonnullos his in temporibus causa quam ignorant iudices audacissimos, qui, ne minus scientes videantur, quecunque nesciunt inutilia predicant aut profana. Iuxta quod Arabes dicunt: 'Nullus maior artis inimicus quam qui eius expers est.' Eoque pertinacius criminandis artibus instant quo ab earum laude imperitie probrum certius sibi conspiciunt imminere. Eos omitto qui honestatis zelo honesta quoque studia persecuntur. Quos pie peccare recte dixerim, dum nocivam curarum putredinem recidere contendentes, a sanarum altrice curarum philosophia manum minime continent indiscretam, sed et eam ipsius partem graviori criminatione persecuntur que ingeniis exquisita clarissimis et exculta quo defecatior ac purior est, eo sapientie vocabulo dignior, eo gratiori quadam compede speculationis iocundissime animos hominum continet alligatos. Horum siquidem error sive coloratus honesto malitioso quoque predictorum testimonio fretus, apud imperitos quorum maxima est multitudo in bonarum neglectum artium efficacissime peroravit, ut iam numerorum quidem mensurarumque scientia omnino superflua et inutilis, astrorum vero studium ydolatria estimetur.

Ita nimirum sentiebat vir religiosus ac prudens cum dicebat: 'Hoc est igitur illud quadruvium quo his viandum est qui a sensibus procreatis nobiscum ad certiora intelligentie perduci volunt'. Eisdem quoque attestatur Remigius dicens: 'Cum omnes artes pessumdate essent apud Egyptios, Abraham eos astrologiam docuisse.' Sed et sanctum Moysen sanctumque Danielem Dominus credo ob astrorum scientiam reprovabit: stultum quippe Creatoris opera contemplari, eorumque speculatione ineffabilem ipsius potentiam ac sapientiam delectabilius admirari; nefarium quoque penitusque liquet illicitum ad Conditoris cognitionem conditorumque cognitione animum sublevare, Creatorem insensibilem sensibilium speculatione sibi quodammodo sensibilem comparare. O mentes cecas viamque philosophandi penitus ignorantes! A creatura siquidem mundi invisibilia Dei intellecta conspiciuntur, nec satis insensibilium veritatem percipere potest mens humana ni ad eam preludio sensibilium sibi viam facultatemque preexcuderit. Hinc a sapientibus institutus est ac diligenter observatus hic studiorum ordo, ut primo quidem ingenite ruditatis nebulas diligenti creatorum disquisitione serenarent, omnibus quidem sed eis potissimum invigilando disciplinis que ipsam sine omni erroris devio, sine omni dubitationis scrupulo veritatem contemplantes, oculum mentis, Boetio teste, rursus illuminant; dehinc vero robore hoc animati in theologica exercitate mentis aciem fiducialiter intendebant. Unde et ab ordine docendi et discendi theologiam metaphisicam nominabant.

Verum nostri, nimirum aquile, hoc quasi quodam molimine giganteo minime indigent sine omni creaturarum adminiculo radiis summe lucis oculos infigere potentissimi atque summe secreta veritatis efficaciter penetrare. Vix rudimentis a puerilibus celum involant terrasque habitare dedignantur. Super nubes eorum conversatio, atque in ipso summe sinu sapientie sese requiescere gloriantur. Mundanam desipiunt sapientiam, eique vacantium deliramenta subsannant. Tibi vero, vir mentis serenissime, longe alia mens est. Tu sacras artes et propter se appetendas, scientibus dulces, insciis adorandas rectissime arbitraris. Nec vero tuum fallit acumen, quoniam perfectio beatitudinis in plenitudine consistit cognitionis, quo sciendo proficimus, hoc accessum ad beatitudinem fieri, presertim cum otio quidem mens corrumpatur, studium vero virtuti sit amicum. Preclarum quoque tibi credo videtur, in quo prestat homo ceteris animalibus hominem homini prenitere. Hinc insurgendum summisque viribus iudicas incumbendum ut omni scientie genere mens illustretur, ad beatitudinem preparetur, suo proprio bono decoretur.

Tui ergo tuique similium gratia presentem hunc laborem ego suscepi, quibus si placeo intentio quoque mihi mea perfecta est. Rideant et insultent artium inimici, ignota iudicent, astrorum studium insaniam predicent. Michi confiteor hec insania dulcis, michi dulce clamare cum Nasone:

'Felices anime quibus hec agnoscere primum
Inque domos superas scandere cura fuit!'

Faveas ergo summisque tibi vigiliis opus elaboratum benignus queso suscipias. Illud tamen unum super omnia moneo ac rogo, ut ea qua et in geometricis usus es edocendis discretione collaudanda ad huius operis lectionem dignos admittas, indignos abigas. Suam quippe rebus dignis adimet dignitatem, siquis eas communicaverit indignis.

Bibliography

Primary Sources

Aristotle, *Analytica Posteriora: Translationes Iacobi, Anonymi sive 'Ioannis', Gerardi et Recensio Guillelmi de Moerbeka*, ed. Lorenzo Minio-Paluello and Bernard G. Dod, Aristoteles Latinus IV/1–4, Bruges/Paris: Desclée de Brouwer, 1968

Aristotle, *Metaphysica, Lib. I–XIV: Recensio et Translatio Guillelmi de Moerbeka*, ed. Gudrun Vuillemin-Diem, Leiden: Brill, 1995

Aristotle, *Meteorologica: Translatio Guillelmi de Morbeka*, ed. Gudrun Vuillemin-Diem, Turnhout: Brepols, 2008

Aristotle, *Meteorologica*, Liber Quartus, tr. Henricus Aristippus, ed. Elisa Rubino, Aristoteles Latinus X/1, Turnhout: Brepols, 2010

Boethius, *De institutione arithmetica*, ed. Godfried Friedlein, Leipzig: Teubner, 1867

Hugo Falcandus, *La Historia o Liber de Regno Sicilie e la Epistola ad Petrum Panormitane Ecclesie thesaurarium di Ugo Falcando*, ed. Giovanni Battista Siragusa, Fonti per la storia d'Italia 22, Rome: Istituto Istorico Italiano, 1897–1904; English translation: *The History of the Tyrants of Sicily by "Hugo Falcandus" 1154–1169*, ed. Graham A. Loud and Thomas Wiedemann, Manchester: Manchester University Press, 1998

John of Salisbury, *Johannes Saresberiensis: Metalogicon*, ed. John Barrie Hall, Turnhout: Brepols, 1991; English translation by John Barrie Hall, Turnhout: Brepols, 2013

Plato, *Meno: interprete Henrico Aristippo*, ed. Victor Kordeuter, London: Warburg Institute, 1940

Plato, *Phaedo: interprete Henrico Aristippo*, ed. Lorenzo Minio-Paluello and Hendrik Joan Drossaart Lulofs, London: Warburg Institute, 1950

Remigius Autissiodorensis Commentum in Martianum Capellam Libri I–II, ed. Cora E. Lutz, Leiden: Brill, 1962–1965

Secondary Literature

Angold, Michael, 'The Norman Sicilian Court as a Centre for the Translation of Classical Texts', *Mediterranean Historical Review*, 35 (2020), 147–167

Berschin, Walter, *Griechisch-lateinisches Mittelalter: Von Hieronymus zu Nikolaus von Kues*, Bern/Munich: A. Francke Verlag, 1980; English translation: *Greek Letters and the Latin Middle Ages: From Jerome to Nicholas of Cusa*, tr. Jerold C. Frakes, Washington DC: The Catholic University of America Press, 1988

Bloch, David, 'John of Salisbury, "John" the Translator and the *Posterior Analytics*', *Classica et Mediaevalia*, 61 (2010), 267–291

Burnett, Charles, *Arabic into Latin in the Middle Ages: The Translators and their Intellectual and Social Context*, Farnham/Burlington: Ashgate Variorum, 2009

Burnett, Charles, 'The Coherence of the Arabic-Latin Translation Program in Toledo in the Twelfth Century', *Science in Context*, 14 (2001): 249–288; reprinted with corrections in *Arabic into Latin in the Middle Ages*, article VII

Carlini, Antonio, 'Vigilia greca normanna: il Platone di Enrico Aristippo', in *Petrarcha e il mondo greco*, ed. Michele Feo, Vincenzo Fera, Paola Magna, and Antonio Rollo, *Quaderni Petrarcheschi*, 12–13 (for 2002–2003), 2007, 50–55

Classen, Peter, *Burgundio von Pisa*, Sitzungsberichte der Heidelberger Akademie der Wissenschaften, philosophisch-historische Klasse, Jahrgang 1974, 4. Abhandlung, Heidelberg: Carl Winter Universitätsverlag, 1974

Camozzi Pistoja, Ambrogio, 'The *Oneirocriticon of Achmet* in the West: A Contribution towards an Edition of Leo Tuscus' Translation', *Studi Medievali*, 55 (2014): 719–758

Collin-Roset, Simone, 'Le *Liber thesauri occulti* de Pascalis Romanus (un traité d'interprétation des songes du XIIe siècle)', *Archives d'histoire doctrinale et littéraire du moyen âge*, 30 (1963): 111–198

Haskins, Charles Homer, *Studies in the History of Mediaeval Science*, 2nd edn, Cambridge MA: Harvard University Press, 1927

Lemay, Richard, 'De la scolastique à l'histoire par le truchement de la philologie: itinéraire d'un mediéviste entre Europe et Islam', in *La diffusione delle scienze islamiche nel medio evo europeo*, Rome: Accademia dei Lincei, 1987, 399–535

Mallette, Karla, *The Kingdom of Sicily, 1100–1250: A Literary History*, Philadelphia PA: University of Pennsylvania Press, 2005

Mavroudi, Maria, *A Byzantine Book on Dream Interpretation: The* Oneirocriticon of Achmet *and its Arabic Sources*, Leiden: Brill, 2002

Minio-Paluello, Lorenzo, *Opuscula: The Latin Aristotle*, Amsterdam: Hakkert, 1972

Murdoch, John Emery, 'Euclides Graeco-Latinus', *Harvard Studies in Classical Philology*, 71 (1966): 249–302

Oxford Dictionary of Byzantium, ed. Alexander P. Kazhdan, 3 vols, New York: Oxford University Press, 1991

Pasnau, Robert (ed.), *The Cambridge History of Medieval Philosophy*, 2 vols, Cambridge: Cambridge University Press, 2011

Schwarz, Werner, 'The Meaning of *Fidus interpres* in Mediaeval Translation', *Journal of Theological Studies*, 45 (1944): 73–78

Translating at the Court: Bartholomew of Messina and Cultural Life at the Court of Manfred, King of Sicily, ed. Pieter De Leemans, Leuven: Leuven University Press, 2014

Williams, Gareth D., 'Cleombrotus of Ambracia: Interpretations of a Suicide from Callimachus to Agathias', *Classical Quarterly*, 45 (1995): 154–169

CHAPTER 10

Why Did Latin Translators Translate from the Greek in the Thirteenth Century and Later?

Pieter Beullens

Part One: Essay

After the translation movement of the twelfth century had brought Aristotle's natural and ethical works, based on Greek and Arabic source texts, to the Latin world, the philosopher firmly established his supreme authority in these fields. Yet the translations transmitted were soon considered unsatisfactory, both for their lack of accuracy and for the incomplete view of the philosopher's output that they offered. Questions were raised about their doctrinal acceptability, and it was thought that more and better translations, made directly from the Greek originals, could resolve the problem.

Three translators, Robert Grosseteste (d. 1253), Bartholomew of Messina (fl. 1258–1266), and William of Moerbeke (d. 1286), each contributed to the correction of existing translations and to increasing the number of works by Aristotle known to the Latin world.[1] Robert Grosseteste's most influential achievement in the field consisted in the compilation of a corpus based on a revision of the existing Latin version of the *Nicomachean Ethics* and the translation of a heterogenous body of ancient and Byzantine commentaries on the same text. Bartholomew of Messina is known to have translated several Aristotelian treatises, including the *Rhetoric*, *On the Parts of Animals*, and some other texts that are now considered pseudo-Aristotelian, as well as two works by Sextus Empiricus. As for William of Moerbeke, in the 1270s the Franciscan philosopher Roger Bacon listed the translators of Aristotelian works whom he knew and reported in particular that William 'the Fleming' 'promised to change all the established translations and hammer out other new ones'. Bacon's overall appreciation of William's endeavour is outspokenly negative, but his evaluation is likely guided by personal preferences, since he considered only the classical

1 For Grosseteste's translations from the Greek, see Mercken, *The Greek Commentaries*, 33–66, and Dionisotti, 'On the Greek Studies'; for Bartholomew of Messina, see De Leemans, *Translating at the Court*, and Beullens, *A Methodological Approach*; and for William of Moerbeke, see Brams and Vanhamel, *Guillaume de Moerbeke*, and Beullens, *De sleutel tot Aristoteles*.

 | DOI:10.1163/9789004472648_011

authority Boethius and his own former teacher Robert Grosseteste adequate translators of Aristotle. Still, Bacon is well informed about William's project to update the Latin versions of Aristotle's works, either by revising the existing translations where they were available, or by making new ones if he found no existing suitable versions (Text 1). An annotation confirms William's ambition to translate the complete works of Aristotle into Latin. It lists titles of Aristotelian treatises, some of which are marked as unknown to the Latin world, and gives guidance about the appropriate order in which to read them. The note is ascribed to 'Willelmus interpres' ('William the translator') in one of the two manuscripts that preserve it (Text 2).[2]

Although the names of these translators are known from Bacon and from other sources, virtually all Latin translations of Aristotelian texts from this period are transmitted in the manuscripts without explicit attributions to specific translators, just as in the previous century. Possibly the towering authority of Aristotle rendered the translators invisible. The only apparent exceptions to the general rule of anonymity are formed by a few pseudo-Aristotelian treatises translated by Bartholomew of Messina at the court of Manfred, king of Sicily 1258–1266, many of which are found in a manuscript preserved in Padua.[3] The colophons mention concisely the names of the translator and of his patron, who is praised as a lover of science (Text 3). In marked contrast, Nicholas of Reggio, who worked in the first half of the fourteenth century, sometimes offers more information concerning the context in which his translations of medical texts originated.[4] In the colophon of his Latin version of Galen's *De compositione medicamentorum secundum locos* of 1335, he explains that the Byzantine Emperor Andronicus[5] had learned that Nicholas's patron Robert, king of Naples and Sicily and titular king of Jerusalem,[6] was eminently learnèd in theology, natural philosophy, and particularly in medicine. He had heard of Robert's interest in acquiring all of Galen's works that were not yet available to the Latin world, and therefore sent him a Greek manuscript as a special gift (Text 4). Obviously, a patron's love for and expertise in science and learning is a recurring theme.

While William of Moerbeke remains invisible behind his Aristotle translations, the translator is more informative when it comes to the circumstances in which his work on some late antique commentaries on treatises by the same

2 Brams, 'Mensch und Natur', 549–551.
3 Padova, Biblioteca Antoniana, MS XVII 370.
4 For Nicholas's works, see Thorndike, 'Translations of Works of Galen', and Fortuna, 'Il corpus'.
5 Andronicus III, emperor from 1328 to 1341.
6 Robert of Anjou (1275–1343).

philosopher originated. In the colophon of his translation of Simplicius's commentary on the *Categories*, completed in 1266, William complains about the bad quality of the Greek model that he had at his disposal. He mentions that about a column of text was missing, and that other passages were inserted in the wrong places. He concludes: 'Yet I did what I could. For it is better to have it so corrupt than to have nothing at all.' (Text 5). The colophon at the end of John Philoponus's commentary on Aristotle's *On the Soul* contains similar laments about the water damage to the Greek manuscript that William had to use. Moreover, he brings up his personal concerns about the accuracy of the commentator's remarks regarding the active intellect, but with a rhetorical flourish leaves the ultimate judgement to the reader's wisdom (Text 6).

The translations of Aristotelian treatises by William of Moerbeke and his contemporaries are characterized by a strict adherence to the word-for-word method. However, the same two colophons that accompany commentaries by later philosophers strikingly contain the term *sensus* ('meaning'). In the first instance, William notes that the corruptness of his Greek model sometimes prevented him from understanding the meaning of the text, in the other he admits that he added missing content in line with the meaning where he was not able to read the Greek manuscript.

William of Moerbeke and his colleagues clearly did not translate in a purely mechanical manner. William's remarks about Philoponus's treatment of the active intellect demonstrate that he was acquainted with the subject matter and could form his own judgement on the topic. Similarly, the words with which he concludes a partial translation of Simplicius's commentary on Aristotle's *On the Heavens*, known as the *Fragmentum Toletanum*, show that he had compared its content with the treatment of the same subject in Ptolemy's works, and that he had thereby acquired a decent background in astronomy (Text 7). Moreover, the wording implies that he was looking for information about a particular problem, maybe because Thomas Aquinas needed that material for the preparation of his commentary on Aristotle's *Metaphysics*. Later, in 1271, William of Moerbeke decided to translate Simplicius's commentary in its entirety, which indicates that he worked independently from Aquinas's direct needs.

Translators and their readers recorded few explicit metatheoretical observations about their activity. In his commentary on Aristotle's *Nicomachean Ethics*, Robert Grosseteste explains why he chose to translate the term *sophrosyne*, one of the four principal virtues, as he did: he translates as *temperantia*, which is well established in the Latin tradition, although the translations of the Pauline epistles and the *Book of Wisdom* offer *sobrietas* for the same Greek word (Text 8). Grosseteste is obviously well aware of the Latin theological and philosophical terminology. As for his linguistic skills in Greek and Hebrew, it is reported that he received assistance from Sicilians Nicholas and Robert to better understand those languages (see also Texts 1 and 11).

Other prologues and colophons written by translators of this period bring up the same narrative themes identified by Réka Forrai in the prefaces to translations in other genres and periods. She distinguishes the *topoi* of (1) poverty, (2) utility, and (3) conflict (Forrai calls the latter the 'bellic' narrative).

(1) The tradition of stressing the poverty of the Latin language, which lacks the means to render the nuances of Greek phrasing, stretches back to authors from antiquity like Lucretius and Cicero.[7] Latin medieval sources focus on the abundance of Greek literature that is unavailable to them rather than on the linguistic aspects (Text 10). Roger Bacon discovered in Aristotle's biography that the philosopher 'wrote a thousand volumes', while his Latin contemporaries only had access to the logical, natural, and metaphysical works. 'Latins have nothing worthwhile', he concludes.[8] Accordingly, when the Greek emperor shares the riches of his culture by sending a manuscript of a treatise unknown to the Latin world, it is considered a special gift (Text 4). In other cases, translators complain that they have to ply their trade using Greek books with deficiencies, with a bad or incomplete text (Text 5), or damaged and difficult to decipher (Text 6).

(2) In spite of the difficulties that they had to face, the translators stuck to their task, because they considered the work useful for their Latin readers. Even if the resulting translation is imperfect, 'it is better to have it so corrupt than to have nothing at all' (Text 5). William of Moerbeke translated a commentary on an Aristotelian text so that its readers could better understand the philosopher's original treatise (Text 6). And if the translator decided to restrict his endeavour to certain passages, he did so because the reader would be able to find the same information elsewhere: to continue the translation effort would be useless (Text 7).

(3) The conflict theme is found less frequently in translations of philosophical texts, nor does it appear explicitly, in the form of military metaphors. Still, the idea subsists that other linguistic communities, like Greeks, Arabs, and Hebrews, hide 'truthful' books from the Latin world and even distort their contents, as Roger Bacon states (Text 1, see also Text 11). One recurring theme forms a logical variation on the previous idea: that translations from the Arabic are deficient and do not offer an adequate and

7 Fögen, *Patrii sermonis egestas*.

8 Maloney, *Roger Bacon: Compendium*, VIII.170, 176–177.

complete view of the Greek original. Such utterances easily combine with the two other standard narratives: the absence of a trustworthy translation bears witness to the poverty of what is available to the Latin world; and, since the conflict theme is often made explicit in the prologue of a new translation, the availability of a better alternative proves that the translation effort will be useful (Text 9). Prologues for translations of philosophical and scientific treatises employ the conflict narrative sparingly. In theology, on the contrary, conflict is more openly stated. In his *Opus tripartitum* (1272), the Dominican Humbert of Romans compiles a list of useful instruments to prepare for the confrontation with the Greeks planned at the Second Council of Lyons (1274). He sees books, obviously in Latin translations, as effective means to find the truth, which is evidently on the side of the Latins. The message is phrased in military metaphors of defence and attack and explicitly compared with the struggle for the truth against the Jews (see also Text 11). Unfortunately, Humbert adds, the translators' focus seems to lie on texts of philosophy and law rather than on works that pertain to the salvation of the soul (Text 10). With this last remark about the insufficiency of what is available to the Latins, the passage from Humbert's treatise that only indirectly addresses the theory of translation complements and confirms the triad of narratives (poverty, utility, conflict) usually found in the accounts given by translators themselves in their prologues or colophons.

Part Two: Texts in Translation

All translations, except for texts 1 and 11, are by the present author.

Text 1. Roger Bacon claims that the existing Latin translations are inadequate.

Roger Bacon: Compendium, 173 and 175; tr. Maloney.

VIII.168 A far greater error occurs in the case of translated philosophy because, if the holy ones erred in their translations, this is much more the case with others who have concerned themselves little or not at all with holiness. So, because we have been given copious translations in every field by Gerard of Cremona, Michael the Scot, Alfred the Englishman, Hermann the German, and William the Fleming, their works contain so much that is false that no one can be surprised. For in order for a translation to be accurate the translator must know the language from which he translates and the language into which he is translating as well as the science which he seeks to communicate. [...]

And so on for the others, especially this William the Fleming, who is now flourishing, even though it is known to anyone literate at Paris that he knew no science in the Greek tongue, about which he makes pretense. So all the things he translates are false, and he corrupts the wisdom of the Latins, for only Boethius sufficiently knew the languages of all interpretations. Lord Robert alone, because of the length of his life and the marvelous methods he employed, knew the sciences more than any other men. Because he did not know Greek and Hebrew sufficiently to translate by himself, he had many helpers. But all the others were ignorant of languages and sciences, and especially this William the Fleming, who knew nothing worthwhile either in the sciences or languages, yet who promised to change all the established translations and hammer out other new ones. But we have seen them and we know them to be full of errors and they are to be avoided.

169. Since in these times the enemies of Christians such as the Greeks, Arabs, and Hebrews have the sciences in their own languages, they do not make available to Christians truthful books but mutilate and corrupt them all, and especially when they see men untutored in languages and the sciences presume to make translations.

Text 2. The translator explains the order in which the Aristotelian treatises have to be read.

Lacombe, *Aristoteles Latinus: Codices*, 83; Smet, *Alexandre d'Aphrodisias*, xii–xiii; Rossi, 'Note', 175; Rossi, 'Problems of Method', 94.

William the interpreter says this. This book, which bears the title *On the Parts of Animals*, immediately follows the book of *Meteorology*, as Alexander says. For Aristotle ordered the science about living beings before the treatises about the other composite and mixed objects. However, the book with the title *History of Animals* must not be counted among the books on natural method, nor the book *On the Anatomy of Animals* [A different hand notes in the margin: 'This one is not yet among the Latins.'], since it is not of the same type. And so that one [i.e., *On the Parts of Animals*] remains the first. After that work, which consists of four books, follows one book *On the Gait of Animals*, after which there are three books *On the Soul*, then one book *On the Senses and What they Perceive*, then one book *On Memory* and one book *On Sleep and Divination in Sleep*. After this there is one book *On the Movement of Animals*. Subsequently *On the Generation of Animals* is discussed in five books. Thereafter is *On Food and the Growth of Animals* [in the same hand: 'This one is not yet among the Latins.'], which Aristotle sometimes calls *On Generation on the Outside of Animals*. Afterwards there is *On the Activity and Passivity and Behaviour of Animals* [in the same hand: 'Neither this one.']. Finally, there are *On Long and Short Lifespans* followed by *On Death and Life*, and *On Youth and Old Age*, and *On Respiration*, (and) lastly *On Sickness and Health* [in the same hand: 'Neither this one.']. And in these the complete science about living beings is contained.

Text 3. Bartholomew of Messina praises his patron as a lover of science.

Giacomelli, 'Le Patavinus Antonianus XVII 370', 65; Padova, Biblioteca Antoniana, MS XVII 370, f.77r.

Here begins the book of Aristotle *On Signs*, translated from Greek into Latin by master Bartholomew of Messina, at the court of the most illustrious Manfred, the most serene king of Sicily, lover of science, at his command.

Text 4. Nicholas of Reggio uses a manuscript sent as a gift to his patron by the Greek emperor.

Lo Parco, 'Niccolò da Reggio', 291–292; Paris, BNF, MS nouv. acq. lat. 1365, f.1r.

Behold, most glorious King Robert of Jerusalem and Sicily, Galen's book on the ailments of each body part and the treatment thereof, which consists of ten treatises. The renowned lord Andronicus, who bears the title of Emperor of the Greeks, heard that you are not only eminently learnèd in theology and natural philosophy, but also distinguished and of outstanding reputation in the science of medicine and the actions pertaining to that science. He also heard that you desire to have Galen's books on that science that are not yet available in the Latin language. Therefore, he sent ⟨this book⟩ as a special gift to your Serene Highness. I, Nicholas of Reggio, faithful physician and your subject, translated it on your command from Greek into Latin without additions, shortenings, or changes. It is needless to explain how much utility physicians can draw from it, since it is clear by merely looking through the very book to whomever is instructed in the medical science.

Text 5. William of Moerbeke complains about the deficiencies of his Greek manuscript.

Pattin, *Simplicius*, 605.

In the Greek model in the preceding chapter *On Movement* stood a few lines that pertain to the last chapter *On Having*. There was no indication where they had to be inserted, they were not continuous, they were full of open spaces without writing, they were corrupt, and therefore I did not translate them. They were about as long as half this column. Whoever looks through this work should also know that the Greek model was very corrupt and in many passages I could not obtain a meaning from the text. Yet I did what I could. For it is better to have it corrupt thus than to have nothing at all.

Text 6. William of Moerbeke complains about the deficiencies in his Greek manuscript (2).

Verbeke, *Jean Philopon*, 119.

I did not judge it necessary to translate the rest of this work. The reader of this work should also know that the Greek model was in many passages so ruined by water that I could not read it at all. And in those cases, I sometimes left open spaces, I sometimes supplemented it on the basis of the meaning, and sometimes I also considered it corrupt on the basis of a false reading. I think that whoever reads this will have more enlightenment to understand Aristotle's text than before. But whether it tells the truth about the active intellect or not, I leave that question to the reader's wisdom.

Text 7. William of Moerbeke explains why he limits his translation to a certain passage.

Bossier et al., *Simplicius*, XXII.

And so Simplicius continues that these events cannot be the causes of those, but neither those that are seen during eclipses. And afterwards he demonstrates that Aristotle too did not consider this thesis true. Subsequently, he proposes hypotheses for eccentric circles and epicycles, and says that their first inventors were the Pythagoreans. Since these hypotheses are more broadly treated by Ptolemy, I did not undertake to translate them. Afterwards he raises some objections against these hypotheses. He claims that he will elsewhere have the opportunity to discuss them and he does not follow them up.

Text 8. Robert Grosseteste justifies a lexical choice in his translation.

Dunbabin, 'Robert Grosseteste', 466; Mercken, *The Greek Commentaries*, 306; Oxford, Balliol College, MS 116, f.68r.

When Greek authors list the four principal virtues, fortitude, temperance, justice, and prudence, they always place the noun *sophrosyne* under the meaning of temperance. I also translate that noun here with the noun 'temperance', while others often translate the same noun with the noun 'sobriety', like those who translated the letters of the blessed Paul and the *Book of Wisdom*, where it is written: 'For it [wisdom] teaches sobriety, and wisdom, and justice, and virtue.'[9]

9 *Book of Wisdom*, 8.7.

Text 9. Nicholas of Reggio illustrates frequently used narratives in a prologue.

Schöne, *Galenus De partibus*, 10; Berlier, 'Niccolò da Reggio', 958–959; *Quarta impressio ornatissima, continens omnes Galeni libros*, Papiae: per Iacob Paucidrapium de Burgofrancho, 1515, I, f.LXVIIr.

Galen's book on the utility of the body parts, which is called *On the Remedies for Limbs* and consists of seventeen treatises according to the Greek text, is only available in the Latin language through the tenth treatise. It is not even a good translation, but one that is acquired from Arabic, which was so deficient that it was considered rather an incompetent compilation or extraction than a faithful and complete translation. You, famous man, master James Pipin of Brindisi, professor of medical science, physician of the illustrious and our common lord Philip, prince of Taranto,[10] and natural philosopher, you desire with a keen mind to have as far as is possible Galen's books that so far have not reached the Latins, or that have been acquired in an unsatisfactory translation, well translated and directly from the Greek language. Therefore, after considering the utility of the present book, you insistently asked and advised me, Nicholas of Reggio, physician, that I translate it from the beginning to the end for you from the Greek language into Latin. Since I feel obliged to serve you most promptly and to obey, on account of both the increase of medical knowledge achieved in me through you and the enormous favours of many kinds, I translated this book for you faithfully and word for word, without additions, shortenings, or changes. At the end of the entire work, I added in a list the terms that are found as unusual in each of the forementioned seventeen treatises with their explanations.

Text 10. Translators from the Greek and Greek books are rare.

Humbertus de Romanis, *Opus tripartitum*, II, 17, in *Secundus tomus conciliorum omnium tam generalium quam particularium* [...], Coloniae Agrippinae: ex officina Ioannis Quentel, 1551, 997; partial translation in Beullens, *A Methodological Approach*, 12.

If there were many good and wise men who knew this language [i.e., Greek] well, they could scrutinize all writings of the Greeks and see on which arguments they rely and deliberate with their wise men. Unfortunately, as it is, there are so few amongst the Latins who know that language, that sometimes even at

10 Philip I of Taranto (1278–1332).

the Roman court hardly anyone is found who is able to read the letters that the Greeks occasionally send. And whenever people are sent to them, they need to have interpreters of whom it is unclear whether they fully understand that language, or, if they know it, whether they translate it faithfully.

Another ⟨requirement⟩ is a huge quantity of Greek books, evidently so that the Latins may have all writings of the Greeks, both of the theologians and the commentators on theology, either in treatises or in commentaries on books, and of their councils and their statutes and the ecclesiastical office, and their histories and chronicles, especially from the times of Christianity. Since the truth is with the Latins, many arguments would without any doubt be found in those writings with which Latins could help themselves against the Greeks, or defend themselves against them, just as we defend ourselves with the writings of the Old Testament against the Jews and attack them.

But alas, there has been much concern to have their philosophical books and laws, yet for those that are related to the salvation and the common good of souls, not so much care has been taken. Although some of these have been translated, we have so few, and we possess no original texts in Greek, from the inspection of which we may gain clearer insight into the truth and stronger arguments to engage with the Greeks.

Text 11. Many texts remain hidden from the Latins by the jealousy of other peoples.

Tr. Giles, *Matthew Paris's English History*, I, 437–438, and II, 484.

Also, in this time, Robert, bishop of Lincoln, a man most skilled in Latin and Greek, translated accurately the Testaments of the twelve patriarchs from the Greek into Latin; which for many years had been unknown and concealed, through the jealousy of the Jews, because of the prophecies concerning our Saviour therein contained. But the Greeks, the most indefatigable investigators of all writings, being the first who learnt about this, translated it from Hebrew into Greek, and kept it to themselves until our own time. Nor in the time of St. Jerome, or of any other holy interpreter, could it in any way whatever come to the knowledge of the Christians, on account of the scheming malice of the Jews. Therefore, the abovenamed bishop, assisted by Master Nicolas, a Greek, and clerk of the abbey of St Albans, translated clearly, evidently, and word for word, into Latin, that glorious treatise, to the strengthening of the Christian faith, and to the greater confusion of the Jews. […]

This Master John had informed Robert, bishop of Lincoln, that when he was pursuing his studies at Athens, he had seen and heard from some learned Greek doctors of certain things unknown to the Latins; amongst which he discovered the testaments of the twelve patriarchs, the sons of Jacob, which were evidently a portion of the Holy Scriptures, but which had long been concealed through the malice of the Jews, in consequence of the manifest prophecies concerning Christ which appear in them. On this the said bishop sent to Greece, and having obtained possession of them, he translated them from the Greek into the Latin tongue, as also some other writings.

Part Three: Original Texts

The spelling and punctuation of the Latin texts have been standardized silently. Minor corrections to the texts as originally published have been introduced silently, where a manuscript or early printed source is cited.

Text 1. Maloney, *Roger Bacon: Compendium*, 172 and 174.

VIII.168 Sed longe maior error accidit in philosophia translata, quia, si sancti erraverunt in suis translationibus, multo magis alii qui parum aut nihil de sanctitate curaverunt. Unde, cum per Gerardum Cremonensem et Michaelem Scotum et Aluredum Anglicum et Hermanum Alemanum et Willielmum Flemingum data sit nobis copia translationum de omni scientia, accidit tanta falsitas in eorum operibus quod nullus sufficit admirari. Nam ad hoc quod translatio fiat vera oportet quod translator sciat linguam a qua transfert et linguam in quam transfert et scientiam quam vult transferre. [...]

Et sic de aliis, maxime iste Willielmus Flemingus, qui nunc floret, cum tamen notum est omnibus Parisius literatis quod nullam novit scientiam in lingua Greca de qua presumit. Et ideo omnia transfert falsa et corrumpit sapientiam Latinorum. Solus enim Boethius scivit de omnibus interpretationibus linguas sufficienter. Solus dominus Robertus, propter longitudinem vite et vias mirabiles quibus usus est, pre aliis hominibus scivit scientias. Quia Grecum et Hebreum non scivit sufficienter ut per se transferret, habuit multos adiutores. Omnes autem alii ignoraverunt linguas et scientias, et maxime hic Willielmus Flemingus, qui nihil novit dignum neque in scientiis neque in linguis, tamen omnes translationes factas promisit immutare et novas cudere varias. Sed eas vidimus et scimus esse omnino erroneas et vitandas.

169. Cum enim istis temporibus inimici Christianorum, ut Greci et Arabes et Hebrei, habeant scientias apud linguas suas, non concedunt Christianis libros veraces sed detruncant et corrumpunt omnes, et maxime quando vident homines indoctos in linguis et scientiis presumere de translationibus faciendis.

Text 2. Lacombe, *Aristoteles Latinus: Codices*, 83; Smet, *Alexandre d'Aphrodisias*, xii–xiii; Rossi, 'Note', 175; Rossi, 'Problems of Method', 94.

Inquit Willelmus interpres. Iste liber, qui inscribitur De partibus animalium, immediate sequitur librum Metheorologicorum, ut dicit Alexander. Aristotiles enim scientiam de animalibus preordinavit tractatibus de aliis complexionatis et commixtis. Liber vero qui inscribitur Hystorie animalium non est numerandus inter libros naturalis methodi, sicut nec liber De anathomiis animalium[11] pro eo quod non sit eiusdem modi. Et sic iste remanet primus. Post istum vero, qui in quatuor libris continetur, sequitur De progressu animalium liber unus. Post quem De anima libri tres. Postea liber De sensibus et sensibilibus. Postea De memoria et sompno et ea que per somnium divinativa, liber unus. Deinde de motu animalium liber unus. Consequenter tractatur De generatione animalium in quinque libris. Deinde De alimento et augmento animalium[12] quem Aristotiles vocat quandoque De generatione ea que deforis animalium. Postmodum De operationibus et passionibus et moribus animalium.[13] Tandem De longa et brevi vitabilitate. Deinceps De morte et vita et iuventute et senectute et de respiratione. Ultimo vero De sanitate et egritudine.[14] Et in hiis completur tota scientia de animalibus.

Text 3. Giacomelli, 'Le Patavinus Antonianus XVII 370', 65; Padova, Biblioteca Antoniana, MS XVII 370, f.77r.

Incipit liber Aristotelis de signis translatus de Greco in Latinum a magistro Bartholomeo de Messana in curia illustrissimi Maynfredi, serenissimi regis Sicilie, scientie amatoris, de mandato suo.

Text 4. Lo Parco, 'Niccolò da Reggio', 291–292; Paris, BNF, MS nouv. acq. lat. 1365, f.1r.

Ecce gloriosissime Ierusalem et Sicilie rex Roberte librum Galieni de passionibus uniuscuiusque particule et cura ipsarum consistentem in decem tractatibus, quem inclitus dominus Andronicus qui intitulatur imperator Grecorum, audiens quod non solum estis in theologicis et physicis eminenter eruditus, verum etiam in medicinali scientia et in operibus secundum scientiam istam estis fama referente precipuus, et quod desideratis habere libros Galieni in ipsa scientia qui nondum habentur in lingua Latina, pro quodam speciali munere vestre serenitati transmisit. Ego Nicolaus de Regio medicus fidelis et subditus vester iuxta mandatum vestrum de Greco in Latinum ydiomate transtuli, nil addens, minuens, vel permutans. Quanta autem sit medicis ex ipso utilitas non expedit exprimere cum pateat quibuslibet edoctis in medicinali scientia a sola inspectione ipsius libri.

---

11 *Iste apud Latinos nondum est* i.m. alia manu.

12 *Iste apud Latinos nondum est* i.m. alia manu.

13 *Et nec iste* i.m. alia manu.

14 *Et nec iste* i.m. alia manu.

Text 5. Pattin, *Simplicius*, 605.

In exemplari Greco in precedenti capitulo de motu stabant quedam pertinentia ad ultimum capitulum de habere, que non erant signata ubi debebant intrare et non erant continua et plena erant spatiis non scriptis et corrupta erant et propterea non transtuli. Erant autem quasi ad quantitatem medie columne huius. Sciat etiam, qui hoc opus inspexerit, exemplar Grecum valde fuisse corruptum et in multis locis sensum nullum ex littera potui extrahere. Feci tamen quod potui. Melius enim erat sic corruptum habere quam nihil.

Text 6. Verbeke, *Jean Philopon*, 119.

Reliqua huius operis non iudicavi oportere transferri. Sciat etiam lector huius operis exemplar Grecum in plerisque locis ab aqua fuisse destructum, ita quod nullatenus legere potui, et ibi spatia quandoque dimisi, quandoque ex sensu supplevi, quandoque etiam corruptum ex falsitate putavi. Puto qui hec legerit, ad intellectum littere Aristotilis plus quam ante lumen habebit. Utrum autem verum dicat de intellectu agente vel non, lectoris prudentie derelinquo.

Text 7. Bossier et al., *Simplicius*, XXII.

Sic prosequitur Simplicius quod isti non possunt ista causare, sed neque que videntur in eclipsibus; et postea ostendit quod nec Aristotiles firmam reputavit hanc positionem; consequenter ponit ypotheses per ecentricos et epiciclos, cuius primos inventores dicit fuisse Pythagoricos, quas quia copiosius tradantur a Ptolomeo non curavi transferre. Deinde movet quasdam obiectiones contra has ypotheses, de quibus dicit alias fore tempus considerandi, et nichil prosequitur de ipsis.

Text 8. Dunbabin, 'Robert Grosseteste', 466; Mercken, *The Greek Commentaries*, 306; Oxford, Balliol College, MS 116, f.68r.

Quando autem Greci auctores enumerant quatuor[15] virtutes principales, scilicet fortitudinem, temperantiam, iustitiam et prudentiam, semper ponunt in significationem temperantie hoc nomen sophrosine. Quod nomen et nos hic transtulimus in nomen temperantie, alii autem idem nomen sepe transferunt

15 *auctores* add. Balliol MS 116.

in nomen[16] sobrietatis, ut qui epistulas beati Pauli transtulerunt, et librum Sapientie, in quo scriptum est: Sobrietatem enim et sapientiam docet, et iustitiam et virtutem.

Text 9. Schöne, *Galenus De partibus*, 10; Berlier, 'Niccolò da Reggio', 958–959; *Quarta impressio ornatissima, continens omnes Galeni libros*, Papiae: per Iacob Paucidrapium de Burgofrancho, 1515, I, f.LXVIIr.

Quia liber Galieni de utilitate particularum, qui dicitur de iuvamentis membrorum, consistens in decem et septem tractatibus secundum litteram Grecam, non nisi usque ad decimum in Latino ydiomate habebatur, non tamen in bona translatione, sed habita ex Arabico, que adeo erat defectuosa, quin potius incompetens compendium seu extractio quam fidelis et completa translatio censeretur. Vos, famose vir, magister Jacobe Pipine de Brundusio [*sic*], scientie medicinalis professor et inclyti et communis domini d. Philippi, principis Tarentini, phisice, qui avido affectatis animo pro posse[17] libros Galieni qui hucusque non pervenerunt apud Latinos aut habiti sunt non in competenti translatione, bene translatos et immediate de lingua Greca habere, considerata presentis libri utilitate me, Nicolaum de Regio de Calabria, medicum, cum instantia requisivistis et monuistis ut ipsum a principio usque ad finem transferrem vobis de Greco ydiomate in Latinum. Ego vero, cum tenear tum propter propagationem doctrine medicinalis in me per vos factam, tum etiam propter immensa multifarie beneficia servire promptissime vobis et obedire, ipsum librum transtuli vobis fideliter et de verbo ad verbum, nihil addens, minuens, vel permutans. In fine autem totius libri vocabula que in singulo predictorum decem et septem tractatuum reperiuntur ut inconsueta cum eorum expositionibus apposui seriatim.

Text 10. Humbertus de Romanis, *Opus tripartitum*, II, 17, in *Secundus tomus conciliorum omnium tam generalium quam particularium* [...], Coloniae Agrippinae: ex officina Ioannis Quentel, 1551, 997.

Si enim essent boni viri et prudentes multi, qui scirent bene hanc linguam, possent omnia scripta Grecorum transcurrere et videre quibus innituntur et conferre cum sapientibus eorum. Nunc autem prohdolor ita pauci sunt inter Latinos qui sciant huiusmodi linguam, quod vix in curia etiam Romana inveni-

16 *temperantie* [...] *nomen* om. Dunbabin.

17 *proponere* Berlier.

tur quandoque qui literas quas Greci mittunt interdum, sciant legere. Et quandocumque mittuntur aliqui ad eos, oportet habere interpretes de quibus nescitur, utrum plene intelligant illam linguam, vel, si sciant eam, utrum fideliter interpretentur.

Aliud est copia librorum Grecorum, ut videlicet haberent Latini omnia scripta Grecorum tam theologicorum quam expositorum theologie, sive in tractatibus sive in expositionibus librorum, necnon et conciliorum eorum et statutorum eorum et officii ecclesiastici, et historiarum et gestorum eorum maxime a tempore Christianitatis. Cum enim veritas sit pro Latinis, indubitanter invenirentur multa in huiusmodi scriptis quibus iuvarent se Latini contra Grecos, vel quibus se defenderent ab eis, sicut per scripta veteris testamenti defendimus nos a Iudeis et impugnamus ipsos.

Sed heu curatum est multum de libris philosophicis et legibus habendis ab eis; de his autem que ad salutem et ad bonum commune pertinent animarum, non est ita curatum. Etsi enim aliqua de his translata sunt, habemus tam pauca, et ipsa originalia in Greco non habemus, ex quorum inspectione veritas magis clareret et fortius possent Greci impugnari.

Text 11. Luard, *Matthaei Parisiensis* [...] *Chronica Majora*, IV, 232–233, and V, 284–285.

Ipsis quoque temporibus, episcopus Lincolniensis Robertus, vir in Latino et Greco peritissimus, Testamenta duodecim Patriarcharum de Greco fideli interpretatione transtulit in Latinum, que per multa tempora incognita et abscondita fuerunt per invidiam Judeorum, propter manifestas prophetias de Salvatore in eis contentas. Sed Greci, omnium scriptorum diligentissimi investigatores, primi in notitiam illius scripti devenientes, illud de Hebreo in Grecum transtulerunt, et penes eos usque in nostra tempora reservarunt. Nec tempore beati Jeronimi vel alicuius sancti interpretis ad notitiam Christianorum, machinante Iudeorum antiquorum malitia, potuit quomodolibet devenire. Illum igitur gloriosum tractatum, ad robur fidei Christiane et ad maiorem Iudeorum confusionem, transtulit plene et evidenter episcopus memoratus de Greco, verbo ad verbum, in Latinum, coadiuvante magistro Nicholao Greco, clerico abbatis Sancti Albani.

Hic magister J[ohannes] intimaverat episcopo Lincolniensi Roberto, quod, quando studuit Athenis, viderat et audierat ab peritis Grecorum doctoribus quedam Latinis incognita. Inter que reperit duodecim patriarcharum, filiorum videlicet Jacob, testamenta; que constat esse de substantia Bibliothece, sed per invidiam Iudeorum dudum fuisse abscondita, propter manifestas, que in eisdem patent, de Christo prophetias. Unde idem episcopus misit in Greciam, et cum ea habuisset, transtulit de Greco in Latinum, et quedam alia.

Bibliography

Primary Sources

Bossier, Fernand, C. Vande Veire, and G. Guldentops (eds), *Simplicius: Commentaire sur le traité du ciel d'Aristote. Traduction de Guillaume de Moerbeke*, I, Corpus Latinum Commentariorum in Aristotelem Graecorum VIII/1, Leuven: Leuven University Press, 2004

Giles, J.A. (tr.), *Matthew Paris's English History from the Year 1235 to 1273*, 3 vols, London: Bohn, 1852–1854

Luard, Henry Richards (ed.), *Matthaei Parisiensis Monachi Sancti Albani Chronica Majora*, 7 vols, London: Longman etc., 1872–1883

Maloney, Thomas S. (ed. and tr.), *Roger Bacon: Compendium of the Study of Philosophy*, Auctores Britannici Medii Aevi 32, Oxford: Oxford University Press, 2018

Mercken, H. Paul F. (ed.), *The Greek Commentaries on the Nicomachean Ethics of Aristotle in the Latin translation of Robert Grosseteste, Bishop of Lincoln*, I, Corpus Latinum Commentariorum in Aristotelem Graecorum VI/1, Leiden: Brill, 1973

Pattin, A. (ed.), *Simplicius: Commentaire sur les Catégories d'Aristote. Traduction de Guillaume de Moerbeke*, Corpus Latinum Commentariorum in Aristotelem Graecorum V/2, Leiden: Brill, 1975

Schöne, Hermann (ed.), *Galenus de partibus artis medicativae, eine verschollene griechische Schrift in Übersetzung des 14. Jahrhunderts*, Greifswald: Abel, 1911

Smet, A.J. (ed.), *Alexandre d'Aphrodisias: Commentaire sur les Météores d'Aristote. Traduction de Guillaume de Moerbeke*, Leuven/Paris: Publications Universitaires/Béatrice-Nauwelaerts, 1968

Verbeke, G. (ed.), *Jean Philopon: Commentaire sur le De Anima d'Aristote. Traduction de Guillaume de Moerbeke*, Corpus Latinum Commentariorum in Aristotelem Graecorum III, Leuven/Paris: Publications Universitaires/Béatrice-Nauwelaerts, 1966

Secondary Literature

Berlier, Stéphane, 'Niccolò da Reggio traducteur du *De usu partium* de Galien. Place de la traduction latine dans l'histoire du texte', *Medicina nei secoli*, 25 (2013), 957–978

Beullens, Pieter, *De sleutel tot Aristoteles: Willem van Moerbeke en de overlevering van antieke wijsheid*, Eindhoven: Damon, 2019 [An English translation, *The Friar and the Philosopher: William of Moerbeke and the Rise of Aristotle's Science in Medieval Europe*, is forthcoming.]

Beullens, Pieter, *A Methodological Approach to Anonymously Transmitted Medieval Translations of Philosophical and Scientific Texts: The Case of Bartholomew of Messina*, Leuven: PhD dissertation, 2020

Brams, Jozef, 'Mensch und Natur in der Übersetzungsarbeit Wilhelms von Moerbeke', in *Mensch und Natur im Mittelalter*, ed. Albert Zimmermann and Andreas Speer, Berlin/New York: De Gruyter, 1992, 537–561

Brams, Jozef, and Willy Vanhamel (eds), *Guillaume de Moerbeke: Recueil d'études à l'occasion du 700e anniversaire de sa mort (1286)*, Leuven: Leuven University Press, 1989

De Leemans, Pieter (ed.), *Translating at the Court: Bartholomew of Messina and Cultural Life at the Court of Manfred, King of Sicily*, Leuven: Leuven University Press, 2014

Dionisotti, A.C., 'On the Greek Studies of Robert Grosseteste', in *The Uses of Greek and Latin: Historical Essays*, ed. A.C. Dionisotti, Anthony Grafton, and Jill Kraye, London: Warburg Institute, 1988, 19–39

Giacomelli, Ciro, 'Le Patavinus Antonianus XVII 370: éléments pour une étude paléographique et textuelle', *Pecia*, 20 (2018), 45–80

Dunbabin, Jean, 'Robert Grosseteste as Translator, Transmitter, and Commentator: the "Nicomachean Ethics"', *Traditio*, 28 (1972), 460–472

Fögen, Thorsten, *Patrii sermonis egestas: Einstellungen lateinischer Autoren zu ihrer Muttersprache. Ein Beitrag zum Sprachbewußtsein in der römischen Antike*, Munich/Leipzig: Saur, 2000

Forrai, Réka, '*Hostili praedo ditetur lingua latina*: Conceptual Narratives of Translation in the Latin Middle Ages', *Medieval Worlds*, 12 (2020), 121–139

Fortuna, Stefania, 'Il corpus delle traduzioni di Niccolò da Reggio (fl. 1308–1345)', in *La medicina nel basso medioevo: tradizioni e conflitti*, Spoleto: Centro italiano di studi sull'alto medioevo, 2019, 285–312

Lacombe, Georgius, A. Birkenmajer, M. Dulong, and A. Franceschini, *Aristoteles Latinus: Codices*, I, Rome: Libreria dello Stato, 1939

Lo Parco, Francesco, 'Niccolò da Reggio. Antesignano del Risorgimento dell'antichità ellenica nel secolo XIV da Codici delle Biblioteche italiane e straniere e da documenti e stampe rare', in *Atti della Reale Accademia di archeologia, lettere e belle arti*, nuova serie, 2 (1913), 241–317

Rossi, Pietro B., 'Note sulla tradizione della *Translatio Guillelmi* del *De partibus animalium*', in *Tradition et traduction: Les textes philosophiques et scientifiques grecs au moyen âge latin*, ed. Rita Beyers, Jozef Brams, Dirk Sacré, and Koenraad Verrycken, Leuven: Leuven University Press, 1999, 167–197

Rossi, Pietro B., 'Problems of Method according to some Medieval Commentators on Book I of the *De partibus animalium*', in *Erfahrung und Beweis: Die Wissenschaften von der Natur im 13. und 14. Jahrhundert. Experience and Demonstration: The Sciences of Nature in the 13th and 14th Centuries*, ed. Alexander Fidora and Matthias Lutz-Bachmann, Berlin: Akademie Verlag, 2007, 89–123

Thorndike, Lynn, 'Translations of Works of Galen from the Greek by Niccolò da Reggio (c. 1308–1345)', *Byzantina Metabyzantina: A Journal of Byzantine and Modern Greek Studies*, 1 (1946), 213–235

CHAPTER 11

Why Translate? Views From Within Judaism: Egodocuments by Translators from Arabic and Latin into Hebrew (Twelfth–Fourteenth Centuries)

Gad Freudenthal

The history of Jewish philosophy
is the story of the receptions of foreign bodies of knowledge.
JULIUS GUTTMANN (1933)[1]

∵

Part One: Essay

Overview

I. Prefatory Note
II. The Arabic and Latin into Hebrew Translation Movement: A Short Survey
III. The Hebrew Translation Movement: A Sociological Perspective
IV. Translators and Patrons: Independent Decisions and Absence of Economic Dependence
V. Towards a History of Medieval Hebrew Translators

I. *Prefatory Note*

Hebrew has been both a source and target language for translations ever since it became the shared language of the Israelite ancestors of the Jewish people (c. 1000 BCE). In the various diasporas in which Jews lived from at least the sixth century BCE, they acquired the diverse vernaculars of their host societies, while maintaining Hebrew as a common written language: the canonical Jewish Bible, the Mishnah (the first major written collection of the Jewish oral

1 'Die Geschichte der jüdischen Philosophie ist eine Geschichte von Rezeptionen fremden Gedankenguts'; Guttmann, *Die Philosophie des Judentums*, 9.

 | DOI:10.1163/9789004472648_012

traditions, redacted at the beginning of the third century CE), and most prayers were in Hebrew. Scholars mastered the 'Holy Tongue', but most 'common folk' knew only a little Hebrew. In Babylonia of the first century BCE, the weekly public reading in the synagogue of a portion of the Hebrew Torah started to be accompanied by an Aramaic translation (or paraphrase) of the text, allowing the uninstructed to follow. In the tenth century, after Arabic replaced Aramaic as the spoken language in the Middle East, the great scholar Saadia b. Joseph Gaon (882–942) translated the Bible into Arabic. The Arabic used by Jews—Judeo-Arabic—was written in Hebrew characters, like all Jewish languages, and had specific linguistic features as well. The regular use of the Aramaic and Judeo-Arabic versions of the Bible made the very idea of translation a household concept in Jewish cultures.

The situation among the Jews in the Islamic empire was one of Arabic–Hebrew bilingualism (or diglossia): for philosophy and science, biblical exegesis, religious law, Hebrew philology, and similar disciplines, Judeo-Arabic was employed; Hebrew was reserved for poetry, as well as the liturgy. Rina Drory has described the situation as one of functional dichotomy: Arabic was used for 'communicative' purposes, Hebrew in 'solemn' contexts.[2] The result was a situation of continuous bilingualism, in which Jewish intellectuals had two written languages at their disposal, in addition to their various vernaculars, whose use as a rule remained oral. This 'division of linguistic labour' was accepted almost universally, although some scholars challenged it, urging that Hebrew be used exclusively, in both solemn and profane contexts.[3] The absorption of non-Jewish lore by Jews living in Arabophone cultures was therefore relatively smooth; numerous Jewish scholars were conversant with Greco-Arabic science and philosophy and drew on it in their writings.

The situation was very different for Jewish cultures living under the Cross, who were divided among three major centres: the Midi ('Provence' in medieval Jewish idiom), the Rhineland and northern France (Ashkenaz/Tzarfat), and the Italian peninsula. The diglossia that prevailed under Islam had no parallel under Christianity: the cultural languages of the Jews in Europe (Andalusia excluded) were Hebrew and Aramaic only. Although Jews spoke the local Romance vernaculars at most time and in most places, they did not know Latin, the cultural language of the host society; this only changed towards the end of the fourteenth century. Thus, unlike their counterparts under the Crescent,

2 See Drory, *Models and Contacts*, esp. 158–177.

3 See, e.g., Halkin, 'The Medieval Jewish Attitude Toward Hebrew'; and Roth, 'Jewish Reactions to the *Arabiyya* and the Renaissance of Hebrew in Spain'.

who had ready access to any Arabic work they wished, Jewish intellectuals in Christian lands could access non-Jewish texts only via translations or Hebrew compositions summarizing them.

Hebrew translations—mainly from Arabic—increased in number and in quality in the late eleventh century and continued to be produced until the end of the fourteenth. Many translated works were philosophical and scientific. It is important to realize that in Judaism, translations of this kind meant not simply 'appropriation' of new lore, to be added to traditional lore, so that both would coexist side by side; rather, it implied a veritable cultural upheaval and the overcoming of high cultural barriers. The principal reason for this is Judaism's traditionally defiant attitude towards non-Jewish lore, sustained by the built-in social conservatism of Jewish culture.[4]

In traditional Jewish cultures, intellectual energy was directed to the study of the authoritative texts of Judaism, especially the Talmud. By virtue of their social values and organization, these cultures had a strong conservative tendency. Consider why. For the Jewish male, studying the Jewish canonical texts, believed to have their source in divine revelation, was a religious obligation. As a result, the symbolic reward was allocated to the Jewish scholar according to his erudition in the Torah and the quality of his contributions to Torah studies.[5] This social system encouraged originality and innovation, but only *within* existing traditions, rendering it a *closed* system. The introduction into Judaism of any alien (non-Jewish) body of knowledge whose authority does not derive from revelation (but from the human mind), *ipso facto* ran against the grain: where it occurred, it needed continual legitimation. This basic hurdle was made even higher by an additional social barrier: as mentioned, in the Arabophone world, Hebrew was employed for 'festive' or ceremonial functions, notably poetry and artistic prose, while communicative functions (Bible exegesis, halakhah, philosophy, grammar, personal and business correspondence) were carried out in Judeo-Arabic. Hebrew thus lacked a tradition of writing on secular topics, such that the translation of philosophy and science into Hebrew required the innovation of appropriate literary models and vocabulary. This again meant a profound change in entrenched socio-linguistic norms.

Nonetheless, a massive cultural transfer got under way, in which scientific, philosophical, and literary bodies of non-Jewish knowledge were 'imported' into the Hebrew cultures north of the Pyrenees, mainly from Arabic, beginning

4 Schacter, *Judaism's Encounter with Other Cultures*.

5 The term 'Torah' denotes the entire corpus of Jewish canonical texts, all ultimately deriving their authority from the Five Books of Moses—the Torah in the narrow, original sense.

in the second half of the twelfth century. During the next two and a half centuries, a Jewish rationalist culture bloomed in the Midi, Christian Spain, and the Italian peninsula (though, it should be noted, hardly at all in Ashkenaz). This cultural transfer had not only to overcome the traditional resistance to all 'alien' or 'secular' thought, but also to do so via the very narrow route of texts (notably translations) written by Arabophone scholars.

In sum, there were powerful factors militating against the introduction of non-Jewish lore into Judaism in Hebrew. These will loom behind all the discussions and often apologetic discourses presented in this chapter.

II. *The Arabic and Latin into Hebrew Translation Movement: A Short Survey*

Between the eleventh and fifteenth centuries, Jewish scholars in Europe produced Hebrew translations of some 570 works.[6] The source language of most of them was Arabic, but a sizable number were translated from Latin (mostly medical works and mostly in the late fourteenth century). In addition, a number of works written in Hebrew by Arabophone scholars made large bodies of Greco-Arabic knowledge available to the Hebrew-reading public.

There is no need to recapitulate the history of the Arabic– or Latin–Hebrew translation movement. One fundamental work continues to constitute the basis for all discussions: the monumental *Die hebraeischen Übersetzungen des Mittelalters und die Juden als Dolmetscher* (Berlin, 1893, hereafter *HÜ*) by Moritz Steinschneider (1816–1907), of which an updated English translation is in progress (see bibliography). This work is usefully complemented by the late Mauro Zonta's 'Medieval Hebrew Translations of Philosophical and Scientific Texts: A Chronological Table',[7] which provides an overview of the gradual development of the corpus with which we are concerned here. Interested readers are also directed to my own recent survey of the translation process, which includes detailed bibliographical indications.[8]

Here let me outline only the historical framework that made cultural transfer possible. Already before the eleventh century, a handful of Arabophone Jewish scholars chose to write in Hebrew, so that their work could be read by Jews living outside their own linguistic region. A few Hebrew writings grounded in Arabic learning were produced, beginning to make non-Arabophone Jews aware of realms of knowledge to which they had no access and initiating a

6 Zonta, 'Medieval Hebrew Translations of Philosophical and Scientific Texts: A Chronological Table'.

7 Freudenthal, *Science in Medieval Jewish Cultures*, 17–73.

8 Freudenthal, 'Science and Medicine'.

transfer of this knowledge. The social context of this early phase of Arabic–Hebrew knowledge transfer is largely shrouded in darkness and will not be considered here.[9]

Next comes the phase of knowledge transfer by those whom I have called in Section One 'harbingers and pioneers'. This phase is characterized by the fact that the works in question are geographically and chronologically scattered, and that together they did not yet constitute a translation movement, which was to emerge shortly after. The phase consists mainly of Arabic–Hebrew knowledge transfer that took place between the mid-eleventh and mid-twelfth centuries, mainly through straightforward translations or Hebrew 'textbooks' that summarized knowledge gathered from Arabic sources. All these works were written by Arabophone Jewish scholars who were making a deliberate effort to share knowledge with those of their co-religionists who were ignorant of Arabic (Section One, Part 1). The 'harbingers and pioneers' phase also includes a small number of Latin– or vernacular–Hebrew translations produced during the second half of the twelfth century (Section One, Part 2). These two segments of early knowledge transfer are characterized by their standalone status and lack of sequels.

Cultural transfer took off seriously following the arrival in the Midi of a number of intellectual families fleeing the Almohad persecutions in al-Andalus, starting c. 1147. These Arabophone scholars quickly became significant cultural intermediaries. Two families of Andalusian origin who played an important role in the cultural transfer were the Tibbonids and the Qimḥis. The first generation—Judah b. Saul Ibn Tibbon (1120–c. 1190), who hailed from Granada and settled in Lunel (north of Marseilles), and Joseph b. Isaac Qimḥi (c. 1111–1170), who settled in Narbonne—translated works on religious philosophy and Hebrew grammar from Judeo-Arabic into Hebrew. As a result of their endeavours, Jewish intellectuals in the Midi came to appreciate that works written by their brethren in Arabic were indispensable for understanding Judaism. The ongoing (informal) religious controversies, specifically the Christian critique of Judaism as theologically unsophisticated, apparently helped persuade Provençal Jewish scholars that they had much to gain from appropriating works in Arabic, notably in grammar and philosophy. Nonetheless, the reception of this new kind of knowledge met with resistance.

The arrival in Provence of the works of Moses Maimonides (1138–1204), and the translation into Hebrew of his *Guide of the Perplexed* (first Hebrew version,

9 Langermann, 'On the Beginnings of Hebrew Scientific Literature'; and Kiener, 'The Hebrew Paraphrase of Saadiah Gaon's *Kitāb al-Amānāt wa'l-I'tiqādāt*'.

1204), produced a true spiritual revolution in certain prominent circles: for the followers of Maimonides—the rationalists—the study of the sciences became legitimate and indeed a religious obligation.[10] Jewish rationalists adopted a *new religious ideal* that extolled the study of scientific and philosophical works in parallel to the study of the Jewish tradition, indeed as a part of it. Beginning in the early thirteenth century, a self-conscious Maimonidean subculture emerged, notably in the Midi and in Italy. The study of the Hebrew translations of Judeo-Arabic works whetted readers' appetite to learn more of the Greco-Arabic sciences underlying them. The resulting demand for scientific and philosophical literature became increasingly independent of its initial trigger. Yet the *Guide of the Perplexed* remained the ideological foundation that legitimized the rationalist stance and it has kept this pivotal role throughout the ages. It was at this stage that the Arabic–Hebrew translation movement, which was to last for at least a century and a half, achieved full stride.

The first significant translators were members of the Ibn Tibbon family, Judah Ibn Tibbon and his son Samuel—the founding fathers of the translation movement (Section Two). They were followed by the next generation of their own family, for whom translation became a routine engagement—Moses b. Samuel Ibn Tibbon, Jacob Anatoli Ibn Tibbon, and Jacob b. Makhir Ibn Tibbon (Section Four). During the course of the thirteenth century and later, other translators appeared, who are not discussed here for lack of space. Most of them were the descendants of families of Andalusian origin, in which Arabic fluency was transmitted from father to son over generations. (Notably, knowledge of Arabic was particularly important among physicians, as it gave direct access to key medical texts; it was only in the fourteenth century, when the new Latin medicine surpassed and replaced Arabic medicine, that Arabic lost this important role. As a result, many translators were also doctors.) In parallel to translations, some thirteenth-century Jewish scholars chose to write general surveys ('encyclopedias') of science and philosophy in order to offer a full picture of these domains in a single volume in Hebrew (Section Three).

It must not be forgotten that throughout this period many Jewish intellectual circles consistently and vehemently opposed the Maimonidean revolution, insisting that only the study of the traditional Jewish disciplines was legitimate. They exerted constant pressure on the rationalists, and a series of clashes over the permissibility of the study of 'alien sciences' punctuated the thirteenth and early fourteenth centuries. This constant traditionalist pressure

10 Davidson, 'The Study of Philosophy as a Religious Obligation'.

on the rationalists is evident in translators' introductions, which often go out of their way to offer religious apologetics for the study of science.

In neighbouring Christian Spain, a command of Arabic was often maintained well into the fourteenth century, reducing the need for translations. In addition, from the early thirteenth century the influence of the Kabbalah increased, leading to decreased interest in science and philosophy and curtailing demand for translations; not even the translations from Provence were much used. Nonetheless, a few translators were active in Spain and a number of works in philosophy and science based on Arabic knowledge were written in Hebrew.

In Italy, things were different. Intellectually and socially, the south of the peninsula has to be discussed separately from the centre and the north. In the Mezzogiorno, Emperor Frederick II (1194–1250), followed by his successors, sponsored the translation of philosophical and scientific works from Arabic into Latin and, to a lesser extent, into Hebrew. A number of Arabophone Jews moved there from the Midi at his invitation, making Sicily and southern Italy a centre for early cultural transfer in the 1230s and 1240s.[11] For Hebrew translators it was exceptional to work in an institutionalized setting and be paid for their efforts, the usual situation for translators into Arabic or Latin. A number of Jewish–Christian collaborations took place at Frederick's court, yielding, among other things, Hebrew translations whose source text was Arabic but for which the Hebrew was checked against the Latin version of the same work. In central and northern Italy, the situation was entirely different, making it wrong to refer to the 'Italian context' as if it were unified. Arabic–Hebrew translations were few and far between, produced by a handful of immigrants from the Midi and a maverick scholar who acquired Arabic during his travels. Here, too, Maimonidean intellectuals depended on Hebrew translations executed in the Midi for their philosophical instruction. Copies of these translations were in high demand but difficult to obtain: the fourteenth-century poet, translator, and social critic Qalonymos b. Qalonymos delightfully describes how scholars in Rome pilfered a store of Hebrew philosophical manuscripts that a traveller scholar put in their custody.[12] In parallel, we encounter a significant phenomenon that has no equivalent in the Midi: Jewish intellectuals who had close and friendly relationships with their Christian counterparts, knew Latin, read Scholastic philosophers, and translated a number of their works from Latin into

11 Steinschneider, 'Robert von Anjou und sein Verhältnis zu einigen gelehrten Juden'; Cohn, 'Jüdische Übersetzer am Hofe Karls I. von Anjou'; Shatzmiller, 'Au service de la Cour de Naples'; and Sirat, 'Les traducteurs juifs à la cour des rois de Sicile et de Naples'.

12 Brener, 'Stealing Wisdom'.

Hebrew. This is a significant social fact that still calls for explanation.[13] These Latin–Hebrew translations do not seem to have circulated widely outside the Italian peninsula however and had only a limited effect on the future course of Jewish philosophy.

The early years of the fifteenth century witnessed the massive translation of Scholastic works from Latin into Hebrew, a cultural phenomenon that Mauro Zonta dubbed 'Hebrew Scholasticism'.[14] This cultural transfer ushered in a new era in the history of Jewish thought that was not a continuation from the preceding centuries; it lies beyond the temporal scope of this chapter.

The quantitative overview of the Hebrew translations produced from the twelfth to fourteenth centuries may be summarized in the following table.

Distribution of Hebrew Translations by Discipline, Source Language, and Century[a]

	Translations from Arabic		Translations from Latin		Total
	Sciences and philosophy	**Medicine**	**Sciences and philosophy**	**Medicine**	
Twelfth century	28	1	0	18	47
Thirteenth century	116	45	12	17	190
Fourteenth century	100	37	34	62	233
	244	83	46	97	470

a Source: Freudenthal, 'Arabic and Latin Cultures as Resources', 76–78. The data in the table there present the distribution by region and period in greater detail.

The number of translations in all domains of secular learning comes close to a respectable 500.[15] The vast majority of those in science and philosophy were translated from Arabic (244), and many fewer from Latin (46), mostly in the fourteenth century (and in Italy, although this cannot be seen here). In medicine, most translations until the thirteenth century were from Arabic, too, but in the fourteenth century Latin outranks Arabic as the source language.

13 See notably the two volumes of *Latin-into-Hebrew*.

14 Zonta, *Hebrew Scholasticism in the Fifteenth Century*.

15 Above (near n. 3) the number given was 570; the discrepancy is because the tables cover different time periods.

III. *The Hebrew Translation Movement: A Sociological Perspective*

From a sociological perspective, the essential quality and hallmark of the cultural transfer sketched here is that it was entirely *decentralized*.[16] Jews lived in countless towns and villages, under many rulers and regimes. They had no central political, economic, or religious authority that was or could have been in a position to initiate and oversee (and fund) the sort of planned and systematic translation movement typical of the Greek–Arabic, Syriac–Arabic, and Arabic–Latin translation movements. Rather, the decision to translate any given work was usually taken by one or two individuals—the translator himself and sometimes his patron. The library of translations into Hebrew from the twelfth to fifteenth centuries is thus the result of some 500 independent decisions, taken by more than 120 translators, as listed by Steinschneider.[17] Each translator (and patron) had his own intellectual and spiritual agenda and preferences that contributed to shaping the total translation movement.

The multicentric social structure of Hebrew translation activity gives the translators' personal statements particular significance, precisely because they were independent actors with a variety of personal motivations. These statements are the principal clue to the reasons behind their choices. In recent decades, such personal statements have been dubbed *egodocuments*. In the present context, these documents are mainly the prefaces and postfaces to translations, though there is the occasional personal letter, or a reflection that the translator has interpolated into the translation.

Certain *topoi* are frequent in these prefaces. In Arabic–Hebrew translations, these include the differences between the two languages, especially with respect to the relative richness of their vocabularies; the resulting challenges facing the translators; observations about translation methods; an exposition of the reasons for Jews' backwardness in matters scientific; an insistence on the alleged Jewish origin of the sciences and their 'theft' by the Gentiles;[18] and arguments for the legitimacy of studying science and philosophy instead of the Torah. While some of these recurrent *topoi* may have become literary conventions,[19] they certainly also reflect shared perceptions and attitudes. The following point, however, seems to be more important: the fact that these

16 For a brief comparative sociological analysis of the main medieval translation movements, see Freudenthal and Glasner, 'Patterns of Medieval Translation Movements'.

17 Steinschneider, *HÜ*, 1067–1068. This list is certainly incomplete, because the many individuals whose translations are anonymous (whether originally or only today) are not included.

18 Roth, 'The "Theft of Philosophy" by the Greeks from the Jews'; and Melamed, *The Myth of the Jewish Origins of Science and Philosophy* [Hebrew].

19 Rothschild, 'Motivations et méthodes des traductions en hébreu'.

topoi keep appearing reveals that the Arabic–Hebrew translation activity was a *socially self-conscious, coherent, and continuous tradition.* The translators may have worked in isolation, but they were usually aware of the work of their predecessors (their role models) and contemporaries, and felt committed to a shared tradition. In particular, the prefaces and translations by the first two Tibbonids, Judah and Samuel, were taken as models by later translators. We may say that the Arabic–Hebrew translators constituted a sort of network extending through time and space. Even if this translation activity was not formally organized by a central authority, it was nonetheless informally structured via the emulation of models taken as canonic and compliance with the norms set by the early translators: this is why the multicentric process of Arabic–Hebrew cultural transmission exhibits a remarkable internal coherence. Put differently, from a sociological point of view, at any moment, the dispersed Arabic–Hebrew translations display a social structure analogous to that of 'normal science' in T.S. Kuhn's *Structure of Scientific Revolutions* (1962), where scientists' work is informally steered by a 'paradigm' consisting of 'exemplars' that gave a unity to a scientific field without any formal control. It is this coherence that warrants our considering the succession of Arabic–Hebrew translations as a translation *movement.*

Looking at many translators' prefaces alongside one another from a sociological perspective uncovers implicit social norms. For example, the prefaces evince a clear reluctance to retranslate a work that has already been translated. As a rule, a translator who is aware that the text he has rendered into Hebrew had already been translated will apologize, explain why he undertook the translation, and assure his readers that it was not his intention to take pride in the 'humiliation' of his predecessor (Document 2.1.1.3) or 'diminish the deference due to the first translator' (Document 4.1.1.3, similarly Document 6.1.1.3).[20] It seems then that Hebrew translators, no matter where and when they lived

20 Another example: in his preface to his translation of a work ascribed to Ibn Zuhr (Avenzoar), whose translation by Jacob ben Makhir had been lost when the Jews were expelled from France, Samuel ben Solomon ben Nathan ha-Me'ati writes: '[…] [Rather than doing this translation,] it would have been appropriate for me to withdraw, donning awe and the cloak of fear. For who am I to follow the king, the great philosopher, and father of knowledge [*te'udah*, after Isa. 8:16], who counted the heavens in his wisdom, R. Jacob [ben Makhir] ben Tibbon of Montpellier? He had translated this book, but owing to (our) sins, he did not have the leisure to complete it. And when (God) decreed to (punish?) the ancient people and destroyed many in their tranquillity, the leaders of the noble communities of France—may God protect them—, the translation of this book was lost in the exile, with the rest of the booty. God knows that I did not venture (to make this translation) in order to aggrandize myself or to boast. Rather, I had a strong desire and attraction

and worked, shared the notion that redoing a translation was an affront to the first translator and best avoided. Only rarely did translators engage in open conflict, as in the case of Judah Alḥarizi and Samuel Ibn Tibbon (Document 2.1.2.6). The fact that Hebrew translators shared this sense of collegiality illustrates the social cohesion of the movement. Historians who are aware of this norm may (cautiously) draw on it. If a work has two (or more) translations, but none of the translators refers to a predecessor, we may prudently posit that they were not aware of the work of their predecessor(s): it is unlikely that a scholar would deliberately redo a translation without a word of apology or explanation.[21]

I may add parenthetically that I doubt that the Hebrew translators from Latin shared similar models and norms, and thus am not sure that the relatively few Latin–Hebrew translations can also be viewed as a translation movement. But this question must be left open here.

The Hebrew translations of medical works constitute a largely distinct tradition (Section Five, Section Seven). I do not have in mind their linguistic characteristics or vocabulary, which often overlap with the Tibbonid tradition. I rather refer to the sociological factors that triggered medical translations and structured the dynamics of the enterprise, and which differ from those that spurred scientific and philosophical translations. In the Midi, until the end of the fourteenth century, Jewish philosophers and scientists were, by and large, not intellectually connected with their Christian environment; their cultural horizon was the body of knowledge translated into Hebrew from the Greco-Arabic and Judeo-Arabic cultures, enriched by what Jewish scholars wrote in Hebrew. (As already observed, in Italy Jewish–Gentile relations were different, and a number of Jewish scholars knew Latin, studied Latin works, and translated a few into Hebrew.) The situation was very different among Jewish physicians. In the Midi, just as in Italy, they were in close contact with their Christian colleagues and thus kept abreast of new develop-

---

to this book, owing to its many treasures, its powerful statements, and its concision. [...]' For the text, see Samuel ben Solomon ben Nathan, 'Aus Samuel b. Salomo's Vorrede zu ibn Zohr', ed. Steinschneider.

21 An example is Judah Ibn Cardinal (Document 1.1.4.1), who in the preface to his translation of the *Kuzari* does not mention Judah Ibn Tibbon's translation. We may assume that had he known of it, he would have apologized for his retranslation; we may thus (prudently) hypothesize that he was not aware of it (just as Judah Ibn Tibbon was not aware of Ibn Cardinal's version). In this specific case, this is all the more likely because Ibn Cardinal mentions in his preface that he had to retranslate part of the *Kuzari* because his first version was lost: had he been aware that the *Kuzrai* had already been translated by someone else, the context would have called for a remark on this.

ments in Latin medicine. This was an existential necessity if they wished to hold their ground in the competition for patients (see §§ 1.2.2 and 7.2.1); it was often also a legal obligation (where licensing was introduced). Jewish physicians were under constant pressure to raise their level of knowledge to that of their non-Jewish counterparts, and this required the translation of medical works, first from Arabic (in the thirteenth century) and later increasingly from Latin. The prefaces by translators of medical works, usually physicians themselves, show clearly that Jewish doctors' main concern was the need for up-to-date medical knowledge. This is what makes medical translations a separate tradition.

IV. *Translators and Patrons: Independent Decisions and Absence of Economic Dependence*

In this section we will look at a number of texts in which translators describe in detail the circumstances of their work, with special attention given to information about who and how the texts to be translated were chosen. A question of particular interest is whether 'patrons' were involved and, if so, what the social relationship between patron and translator was. As we will see, it is sometimes possible to reconstruct the full chain of events that led to the decision to translate a given text. For example, we can follow in minute detail the discussions between R. Judah Ibn Tibbon and R. Meshullam b. Jacob that induced the former to agree, in 1161, to translate the *Duties of the Heart* (§§ 2.1.1–2.1.3). We can observe in similar detail the process whereby R. Jonathan ha-Kohen of Lunel persuaded R. Samuel Ibn Tibbon to translate Maimonides's *Guide of the Perplexed* (§§ 2.2.1–2.2.3). We will see that the texts and translators were chosen through a decision-making process in which both the future translator and his 'patron' (where there was one) took part.

The term 'patron' should not be misunderstood. Nowhere in these extended discussions is there any reference to financial remuneration for the translator's labour. In all the egodocuments that I have seen where recompense is mentioned, it is always understood as a reward of divine origin. Steinschneider already drew this conclusion in 1893: '[although] translators by training supposedly were working as professionals, [...] no reference to a honorarium has been found so far, [...] whereas it is part of the business of copyists of all kinds of Hebrew manuscripts to acknowledge receipt of their "fee".'[22] Thus, contrary to the situation in other translation movements, in our case 'patrons' offered

22 Steinschneider, *HÜ*, XVI; quoted from Moritz Steinschneider, *The Hebrew Translations of the Middle Ages*, ed. and tr. Manekin et al., I, 53.

translators moral support only, symbolic capital if you will. It should therefore be remembered that the use in this context of the term 'patron' does not denote any formal or economic relationship with the translator. An apparent exception to this rule is Jacob Anatoli, who explicitly refers to a payment he received from Emperor Frederick II (§ 4.1.2). But this exception in fact confirms the rule. Anatoli was engaged by the imperial court and was paid for his work, just like the other translators (into Latin) employed by the court (see further above). But only rarely were the rulers or ecclesiastical authorities, who funded most translations into Latin, interested in translations into Hebrew. With very few exceptions, Hebrew translators did not work for a secular or religious institution and were not paid.

Money does seem to have changed hands in a different context, namely when a translator had to buy a source text or gain access to it. Manuscripts were expensive (they were often used as pawns for loans). Accordingly, they could be rented out and produce revenue for their owner: an interested party would pay handsomely for the opportunity to consult or hold on to a manuscript for a while. Not surprisingly, this practice is attested mainly with regard to medical treatises. We know of only a few such rental contracts, but they are likely to be only the tip of an iceberg.[23] The economics of manuscript rental thus made a significant contribution to the structure of the *Hebraeische Übersetzungen des Mittelalters* corpus.

v. *Towards a History of Medieval Hebrew Translators*

Steinschneider's monumental *Die hebraeischen Übersetzungen des Mittelalters und die Juden als Dolmetscher* is primarily, as the title states, a history of medieval Hebrew *translations*. It concentrates on the genealogy of texts and accords only secondary importance to the *translators* who produced them, although it does contain much material on this topic as well.[24] The time seems

---

23 Shatzmiller, 'Livres médicaux et éducation médicale'; Beit-Arié, *Hebrew Codicology*, 77, n. 121. In the postface to his revision of the translation (by Jacob ben Makhir) of Jābir Ibn Aflaḥ's *Epitome of the Almagest* (Document 6.1.1.5), Samuel ben Judah of Marseilles describes the difficulties he and his learnèd brother faced in obtaining a copy of Ibn Aflaḥ's work. He writes that when they finally did locate a copy, they were given access to it for only two days. Samuel does not explain why this was so, and we must content ourselves with a conjecture. Possibly this restriction may have been economic. I surmise that the two brothers had only two days to inspect the precious Ibn Aflaḥ manuscript, because its owner limited their access to it and/or charged a hefty fee for each day of use.

24 See Freudenthal, 'The Aim and Structure of Steinschneider's *Die hebraeischen Übersetzungen des Mittelalters*'.

ripe to complement Steinschneider's work with a *History of Medieval Hebrew Translators*, and this section hopefully offers a first step in the direction of such a work. I have chosen to organize the material by applying the following criteria: first, the distinction by period, which largely overlaps with that of the social organization of the translation activity; second, the distinction by source language—Arabic and Latin (eventually also vernaculars); and third, the distinction between philosophical/scientific, medical, and literary translations. Implicitly, the distinction by region—the Midi, Christian Spain, and the Italian peninsula—will be borne in mind.

Each historical period is dealt with in a separate section, and within each section, individual translators are numbered consecutively, as is each of that translator's translations. Each translator thus receives a three-digit number (indicating period, source language, and translator number), and each translation a four-digit one. Clearly, not all these theoretical rubrics will be populated, so the numbering is not always sequential—but the lacunae are themselves a significant part of the story. This ideal structure served me as a heuristic when writing the present chapter, although constraints of space and competence did not allow me to try to offer a full account along these lines. It is to be hoped that a fuller treatment will follow in the future.

As a result of these considerations, Parts Two and Three (which present the translators' egodocuments in English and in Hebrew) are structured as follows.

Section One. Harbingers and Pioneers, Eleventh–Twelfth Centuries

Part 1. Arabic–Hebrew Translations (and Knowledge Transfer by Other Kinds of Texts)

1.1.1. Moses Ibn Giqatilla (c. 1020–1080, Saragossa)
1.1.2. Abraham bar Ḥiyya ha-Nasi (c. 1065–c. 1136, Barcelona)
1.1.3. Abraham Ibn Ezra (1089, Tudela–c. 1167, London)
1.1.4. Judah Ibn Cardinal b. Isaac (fl. second half twelfth century?)
1.1.5. Judah Alḥarizi (1165, Toledo–1225, Aleppo)

Part 2. Latin/Vernacular–Hebrew Translations

1.2.1. Berechiah b. Naṭronai ha-Naqdan (fl. 1150–1180, Normandy and the Midi)
1.2.2. Anonymous Translator of Gundissalinus's *Tractatus de anima* (first half of the thirteenth century, Catalonia?)
1.2.3. 'Doeg the Edomite' (fl. 1197–1199, Montpellier?)

Section Two. Founding Fathers

Part 1. Arabic–Hebrew Translations (The Tibbonid Dynasty)
2.1.1. Judah b. Saul Ibn Tibbon (c. 1120, Granada–c. 1190, Lunel)
2.1.2. Samuel b. Judah Ibn Tibbon (c. 1165, Lunel–1232, Marseilles)

Part 2. Latin–Hebrew Translations

Section Three. 'Encyclopedias' of Science and Philosophy
Void

Part 1. 'Encyclopedias' Drawing on Sources in Arabic
3.1.1. Shem-Tov Ibn Falaqera (c. 1225–after 1290, Spain?)

Part 2. 'Encyclopedias' Drawing on Sources in Latin/Vernacular
Void

Section Four. Routine Philosophical and Scientific Translations: The Thirteenth Century

Part 1. Arabic–Hebrew Translations
4.1.1. Moses b. Samuel Ibn Tibbon (fl. 1244–1283, Provence)
4.1.2. Jacob b. Abba b. Samson Anatoli (c. 1194–1256, Midi and Southern Italy)
4.1.3. Jacob b. Makhir Ibn Tibbon (c. 1236–1305, Provence)

Part 2. Latin–Hebrew Translations
Void

Section Five. Routine Medical Translations: The Thirteenth Century

Part 1. Arabic–Hebrew Translations
5.1.1. Nathan b. Eli'ezer ha-Me'ati (fl. 1275–1300, Rome)

Part 2. Latin–Hebrew Translations
5.2.1. Jacob, known as *ha-qaṭan* [= *minimus*] (fl. c. 1280, Provence)

Section Six. Routine Philosophical and Scientific Translations: The Fourteenth Century

Part 1. Arabic–Hebrew Translations

6.1.1. Samuel b. Judah b. Meshullam of Marseilles (1294–after 1343, Marseilles)

Part 2. Latin–Hebrew Translations

Void

Section Seven. Routine Medical Translations: The Fourteenth Century

Part 1. Arabic–Hebrew Translations

Void

Part 2. Latin–Hebrew Translations

7.2.1. Leon Joseph of Carcassonne (1365, Carcassonne–1418, Perpignan)

In what follows, the chosen texts are first presented with short prefaces in English translation (Part Two); the Hebrew originals of these texts follow (Part Three).

Part Two: Egodocuments in English Translation

A Word on the Translations Presented Here

Wherever possible I have drawn on existing translations. Only rarely are they quoted unchanged, however; I have nearly always modified them (including my own), some more, some less. Where an existing translation has been used, its source is indicated; however, to avoid repetition, the mention 'quoted with modification' has been systematically avoided. Where no source is indicated, the translation is mine.

'Mine' with tongue in cheek, however. Lenn Schramm, a collaborator for almost three decades, did wonders in improving my translations and revisions of my predecessors—both in upgrading the English style and very often also in improving the understanding of the Hebrew texts. As always, his involvement in his work went far beyond the normal call of duty.

The translated documents themselves call for an apology, not unlike those offered by the medieval Hebrew translators studied here. The Hebrew egodocuments are often written in part in rhymed prose, much in the style of the *maqāmah*, a literary genre that entered Hebrew letters in the twelfth century, notably through Judah Alḥarizi's translation of al-Ḥarīrī. In addition, long stretches of the texts often consist of fragments of biblical verses that were lifted out of their original context and used to convey a metaphor or embellish the prose. Translators often despair of conveying such texts faithfully and accurately (unless they are gifted with talents like Alḥarizi's!). Like my medieval forebears, I apologize for having often failed to reproduce the beauty of the source texts and humbly assure readers that, assisted by Mr Schramm, I have done my best to convey at least their prosaic, factual content—i.e., the medieval translators' ideas on 'why we translate'.

Section One. Harbingers and Pioneers: The Eleventh and Twelfth Centuries

Part 1. Arabic–Hebrew Translations, and Knowledge Transfer by Other Kinds of Text

1.1.1. Moses Ibn Giqatilla (c. 1020–1080, Saragossa)

Judah b. David Ḥayyūj (Abū Zakariyā' Yaḥyā ibn Dāwūd Ḥayyūǧ, c. 945–1000) is considered the father of scientific Hebrew grammar, which he based on Arabic grammar. He posited that all Hebrew roots consist of three letters and devoted special attention to the problem of the 'weak letters'. About a century later (c. 1080), Moses Ibn Giqatilla, a noted poet, grammarian, and Bible commentator, translated Ḥayyūj's first two grammatical treatises into Hebrew at the request of a dignitary of a Jewish community north of the Pyrenees.

Document 1.1.1.1. Moses Ibn Giqatilla, preface to his translation of *Three Treatises in Hebrew Grammar* by Judah b. David Ḥayyuj.

Ḥayyuj, Judah b. David, in Nutt, *Two Treatises*, Hebrew part, 1–2; tr. Freudenthal, with occasional borrowings from Nutt, *Two Treatises*, English part, 1–3.

This is one of the earliest Arabic–Hebrew translations.[25] Giqatilla was an erudite translator with high professional standards. Throughout his translation he interpolated sentences or short remarks of explanation, added examples, and at times even corrected the translated text; his objective was to produce a Hebrew work that would make Hebrew grammar accessible to readers of Hebrew.[26]

This book was composed by a learned man [R. Judah Ḥayyuj], knowledgeable in the grammar of the Jewish language and erudite in the rules of the Hebrew tongue. His and his father's names will be made known at the end of this text. He was the head of those who speak about and are experts in the nature of the Hebrew language.

But a 'people speaking a foreign tongue' [cf. Ps. 114:1] overpowered us and we were swallowed up among peoples 'of a strange speech and a hard language' [Ezek. 3:5], […] and the Holy Tongue was lost among the profane languages, whose speakers are as plentiful as the sand[27] [after Gen. 22:17], and we 'are left but a few of many' [Jer. 42:2]; the wisdom of our wise men has perished and the understanding of our sages has been concealed [after Isa. 29:14]. No one is left from whom we could learn the nature of the language and no one remains from whom we could learn all its topics. We can only follow what is found in the Holy Scriptures and learn from the prophetic books, but this is a tiny portion of the whole. For the prophets did not aim to investigate the language in full, but only as much as they needed for their prophecies and visions.

For this reason, the authors of grammars of the Holy Tongue and the expounders of the secrets of the Hebrew language had to compose their works in Arabic: for it is current among the victorious nation. It is comprehensible, whereas the Hebrew language is obscure. The Arabic tongue is straightforward and explicit, but the Holy Tongue is cryptic. […]

25 About half a century later, Abraham Ibn Ezra retranslated these works, as well as the treatise on punctuation (1040–1042).

26 Steinschneider, *HÜ*, § 549, 915–917; Olszowy-Schlanger, 'The Science of Language among Medieval Jews'; and Delgado, 'Moshe ben Shemuel ha-Kohen ibn Chiquitilla'.

27 An elegant pun: sand = *ḥol*, but *ḥol* also has the meaning 'secular'.

However, most of the [Jews] of France, who dwell in the territory of our brethren the sons of Esau [after Deut. 2:8] [= Christians],[28] do not know the Arabic language, [whereas] they are very fond of the Holy Tongue and are accustomed to speaking it. [Thus,] God has stirred up the spirit of the man of reason, a 'handsome young man' [after Ezek. 23:6], who delights in the law of the Lord [Ps. 1:2] [...], [namely,] the agreeable R. Isaac son of R. Solomon the Nasi [the Prince]. He wished to understand the topics of the book composed by R. Judah son of R. David from the city of Fez, known as Ḥayyuj, and fathom its secrets.

[Now,] in the entire [Hebrew] language there is nothing more profound and wondrous than the verbs and nouns that contain the letters *alef*, *heh*, *waw*, and *yod*. At times they appear at the beginning [of the verb or noun], at times in the middle, and at times at the end. These letters are known in Arabic as *ḥurūf al-līn* [weak letters]. And also the verbs whose middle and last letters are double—the same as each other—and are called in Arabic *ḏawāt al-maṯalayn* [verbs with double letters].[29] Accordingly, the author [Ḥayyuj] chose to uncover their mysteries and expound their secrets: for when these arcana have been revealed, the other arcana of the Hebrew language also become plain to anyone who understands one thing from another but is careful not to extrapolate from the part to the whole.[30] As it is said: 'Give instruction to the wise man and he will be yet wiser' [Prov. 9:9].

[However,] we did not find all the terms of the Arabic in the Holy Tongue. This may be because we do not possess it [Hebrew] in its totality, or because their language is not the same as ours in all words and does not resemble it in all matters. Therefore, it would be incorrect to [render] word for word. Instead, an Arabic word in the author's book would be the cause and reason for a multitude of words in the Hebrew tongue [i.e., an Arabic word would trigger a phrase rather than a single word], making the matter clear and settled in the reader's heart, just as it is in the first [i.e., source] language.

The aforesaid R. Isaac, may he be remembered for good, asked me, Moses ha-Kohen, son of R. Samuel ha-Kohen from Cordoba, to translate this book into Hebrew for him, and so I did. [...]

---

28 Usually Christianity is identified with Edom, but the Bible identifies Edom with Esau (e.g., Gen. 36:1).

29 The Arabic title of the second treatise translated here is *Kitāb al-Afʿāl ḏawāt al-maṯalayn* (*The Book on the Verbs with Double Letters*).

30 Giqatillah has in mind the famous talmudic definition of the perspicacious man as one 'who understands one thing from another' (B *Ḥagigah* 14a), but who 'is careful not to extrapolate from the part to the totality': the imprint of Aristotelian logic is perceptible.

1.1.2. Abraham bar Ḥiyya ha-Nasi (c. 1065–c. 1136, Barcelona)

Abraham bar Ḥiyya Savasorda was a philosopher, mathematician, and astronomer who also dabbled in astrology. Although Barcelona was under Christian rule, his culture was Arabic and Judeo-Arabic, in addition to traditional Jewish. Bar Ḥiyya wrote—exclusively in Hebrew—on scientific topics that had never been discussed in this language, creating along the way many new Hebrew scientific terms. Unlike his younger contemporary Abraham Ibn Ezra, who favoured linguistic purism (as befits a Hebrew poet and grammarian), bar Ḥiyya tended to coin Arabisms.

In addition to original monographs on religious philosophy (notably on repentance and on the Redemption), Abraham bar Ḥiyya wrote a general survey of all branches of science (an 'encyclopedia') and introductory works on mathematics and astronomy. In all three, bar Ḥiyya presented in Hebrew—usually for the first time—knowledge he had gained from Arabic works. They can therefore be seen as comparable to translations, in that they pursue intercultural transfer of knowledge. In two works, bar Ḥiyya declares that he is responding to requests for instruction addressed to him by his coreligionists north of the Pyrenees (in Tzarfat), who could not read books in Arabic; of the third, he says that he wrote it on his own initiative, having become aware that rabbis of Tzarfat were transgressing halakhah as a result of their ignorance of mathematics.

Document 1.1.2.1. Abraham bar Ḥiyya, author's preface to *Yesod ha-tevunah u-migdal ha-ʾemunah* (*The Foundation of Reason and the Tower of Faith*) (concluding passage).

Millás-Vallicrosa, ed., *La obra enciclopédica*, 10 (Hebrew pagination).

Abraham bar Ḥiyya planned his *Yesod* (sometimes *Yesodei*) *ha-tevunah u-migdal ha-ʾemunah* as a comprehensive and systematic presentation of all branches of knowledge, of the sort often called an 'encyclopedia'. Only a small fraction of the planned work is extant, and it was probably never completed. Bar Ḥiyya seems to have based it on several Arabic sources.

The long preface, which I translate only in part, essentially seeks to legitimize the study of the 'alien' sciences in parallel to traditional Jewish learning. It discusses several topics. The first is epistemology: bar Ḥiyya argues for the superiority of the science that studies the Holy Scriptures over those that investigate 'existents' with the power of reason. This brings him to psychology: he discusses the faculties of the soul involved in the different kinds of cognition. Next bar Ḥiyya comes to the rational soul's rule over the vegetative and appetitive souls, a discussion that leads him to ethics, the discipline that can guide individuals to the World to Come. Lastly, bar Ḥiyya gives an overview of the subjects to be presented in the various parts of his encyclopedia. Only the very last paragraph of the preface, which presents bar Ḥiyya's motives for writing the work, is an egodocument and is translated here.

My masters, who review these statements of mine, should not blame me, perceiving me as someone who has got himself into a confined and narrow place where one should not go, putting myself in danger by going out of my (customary) way. I will divulge my secret to them. It is not out of my own will that I embarked on (this inquiry) and not in order to be honoured and be due gratitude. Rather, many of the great (scholars) of my generation, whose counsel I am obliged to accept, drove me into this. For in the entire land of Tzarfat [= north of the Pyrenees] there was not a single book on these sciences written in the Hebrew tongue. At their behest, I translated (the sciences) from the books of Ishmael into the Holy Tongue, as best I could. As it is written: 'Every man shall give as he is able, according to the blessing of the Lord thy God which he hath given thee' [Deut. 16:17]. Each individual is obligated to teach what he has learned, to the measure of his understanding, one more, the other less, directing his heart to heaven [i.e., doing it out of a religious motivation, not to gain honour].

And I beseech every reader of this composition to judge me leniently—as it is written: 'in righteousness shalt thou judge thy neighbour' [Lev. 19:15]—and kindly rectify any error or mistake that he may find in my words, be it in matters of language or of substance. May God multiply his reward.

Document 1.1.2.2. Abraham bar Ḥiyya, author's preface to *Ṣurat ha-ʾareṣ we-tavnit kaddurei ha-raqiʿa we-seder mahalaḵ koḵavehem* (*The Form of the Earth and the Shape of the Celestial Globes and the Order of the Movements of Their Stars*).

Ṣurat ha-ʾareṣ, ed. Offenbach, 1720, 1a–1b; *Ṣurat ha-ʾareṣ*, ed. Lasri, 37–38.

This is an elementary treatise on astronomy, intended for the educated layman. It is complemented by another, more mathematical work, and by tables. The three compositions are based on Arabic sources, notably al-Farghānī and al-Battānī, but are not straightforward translations. Nearly 70 manuscripts of *Ṣurat ha-ʾareṣ* are extant, attesting to the work's great popularity.

The author's short preface seeks to legitimize the study of astronomy. It essentially argues that the study of the heavens leads one to appreciate the glory of God's creation and His 'wonders'. True, Jewish law requires one to devote his time to the study of the Torah, but the study of heaven, which is divinely created, is equally a route to the 'secrets of faith', bar Ḥiyya reasons. He peppers his text with numerous biblical verses that express the sense of wonder a man feels when contemplating the sky: he thereby makes the study of astronomy into a religious activity, a point Maimonides and his followers would later emphasize explicitly. It is notable that bar Ḥiyya does not seek to ground the legitimacy of astronomy in its practical value, including its role in establishing the Jewish calendar (which is mentioned en passant in a single sentence).

[...] David, king of Israel, said: [...] 'When I consider Thy heavens, the work of Thy fingers, [the moon and the stars, which Thou hast ordained]' [Ps. 8:4]. Here the action of 'seeing' is seeing with the heart, i.e., understanding [*binah*]. He [David] thus indicated that one who observes the fabric of the heaven and the movements of the moon and stars, which the Deity has put in place to exhibit His wonders, will be cognizant of the praise of His great Name far and wide and he will proclaim [*yetanneh*] and interpret it, [namely as pointing to] His splendour, which is spread over the heavens [after Ps. 148:13]. [...]

From this verse it will become clear to you that whoever observes the fabric of heaven and earth and their shape and observes the motions of the stars and the wisdom of their shapes will be aware of the might and splendour of his Creator. Of such men our rabbis of blessed memory said: 'Anyone who knows how to calculate the astronomical seasons and the movement of constellations and does not do so, the verse says about him: 'They do not take notice of the work of God, and they do not see His handiwork' [Isa. 5:12]' [B *Shabbat* 75a]. This implies that whoever does calculate them thereby takes notice of God's labour and sees His handiwork.

You should thus hold it as certain from every side that one who studies the Torah from early on according to his ability, and who attends to it and observes its commandments, and concomitantly inquires into the science of the stars, studies it and attends to it, is not doing something vain; rather he is embracing a science that leads one to the belief in mysteries of faith, and it straightens for him the ways toward true piety, and is also needed for the observance of many commandments.[31]

Throughout my life I have searched and sought for a composition in the Holy Tongue that expounds this science systematically, but found none. I sensed that I was obligated to produce one, to the limits of my abilities, but I was negligent

31 The opening lines seek to legitimize the study of astronomy. Bar Ḥiyya adduces proof texts to show that King David and others approved of this study as a contribution to one's religiosity. The last half-sentence is the only one in which bar Ḥiyya very briefly legitimizes the study of astronomy by the argument that it is required for the observance of certain religious obligations.

and indolent about doing this, because I did not believe it appropriate to rely on my own knowledge and carry out my intention of my own accord. But then I received precious lines and columns[32] from the leader and chief of this generation, the sun and light of science, the man of God and His elect,[33] may he be well for ever, and may his name endure as long as the sun [Ps. 72:17]. And he encouraged me to compose a work on the science of the stars[34] that would explain its subject-matter and explicate properly its methods and reasoning. When I received his pleasant words, and since it is appropriate to fulfil a request from him, at once I endeavoured to follow his wish: this was incumbent upon me owing to my respect and affection for him, as is the case for all his lovers and friends and *a fortiori* for his students who attend on him [*lit.* pour ⟨water⟩ on his hands] [...]

Document 1.1.2.3. Abraham bar Ḥiyya, author's preface to *Ḥibbur ha-Meshiḥah we-ha-tishboret* (*The Book on Measurement and Geometry*).

Chibbur ha-Meschicha we-ha-Tischboreth, ed. Guttmann, 1–5.

In his *Ḥibbur ha-Meshiḥah we-ha-tishboret*, Abraham bar Ḥiyya again pursues the goal of teaching science to the Jews north of the Pyrenees, whose only cultural tongue is Hebrew. This time the science in question is mainly practical geometry, and the motive for teaching it derives from its use for observing certain religious commandments as well as its utility in many crafts. Bar Ḥiyya explains that one who does not know arithmetic necessarily makes mistakes in calculations, treating his fellows unfairly and hence sinning. Similarly, geometry is necessary to fulfil commandments that require the measurement of land. Bar Ḥiyya next explains that arithmetic is quite easy and most people know enough of it, so there is no need for a Hebrew work on the subject. By contrast, geometry is 'hard to understand and its details are beyond most people'.

Bar Ḥiyya therefore resolved to compose a Hebrew treatise on geometry that, he implies, could be read and understood by all Jews. Its purpose is to teach the measurement and division of land 'without any approximation or doubt [...] [so as to] eradicate from those lands [Tzarfat] the error[s] that are prevalent among them'. It is noteworthy that bar Ḥiyya does not refer to the possible soteriological value of the study of geometry (as a means to train the mind to apprehend abstract entities, say), which was an important justification for the study of geometry in post-Maimonidean generations.

32 Possibly: lines = prose text, columns = poetical text.

33 One manuscript adds here a name: R. Abraham b. R. Solomon, otherwise unidentified.

34 In the sequel (not translated here), bar Ḥiyya clearly distinguishes between astronomy and astrology and explains that the expression 'science of the stars' refers to both.

Ḥibbur ha-Meshiḥah we-ha-tishboret is an introduction to practical geometry and provides only practical methods, without proofs. Its four sections are devoted to general notions of the mathematical sciences, the measurement of the area of various figures (triangles, rectangles, circles, etc.), the division of these figures, and the measurements of various heights and depths. Bar Ḥiyya was successful: it is notable that his work was used by the Jews of northern Europe, where the 'foreign sciences' were usually rejected. It could be accepted there precisely thanks to its practical approach and, hence, the absence of any possibly objectionable theological implications. Bar Ḥiyya's work was translated into Latin by Plato of Tivoli under the title *Liber embadorum*.

[...] It is written: 'I am the Lord thy God who teacheth thee to profit, who leadeth thee by the way that thou shouldest go' [Isa. 48:17], [i.e., He] teaches you everything that is profitable to you and leads you on the way that you should go—and this is the way of the Torah. And you infer from this verse that it is appropriate for him to study and engage with every occupation and every science that is profitable to man in his daily life and in [fulfilling] a commandment.

I observed that arithmetic [*lit.* the science of numeration] and [practical] geometry [*lit.* the science of measurement] go this way: they are profitable in many crafts that are used in [fulfilling] the commandments and laws of the Torah. We find that in many places Scripture requires them, as for example in [there follow quotations from Lev. 25:15–16, which show that the issue here is one of ratio and proportion, not just addition and subtraction]. And then it adds: 'You shall not wrong one another, but you shall fear your God' [Lev. 25:17]. Now you cannot perform these calculations correctly so as to avoid doing wrong unless you are well versed in the science of arithmetic. And it is written: 'And he shall reckon with him that bought it from the year that it was sold to him until the year of Jubilee: and the price of his sale shall be [according to the number of years]' [Lev. 25:50]. Now 'he shall reckon with him that bought it' [means] that he should be scrupulous with the buyer so as not to defraud him. You will not be able to perform these calculations correctly, to set the payment for the field and its harvests, or the rent, to be proportional to the number of years remaining and divide this number by that number if you are not well versed in arithmetic and have practised its use.

The Torah also requires the science of geometry in order to measure the land and divide it, and to [mark] the Shabbat boundary[35] and for other commandments. [There follow quotations of Num. 35:5, 26:53–54.] Now whoever is not

35 This is the distance a Jew is permitted to walk beyond the place he resided when Shabbat began.

expert in geometry and trained in it will not measure the land and divide it truly and justly, without fraud, so as to give each party its due part: more to the many and less to the few, as commanded by the Torah; he must first study this science so as to rely on it in his practice. From this you learn that the Torah requires these sciences and that they deserve to be studied and taught.

It is sufficient for [recommending] this science that the Holy One, may He be blessed, Himself boasts [of using it], as it is said: 'He stood, and measured the earth' [Hab. 3:6]; and 'Who hath measured the waters in the hollow of His hand, and meted out heaven with the span?' [Isa. 40:12]. From these verses you see that the Holy One, may He be blessed, created His world by means of measurement and calculation. Now a man should [seek] to resemble his Maker to his utmost power and capacity; this is a man's praise, according to all the sages. And from all these considerations you see the dignity of these sciences.[36] One who occupies himself with them is not occupying himself with something vain, but with something that is needed for worldly matters and for [observance of] the commandments.

As I see it, arithmetic, though needed for worldly matters and useful for men's activities and many commandments, is not hard to comprehend, and most people understand something of it and are accustomed to using it. So, no one needs to compose a book on it in the Holy Tongue. As for geometry, there are as many matters that require it as require arithmetic, both mundane matters and many Torah commandments. But it is hard to understand, and its details are beyond most people. Therefore, one has to set one's mind to it and study it, for there is a very great need for measuring land and dividing it among heirs and partners, to the point that one cannot measure them and divide them fairly and justly if he does not employ this science.

I observed that the majority of the sages of our generation in the land of Tzarfat are not proficient in surveying and are unskilled in parcelling land out, but they strongly disdain this subject, and apportion land among heirs or partners by estimation and guessing and thereby bring on themselves sin and guilt. […] Do not think that they measure and calculate; rather, they estimate and mislead. […]

There is no greater fraud than the fraud occasioned by them. In their mode of calculation, it may happen that he who [in truth] owns a third receives a quarter, and the owner of a quarter takes a third—and there is no greater theft

36 To legitimize the study of mathematics, bar Ḥiyya adduces the theological-metaphysical argument that the world was created according to mathematical principles, so that whoever strives to resemble the Creator needs to master mathematics.

and fraud than this. [Bar Ḥiyya next presents a possible halakhic objection to his argument, which he quickly refutes.]

Having become aware of the great damage and losses that are occasioned by them [our contemporary rabbis], and although I am well aware that I am not worthy of it and not qualified for it, I took upon myself the task of writing a work about surveying and parcelling land, one that explains this entire subject and teaches the method of measuring and dividing. [This I will do] adducing justifications and proofs derived from geometry and arithmetic, without any approximation or doubt, to the best of my ability. Perhaps someone will learn them and eradicate from those lands [Tzarfat] the error[s] that are prevalent among them. [...]

1.1.3. Abraham Ibn Ezra (1089, Tudela–c. 1167, London)

Abraham Ibn Ezra was a bicultural polymath, raised in the Arabic and Hebrew cultures of al-Andalus. During the five decades he lived there, he wrote mostly or exclusively poetical works—both profane and liturgical. After he relocated to Italy in 1140, he composed numerous Hebrew treatises in various disciplines: philosophy, biblical exegesis, Hebrew grammar, mathematics, astronomy, and astrology.[37] His name is associated with the introduction of the 'Arabic numerals' and the decimal system to Latin Europe. Ibn Ezra's writings, while highly original and personal, drew on wide knowledge acquired from Arabic texts, which they communicate to Hebrew readers.

Document 1.1.3.1. Abraham Ibn Ezra, preface to his Hebrew version of Ibn al-Muṯannā's commentary on the astronomical tables of al-Ḫwārizmī.

Goldstein, ed. and tr., *Ibn al-Muthannâ's Commentary on the Astronomical Tables of al-Khwârizmî*, 147–150 (with added annotations); Milan, Biblioteca Ambrosiana, MS X 138 Sup.

Ibn Ezra rarely translated entire works from Arabic into Hebrew. An exception is Ibn al-Muṯannā's *Commentary on the Astronomical Tables of al-Ḫwārizmī*.[38] In the long preface to this translation, dated 1160, he describes how Indian literary lore (*Kalilah wa-Dimnah*) and Indian mathematics (the decimal system) were introduced into Arabophone culture. En passant, the preface makes a few remarks that (implicitly) legitimize

37 For a list of Ibn Ezra's works and bibliography, see Sela and Freudenthal, 'Abraham Ibn Ezra's Scholarly Writings'.

38 Another exception is his translation of Judah ben David Ḥayyuj's three grammatical treatises.

the study of this 'alien wisdom' by Jews, but this issue was certainly not his main concern there. His account of the transfer of Indian lore to Arabic culture differs from what we hear from other sources, but we need not go into this problem here.[39] It is worth mentioning that Ibn Ezra's account is not part of the legendary tradition of the Indian origin of knowledge.[40]

Ibn Ezra's purpose in describing the history of the transfer of knowledge from India to the West is not clear. He seems to have had a special interest in Indian thought and practices, in particular astrology, to which he occasionally refers in his biblical commentaries. In his scientific works he refers to the 'sages of India' as valid sources of astrological knowledge and political counsel and mentions *Kalilah wa-Dimnah* among the works that were made accessible in Arabic (see below). It was translated into Hebrew some fifty years later by Jacob b. Eleazar; this suggests that some Arabophone Jewish circles in al-Andalus were interested in 'wisdom' of Indian origin.[41] Nor is it clear who commissioned Ibn Ezra's translation of al-Muṯannā or whom he expected to be interested in it: in his day all the Jewish astronomers familiar with advanced mathematical astronomy were, like himself, fluent in Arabic, and they had no need for a Hebrew version of the work.

Here I reproduce only the end of the long preface, where Ibn Ezra legitimates the (non-Jewish) tradition of mathematical astronomy and explains why he chose to translate Ibn al-Muṯannā's work.

Abraham Ibn Ezra the Spaniard said:

[…] After him arose [another] great scholar who translated the venerable book about the stars [*lit.* constellations] written by Ptolemy, the king of Egypt. He was a Greek who lived a thousand years ago.[42] This book is perfect. There is none superior to it in the science of the spheres, their secrets, their motions [as seen] from the earth, and the magnitudes of the upper [fixed] stars, which are in the sphere of the zodiac. […]

Many Ishmaelite scholars have composed commentaries on this book, among them a scholar more eminent than the others in the sciences of geometry and astronomy, whose name was Muḥammad b. Muṯannā. For one of his relatives, he composed a most venerable book on the rules of planetary

39 For a full discussion, see Goldstein, 'Astronomy and Astrology in the Works of Abraham Ibn Ezra', 11–14. Steinschneider studied Ibn Ezra's preface in great detail in his 'Uebersetzungen aus dem Indischen in's Arabische'.

40 For a recent overview and bibliography, see Yoeli-Tlalim, *'Book of Asaf'*, 14–17.

41 Marks, 'Hindus and Hinduism in Medieval Jewish Literature'.

42 The misidentification of Claudius Ptolemy the astronomer (100–170), author of the *Almagest*, with the kings of the Ptolemaic dynasty was common.

motions, which explains the reasons underlying the tables of al-Ḫwārizmī.[43] He [also] included short proofs and diagrams whose principles are taken from the *Almagest*. In some passages, he elaborated on the explanation given by King Ptolemy. Occasionally he even added geometrical proofs of his own. [...] He arranged his book in the form of questions and answers for greater clarity.

1.1.4. Judah Ibn Cardinal b. Isaac (fl. second half twelfth century?)

Nothing definite is known about Judah Ibn Cardinal (or Cardenal, or Qardanel), save that at some point in time he translated Judah Hallevi's *Kuzari* into Hebrew. Only fragments of his translation (including the preface) are preserved. Some scholars conjecture that Cardinal produced his translation before Judah Ibn Tibbon published his own (in 1167, according to scholarly opinion); others think he worked later, at the beginning of the thirteenth century.

With regard to our present concern, the following observation can be made. Beginning with Judah Ibn Tibbon, translators' prefaces gradually took shape as a distinct literary genre, in which certain *topoi* appear regularly: the respective merits of the Arabic and the Hebrew languages; the translator's insufficient competence; the Hebrew origin of the secular sciences; etc. None of these themes is broached in Ibn Cardinal's (admittedly short) preface. Moreover, when Ibn Cardinal briefly refers to his method of translating (*ad verbum* or *ad sensum*), he seems unaware of the early and thorough discussions of the issue. Leaving aside other considerations, it seems very unlikely that a scholar living at the turn of the twelfth century or later (that is, a few decades after the first translators' prefaces were published), in a region in which Judeo-Arabic writings circulated (the Midi or Christian Spain), would have been entirely unaware of these discussions. His apparent ignorance of Tibbonian terminology points in the same direction. I thus concur with Sonne (among others) that Ibn Cardinal's translation predates (or is roughly contemporary with) Judah Ibn Tibbon's, and that the wide circulation of the latter led to the disappearance of Ibn Cardinal's (just as Samuel Ibn Tibbon's translation of the *Guide* overshadowed Judah Alḥarizi's). This is why Ibn Cardinal's preface is included in the present section.[44]

---

43 This is the work Ibn Ezra translates here.

44 Steinschneider, *HÜ*, § 235, 404–405; Sonne, 'Ein Beitrag zu der verschollenen Kuzariübersetzung des Jehuda ben Cardinal'; and Baneth, 'Different Translation Methods in the Hebrew Versions of the *Kuzari*' [Hebrew].

Document 1.1.4.1. Judah Ibn Cardinal b. Isaac, preface to his translation of Judah Hallevi's *Kuzari*.

The Kuzari, ed. Brecher, II, 4b (Hebrew); Parma, Biblioteca Palatina, Cod. Parm. 1936 (= De Rossi 625), f.303a.

The short preface consists of two parts. First, Ibn Cardinal describes the purpose of the *Kuzari* and gives a brief and unsophisticated characterization of his translation method. Second, in a few lines that seem to have been added later, he recounts that a certain Joseph b. Baruch (apparently the person who had commissioned the translation in the first place) took the translation of the two first sections with him to England; there, the manuscript was confiscated—apparently Joseph was arrested and held in custody for a while—and never returned to Ibn Cardinal, who had to retranslate the two sections again.

Judah b. R. Isaac b. Cardinal of blessed memory [said]:

I translated the book *al-Kuzari* by R. Judah Hallevi, may his soul be bound up in the bundle of life, from Arabic into Hebrew. R. Judah's intention was to teach [*or* point out] the praise of the Torah and to answer the philosophers, the theologians of the nations, and the Karaites.

When translating it, I will not alter the gist of its contentions and I will not modify its contents. And if I add to what is said or subtract from it—as far as this is within my capability and intellect—I will not corrupt the contents. I ask God for assistance.

A while after having begun to translate it, following the directive of the erudite [*var.* Nasi] Joseph b. Baruch, may the Lord protect him, [the latter] took its first two sections with him to England. There, those who captured [*or* imprisoned] him kept them [the two sections], and he could not return them. [Thus,] I was obliged to translate them anew, so that we may have a full [translation], with God's help. […]

1.1.5. Judah Alḥarizi (1165, Toledo–1225, Aleppo)

Judah Alḥarizi (al-Ḥarīzī) was a Hebrew poet and translator, one of the central figures of the Hebrew Golden Age in medieval Spain. He was born in Toledo where, though Castile was under Christian rule, Arabic was still alive, and acquired a mastery of both Arabic and Hebrew. In c. 1195, he moved north, to the Midi, where he resided for a number of years. Alongside his original literary composition of poetry and *maqāmāt*, he devoted considerable time and talent to Arabic–Hebrew translations, both literary and philosophical. He journeyed from one patron to the next, dedicating works to his benefactors, on whom he heaped praise, while pouring scorn and mockery on those who refused to sponsor him. This was an established social pattern in Spain, and Alḥarizi practiced it artfully: he went so far as to dedicate one work to multiple patrons, each of course unaware of the others.[45]

The pinnacle of his achievements as a translator was the artful Hebrew translation of the celebrated *Maqāmāt al-Ḥarīrī*, executed at the request of several high-ranking and wealthy intellectuals in Christian Spain. This translation was a sort of trigger for the composition a few years later of his greatest original Hebrew work, the *Taḥkemoni*.

Alḥarizi's scientific and philosophical translations do not seem to follow any agenda, other than to satisfy commissions and please patrons. He translated sections of Maimonides's commentary on the Mishnah at the behest of the influential leader and scholar R. Jonathan ha-Kohen of Lunel (c. 1135–c. 1211).[46] Subsequently, Alḥarizi translated the *Guide of the Perplexed* (c. 1213), an unsuccessful alternative to Samuel Ibn Tibbon's rigorous but difficult translation, which soon became canonical.[47] Alḥarizi also prepared a Hebrew version of Maimonides's *Treatise on Resurrection*. Only some of his many translations have prefaces.

Sometime around 1216, Alḥarizi decided to travel to the east—visiting Fustat, Jerusalem, Baghdad, Mosul, Damascus, and Aleppo (where he died). This long journey supplied much of the inspiration and material for his literary masterpiece *Taḥkemoni*, which sheds much light on Alḥarizi's ideas about translation and on the social context of Arabic–Hebrew translation in the early thirteenth century. Alḥarizi explains that he now regretted having translated al-Ḥarīrī into Hebrew: he had come to believe that Arabic–Hebrew translations present Hebrew, the Holy Tongue, as subservient and inferior to Arabic, a view he now wished to combat. This is why he resolved to compose an original Hebrew work of *maqāmāt*, modelled after al-Ḥarīrī's, which would (he writes) salvage the honour of the Hebrew language.

45 For an up-to-date biographical account in English of Alḥarizi's life and œuvre, see Blau et al., Judah Alḥarizi, *Kitāb al-Durar*, English introduction, 9*–51*. See also Sadan, 'Un intellectuel juif'. Judah Alḥarizi's period of activity parallels that of Samuel Ibn Tibbon, but socially he is a 'free atom', closer to the 'harbingers' than the Tibbonids were; this is why he has been placed in Section One.

46 Steinschneider, *HÜ*, § 554, 923; and Schirmann, *The History of Hebrew Poetry*, 147.

47 Only a single MS of Alḥarizi's translation, which was the exemplar for the anonymous Latin translation of the *Guide* made in the 1230s, is extant.

Document 1.1.5.1. Judah Alḥarizi, preface to his Hebrew translation of Maimonides's *Guide of the Perplexed.*

Maimonides, *Moreh ha-nevukhim*, tr. Judah Alḥarizi, ed. Schlossberg, 2.

We hear from Judah Alḥarizi that he wrote his translation of the *Guide of the Perplexed* (between 1205 and 1214) at the request of 'several of the noblemen and scholars of Provence' who were baffled by the Hebrew style used by Samuel Ibn Tibbon in his scholarly translation (first published in 1204) and wished to have a more accessible Hebrew version of the venerated book. Alḥarizi's translation is indeed written in elegant and readable Hebrew, bought however at the price of philosophical imprecision.

Judah b. Solomon Ḥarizi said:[48]

The authority [*lit.* hand] of several of my masters, the dignitaries and scholars of Provence, descended upon me that I should translate for them the book *Guide of the Perplexed*, in simple and clear words, comprehensible to the intelligent person [after Prov. 8:9], guiding those who err along a straight path, so that 'the crooked shall be made straight, and the rough places made plain' [Isa. 40:4]. For, [they stated,] all its ideas are hidden and its secrets are locked up and sealed.[49]

I [first] declined to accede to their request because [the *Guide*] is a vast ocean, whose crossings are deep; and who knows its [the sea's] boundaries [cf. Num. 34:4f.; 1Chron 5:16]? Moreover, a gifted and educated man translated it before me, although he intended for his words to conceal (its meaning) and make it even deeper. But when I pleaded with them to relieve me of that burden, they did not accede to my request. I rowed hard to return to dry land, but could not do so [after Jon. 1:13].

When they pressed me [...] and compelled me to translate it and to pour it from one vessel to another, I applied all my might to bring its benefit closer, as far as my force and my ability reached. I put its matters in order, and I erected its building on a level foundation. Nonetheless, I did not aim to divulge any of the book's secrets, or to interpret any of its themes. Rather, I wished only to put it in order, as does a person who clears stones from the road.[50]

48 The preface is preceded by a poem in praise of Maimonides, which is not translated here.

49 In these early years of the thirteenth century, scholars in Provence still lacked a philosophical education, so Samuel Ibn Tibbon's demanding translation of the *Guide of the Perplexed* was 'locked up and sealed' for them.

50 Alḥarizi defends himself against the possible charge (soon to be levelled by Samuel Ibn Tibbon) that his translation made the *Guide* too easy to access, in violation of Maimonides's explicit injunction.

In making this translation I have [indeed] seen fit to add two chapters [*or* 'gates'], through which one may access the [book's] innermost chambers.

The first [new] chapter glosses every unfamiliar word that I used in this book; I arranged them alphabetically so that the reader can peruse it easily.

The second [new] chapter enumerates the parts and the chapters in this book: for each chapter I indicate its contents and its intention.

This is so that one can find, in the first [new] chapter the meaning of any word he does not know; and in the second [new] chapter he will find the book's purpose and the contents of its chapters and parts. Thus, everything is arranged in its appropriate place, 'every man by his own camp, and every man by his own standard' [Num. 1:52].

Part 2. Latin/Vernacular–Hebrew Translations

1.2.1. Berechiah b. Naṭronai ha-Naqdan (fl. 1150–1180, Normandy and the Midi)

Starting in the eleventh century, Jewish culture in northern France and the Rhineland (the 'Ashkenaz' of Jewish historiography) was unified by the primary role accorded to the study of the Babylonian Talmud and other traditional texts. 'Alien' bodies of knowledge—namely, those that do not derive their authority and legitimacy from the Jewish traditional texts and ultimately from revelation—were usually treated with indifference or rejected outright. Even the Hebrew philosophical culture that emerged in the Iberian Peninsula and the Midi never took root in Ashkenaz.[51]

Berechiah b. Naṭronai ha-Naqdan (Berechiah the Punctuator) is associated primarily with northern France (Rouen). Remarkably and exceptionally, he was familiar with a number of non-Jewish texts of Anglo-Norman culture. A few traces of Arabic–Hebrew translations, especially by Judah Ibn Tibbon, are also discernible in his writings, allowing the conclusion that around 1160 he lived for some time in Lunel in southern France, the capital of Jewish translation activity during the second half of the twelfth century.[52] In addition to composing a number of original works (animal fables or 'fox tales'), he translated or (mostly) paraphrased in Hebrew two works on natural science: (i) Marbode's *Lapidary*, which he read in one of its numerous reworkings in Old French; and (ii) Adelard of Bath's *Questiones naturales*, which he read in Latin or Old French. Berechiah's brief preface to *Dodi we-nekhdi* (*Uncle and Nephew*), his Hebrew version of *Questiones naturales*, is his only extant egodocument.

51 See Schacter, *Judaism's Encounter with Other Cultures*; and Freudenthal, *Science and Philosophy in Ashkenazi Culture*.

52 See Visi, 'Berechiah ben Naṭronai ha-Naqdan's *Dodi ve-Neḵdi*'.

Document 1.2.1.1. Berechiah b. Naṭronai ha-Naqdan, preface to *Dodi ve-nekhdi*, the Hebrew adaptation of Adelard of Bath's *Quaestiones naturales* (c. 1150).

Visi, 'Berechiah b. Naṭronai ha-Naqdan's *Dodi ve-Neḵdi*', 45–46.

I will now begin the book of the questions posed by a nephew to his uncle.

I, Berechiah, son of R. Naṭronai, was pressed by my thoughts [and said to myself]: 'Let me gird my loins and translate into the Hebrew tongue the following discussions' [after Jer. 1:17; Prov. 31:17]. For I found them in a text [composed] by non-Jews, who had translated [*or* copied] them from the Hebrews [*or* from the Arabic]. They include excellent matters unknown to the scholars of our age, magnificent sciences, [destined] to the learnèd, unsuitable for [*lit.* never seen by] the eye of the unlearnèd. When I saw such magnificent science in an ugly vessel, pearls before swine,[53] I purified it from the touch [*lit.* hand] of the Gentile, and wrote them in the Holy Tongue, which is superior to any tongue.

This book is a [dialogue] between two scholars of the generation, the Uncle and the Nephew, who refine and clarify their statements. As it is said: '[Iron sharpens iron, so also] one man sharpens another' [Prov. 27:17]. The learnèd [men] will comprehend.

1.2.2. Anonymous Translator of Gundissalinus's *Tractatus de anima* (first half of the thirteenth century, Catalonia?)

Document 1.2.2.1. An anonymous translator's interpolation in the Latin–Hebrew translation of Gundissalinus's *Tractatus de anima* (first half of the thirteenth century)

Schwartz, 'Dominicus Gundissalinus: *Sefer ha-Nefesh* (*Tractatus de anima*)', 228.11–14; tr., Schwartz, 'The Medieval Hebrew Translations of Dominicus Gundissalinus', 22–23.

The *Tractatus de anima* by Dominicus Gundissalinus (c. 1110–c. 1190) was translated into Hebrew by an anonymous scholar as *Sefer ha-Nefesh*. Although the translation survives only in a single Cambridge manuscript, it attracted a wide readership, because large parts of it were incorporated in the late thirteenth-century *Sha'ar ha-shamayim* (*Gate of Heaven*), the popular encyclopedia by Gershom b. Solomon of Arles, where, however,

53 This implicit quotation from Matt. 7:6 was known to and used by several Jewish medieval writers.

the text is attributed to Aristotle. The date of the translation cannot be determined with certainty, but terminological considerations led its editor, Yossef Schwartz, to suggest that it was written in the first half of the thirteenth century by a Jew whose vernacular was Catalan.

Sefer ha-Nefesh has no translator's preface. But soon after the beginning, the translator inserted an interesting personal statement of a few lines. He tells us that he started to translate the Latin text of Aristotle's *De anima* (he must mean the *translatio vetus* by James of Venice), but soon despaired of the task ('it was hard and difficult') and switched to translating Gundissalinus's *Tractatus de anima* instead. His short statement follows.

> Until here I have translated the text of Aristotle's book itself, and it was hard and difficult. [So] I gave up translating it and took [another work], one written by a wise man from Toledo who explicated the meaning of the wise man's [= Aristotle's] book. It begins with the words 'All humans'—this is a Proemium to the book—and it extends up to the words 'It is proper', which are the beginning of the book itself.[54] And in the Proemium he said that the examination of the soul consists of ten investigations, which I will list.[55]

1.2.3. 'Doeg the Edomite' (fl. 1197–1199, Montpellier?)

In 1197–1199, an anonymous Jewish physician, who referred to himself as 'Doeg the Edomite', translated a set of no fewer than twenty-four medical works from Latin into Hebrew.[56] This is a most exceptional cultural phenomenon. The translation of non-halakhic works into Hebrew had begun fifty or sixty years earlier, but until the closing years of the twelfth century all translated works were distinctly Jewish (theology, Hebrew linguistics, etc). 'Doeg's' translations were the first of secular works, of non-Jewish authors, and from Latin.

All we know about 'Doeg' comes from the preface to his collection of translations and from a few short poems he affixed to some of them. As already recognized by Steinschneider, the translator had converted to Christianity but later repented of this step.[57] As a Christian, he held a high social position and enjoyed the esteem of the powerful and the noble, yet he came to feel that his success rested on falsehood and treason. He

54 In the Hebrew translation the two phrases are at 228:15 and 229:14 respectively.

55 The list follows at 229:1–11.

56 The following introduction draws on Freudenthal, McVaugh, and Mesler, 'Twelfth-Century Latin Medicine in Hebrew Garb'. The name 'Doeg the Edomite' derives from 1 Sam. 21:8, 22:9, 18; Ps. 52; and the talmudic tradition.

57 Steinschneider, *HÜ*, 712–713.

refers to himself by the epithet *aḥer* (a pejorative term denoting a Jewish heretic) and perceives his new clothes, in which the Christians respected him, as the 'clothes of his betrayal'. He calls himself 'Doeg the Edomite', a highly evocative pseudonym: Doeg the Edomite, as we know him from the Bible and from talmudic sources, was erudite in the Torah, but also a foreigner and a villain—an informer and killer. The pseudonym thus reflects a rather gloomy self-image of profound remorse and repentance. The presence of numerous Occitan terms in Doeg's translations establishes that he lived in the Midi; the exceptional richness of his bookshelf suggests that he was associated with a university, presumably that of Montpellier.[58]

Document 1.2.3.1. 'Doeg the Edomite', preface to his Latin–Hebrew translation of a collection of twenty-four medical books.

Freudenthal, McVaugh, and Mesler, 'Twelfth-Century Latin Medicine in Hebrew Garb'.

While still wallowing in his schemes, [and] in the height of his deceptions, [the writer[59]] said to himself: I observed the holy seed [i.e., the Jews] early in the morning crowding at the doors of the Gentile sages [to consult] about their illnesses, and the illnesses of their sons and daughters. They take from them any stuff, without any understanding, without knowing what it is. They do not distinguish between the impure and the pure; in most of their concoctions they mix sorts of resin with wine, and introduce impure meat, suet, and blood. Not one of them escapes [the category] of prohibited or abominable.[60]

I therefore undertook to translate into the Holy Tongue all the books of medical theory [*ḥokmah*] that are in their [the Christians'] possession and that I could get, as well as the best books of [medical] practice [*melakah*]. These are the names of the books whose translation I completed in the year(s) 17 and 18 and 19 of the 231st cycle [of the moon] [= 1197–1199].

[A list of twenty-four medical books follows. Sometimes Doeg transliterates the Latin/vernacular title, and sometimes gives the book a Hebrew title.[61]]

58 For Moses Ibn Tibbon's balanced critique of one of Doeg's translations, see below, Document 4.1.1.3.

59 The text opens with a short introductory poem and a short passage written in rhymed prose (not translated here) in which Doeg refers to himself in the third person and describes himself as a renegade to his people, one who repents having held a high and honoured position in Christian society.

60 These ingredients—wine, impure meat, suet, and blood—are not permissible according to the Jewish dietary regulations.

61 The books are identified in Freudenthal et al., 'Twelfth-Century Latin Medicine in Hebrew Garb'.

Thus, the number of the works on [medical] practice that I translated is seventeen; as well as seven books of [medical] theory. [Altogether,] they are twenty-four.

Now,[62] I prostrate myself before all the sages of the people, so they may acquit me and find in my favour in this matter, for my intention was good. [Indeed,] I relied on the words of our sages, who said: 'One must teach his son Torah and teach him a craft [*or* skill, profession: *ʾummanut*].'[63] There is no craft in the world that is cleaner than this one [medicine].[64] [...]

Section Two. Founding Fathers

Part 1. Arabic–Hebrew Translations (The Tibbonid Dynasty)

2.1.1. Judah b. Saul Ibn Tibbon (c. 1120, Granada–c. 1190, Lunel)

Born in Granada, Judah Ibn Tibbon left Spain c. 1150, in the wake of the Almohad persecutions, and settled in Lunel in southern France. A physician by profession, he initially had no intention of producing translations. However, an exchange with the charismatic and highly influential R. Meshullam b. Jacob (d. 1170), reconstructed in detail below, persuaded him otherwise. Judah thus became the first scholar to engage in systematic translation from (Judeo-)Arabic into Hebrew.

Judah Ibn Tibbon translated the most important treatises of Judeo-Arabic thought of his time: ibn Paqūda's *Duties of the Heart* (or *of the Hearts*; 1161, first chapter; after 1180, chapters 2–10); Solomon Ibn Gabirol's *Improvement of the Moral Qualities* and *Choice of Pearls* (a collection of apothegms; c. 1161 and 1167); Jonah Ibn Janāḥ's Grammar and Lexicon (c. 1171); Judah Hallevi's *Kuzari* (1171?); and Saadia Gaon's *Book of Beliefs and Opinions* (1186).

---

62 This opens the second part of Doeg's preface. The introduction of 'alien wisdom' into the Jewish cultural system always met with opposition. Doeg was writing just before Maimonides's writings reached the Midi, and thus before scholars became aware of the halakhic legitimization of profane studies contained in them. When Jewish religious authorities took Doeg to task for introducing 'profane sciences' into Judaism, he defended himself with arguments showing that he was well versed in halakhic dialectics.

63 Paraphrasing B *Qiddushin* 29a. The Gemara adds: 'And some say, [the father must teach him] to swim, too', a point mentioned by Doeg in the sequel, showing that he had precisely this text in mind. The obligation to teach one's son to swim is mentioned also in T *Qiddushin* 1:11 (Lieberman).

64 Doeg here implicitly refers to M *Qiddushin* 4:14: 'Rabbi Meir says: one must always teach his son a clean craft [*ʾummanut neqiyyah*]'.

Judah created a distinctive style of translation that became a model for most later translators. This style is characterized above all by faithfulness to the *Vorlage*, which ensures great rigour in the transmission of ideas. He was often compelled to coin terms for notions that had no names in Hebrew; this, too, he did under the strong influence of Arabic. Many members of the Ibn Tibbon family became translators as well, and one can justly speak of a 'Tibbonid dynasty of translators' who followed roughly the same translation methods. Owing to Judah's momentous role in launching the Hebrew translation movement, his son Samuel referred to him as 'the father of the translators', a fitting designation that has been widely accepted.

The relatively large number of egodocuments by Judah that have come down to us allow a fairly detailed reconstruction of his career as a translator under the benevolent patronage of R. Meshullam b. Jacob, the head of the yeshivah in Lunel. The documents connected with Judah's first translation, in particular that of *Duties of the Heart*, offer us a rare and precious insight into how the decision to translate a specific work was taken as a result of protracted exchanges between translator and patron. In the preface to his translation of Ibn Janāḥ's Grammar, Judah discusses his translation method in detail. Last but certainly not least is the 'Ethical Will' that Judah wrote for his son Samuel. When he wrote it, Samuel, initially a physician like his father, had not yet become a translator, so the document does not broach matters specifically related to that activity; however, Judah advises Samuel about matters related to books, the study of Arabic, calligraphy, and other skills a bilingual scholar should possess. Considerations of space kept this text from being included here.[65]

Document 2.1.1.1. Judah Ibn Tibbon, first preface to his translation of *Duties of the Heart* (to the first chapter, on the unity of God, 1161).

Ibn Paqūda, *Sefer Ḥovot ha-levavot*, tr. Judah Ibn Tibbon, ed. Zifroni, 1–5.

Baḥyā b. Joseph ibn Paqūda, who lived in Zaragoza, wrote his ethical-moral work with Sufi tendencies, the *Book of Direction to the Duties of the Heart*, around 1040. Judah Ibn Tibbon translated it into Hebrew in two instalments. In 1161, he presented the work's first chapter orally (in Hebrew) to R. Meshullam b. Jacob, whereupon the latter asked him to translate it for him. As Judah informs us, this is how he became a translator, despite his determination not to pursue this occupation. However, at the behest of his close friend R. Asher, one of R. Meshullam's sons, Judah abandoned the translation of the *Duties of the Heart* after completing only the first (and most philosophical) part and took up Solomon Ibn Gabirol's *Improvement of the Moral Qualities* instead. He did not

65 See Judah Ibn Tibbon, 'Ethical Will'.

translate the remaining chapters, 2–10, until about two decades later, at the request of R. Abraham b. David.

Judah Ibn Tibbon wrote separate prefaces for the translation of the first chapter and for that of the remaining nine. In addition, he discussed this translation and its history in a letter to R. Asher b. Meshullam. Together, these three texts permit a detailed reconstruction of the motives behind the translation and the interactions between the three persons involved: Judah Ibn Tibbon the translator, and his two (competing) patrons, father and son.

Judah begins his first preface by discussing why the Jews, although the chosen people, are mostly lacking the philosophical sciences. He explains that the little philosophy contemporary Jews possess they access in Arabic, the language of the Islamic culture in which they live. He notes the 'narrowness' of the Hebrew language as compared with Arabic—a well known *topos* of the time. Commenting apologetically on the dearth of philosophical studies among the Jews under the Cross—they have devoted themselves exclusively to the study of the Talmud, he explains—Judah takes notice of the recent epoch-making cultural change brought about by R. Meshullam b. Jacob of Lunel, a man who excels in both piety and wisdom. Judah refers to books R. Meshullam had obtained and had had translated in the past, but we have no independent information on this. Judah goes on to recount how R. Meshullam little by little got him involved in the translation of Baḥyā Ibn Paqūda's *Duties of the Heart*, although he (as a physician) had intended to steer clear of translation and had refused similar requests in the past. Judah also allows us to understand why he had shunned translation: 'At the present time, most people are prone to abuse and lash out at anyone among their contemporaries who innovates anything, be it in a translation, in a composition, in liturgical poetry, or in any similar matter. The sagacious person will therefore avoid entering this path and similar ones, and will [thereby] escape being the object of people's tongues and their vilifications.'

Judah next reflects on the method of good translation and on the qualities required of a good translator. These are old *topoi* in Syriac and Arabic literature, but this preface seems to be the first time they were stated in Hebrew.

Said the translator:

[....] After our sins—that we deviated from the path of the Torah and of wisdom—caused the Divine Presence to depart from our midst and prophecy and the Holy Spirit to cease among us, nonetheless our God's mercy was not exhausted and He left us a remnant. In each generation He left us in His vineyard some gleanings and two or three berries [after Isa. 17:6], so as to enlighten the eyes of the remnant with the light of wisdom: to give strength to the weary and force to the weak [after Isa. 40:29]. They [the sages of the Mishnah and the Talmud] investigated the secrets of the sciences and illuminated mysteries.

They clarified the intentions of the Torah and the Mishnah and the Talmud; they laid foundations and erected pillars to strengthen the edifice of our divine Torah, upon which the House of Israel is firmly emplaced. In matters of appropriate behaviour and of morals, and the improvements of moral qualities, too, they composed treatises [so precious that] their value cannot be measured by gold and crystal [Job 28:17]. […]

After them [came] the Geonim, most of whom lived in exile of the kingdom of Ishmael, in Babylonia, in Palestine, and in Persia. They spoke Arabic; and all Jewish communities residing in these countries spoke that language. [Consequently,] most of their exegetical writings—on the Bible, on the Mishnah, and on the Talmud—were in Arabic. They did the same in the majority of their works, as also in the response to [halakhic] queries addressed to them, too. They did this because everyone understood that language. [In addition,] Arabic is a vast language, abounding in all domains, whatever the needs of each speaker or writer. Its rhetoric is straightforward and clear and captures the gist of each subject more than is possible in Hebrew. This is so because what we have today of the Hebrew language is only what we find in the biblical books, and this is insufficient to meet all of a speaker's needs. Moreover, [they used Arabic because] their intention was to benefit the common folk, who are not versed in the Holy Tongue. For this reason, most of their compositions were in Arabic, whatever the topic—whether Torah scholarship or other disciplines.

A remnant of our people resides in the lands of Edom [Christendom], too. From ancient times, there were great scholars in Torah and Talmud scholarship among them. However, they did not engage in other disciplines, because the Torah was their [sole] occupation and also because they did not have books in the other sciences.

[This situation continued] until was set among them the pure candelabrum, the lamp of the precepts and of the Torah [after Prov. 6:23], the great, pious, and holy rabbi, our Master Meshullam, may the Merciful One guard and deliver him, son of the elderly scholar, R. Jacob, of blessed memory. The oil of his understanding is a pure pressing for kindling the perpetual lamp of wisdom. His soul was drawn to the Torah of his God and to the fear of God. He made learning his cup and his portion and longed for the books of the sciences composed by the Geonim. He did his utmost to gather, spread, and [cause to be] copied [*he'etiq*][66] [works], whether in the discipline of the Torah, the science of language [grammar], the knowledge of [true] belief [theology], the books of

66 The verb *'a.t.q.* is often used in the sense of 'translate'. But in this text, Judah Ibn Tibbon mostly conveys the meaning of 'translate' with the (talmudic) verb *t.r.g.m.*, so *he'etiq* here has the meaning of 'copied'. (Nonetheless, it is used once below in the sense of 'translate'.)

rhetoric, and treatises on ethics and the parables of the sages. His hand became like a nest for all their treasures. Through his potent intellect and his power, he understood one thing from another and inferred one matter from another. He also produced works [*lit.* bore children] in the [domains of the] sciences and of ethics. He opened new portals on the paths to the fear [of God] and to ethics, illuminating our darkness by them and straightening our crooked ways. And he is still engaged in gathering and disseminating [knowledge].[67]

Having now learnt that one of the scholars of Sefarad, namely R. Baḥyā the judge, son of R. Joseph of blessed memory, composed a treatise on the doctrine [*torah*] of the duties of the heart, grounded in the idea of the unity [of God], he craved to see it. As soon as it reached him, he instructed me to interpret [translate orally, *le-targem*] its content into Hebrew for him. And when he heard its statements and its discourse, he did not relent until he had it copied and instructed me to translate [*le-targem*] for him its first chapter, namely the chapter on the unity [of God].

Although his command was precious to me, I realized that he was getting me involved in a matter that I did not wish to enter and that I was always careful to avoid. [Indeed,] several times, a few of the magnates who love us [i.e., me] among the scholars of these places beseeched me to translate [*le-haʿatiq*] for them some of the works of the Geonim that are in the Arabic language into the Holy Tongue. But I did not yield to them, due to several concerns that one must take into account. [...]

I have said all this so as to inform all men in the preface to this chapter that I did not undertake this endeavour [of translating] of my own desire and my own will. Nor am I unaware of all I have mentioned. However, because I am under the obligation to annul my own wishes in favour of the wishes of the [person] who has commanded me and must subordinate my honour to his, I obeyed his command and hastened to do as he wished and desired. I 'took my life in my hand' [after 1 Sam. 28:21] and did not put my hand on mouth [i.e., did not keep silent]; I made myself the target of the archers' arrows and 'I gave my back to the smiters' [Isa. 50:6]. [...]

I noticed that this author, his memory for a blessing, left undone one thing that he should have done at the beginning of his book or else in the appropriate places: namely, he should have explained the words and terms that he took

---

Thus, the three verbs used here by Judah Ibn Tibbon all refer to R. Meshullam's collection and dissemination of knowledge already available in Hebrew, not to his translating any works into Hebrew.

67 Unfortunately, we have no information about the translations Judah Ibn Tibbon had in mind.

from the books of the secular sciences for the purposes of his discussions and should have [clarified] what they mean in those sciences. This is indeed the usual practice of scholars in all works of science, because most people are not familiar with those words and terms and do not understand them; and they use some of them in a different way [than in the sciences] and apply them to different matters or interpret them differently. Perhaps [the author] acted as he did because, as he said, he composed the book to serve as an aide-mémoire for himself and his friends [who know the sciences and the correct meanings of all terms]. But since he said later that he wished to be helpful to other people as well, he should have been aware of the issue I raised. Therefore, whoever is not at home with these words and terms and with those sciences will have to inquire about them. […]

Document 2.1.1.2. Judah Ibn Tibbon, epistle to R. Asher b. Meshullam (1161).

Freudenthal, 'R. Judah Ibn Tibbon's Epistle to R. Asher b. Meshullam of Lunel', 668–670; tr. Freudenthal, 'Abraham Ibn Ezra and Judah Ibn Tibbon as Cultural Intermediaries', 80–81.

From the following letter, which Judah Ibn Tibbon addressed to his 'comrade' R. Asher b. Meshullam, we learn that the latter was dissatisfied with the text his father had chosen for translation, on the grounds that studying this long and difficult text would occupy much of the time of his students, who, he thought, should devote themselves exclusively to the Talmud. R. Asher consequently asked Judah Ibn Tibbon to drop his translation of *Duties of the Heart* and instead translate a short and fairly simple work on ethics. They chose Solomon Ibn Gabirol's *Improvement of the Qualities of the Soul*. When Judah had completed the translation, he sent it to R. Asher b. Meshullam, with the covering letter translated below (c. 1161). We learn en passant that while translating the *Duties of the Heart* Judah Ibn Tibbon also 'read', i.e., taught, the translated text 'in the presence of' R. Meshullam, presumably in the latter's yeshivah, where many scholars attended this 'work-in-progress' seminar.

I eased myself of the burden of translation that I had shunned for the reasons I mentioned in the preface to my translation of the chapter on the unity [of God], which I translated for our master, your father, the holy rabbi, may God protect him, from the *Duties of the Heart*. For recall: you and I were studying that chapter in his presence. During our conversation, I told [the two of you] that I had found a small composition on the *Improvement of the Moral Qualities of the Soul* by the wise philosopher R. Solomon son of R. Judah Ibn Gabirol, which encompasses all the themes discussed in the nine [remaining] chapters of that book [*Duties of the Heart*], and even adds to them. Its words are spoken appropriately, presented according to the method of wisdom, they rely on Scripture and are grounded in the sayings of the sages, the parables of the ancients, and the ethical sayings of the philosophers. [...]

When you heard this, your soul craved to see [the book] and learn its content thoroughly.

You said that if we had that work in our hands, together with the present chapter [on the unity of God], it would suffice us as a substitute for the entire book of the *Duties of the Heart*. It could supplant it and ease us of the burden of excessive length; we would better be able to commit to memory its words according to the order of its chapters. For no man can remember things that are long and drawn out.

And at that time, you instructed me to translate it for you after I completed [the translation of] that chapter [on the unity of God].

I have now translated it for you as best I could, to the utmost of my abilities. [I translated] all its statements, phrases, parables, ethical maxims, and riddles. I have, however, left out the poems that are inserted in each of its chapters, for it was imperative that I [first] fulfil the request [for a further translation] of our master, the rabbi, your father, may God protect him, before I have leisure to render their content in poems as well. [Then,] if I find in the writings of the earlier poets—as, for example, in those of the author of this treatise [Solomon Ibn Gabirol], those of the Prince [Samuel ha-Nagid], and those who trod in their path—something [viz., Hebrew poems] that resembles them and is close to them [the original poems in *Improvement of the Moral Qualities of the Soul*], I will insert it [*lit.* write it down] in the name of its author. For those [poems] for which I do not find [an appropriate equivalent], I will do [i.e., translate] according to my understanding and to the best of my capacities. I will [then] insert each poem in its appropriate place in the folio. [...]

Document 2.1.1.3. Judah Ibn Tibbon, second preface to his translation of *Duties of the Heart* (to chapters 2–10).

Bahya Ibn Paqūda, *Sefer Ḥovot ha-levavot*, tr. Judah Ibn Tibbon, ed. A[braham] Zifroni, 162–164.

In 1166, Ibn Tibbon discontinued the translation of the *Duties of the Heart*. The disappointed R. Meshullam thereupon turned to another exile from Andalusia who was living in Narbonne, R. Joseph Qimḥi (c. 1105–c. 1170), and asked him to translate the rest of the book. Qimḥi complied, and eventually translated chapter 1 as well. Only fragments of Qimḥi's translation are extant. A few years later, however, although a complete translation of *Duties of the Heart* now existed, the noted R. Abraham b. David (c. 1125–1198) asked Judah Ibn Tibbon to complete his own translation of the work, and he did so.[68] He inserted an additional 'translator's preface' in which he recounted the history of the translation.

[...] The learned sage, our rabbi Meshullam son of Jacob of blessed memory, asked me to translate for him the first chapter [of the *Duties of the Heart*], on the unity [of God]. I translated it for him, although I was upset that he involved me in this matter, on account of the concerns that I described in my preface to that chapter. Nonetheless, I suppressed my own wishes in favour of his wishes.

Later, R. Joseph Qimḥi the Spaniard translated the rest of the book for him, from chapter two on. And now, the great sage R. Abraham b. David asked me to bring my own translation to completion. This, too, was problematic for me, for someone else preceded me in translating it and [I did not wish him] to think that I take pride in his humiliation. I beseeched him [R. Abraham] to leave me, so that he would not lead me to do that [viz., take pride in his predecessor's humiliation]. I revealed to him that my [decision] was motivated by respect for R. Joseph, who had preceded me in translating the book. But we could not agree. When I saw that he was decided, I undertook to annul my own view in favour of his view, so as to satisfy his wishes and desire.

I saw fit to mention this at the beginning of this book, so that no one may think that it was by my own will that I trod in the wake of that translation. He who knows the thoughts of all men knows that this was not the reason for my embarking upon this enterprise, and that I did not begin it out of my own will. Rather, I let the wishes of the person whose wishes I undertook to satisfy [viz., R. Abraham b. David] overpower my own wishes. He is my witness for that, and this is enough for me to absolve myself.

After I began translating, I could not forsake my convictions and leave the path of truth and correctness and do God's work dishonestly, [wherefore I would] regard myself as stupid and unwise, for the sake of [saving] the honour of others [viz., R. Joseph Qimḥi]. I translated according to what seemed to me appropriate and according to my understanding.

This book now exists in two translations and in three forms.[69] I therefore beseech anyone who copies it to choose whichever he wishes; I make him take

68 R. Abraham ben David of Posquières (present-day Vauvert) was a very influential teacher of halakhah, the head of a yeshiva, a critic of Maimonides, and favourably inclined toward the nascent Kabbalah.

69 Namely: Ibn Tibbon's translation, Qimḥi's, and a mixed form (with the first chapter translated by Ibn Tibbon and the rest by Qimḥi).

an oath by God that he will write the name of the translator at the beginning of the book, so that the faults of others will not be charged to us, inasmuch as we have neither acted foolishly nor sinned [after Num. 12:11]. Our own errors are enough, and we need not burden ourselves with those of others and add the offences of others to our own. Our rabbis of blessed memory have warned of this, saying: 'Whoever repeats a remark in the name of its author brings redemption to the world' [*Avot* 6:6].

Apropos of what I said, that the book now exists in two translations and in three forms: this is because a few persons have copies [i.e., manuscripts] in which only the first chapter is in my translation [and the rest in that of R. Joseph Qimḥi]. Now R. Joseph later translated the first chapter as well, so as to have translated the entire work, and [so] other persons have the entire work in his translation. And I, too, have completed the translation, as the aforementioned wise rabbi, may the Merciful One guard and deliver him, asked of me [and some persons have that version].

And whoever corrects its words in the appropriate place and judges me leniently where I came up short—may his Creator allow him to witness His kindness.

Document 2.1.1.4. Judah Ibn Tibbon, translator's preface to the translation of Jonah Ibn Janāḥ's *Sefer ha-Riqmah* (1171).

Sefer ha-Riqmah, ed. Wilenski et al., I, 1–7.

Jonah Ibn Janāḥ (Abū l-Walīd Marwān ibn Ǧanāḥ, c. 990–c. 1050), a physician by training and (by his own account) poet manqué, was a central figure of the school of Hebrew grammar that flourished in Andalusia in the tenth and eleventh centuries. Its major innovation was to introduce the notion of the triliteral root (borrowed from Arabic grammarians) into the study of Hebrew. Ibn Janāḥ engaged in constructive criticism of his great predecessor Judah b. David Ḥayyuj (c. 945–1000), the founder of systematic Hebrew grammar.

Ibn Janāḥ's major work is *Kitāb al-Tanqīḥ* (*Book of Minute Investigation*), in two independent parts: *Kitāb al-Lumaʿ* (*Book of Multicoloured Flowerbeds*), a grammar of Hebrew, and *Kitāb al-Uṣūl* (*Book of Roots*), a Hebrew lexicon. Judah Ibn Tibbon translated both in 1171, as *Sefer ha-Riqmah* and *Sefer ha-Shorashim*.[70]

Judah Ibn Tibbon added two texts to his translation: a fairly substantial preface to the entire project, included in *Sefer ha-Riqmah*, and a short poem at the start of *Sefer ha-Shorashim*. Here I offer an English translation of the former.[71]

70 The latter was published as *Sefer ha-Shorashim*, ed. Bacher.

71 Steinschneider, *HÜ*, § 550, 918–919.

Blessed be the God of Israel, who has not left science without a redeemer, one who seeks to understand its ways and seeks knowledge of its path. […] Nonetheless, day after day, folly conceives and bears children, whereas wisdom is barren and does not conceive. Ignorance increases and the sciences decrease. […]

More than all other [sciences], it is the science of language, which is the foremost among the sciences and a ladder to all the higher ranks, [that has diminished in our times]. Even though it is not the first in the enumeration [i.e., classification] of the basic sciences, it is the base of the pillar and comes first in the curriculum. Fools have broken through its fences, […] deformed its truths, twisted its paths, contorted its words and terms; they have abused it as they pleased; in their discourse they have annihilated and debased its pleasant statements. They have not done justice to it, for they have not understood its workings or recognized its merits. They used to poke abuse at its scholars and deride those who taught and studied it. They thought its statements were vain, not useful and offering no succour [after 1 Sam. 12:21]. This is why they turned away from it and did not study its ways.

[So it was] until the Lord awakened the spirit of its [the science of language's] warriors and the heroes of its battles, namely the chiefs of the hosts of Israel, the sages of the exile of Jerusalem in Sepharad [Obad. 1:20]. When they saw that [this science] had vanished and become non-existent, they became jealous for it, because it became concealed. […] Until there arose the great teacher, the Teacher of Justice, the restorer [of the building], who uncovered the profundities of the science, a valiant warrior and fighter, knowledgeable in holy matters, a counsellor and cunning artificer [Isa. 3:3], R. Judah b. David, known as Ḥayyuj. Great wisdom was found in him [after Dan. 5:11–12, 14], a spirit of counsel and might [Isa. 11:2]. He understood their [the verbs'] ways and he knew their place. He did to them as he had intended, fought them, and overcame them. And he excelled in wisdom all who preceded him, led them by the cords of his wisdom and brought them into his covenant. […]

After him [Ḥayyuj] came two commanders of the hosts of wisdom, with great power and an upraised hand: our Rabbi Samuel Halevi, the Nagid [= prince], the commander of Israel's host, of blessed memory;[72] and the sage and teacher, R. Jonah, known as Marwān Ibn Ǧanāḥ. They followed in his footsteps and diffused his statements, and his light illuminated them. They became the

72 Samuel ha-Levi ben Yosef ha-Nagid (Abū Isḥāq Ismāʿīl ibn al-Naġrīla, 993–after 1056) was an important grammarian and noted poet, who rose in the administration until he became vizir to the king of Granada and commander of his armed forces (whence the military metaphors); he was a very influential patron of Hebrew learning.

strong heroes who did his behest, girding their might for war, to bring what was hidden into the light, and were victorious in the campaign, and drew up battle lines against battle lines. And the substance of their wars and words, behold they are written in their books and mentioned in their compositions.

When the battles ceased, in which they showed [their] might and power [Ps. 68:36] in the art, the sage R. Jonah composed a work on the grammar of the Holy Tongue. He divided it into two parts: he called one *Sefer ha-Riqmah* [*Book of Embroidery*] and he called the other *Sefer ha-Shorashim* [*Book of Roots*]. He included in them most of the rules of the language, its topics, structures, causes, and stratagems, its highways and byways. Apropos of the two mentioned families,[73] he relied on the books by R. Judah, but in his own books he mentioned only a few [of these] matters, where necessary.

He wrote these books in Arabic, the language of the people among whom he dwelt. For this was the case of the majority of the compositions by the Geonim and by the [Jewish] sages throughout the kingdom of Ishmael. For it [Arabic] is a vast and eloquent language, in which nothing is lacking for a speaker. [By contrast,] of the Holy Tongue we have only what is found in the biblical books, and this is not sufficient for all of a speaker's needs. Moreover, the masses do not understand it, but only the few, whereas all their contemporaries [the sages in the kingdom of Ishmael] knew Arabic; therefore, they chose to write their compositions in it.

However, the sons of this exiled host [*ha-ḥel*] who are in Tzarfat [see Obad. 1:20] and in the entire domain of Edom [= Christendom] do not know Arabic, so these books are like a sealed book to them; they could not comprehend them until they had been translated into the Holy Tongue.

Thus, a number of years ago, the spirit of the sage R. Moses son of R. Samuel ha-Kohen ben Giqatilla prompted him to translate the two aforementioned books by R. Judah b. David [Ḥayyuj] into the Holy Tongue for one of their young men and a few of the sons of their leaders.[74] They kept them and so preserved them, keeping them in memory. Then the sage R. Abraham Ibn Ezra came to their countries and he too helped them in this subject with short treatises, in which he included fine and valuable discussions.[75] R. Isaac ha-Levi joined him and wrote *Sefer ha-maqor* for them, in which he collected helpful rules for the

73 'Families' of problematic roots mentioned earlier—those containing the letters *aleph*, *heh*, *waw*, and *yod*, and double letters.

74 Moses Ibn Giqatilla's preface to his translation of these grammatical treatises by Ḥayyuj is translated above, § 1.1.1.

75 Although Judah Ibn Tibbon does not indicate this here, Abraham Ibn Ezra also translated Judah Ḥayyuj's three grammatical treatises.

use of the language and the *binyanim* [verbal paradigms or patterns].[76] Since then, some of them have engaged in this art and occupied themselves with it a little.

I met some of those who crowd around its portals and who appropriated its insights, namely people who began to inquire into it and savoured the taste of its honey. And when they realized that [through this study] their eyes were lit up and their ears opened, they were drawn to it and yearned to penetrate its books. They addressed themselves to me [with the request] that I translate these two books for them. With much friendliness they urged me to fulfil their request and accede to their wish. They beseeched me until they changed my mind.

I knew that the work is heavy and the labour hard, on account of the points that confront the translator and that I already mentioned in the preface to the translation of the *Duties of the Heart*. Given the deficiency of the [Hebrew] language, those difficulties are difficult to overcome, except with many stratagems and long reflection, turning the matter round and round until the two ends are joined [after Exod. 28:7]. [...]

2.1.2. Samuel b. Judah Ibn Tibbon (c. 1165, Lunel–1232, Marseilles)

Samuel Ibn Tibbon is arguably the finest medieval Arabic–Hebrew translator and certainly the most influential. His translation of Maimonides's *Guide of the Perplexed* (1st edn 1204, 2nd rev. edn 1210), his most ambitious and prominent project, almost immediately became the foundational text of all Hebrew writing on religious philosophy. It is difficult to exaggerate the importance of Samuel's role in the formation of the Hebrew philosophical tradition.

Samuel also translated other works by Maimonides and produced the first Hebrew translations of non-Jewish writers on philosophy and science (Galen, ʿAlī b. Riḍwān, Aristotle, Averroes). In addition, he wrote three philosophical works of his own, construed as contributions to Maimonides's philosophical programme. Samuel Ibn Tibbon is thus much more than a translator: he is an independent thinker who set himself the mission of winning over his brethren to the Maimonidean interpretation of Judaism.

Samuel lived in the Midi (chiefly in Lunel, his birthplace, and in Marseilles), where his father and his family had found refuge after fleeing their hometown of Granada and the Almohad persecutions in Islamic Spain in 1148. In exile, circumstances gradually led

76 Steinschneider (*HÜ*, § 510, 918–919) identifies R. Isaac as a scholar who also translated part of Ibn Janāḥ's *Kitāb al-uṣūl*.

Judah, originally a physician, to devote himself entirely to the translation of works of Jewish religious philosophy from Judeo-Arabic into Hebrew (§ 2.1). Two further generations of Ibn Tibbons continued in this craft.[77]

Samuel Ibn Tibbon throws much light on the evolution of his thought, including his path to becoming a translator, in the long preface to his commentary on Ecclesiastes. Unfortunately, lack of space prevents even samples of this text from being offered here (see Robinson, *Samuel Ibn Tibbon's Commentary on Ecclesiastes*).

Two kinds of egodocuments by Samuel have reached us: I. Letters by him, to him, and letters in which he is mentioned; II. Prefaces to his translations.

I. Triangular Correspondence between Samuel Ibn Tibbon, R. Jonathan ha-Kohen of Lunel, and Maimonides on the Preparation of the Hebrew Translation of the *Guide of the Perplexed* (1194–1204)

Samuel Ibn Tibbon became the translator of the *Guide of the Perplexed* through a series of stages that can be reconstructed in detail. Sometime in the early 1190s, the foremost scholars of Lunel and Montpellier, headed by R. Jonathan ha-Kohen, wrote to Maimonides with queries on various halakhic matters, as well as on the validity of astrology.[78] This letter and those that followed are a vivid indication of Maimonides's prestige and unquestioned authority throughout the Jewish world. Eventually, the scholars asked for the text of the *Guide of the Perplexed*, of which they knew of by hearsay: its 'renown' kindled their vivid interest.[79] Later, Samuel Ibn Tibbon would recall that the Provençal scholars asked him for an oral summary of the book—they were obviously unable to access it directly—and that he fulfilled their request, allowing them to become 'cognizant of some of the book's discussions' (see below, Document 2.1.2.5). (It should be recalled that forty years earlier, his father, Judah, had provided Meshullam b. Jacob with an oral summary of the first chapter of the *Duties of the Heart*, after which the latter decided to have it translated into Hebrew; see above, Document 2.1.1.2.) Having thus tasted of the *Guide*, the Lunel scholars' interest in it redoubled, and they commissioned Samuel Ibn Tibbon to translate it. Other letters followed.

The triangular correspondence across the Mediterranean that ensued, which lasted until Maimonides's death in 1204, is a rare example of sustained exchanges among author, patrons, and translator, in which texts and translators were selected. Below I present excerpts from four letters that are germane to issues of translation.[80]

77 For synthetic accounts of the external circumstances of Ibn Tibbon's life and works, see Robinson, *Samuel Ibn Tibbon's Commentary on Ecclesiastes*, 3–23; id., 'Samuel Ibn Tibbon'; and Fraenkel, *From Maimonides to Samuel Ibn Tibbon* [Hebrew].

78 See Sela, 'Queries on Astrology'.

79 Lichtenberg, *Qovetz Teshuvot ha-Rambam*, I, 6b.

80 Steinschneider, *HÜ*, § 243, 415–420; Marx, 'The Correspondence between the Rabbis of Southern France and Maimonides', 325–335; Stern, 'Maimonides' Correspondence with

Document 2.1.2.1. R. Jonathan ha-Kohen and Colleagues to Maimonides (c. 1199).

'Letter of R. Jonathan of Lunel to Maimonides', ed. Wertheimer.

Blessed be the God of our fathers, who encouraged the king of the sciences [= Maimonides] to glorify our faith and fortify the structure of our Torah. [...] By God's benevolence, we resolved to collect and gather his [Maimonides's] books, which are like sapphires,[81] and his choicest compositions, which are incomparably more valuable than hidden treasures. [...]

We have seen and studied the *Mishneh Torah*, which solves doubts and uncovers profundities. [...] You have given us life and the good.[82]

You have heaped upon us wisdom and the good by sending us the *Guide of the Perplexed*.[83] [...] Its [the *Guide*'s] words are the words of the Living God. He who discovers it [the *Guide*] has discovered life itself [after Prov. 8:35]; and all who turn away from it are as if turning away from life.

For the wise, the book is just as its name indicates: it is a guide for the perplexed. But for the masses it has no utility. For us [too] it would have been like a stone tied up in a sling,[84] like a rose in the midst of thorns [after Song 2:2], a book put in front of the illiterate, were it not that our Creator brought us a wise son, learned in all the sciences, whom his father, the rabbi, taught to read and understand Arabic. [This is] the son of the outstanding scholar, the excellent physician, the Rabbi Judah Ibn Tibbon the Spaniard, who benefited and instructed and taught us from the books of the [philosophical] sciences by translating for us the *Book of Doctrines* [*and Opinions*], and the *Book of the Duties of the Heart*, and [Solomon Ibn Gabirol's] *Improvement of the Moral Qualities* and *The Choice of Pearls*, and the *Kuzari*, as also the grammatical

---

the Scholars of Provence'; and Shailat, ed. and tr., *Iggerot ha-Rambam*. (The latter edition, which has become standard, is not strictly scientific. It nonetheless gives a sufficiently reliable text to be used for the present purposes. It includes only Maimonides's letters, not those of his correspondents. Shailat discusses the chronology of the correspondence is in his introduction to each letter.)

81 The Hebrew word for 'book' is *sefer*; that for 'sapphire' is *sappir*, yielding an intended pun.

82 The authors are making an intertextual reference to the verse: 'I have set before thee this day life and good, and death and evil' (Deut. 30:15).

83 As a halakhic work, the *Mishneh Torah* brings with it 'life and the good', according to the verse in Deuteronomy. The *Guide of the Perplexed*, for its part, offers 'wisdom', which similarly, but differently, opens the door to life ('he who discovers it [the *Guide*] has discovered life itself', alluding to Proverbs, as noted in the text), namely eternal life of the soul in the world to come.

84 After Prov. 26:8: 'As he that bindeth a stone in a sling, so is he that giveth honour to a fool.'

works by Ibn Janāḥ: the *Book of Roots* [*Sefer ha-Shorashim*] and the *Book of Embroidery* [*Sefer ha-Riqmah*]. Each of them brings glory to its author [after Esth. 1: 20].

Before the lamp of Eli [i.e., Judah Ibn Tibbon] was extinguished, the sun of Samuel of Ramah [= Samuel Ibn Tibbon] shone forth.[85] [...] [Samuel] enlightened our eyes, allowing us to see hidden things—our eyes could not be satisfied, nor our ears hear enough. He whetted our initial appetite [for instruction], [and we are now] like an exhausted and thirsty person who began drinking but the glass was snatched away from him, or like a suckling who has been removed from his mother's breast. The rivers of our desire burned in our bosom and heated us ever more. [...]

Therefore, if it pleases our master, the rabbi and outstanding scholar, and if he approves of it, would he complete our meal [...] and send us the *Guide of the Perplexed* up to the end? May the master listen with kindness and sympathy to the plea of these scholars resident in Lunel, [...] so that we may know the truth and be sure that we have not spent our money for nothing.[86] [...]

[Signed by] Jonathan ha-Kohen b. David, of blessed memory

Document 2.1.2.2. Maimonides to Samuel Ibn Tibbon (1199).

Shailat, *Iggerot ha-Rambam*, II, 530–533, 554; tr. Stitskin, *Letters of Maimonides*, 130–136.[87]

I, Moses son of R. Maimon the Spaniard of blessed memory, received all the letters of the esteemed, learnèd, and perceptive student who is the crown of disciples, the glory of the learned, the revered Rabbi Samuel, may the Lord preserve him, son of Rabbi Judah Ibn Tibbon of blessed memory. [...]

The questions you raised are all pertinent. All your conjectures concerning the [copyist's] omissions of a word or more are correct. I have explained to you all these things in detail in Arabic at the end of this letter and advised you about

85 The authors draw on *Genesis Rabbah*, 58:2.

86 This is a significant offhand remark. Maimonides had the two first parts of the *Guide* copied for the scholars of Lunel, but presumably they covered the costs. The costs include the writing materials, the copyist's honorarium, and perhaps also a payment to the owner of the copied manuscript. The scholars seem to refer to these costs, which is why they suppose that Maimonides would understand what they meant (and also feel under moral pressure to fulfil their request). It is unlikely that they have a payment to Samuel Ibn Tibbon in mind.

87 I also drew on Kraemer, *Maimonides*, 711–722 (only part of the text).

all your undertakings [*variant*: as well as the books you should read and those that should be left out]. To you it is certainly advisable to engage in translation from one language to another. For the Creator has endowed you with an understanding heart to comprehend all texts [*lit.* to comprehend proverbs and elegant language]. I recognized from your correspondence that you have the capacity to delve into the depth of a subject and reveal hidden meanings.

Now, I have explained to you in Arabic [in the second part of this letter] how you should go about any translation. 'Give [instruction] to the wise and he will become yet wiser' [Prov. 9:9]. Become wiser, my son, and gladden my heart! Verily, when I saw your correspondence, I was exceedingly happy and puzzled: I was happy about the discovery of such a wise son; and I was puzzled—how can the nature of one who was born among the 'babblers' be such that he pursues the sciences and displays such proficiency in the Arabic language (which surely is only somewhat corrupted Hebrew[88])? Moreover, how is it possible that he has mastered the fine points of the language and has also probed into profound subjects? Surely this is as [wondrous as] 'a root out of dry ground!' [Isa. 53:2]. [...]

[In the manuscripts, the following note (in Hebrew) is inserted at this point: 'Until here the words of our Rabbi Moses were in the Hebrew language; from here onward they were in Arabic.' The text has survived only in Hebrew translation, and the above note was presumably added by a translator of the Arabic part of the letter. The following translation is based on one of the medieval Hebrew versions.]

The writings of the exalted, esteemed, and learned scholar [Samuel Ibn Tibbon] have reached me. May the Lord maintain his excellence and increase his success!

I checked all the points [in the translation of the *Guide*] about which you raised doubts and I examined all the places where the scribe erred. I took note of the premises and the chapters that you did not entirely comprehend and on which you sought guidance. I shall explain everything to you. [...]

I now begin to answer you about every chapter and to explain what you asked me to explain, and inform you of the correct readings against which you should check the mistaken version, in the order they are listed in your letter.

I have set you straight and instructed you concerning the books you should study and with what [subjects] you should occupy your precious soul. [...]

---

88 In chapter 25 of his *Aphorisms*, Maimonides states that 'the Hebrew and the Arabic languages are indubitably a single language, as anyone who knows them both has verified'. It is unclear why he inserted this offhand remark here.

Document 2.1.2.3. Maimonides to the Community of Lunel (1200).

Shailat, *Iggerot ha-Rambam*, II, 555–559; tr. Twersky, *A Maimonides Reader*, 481–482.

[…] With the present text I enclose the third part of the *Guide of the Perplexed* in Arabic. With regard to your request that I translate it into the Holy Tongue for you—would that I were young enough to be able to comply with your request with regard to this book and also the other works that I have composed in the language of Qedar, whose sun has darkened,[89] because I dwelled in their tents.[90] I should be very pleased thus to 'extract the precious from the vile' [Jer. 15:19] and to give back the lost item to the rightful owner.[91] But the eddies of time have surrounded me; even some commentaries I wrote and other works composed in the rabbinic language,[92] which contain various obscurities, I do not have the time to correct carefully so that they may be published; *a fortiori* I do not have the time to make translations from one language into another. Yes, my honoured masters and friends, I have no leisure even for writing a short chapter, neither during the day nor during the night. It is only out of respect for your honour that I took the pain and trouble to write this entire letter with my own hand.

But you have in your midst a 'wise son [who] maketh a glad father' [Prov. 1:10], his mother's joy, [namely,] the learnèd fellow-student, the crown of the learned, R. Samuel son of the wise Judah Ibn Tibbon of blessed memory, who is wise and erudite, of vast understanding and a sound intellect,[93] on whom the Lord has bestowed skill in language and eloquence. Nothing of what you may wish him to translate will be beyond him. I have already written to him about issues of the translation in my own hand, in Hebrew and in Arabic.

89 Wordplay: *qedar* = Arabic; *qadar* = darken.

90 Alluding to Ps. 120:5: 'Woe is me, that I sojourn in Meshech, that I dwell in the tents of Kedar!'.

91 This is a passing allusion to the widespread view that the sciences originated with the Jews and in Hebrew and were subsequently 'stolen' by the Greeks, from whom they passed to Arabic culture.

92 Hebrew mixed with Aramaic.

93 *De'ah nekonah*; cf. *Mishneh Torah*, Foundations of the Torah, chapter 7:1.

Document 2.1.2.4. Samuel Ibn Tibbon to Maimonides (1199/1200).

Diesendruck, 'Samuel and Moses Ibn Tibbon on Maimonides' Theory of Providence', 352–353.[94]

After Samuel Ibn Tibbon had seen all three parts of the *Guide*, he wrote to Maimonides again to inform him that the text of the third part of the *Guide* that reached him was extremely faulty: he suspected that the problem derived from a copy that was written in Arabic letters. Samuel enclosed a copy of the text he had and asked Maimonides to have it checked and sent back to him corrected. Also, in the third part of the *Guide* he had encountered a major philosophical issue concerning the nature of divine providence and wishes to discuss it with Maimonides.

[...] The youngest of his servants [...] writes to inform your honour that his epistle has reached his servant.[95] It is more precious to me than anything else that was written at midday.[96]

[...] I almost fainted when I read its first lines, but the last lines comforted me when I learnt that our master has recovered from his long illness that he bore on account of the sins of us all [after Isa. 53:6]. [...]

I now yearn for the late rains of his clear answers and his pure statements concerning everything I asked the honourable sage in my earlier letters concerning the three parts of the esteemed treatise, the *Guide of the Perplexed*. I am afraid to ask more, lest I be a burden on his honour. But there is one question that is like a raging fire in my heart, shut up in my bones [after Jer. 20:9], occasioned by one of the last chapters of the third part, which arrived recently. It causes me great perplexity, and I will address it to his honour, in addition to the earlier questions. And it will be his merit before the Lord his God.

I am also sending [the Arabic text of] the three parts of the treatise as we have it, which reached us from there [Egypt]. This is [the text] from which I made my translation. I send them to your honour with the request that he order a few of his students to proofread them with extreme rigour, once and twice, until they are sure that no error remains. May our Master sign these copies after having verified the correctness of the text. May he then have them sent to the wise and erudite student, R. Abraham ha-Kohen, may the Lord protect him, in

94 Also Lichtenberg, *Qovetz Teshuvot ha-Rambam*, II, 26b–27a.

95 Samuel refers to the letter in which Maimonides responded to queries about the *Mishneh Torah* (and which therefore has not been translated here).

96 Samuel stresses this because he had heard from Maimonides that as a rule he was very busy during the morning.

Alexandria. I know that he, by his kindness, will endeavour to send them to us by our Jewish brethren who travel there regularly. He was already so kind to do whatever I asked of him there, and *a fortiori* he will do his best if asked by our master.

I did not refrain from troubling our master in this matter, too, because I know that his honour will favour eliminating all offences from our midst [i.e., eliminate errors in the text of the *Guide*]. I myself will not rest until I know that this treatise [the *Guide*] has been corrected. For who can trust himself in [detecting] errors in a manuscript whose errors are innumerable? Even if the reader were wise, he would miss many spots—*a fortiori* if he is someone who has not inhaled the perfume of wisdom.[97] [Samuel Ibn Tibbon's query concerning Maimonides's theory of providence follows.]

II. Samuel Ibn Tibbon's Prefaces to his Translations

Document 2.1.2.5. Samuel Ibn Tibbon, preface to his translation of Maimonides's *Guide of the Perplexed.*

Samuel Ibn Tibbon, 'Preface [to his Translation of the *Guide of the Perplexed*]', 117–122.

The translation of Maimonides's *Guide of the Perplexed* was doubtless Samuel Ibn Tibbon's major translation project; Moritz Steinschneider called it 'one of the most significant documents in the history of translations'. Samuel viewed Maimonides as an exceptionally inspired individual, whose thought, as expressed notably in the *Guide*, played a pivotal role in Jewish history. One can imagine the historical weight Ibn Tibbon felt on his shoulders when he agreed to reformulate the Master's message in Hebrew. This acute sense of historical consciousness, let me add in passing, explains the emotional vehemence of his attack on Alḥarizi for what he viewed as a betrayal of Maimonides's legacy (Alḥarizi retranslated the *Guide* to make it easier for readers to understand; see Document 1.1.5.1).

In his extensive but conventional preface, Samuel Ibn Tibbon discussed his approach to translation. It has become a classic of Hebrew letters and guided later translators. Only a few passages are translated here.

Samuel b. Judah Ibn Tibbon of blessed memory, the translator of this treatise from Arabic into Hebrew, said, apologetically: [Samuel Ibn Tibbon refers to all the reasons why one should shun translating and continues:]

97 In the last sentence, Samuel Ibn Tibbon modestly refers to himself.

My father and teacher of blessed memory has already mentioned all this in the preface he added to his translation of the *Duties of the Heart* [see Document 2.1.1.1]. [...] Nothing of this escapes me! I am aware that my mastery of both languages, Hebrew and Arabic, is very imperfect. My knowledge of Arabic is especially sparse, because I did not grow up among its people and lands and did not acquire it thoroughly. I am also aware of my great imperfection in understanding the esteemed treatise, the *Guide of the Perplexed* and its themes, because most of them are very profound. It involves [*lit.* comprises] many high sciences that are unknown to the great majority of our people in our country, perhaps even to all of them, because they do not study these [sciences] and have no access to them. I myself understood a very tiny part of some of them, because I had some of the relevant books in Arabic, which I know a little, and my heart incited me to learn from them as far as my limited capability permits.

Owing to all these considerations, I should have abstained from translating this esteemed book. [What convinced me to translate it is as follows:] the sages of this land, headed by R. Jonathan ha-Kohen, may God protect him, and the other scholars of Lunel, my hometown, heard the fame of that book, and their desire led them to send to Egypt for it. In their letters they implored the great rabbi, the divine philosopher, the pure Crown of the Exile, our teacher and master, R. Moses, may God protect him, son of the great and honourable Rabbi Maimon of blessed memory, the author of this book, asking him to send it to them translated [into Hebrew], if possible, or otherwise untranslated. They made great efforts in this matter. And when the work came to their hands, untranslated, their desire [for the book] similarly led them to make great efforts to have it translated. Moreover, when they became cognizant of some of the book's ideas, their desire increased even more.[98]

They summoned me, Samuel, son of Judah b. Tibbon of blessed memory, and turned to me, for they knew I had some knowledge of Arabic and because there is no native speaker of this language in their district.[99] They strongly implored me to translate it for them the best I could. I explained to them my incompetence in translating, for the aforementioned considerations, but they retorted: 'It is impossible that we will not derive some benefit from it; even if the trans-

98 As already noted above (introduction to § 2.1.2, *Triangular Correspondence*), this short sentence seems to indicate that Samuel orally summarized for the sages of Lunel 'some of the book's discussions', much the same way as his father had done in 1161. (He explained the first chapter of the *Duties of the Heart* to R. Meshullam ben Jacob, whereupon the latter asked him to translate it; see Document 2.1.1.2.)

99 Quite remarkably and surprisingly, Samuel does not connect his calling to his father's reputation as the 'father of the translators'.

lation is feeble and of little benefit for the momentous topics discussed in this book, in our eyes the benefit will be considerable.' [Thereupon,] I could not refuse them: I translated it despite my shortcomings, making myself into a 'fool who behaves overbearingly and is confident' [after Prov. 14:16]. [...]

Document 2.1.2.6. Samuel Ibn Tibbon, *Glossary of Unfamiliar Terms* (*Perush ha-millot ha-zarot*, 1213).

Samuel Ibn Tibbon, *Perush ha-millot ha-zarot*, 11:1–12:6.

Samuel Ibn Tibbon was fully aware that simply putting a Hebrew translation of philosophical works in the hands of his contemporaries would not suffice to introduce them to philosophical thinking: they had little or no prior acquaintance with the concepts of philosophy and could not be expected to know what meanings to assign the Hebrew terms. To the second edition of his translation of the *Guide of the Perplexed* (1213), Ibn Tibbon therefore appended *Perush ha-milim ha-zarot* (*Glossary of Unfamiliar Terms*), where *zar* (pl. *zarim/zarot*) refers simultaneously to that which is strange because unfamiliar, and to ideas of foreign origin. (In what follows, I will write 'unfamiliar', but the connotation of 'foreign' should be borne in mind.) We recall that the idea had already been countenanced by Samuel's father (see above, Document 2.1.1.1).[100]

The *Glossary* is almost entirely devoted to explaining philosophical terms and concepts. It can easily be qualified as a philosophical *vade mecum*.[101] Here I will translate only one brief passage, in which Samuel Ibn Tibbon tells us something about himself as a translator and his methods, including the vehement (but understandable) critique of his colleague Judah Alḥarizi, who had translated the *Guide* at the request of Provençal scholars who found themselves unable to read Ibn Tibbon's scholarly translation (see Document 1.1.5.1).

Samuel, son of R. Judah Ibn Tibbon said: Having completed the translation of this esteemed treatise, *The Guide of the Perplexed*, I remarked that I was unable to avoid using unfamiliar terms that most of those who study it will not understand: [this is so] owing to the narrowness of our language and to our nation's lack of works in the philosophical [*lit.* demonstrative] sciences, as a result of which our language does not include the unfamiliar words employed by the masters of the specialized philosophical sciences. For this reason, I have ventured to compose a treatise in which I explain those terms.

100 Steinschneider, *HÜ*, § 244, 420–423; and Goshen-Gottstein, 'Samuel Ibn Tibbon's *Explanation of Unfamiliar Terms*' [Hebrew].

101 Fraenkel, *From Maimonides to Samuel Ibn Tibbon*, 103 [Hebrew].

All the more so [was this needed] because the poet R. Judah Ibn Alḥarizi, who translated this book after I did, anticipated me in this and prefaced two treatises to his translation. The first, in which he explains the terms, is replete with nonsense, mistakes, and obstacles. The second, a treatise in which, he said, he would enunciate the intention of each chapter of the esteemed treatise [the *Guide*], is full of invective, obstacles, and stumbling blocks.

To remove the obstacles created by the first treatise from our people's path, and to be useful to those who will study my translation, I made up my mind to compose a treatise in which I would explain all the unfamiliar words [found in it], as well as those that I will use in the present section that I preface to the treatise in question.

Document 2.1.2.7. Samuel Ibn Tibbon, preface to his translation of Maimonides's *Eight Chapters* (1202).

Slutzky, *Shemonah Peraqim le-ha-Rambam*, 1a–b, 2b–3a.

Tractate *Avot* (= Fathers) of the Mishnah bears on ethics and is a relatively easy read. Maimonides prefaced his commentary on this tractate with a short treatise on ethics, known as the *Eight Chapters*. This was often transmitted as an independent work and became very popular. As part of translations of some of Maimonides's *opera minora*, Samuel Ibn Tibbon translated it as well. He states that he did so at the request of Lunel scholars, after completing the *Guide of the Perplexed* (first edition). The translation is dated Tevet 4963 (= mid-December 1202 to mid-January 1203).[102] The translator's preface has come down to us in two versions; evidently Samuel revised it and produced a second edition, as he did with his translation of the *Guide*. I use David Slutzky's edition of 1864, in which the two versions are clearly distinguished. Below I translate only the passages bearing on the circumstances and considerations that led Samuel Ibn Tibbon to undertake the translation.

Passage 1: 1a–b

Samuel son of R. Judah Ibn Tibbon said:

This tractate, viz. *Avot* [*Fathers*], of the *Seder Yeshu'ot* [*Order Salvation*[103]], is placed between the tractates *Avodat kokhavim* [*Idolatry*] and *Horayot* [*Erroneous Rulings*]. It has no *gemara*.[104] The great rabbi, our R. Moses ben Maimon

102 Steinschneider, *HÜ*, § 254, 437–438.

103 The Mishnah consists of six orders, each divided into tractates. Ibn Tibbon here uses an alternate name for the order (see B *Shabbat* 31a) that is usually called *Neziqin* [= *Damages*].

104 The Talmud consists of the Hebrew text of the Mishnah and discussions related to it

of blessed memory, commented on it in Arabic, as part of his commentary on the six orders of the Mishnah, all of which is in that language. His commentary on this tractate is particularly broad, because its subject-matter made it precious in his eyes, inasmuch as it discusses the human qualities—the noble and righteous ones as well as the lowly and vile ones. The noble and righteous ones clear the way and 'remove the stones from the road' [after Isa. 62:10] that leads upward to the intellectual virtues. As the Wise Man [King Solomon] said: 'Hear counsel, and receive instruction, that thou mayest be wise in thy latter end' [Prov. 19:20]. It is via the intellectual virtues that one can attain the knowledge of his Creator.

Passage II: 2b–3a (end of the preface)

The philosophers have composed books about the human qualities [*or* about ethics] and discussed them at great length, explaining how a man can habituate himself in the virtues until he acquires them and they become fixed in his soul, and how someone whose qualities are perverted can correct what is perverted in them. This is why the rabbi of blessed memory [Maimonides] saw fit to preface the present tractate [*Avot*] with chapters [i.e., *Eight Chapters*] in which he reviews briefly the topics discussed in detail by the philosophers, as he explained in his own preface [to *Eight Chapters*].

And when the scholars of Lunel, the town of my birth, saw [Maimonides's] commentary on this tractate and also these chapters he prefaced to it in order to explain the matters discussed [in the Mishnah], they asked me to translate [the preface and the commentary] for them, as they had done with respect to the *Guide of the Perplexed*.

Document 2.1.2.8. Samuel Ibn Tibbon, preface to *Otot ha-Shamayim* (his translation of al-Biṭrīq's Arabic version of Aristotle's *Meteorology*).

Fontaine, ed. and trans, *Otot ha-Shamayim*, 4–9.

Samuel Ibn Tibbon chose to translate Aristotle's *Meteorology*—certainly not a centrepiece of the Stagirite's work—for philosophical reasons. Following a lead provided by Maimonides, he sought to understand how the account of creation in Genesis can be reconciled with what the science of his day had to say about physical reality: the

(mostly in Aramaic), called 'Gemara'. Samuel here notes that there is no *gemara* for Tractate *Avot* of the Mishnah. This statement prepares the ground for readers to understand why Maimonides devoted particular attention to commenting on this tractate.

Meteorology was to serve him as a 'semantic key' for discovering the secrets of Creation. Ibn Tibbon writes that he made the translation at the request of a friend (he had around him a circle of students, who read books of science and philosophy with him); it was apparently one of them who requested that he translate the *Meteorology*. Samuel accessed the Aristotelian treatise in the Arabic version of Yaḥyā Ibn al-Biṭrīq (c. 775–c. 830), which was notoriously unsatisfactory. Ibn Tibbon's aim was twofold: to produce a useful translation of the Aristotelian text and at the same time to delve into Aristotle's thought on sublunar phenomena and understand them scientifically. To undo the damage done by Ibn al-Biṭrīq, Ibn Tibbon engaged in a systematic study of the scholarly literature on the topic; above all he drew on the commentary by Alexander of Aphrodisias (in Yaḥyā Ibn 'Adī's Arabic translation), complemented by the meteorological section of Avicenna's *Shifā'* and Averroes's *Epitome*.

Samuel Ibn Tibbon, the translator of this work, entitled *Otot ha-Shamayim* [*The Signs of Heaven*], from Arabic into Hebrew for one of the erudite lovers of science, R. Joseph, son of the pious R. Israel, blessed be his memory, says:

In his first entreaties, the learned scholar, my dear friend, expressed the wish that I translate for him books on natural science composed by the philosophers. But because of the difficulty of what he asked me do do, which required much arduous labour, and because [my] time is short—I have very few days free—and because of other reasons that there is no need to detail, I could not satisfy his wishes regarding this request. When he saw that this was so, and given that he ardently longed for wisdom and that even a small amount of it was precious in his eyes, he replaced his request covering all parts [of natural science] with a request for only one part of it. He repeatedly asked me and beseeched and implored me, by virtue of the perfect love between us, that I translate *Otot ha-Shamayim* for him. […]

But [even] this entreaty, small as regards quantity but huge as regards its quality and its other accidents, was difficult for me [to comply with]. It became ever larger in my eyes owing to the aforementioned reasons, both manifest and hidden. For none of them was missing except for the first one [translating a single book is less 'burdensome' than translating the entire corpus of natural philosophy]. [However,] if one reason dropped, another one was added, namely that I have this work only in the translation of al-Bitrīq. In fact, I have never seen any other translation than his. But his translation is very confused, [just] as the great Rabbi, the divine philosopher, our master and teacher, Moses [Maimonides], the servant of God, wrote me in his first letter that reached me.[105]

---

105 Forte, 'Back to the Sources', 89 (see also 77).

Moreover, errors and omissions are discernible in this translation. Some are indicated by a locution and phrasing that are incompatible with that particular passage; others are indicated by what is found in Alexander's commentary on this book, either by what he quotes there from Aristotle or by what he writes in his own name.

Even for this [latter] reason alone, it would have been appropriate to abstain from translating this book, until it became clear, by God's will, whether another translation can be found, or a better-corrected copy of this one, although I myself have already corrected it against [copies] I found in Toledo and Barcelona. However, I could not go against the wish of the man who asked for [the translation] or refuse him, because of his intense desire, his great insistence, and his precious friendship.

So I will begin to translate the book, according to my limited intellectual capability. I ask God to direct me in my translation and to save me from error. I also ask him to indicate to those who study the book's truthful statements and 'upper ways'[106] the path they should follow, so that the verse of Scripture: 'for the ways of the Lord are straight, and the just walk in them' [Hos. 14:10], may be fulfilled. […]

Document 2.1.2.9. Samuel Ibn Tibbon, preface to his translation of *Three Treatises on Conjunction with the Active Intellect*, by Averroes and his son.

Hercz, ed. and tr., *Drei Abhandlungen*, Appendix IV, 23 (Hebrew numbering); Robinson, 'Samuel Ibn Tibbon's *Commentary on Ecclesiastes*', Appendix A, 501–508.

Samuel Ibn Tibbon translated the three short treatises on conjunction with the active intellect, written by Averroes and his son 'Abd Allāh, after he completed his commentary on Ecclesiastes. In his short preface, he explains why he chose these texts for translation: he wished to demonstrate that both Solomon and Averroes defended the thesis of the possibility of conjunction with the active intellect against its deniers. The commentary on Ecclesiastes and the translation of the 'three treatises' are transmitted together in several manuscripts, showing that Samuel viewed them as complementary works. It was thus a philosophical problem that moved Samuel Ibn Tibbon to translate these three treatises—the same kind of consideration that had led him to translate the *Meteorology* (Document 2.1.2.8). This was the first Hebrew translation of a work by Averroes.

106 I.e., sublime discussions. Samuel makes a small pun: one of the titles he considered for the *Meteorology* was (following the Arabic) 'the upper ways'.

[What follows] is [a text on] the survival of the soul by one of the philosophers. It seems that its proper place was after the *Book on the Soul.* [It treats] ancient theories of ancient philosophers.

I realized from this esteemed work that its main purpose is to expose all the opinions of the sages concerning the rising upward of the human soul. That the human soul 'rises upward' means that it becomes perfect and rises [in rank] until it achieves conjunction with the separate intellect and unites with it, i.e., they become one thing. Now the separate intellect is surely of the highest rank [among existents]. For it is known that the existents are of three kinds. The sublunar matter and what comes to be from it, is the lowest in rank and the lowest in place. Above it, both in rank and in place, are the celestial bodies, whose forms persevere in their matter eternally. (Indeed, the words 'matter' and 'form' are applied to them and to the things that come to be of the sublunar matter by homonymity [only].) Above them all in rank and dignity are the separate intellects, which are forms that do not inhere in a substrate. This is why, when [Solomon] wanted to intimate that the human soul conjoins with one of the separate intellects, he said that it 'rises upward' [Eccl. 3:21].

And when I realized that these notion[s]—the conjunction and the rising upward [of the soul]—are one and the same and are the main objective of this book [Ecclesiastes], I resolved to add to my commentary on it short treatises by recent scholars about the conjunction of the separate intellect with man.

Section Three. 'Encyclopedias' of Science and Philosophy

Three so-called 'encyclopedias', works presenting overviews of the sciences and of philosophy, were written in Hebrew in the thirteenth century: Judah b. Solomon, *Midrash ha-Ḥoḵmah* (c. 1245), written in Arabic and translated into Hebrew by the author himself, at the court of Frederick II in Sicily; Shem-Tov Ibn Falaqera, *The Doctrines of the Philosophers* (c. 1245); and Gershon b. Solomon, *The Gate of Heaven* (c. 1280). The first two are based on their authors' reading in Arabic, the third on the author's reading in Hebrew and (it seems) extended discussions with Christian scholars. Here I will discuss only Shem-Tov Ibn Falaqera, the most erudite and knowledgeable among these three encyclopedists.

Part 1. 'Encyclopedias' Drawing on Sources in Arabic

3.1.1. Shem-Tov Ibn Falaqera (c. 1225–after 1290, Spain?)

Shem-Tov Ibn Falaqera, of whose life next to nothing is known, composed didactic works for the 'educated layman', whom he wished to introduce to philosophy in the spirit of Maimonides. His most ambitious work is certainly the huge *De'ot ha-pilosofim* (*The Doctrines of the Philosophers*), composed between 1263 and 1280. The late Mauro Zonta characterized it concisely as follows: 'It consists of an anthology of texts (mostly by Averroes, but also by Avicenna, al-Farabi, Alexander of Aphrodisias, Themistius and other authors, including Aristotle himself) about physics and metaphysics, directly translated from Arabic into Hebrew, so providing a complete and detailed commentary on Aristotle's *Physica, De generatione et corruptione, Meteorologica, De animalibus, De anima, De caelo* and *Metaphysica.* [...] Ibn Falaqera did not restrict himself to a mere reproduction of the contents of Averroes' commentaries: in his encyclopaedia, he [also] deals with non-Aristotelian subjects, such as mineralogy and botany.'[107]

Document 3.1.1.1. Shem-Tov Ibn Falaqera, preface to *The Doctrines of the Philosophers* (excerpts).

The Hebrew text given here is its *editio princeps* (see Part Three). My translation is based on this edition.

Ibn Falaqera's preface to *De'ot ha-pilosofim* is rather long. Below I translate the sections in which the author discusses his reflections about 'what I should translate' and justifies the study of philosophy by Jews and thus his decision to write *De'ot.*

[2] Having given Him praise, let me say that it is well known and celebrated and agreed upon, both by the Torah and by the sages [i.e., the philosophers], that man's true realization [*or* perfection; *haṣlaḥah*] consists in the [intellectual] apprehension [*haśśagah*] of the Creator, blessed be He, and knowing Him as accords with the power of the human intellect. The true sages [i.e., the philosophers] also agree that man acquires that apprehension by knowing His works and by forming an intellectual conception of them. For man can apprehend something immaterial [*lit.* separate from matter] only via its works. This is why they [the philosophers] said that whoever wishes to inquire into the intellect must begin by inquiring what its works are. For the works are better known by

107 Zonta, 'Mineralogy, Botany and Zoology', 275.

us than are the forces of which these works are the result. Further, if we wish to know the works, we must first know the substrates [of those actions], for we know the substrates better than the works.

[3] R. Moses [ben Maimon] said: 'I have already let you know that there exists nothing except God, may He be exalted, and this existence [i.e., the world]. And there is no demonstration of His existence, may He be exalted, except that deriving from this existent [world], taken as a whole and from its details. Accordingly, it necessarily behooves one to consider this existent [world] as it is, and to derive premises from what is perceived of its nature. It follows that it is necessary to know its form and its perceptible nature, and then it will be possible to infer from it concerning what is other than it'.[108] End quote.

It seems to me that his words imply that this [mode of scientific inquiry] is necessary for anyone whose [intellectual] apprehension is based upon inquiry; but someone, like a prophet, who receives intellectual emanation, will apprehend the truth without engaging in inquiry.

[4] [Hence,] whoever employs [scientific] inquiry in order to achieve [intellectual] apprehension will have to investigate [scientifically] the created beings and study the words of the sages [viz. the philosophers], who devoted their lives to their investigations, seeking to establish the truth of their essences [i.e., of the created beings]. Now our [own] sages of blessed memory did not compose books on these subjects, or perhaps they composed such books and these were lost during the long exile and [owing to] our subjugation by the nations. As [the prophet Isaiah] said: 'the wisdom of their wise men shall perish' [Isa. 29:14]. [Therefore,] whatever is found among the statements of the [Gentile] philosophers [*lit.* sages {who follow the method of} inquiry] that is true and is consistent with our faith, one should accept and praise them for it; and whatever is untrue, relinquish and reject.

[5] Aristotle said: 'We must praise the ancients for their inquiries, for their inquiries are a cause [allowing us] to arrive easily at the truth. And if the earlier [sages] had not existed, the later [ones] would have been unaware of much of the truth; [indeed,] if the earlier had not existed, the later would not have existed at all.'[109] End quote. It is improper for the inquirer [*maśkil*] who desires to know the truth to reject the many things that the philosophers have discovered and made evident by true demonstration and which are indubitably as they have stated, on the grounds that they [the philosophers] are not of our

108 *Guide*, 1:71, in Samuel Ibn Tibbon's translation; English translation adapted from Maimonides, *Guide of the Perplexed*, tr. Pines, 183. See Falaqera's *Moreh ha-moreh*, 1:71, ed. Schiffman, 173.54–59.

109 Aristotle, *Metaphysics*, 2.1, 993a31–b14; and *Sophistical Refutations*, chap. 34.

nation. Rather, one should hear the truth from one who says it and should not focus on the speaker, but on the truth of his words.[110]

[6] Rabbi Moses [ben Maimon] said: 'With respect to anything whose truth has been established by sound proofs—we do not rely on the person who stated it or taught it, but rather on the proof that was discovered and the reason that became known.'[111] End quote. Clearly there are many things that are true and whose truth is indubitable, but which, being unfamiliar to a man, his mind will not tolerate when he hears them. This will happen either owing to the shortcoming of his intellect, or because he lacks familiarity with these things, or because he was accustomed to believing their contrary. For it is human nature that it is difficult for a man to abandon long-held beliefs in a short time. This is why when one hears such [new] things, he hastens to deny them and to make up his mind that they are the opposite of the truth. However, when he examines them carefully, and probes into them without heeding what he was accustomed to regard as true since his youth, then he will discover their truth and will realize that things are not as he had thought.[112]

[7] Porphyry of Tyre said: 'When one hears something strange to which he is unaccustomed and whose contrary he had believed for a long time, he should not push it off and reject it without inquiring into it; nor should he inquire into it as if it were a falsehood and untrue. Rather, after he hears it, he should inquire into it and, if he finds that it is true, he should scrutinize it [further]. For there are in those [unaccustomed] ideas many things that require the student to make additional investigation, or to have doubts about some of them; however, this does not happen only with this kind of [unaccustomed] doctrine, but is a universal feature of all doctrines, in most cases. For in all these matters there are things that are impossible or difficult to grasp, and oftentimes the error is due to our intellect's incapacity to grasp them. Therefore, it is inappropriate for us to relinquish any doctrine we have not grasped or that is beset with doubts. For if we do this, we will not acquire solid knowledge. Consequently, we must inquire into all ideas that have been advanced or that can be advanced, and investigate them. And we will choose those [doctrines] that we find clearly excel the others and follow with necessity from the intellectual [*or* intelligible, intellected] principles that are accessible to the senses, even if there remain some enigmatic things in them.'[113] End quote.

---

110 Falaqera here paraphrases a well-known statement by Maimonides (found, for example, in his foreword to the *Eight Chapters*).

111 Maimonides, *Laws of the Sanctification of the Moon*, 17:24.

112 Falaqera here directly draws on Maimonides, *Guide*, 1:31.

113 Maimonides discusses this topic in *Guide*, 1:31, but when Falaqera comments on the pas-

[8] This is why I, Samuel son of R. Judah Ibn Tibbon [*sic*[114]] saw fit to explicate in this composition the ideas of the men of [philosophical] inquiry concerning everything that exists. These are the sages who are called 'true philosophers'. [And I refer to] the ideas that they take to be those of the true philosopher. For the word 'philosopher', whose meaning is 'love of wisdom' [*or* 'of knowledge'], denotes two kinds of perfection: the first species is that of ethical perfections,[115] and the second species is that of intellectual perfections. Therefore, among them, the true philosopher is someone whose goal is to achieve knowledge of truth and to practice kindness and truth—[who] in all his deeds aims to achieve these two kinds of perfection, and all his efforts in his life are to attain them as completely as humanly possible and as far as he is able to. This is what the noun 'philosopher' denotes according to them, and this is the meaning of the noun *ḥasid* in our language, [that is,] someone who has acquired the ethical and the intellectual qualities. [...]

[9] The reasons that made it necessary to write down these doctrines in this composition are two.

[10] The first reason, which is of greater necessity, is that any of our people who desires to understand these doctrines will know them with true knowledge, which is demonstrative knowledge, so that he will clearly distinguish the true from the false and the impossible from the possible. For one of the things that cause those who study the statements of the philosophers to deny the truth contained therein is that they did not understand their truth and did not fully grasp the depth of their demonstrations and reasons. This is one reason leading [people] to dispute them on many points. Furthermore, one of the reasons leading [people] to believe anything they [the philosophers] say is that they assume that there is a proof for all their doctrines. However, recent [philosophers] have stated that while in matters of natural science only few of their statements are subject to doubt, there are many doubts about what they say in metaphysics. Therefore, in this composition it will be made clear which of their doctrines have been proven, according to their view, and which have not been proven and were asserted by them [solely] according to what seemed right to them.

---

sage in *Moreh ha-moreh* he draws on Aristotle, with no mention of Porphyry (see Schiffman's edition, 131, with note). The Porphyrian quotation itself does not seem to be known from other sources; see Johnson and Freudenthal, 'A New Porphyry Fragment?'.

114 This statement of authorship appears in both manuscripts. However, it has been conclusively established that the author of *De'ot* is Shem Tov Ibn Falaqera (see *HÜ*, § 2) and that the ascription to Samuel Ibn Tibbon is a scribal error, possibly due to similar acronyms of the two names.

115 *Šelemut yeṣirit*, a term that is frequent in Falaqera's writings.

[11] For this reason, I sought to translate these doctrines from Arabic into Hebrew and to assemble them from the books in which they are scattered. Thus, anyone who wishes to become acquainted with them may find them in a single work and will not have to fatigue himself by reading all those books. For this composition includes all their doctrines in matters of natural science and of metaphysics, both the general principles and the particulars.

[12] The second reason [for composing this book] is this. Most of those doctrines that have been translated into our tongue are muddled and not rendered correctly. [Consequently,] one who reads these translated books is unable to gauge their truth. Many of these [translated] doctrines are from the ancient philosophy that has been rejected by recent [philosophers]. This is not the case only with books translated into our language; for there are compositions that have been translated into Arabic that ascribe to the philosophers many doctrines that are not their true ideas. [...]

[15] I say: the philosophers [*lit.* sages {who follow the method of} inquiry] have agreed that [all] the true ideas of the philosophers are included in the ideas of Aristotle, the Greek philosopher. They said that he 'pondered, and sought out, and set in order' [after Eccl. 12:9] the ideas of the ancients who had lived before him. He pointed out their mistakes and separated the truth contained in their statements from the falsehoods immixed in them, and made human wisdom [*or* science] known. [...]

[17] For this reason, my statements in this composition [consist] of his [Aristotle's] words as interpreted by the philosophers who share his ideas. For understanding the actual text of everything he said would be very difficult, even for philosophers, all the more so for other readers. The sage Averroes said: 'It took a long time for the commentators to grasp the meaning of Aristotle's works, almost as long as the time in which philosophy was practised [*lit.* perfected] [up to Aristotle], or nearly so.'[116] For this reason, I will not reproduce his actual state-

116 This is possibly a paraphrase of a well-known sentence in Averroes's Proemium to his *Long Commentary on the Physics*: 'The author of this book is Aristotle, the son of Nicomachus, who is well known in science among the Greeks, and who is the author of other books on this art and books on the art of logic and treatises in the divine art. He is the one who originated these three arts, i.e., the art of logic, natural science, and divine science, and it is he who completed them. Indeed he originated them, for what one finds of others in these things should not be posited as a principle of these arts, nor [even] as a part of them. Further, what one finds of others should not even be considered as difficulties in these arts, let alone principles of them. This is clear from what he relates in his books of their arguments in these things. This is also clear from the extant books that are attributed to them. But these are few since men kept away from the books of those who preceded him [Aristotle] after seeing the books of this man. [...] As for Aristotle's having completed them [logic,

ments everywhere. Moreover, those who came after him have explained these doctrines and resolved the greater part of the doubts that were expressed with respect to them, and answered most of the problems that were raised about them.

[18] There is nothing in this entire composition that I will say out of my own mind. Rather, everything I write will be Aristotle's words, interpreted by the sage Averroes. For he was the last of the commentators and selected the best of their commentaries and those that [best] fit Aristotle's doctrine. And Aristotle's doctrines are the doctrines one should study, in preference to those of the other ancients. All the more so because, in our own time, the doctrines of the other sages of the nations have been forgotten, and are remembered only insofar as they are mentioned in his books. Therefore, I will set down of these doctrines only what is necessary for this composition. […]

[19] The sage Averroes said: 'Aristotle's inquiry in these subjects is very much superior to [*lit.* distant from] that by any other sage. And what becomes clear to him [Aristotle] swiftly, other people need long and intensive investigation to clarify, from [i.e., drawing on] his words. This is why Aristotle said that had his statements on these matters not been put down in writing, it would have been impossible to discover them, or it would have been very difficult and taken a long time. Therefore, it happens that commentators have doubts about his statements, but then, after a long time, the correctness of his statements becomes clear, as well as the shallowness of the inquiry of the others when compared with his. Thus, with the divine power that was in him he brought philosophy into existence and perfected it. And this is something that happens only infrequently in the arts—in all arts, but *a fortiori* in this great art. Indeed, we have said that he is the one who brought philosophy into existence and who perfected it, because all that can be found in these matters by anyone else among the ancients does not [even] deserve to be the subject of doubts, much less to be posited as premises. Owing to this perfection with which God, blessed be He, endowed him, the ancients called him "the divine".'[117]

---

natural science, and divine science], no one who has come after him to this our time—and this is close to fifteen hundred years later—has been able to add a word worthy of attention to what he said.' Quoted from Averroes, 'The Hebrew Translation of Averroes' Prooemium', ed. Harvey, 83.

117 The entire passage appears also in Falaqera's *Moreh ha-moreh*, 2:22, ed. Schiffman, 278.177–190. For Falaqera's source, Schiffman refers to Averroes's preface to his *Long Commentary on the Physics*, as reproduced by Munk in his *Mélanges*, 361. An independent Hebrew translation is published in Averroes, 'The Hebrew Translation of Averroes' Prooemium', ed. Harvey, 70 (English translation on 83). Both texts deviate from Falaqera.

[20] He [Averroes] further said that 'none of those who lived after him [Aristotle], through our time, which is nearly 1,500 years, could add anything to what he had said; and it is extremely rare for such [accomplishment] to be found in a single man. And if it is found in a man, it is more reasonable to ascribe it to the divine rather than to the human. Therefore, there is nothing in Aristotle's statements that needs to be supplemented, as al-Ṣāʾigh [i.e., Ibn Bājja] thought. For he had many ideas about things [in Aristotle] that he did not understand; nor do we yet understand them, especially in the books on which the commentators' discussions have not reached us.' End quote.[118]

[21] Rabbi Moses [ben Maimon] said: 'Everything that Aristotle has said about all that exists from the sphere of the moon to the centre of the earth is indubitably correct, and no one will reject it, unless he does not understand it, or has preconceived convictions that he wishes to defend or that lead him to deny a thing that is manifest. But what Aristotle expounds with regard to the sphere of the moon and above it is very partial [i.e., is uncertain].'[119]

[22] He further said: 'What Aristotle said is the pinnacle of what man can apprehend by [philosophical] inquiry, as distinguished from what can be apprehended by prophecy.'[120]

[23] I [Ibn Falaqera] say that these sages exaggerated in what they said in his favour. The truth is what Alexander the Commentator said, namely, that there are fewer doubts about his ideas than about those of any other [thinker].[121]

[24] And here I plead *pro domo sua* and say: may no one suspect that when I write down these doctrines I believe any of them that contradicts our faith, for this would be an instance of suspecting blameless persons. In this composition, I will not write anything of my own, but will only set forth the view[s] of the philosophers. And I do not refer to knowledge acquired through divine inspiration or prophecy, but to knowledge acquired through [philosophical] inquiry and logical reasoning [*lit.* syllogism].

[25] I have divided this composition into two parts: in the first part I will discuss the natural existents; and in the second I will discuss divine existents.

118 As last note.

119 Maimonides, *Guide*, 2:22 (where the received text of Ibn Tibbon's translation is a little different); English translation adapted from Pines, 319–320.

120 The sentence also appears in Falaqera's *Moreh ha-moreh*, 2:22, ed. Schiffman, 273.91–92. The source of this statement (indicated by Schiffman) is Maimonides's letter to Samuel Ibn Tibbon, published in Marx, 'Texts by and about Maimonides', 380.

121 A similar sentence appears in Falaqera's *Moreh ha-moreh*, 2:22, ed. Schiffman, 278.193–194. Schiffman found the source of this statement in Maimonides, *Guide*, 2:3 and gives its source as Alexander's *On the Principles of the All*; similarly also in *Guide*, 2:22.

Section Four. Routine Philosophical and Scientific Translations: The Thirteenth Century

Part 1. Arabic–Hebrew Translations

4.1.1. Moses b. Samuel Ibn Tibbon (fl. 1244–1283, Provence)

Moses, Samuel Ibn Tibbon's son, who resided in Montpellier most of his life, belongs to the third generation of Tibbonid translators. Like his father, he followed two parallel intellectual agendas: he translated sundry books from Arabic into Hebrew; and he wrote a number of monographs on topics in religious philosophy. The latter are more conservative, and also less original, than his father's. Unlike Samuel Ibn Tibbon, he does not seem to have chosen the titles to be translated on the basis of his own intellectual interests as reflected in the monographs. This is easily understandable: when Moses embarked upon his career (1240s), there was already a sizeable group of philosophically and scientifically minded scholars in the Midi who needed Hebrew versions of advanced works in mathematics, science, and philosophy. Moses stepped into the breach and translated what was to form the nucleus of the Provençal Hebrew scientific library.

Moses began by translating a medical text by Maimonides (*Regimen of Health*, 1244), and then Maimonides's *Book of Commandments*, a sequel to his father's project of translating the Arabic writings of the 'Great Eagle' into Hebrew. But most of Moses's translations were of non-religious and non-Jewish works. They include Euclid, *The Elements*; al-Ḥaṣṣar, *Arithmetic*; Geminus, *Introduction* [to Ptolemy's *Almagest*]; Averroes, nearly all the epitomes of Aristotle's works; Baṭalyusi, *Imaginary Circles*; al-Fārābī, *On the Principles of Existing Things*; the introductions to Euclid's *Elements* by Ibn al-Haytham and al-Fārābī; as well as a number of medical works. Moses Ibn Tibbon rarely reveals his motivation for choosing a text for translation; as already observed by Steinschneider, almost none of the extant manuscripts of Moses's translations carries a translator's introduction.[122] The rare exception is his translation of al-Fārābī's *Short Syllogism*.

Document 4.1.1.1. Moses Ibn Tibbon, a note on his translation of al-Fārābī's *Short Syllogism*.

Paris, BNF, MS héb. 917, ff.210a–b; Jena, Thüringer Universitäts- und Landesbibliothek, MS Rec. adj. fol. 10, f.43a.

---

122 Steinschneider, *HÜ*, § 312, 504.

I translated this short abridgement [*or* epitome] about the syllogism from Arabic into Hebrew, I, Moses son of R. Samuel, of Granada in Spain, may his memory be for a blessing. The translation was completed in the year 5015 of the Creation [1254/55]. And I was aroused to translate it because I discovered that my grandfather, the sage, translated the composition that follows this work by Abū Naṣr al-Fārābī.[123]

Document 4.1.1.2. Moses Ibn Tibbon, a note on his translation of al-Fārābī's *On the Principles of Existing Things*.

Paris, BNF, MS héb. 893, f.65a.

From his father, Moses learned a number of lessons about the standards to be followed in translation. At the end of one copy of his translation of al-Fārābī's *On the Principles of Existing Things* we find a rare personal remark concerning the need for rigour in translating.

[Here] ends the book *The Principles of Existing Things* by the sage, the philosopher Abū Nasr Muhammad Ibn Muhammad al-Fārābī. [...] And I translated it from the Arabic language, I, Moses son of R. Samuel, son of R. Judah, son of R. Saul b. Tibbon the Spaniard. Its translation was completed in the month of Marheshvan of the year five thousand and nine of Creation [= Oct./Nov. 1248]. Praise be to God. And I checked [i.e., revised] it against another copy that reached me in the month of Av, the year five thousand eleven [July/Aug. 1251]. Praise be to God.

Document 4.1.1.3. Moses Ibn Tibbon, preface to *Ṣedat ha-derakhim*, his Arabic–Hebrew translation of Ibn al-Ǧazzār's *Zād al-musāfir wa-qūt al-ḥāḍir* (*Provisions for the Traveller and Nourishment for the Sedentary*).

Ibn Tibbon, Moses, Ibn al-Djezzar's *Viaticum* [...]. a) 'Vorrede des Moses ibn Tibbon', ed. Steinschneider; Munich, Bayerische Staatsbibliothek, Cod. hebr. 19, fol. 1a–b; tr. Freudenthal and Fontaine, 'Philosophy and Medicine in Jewish Provence', 17 (partial).

Zād al-musāfir wa-qūt al-ḥāḍir (*Provisions for the Traveller and Nourishment for the Sedentary*) by Ibn al-Ǧazzār of Qayrawān (c. 895–979), a disciple of Isaac Israeli's, is one of the most influential medical compendia of the Middle Ages. It was (freely) trans-

123 See also Steinschneider, *HÜ*, §15, 47, and §71, 149.

lated into Latin by Constantine the African as *Viaticum*, and it is best known under this title. This Latin version in turn served as the *Vorlage* for 'Doeg the Edomite', the pseudonymous physician who translated twenty-four medical works from Latin into Hebrew around 1197–1199.[124] In 1259, Moses Ibn Tibbon retranslated the work, this time directly from the Arabic original. Aware of the earlier translation, he justified his decision to produce a new version in the short preface to his translation.

Two points deserve particular attention. First, the preface makes it clear that Moses Ibn Tibbon compared Doeg's version with the Arabic original and found that the former was imprecise and condensed, but that the Arabic–Latin translator, not the Hebrew translator, was to be blamed for this: the latter simply followed his Latin model. Since Moses himself knew no Latin, he must have taken the trouble to compare the Arabic and Latin texts with the help of a Gentile confrère. Second, Moses wrote that the Hebrew translator's extensive use of technical terms in 'the language of the Gentiles' (Latin or Romance terms transcribed in Hebrew letters, called *leʿazim*) makes the work useless to some readers; he may have had in mind émigrés from al-Andalus who (unlike the Jews who were natives in the Midi) were not fluent in the Provençal vernacular. These points illustrate how attentive translators were to the needs of their intended readership. In the present case, the translation was not intended specifically for physicians, but rather for 'general readers' who travel and may require self-medication.

Moses Ibn Tibbon, the translator of this book from Arabic into Hebrew, says, as an apology to any understanding [reader]:

It is not because I view myself as erudite or knowledgeable in the two languages and expert in this science [medicine] that I am qualified to translate this lovely book. Nor did I dare to do this work in order to diminish the deference [due to] the first translator [Doeg the Edomite], or to cause him and the good he did to be forgotten. Rather, I ventured to try my hand in this work when I saw that he who had translated this work into the tongue of the Gentiles [i.e., Latin, viz. Constantine the African] abridged it throughout, obscured what is clear and obfuscated what is evident, and altered the contents. With respect to most of the prescriptions, he abridged the accounts of their utility, of their powers, of their composition, and of their modes of preparation. He increased or reduced the weights, omitted some of the ingredients, separated what belongs together and combined what is separate.

All this is not the fault of the translator from the tongue of the Gentiles into the Holy Tongue. However, he specified most of the names of the illnesses in

124 See §1.2.3

the tongue of the Gentiles, and these names are understood only by someone learned in this art, who learned them [here one or more words seem to be missing]. I therefore resolved to translate it from the language in which it was written [Arabic], in a clear style that can be understood by any literate person. I paid no attention to the purity of the style or to the concision of the language. In this translation I followed my customary manner in the other translations, [namely] to translate word-for-word as far as I can; or, when our language makes this impossible for me, to leave out a little, or add, or postpone or advance [something]—all this in order to avoid any error and not miss [the author's] intention. For I did not wish to rely on my apology and [then] depend on my own understanding and express what I understood in a concise and elegant language. I thus chose to try to stick to the author's intention and not [strive] for elegant language. I observed that when I follow this method, if I err in a word, it can be corrected quickly; this is not possible if I use the language to express what I [myself have] understood of the text. [...]

4.1.2. Jacob b. Abba b. Samson Anatoli (c. 1194–1256, Midi and Southern Italy)

Jacob Anatoli was primarily a translator from Arabic into Hebrew, indeed a particularly expert and prolific one. He was the son-in-law of Samuel Ibn Tibbon, with whom he also studied, notably mathematics and astronomy. Around 1230, he was invited to the imperial court in Naples, which in those years was investing heavily in the translation of Arabic philosophical works into both Latin and Hebrew.[125] Jacob Anatoli confirms that he, too, was generously paid by Emperor Frederick II, as were the Latin translators (Document 4.1.2.1). This contrasts with the lack of similar reports of remuneration by other Hebrew translators in Provence, who rarely worked for a state or a church institution: as a rule, translators were associated with 'patrons' from whom they received symbolic reward only. At the court, Jacob 'associated himself' (as he says) with Michael Scot, whose views on some matters he reports in his writings.

Jacob Anatoli translated Averroes's Middle Commentaries on Aristotle's *Organon*, and (later) astronomical works. Most noteworthy is the fact that with the help of a Christian colleague Anatoli checked his Hebrew translations of the *Almagest* and al-Farghānī against Gerard of Cremona's Latin versions of these works.[126]

Along with his intensive activity as a translator, Anatoli was also a preacher who set out to diffuse Maimonides's 'enlightened' conception of Judaism beyond the narrow

125 See Gauthier, 'Le début du premier "averroïsme"'; Sirat, 'Les traducteurs juifs'; Cohn, 'Jüdische Übersetzer'; and Shatzmiller, 'Au service de la Cour de Naples'.

126 Zonta, 'La tradizione ebraica dell'Almagesto di Tolomeo'.

circles of scholars who could read the *Guide of the Perplexed*. He collected his sermons in his only original work, *Malmad ha-talmidim* (*Goad for Students*).

Document 4.1.2.1. Jacob Anatoli on the support he received from Emperor Frederick II.

Leopold Dukes, 'Literaturhistorische Miscellen', *Literaturblatt des Orients*, 13 (1848), annexed to *Der Orient*, 9. Jahrgang, 13 (25 March 1848), cols 193–197, at col. 196; Paris, BNF, MS héb. 930, f.60a, and MS héb. 994, f.211b.

Said the translator:

Blessed be God, who has not turned away my prayer or His steadfast love from me [Ps. 66:20], Jacob son of R. Abba Mari son of R. Samson Anatoli of blessed memory. He has given me strength to complete, in [the month of] Second Adar in the year [4]992 [February–March 1232], in the city of Naples, the translation of the books on logic composed by the Andalusian philosopher Ibn Rushd in a clear and elegant style. [They are] commentar[ies] on the books of the master of this discipline, the finest of the sons of Japhet, the paladin of the philosophers, Aristotle.

I have translated five books: four are by Aristotle, namely the *Categories*, the *Rhetoric*, the *Prior Analytics*, and the *Posterior Analytics*; and Porphyry's preface to them, the *Isagoge*.

Before I begin working on [i.e., translating] the remaining books in this discipline, I will review and study the translations of the aforementioned books, in order to correct errors as far as I can. And after I am done with this, I will undertake to complete the task with the aid of Him who giveth aid to all that stand in need of aid, and who has led our lord, the Emperor Frederick [II], the lover of science and of the students of science, to provide for me and support me and my family in full. May the Lord increase His grace to him and raise him above all the kings of the nations. And may the King Messiah come in his lifetime, and we witness it.[127]

Document 4.1.2.2. Jacob Anatoli, preface to the translation of Averroes's *Middle Commentary on Porphyry's 'Isagoge'*.

Averroes, *Middle Commentary on Porphyry's 'Isagoge'*, ed. Davidson, 1–3 (text), 93–94 (notes); tr. Averroes, *Middle Commentary on Porphyry's 'Isagoge'*, 1–3 (text) and 93–94 (notes).

---

127 Averroes, *Middle Commentary on Porphyry's 'Isagoge'*, ed. Davidson, x.

In the following document, Jacob Anatoli defends the study of logic by Jews, arguing that this discipline is an indispensable tool for controversies with opponents, notably Christians. He also notes and rebuts two talmudic passages that were occasionally viewed as banning such study and shows that, properly understood, nothing in these passages opposes the study of logic.

Those who seek understanding as men seek silver ask: 'Where is the crucible that will refine our dross?'

The science of logic makes reply: 'The furnace for refining understanding has been mine of old, wherein one may remove dross from silver so that there will come forth a vessel for the refiner.'

A crucible for refining the silver of understanding and a furnace for refining the gold of belief [cf. Prov. 27:21] are the portion of the science of logic. For therein every statement may be tested to determine whether it is sweet or bitter, evil or fair, strong or weak, whether it is proper and acceptable, small or great.

Speech designates the conceptions that are present in the human mind; those conceptions have their referents in things that exist outside the mind; and the totality of existent things proclaims the cause of their existence and professes that He created them. Therefore, everyone who truly desires to seek God stands in need of the science of logic.[128] For although it is not itself a science,[129] it nevertheless is very useful for refining each of the sciences, since no philosopher can possess perfection without it. Logic serves the philosopher as a tool serves the artisan.[130] This is true not merely in theoretical [= philosophical] science, but can also be very helpful in solving numerous difficulties in many places in the Pentateuch, in the rest of the Bible, and in the Talmud. Such is the first object of the science of logic.

The second object [of logic] is that it sharpens the mind and prepares [it for] study, like condiments, which season food and arouse one's appetite. This is of great utility to anyone who pursues knowledge.

There is an additional consideration here: logic is indispensable if one is to stand up to opponents who employ sophistic arguments. Our rabbis already

---

128 Implicitly addressing the traditionalists, Anatoli recapitulates a proof for the existence of God. Inasmuch as this proof appeals to logic, it demonstrates the latter's usefulness for bolstering arguments to which the traditionalists ascribe great importance.

129 Namely because it is only an 'instrument' (*organon*) required for the practice of science and philosophy.

130 The word כלי is used here in the sense of 'tool', but in the opening lines it was used in the sense of 'vessel'.

warned about that, saying: 'Be diligent in studying the Torah [so that you will know] what to answer the heretic' [*Avot* 2:14]. And in their warning, they meant not merely the heretic, but everyone who opposes the truth. It is well known that without studying the science of logic none of us [Jews] will be able to hold his ground against clever representatives of the other nations who oppose us.

Since I, Jacob, the son of Abba Mari b. Samson b. Anatolio of blessed memory, saw how numerous the wicked fools who presume against us in disputation and dialectic are, I became irritated with them, and my desire arose to translate this science as far as lay in my power.

The truth is that initially I had planned to unburden myself by first completing the work on the books of astronomy, a science difficult both to translate and to study, requiring, as it does, numerous books and great concentration. However, I was pressed by my brothers and friends, my companions and intimates, the scholarly and educated men of Narbonne and Beziers, who were eager to approach this subject [after Ex. 36:2] and out of affection for them I shall bend my shoulder and assume an added burden. With the aid of Him who giveth power to the faint, I shall sow in the morning in the science of astronomy, and in the evening my hands will be steady in the science of logic.

If someone should object that our rabbis prohibited this science, inasmuch as they said 'Restrain your sons from logic [*higgayon*]' [B *Berakot* 28b], I will reply: 'Listen to what your own mouth says!' The rabbis said, 'Restrain *your sons*', not 'Restrain *yourselves*.' [...]

Considering all this, I decided to make a mighty effort in the present discipline, although I realized that it is improper for a person like myself to approach the mountainous task of translation rashly [alluding to Ex. 19:24]. The way is too long for me, and the stumbling blocks are too numerous, both because of my deficient knowledge of Arabic and because of my imperfect understanding. Nevertheless, I thought that even the small amount of honey to be found in my translations might enlighten the eyes of the reader so that he will be able to set straight whatever might be awry. And in science, even a small benefit has value for those who love it. [...]

4.1.3. Jacob b. Makhir Ibn Tibbon (c. 1236–1305, Provence)

Jacob b. Makhir (known in the Latin world as Don Profeit Tibbon), a grandson of Samuel Ibn Tibbon, was active as an astronomer and as a translator of mathematical, astronomical, and philosophical works. In 1306, a heated large-scale epistolary controversy about the legitimacy of the study of philosophy and science flared up among the Jews of Provence and Christian Spain. As a scientist and scion of the Tibbon family, Jacob b. Makhir was naturally an ardent defender of the legitimacy of the study of the rationalist sciences, and even the obligation to engage in it. He contributed to the controversy a single epistle in which he expressed his beliefs, including the following popular justification for the translation of books of science.

Document 4.1.3.1. Jacob b. Makhir on the value of science for the defence of Judaism.

Aderet, *Teshuvot ha-Rashba*, ed. Dimitrovsky, 510.54–57.

[We must demonstrate] to the Gentiles our knowledge and our understanding [in the books of science and philosophy], so that it not be said: 'they are devoid of all intelligence and science.' We have to follow the ways of the most enlightened of the Gentiles, who have translated the books of science into their various languages, even when their statements and proofs contradict their own law and faith. They respect the sciences and those who master them, without asking what their faith is.

Document 4.1.3.2. Jacob b. Makhir, preface to the translation of Euclid's *Elements.*

Krafft and Deutsch, *Die handschriftlichen hebräischen Werke*, 173–174 (no. CLXXIII).[131]

Scholars are as yet undecided as to whether the *Elements* was translated into Hebrew by both Moses Ibn Tibbon and Jacob b. Makhir, or whether Moses produced a translation that b. Makhir subsequently revised. In any event, a preface to the translation, by Jacob b. Makhir, dated 1289, is appended to some manuscripts and is translated below.

At the time, the legitimacy of the study of the sciences was under attack by traditionalists. Jacob b. Makhir defends it. He asserts that the ancient Israelites mastered the very same sciences that at present they have to learn from the Gentiles. Thus, scientific study amounts to 'return[ing] a lost property to its [Israelite] owner' and cannot possibly be religiously illegitimate.

131 I am grateful to Dr Ofer Elior for permission to consult his unpublished edition, forthcoming in *Mada'ey ha-Yahadut.*

Glory to God, who created man in His image [Gen. 1:26]. He has made [man] more intelligent and wiser than all living beings on earth. He has made man stand erect, so that he may seek numerous calculations [Eccl. 7:29] and frame thoughts stretching from the lowest point of the earth to above the Cherubim.[132] [...]

However, because our fathers sinned, we have forsaken the land and we were cast out of our dwellings [after Jer. 9:18]. The wisdom of our wise men has perished, and the understanding of our scholars has been lost.[133] Not one in a thousand of [their] books of the sciences is extant. [Hence,] we must seek them [the sciences] in the [works of] the ancient scholars of the nations, who [contrary to us] have been tranquil from their youth and undisturbed and settled on its lees [after Jer. 48:11]; their cleverness [*lit.* sense] has remained intact [after Jer. 48:11] and [their wisdom] has not been lost.

Geometry [*lit.* the science of measure, referring in fact to all mathematical sciences] is the foundation of all the sciences, especially the mathematical sciences. And this book [*The Elements*] is the foundation and root and starting point for all other books in this science. Wherefore my heart stirred me up [after Ex. 35:21], me, Jacob son of R. Makhir, to translate it from Arabic into our tongue, [so as to] return a lost property to its owner and to make us a name like the name of the great [after 2 Sam. 7:9]. [Thereby we will] rebuff the reproach of the Christians [*lit.* the uncircumcised], who allege that we are devoid of all science.

In my heart I know that I have not studied the sciences and that I lack human understanding [after Prov. 30:2–3]; also, my knowledge of Arabic is slight. But [in writing this translation] I relied on two things: first, this science is not verbose, and a few words embrace it; second, I was reinvigorated by the Name of the One who gives power to the faint [Isa. 40:29] and who gives subtlety to the simple, to the young man knowledge and discretion [Prov. 1:4].

---

132 Jacob apparently alludes to the Aristotelian view that man's upright posture endows him with an intelligence greater than all other living beings. See Brague, *Aristote et la question du monde*; and Freudenthal, *Aristotle's Theory of Material Substance*, 56–60.

133 The Prophet Isaiah formulated this verse as a prophecy for the future; Maimonides (introduction to the *Mishneh Torah*), whom Jacob presumably quotes, had already put the verse in the past tense, making this purported situation into the fulfilment of a prophecy.

Section Five. Routine Medical Translations: The Thirteenth Century

Part 1. Arabic–Hebrew Translations

5.1.1. Nathan b. Eli'ezer ha-Me'ati (fl. 1275–1300, Rome)

In Rome, in 1279, the physician Nathan ha-Me'ati (i.e., of Cento) translated Avicenna's (Ibn Sīnā) *Canon* (the *Kitāb al-Qānūn fī l-Ṭibb*) into Hebrew.[134] He produced other translations as well, but this was the most important and also the longest. Unlike his confrères in Provence, who usually came from families in which fluency in Arabic was passed down from father to son, Nathan, like all scholars native to central or northern Europe, had no roots in Arabic culture. (The exceptions are some immigrants from the Midi who lived in Italy for a time, such as Zeraḥiah b. She'altiel Ḥen of Barcelona, who lived in Rome between 1277 and 1290.) Nathan acquired Arabic as an adult, during the wanderings to which he alludes in the preface to his translation of the *Canon* (below), but without giving any details.

Nathan's translation of the *Canon* raises two questions. First, why was the translation of this central text of medical education delayed so long—the first Arabic–Hebrew translations of a medical work date from 1199—long after many other medical works had made the linguistic passage? Second, why was this fundamental text not translated in Provence, the heart of Hebrew translations from Arabic? Chaim Rabin answered both questions.[135] Most Jewish physicians in the Midi knew Arabic and could read the *Canon* in the original; in Italy, by contrast, very few Jews knew Arabic, or for that matter Latin. So in the thirteenth century, when Christian medicine in Italy came to rely increasingly on Avicenna, Jewish physicians there, but not in Provence, felt the need for a Hebrew version of the work in order to keep pace with their Gentile colleagues. In his preface, Nathan alludes to the fact that the Jewish physicians were derided and scorned by their Christian confrères, so by translating the *Canon* he was responding to a pressing *local* social need. Nathan basically followed the lead of the Tibbonids in his translation style and method.

134 In 1280, while living in Rome, Zeraḥiah ben She'altiel Ḥen revised this translation partially, but this has no bearing on our topic.

135 Rabin, 'The History of the Translation of the *Canon* into Hebrew'.

Document 5.1.1.1. Nathan ha-Me'ati, preface to his Hebrew translation of Avicenna's *Canon*.[136]

Nathan ha-Me'ati, 'Nathans Vorrede zur Übersetzung des *Kanons*', ed. Steinschneider; Leibowitz, 'The Preface by Nathan ha-Me'ati', 4–6; tr. Leibowitz [and Bodian], 'The Preface by Nathan ha-Me'ati', I–IV.

Said Nathan of Cento Ba-Gad [= the Fortunate[137]], son of Eliezer, may God protect him:

I am the man who has known wanderings; a stranger to my brethren,[138] bereft of comrade and friend. From my youth, my heart has never been still and has found no rest; it has gone to and fro, from one misfortune to another; my eyes have wandered to the end of the earth; my feet have almost failed me [Ps. 73:2] at the corners of the world; my fate has hurled me like a ball; with a mighty toss I have been displaced,[139] but now I have become weary;[140] it has exiled me among various nations and peoples; and among various languages it has made me roam; I have learned the tongues of these foreign nations among which I have dwelled for a time and have lingered until now [echoing Gen 32:5].

I have acquired a pure speech [= Hebrew] and the tongue of the learnèd [Arabic].[141] The lovely and the pleasant:[142] the tongue of Eber [= Hebrew] and that of Arav [= Arabic] joined together. I have stowed them in my heart, and they have become one in my hand [Ezek. 37:17].

I saw that every day Christian scholars curl their lips at us [Ps. 22.8] and increase the force of their despising words. For we are dispossessed of the medical works that were composed in the days of [King] Solomon and of other sages of our nation who lived before us: they possessed them, but we have lost them; our forefathers have told us of them, but we have not seen them.[143] I therefore put on the cloak of anger and girded up my lazy loins; and I said to my mind

136 The text is given almost *in extenso*, because the periodical in which it was published is not widely available.

137 'Ba Gad'; see Gen. 30:11.

138 See Ps. 69:9.

139 Isa. 22:17.

140 Job 16:7.

141 The locution *śafah berurah* is taken from Zephaniah 3:9, but it was also the title of a book on Hebrew grammar by Abraham Ibn Ezra. *Lešon 'arumim* (Job 15:5) was taken by traditional exegesis to denote the refined and polite speech of the learnèd.

142 2 Sam. 1:23.

143 The idea that secular wisdom was known to King Solomon and the talmudic rabbis but was lost during the hardships of the Exile is a frequent trope. See n. 18.

[*lit.* thoughts]: let us be strong and powerful for the sake of our people [2 Sam. 10:12], the remnant of the Exile, so as to revive wisdom as a great deliverance [after Gen. 45:7].

[Moreover,] I witnessed those 'pleasant scions' [Isa. 17:10], the excellent sons of Tibbon; they are in harmony with us [Gen. 34:21], for their purpose was to raise the horn of wisdom, to draw and pull it out of the miry bog of Arabic;[144] they have delivered it from sombre depths of darkness [= Arabic culture].[145]

In this matter I have followed in their footsteps, to serve my people as they did, to translate this great book for them [viz., readers of Hebrew]. [In translating] most terms I have followed the way they have paved in their versions of other books of medicine and used their phrases and terms.

Besides,[146] in his use of language and reasoning this author [Avicenna] deviated from the way of other authors. For he was a Persian by birth. He was excessively brief in his explanations, and in his book his language is awkward [*lit.* heavy]. I was therefore compelled to introduce new words in many places—they are newly coined and lay deserted until today. For the most part, I derived them as far as I could from [other] Arabic scholars. Owing to the difficulty of his prose, they, too, had difficulties catching the content and derivation of some of the words he used, except by [gauging] his intentions from the context.

Moreover, I was occasionally obliged to leave blanks, for the copy [from which I translated] was faulty in parts. But I entreat every reader of this book not to modify anything except on the authority of a skilled and famous linguist. However, if God preserves me and prolongs my days I will search in other copies and inquire of masters and then understand these words correctly.

After this [i.e., when the translation was completed], I conceived the idea to compose another chapter, with which to complete this work: in it I will indicate most of the difficult Arabic and Hebrew words occurring in this work; and also the different names of drugs, along with those used in our language, so that the reader will not mistake their meaning.[147]

My worthy intention in translating this book was to protect and guard [us] from the stream of devious talk [Prov. 4.24] of the proud [i.e., the Christians; cf. Ps. 123:4]; [it is] an action [aiming] to give the lie to their words of humilia-

144 See Ps. 40:3. Nathan plays on the equivocation of the word יון, which can be read as *yaven* [= mire or miry] or as *yavan* [= Greece]

145 For 'darkness' the author uses *qadrut*, derived from the same root as *qedar*, meaning Arabic.

146 See Gen. 20:12.

147 This small lexicon of medical terms has been edited by Gerrit Bos in 'Medical Terminology in the Hebrew Tradition'.

tion and mockery. Hence, every reader shall bless me, along with all the other authors [*var.* providers]. The reader shall not wish ill to me on account of its deficiency; rather he shall judge in my favour.

May the youngest in his father's house, Nathan ha-Me'ati, the poorest in the house of Israel [after Judg. 6:5] be blessed with the 'Benediction of the Wise'.[148] [...]

Document 5.1.1.2. Nathan ha-Me'ati, preface to his translation of Maimonides's *Aphorisms*.

Maimonides, (*Medical*) *Aphorisms of Moses*, ed. Muntner, 9–10.

Nathan ha-Me'ati said:

Praised be the Lord, for so He deserves on account of His Glory and Greatness, and we praise Him as far as humanly possible, (although in truth He is) above every blessing and praise.

One of my dear friends, who wishes me well, asked me to translate this book, for he desires and has pleasure from the books of the rabbi, the righteous teacher, and thirsts for his texts. I complied with his wish and translated it for him and for others like him who wish to follow the wise men [*or* the sages]. [...]

Document 5.1.1.3. Nathan ha-Me'ati, preface to his translation of Hippocrates's *Aphorisms* with Galen's commentary.

Maimonides, (*Medical*) *Aphorisms of Moses*, ed. Muntner, 9, n. 1.

Nathan ha-Me'ati said:

When I heard the Christian scholars abusing and decrying us on account of the many books we do not have, I was seized with great zeal and I took my inkstand and translated this book, so that it may shelter the sons of our people and [provide] them with honour and fame. Today, I removed from us the Christians' reproach [Josh. 5:9], when they claim that we have lost our science.[149]

---

148 According to the Talmud, when one sees a Jewish scholar one must recite the benediction 'blessed be He Who has imparted of His wisdom to them that fear Him' (B *Berakhot* 58a).

149 The last sentence is not in all MSS.

Part 2. Latin–Hebrew Translations

5.2.1. Jacob, known as *ha-qaṭan* [= *minimus*] (fl. 1280, Provence)

One Hebrew translation or adaptation of an *antidotarium*, possibly that of Nicolai,[150] includes a preface by the translator, who calls himself *Ya'aqov ha-qaṭan*, 'little Jacob' (or *minimus*). His identity has been much discussed, but I will not enter into that issue here. The translator was certainly a physician, and it seems that his *floruit* was the last two decades of the thirteenth century. This translation is an instance of a routine Hebrew translation from Latin: towards the close of the thirteenth century, an increasing (although still small) number of Jewish physicians knew Latin and could produce Hebrew translations of Latin texts. At the same time, however, most of their Jewish colleagues remained literate only in Hebrew. Those with Latin thus felt it their duty to translate important medical texts for those who were without, so that they could compete in their field. Note the absence of apologetic rhetoric in the preface.

Document 5.2.1.1. Jacob *ha-qaṭan*, translator's preface to the *Antidotarium Nicolai.*

Steinschneider, 'The Translator's Preface to the *Antidotarium*'.

Thus spoke Jacob *minimus*, the fortunate:

This book is called in their tongue *Antidotarium*, meaning 'the comprehensive [work]'. The scholars of Montpellier gathered around it [i.e., around the project of producing it] and appointed Nicolas as their chief. They then assembled a pharmacopeia, so as to allow a physician to find compounds that are regularly prepared by the druggist for any ailment and any temperament. The physician can then obtain as much as he needs of them [the medications] and according to what the patient can afford. He will not have to expend a large amount on preparing the compounds for him [the patient]. When he finds ready-made compounds, he will purchase of the one that is beneficial for him [the patient] a little or much, according to his judgment, and [the patient's] needs and [financial] capacity. If it is within his power [i.e., if he can afford it], he will find in this book good and important compounds, some hot and some cold, some tonic, some purgative, and some constipating. You will also find therein syrups, balms, oils, powders, and bandages. Each of them has

150 A number of Hebrew translations of *antidotaria* are known. See Steinschneider, *HÜ*, § 508, 811–816.

well-known effects and particular powers, and its virtue and effect are detailed [alongside its name]. All of these compounds or most of them can regularly be found at the druggist's. And if one of them is not in stock, then, because everyone knows them, they [the druggists] know how to prepare them when they are needed. One has only to refer to them by their name, for every druggist has this book, with its exact text.

In my opinion, when someone has the books of [medical] theory, then this book [the *Antidotarium*] is almost sufficient for the practice, and he can act well following it. And because I found that it has great utility for life and was a great succour for medicine, I made up my mind to translate it and transfer it from their language into ours. And I made the effort to arrange it according to our alphabet.

I did not seek to translate it word-for-word, but only [keep] most of the meaning. However, the names of the drugs and their weights I will not alter.

Because all people are in need of medical science, every learned man should learn something about medications. For no one knows when he may happen to go on a long voyage and not find a physician there [at his destination], or when he may fall ill or recover and have to draw on his own knowledge.[151] [Medicine] is the most important and most pleasant of all sciences, save the Torah, and is a magnificent science.[152]

Section Six. Routine Philosophical and Scientific Translations: The Fourteenth Century

Part 1. Arabic–Hebrew Translations

6.1.1. Samuel b. Judah b. Meshullam (1294–after 1343, Marseilles)

The life and work of Samuel b. Judah can be reconstructed from the postfaces to his translations, in which he recounted the circumstances of each of them. The short texts provide an unusually detailed insight into his biography, his motivations for translating, and his methods of translation. These rich materials were thoroughly examined by

151 This argument also appears in the introduction to Ibn al-Ǧazzār's *Zād al-musāfir wa-qūt al-ḥāḍir* (*Provision for the Traveller and Nourishment for the Sedentary*), which Moses Ibn Tibbon translated into Hebrew (see Document 4.1.1.3.). But the wording there is different (see, e.g., Munich, Bayerische Staatsbibliothek, MS Cod. hebr. 19, f.2a), and there is no reason to think that our Jacob was acquainted with that translation.

152 A second short text follows in the manuscripts, but, as Steinschneider remarked, it is a late interpolation.

A. Neubauer and E. Renan, followed by Lawrence V. Berman. Here I recapitulate only the most basic facts.

At age eighteen (1312), Samuel b. Judah began studying philosophy and science, notably under Abba Mari b. Eligdor (Sen [= Senior] Astruc de Noves) in Salon-de-Provence. He completed his first translation, of Averroes's *Questions on Logic*, in 1320. He then moved to another field, in which he was pioneer in medieval Hebrew letters—ethics and political philosophy: he translated Averroes's *Middle Commentary on the Nicomachean Ethics* and his epitome of Plato's *Republic*. In the following years, he translated or revised the following works: Averroes's *Epitome of Logic* (1329, a revision of Jacob b. Makhir's translation); Jābir Ibn Aflaḥ's *Epitome of the Almagest* (1335, a revision of Jacob b. Makhir's translation); and Alexander of Aphrodisias's *De anima* (1324, revised and published 1340).

Samuel's postfaces to his translations give us true insight into their author's mindset. What emerges from them is the picture of a scholar who 'shows a great deal of idealism in his pursuit of philosophy, which makes his character attractive'.[153]

Document 6.1.1.1. Samuel b. Judah of Marseilles, postface to his translation of Averroes's *Middle Commentary on the Nicomachean Ethics*.

Berman, 'Greek into Hebrew', 303–304; tr. Berman, 'Greek into Hebrew', 304–306.

The commentary of Averroes on the *Ethics* of Aristotle, called in Greek *Nicomachean*, is finished. It is the first part of the two parts of political science known as 'the theoretical part'. Its relationship to the second part, which is called 'the practical', is the same as the relationship in the medical art of the contents of the *Book of Health and Sickness* to the contents of the *Book on the Preservation of Health and the Removal of Sickness*.

The second part is contained in the book known as the *Politics* by the Philosopher, but that book did not come into the hands of the sage Averroes, and he apologized for not commenting on it. Instead, he explained the scientific statements of that second part, which he found in the *Republic* of Plato. For this reason, I, the translator, went on to the first part [i.e., the *Politics*], even though one author did not compose them, since they are parts of one science and the same commentator explained them.

I, Samuel the son of Judah, may God preserve him, the son of Meshullam the son of Isaac the son of Solomon, may he rest in paradise, of Marseilles, translated this book, which is the first part of this science—that is, political

153 Berman, 'Greek into Arabic', 301.

science—in the twenty-seventh year of my life in the citadel of Beaucaire. Its translation was completed on Monday the tenth of Adar I in the year 5081 of the Creation [= 9 February 1321]. May the Helper be praised and blessed forever, Amen. Amen. Selah. [...]

Document 6.1.1.2. Samuel b. Judah of Marseilles, postface to his translation of Averroes's *Commentary on Plato's Republic.*

Averroes' Commentary on Plato's Republic, ed. Rosenthal, 106–107; tr. Berman, 'Greek into Hebrew', 307–311.

The commentary of Averroes on the scientific statements found in the *Republic* of Plato is complete. It is the second part of political science. Its translation was completed, in the city of Uzès, on the twenty-second of Kislev in the year 5081 of the Creation [= 24 November 1320]. I, Samuel son of Judah son of Meshullam son of Isaac son of Solomon Barbevaire of Marseilles, translated it.

When I translated it, I still did not have the commentary by Averroes on the *Ethics* of the philosopher Aristotle, which is the first part of this science that deals with matters dependent on the will [...]. I was unable to translate them on account of their profundity and difficulty. I therefore put myself to much effort until I procured Averroes's commentary on that part, written in a pure and clear style, as is his habit in all of his commentaries, and I roused myself to translate it. God in His kindness backed me, until the translation of the commentaries on the whole of political science was complete.

I revised the whole translation bearing on this science and corrected its mistakes in so far as I was able. The work of revision and correction was completed on the twenty-sixth of Elul of that year [= 20 Sept. 1321] in the citadel of Beaucaire, shut up and abandoned with the rest of our brethren, imprisoned in one of its fortresses called 'Redorta' [= Château de la Redorte]. [...]

Because of my liking for, and diligent attention to this science, I thought and conceived the desire to correct this translation again together with the Christian scholars, especially the first part of it, because they have the treatises of the Philosopher on that part, together with the commentary by Abū Naṣr al-Fārābī. By the Law and the Testimony [Isa. 8:20], as I had planned so would it have come about [after Isa. 14:24], if the aforementioned long and arduous imprisonment which has us in its grasp at this time had not prevented me from performing this task. But if God gives me longer life, takes me out of confinement and prison, and gives me a little free time, I will make an effort and will study and investigate it very closely, from the beginning of the science till its end, in order [i.e., the *Ethics*, followed by the *Republic* or *Politics*], so that the translation will be as correct as possible. […]

And in general, I say: O you who follow rationalist inquiry, the congregation of believers, truth blazes its own path, so show your gratitude to whomever deserves it. For, until this day, no part of this science was translated or came into our possession, neither by the philosopher nor by anyone else, except what is to be found in the *Book of the Principles of Existing Things* of Abū Naṣr al-Fārābī, which many of our nation have. It contains a little of the second part of this science, but nothing of the first part. I took courage and began to show the splendour of the greatness [Esth. 1:4] of this science and to present it in our language. Although, because of my deficiencies, the good that is in it is small, nevertheless the [present] treatise is a beginning; although it is meagre, yet is it great potentially.

In any case, awake, O you sleepers, from the deep sleep of laziness and ignorance, devoid of all [intellectual] possessions, [a sleep] that causes you to descend into the lowest pit, which is transitory, and which destroys absolute happiness. Awake, O you who are drunk with calamities, rejoice, take hold of this fair portion that God has allotted you. Hear, O you deaf, what these two divine kings have said. Look and behold, O you blind, the strong light that is in the heavens. Ask who created these wonders, which are more exalted and excellent than can be praised! Contemplate this vision! Your reward is with you and your recompense is before you [Isa. 40:10 and 62:11]. […]

Document 6.1.1.3. Samuel b. Judah of Marseilles, postface to his revision of Jacob b. Makhir's translation of Averroes's *Epitome of Logic*.

Neubauer-Renan, *Écrivains*, 559; Paris, BNF, MS héb. 956, f.350a; tr. Berman, 'Greek into Hebrew', 311–312.

The epitome of the *Logic*, which was composed by Averroes the Andalusian, the great philosopher and chief of the commentators, is complete—praise be to God—and its translation [too] is also complete. I, Samuel Marsilius son of Judah, may God preserve him, son of Meshullam son of Isaac son of Solomon, may he rest in Eden, translated it from Arabic into Hebrew.

It had already been translated before, but there were many mistakes in that translation, until the book became useless. Afterwards, outstanding contemporaries, knowledgeable in Arabic, came and tried to correct that translation, but they did not complete what they attempted to do. I might almost say that their correction was not worth consideration. Thus, the book remained full of mistakes, confused, and unintelligible to us, the congregation of [those] who follow rationalist inquiry.[154]

When I saw this, I recognized the great benefit that would accrue from this text, inasmuch as it is short, small in quantity, but of exceedingly great quality: it contains the choicest and most precious [material] from Abū Naṣr's book on logic, which is of high value and great excellence. For in this book Averroes collected the pearls upon which Abū Naṣr founded his book on logic, that is, [only] what seemed to him [Averroes] to be necessary and indispensable in accordance with his intention; he [Averroes] followed in his [al-Fārābī's] footsteps, deviating neither to the right nor the left. For this reason, I inconvenienced myself, although time pressed me, and I translated it afresh from the beginning.

And God is my witness if my intention was to disparage and criticize the first translator and his correctors in order to make a name for myself in this art. Not at all, by God's life! My only motive was the search for truth, which is its own witness and entirely consistent.

[Only] from God the Rewarder I ask a [spiritual] reward for my trouble. I completed this work on the twentieth of Tevet in the year 5090 of the Creation [= 13 December 1329] in the citadel of Tarascon. May the Helper be praised and exalted. Amen.

Document 6.1.1.4. Samuel b. Judah of Marseilles, postface to his translation of Alexander of Aphrodisias's *De anima* (from the Arabic of Ishāq ibn Ḥunayn).

Neubauer-Renan, *Écrivains*, 565–566; Steinschneider, 'The Translator's Epilogue at the End of Alexander of Aphrodisias' *De Anima*'; tr. Berman, 'Greek into Hebrew', 317–320.

The treatise of Alexander of Aphrodisias entitled *De anima* is complete and its translation is complete, thanks be to God. It was composed in Greek, and as far as I know, Ishāq ibn Ḥunain translated it from that language into Arabic, as you may see at the beginning of the treatise. After that, I translated it from Arabic into Hebrew.

It is profound and difficult to understand, because this is the manner of the author in general, and how much more so in treating this science, the science of the soul, which by its very nature is difficult and profound. The fact we mentioned, that it has been translated twice, from one language into another, increased the difficulty and obscurity. But although [the work] is difficult and obscure on account of these two reasons—by the nature of the science and on

154 See Steinschneider, *HÜ*, § 17, 54–56.

account of the [double] translation, as we have explained—nonetheless whoever was correctly schooled in science will be able to understand it completely.

Happy will be the one who is patient enough to finally understand its statements! For this treatise is utterly perfect, honoured, and precious, too great for praise. There are two reasons for this: first, its author is the most perfect of the philosophers, no one is like him, a sage and true prophet, as it were, in the camp of the philosophers. Even Aristotle, the son of Nicomachus, the most famous among the Greeks in science, who invented the sciences according to what has been related and was the first chief, did not attain the stature of this man in wisdom. His statements in this treatise and elsewhere bear witness to the truth of my words. For indeed, not one of all the philosophers after Aristotle to this day, of those whose words have reached us, has attained his degree and stature in philosophy. The second reason is that this treatise has as its subject a precious science, more elevated than all the sciences, namely, natural science, and the treatise is concerned with the most noble part of all of its parts, namely, the psychological part, for it does not pertain to natural science alone, but has a strong bearing on metaphysics. The other branches of natural science, too, have much in common with metaphysics, but this part—that is, the psychological part—is superior to all of them in the commonality and closeness that exist between it and metaphysics.

For these two reasons, this treatise is the most noble of treatises. Examine it carefully, [making it the object of] a complete investigation and perfect study. It [the treatise] will then emerge like gold. […]

Document 6.1.1.5. Samuel b. Judah of Marseilles, postface to his revision of Jacob b. Makhir's translation of Jābir ibn Aflaḥ's *Epitome of the Almagest.*

Neubauer-Renan, *Écrivains*, 562–563; tr. Berman, 'Greek into Hebrew', 312–317.

Samuel, son of Judah, may God protect him, son of Meshullam son of Isaac son of Solomon, may he rest in Paradise, known commonly as Miles of Marseilles, said:

[…] My ear heard and realized the exalted character of this book [on astronomy], composed by someone who was a prince of God, a well-known Ishmaelite philosopher commonly called Jābir ibn Aflaḥ of Seville. And while still a student, the wise Averroes, who was a disciple of Ibn Aflaḥ's, also wrote on this science. [From him] I gauged that [Ibn Aflaḥ] was divinely inspired [*lit.* sat in the seat of God, after Ezek. 28:2].

[But] at that time, I was unable to acquire it [Ibn Aflaḥ's treatise], because it was to be found only in a translation by the sage Nathan ha-Me'ati, may he rest in Eden, as I believe,[155] and his translation was very defective, according to what was said. Moreover, it was to be found only in the hands of a few people. I thus quenched the fire of my great burning desire and exchanged a cow for an ass,[156] that is, the *Almagest* by Averroes, which I mentioned. For I thought that since he was the most recent [author] and renowned for his knowledge, his book would be sufficient without [my] having need for another composition. For these reasons, I did not pursue this precious book [by Ibn Aflaḥ] with ardour and comforted myself with its substitute. For my judgment at that time was not [yet] ripe—I was [still] being educated in the sciences and at the beginning of my study Averroes's discussions in his aforementioned book were very delectable to me. I believed that there was no one like him in excellence and intellectual distinction.

However, with the passage of long time, when I reached the age of thirty, I ardently desired to return to this lofty science. I directed myself to the first book, the great ass, composed by the first chief, Ptolemy Claudius,[157] which is known as the *Almagest*. It is the book that I studied first, in my youth. Although I knew how profound the book was, I longed, nevertheless, to read it again from the beginning. I studied it at that time with the philosopher, my brother En Bondavi of Marseilles, who is perfect in every science, younger than I in years, may God protect and preserve him. We were able to read all of it, except the last treatise, because of impediments, which overtook us frequently. At that time, I also was able to complete a commentary on the first three treatises, but I was unable to comment on the rest of the treatises of the book to the end as I wished, because of the immensity and frequency of the calamities that came upon me. This period of study took place in the citadel of Tarascon, because our home was there at that time. I was about thirty-five years old.

And when, at that time, my eyes were opened to this noble science and all or nearly all of the other sciences, I realized from the discussions of Averroes in his book on this science that whatever is valuable in them [i.e., in those discussions] was gleaned from the book by Ibn Aflaḥ, which is more precious than gold, and that most of the wonderful and excellent statements found in this book [by Ibn Aflaḥ] were omitted in that book [by Averroes], so that there was almost nothing good in Averroes's entire book, except what he gleaned from this one.

155 Wrongly; in fact the translation was by Jacob ben Makhir (Steinschneider, *HÜ*, § 337, 544).

156 This fanciful metaphor is inspired by B *Ketubbot* 76a and B *Bava Metzi'a* 100a.

157 On אלפלודיי = Claudius, see Steinschneider, *HÜ*, § 321, 519.

Therefore, my brother, the aforementioned philosopher, and I girded up strength and courage and stripped off the mantle of haughtiness and grandeur in order to acquire knowledge. We got up and set out and journeyed from our city to Trinquetailles. We exerted ourselves until a very accurate copy of the Arabic version of this book [by Ibn Aflaḥ] reached our hands there, very carefully proofread in the usual manner. We copied from it at that time very quickly, shut up two days in one of the houses of the seekers of wisdom who lived there, nourished on meagre bread and scanty water. We then had to return the Arabic original to its owner and return to our home because of mundane affairs. Therefore, we were forced [only] to glean from the book discontinuous passages, namely those that were most doubtful to us. The excerpted material did not amount to an eighth of the book. In my haste I said [to myself], this [will have to] solace us.

One day, a mighty and strong wind, which split my strength and broke my bones [after 1 Kgs 19:11], slung me around, exiled me, and carried me from the nests of my place of rest to this place. Here, I exerted myself and found this book translated by the sage R. Jacob b. Makhir. My heart did not rest. I used cunning and I laid low my former arrogance for the sake of this precious book until I came to an agreement with its owner and he permitted me to copy it. I wrote it down, and the handwriting in the book from which I copied was that of the aforementioned sage, the translator [himself]. However, while I was copying, I always felt that he had made many mistakes. Then I said: I have busied myself for nothing and I have come into the possession of falsehood.

However, 'The end of the matter, all having been heard' [Eccles. 12:13], I did not rest until I received divine aid: today I received the first Arabic copy, the one from which I had previously excerpted what I could. I have checked this translation against the Arabic text twice. I also found many readings in the Arabic text that were not in the translation by the aforementioned sage. In addition, the author's discussions of the Milky Way, the solid sphere, and the figures appropriate for the fixed stars were missing from that translation. All that was lacking I completed and what was incorrect I corrected, insofar as I was able. As I live, I expended more effort in this matter and I had more expense of spirit than if I had translated it from the Arabic afresh, owing to the many mistakes that I found in the translation. I was told that the sage R. Moses Ibn Tibbon also translated it, but his translation has not reached me, nor have I seen it.[158]

These [the heavenly bodies] are wondrous, exalted existents. 'Whoever is wise will understand these things; whoever is discerning will know them' [Hos.

158 Steinschneider, *HÜ*, § 337, 544.

14:10]. 'O house of Jacob, come ye, and let us walk' in this great and terrible light of the Lord [after Isa. 2:5], for there was never anyone like this author before him, a king of this science and a true prophet. After him, who can find another like him? He criticized the high priest of this science, Ptolemy Claudius. He mentioned great criticisms and explained his discussions, revealed his secrets, and shortened his lengthy passages. Everything is true and valid, as you will see in the book if you study it. Who is like him, teaching, ruling, giving close attention to a true matter [after Prov. 29:12]?

Thus far God, His name be blessed, has helped us. This took place in the forty-second year of my life, the first of Tevet, 5096 of the Creation [17 December 1335], in the city of Aix. [...]

Section Seven. Routine Medical Translations: The Fourteenth Century

Part 2. Latin–Hebrew Translations

7.2.1. Leon Joseph of Carcassonne (1365, Carcassonne–1418, Perpignan)

The physician Leon Joseph of Carcassonne wrote a long preface to his Hebrew translation of Gerard de Solo's *Pratica super nono Almansoris* (c. 1394). This is an exceptional egodocument that is revelatory of the social position of Jewish physicians in his time and place and of the intellectual and spiritual dilemmas that faced them. As the fourteenth century progressed, Arabic medical knowledge became outdated. Under the pressure of the Gentile environment, Jewish physicians increasingly turned to Latin medicine. Many Jewish physicians in Provence and elsewhere now knew Latin and some of them translated Latin medical literature into Hebrew for their still numerous Jewish colleagues who had not mastered it. One of these translators was Leon Joseph. He was acquainted with the Latin intellectual world and offers us highly perceptive comparative observations on the study of medicine—and of profane science in general—in the Jewish and Christian cultural systems. He recounts his dissatisfaction with the Jewish physicians and intellectuals of his time, who lack curiosity and open-mindedness. As the fourteenth century advanced, Latin medicine increasingly progressed, notably through the use of the new *quaestio disputata* research and teaching method introduced in the universities. New textbooks were being written and used. The competition between physicians was conducted mainly through oral disputations, in which the Christians understandably had the upper hand—both because they were better trained in disputations, and because they had access to the most recent literature. Leon Joseph perceptively identifies the Jews' lack of institutions of higher learning comparable to the universities and especially their ignorance of the technique of disputations as a major cause of Jewish intellectual backwardness. This gap in the proficiency in medical disputations gave Christian physicians a strong competitive edge over the Jews; Leon Joseph wishes to allow them to compete on an equal footing. Like many other Jewish physicians of the period, his close contacts with the Christian world eventually led him to be baptized.[159]

I published a new edition and translation of this text recently and therefore reproduce here only the sections with a direct bearing on 'why we translate'.

Document 7.2.1.1. Leon Joseph, preface to his Hebrew translation of Gerard de Solo's *Pratica super nono Almansoris* (c. 1394).

Freudenthal, 'Hebrew Science in the Medieval Midi', 39–52 (text edition with H. Yishai); tr. Freudenthal, 'The Brighter Side of Medieval Christian-Jewish Polemical Encounters', 48–61.

[1] Thus says Leon Joseph, the translator, a resident of Carcassonne.

[2] Ever since I opened my eyes to study and explore the foreign [i.e., profane] sciences—which are many both in species and number—equal in number to the days of the week, so that they cannot all be subsumed under a single definition—and put my hope in them, so that [I might] acquire knowledge and undertake various inquiries, [ever since then] the merits of these sciences grew in my eyes above all praise. Thus, owing to my great yearning and desire to study them and my craving to investigate them, I followed in the footsteps of our learnèd sages, those who live today and those of the past, so that I may frame thoughts that would illuminate my eyes by the light of the intellect and of the heart. [...]

[3] [However,] I discovered that the deficiency in them ['our learned sages'], as also in some of our nation in the present generation, is great and huge and what they say of the profane sciences is like 'a sealed book' [after Isa. 29:11] [i.e., meaningless]. [But] I said [to myself]: 'Perhaps my desire [for knowledge] is greater than my intellectual capacity, and the [perceived] deficiency is my own, on account of the profundity of what is to be apprehended, and of the shortcomings of my intellect and my understanding, and [on account] of my [intellectual] poverty and weakness, and my incapacity to grasp their idea[s], and it is not [due] to their ignorance or lack of knowledge. For [surely] they are wiser than any man [after 1 Kgs 5:11], whereas I am only a remnant of their company. And I was perturbed.' [...]

159 On the particular proneness of Jewish physicians to convert to Christianity, see Emery, 'Jewish Physicians in Medieval Perpignan'.

[6] Even those few, who, by God's grace, inquired into [the sciences] and picked the food out of the husks, were forced to do this in secret and hidden away, 'in the clefts of the rocks and in the hiding places' [Song of Solomon 2:14]. They were not allowed to teach rationalist science [*ha-ḥokmah*] in the market-places and in the streets, or to discuss it and explain its rationale, or set up a public place of study [*le-hoshiv yeshivah ba-rabbim*], so as to bring forth the truth properly: for the truth can be known only through its contrary.

[7] [This was so] because they feared the tongues of the multitude of ignoramuses, for they [the true scholars] are few, but the others are legion. In addition, they dreaded some of the Torah-scholars [*toriyyim*], who are bare of all other sciences and who intimidate the rationalist inquirers—not by virtue of their [intellectual] power and the breadth of their knowledge, but only on account of the strength of their arms and their many frauds. [...]

[8] As a result [of all the above], the deficiency of some of the nation grew and the absence [of learning] increased. On top of this comes another cause: most of the books in those sciences were written by wise men of the nations, such as the Greeks, the Ishmaelites, or the Christians, and other believers whose religions are different from the religion of the Jews, [so that] their books reached us through the intermediary of the translators. [Now] they or some of them may not have been perfectly proficient in that [source] language or languages, and may have mistaken one word for another, resembling it in its intellectual content [i.e., meaning] or in its external shape [i.e., verbal representation]. Or they may have left out or added words, on account of their defective knowledge of the essence of that [source] language; or there may have been a mistake in the copies from which they translated, and consequently the mistake is in those [translated] books, so that [the readers] fail to make sense of them and exhaust themselves [trying to find the sense]. [...]

[12] When I recognized that these are the hindering causes, and knowing that these sciences are known among the Christians and their study is formidable and wondrous—[indeed,] who can embrace it all, understand their subject-matter and encompass it?!—I said to myself: let me learn something of their language, and sit in their universities [*yeshivotehem*] and their house of study [*beit 'iyyunam*] and follow in their footsteps, so that I may benefit myself and others by grasping their statements and some of the truth of their books. I will apprehend what the perfect ones among them have apprehended, provided that time will be sufficient, that there are no impediments or accidents, and that the mind can grasp it without being distracted by [sense perceptions?] to and from it. [...]

[15] Thus, we have been prevented from having knowledge of the sciences and of similar [lore]. [However,] this is not because our intellect is inferior to

their intellects, for we have minds [*lit.* a heart] just like theirs. Rather, we have been hindered by the aforementioned causes, on top of which come our exile and the oppression, [as against the Christians'] pleasures, splendour, and large numbers. This is why they [the Christians] rose ever higher in the profane sciences, whereas we descended on account of the pressing annoyances. [...]

∴

[20] Having perceived all this, I said [to myself]: there is 'no one discerning and wise' [after Gen. 41:39] in this climate of ours. I [thus] followed in the footsteps of the Christian sages, and they directed me to the right path, showed me the ways, and revealed to me the books, both new and old. Among them I saw the discourses of the wise physician Gordon, who in his numerous discourses provides appropriate responses. He excelled in [medical] theory and practice, and earned the crown of good reputation [see *Avot* 4:13]. His books are many, so too his magnificence and splendour, and he excelled all his brethren.

[21] Now some of his books have reached us, namely *Shoshan ha-refu'ah* [*Lilium medicinae*] and *Haqdamat ha-yedi'ah* [*Tractatus de prognosticis*].[160] However, they reached us in a very faulty translation, indeed a shaky ladder. For the Jews who translated it [*Shoshan ha-refu'ah*][161] in those countries[162] accessed it only through the vernacular: they vernacularized the Latin [text], heaping up errors.[163] [...]

[22] When I searched through the satchels of their books, I found two new books, comparable to 'two golden pipes' [Zech. 4:12]. They are very good and worthy, but the sons of our nation lack them. These are: the book of the learnèd Gerard de Solo and the book by the wise Master Jean de Tournemire. It has been some ten years since I heard about them, but I could not obtain them—neither in Montpellier, although this is the place from which they hailed [*lit.* were

160 These are two well-known works by Gordon; see Steinschneider, *HÜ*, § 490 (1b), 785–786, and § 490 (4b), 787.

161 Now the singular form is used, referring to only one translation. Leon Joseph discusses the two bad translations in succession, relating to *Haqdamat ha-yedi'ah* later in the paragraph.

162 Yekutiel ben Solomon, the translator of *Shoshan ha-refu'ah*, was from Narbonne, which is near Carcassonne, Leon Joseph's hometown. Leon Joseph nonetheless writes '*those* countries' (referring to Narbonne, and more generally, to the Midi), because after the Jews were expelled from France in 1394, he settled in Perpignan, which was then subject to the Crown of Aragon.

163 Leon Joseph alludes to a common practice ('traduction à quatre mains'): a Christian scholar turned the Latin text into Provençal on the fly, and the Jew listened and set it down in Hebrew—a practice that naturally produced mistakes.

hewn], nor in Avignon, nor in any of the other illustrious places. For they exist in few [copies] only and, moreover, the scholars of Montpellier used to boycott and ostracize whoever sold them to non-Christians. In order to obtain them, I spent fortunes, to the utmost of my possibilities, in disregard of everything. And this very year, which is the year ninety-four according to the Christian calendar and reckoning, they came into my possession. Forsooth! I bought them for double their value, out of my desire for them, my yearning, and my wish to get them.

[23] I [indeed] recognized how exceedingly fine they are; they make all topics comprehensible and account appropriately for all matters. One who acts according to them will stand like a fortified pillar, not fearing anything, not even the myriads of the company of physicians, if they were to confront him with trials and disputations. For [these books] are fine flour, as far as memory reaches there has been nothing like them. I believe they [the two authors] were not sluggish and chose the very best [available] in medicine.

∴

[24] I now begin with the book by Gerard, which is based on part IX of the *Almansor*. It is small in quantity, but great in quality; it is 'harder than flint' [Ezek. 3:9]. It provides appropriate replies and gives correct advice. Its art is safe, and it provides good reasons for all its statements, which is a reason to praise it. Nothing was concealed from him, nor was anything forgotten. Therefore, the decision to translate it was taken in all tranquility.[164]

[25] If God decrees that I complete this translation, I will thereafter begin [the translation of] Tournemire's book.[165] In his days he was the head [i.e., chancellor] of all the scholars of Montpellier, and they were subordinate to him. I saw him with my own eyes and spoke with him. In his conversation he was gracious, and his manner was not like that of the scholars of his generation, namely to humiliate the Jews who practise the art of medicine. For he used to steer them in the right direction with his [intellectual] power. May his soul be bound in the bundle of life of the righteous of the nations.[166] No other book has been comparable to it in the beauty of its wisdom. It speaks [*lit.* spoke] of the organs and their faculties and of the maladies that afflict them, questioning

164 See Steinschneider, *HÜ*, § 496 (1b), 794–797.

165 Jean of Tournemire (c. 1330–1390/96) was physician to Pope Gregory XI and became chancellor of the University of Montpellier in 1384.

166 The traditional formula for the dead (derived from 1 Sam. 25:29), here applied to Tournemire; see Steinschneider, *HÜ*, 834, n. 376.

and answering, clarifying and elucidating, like a father teaching his son.[167] He [Tournemire] produced new statements that are accurate and correct, pleasing and agreeable to listeners.

[26] I came up with the idea to translate these books, not for myself but for [the benefit] of others of the people, both contemporaries and those that will come after me, who know nothing of their [the Christians'] language. For when they peruse them and read them and perceive their perfection and the beauty of their arrangement, they will bless me on account of them and will remember me in their hearts, inasmuch as I will be the cause of their vitality and their ability to hold their ground when confronted by the Christian physicians [who challenge them] in trials and disputations.[168] This [will be] my second life. I know that with these books they will be able to gainsay the Christians' words—unless they anoint their flesh with wine, eating and drinking and celebrating, for then the translations will be of no avail to them; it will be as if [the books] were not translated and were removed from their minds. [By contrast,] if they read them repeatedly, they will derive multiple benefits, they will live in 'quietness and in confidence' [Isa. 30:15], and sleep peacefully in their beds; they will not fear the 'the voice of the oppressor' [Job 3:18] that beats them with the rod of medicine. For they will be able to fight their battles—they will not dread the crowd of the [Christian] doctors, even though they are shrewd. They [the Christian doctors] may [then] hide and conceal all other books of the art of medicine—but with these two books they [the Jews] will utterly destroy them. [...]

[30] I will write down all the medicinal compounds in Latin [in fact: vernacularized Latin], after first translating them into our tongue, God willing, so as to be helpful to the neophytes in this art: they will be able to organize [their knowledge] without toil or fatigue, to view themselves as learnèd and wise, as elders of the art and perfect [men], even though they are beginners who had never before seen the lights in the art. The pharmacists will not despise them, and they will not have to consult pharmacopeias if they know the names of the medicines by heart. As is written: '[God] shall keep your foot from being snared' [Prov. 3:26]. He will then be considered a distinguished physician. [...]

---

167 Two of Tournemire's works were translated into Hebrew, and Leon Joseph may have been responsible for both of them; see Steinschneider, *HÜ*, § 516, 833–834.

168 Cf. § 23.

Part Three: Egodocuments—the Original Hebrew Texts

Section One. Harbingers and Pioneers: The Eleventh and Twelfth Centuries

Part 1. Arabic–Hebrew Translations (and Other Means of Knowledge Transfer)

1.1.1. Moses Ibn Giqatilla (c. 1020–1080, Saragossa)

Document 1.1.1.1. Moses Ibn Giqatilla, preface to his translation of *Three Treatises in Hebrew Grammar* by Judah b. David Ḥayyuj.

Ḥayyuj, Judah b. David, in Nutt, *Two Treatises*, Hebrew part, 1–2.

זה ספר אשר חברו איש שכל יודע מבין בדקדוק לשון יהודית ומשכיל בהגיון שפת עברית. עוד ינקב בשמו ושם אביו ויודע בסוף הדברים. כי כן היה בראש המדברים ומביני תכונת לשון עברית.

אך מפני אשר גברו עלינו עם לועז ונבלענו בגוים עמקי שפה וכבדי לשון. [...] ושקעה לשון הקדש בין לשונות החול. אשר רבו בעליהן כחול. ונשארנו מעט מהרבה ואבדה חכמת חכמינו. ונסתתרה בינת נבונינו. ולא נשאר ממי נשמע תוכן הלשון. ולא נותר ממי נלמוד כל עניניה. כי אם מה שנזהר מן הנמצא בכתבי הקדש ונלמוד מן הנכתב בספרי הנבואה. והוא מעט מזער מן הכל. לפי שלא באו הנביאים לחקור כל הלשון. כי אם לפי צרכם בדברי נבואתם ועניני חזיונם.

ולזאת הצרכו מחברי דקדוקי לשון הקדש ומבארי רזי שפת עברית לחבר ספריהם בלשון ערבית מפני שהיא עוברת בפי אומה גוברת. והיא מבוארת. ולשון עברית עלומה. ולשון ערבית גלויה ומפורשת. ולשון הקדש סתומה. [...]

רק אנשי צרפת היושבים בגבול אחינו בני עשו אין רובם מכיר בלשון ערבי והרבה מחבבים לשון הקדש ורגילים לדבר בה. אז העיר ה׳ את רוח איש תבונה בחור חמד בתורת ה׳ חפצו. [...] ר׳ יצחק הנעים ברבי שלמה הנשיא. [...] ובקש להבין עניני הספר אשר חבר ר׳ יהודה ברבי דוד דממדינת פאס המכונה חיוג ולעמוד על סודו.

ולא נמצא בכל הלשון עמוקים ונפלאים מן הפעלים והשמות אשר יהיו בהם אותיות א׳ה׳ו׳י׳. יש כשיהיו בתחלתם. ויש כשיהיו בתוכם. ויש כשיהיו בסופם. והאותיות האלה ידועות בלשון ערבית חרוף אללין. וגם הפעלים אשר יהיה תוכם וסופם שתי אותיות כפולות דומות זאת לזאת וקרואים בלשון ערבית ד֗ואת אל מת֗לין: ועל כן בחר המחבר לגלות רזיהם ולבאר סודיהם, כי בהגלות רזי אלה יגלו שאר רזי לשון עברית למבין דבר מתוך דבר ולנזהר מן הקצת על הכל. כענין שנ׳ תן לחכם ויחכם עוד.

ולפי שלא מצאנו כל כנויי לשון ערבית בלשון הקדש. אם מפני שהיא אין עמנו כולה, ואם מפני שלא שותה לשונם ללשוננו—בכל המלים ולא דמתה אליה בכל הדברים לא נכון שתהיה מלה כנגד מלה, כי אם בסבה ועלילה—שיסבבו למלה ערבית שהיתה בספר המחבר הרבה מלים בלשון עברית, עד שיתבאר הענין ויכון בלב הקורא כמו שהוא בלשון ראשון.

ובקש ר׳ יצחק זה זכור לטוב ממני אני משה הכהן ברבי שמואל הכהן נ״ע דממדינת קרטבה שאתרגם לו זה הספר בלשון עברית וכן עשיתי. [...]

1.1.2. Abraham bar Ḥiyya ha-Nasi (c. 1065–c. 1136, Barcelona)

Document 1.1.2.1. Abraham bar Ḥiyya, author's preface to *Yesod ha-tevunah u-migdal ha-ʾemunah* (concluding passage).

Millás-Vallicrosa, *La obra enciclopédica*, 10 (Hebrew pagination).

ואל יאשימוני רבותי המסתכלים בדברי אלה ואהיה בעיניהם כמו שהכניס עצמו במצוק [בדפוס: ממצוק] ובמקום צר שאינו ראוי ללכת בו ומסתכן בעצמי בצאתי מדרכי. ואני מגלה להם את סודי. ולא לרצוני נכנסתי ולא לכבודי להחזיק לי טובה. אבל רבים מגדולי דורי שאני חייב לקבל עצתם הביאוני בכל זה. מפני שלא היה בכל ארץ צרפת מהחכמות האלו ספר כתוב בלשון עברית ועל פיהם העתקתים מספרי ישמעאל אל לשון הקדש כהשגת ידי. ככתוב ׳איש כמתנת חלקו כברכת יהוה אלהיך אשר נתן לך׳. וכל אדם חייב להורות מתלמודו כפי בינתו, אחד המרבה ואחד הממעיט, ושיכוון לבו לשמים.

ואני מפייס מכל הקורא בחבור הזה שידינני לכף זכות, ככתוב ׳בצדק תשפט עמתך׳, ויגיה בחסדו כל טעות ושגגה שיראה בדברי אלה, הן בלשון הן בענין לכשיתברר לו. האלהים יכפיל שכרו.

Document 1.1.2.2. Abraham bar Ḥiyya, author's preface to *Ṣurat ha-ʾareṣ we-tavnit kaddurei ha-raqiʿa we-seder mahalaḵ koḵavehem* (*The Form of the Earth and the Shape of the Celestial Globes and the Order of the Movements of their Stars*).

Ed. Offenbach, 1720, 1a–1b; *Sefer Ṣurat ha-ʾareṣ*, ed. Lasri, 37–38.

[...] אמר דוד מלך ישראל: ׳כי אראה שמך מעשה אצבעותיך וגו״ [תהל׳ ח ד]. ופעולת ׳אֶראֶה׳ בכאן מן לשון ראות הלב אשר היא הבינה. ואמר כי המתבונן במעשה שמים ומהלכות הירח וכוכבים, אשר כונגם המקום להראות נפלאותיו, יהיה יודע שבח שמו האדיר בכל הארץ, אשר יתַנֶה אותו ויפרשֵהו, הודו אשר על השמים. [...]

ויתברר לך מן הכתוב כי המסתכל במעשה שמים וארץ ותוכן צורתם, ויהיה מתבונן במהלכות הכוכבים וחכמת תבניתם, יראה גבורות בוראו והודו ויבין חכמת קונו ויוצרו. ועליו ארז״ל [שבת עה ע״א]: ׳כל היודע לחשב בתקופות ובמזלות, ואינו חושב עליהן, עליו הכתוב אומר: ״את פועל יי לא יביטו ומעשה ידיו לא ראו״ [ישע׳ ה יב]׳. ונמצא, החושב אותן מביט את פועל יי ורואה את מעשה ידיו.

ויהיה מוחזק בידך מכל צד כי המקדים ללמוד תורה כפי יכולתו ויהיה עוסק בה ומחזיק במצוותיה, ויהיה עם מעשיו אלו דורש את חכמת הכוכבים ולומד אותה ועוסק בה, אינו מתעסק בדבר ריק, אבל הוא מחזיק בחכמה שהיא מסדרת להאמין סתרי אמונה, ומיישרת לפניו דרכי היראה הנאמנה. ובמקומות רבות הוא מצורך המצוֹת.

וכל ימי דרשתי וחפשתי למצוא חבור בלשון הקודש מפרש את החכמה הזאת על סדרה, ולא מצאתי. ונראיתי בעינַי כאלו מחויב להמציא אותו כפי יכולתי. והייתי מתרשל ומתעצל מעשות זאת כי לא ראיתי לסמוך על דעתי, ולא לקיים דברי בזה מעצתי ורשותי. וכשבאו אלי שורות ישרות וטורים יקרים, יוצאין מלפני קצין הדור וגבירו, ושמש המדע ומאורו, והוד השכל והדרו, הנקוב בשם איש האלהים ובחירו [בכת״י אחד נוסף כאן: ר׳ אברהם ב״ר שלמה], יסגא לעד שלומו ולפני שמש ינון שמו, ויועץ אותי לחבר בחכמת הכוכבים חבור שיהיה מבאר את ענייניה ומפרש כהוגן דרכיה ופניה. וכעמדי על דבריו הנעימים, ומצותו ראויה להשלים ולהתמים מיד השתדלתי לשמוע לעצתו, כפי כבודו עלי וחבתו, וכן הוא הדין על כל אוהביו וידידיו, כל שכן על תלמידיו המציקים על ידיו.

[...]

Document 1.1.2.3. Abraham bar Ḥiyya, author's preface to *Ḥibbur ha-Meshiḥah we-ha-tishboret* (*The Book on Measuring and Geometry*).

Chibbur ha-Meschicha we-ha-Tischboreth, ed. Guttmann, 1–5.

[...] וכתיב: ׳אני ה׳ מלמדך להועיל מדריכך בדרך תלך׳ [ישע׳ מח יז], מלמדך כל דבר שהוא מועיל לך ומדריכך בדרך תלך והוא דרך התורה. ואתה דורש מן הפסוק הזה, כי כל עסק וכל חכמה שהיא מועילה לאדם בדרכי החיים ובדבר מצוה ראוי לו ללמדה ולהתעסק בה.

וראיתי חכמת המנין וחכמת השיעור נוהגות המנהג הזה ומועילות במלאכות רבות שהן משמשות בחקות התורה ומצותיה. ומצאנו הכתוב במקומות הרבה מצריך אליהן, ככתוב ׳במספר שנים אחר היובל תקנה מאת עמיתך׳ [ויק׳ כה טו], וכתיב: ׳לפי רוב השנים תרבה מקנתו ולפי מעוט השנים תמעיט מקנתו׳, ואחריו כתיב: ׳לא תונו איש את עמיתו ויראת מאלהי׳ [ויק׳ כה טז]. ואין אדם יכול לדקדק בחשבון עד שלא יבוא בו לידי הונאה אם לא ילמד חכמת החשבון. וכן כתוב: ׳וחשב עם קונהו משנת המכרו לו עד שנת היובל והיה כסף ממכרו׳ [ויק׳ כה נ] וגו׳. וכתיב: ׳וחשב את שני ממכרו והשיב את העודף לאיש אשר מכר לו׳ [ויק׳ כה כז], וכתיב: ׳וחשב עם קונהו׳—ידקדק עם קונהו שלא יגלום עליו. ולא תוכל לתקן החשבונות האלה ולהקיש מנין דמי השדה ותבואתה או דמי השכירות אל השנים הנותרות ולהוציא קצב מנין זה ממנין זה אם לא תהיה בקי בחכמת החשבון ורגיל בה.

וכמו כן הצריכה התורה אל חכמת השיעור במדידת הארץ וחלוקתה ובתחומי שבת ובמצות אחרות, ככתוב: ׳ומדותם מחוץ לעיר׳ [במד׳ לה ה], וכתיב: ׳ומדדו אל הערים אשר סביבות החלל׳ [דבר׳ כא ב], וכתיב: ׳לאלה תחלק הארץ בנחלה במספר שמות׳ [במד׳ כו נג], וכתיב: ׳לרב תרבה נחלתו ולמעט תמעיט נחלתו׳ [במד׳ כו נד]. ומי שאינו בקי ורגיל בחכמת השיעור אינו מודד את הארצות ולא חולק אותה באמת ובמשפט בלא הונאה, עד שיהא נותן לכל אחד חלקו הראוי לו—ולרב ירבה, ולמעט ימעיט, כמצות התורה עד שילמד החכמה הזו וישען במעשיו עליה. ומכאן אתה למד שהחכמות האלה מצורך התורה הן וראויות ללמוד וללמד.

ודי לחכמה הזאת שהקב״ה משתבח בה, ככתוב: ׳עמד וימודד ארץ׳ [חבקוק ג ו], וכתיב: ׳מי מדד בשעלו מים ושמים בזרת תכן׳ [ישע׳ מ יב]. ואתה רואה מהכתובים האלה שהקב״ה

ברא את עולמו במדידה ושיעור מתוכן ושקול. וחייב אדם להמשל לקונו בכל כחו ובכל מאודו, וזהו שבחו של אדם לדעת כל החכמים. ומכל הענינים האלה אתה רואה כבוד אלו החכמות, והמתעסק בהן אינו מתעסק בדבר ריק אלא בדבר שהוא מצורך העולם והמצות.

ואני רואה חכמת המנין ואם שהיא מצורך העולם ומועילה בעסקיו ובעסק מצות רבות אינה קשה להבין, ורב בני אדם מבינים ממנה מעט ורגילים להשתמש בה. ומפני זה אין אדם נזקק לחבר בה חבור בלשון הקדש. וחכמת השיעור כמו כן הענינים המצריכים אליה רבים כענינים המצריכים לחכמת המנין בעסקי העולם הזה ובמצות רבות מן התורה, והיא קשה להבין וענייניה נפלאים מרוב בני אדם. ומפני זה צריך אדם ליתן דעתו עליה ולפרשה, כי הצורך אליה גדול הוא במדידת הארצות וחלוקתן בין היורשין והשותפים, עד שאין אדם יכול למדוד אותן ולחלקן בדרך ישרה ומשפט אמת אם לא ישען על החכמה הזאת.

וראיתי רוב חכמי דורנו בארץ צרפת אינן בקיאין במדידת הארצות ולא זריזין בחלוקתן, אבל הם מזלזלין בדבר הזה זלזול גדול ומחלקין את הקרקעות בין היורשין והשותפין לפי אמידה וגוזמא ובאין במעשה הזה לידי עוֹן ואשמה. ורובם מזידים וחוטאים וכל אחד מהם חטאו בטעותן כפי שבוש אומדנותן. [...] ואל יעלה בלבך שהם מודדים וחושבים, אבל הם מאמדים וכוזבים. [...] שאין בעולם הונאה גדולה מן ההונאה הבאה על ידם, עד שיכול שיקרה בחשבונם שבעל השליש נוטל רביע ובעל הרביע נוטל השליש, ואין לך גזילה ואונאה גדולה מזו. [...]

ומפני שראיתי הנזק הגדול וההפסד הבא על ידם, הטרחתי את עצמי, ואע״פ שאיני כדאי וראוי לכך, לחבר במדידת הארץ וחלוקתה חבור שיהיה מפרש את כל הענין הזה ומלמד דרך המדידה והחלוקה ומביא עליה אותות וראיות מחכמת השיעור וחכמת המנין, אשר אין בהם שום קירוב ולא ספיקא כפי יכלתי, אולי ילמד אדם אותם ויסיר מן הארצות האלה הטעות הזה הנוהג ביניהם. [...]

1.1.3. Abraham Ibn Ezra (1089, Tudela–c. 1167, London)

Document 1.1.3.1. Abraham Ibn Ezra, preface to his Hebrew version of Ibn al-Muṯannā's commentary on the astronomical tables of al-Ḫwārizmī (1160)

Goldstein, *Ibn al-Muthannâ's Commentary on the Astronomical Tables of al-Khwârizmî*, 302.

נאם אברהם א״ע הספרדי.

[...] ואחריו קם חכם גדול שהעתיק ספר נכבד במזלות שחבר בטלמיוס מלך מצרים. הוא היה יוני ויש לו אלף שנים. וזה הספר הוא שלם, אין למעלה ממנו בחכמת הגלגלים וסודם ומהלכם מהארץ ומדת הכוכבים העליונים שהם בגלגל המזלות [...]

וזה הספר פירשוהו חכמים רבים בישמעאל, ומהם חכם מופלא מכולם במדות ובמזלות ששמו מחמד בן מתני. וחבר לאחד מקרוביו ספר נכבד מאד בתקוני המשרתים וטעם כל מעשה הלוחות הכתובים בספר אלכורזמי. והזכיר הראיות והדמיונות בצורות בדרך קצרה, ועקריהם לקוחים מספר אלמגיסטי. ויש מקומות שיוסיף לבאר יותר מתלמי המלך, וגם יש

מקומות שהוסיף מלבו ראיות מחכמת השיעור. [...] ועשה ספרו על דרך שאלה ותשובה למען היותו מבואר היטב. [...]

1.1.4. Judah Ibn Cardinal b. Isaac (fl. second half twelfth century?)

Document 1.1.4.1. Judah Ibn Cardinal b. Isaac, preface to his translation of Judah Hallevi's *Kuzari.*

The Kuzari, ed. Brecher, vol. 2, 4b; Parma, Biblioteca Palatina, MS Cod. Parm. 1936 (De Rossi 625), f.303a.

[אמר] יהודה ברבי יצחק בן קרדנאל זצ"ל: העתקתי ספר אלכוזרי לר' יהודה הלוי תנצב"ה מלשון הגרי אל לשון עברי. ורבי יהודה היה מתכוין להורות שבחה של תורה ולהשיב תשובות לפילוסופים ולחכמי דת האומות ולקראים. ואני בתרגמי אותו לא אשנה עקר דבריו ולא אחליף עניניו, ואם אוסיף ואגרע במדבר לא אשחית הענין כפי יכלתי ותבונתי, ומאת ה' אשאל עזרתי.

וכאשר התחלתי להעתיק אותו במאמר החכם [כת"י: הנשיא] ר' יוסי [כת"י: יוסף] בן ברוך נר"ו [= נטריה, רחמנא ופרקיה] הוליך ממנו שני השערים הראשונים אל אנגלטירא. והתופשים אותו החזיקו בם ולא יכול להחזירם, והוצרכתי לחזור להעתיקם כבתחלה, כדי שיהיה שלם אצלנו כעזרת האל.

1.1.5. Judah Alḥarizi (1165, Toledo–1225, Aleppo)

Document 1.1.5.1. Judah Alḥarizi, preface to his Hebrew translation of Maimonides's *Guide of the Perplexed.*

Maimonides, *Moreh ha-nevukhim*, tr. Judah Alḥarizi, ed. Schlossberg, 2; Paris, BNF, MS héb. 682, f.1b.

אמר יהודה ב"ר שלמה חריזי:

היתה עלי יד רבותי קצת אצילי פרובנצה וחכמיה להעתיק להם **ספר מורה הנבוכים**, במלים פשוטים וצחים ולמבין נכוחים, ולהנחות בדרך מישור אנשי תועה, והיה העקוב למישור והרכס לבקעה. כי כל ענייניו נעלמים ורזיו סתומים וחתומים.

ונמנעתי לתת שאלתם בעבור היותו ים גדול, עמוקים מעברותיו, ומי היודע תוצאותיו? אף כי הקדימני איש נבון ומשכיל להעתיקו, אע"פ שהתכון בו להסתירו בדבריו ולהעמיקו. וכאשר חליתי פניהם להקל מעלי עלו לא השאילוני כאשר שאלתי. וחתרתי להשיב אל היבשה ולא יכולתי.

וכאשר לחצוני [...] והכריחוני להעתיקו, ומכלי אל כלי להריקו, שמתי כל מאדי להקריב התעלת כפי האון והיכולת, וסדרתי כל עניניו, ויסדתי על קו המישור בניניו. ולא באתי לגלות סוד מסודות הספר ולא לפרש ענין מענייניו, כי אם לסדר בו הענינים כמסקל המסלה מהאבנים.

וראיתי לחדש מלבי בזאת ההעתקה שני שערים אשר מהם יגיע אדם לחדרי חדרים:
השער הראשון. בפיר[ו]ש כל מלה זרה אשר הבאתי בזה הספר. ואסדר אותם על דרך אל״ף בי״ת למען ירוץ הקורא בה;
השער השני. במנין החלקים והפרקים אשר בזה הספר, ואזכר בכל פרק ופרק ענינו וכוונתו;
כדי שיהיה אדם מוצא בשער הראשון פירוש כל מלה אשר לא ידעה, וימצא בשער השני כל כוונת הספר וענין פרקיו וחלקיו, למען יהיה כל דבר מסודר במקומו הראוי לו, איש על מחנהו ואיש על דגלו.

Part 2. Latin/Vernacular–Hebrew Translations

1.2.1. Berechiah b. Naṭronai ha-Naqdan (fl. 1150–1180, Normandy and the Midi)

Document 1.2.1.1. Berechiah b. Naṭronai ha-Naqdan, preface to *Dodi ve-nekhdi*, the Hebrew adaptation of Adelard of Bath's *Quaestiones naturales* (c. 1150).

Dodi ve-nechdi, ed. Gollancz, Hebrew part, 10; Visi, 'Berechiah b. Naṭronai ha-Naqdan's *Dodi ve-Neḵdi*', 45–46.

כאן אתחיל ספר השאלות נכד לדוד:
אני ברכיה בר נטרונאי, לחצוני רעיוני, אשנס מותני להעתיק אלה הדברים בלשון עברית, כי מצאתים במכתב של גוים אשר העתיקום מן העבריים [נ״א: מלשון ערבית] ובם דברים טובים נעלמו מחכמי דורינו, וחכמות מפוארות למשכילים, לא שזפתם עין סכלים. ובראותי חכמה מפוארת בבלי מכוער, פנינים לפני חזירים, טיהרתים מיד בן נכר וכתב[ת]ים בלשון הקודש אשר נעלה על כל לשון.
והספר הזה הוא בין שני חכמי הדור, הדוד והנכד, לצרף וללבן מליצותם, כמו שנאמר: 'ואיש יחד פני רעהו' [משלי כז יז]. והמשכילים יבינו.

1.2.2. Anonymous Translator of Gundissalinus' *Tractatus de anima* (first half of the thirteenth century).

Document 1.2.2.1. Anonymous translator's interpolation in the Latin–Hebrew translation of Gundissalinus's *Tractatus de anima* (first half of the thirteenth century).

Schwartz, ed., '*Sefer ha-nefeš* (*Tractatus de anima*)', 225–279, at 228.11–14.

עד כאן היא העתקת לשון ספר ארסט״ו עצמו, והיתה חזקה וקשה. והנחנו העתקתו ולקחנו [ספר אחר], והוא לחכם מטולטולא פרש ענין ספר החכם, והוא מתחיל: 'כל בני האדם', וזה פתיחה לספר, עד 'הדין נותן', שהוא תחלת הספר. ובפתיחה אמר שיש בחקירת הנפש עשר חקירות אעשה להם סמנין.

1.2.3. 'Doeg the Edomite' (fl. 1197–1199, Montpellier?)

Document 1.2.3.1. 'Doeg the Edomite', preface to his Latin–Hebrew translation of a collection of twenty-four medical books.

Freudenthal, McVaugh, and Mesler, 'Twelfth-Century Latin Medicine in Hebrew Garb', Appendix 1.

[...] וְעוֹדֶנּוּ מִתְהַפֵּךְ בְּתַחְבּוּלוֹתָיו / בְּמַעֲלוֹת מַהֲתַלּוֹתָיו, / אָמַר בְּלִבּוֹ: 'הֵן רָאִיתִי זֶרַע קֹדֶשׁ מַשְׂכִּימִ[ים] עַל פִּתְחֵי חַכְמֵי הָאֻומּוֹת עַל אוֹדוֹת תַּחֲלוּאֵיהֶם / וְתַחֲלוּאֵי בְּנֵיהֶם וּבְנוֹתֵיהֶם, / וְלוֹקְחִים כָּל דָּבָר מִיָּדָם בְּאֵין מֵבִין, / אֲשֶׁר לֹא יָדְעוּ מַה הוּא; / וּבֵין טָמֵא לְטָהוֹר לֹא הִבְדִּילוּ, / וּבְרוֹב מֶרְקָחֵיהֶם יִמְזְגוּ מִינֵי הַשָּׂרָף בַּיַּיִן, וּבְשַׂר טָמֵא וְחֵלֶב וָדָם יָבֹא בָהֶן. וְאֵין אֶחָד מֵהֶם נִמְלָט אוֹ מֵאִסּוּר אוֹ מִגֵּעוּל'.

עַל כֵּן בָּאתִי לְהַעְתִּיק בִּלְשׁוֹן הַקֹּדֶשׁ כָּל סִפְרֵי הַחָכְמָה, אֲשֶׁר הִשַּׂגְתִּי בְיָדָם מֵחָכְמַת הָרְפוּאוֹת וְכָל הַנִּבְחָר מִסִּפְרֵי הַמְּלָאכָה. וְאֵלֶּה שְׁמוֹת הַסְּפָרִים אֲשֶׁר הִשְׁלַמְתִּי הַעְתָּקָתָם שְׁנַת י"ז וי"ח וי"ט לְמַחְזוֹר רס"א.

[...]

וְהִנֵּה מִסְפַּר סִפְרֵי הַמְּלָאכָה אֲשֶׁר הֶעְתַּקְתִּי—שִׁבְעָה עָשָׂר, וְשִׁבְעָה סִפְרֵי הַחָכְמָה, וְהֵם כ"ד.

וְהִנֵּה הִתְנַפַּלְתִּי לִפְנֵי כָל מַשְׂכִּילֵי עָם, אֲשֶׁר יְזַכּוּנִי וְיַצְדִּיקוּנִי בַּדָּבָר הַזֶּה, כִּי לְטוֹבָה נִתְכַּוַּנְתִּי / וְעַל דִּבְרֵי רַבּוֹתֵי[נוּ] נִשְׁעַנְתִּי / אֲשֶׁר אָמְרוּ: חַיָּיב [אָדָם] לְלַמֵּד אֶת בְּנוֹ תוֹרָה וּלְלַמְּדוֹ אֻומָּנוּת, וְאֵין לְךָ בָעוֹלָם אֻומָּנוּת נְקִיָּה כָּזֹה. [...]

Section Two. Founding Fathers (Arabic–Hebrew Translations): The Tibbonid Dynasty

2.1.1. Judah b. Saul Ibn Tibbon (c. 1120, Granada–c. 1190, Lunel)

Document 2.1.1.1. Judah Ibn Tibbon, first preface to the translation of *Duties of the Heart* (to the first chapter, on the unity of God, 1161).

R. Bahya b. Joseph Ibn Paqūda, *Sefer Ḥovot ha-levavot*, tr. Judah Ibn Tibbon, ed. A[braham] Zifroni, 1–5.

[...] ואחר שגרמו עונותינו בנטותנו מדרך התורה והחכמה, ונסתלקה השכינה, ונפסקה הנבואה ורוח הקודש מבינינו, לא כלו רחמי אלוהינו והותיר לנו שריד. והשאיר בכרמו עוללות שנים שלשה גרגרים בכל דור ודור להאיר עיני הנשארים הנמצאים באור החכמה, 'לתת ליעף כח ולאין אונים עצמה' [ישע' מ כט]. וחקרו על צפוני החכמות, והוציאו לאור תעלומות, ובארו טעמי התורה במשנה ובתלמוד, ויסדו יסודות, והעמידו עמודים לחזק בדק תורת אלוהינו, אשר בית ישראל נכון עליהם. וגם בדרך ארץ ובשערי המוסר ובתיקון המידות חיברו אמרים אשר לא יערכם זהב וזכוכית [עפ"י איוב כח יז], מהם מחוברים במסכת מיוחדת להם, במסכת אבות והלכות דרך ארץ, ומהם מפוזרים בסדרי המשנה והתלמוד בכמה מקומות.

ואחריהם היו רוב הגאונים בגלות מלכות ישמעאל, ובבבל, ובארץ ישראל, ובפרס. ומדברים בלשון ערבית, וכל קהלות ישראל, אשר היו במקומות ההם, מדברים בלשון ההיא. פירשו רוב מה שפירשו מספרי המקרא וסדרי המשנה והתלמוד בלשון ערבית. וגם ברוב חיבוריהם אשר חיברו ותשובות השאלות הנשאלות מהם עשו כן. מפני שכל העם היו מבינים בלשון ההיא, ועוד כי היא לשון רחבה וממולאה בכל ענין וכפי הצורך לכל מדבר ומחבר, המליצה בה מיושרת ומבוארת ומגעת לתכלית כל ענין יותר ממה שייתכן בלשון העברית, מפני שלשון העברית אין ממנה נמצא בידינו כי אם מה שמצאנו בספרי המקרא, ואינו מספיק לכל צורך המדבר. וגם כוונתם היתה להועיל בחיבוריהם לעמי הארץ, שאינם בקיאים בלשון הקודש. על כן היו רוב חיבוריהם בלשון ערב בכל הענינים אשר חיברו, בין בחכמת התורה בין בחכמות אחרות.

וגם בארצות אדום היתה פליטה לשאר עמנו. היו בהם חכמים גדולים בחכמת התורה והתלמוד מימים קדמונים, אך לא היו מתעסקים בחכמות אחרות, מפני שהיתה תורתם אומנותם ומפני שלא היו ספרי חכמות אחרות נמצאים אצלם. עד שנקבעה בתוכם המנורה הטהורה, נר מצוה ותורה, הרב הגדול החסיד הקדוש רבנא משולם, נטריה רחמנא ופרקיה, בן החכם הישיש רבי יעקב, זכרונו לברכה, אשר שמן תבונתו זך כָּתִית להעלות נר החכמה תמיד, ותדבק נפשו בתורת אלוהיו ובירא‍תו, ושם החכמה כוסו ומנתו, ויכסוף לספרי החכמות אשר חיברו הגאונים, וכפי יכלתו קבץ ורבץ והעתיק בין מחכמת התורה, וחכמת הלשון, וחכמת האמונה, וספרי המליצות, וחיבורי המוסר, ומשלי החכמים. ותמצא כקן ידו [עפ״י ישע׳ י יד] לכל חמודותיהם, ובגבורת שכלו ועוצם כחו, הבין דבר מתוך דבר והוציא ענין מענין. והוא גם הוליד ילדים במשלי החכמות ועניני המידות, ופתח שערים בדרכי היראה, והמוסר, הגיה בהם חשכנו ויישר מעקשות דרכנו, ועוד יד גבורתו נטויה לקבץ ולרבץ.

וכשמעו עתה, כי לאחד מחכמי ספרד, הוא רבנו בחיי הדיין ברבי יוסף, זכרונו לברכה, חיבור בתורת חובות הלבבות, יסודתו על ענין הייחוד, כלתה נפשו לראותו. ובהגיעו לידו, צִיוָּנִי לתרגם לו עניניו בלשון עברית. וכאשר שמע את דבריו ואת שיחו, לא רפתה רוחו עד אשר ציוה להעתיקו, ואותי ציוה לתרגם לו ממנו השער הראשון, שהוא שער הייחוד.

ואף על פי שיקרה בעיני מצוותו, ראיתי שהכניסני בענין שלא היה רצוני להיכנס בו ובכל עת הייתי נשמר ממנו. וכמה פעמים פייסו אותי קצת הנדיבים אוהבינו מחכמי המקומות האלה להעתיק להם קצת מחיבורי הגאונים אשר בלשון ערב אל לשון הקודש, ולא נתפייסתי להם מפני כמה פנים, שצריך האדם להזהר בהם בנין הזה. [...]

והגדתי כל זה להודיע לכל בני העולם בהערת השער הזה, כי לא מלבי ולא ברצוני נכנסתי בענין זה, וכי לא נסתר מנגד עיני כל אשר זכרתי. אבל מפני שאני חייב לבטל רצוני מפני רצון המצוה אותי ולמחול כבודי מפני כבודו, שמעתי למצוותו, ואצתי לעשות כחפצו ותאוותו. ואשים נפשי בכפי, ולא שמתי יד לפי, והצבתי עצמי כמטרה לחצי המורים, וגֵוִי נתתי למכים. [...]

וראיתי, כי המחבר הזה, זכרונו לברכה, הניח דבר אחד שהיה צריך להקדימו בתחילת ספרו, או לבאר אותו במקומו, והוא שיפרש ענין המלות והשמות, אשר הביא מספרי החכמות החיצונות לצורך דבריו, ועל מה הם אמורים בחכמות ההן. כי כן מנהגי החכמים בכל חיבורי החכמות, מפני שהם שמות ומלות שאין המון בני אדם רגילים ואינם מבינים בהם, ומשתמשים במקצתם בדרך אחרת, ומפילים ומפרשים אותם על ענינים אחרים. ושמא עשה זה מפני שאמר כי חיברו לזכרון לעצמו ולחבריו. אבל מפני שאמר אחר כן כי גם רצונו היה להועיל

בו לשאר בני אדם, היה לו להרגיש למה שאמרתי. ועל כן מי שאינו בקי במלות ובשמות ההם ואינו רגיל בחכמות ההן, צריך לשאול עליהם. [...]

Document 2.1.1.2. Judah Ibn Tibbon, epistle to R. Asher b. Meshullam (1161).

Freudenthal, 'R. Judah Ibn Tibbon's Epistle to R. Asher b. Meshullam of Lunel', 668–670.

[...] הקלתי על עצמי טורח ההעתקה שהייתי נזהר ממנה בעבור הדברים אשר זכרתים בפתיחת העתקת **שער הייחוד** אשר העתקתי לאדונינו אביך הרב הקדוש נר״ו [= נטריה, רחמנא ופרקיה] מ**ספר חובת הלבבות**. כי זכור: אני ואתה לומדים השער ההוא לפניו, ומתוך דברינו ספרתי לכם שמצאתי לחכם הפילוסוף ר׳ שלמה ב׳ר יהודה בן גבירול ז״ל חיבור קטן **בתיקון מדות הנפש**, כולל כל ענייני התשעה שערים מן הספר ההוא, ועוד מוסיף עליהם. ודבריו דבורים על אופניהם, מובאים על דרך החכמה, וסמוכים על המקרא, ומיוסדים על דברי החכמים ומשלי הקדמונים ומוסרי הפילוסופים.

[...] וכאשר שמעת, נכספה נפשך לראותו ולבא עד תכונתו. ותאמר: ׳אם היה בידינו החיבור ההוא עם השער הזה היה מספיק לנו בכל **ספר חובת הלבבות**, וממלא את מקומו, ומקל מעלינו טורח האריכות. והיינו יכולים לזכור דבריו על סדר ענייני שעריו. מפני שהדברים הארוכים אין אדם יכול לזוכרם׳. וייעדת אותי בעת ההיא להעתיקו אחרי השלמת השער ההוא.

והנה העתקתיו לך כפי יכולתי והשגת ידי, כל דבריו ומליצותיו ומשליו ומוסריו וחידותיו. אך הינחתי מקום השירים המובאים בו בכל שער משעריו, מפני שהיה דבר אדונינו הרב אביך נר״ו נחוץ להשלימו עד שאפנה להפוך ענייניהם בשירים כמוהם. ואם אמצא בדברי אנשי השיר הקדמונים כדברי בעל החיבור ודברי הנגיד ז״ל ומי שנהג מנהגם מה שהוא דומה להם וקרוב אליהם אכתבהו בשם בעליו. ומה שלא אמצא אעשה כפי דעי והשגת ידי, ואכתוב כל שיר במקומו מן הגליון. [...]

Document 2.1.1.3. Judah Ibn Tibbon, second preface to the translation of *Duties of the Heart* (to chapters 2–10).

Bahya Ibn Paqūda, *Sefer Ḥovot ha-levavot*, tr. Judah Ibn Tibbon, ed. A[braham] Zifroni, 162–164.

[...] החכם רבנא משולם בר״י זצ״ל שאל ממני להעתיק לו ממנו [= מן הספר הזה] השער הראשון שהוא שער היחוד. והעתקתיו לו, אע״פ שהיה קשה בעיני שהכניסני בזה, מפני הענינים אשר זכרתים בפתיחה אשר כתבתי לו על השער ההוא, אך בטלתי רצוני מפני רצונו.

ובא אחר כן ר׳ יוסף קמחי הספרדי להעתיק לו שאר הספר והעתיק מן השער השני והלאה. ועתה שאל ממני החכם הגדול הרב ר׳ אברהם ברבי דוד זכר צדיק לברכה להשלים לו כל הספר מהעתקתי. ויהי גם כן קשה בעיני, בעבור שקדם זולתי להעתיקו—שלא יחשב אותי כמתכבד בקלונו. ופייסתי אותו להניח לי, שלא יביאני אל הדבר הזה, וגליתי את אזנו, שבעבור כבוד ר׳ יוסף שקדם להעתיקו אני עושה. ולא שוה לי עמו. וכאשר ראיתי הסכמתו, התחייבתי לבטל

דעתי מפני דעתו, כדי להפיק רצונו ותאותו.

ונצרכתי לזכור את זה בתחלת הספר הזה, שלא יחשב אדם עלי, כי מרצוני באתי בעקבי ההעתקה ההיא. ויודע מחשבות אדם ידע, כי לא לאחת מאלה נכנסתי בדבר וכי לא לרצוני החלותי אותו, אך הגברתי רצון מי שנתחייבתי לעשות רצונו על רצוני והוא עדי ושהדי על זה, ודי בזה לנקות את עצמי.

ואחרי אשר החִלותי להעתיק, לא יכלתי לשקר באמונתי ולנטות מדרך האמת והנכונה ולעשות מלאכת ה׳ ברמיה. ואחזיק עצמי כפתי וסכל בעבור כבוד אחרים. והעתקתי כאשר נראה לי וכפי הבנתי.

ומפני שהספר הזה עתה בשתי העתקות ועל ג׳ ענינים, אני מפייס כל מי שיבא לכתוב מאיזו מהם שירצה ומשביעו באלהים, שיכתוב שם המתרגם תחלה על הספר, כדי שלא תושת עלינו חטאת באשר לא נואלנו ולא חטאנו. ודי לנו בשגיאותינו מעמס עלינו משל אחרים ומהוסיף על חטאותינו פשעי זולתנו. ורבותינו ז״ל הזהירו על זה ואמרו: ׳כל האומר דבר בשם אומרו מביא גאולה לעולם׳ [אבות ו ו].

ומה שאמרתי, כי הספר הזה בשתי העתקות ועל שלשה ענינים, מפני שיש ממנו ביד קצת בני אדם ספרים שהשער הראשון בהם מהעתקתי בלבד, ואח״כ העתיק ר׳ יוסף גם הוא השער הראשון והשלים הספר משלו; וישנו ביד אחרים כלו מהעתקתו. וגם אני השלמתיו כלו בהעתקתי כאשר שאל ממני הרב החכם הנזכר נר״ו. ומי שמתקן במלותיו במקום הראוי לתקן וידינני לזכות במה שקצרתי, יזכהו יוצרו לראות בטובו.

Document 2.1.1.4. Judah Ibn Tibbon, translator's preface to the translation of Jonah Ibn Janāḥ's *Sefer ha-Riqmah* (1171).

Sefer ha-Riqmah, ed. Wilenski et al., I, 1–7.

ברוך יי אלהי ישראל, אשר לא השבית לחכמה גואל, דורש להבין דרכה, ודעת נתיבה שואל. [...] ויום יום האולת הרה ויולדת, והחכמה עקרה ואיננה נפקדת. [...] ותלך האולת הלוך וגבור, והחכמות היו הלוך וחסור. [...]

ויותר מהמה [החכמות] חכמת הלשון אשר היא ראש לכל החכמות, וסלם אל כל המעלות הרמות. ואם איננה קודמת במנין היסודי, היא האדן העמודי, וקודמת בסדר הלמוד. פרצו אוילים גדריה [...] ועותו נכוחותיה, ועקלו נתיבותיה, ועבתו מלותיה ותיבותיה; ויתעללו בה כרצונם, ויכחידוה תחת לשונם, ושחתו דבריה הנעימים. ולא עשו עמה באמת ובתמים, כי לא הבינו אל פעולותיה, ולא הכירו מעלותיה. ויהיו מלעיבים באנשיה, ומלעיגים לנוחליה ויורשיה. ויחשבוה כדברי התהו אשר לא יועילו ולא יצילו, אשר על כן סרו מאחריה ואת דרכיה לא השכילו.

עד אשר העיר האלהים את רוח אנשי מלחמותיה, וגבורי מערכותיה, המה שרי צבאות ישראל חכמי גלות ירושלם אשר בספרד. וכראותם כי נעלמה ונבצרה, ויקנאו לה כי נסתרה [...] עד בא המורה הגדול מורה צדק, ומחזיק הבדק, מגלה עמוקות החכמה, גבור חיל ואיש מלחמה, יודע דעת קדושים, יועץ וחכם חרשים, ר׳ יהודה בן דוד המכנה חיוג ז״ל. ואשתכחת ביה חכמה יתירה, ורוח עצה וגבורה. הבין דרכם וידע מקומם, ועשה להם כאשר זמם, נלחם אליהם ויוכל, ומכל אשר היו לפניו שכל, משכם בעבותות תבונתו, ויביאם במסורת בריתו.

[...]

ואחריו באו שני שרי צבאות החכמה, בכח גדול ויד רמה, רבנא שמואל הלוי הנגיד, שר צבא ישראל ז״ל, והחכם המורה ר׳ יונה המכֻנה מרון בן גנאח, ויצאו בעקבותיו, וישאו מדברותיו, ויאור להם באורו. ויהיו גבורי כח עושי דברו, ויתאזרו חיל למלחמה, להוציא לאור תעלומה, ויהיו מנצחים במלאכה, ויערכו בה מערכה לקראת מערכה. ותוכן מלחמותם ודבריהם, הלא הם כתובים על ספריהם, ונזכרים בחבוריהם.

ואחר אשר שבתו המלחמות, ונראה להם במלאכה עוז ותעצומות, חבר החכם ר׳ יונה מחברת בדקדוק לשון הקדש וחלק אותה לשני חלקים. האחד קרא: **ספר הרקמה**, והשני קרא: **ספר השרשים**, כלל בהם רוב הליכות הלשון ועניניו ובניניו וסבותיו ועלילותיו ומוצאיו ומובאיו. וסמך על ספרי ר׳ יהודה בשתי המשפחות הנזכרות ולא זכר מהם בתוך ספריו כי אם קצת [בנדפס: קצות] דברים במקום הצרך.

והספרים האלה חברם בלשון ערבית, כלשון העם אשר היה יושב בקרבם. כי כן היו רוב חבורי הגאונים והחכמים בכל מלכות ישמעאל. בעבור שהיא לשון רחבה וצחה ולא יחסר המדבר כל בה, ולשון הקדש אין בידינו ממנה כי אם הנמצא בספרי המקרא, ואיננו מספיק לכל צרך המדבר. ועוד שאין המון העם מבינים אותה כי אם היחידים. וכל בני דורם היו מכירים בלשון ערבית. ועל כן בחרו לחבר את דבריהם בה.

אך בני גלות הַחֵל הזה אשר בצרפת ובכל גבול ארץ אדוֹם אינם מכירים בלשון הערבית, ויהיו להם הספרים האלה כדברי הספר החתום. ולא יכלו להגיע אליהם, כי אם אחרי העתקתם אל לשון הקדש.

וזה כמה שנים נדבה רוח החכם ר׳ משה בר׳ שמואל הכהן בן ג׳קטילה אותו והעתיק את שני ספרי ר׳ יהודה בן דוד הנזכרים בלשון הקדש לאחד מבחוריהם, ומבני נשיאיהם, ויהיו בידם למשמרת, לאות ולמזכרת. עד אשר בא החכם ר׳ אברהם בן עזרא בארצותם, ועזר להם גם הוא בענין הזה בחבורים קצרים, כלל בהם דברים נחמדים ויקרים. וגם החכם ר׳ יצחק הלוי נלוה אליו וחבר להם **ספר המקור**, אסף בו כללים מועילים, בשמוש הלשון ובניני הפעלים. ומאז נטו קצתם אחרי המלאכה הזאת והתעסקו בה עסק מעט.

אך פגעתי מהשוקדים על דלתותיה, אשר שמו אותות אותותיה, אנשים החלו לדרשה, וטעמו טעם דבשה. וכאשר ראו כי אורו עיניהם, ונפקחו אזניהם, נמשכו אחריה, ונכספו לעמוד על ספריה. ועלי שמו פניהם, להעתיק להם את שני הספרים האלה, והרבו עלי רעים למלאת את שאלתם, ולעשות את בקשתם. והם פגעו בי, עד אשר הטו את לבבי.

ואני ידעתי כי העבודה כבדה והמלאכה קשה, בעבור הדברים אשר זכרתים כבר בפתיחת העתקת **ספר חובות הלבבות** אשר יצטרך אליהם המעתיק. והדברים ההם עם קוצר הלשון קשים להעשות כי אם על ידי תחבולה גדולה ומחשבה ארוכה, לבעבור סבב את פני הדבר, משני קצותיו וחבר. [...]

2.1.2. Samuel b. Judah Ibn Tibbon (c. 1165, Lunel–1232, Marseilles)

I. Triangular Correspondence between Samuel Ibn Tibbon, R. Jonathan ha-Kohen of Lunel, and Maimonides in Preparation of the Hebrew Translation of the *Guide of the Perplexed* (1194–1204)

Document 2.1.2.1. R. Jonathan ha-Kohen and colleagues to Maimonides (c. 1199).

'Letter of R. Jonathan of Lunel to Maimonides', ed. Wertheimer.

[...] ברוך יי אלהי אבותינו אשר נתן כזאת בלב מלך החכמות לפאר את דתינו ולחזק את בדק תורתינו [...] ואנחנו החזקנו כיד אלהינו עלינו לקבץ ולאסוף ספירי ספריו ומבחרי חבוריו לא יערכום מטמונים ולא ישוו [...].

למדנו ראינו את **ספר משנה תורה** פשר הספקות מגלה העמוקות. [...] הנה נתת לפנינו את החיים ואת הטוב.

והוספת עלינו חכמה וטוב כי שלחת אלינו **הספר מורה הנבוכים**. [...] דבריו דברי אלהים חיים. מוצאו מצא חיים וכל הפורש ממנו כפורש מחיים.

כשמו כן הוא—**מורה** לחכמים **הנבוכים**, ולפני ההמון כאבן שאין לו הופכים. ואמנם היה בידנו כצרור אבן במרגמה וכשושנה בין החוחים—נתן הספר לאשר לא ידע הספר. לולא אשר הקרה בוראנו לפנינו בן חכם ומשכיל בכל חכמה, למדו הרב אביו ספר ולשון הערב, בן לחכם המופלא הרופא המעולה הרב יהודה אבן תיבון הספרדי, אשר הועילנו והשכילנו ולִמְדָנו מספרי החכמות בהעתיקו אלינו **ספר האמונות וספר חובות הלבבות ומדות הנפש ומבחר הפנינים וספר הכוזר** וספרי הדקדוק: **ספר השרשים** לאבן ג׳נאח אשר לו **הרקמה**, כל איש מהם יתנו יקר לבעליהם.

ועד שלא כבתה נרו של עלי [זרחה שמשו] של שמואל הרמתי. [...] ויאורו עינינו לראות בנסתר, לא תשבענה לראות עינינו ולשמוע אזנינו. והגדיל תשוקתנו [וחמדתינו] הראשונה כעיף הצמא אשר החל לשתות והסירו הכוס מפיו, וכיונק העתיקוהו משדי אמו. [בערו בלבנו] נחלי תשוקתנו והוסיפו לחמם. [...]

לכן אם [טוב] הדבר לפני אדונינו הרב הגאון וכשר הדבר לפניו [...] ישלים חקנו וימלא טרפנו [...] וישלח לנו מערכות [...] **ספר מורה הנבוכים** עד תֻּמה. [...] ויקשיב באזן אהבה ורצון לתחינת חכמינו אלה יושבי ירח יקר [...] למען נעמוד על האמת ונדעה כי לא לחנם פזרנו מעותינו. [...]

יהונתן הכהן בן דוד זצ״ל

Document 2.1.2.2. Maimonides to Samuel Ibn Tibbon (1199).

Shailat, *Iggerot ha-Rambam*, II, 530–533, 554.

הגיעו אלי, אני משה בר׳ מימון הרב הספרדי זצ״ל, כל כתבי התלמיד היקר, המבין והמשכיל, נזר התלמידים, צבי החכמים, ר׳ שמואל ש״צ [= שְׁמָרוֹ צוּרוֹ] בר׳ יהודה זצ״ל. [...]

יפה שאלת בכל אשר שאלת, וכל המקומות ששערת בהם חסרון תבה או תבות כך הוא. וכבר פרשתי לך כל הדברים בלשון ערבי בסוף כתב זה, והודעתיך כל עניניך [נ״א: + וגם הספרים שתתעסק בקריאתם ותניח זולתם]. לך בודאי ראוי להעתיק מלשון ללשון, לפי שלב נבון נתן לך הבורא להבין משל ומליצה. והכרתי מן דבריך שלבך ירד לעמק הענין ויגלה מצפון הסודות.

והנה בארתי לך בלשון ערבי היאך תעשה בכל העתקה, ו׳תן לחכם ויחכם עוד׳ [משלי ט ט], ׳חכם, בני, ושמח לבי׳ [משלי כא יא]. חייך, שבעת ששמעתי דבריך—שמחתי ותמהתי: שמחתי למציאת בן חכם, ותמהתי היאך יהיה טבע בן שנולד בין העלגים כך, וירדוף אחר החכמות, ויהיה מהיר בלשון ערבי—שהיא ודאי לשון עברי שנשתבשה מעט—ועוד היאך יבין דקדוקי הלשון וענינים עמֻקים. אין זה אלא כשרש מארץ ציה [ישע׳ נג ב]. [...]

הגיעו כתבי הישיבה הנשאת, הנכבדת, המשכלת, יתמיד ה׳ מעלתה ויוסיף הצלחתה. ועמדתי על מה שסֻפֵּק לך העתקתו, וראיתי המקומות אשר טעה בהם הסופר, וידעתי ההקדמות והפרקים אשר לא התבאר לך ענינם הבנה שלמה, ובקשת פרושם. ואני אבאר לך הכל. [...]

והנני מתחיל עתה לענות אותך על כל פרק ופרק, ולפרש לך מה שבקשת פרושו, ולהזכיר לך את הנסחא האמתית אשר תגיה בה הנסחא ההיא המַטעית, על סדר שבא בכתבך. [...]

הנה הישירתיך והוריתיך למה שתעיין בו, ולמה שתטריד את נפשך היקרה. [...]
כתב משה בר׳ מימון זצ״ל ביום שמיני לתשרי אתקי״א לשטרות. ושלום.

Document 2.1.2.3. Maimonides to the Community of Lunel (1200).

Shailat, ed., *Iggerot ha-Rambam*, II, 557–559.

[...] ושלחנו לכם עם זה הכתב החלק השלישי מ**מורה הנבוכים** בלשון ערבי. ואשר שאלתם שאעתיקנו לכם בלשון הקדש—מי יתנני כירחֵי קדם לעשות שאלתכם, בזה הספר ובשאר הספרים שחברתי בלשון קֶדָר אשר הקדיר שמשה, כי עם אהליהם שכנתי. ושמחה גדולה היה לי בזה, להוציא יקר מזולל ולהשיב האבדה לבעליה. אבל סבות הזמן סבבוני, ואפילו הפֵּרושים שעשיתי, וכמה ענינים שחברתי בלשנא דרבנן, שהן עדין אפלות—לא נשאר לי פנאי לדקדק אותן ולהגיהן עד שיֵּצאו לאור העולם, וכל שכן להעתיק מלשון אל לשון. חיֵי נפשכם, אלוּפַי ומיֻדָּעַי, אפילו לכתב פרשה קטנה אין לי פנאי, לא ביום ולא בלילה, ומפני כבודכם צערתי עצמי וטרחתי לכתב כתב זה כלו בכתב ידי.

ומכל מקום הנה יש אצלכם בן חכם משמח אב [עפ״י משלי י א], אשרי יולדתו, הוא התלמיד החבר היקר, נזר המשכילים, ר׳ שמואל בן החכם ר׳ יהודה אבן תיבון ז״ל, נבון ומשכיל

ורחב לב הרבה, ובעל דעה נכונה, ולשון למודים לדבר צחות נתן לו האל ית׳, ולא יבצר ממנו כל אשר תרצו להעתיק לכם. וכבר כתבתי לו כתב בכתב ידי בלשון עברי והגרי על עסקי ההעתקה.

Document 2.1.2.4. Samuel Ibn Tibbon to Maimonides (1199/1200).

Diesendruck, 'Samuel and Moses Ibn Tibbon on Maimonides' Theory of Providence', 352–353.

[...] ישגה כרצון הצעיר בעבדיו הנכסף להתחבר לאחד מתלמידיו הכותב להודיע לכבודו כי הגיע ליד עבדו כתבו היקר עלי מכל יקר הכתוב בחצי היו׳. וכמעט יצאה נפשי בשורותיו הראשונות, רפדוני שורותיו האחרונות במצאי בפיהם רפואת אדוננו מחוליו הארוך אשר נשא בעון כלנו. [...]

והן אפער פי מיחל למלקוש תשובותיו הברורות ואמרותיו הטהורות על כל אשר שאלתי מכבוד חכמתו בכתבֵי הקדומים בשלשת חלקי המאמר הנכבד **מורה הנבוכים**. ואירא מהוסיף שאלותי פן אטריח מאד על כבודו. רק שאלה אחת היא בלבי כאש בוערת עצורה בעצמותי, עוררה לי פרק אחד מן הפרקים האחרונים מן החלק השלישי הבאים באחרונה. אוסיף לשאול מכבוד אדוננו גם בה מבוכתי הגדולה עם יתר השאלות הקדומות. וצדקה תהיה לו מעם יי׳ אלהיו.

וכן אני שולח את שלשת חלקי המאמר אשר אתנו העלולים המגיעים משם אשר מהם העתקתי. אשלחם אל כבוד אדוננו, מבקש מכבודו לצוות לקצת תלמידיו להגיהם בדקדוק גדול פעם ושתים עד שיהיו בטוחים שלא נשאר בהם טעות. ויחתום אדוננו בהם אחר שיתאמת לו טוב הגהתם ויצוה אז לשלחם אל התלמיד הנבון המשכיל רבי אברהם הכהן ש״צ [= שְׁמָרוֹ צוּרוֹ] אל אלכסנדריא. וידעתי כי הוא בחסד ישתדל לשלחם אלינו על ידי אחינו היהודים ההולכים לשם תמיד. כי גם כן גמלני חסד להשתדל בכל אשר שלחתי לעשות שם, וכל שכן שישתדל אם יבא לו כמו מצות אדוננו ירום הודו.

לא נמנעתי להטריח אדוננו גם בזה כי דעתי כי ייטב בעיני כבודו להסיר עולה מאהלינו. וגם לבבי לא ישקוט ולא ינוח עד דעתי מאמר ההוא מוגה. כי מי יוכל לבטוח בעצמו בשגיאת ספר שרבו שגיאותיו מספור. ואפילו אם היה המעיין חכם היו נשמטין לו מקומות רבים, כל שכן למי שלא הריח ריח חכמה.

ואשר יראה אדוננו בחלק השלישי קצת הגהות בקצתו—הגהתים מנוסחא שנית שהגיע לידי מקצת החלק ההוא, לא היתה מוטעית כראשונה. כי הראשונה, כאשר הודעתי אל כבוד אדוננו, נראה ממנה שנכתבה מספר כתוב ערבי או מספר שנכתב הוא מספר כתוב ערבי, ולזה רבו טעיותיו כאשר יראה בה אדונינו. ואני בקצת השגיאות ששיערתי בהן הייתי מציין בחוץ בדיו, ובקצת הייתי מעביר צפרני בגליון כנגד השטה שבה הטעות שבקצתה אין בה ציון כלל. ומחסד אדוננו לצוות על המגיהים שלא יגרדו מכל מה שימצאו בספרים אפילו אות אחת, אך יגיהו מבחוץ, לא על ידי גרד, ויציינו היטב המקומות כדי שלא ישאר לנו ספק במה שיגיהו.

II. Samuel Ibn Tibbon's Prefaces to his Translations

Document 2.1.2.5. Samuel Ibn Tibbon, preface to his translation of Maimonides's *Guide of the Perplexed.*

Samuel Ibn Tibbon, 'Preface to his Translation of the *Guide of the Perplexed*', 117–122.

דברי שמואל בן יהודה אבן תיבון ז״ל, מעתיק המאמר הזה מלשון הגרי אל לשון עברי. אמר, מתנצל:

[...] וכבר גילה אבא מורי ז״ל כל זה בפתיחתו אשר עשה להעתקתו **ספר חובת הלבבות**.

[...] ואני, לא נעלם ממני מהם מאומה, ויודע אני ומכיר במיעוט שלימותי בשתי הלשונות העברית והערבית, וכל שכן בלשון הערבית שידיעתי בה מעוטה, כי לא גדלתי בין אנשיה ובארצותיה, ולא באתי עד תכונותיה. ועוד כי ידעתי חסרוני הגדול בהבנת דברי הספר הנכבד הזה **מאמר מורה הנבוכים** ועינייניו, מפני שרובם עניינים עמוקים מאוד והוא כולל חכמות רבות ורמות, מעיני רוב אנשי עמנו אשר בגבולנו זה נעלמות, ואולי מעיני כולם מפני שאינם מתעסקים בהם ואינם נמצאות איתם, ומעט מזער פקחתי עין בקצתם כי היו אתי קצת ספריהם בלשון ערבי אשר אני יודע בה מעט, והעירני לבבי ללמוד מהם כיכולתי הקצרה.

ומכל אלה הפנים אשר זכרתי היה לי להימנע מהעתיק הספר הנכבד הזה, אבל כמו שהביאה תשוקת חכמי הארץ הזאת ונבוניה אותם, ובראשם החסיד הכהן רבי יהונתן נר״ו [= נטריה, רחמנא ופרקיה] ויתר חכמי בקעת יריחו עיר מושבי, לשלוח אחרֵי הספר הנזכר אל מצרים בשמעם את שִׁמעו, ויתחננו על פי כתביהם אל הרב הגדול הפילוסוף האלהי הנזר הטהור על ראש גליותנו מרנא ורבנא משה נר״ו בן כבוד הרב הגדול רבי מיימון ז״ל, הוא המחבר את הספר הזה, ויבקשו מאתו לשלחו אליהם מועתק אם יוכל או בלתי מועתק, והשתדלו בזה השתדלות גדולה, כן הביאה אותם תשוקתם להשתדל בהעתקתו בהגיעו אליהם בלתי מועתק, ועוד כי הוסיפו כוסף על תאותם בעמדם על קצת מעניני הספר.

ולי אני שמואל ב׳ר יהודה ן׳ תיבון ז״ל קראו, ועלי שמו פניהם, מפני שידעוני שאני מבין קצת בלשון הערב, ומפני שאין בגבולם איש מאנשי הלשון ההוא. ויפצירו בי מאוד להעתיקו להם כיכולתי וכאשר תשיג ידי. ואע״פ שהודעתים את קצורִי בהעתקה מן הפנים שזכרתי, אמרו לי אי אפשר בלתי שנקבל בה קצת תועלת, ואם תהיה ההעתקה חלושה ומעוטה מן התועלת בענינים הנכבדים אשר יכללם הספר הזה—הוא הרבה בעינינו. ולא יכולתי להשיב פניהם ושמתי עצמי ככסיל מתעבר ובוטח, והעתקתיו עם קוצר השגתי. [...]

Document 2.1.2.6. Samuel Ibn Tibbon, *Glossary of Unfamiliar Terms* (*Perush ha-millot ha-zarot*, 1213).

Samuel Ibn Tibbon, *Perush ha-millot ha-zarot*, 11.1–12.6.

בהשלימי העתקת המאמר הנכבד הזה, **מאמר מורה הנבוכים**, וראיתי שלא יכולתי להימנע מלהשתמש בו במלות זרות לא יבינו אותם רוב המעיינים בו מפני קוצר לשוננו, ובהעדר חבורי

החכמות המופתיות מאומתנו, עד שלא יימצאו בלשוננו המלות הזרות שמשתמשים בהם בעלי החכמות המיוחדות המופתיות, על כן מלאני לבי לחבר שער אחד אפרש בו המלות ההם.

וכל שכן שכבר הקדימני לחבר כיוצא בכוונתי המשורר ר׳ יהודה אל חריזי, אשר העתיק המאמר הזה אחרי, והקדים להעתקתו שני שערים: השער האחד—שער יבאר בו המלות, מלאהו הבלים וטעויות ומכשולים; והאחר—שער אמר שיודיע בו כוונת כל פרק ופרק מפרקי המאמר הנכבד, כולו נאצות ומכשלות ואבני נגף.

ולהסיר מכשולי השער ההוא הראשון מדרך עמנו ולהועיל המעיינים בהעתקתי, אמרתי לחבר שער אחד, אבאר בו המלות הזרות, וכן כל מלה זרה שאשמש בה בפתיחתי זאת אשר אקדים לשער הנזכר. [...]

Document 2.1.2.7. Samuel Ibn Tibbon, preface to his translation of Maimonides's *Eight Chapters* (1202).

Slutzky, *Shemonah Peraqim le-ha-Rambam*, 1a–b, 2b.

Passage 1: 1a–b

אמר שמואל בר יודא אבן תיבון.

זאת המסכתא—ר״ל מסכת אבות, ידועה מסדר ישועות—סדורה בין מסכת עבודת כוכבים ומסכת הוריות. ולא נמצא לה גמרא. והרב הגדול רבינו משה בן הרב רבינו מימון ז״ל פירשה בלשון הגרי, בכלל פירושו למשנת שיתא סדרי שפירשה כולה בלשון ההיא. והרחיב במסכתא הזאת להיותה יקרה בענייניה, מפני שהיא מדברת במדות האדם החשובות והישרות, והפחותות והמעוותות, אשר החשובות והישרות מהן מפנות הדרך ומסקלות המסילה העולה אל המעלות השכליות, כמ״ש החכם: ׳שמע עצה וקבל מוסר למען תחכם באחריתך׳ [משלי יט כ]. ובמעלות השכליות ישיג המשכיל את ידיעת בוראו.

Passage 2: 2b

ומפני שחכמי המחקר חברו ספרים במדות האדם, והאריכו לדבר בהן ולהודיע איך אפשר לאדם להרגיל עצמו במידות הטובות עד שישובו קנין חזקה ויהיה קביעות בנפשו, ואיך יוכל בעל מדות מעוותות לתקן המעוות בהן, ראה הרב ז״ל להקדים למסכתא זאת פרקים יזכור מהם בקצרה הצריך מן העניינים אשר האריכו בהם חכמי המחקר, כמו שזכר בפתיחתו.

וכאשר ראו חכמי לוניל עיר מולדתי פירוש זאת המסכתא והפרקים אשר הקדים הרב בביאור עניינים, בקשו ממני להעתיקם להם כאשר עשו במאמר **מורה הנבוכים**.

Document 2.1.2.8. Samuel Ibn Tibbon, preface to *Otot ha-Shamayim* (his translation of al-Biṭrīq's Arabic version of Aristotle's *Meteorology*).

Fontaine, *Otot ha-Shamayim*, 4–9.

אמר שמואל בן תבון המעתיק הספר הזה הנקרא **ספר אותות השמים** מלשון הגרי אל לשון עברי אל אחד מחומדי החכמה ומן המשכילים בה, רבי יוסף בן החסיד רבי ישראל זצ״ל.

בתחילת תחנוּנֵי המשכיל הנבון החבר היקר רצה דבר להעתיק לו ספרי חכמת הטבע אשר חיברו חכמי המחקר. ומקשי שאילתו עלי, לרוב המלאכה ולכבידותה, ולקוצר הזמן ולמיעוט ימי הפנאי בו ולסבות אחרות אין צורך לזוכרם, לא הפקתי רצונו בשאילה ההיא. ובראותו כי כן, עם רוב כֹספו לחכמה וליקרת המעט ממנה בעיניו, העתיק שאילתו הכוללת את כל חלקיה אל חלק אחד ממנה, ושָׁנָה לבקש ממני, ופצר בי, וגזר עלי בגזירת האהבה התמימה אשר בינינו להעתיק לו **ספר אותות השמים**. [...]

ותִקשֶׁה גם השאילה הזאת, הקטנה בכמותה הגדולה באיכותה וביתר מקריה, ותגדל גם היא בעיניי בגלל הסבות הנזכרות, הגלוייות והנסתרות. כי לא חסר מהם אחת, רק הראשונה שבהם. ואם חסרה מהם אחת נוספה עליה אחרת, והיא: שזה הספר אינו אצלי רק מהעתקת הבטריק. ולא ראיתיו במקום אחר רק מהעתקתו. והעתקתו מבולבלת מאד, כאשר כתב אלי הרב הגדול, הפילוסוף האלהי, מרנא ורבנא משה עבד האלהים בכתבו הראשון הבא אלי.

ועוד, כי נראה בו טעיות וחסרונות: יורה על קצתם מקומותם ולשונותם שאינן סובלין העיניין הראוי במקום ההוא; ועל קצתם יורה הנמצא בפירוש אלכסנדר לספר הזה, ומה שמביא מלשון אריסטוטליס בפירושו ומה שכותב בשמו.

ואפילו אם לא היה שם רק זאת הסבה לבדה היה ראוי להמנע מהעתיק הספר הזה עד שנראה בגזירת השם אם נמצא העתקה אחרת, או אם נמצא זאת מוגהת יותר, ואף על פי שאני הגהתי אותה כאשר מצאתי בטוליטולה ובברצלונה. אך לא יכולתי לעבור על רצון השואל ולהשיב פניו לפי עוצם תשוקתו ורוב הפצרו עם יקרת אהבתו.

ובגזירת השם ובעזרו הטוב אתחיל להעתיקו כפי השגת יד שכלי הקצרה. וממנו אשאל להישיריני בהעתקתי ולהצילני ממכשול, כאשר אשאל ממנו להורות המעיינים בדברי הספר הישרים ובדרכיו העליונים את הדרך ילכו בה. ויתקיים בהם מקרא שכתוב ׳כי ישרים דרכי ה׳ צדיקים ילכו בם׳. [...]

Document 2.1.2.9. Samuel Ibn Tibbon, preface to his translation of *Three Treatises on Conjunction with the Active Intellect*, by Averroes and his son.

Hercz, ed. and tr., *Drei Abhandlungen*, Appendix IV, 23 (Hebrew numbering); Robinson, 'Samuel Ibn Tibbon's *Commentary on Ecclesiastes*', Appendix A, 502, 503, 505.

מה שנמצא לאחד מן הפילוסופים בענין השארת הנפש האדם. ונראה שהיה מקומו המיוחד לו אחר **ספר הנפש**, והם דעות קדומות לפילוסופים הקדומים.

כאשר הכרתי מדברי זה הספר הנכבד שעיקרו להודיע דעת החכמים בענין עלות הנפש האדם למעלה. וענין עלותה למעלה הוא שתשלם ותתעלה עד שתדבק בשכל הנפרד ותתאחד עמו, כלומר שישובו דבר אחד. והשכל הנפרד הוא בלא ספק במעלה העליונה. כי ידוע שהנמצאות שלשה מינים: החמר השפל והמתהוה ממנו, והוא הפחות במעלה והשפל במקום; ולמעלה ממנו במעלה ובמקום הגופות הרקיעיים שצורותיהם עומדות בחמריהם תמיד, עד ששֵם ה׳חומר׳ וה׳צורה׳ נאמרים עליהם ועל ההוים בעולם השפל בשתוף השם; ולמעלה מכולם מעלת גדולה וכבוד השכלים הנפרדים שהם צורות לא בנושא. ועל כן אמר בנפש האדם כשרצה לרמוז בדבקה באחד השכלים הנפרדים שהיא ׳עולה למעלה׳ [קהלת ג כא].

ומפני שראיתי שענין זה—הדבקות והעליה למעלה—הם דבר אחד, והם היו עקר כונת זה הספר, ראיתי לחבר אל פרושי בו מאמרים קטנים שנמצאו לחכמים האחרונים בענין הדבקות השכל הנפרד באדם.

Section Three. 'Encyclopedias' of Science and Philosophy

3.1.1. Shem-Tov Ibn Falaqera (c. 1225–after 1290, Spain?)

Document 3.1.1.1. Shem-Tov Ibn Falaqera, preface to *The Doctrines of the Philosophers* (excerpts).

The Hebrew text given here is its *editio princeps*, based on the two extant manuscripts: Parma, Biblioteca Palatina, 3156 (= De Rossi 164), f.3a (P); Leiden, Universiteitsbibliotheek, Cod. Or. 4758, f.106b (L). The division into paragraphs is mine.

[2] ואחר תהלתו אומר כי מהידוע והמפורסם והמוסכם עליו מן התורה והחכמי׳ כי הצלחת האדם האמיתית היא השגת הבור׳ ית׳ וידיעתו כפי כח השכל האנושי. ומהמוסכם עליו כמו כן אצל כל החכמי׳ האמיתיים כי אותה ההשגה תגיע לאדם בידיעת מעשיו ובציירו אותם ציור שכלי, כי אין הנבדל מהחומר מושג לאדם אלא מצד פעליו. ועל כן אמרו כי הרוצה לחקור השכל צריך לחקור בתחילה פעליו מה הם. כי הפעלים יותר נודעים אצלינו מהכוחות שהפעלי׳ הווים מהם. וכמו כן כשנרצה לדעת הפעלים צריך שנקדים ונדע הפעולים, כי הפעולים יותר נודעים אצלינו מהפעלים.

ועל כן אמרו ... בתחילה פעליו] ליתא P

[3] **אמר** רבי׳ משה ז״ל: ׳וכבר הודעתיך כי אין שם זולתי האלוה יתע׳ וזה הנמצא; ואין ראייה עליו ית׳ אלא מזה הנמצא מכלליו ומפרטיו. ועל כן יתחייב בהכרח לחפש ולחקור זה הנמצא כפי מה שהוא עליו ולקחת המוקדמות ממה שיראה מטבעו. ועל כן יתחייב לדעת צורתו וטבעו הנראה ואז איפשר לקחת [P fol. 3b; L 107a] ראיה ממנו על זולתו׳. ע״כ. והנראה לי מדבריו כי זה הכרחי למי שהשגתו על דרך החקירה, אבל מי שישפע עליו השפע השכלי כמו הנביא ישיג האמת מאין חקירה.

לי] ליתא P מדבריו] ליתא L

[4] ומי שהשגתו על דרך החקירה יצטרך לחקור הנבראים ולעיין בדברי החכמי׳ שכלו ימיהם בחקירתם והשתדלו לעמוד על אמתת מהותם. מאחר שלא חברו חכמינו ז״ל ספרים בזה העניין, או שחברו ואבדו באורך הגלות ובשעבוד האומות אותנו, כאמרו ע״ה ׳ואבדה חכמת חכמיו׳ וגו׳ [ישע׳ כט יד]. ומה שימצא מדברי חכמי המחקר שהוא אמת ומסכים עם אמונתינו יקבל אותו וישבח אותם עליו, ומה שאינו אמת יטוש אותו וירחיקהו.

[5] **אמר** אריסטו: וצריך שנשבח לקדמונים במה שחקרו כי חקירתם היא סבה שנעמוד על האמת בקלות. ואילו לא היה המוקדם היה נעלם מהמאוחר הרבה מהאמת, אבל אילו לא היה המוקדם לא היה המאוחר כל עיקר. ע״כ. ואין ראוי למשכיל המבקש לדעת האמת להרחיק דברים רבים שנגלו לחכמי המחקר ובארום במופת אמיתי והם בלי ספק כמו שאמרו, בעבור שאינם מאומתינו. אלא ישמע האמת מהאומרו, ולא יביט אל האומר אלא לאמתות דבריו.

לא היה המוקדם] היה מוקדם L

[6] **אמר** רבי׳ משה ז״ל: ׳וכל דבר שנודעה אמתתו בראיות ברורו׳ אין אנו סומכין על אותו האיש שאמרו או למדו אלא על הראיה שנגלתה ועל טעמו שנודע׳. ע״כ. ומהמבואר כי יש דברים רבים אמיתיים אין ספק באמיתתם והם זרים לאדם וכשישמעם לא תסבלם דעתו. ויקרה לו זה או לחסרון שכלו או למיעוט הרגלו באותם הדברים או מפני שהורגל להאמין הפכם. כי מטבע האדם שיקשה עליו לעזוב בזמן מועט מה שהורגל להאמינו בזמן ארוך. ועל כן כשישמע באותן הדברים ימהר לסתור אותן ויגזור עליהן שהם הפך האמת. אבל כשידקדק אותן ויעיין בהן כראוי מאין פנות למה שהורגל להאמינו מזמן הנערות אז ימצא אמתתם ויראה׳ כי אין הדבר כמו שהיה חושב.

[7] **אמר** פורפירוס הצורי: ׳ואין צריך האדם כשישמע דבר זר לא הורגל בו ויאמין הפכו זמן ארוך שיטה ממנו וירחיקהו מבלתי שיחקור אותו, או שיחקור אותו כאלו הוא דבר בטל ואינו אמיתי. אלא אחר שישמע אותו ישתדל לחקרו ואם ימצא [4a P] כי הוא אמת יבקר בו. כי אף על פי שיש באילו הדיעות דברים רבים יצטרך המעיין בהם לחקירה יתירה או שיתחדשו לו במקצת הדברים ספיקות, לא יקרה זה באלו הדיעות בלבד אלא זה הוא דבר כולל כל הדיעות ברוב. כי יש בכל אילו דברים לא יושגו או תקשה השגתם, וברבים מהדברים יקרה הטעות מפני חולשת דעתינו להשיגם. ועל כן אינו מן הנכון שנעזוב כל דעת לא השגנוה או יש בה ספיקות, כי כשנעשה זה לא תגיע לנו דעה קיימת. ועל כן צריך שנעיין בכל הדיעות שדברו בהן או איפשר לדבר בהן ונחקור אותן, ומה שנמצא מהן שיש לו יתרון על השאר בביאור ומתחייבת מהשרשים המושכלים הנופלים תחת החושים נבחר בה, ואע״פ שיש בה דברי׳ נעלמים׳. ע״כ.

פורפירוס] פורפוריאוס L

[8] על כן ראיתי אני שמואל בר׳ יהודה בן תבון [!] ע״ה לבאר בזה החיבור דעות אנשי המחקר בכל הנמצאים, והם החכמים הנקראי׳ פילוסופי׳ האמיתיים [107b L], והם דעות הפילוסוף האמיתי אצלם. כי מלת ׳פילוסוף׳, שעניינה אהבת חכמה, היא מורה על שני מינים מהשלמיות: הראשון מין השלמויות היצירויות, והשיני מין השלמיות השכליים. ועל כן הפילוסוף האמיתי אצלם מי שכוונתו להשיג ידיעת האמת ולעשות מעשה החסד והאמת, ומכוין בכל מעשיו שיגיעו לו אלה השני מינים מהשלמיות ושם כל יגיעו בחייו להשיגם על השלמות כפי הכח האנושי והגיעו לו. ועל זה העניין מורה שם ׳פילוסוף׳ אצלם, ועל זה העניין מורה שם ׳חסיד׳ בלשונינו, והוא מי שיקנה מעלות היצירויות והשכליות. [...]

הראשון מין השלמויות היציריות] האחד מן השלמיות הישרים L // הראשון מין השלמויות היצורים P

[9] והסבו׳ המצריכות לכתוב אלו הדיעות בזה החיבור הם שתים:

[10] **הסבה הראשונה**, והיא מההכרחית יותר, כדי שידע אלו הדיעות הרוצה לעמוד עליהם מבני עמנו ידיעה אמיתית, שהיא ידיעה מופתית, ותתבאר לו האמת מן הכזב, והנמנע מן האיפשרי. מפני שהאחד מהדברים הגורמים למעיינים בדברי הפילוסופים להכחיש האמת הנזכרת בהם הוא שלא עמדו על אמתתם ולא ירדו לעומק ראיותיהם וטעמיהם. וזו היא סבה לחלוק עליהם בדברים רבים. וכמו כן אחד מהם [4b P] מהדברים הגורמים להאמין כל מה שאומרים הוא מה שחושבים שיש מופת על כל דיעותיהם. וכבר זכרו אחרוניהם כי הספיקות בדבריהם בחכמת הטבע מועטים ובדבריהם בחכמת האלהות רבים. ועל כן יתבאר בזה החיבור על אי זו דעת מדעותיהם יש עליה מופת לדעתם, ואי זו מהם אין עליה מופת, אלא אמרוה כפי מה שנראה להם.

מופת] ספק P

[11] על כן השתדלתי להעתיק אלו הדיעות מלשון הערבי אל לשון העברי ולחברם מהספרים שהן מפוזרות שם כדי שימצאם הרוצה לעמוד עליהם מקובצים בספר אחד ולא יצטרך ליגע עצמו בקריאת כל הספרים. כי כל דעותיהם בחכמת הטבע והאלהות נכללות בזה החיבור כלל ופרט.

להעתיק] ליתא P

[12] **הסבה השנית** מפני כי רוב מה שהועתקו מאלו הדיעות ללשונינו הן משובשות ואינן מבוארות כראוי, ולא יוכל המעיין באותן הספרים המועתקים לעמוד על אמתתם. והרבה מאותן הדיעות הן מהפילוסופיא הקדומה שדחו אותה האחרונים. ולא קרה זה בספרים המועתקים ללשונינו בלבד, אלא גם כן יש ספרי׳ מחוברי׳ מועתקים בערבי ומיחסים דעות רבות אל הפילוסופים שאינן דעותיהם על האמת.

הקדומה] הראשונה P

עת] אם P

[15] **ואומר** כי חכמי מחקר הסכימו כי דעות חכמי המחקר האמיתיות נכללות בדעות אריסטוטלוס הפילוסוף היוני. ואמרו שהוא איזן וחקר ותקן דעות הקדמוני׳ שהיו לפניו, והעיד על טעותם והבדיל האמת הנזכרת בדבריהם מהכזב המעורב בו והוציא לאור החכמה האנושית.

[17] על כן דברי בזה החיבור הם דבריו המפורשים בפירושי הפילוסופים שהם על דעתו. כי הבנת טופס דבריו בכל מה שדיבר תקשה מאד אפילו על הפילוסופי׳, כל שכן על שאר המעיינים. אמר החכם בן רשד: ׳ולא עמדו המפרשים על פירוש ספרי ארסטו אלא בזמן ארוך, כמעט שיהיה שוה לזמן שנשלמה בו החכמה או קרוב ממנו׳, ומפני זה לא אזכור טופס דבריו בכל המקומות. ועוד, כי הבאים אחריו מן הפילוסופי׳ בארו אלו הדיעות והתירו רוב הספקות שנסתפקו בהם ותרצו רוב הקושיות שהקשו עליהם.

[18] ואין בכל זה החיבור דבר שאומר אותו מעצמי. אלא כל מה שאכתוב הם דברי ארסטו׳ המפורש׳ בדברי החכם בן רשד. כי היה אחרון המפרשי׳ ולקח הנבחר שבפירושיהם והנאות

לדעת ארסטוטלו׳. ודעת ארסטו׳ הן הדעות שצריך לעיין בהם מבין שאר דעות הקדמונים. וכל שכן כי דעות זולתו מחכמי האומות כבר נשכחו בזמנינו זה, ואינן נזכרות אלא מצד שהן כתובות בספריו. על כן לא אכתוב מאותן הדעות אלא הצריך בזה החיבור.

בדברי] בפירושי P

לדעת] לדברי P

[.....]

[19] [108b L] אמר החכם אבן רשד: ׳ועיון ארסטו׳ באלו העניינים רחוק מאד מעיון זולתו מהחכמים, ומה שיתבאר לו במהירו׳ יתבאר לבני אדם מדבריו אחר חקירה חזקה וזמן ארוך. על כן אמר ארסטו׳ כי אילו לא היו דבריו באלו העניינים כתובים היתה העמידה עליהם כמו הנמנע או הקשה המציאות אלא בזמן ארוך. ועל כן יקרה למפרשי׳ ספיקות בדבריו, ואחרי כן יתבאר בזמן ארוך יושר דבריו וקוצר עיון זולתו בצרוף אל עיונו. ובזה הכח האלהי שהיה בו הוא היה הממציא לחכמה והמשלימה. וזה דבר ימעט מציאותו במלאכות, אי זו מלאכה שתהיה, כל שכן בזו המלאכ׳ הגדולה. ואמנם אמרנו כי הוא הממציא החכמה ומשלימה מפני כי כל מה שימצא לזולתו מהקדמוני׳ באלו העניינים אינם ראויים שיושמו ספיקות באלו הדברים, כל שכן שיושמו התחלות. ומפני זה השלימות שהשלימו האל ית׳ היו קוראים אותו הקדמונים ״האלהי״׳.

[20] ואמר עוד כי ׳לא יכול שום אדם מהבאים אחריו עד זמנינו זה שהוא קרוב מאלף וחמש מאות שנה להוסיף על דבריו, ושימצא זה באדם אחד הוא בתכלית הזרות. וכשימצא זה העניין באדם מה יהיה יותר הגון שייוחס אל המציאות האלהי משייוחס אל המציאו׳ האנושי. ועל כן אין בדברי אריסטו׳ דבר שצריך להשלימו, כמו שחשב בן אלצייג. כי חשב בהם דברי׳ רבים לא הבינם הוא, ולא אנחנו עדיין, ובלבד בספרים שלא הגיעו לנו בהם דברי המפרשים׳. ע״כ.

וחמש מאות] ותק׳ L

אלצייג] אל צאיג L

[21] אמר רבי׳ משה ז״ל: ׳וכל מה שאמר ארסטו׳ בכל הנמצא שמגלגל הירח עד מרכז הארץ הוא אמת בלי ספק ולא ירחיק דבריו אלא מי שלא יבינם, או מי שקדמו לו דיעות וירצה להגן אותם, או ימשכו אותו אותן הדיעות להרחיק דבר נראה. אבל כל מה [6a P] מה שדיבר בו ארסטו מגלגל הירח ולמעלה, כולו כמו אומר אלא מקצת דברים׳.

[22] ואמר עוד ז״ל: ׳ומה שאמר ארסטו׳ הוא תכלית מה שיוכל האדם להשיג מצד העיון, מלבד מה שיושג מצד הנבואה׳.

[23] ואני אומ׳ כי אלו החכמים הפליגו במה שאמרו בעדו. והנכון מה שאמר אלכסנדר המפרש: כי דעותיו מועטות הספיקות מדעות זולתו.

[24] ואני מתנצל ואומר שלא יחשדני חושד בכתבי אלו הדעות שאני מאמין מהם דבר שיסתור דבר מאמונתינו, כי יהיה חושד בכשרים. כי אני לא אכתוב בזה החיבור דבר מעצמי, אלא כל מה שאזכור בו הוא דעת הפילוסופי׳. ואני איני מדבר בחכמה המושגת בהתעוררו׳ [109a] אלהי[ת] או בנבואה, אלא בחכמה המושגת מצד החקירה וההיקש.

[25] וחלקתי זה החיבור לשני חלקים: אדבר בחלק ראשון בנמצאים הטבעיים, ובחלק השני בנמצאים האלהיים.

Section Four. Routine Philosophical and Scientific Translations: The Thirteenth Century

Part 1. Arabic–Hebrew Translations

4.1.1. Moses b. Samuel Ibn Tibbon (fl. 1244–1283, Provence)

Document 4.1.1.1. Moses Ibn Tibbon, a note on his translation of al-Fārābī's *Short Syllogism*.

Paris, BNF, MS héb. 917, ff.210a–b; Jena, Thüringer Universitäts- und Landesbibliothek, MS Rec. adj. fol. 10, f.43a.[169]

העתקתי הקצור הקצר הזה מן ההקש מלשון ערבי ללשון עברי, אני משה ב״ר שמואל מרמון ספרד זצ״ל. ונשלמה שנת העתקתו שנת חמשת אלפים וחמש עשרה ליצירה. והתעוררתי להעתיקו מפני כי מצאתי שהעתיק זקני החכם הספר הבא אחר זה מן החבור הזה לאבו נצר אלפרבי.

Document 4.1.1.2. Moses Ibn Tibbon, a note on his translation of al-Fārābī's *On the Principles of Existing Things*.

Paris, BNF, MS héb. 893, f.65a.

תם ספר **התחלות הנמצאות** לחכם פילוסוף אבונצר מוחמד אבן מוחמד אלפרבי. [...]
והעתקתיו מלשון הערב, אני משה בר׳ שמואל בר׳ יהודה בר׳ שאול בן תיבון הספרדי. ונשלמה העתקתו בחדש מרחשון שנת חמשת אלפים ותשע לבריאת העולם ת״ל. והגהתיו מספר אחר הגיע לידי בחדש אב שנת חמשת אלפים ואחת עשרה. שבח לאל.

Document 4.1.1.3. Moses Ibn Tibbon, preface to *Ṣedat ha-derakhim*, his Arabic–Hebrew translation of Ibn al-Ǧazzār's *Zād al-musāfir wa-qūt al-ḥāḍir* (*Provisions for the Traveller and Nourishment for the Sedentary*).

Ibn Tibbon, Moses, Ibn al-Djezzar's *Viaticum* [...] a) Vorrede des Moses ibn Tibbon; Munich, Bayerische Staatsbibliothek, Cod. hebr. 19, f.1a–b.

אמר משה אבן תבון מעתיק הספר הזה מלשון הגרי ללשון עברי, מתנצל לפני כל חכם לב:
לא מפני היותי חכם בעיני ובקי בשתי הלשונות ויודע בחכמה הזאת אני ראוי להעתיק

169 See Steinschneider, *HÜ*, § 15, 47, and § 71, 149.

הספר החמוד הזה, ולא לחסֵר כבוד המעתיק הראשון, ולשַׁכחוֹ וטובתו וזכר פעולתו מלאני לבי לשלוח יד במלאכה הזאת. אך מלאני לבי לשלוח יד במלאכה הזאת בראותי כי מי שהעתיק הספר הזה אל לשון הגוים קצר ברוב המקומות, וסתם המפורש, והסתיר המבואר, והמיר הענינים, ורוב הנוסחאות קצר זכר תועלתם וסגולתם ודרך מתכונתם ועשייתם, והוסיף וגרע במשקלות, וחסר קצת הסמים, והפריד המחובר וחבר המפוזר.

ואין על כל זה אשם על המעתיקו מלשון הגויים אל לשון הקודש. אמנם עשה רוב שמות החלאים בלשון הגוים, והם שמות לא יבינום כי אם חכם בחכמה הזאת שקבלם מרוב חכמתו. אמרתי אני בלבי להעתיקו מהלשון אשר חובר בו בלשון מבואר ומובן לכל יודע ספר. ולא שמתי לבי לצחות המליצה וקצור הלשון, אבל הלכתי בהעתקה הזאת כמנהגי בשאר ההעתקות: להעתיק מלה במלה ככל אשר אוכל, או לחסר מעט, או להוסיף, או לאחר, או להקדים, כאשר לא אוכל בלעדי זאת, כפי לשוננו. וכל זה כדי להשמֵר מן השגיאה ושלא להחטיא הכוונה. כי לא רציתי להשען על התנצלותי [נ״א: + ולסמוך בהבנתי] ולעשות לשון קצר וצח לפי המובן לי. אבל בחרתי לשום כוונתי להשאיר כוונת המחבר לא לצחות הלשון. וראיתי כי בדרך הזאת אם אטעה במלה יתוקן מהר. ולא כן אפשר אם אעשה הלשון לפי מה שאבין מן המאמר. [...]

4.1.2. Jacob b. Abba b. Samson Anatoli (c. 1194–1256, Midi and Southern Italy)

Document 4.1.2.1. Jacob Anatoli on the support received from Emperor Frederick II.

Leopold Dukes, 'Literaturhistorische Miscellen', *Literaturblatt des Orients*, 13 (1848), annexed to *Der Orient*, 9. Jahrgang, 13 (25 March 1848), cols 193–197, at col. 196; Paris, BNF, MS héb. 930, f.60a, and MS héb. 994, f.211b.

אמר המעתיק:

ברוך אלהים אשר לא הסיר תפלתי וחסדו מאתי, יעקב בר אבא מרי בר שמשון בר אנטולי זצ״ל. והחליף כחי להשלים באדר שני בשנת תתקצ״ב בעיר נאפולי העתקת ספרי חכמת הדבר שחבר הפילוסוף האנדלוסי אבן רשד במליצה צחה ומבוארת על דרך באור לספרי בעל החכמה הזאת, היפה שבבני יפת, אביר הפילוסופים אריסטוטליס.

והספרים שהעתקתי חמשה, ארבעה מהם לאריסטוטלי׳, והם ספר המאמרות, וספר המליצה, וספר ההיקש, וספר המופת. והקודם להם לפורפיריוס, הוא ספר המבוא.

וטרם החלי במלאכת הספרים האחרים בחכמה הזאת [חכמת הדבר] אשוב בג״ה בדרך לימוד על העתקת הספרים הנזכרים לתקן השגיאות כפי אשר אשיג. ואחר השלימי זה אשלח ידי להשלים המלאכה בעזר העוזר לכל נעזר, אשר נתן בלב אדוננו האנפרדור פרדריק [נ״א: האנפרדוק פרי דרקו] אוהב החכמה ודורשיה לזון אותי ולכלכל אותי ואת ביתי לשבעה. יגדיל השם חסדו אתו לנשאו על כל מלכי האומות. ויבוא המלך המשיח בימיו ועינינו רואות.

Document 4.1.2.2. Jacob Anatoli, preface to the translation of Averroes's *Middle Commentary on Porphyry's 'Isagoge'*.

Averroes, *Middle Commentary on Porphyry's 'Isagoge'*, ed. Davidson, 1–3 (text), 93–94 (notes).

קול חופשִׂי בינה ככסף: איה מצרף מסיג שלי?
חכמת דבר תשיב אומר: כור לַבִּין מאז נִתן לי,
בו הגו סיגים מכסף ויצא לצורף כלי.

מצרף לכסף התבונה וכור לזהב האמונה לחכמת הדבר היה למנה, כי בו יבחַן כל מאמר— המתוק הוא אם מר הרע הוא? אם יפה? החזק הוא הרפה? אם יכשר ואם יערב? המעט הוא אם רב?

והמאמרים מורים על הענינים אשר בשכל האדם מצויירים והציורים ההם נשואים על הדברים שהם חוץ משכלו נמצאים וכל הנמצאים יודיעו ממציאם ויודו כי הוא בראש. לכן כל הבא לדרוש את יי באמת הוא צריך לחכמה הזאת. כי אם איננה חכמה לעצמה אמנם תועלתה גדולה לצרוף כל חכמה, שאין שלמות לחכם זולתה, והיא אצלו ככלי האומנות אצל האומן. ואין זה בחכמה העיונית לבד, כי גם במקומות רבים בתורה ובמקרא ובתלמוד יש לה מבוא גדול להתיר ספקות רבות. וזאת היא הכונה הראשונה בחכמת הדבר.

והכונה השנית כי היא מחדדת השכל ומכשרת הלמוד, כסמים המכשירים האוכל ומעירים תאות המזון. והיא תועלת טובה לכל מבקש חכמה.

מצורף אל זה שהיא צריכה מאד לעמוד כנגד החולקים הטוענים בטענות השָלִי ר״ל ההטעאה. וכבר הזהירו רבותינו ז״ל על זה ואמרו ׳הֱוֵי שקוד ללמוד תורה מה שתשיב את אפיקורוס׳ [אבות ב יד]. ולא על אפיקורוס בלבד היתה כונת הזהרתם, כי אם על כל החולקים על האמת. וידוע שאין כח באדם ממנו לעמוד כנגד הפקחים משאר האומות החולקים עלינו אם לא ילמד החכמה הזאת.

ומפני שראיתי כי רבו ההוללים הרעים המתפארים עלינו בדרך המחלוקת והנצות, קנאתי בהם, אני יעקב בר׳ אבא מרי בר׳ שמשון בר׳ אנטוליו ז״ל, והתעוררה תשוקתי להעתיק החכמה הזאת כפי אשר תשיג ידי.

ואמת כי נתתי אל לבי להקל מעלי תחלה עבודת ספרי חכמת התכונה, הכבדה בהעתקתה ולמודה, לפי שהיא צריכה לספרים רבים ולהתבוננות רב. אכן הציקתני רוח אֶחַי וְרֵעַי אלופַי ומיודעַי החכמים והמשכילים אשר בעיר נרבונה ובדרש, אשר מלאם לבם לקרבה אל המלאכה הזאת, ולאהבתם אטה שכמי לשאת משא על משא, ובעזרת הנותן ליעף כח אזרע בבקר בחכמת התכונה ולערב בחכמת ההגיון תהיינה ידי אמונה.

ואם יאמר אומר: הלא רבותינו ז״ל מנעו החכמה הזאת, כי הם אמרו: ׳מנעו בניכם מן ההגיון׳ [ברכות כח ע״ב], נשיב לו: השמע לאזניך מה שאתה מוציא מפיך! כי הם אמרו ׳מנעו בניכם׳ ולא אמרו ׳מנעו עצמכם׳. [...]

ועל זה אמרתי להתאמץ ולהתחזק במלאכה הזאת, ואם ידעתי כי לא יאות לאיש כמוני להרוס ולעלות אל ההר הגדול הר ההעתקה, כי רב הדרך ממני ורבים המכשולים לפני, מחסרון ידיעתי בלשון הערב ומקוצר יד הבנתי. אמרתי אולי במעט הדבש אשר ימצא בספרי העתקתי יאורו עיני המעיינים בהם ויתקנו המעוות בם, ומעט התועלת בחכמה הוא גדול לאוהביה. [...]

4.1.3. Jacob b. Makhir Ibn Tibbon (c. 1236–1305, Provence)

Document 4.1.3.1. Jacob b. Makhir on the value of science for the defence of Judaism.

Aderet, *Teshuvot ha-Rashba*, ed. Dimitrovsky, 510.54–57.

גם לעיני הגוים ידיעתנו בהם [ספרי היונים] חכמתנו ובינתנו. לבל יֵאָמֵר, מכל שכל וחכמה רֵק לבבנו. והיה לנו ללמוד ממשפטי הגוים: המתֻקָנים שבהם העתיקו ספרי החכמות ללשונותם בגוייהם, ואף אם דבריהם וראיותם הם כנגד דתם ואמונתם. כִּבדוּ החכמה ויודעיה ולא שאלו מאי זה דת הם.

Document 4.1.3.2. Jacob b. Makhir, preface to the translation of Euclid's *Elements*.

Krafft and Deutsch, *Die handschriftlichen hebräischen Werke*, 173–174 (no. CLXXIII).[170]

תהלה לאל אשר ברא את האדם בצלמו, ומכל החי אשר על פני האדמה השכילו והחכימו. עשה אותו ישר לבקש חשבונות רבים, לחשוב מחשבות משפל הארץ עד ממעל לכרובים. [...]

ומאשר אבותינו חטאו, עזבנו ארץ והשליכו משכנותינו, אבדה חכמת חכמינו ונסתרה בינת נבונינו. לא ימצא אחד מאלף מספרי החכמות, וצריך שנבקשם מקדמוני חכמי האומות, היו שאננים מנעוריהם ושקטים על שמריהם, עמד טעמם בם ולא נפקדה מהם.

ומאשר חכמת השעור הוא יסוד לכל החכמות ובפרט לכל החכמות הלמודיות, וזה הספר הוא יסוד ושרש והתחלה לכל שאר הספרים מזאת החכמה, נשאני לבי, אני יעקב ב"ר מכיר, להעתיקו מלשון הגרי אל לשוננו, ולהחזיר אבדה לבעלים, ולעשות לנו שם כשם הגדולים, להסיר מעלינו חרפת הערלים האומרים כי מכל חכמה אנו משוללים.

ואם ידעתי בנפשי כי לא למדתי חכמה ולא בינת אדם לי, וידיעתי גם כן בלשון הערב היא מעט מזער, נשענתי בזה על שני דברים: האחד, כי אין בחכמה הזאת רבוי דברים ומלות מעטות יכללו אותה. והשנית, כי התחזקתי בשם אשר לאין אונים ירבה עצמה, ונותן לפתאים ערמה, לנער דעת ומזמה.

170 I am grateful to Dr Ofer Elior for permission to consult his unpublished edition.

Section Five. Routine Medical Translations: The Thirteenth Century

Part 1. Arabic–Hebrew Translations

5.1.1. Nathan b. Eli'ezer ha-Me'ati (fl. 1275–1300, Rome)

Document 5.1.1.1. Nathan ha-Me'ati, preface to his Hebrew translation of Avicenna's *Canon*.[171]

Nathan ha-Me'ati, 'Nathans Vorrede zur Übersetzung des *Kanons*', ed. Steinschneider; Leibowitz, 'The Preface by Nathan ha-Me'ati', 4–6.

אמר נתן המאתי בא גד בנו אליעזר יצ״ו. אני הגבר ראה נדוד, מוזר הייתי לאחי, נזיר רע ודוד. מנעורי לא שקט לבי ולא חנה, מדחי אל דחי הלך אנה ואנה. עיני בקצה הארץ היו, רגלי בפאתי תבל כמעט נטיו. כדור צנפני, זמני טלטלה גבר טלטלני. אך עתה הלאני, בין עמים ולאומים שונים הגלני, ובין לשונות שונות הנידני. ואבין בלשונות הגויים ההם אשר ביניהם משך גרתי, עד עתה אחרתי.

ואבחר בשפה ברורה ולשון ערומים, הנאהבים והנעימים, שפת עבר וערב יחד צמודים, בלבי צפנתים והיו בידי לאחדים.

ויהי בראותי יום יום כי חכמי הנוצרים שפה עלינו יפטירו, וחיילי לעגי שפתים יגבירו, בהפקד מאתנו ספרי הרפואה אשר חוברו בימי שלמה ושאר חכמי אומתנו, אשר היו לפנינו, הנמצאים אתם ואבדנום, אבותינו ספרו לנו אך לא ראינום. לכן לבשתי מעיל קנאה, ואשנס מתני העצלה, ואומרה אל סעיפי: חזק ונתחזק בעד עמנו שארית הגולה, להחיות החכמה, לפליטה גדולה [כצ״ל].

וגם כי ראיתי נטעי נעמנים בני תבון היקרים, שלמים הם אתנו בהיות את נפשם קרן החכמה להרים, אשר מטיט יון הערב משוה ומשכוה, משחת קדרות הקודרת גאלוה. ואלך אני בזה בעקבותיהם, להטיב לעמי כמוהם, להעתיק הספר הזה הגדול אליהם, וברוב הלשונות בשאר ספרי הרפואה הלכתי בדרך הסלולה להם, בשימוש לשונותיהם ומלותיהם.

וגם אָמנה כי המחבר הזה נטה בלשונו מדרך שאר המחברים והגיונו, כי פרסי היה ממולדתו. קצר דברים בכובד לשון חבר במחברתו, ולכן הוצרכתי במקומות רבים להוציא לשונות חדשים, מקרוב באו, עד היום היו נטושים. ורובם דקדקתי כפי יכלתי מחכמי הערב. ומרוב כובד מליצותיו, גם המה ייגעו מהשיג תוכן גזרת קצת מלותיו, כי אם כפי כונת בחינותיו.

וגם בקצת הספר הוצרכתי להניח חלק בקצתם כי היה חלק מהספר מוטעה. ואני משביע כל קורא בו לבל יגיה בו דבר אם לא מפי מומחה ומפורסם בלשון. ואולם אם השם יספיק בחלדי וצבאי יאריך, אחקור אחריהם בספרים אחרים ומפי חכמים ואבא להבינם עלי מתכונתם.

171 The text is reproduced almost *in extenso*, because the periodical in which it was published is difficult to find.

אחרי זאת עלה בלבי לעשות מאמר אחד לחתום הספר בו, ואזכיר בו רוב מלות הכבדות, ערביות ועבריות, הנמצאות בספר זה, גם שמות הסמנים המשתנים ושמותם הנזכרים בלשוננו, למען לא יטעה הקורא בהבנתם.

ועל כונתי הרצויה בהעתיק הספר הזה להיות למחסה ולמסתור מזרם לזות שפתי גאיונים, ומעשה לסתור דברי גדופותם ומהתלותם, כל קוראו בנמצא יברכני כשאר המחברים [נ"א: משבירים], ובחסרון אל יקללני, ולזכות ידינני.

ומברכת החכמים יבורך הצעיר בבית אביו, נתן המאתי, הדל באלפי ישראל. [...]

Document 5.1.1.2. Nathan ha-Me'ati, preface to his Translation of Maimonides's *Aphorisms*.

Maimonides, (*Medical*) *Aphorisms of Moses*, ed. Muntner, 9–10.

אמר נתן המאתי.

תהלה לאל כי לו יאות לרוממות גדלו והשבח לו מאתנו, כפי היכלת האנושי[ת], אמנם [הוא מ]על כל ברכה ותהלה.

אחרי אשר שאלני אחד מיקירי מי[ודעי], דורש שלומי וטובתי, להעתיק לו זה הספר, בהיותו חומד ומתענג בספרי הרב מורה צדק ומשתוקק במאמריו. ואני מלאתי את דברו והעתקתיו לו, וכיוצא בו מהאוהבים ללכת בעקבות החכמים. [...]

Document 5.1.1.3. Nathan ha-Me'ati, preface to his translation of Hippocrates' *Aphorisms* with Galen's commentary.

Paris, BNF, MS héb. 359.

אמר נתן המאתי:

בשמעי חכמי הנוצרים חרפונו ויגנו אותנו בחסרון ספרים רבים אשר נעדרו מאתנו קנאתי קנאה גדולה ואמשוך בקסת הסופר והעתקתי הספר הזה להיות לבני עמינו למחסה ולמסתור ולכבוד ולתהלה. והיום גלותי חרפת הנוצרים מעלינו באמרם אבדה חכמתנו.[172]

Part 2. Latin–Hebrew Translations

5.2.1. Jacob *ha-qaṭan* [= *minimus*] (fl. 1280, Provence).

Document 5.2.1.1. Jacob *ha-qaṭan*, translator's preface to the *Antidotarium Nicolai*.

Steinschneider, 'The Translator's Preface to the *Antidotarium*'.

172 The last sentence in not in all MSS.

נאם יעקב הקטן ר׳ ב׳ מ׳ ז׳ ל׳. זה ספר נקרא בלשונם אנטיאוטריאוס, פי׳ הכללי. אשר נתקבצו עליו חכמי מונפלירי, ומינו עליהם ניקולא, וקבצו ספר של מרקחות למען ימצא הרופא מרקחות מתוקנים אצל הבַשָׂם תדיר לכל חולי ולכל טבע. ויקח מהם לכל צרכו וכפי אשר תשיג יד החולה, ולא יצטרך לעשות הוצאה מרובה לתקן מרקחות בעבורו, כי ימצא מרקחות מתוקנות—הטובה אליו יקנה ממנה מעט או הרבה, כפי דעתו וצרכו, כפי השג ידו. ואם יש לאל ידו ימצא בזה הספר מרקחות טובות וחשובות, מהם חמים מהם קרים, מהם מחזקים מהם משלשלים מהם עוצרים. וגם תמצא בתוכן שירופי ומשיחות, ושמנים ואבקים ותחבושות. וכל אחת מהם יש לה פעולות ידועות וכוחות בפני עצמן, וסגולתה ופעולתה מפורשת בצדה. ומהם יש למצוא כל המרקחות או רובם תדיר אצל הבשם, או אם לא ימצֵא אחת מהן, כיון שהם שגורים בפי כל, יודעים לתקן בעת הצורך, רק שיקרא להם שמם, כי ביד כל בשם תמצא זה הספר בכתיבתו ובלשונו.

ודעתי בזה: אחר שיהיה לאדם ספרי הטיאוריקה כמעט שיספיק לו זה בפראטיקה ויוכל לפעול בו בטוב. ולפי שראיתי בו תועלת גדולה לחיים וסיוע גדול לרפואה נמלכתי להעתיקו ולהוציאו מלשונם ללשוננו. וטרחתי בו לסַדרו על פי האלפא ביתא שלנו.

ולא כוונתי להעתיק מלה במלה, אלא רוב המשמעות. אבל שמות הסמים ומשקלם לא אשנה.

ולפי שכל העולם צריכין מחכמת הרפואות צריך לכל אדם משכיל לדרוש מלמוד קצת הרפואות. כי אינו יודע מתי יזדמן לו לצאת לדרך רחוקה שלא ימצא שם רופא, או מתי יחלה או יבריא ויצטרך מחכמתו. כי היא החשובה והנעימה שבכל החכמות אחר התורה וחכמה מפוארה.

Section Six. Routine Philosophical and Scientific Translations: The Fourteenth Century

Part 1. Arabic–Hebrew Translations

6.1.1. Samuel b. Judah b. Meshullam of Marseilles (1294–after 1343, Marseilles)

Document 6.1.1.1. Samuel b. Judah, postface to his translation of Averroes's *Middle Commentary on the Nicomachean Ethics*.

Berman, 'Greek into Hebrew', 303–304.

נשלם באור בן רשד **לספר המדות** לאריסטו, הנקרא בלשון יון **ניקומאכיה**. והוא החלק הראשון משני חלקי החכמה המדינית, הנקרא ׳חלק החכמה׳. ויחסו אל החלק השני ממנה, הנקרא ׳חלק המלאכה׳, הוא יחס מה שבספר הבריאות והחולי אל מה שבספר שמירת הבריאות והסרת החולי במלאכת הרפואה.

והחלק השני הוא בספר אשר ידוע **בספר ההנהגה** לפלוסוף. אך לא נפל אל החכם בן רשד הספר ההוא, וכבר התנצל בזה על כי לא בארו, ובאר תמורתו מה שמצא מן החלק ההוא השני

בספר הנהגת המדינה לאפלאטון ממאמריו המדעיים. ולזאת הכונה המשכתי, אני המעתיק, אל החלק הראשון, אעפ״י שלא חברם מחבר אחד, אחר שהם חלקים מחכמה אחת ובארם מבאר אחד.

והעתקתי הספר הזה, אשר הוא החלק הראשון מחכמה הזאת, רצוני החכמה המדינית, אני שמואל בן יהודה י״צ [= ישמרו צורו] בן משלם בן יצחק בן שלמה נ״ע ממרשלייא בשנת שבע ועשרים משנותי, וזה במגדל בלקיירי. ונשלמה העתקתו יום ב, י אדר ראשון משנת השמנים ואחד לפרט האלף הששי ליצירה. ישתבח העוזר ויתברך לנצח אמן אמן סלה. [...]

Document 6.1.1.2. Samuel b. Judah of Marseilles, postface to his translation of Averroes's *Commentary on Plato's Republic*

Averroes, *Commentary on Plato's Republic*, ed. Rosenthal, 106–107.

נשלם באור בן רשד מן המאמרים המדעיים הנמצאים **בספר הנהגת המדינה** לאפלטון, והוא תחלת החלק השני מן החכמה המדינית. ונשלמה העתקתו כ״ב כסליו משנת פ״א לפרט האלף הששי ליצירה, בעיר אוריץ. והעתקתיו אני, שמואל בן יהודה בן משלם בן יצחק בן שלמה ברבא ויירי ממרשילייא.

ובעת העתיקי אותו לא היה מגיע לידי באור בן רשד **מספר המדות** לפילוסוף ארסטו, אשר הוא החלק הראשון מן החכמה הזאת הרצוניית זולת מאמרי הפילוסוף עצמם. ולא יכולתי להעתיקם לעמקם וקושיים. והשתדלתי בזה וחתרתי חתירה מופלגת עד שהגיע לידי ביאור החלק ההוא לחכם הנזכר בן רשד בלשון צחה ומבוארת כמנהגו הטוב לכל באוריו. והתעוררתי להעתיקו והשם הנכבד בחסדו הסכים על ידי לטוב, עד שנשלמה העתקת ביאורי החכמה המדינית כלה.

עוד עברתי על העתקתי בזאת החכמה כלה ותקנתי השגיאות כפי כחי. ונשלמה מלאכת הדיוק והתקון כ״ו אלול מהשנה ההיא במגדול בלקיירי, עצור ועזוב עם יתר אחינו, וכלוא באחד ממבצריו יכונה רידורטא. [...]

ואני לאהבתי אותה וחריצותי בה חשבתי ודמיתי להפליג עוד בתקון ההעתקה הזאת עם חכמי הנוצרים, ובפרט החלק הראשון ממנה, כי מאמרי הפילוסוף בחלק ההוא נמצאים אצלם, גם פרישתם לאבונצר אלפראבי. לתורה ולתעודה—כאשר דמיתי כן היה, לולא הסבה הרמוזה מן המאסר הארוך והחזק השיגנו בעת הזאת מנעתני מזה. ואם יאריך השם חיי ויוציאני ממסגר אסיר מבית כלא ויכין לי הפנאי המועט אשתדל בזה ואפליג בעיון והמחקר מראשית החכמה עד אחריתה על סדר, עד שתתוקן ההעתקה ביותר שלם.

[...] ובכלל אומר: אתם המעיינים קהל המאמינים האמת יעשה דרכו, החזיקו טובה לראוי לה. כי עד היום לא הועתק דבר או לא נפל אלינו כלל מזאת החכמה—לא מהפילוסוף ולא מזולתו, זולת מה שנמצא מזה **בספר התחלות הנמצאות** לאבונצר אלפראבי הנופל ביד רבים מאנשי אומתינו, והוא קצת מהחלק השני מזאת החכמה, ומהחלק הראשון ממנה לא נמצא בו דבר. ואני התחזקתי והחלותי להראות את יקר תפארת גדולת החכמה הזאת ולהמציאה אל לשונינו. ואם הטוב הנמצא בה מצד חסרוני הוא מעט הנה המאמר בהתחלה, ואם היה מועט הוא גדול בכח.

איך מה שהיה—עורו ישנים מתרדמת העצלה והסכלות בלא קנין, המביאה המורידה את בעליה לבאר שחת העוברת, המפסידה ההצלחה הגמורה! הקיצו שכורי הצרות! שוש

תשישו! החזיקו בזאת המנה היפה חלק השם לכם! החרשים—שמעו מה דברו שני המלכים האלה האלקיים! העורים—הביטו וראו אור בהיר הוא בשחקים! שאול תשאלו מי ברא אלה הפליאות, הנשגבות המעולות ביותר משישובחו, ובינו במראה! הנה שכרכם אתכם ופעולתכם לפניכם.

Document 6.1.1.3. Samuel b. Judah of Marseilles, postface to his revision of Jacob b. Makhir's translation of Averroes's *Epitome of Logic*.

Neubauer-Renan, *Écrivains*, 559; Paris, BNF, MS héb. 956, f.350a.

נשלם קצור ההגיון ת״ל אשר חברו החכם הגדול, ראש המפרשים בן רשד האנדלוסי, ונשלמה העתקתו. והעתקתיו מערבי לעברי אני שמואל מרשיליי בן יהודה י״צ בן משולם בן יצחק שלמה נ״ע.

וכבר הועתק לפנים ונפלו טעויות רבות בהעתקה ההיא, עד שנפסד הספר. ובאו אחר זה חשובי זמנינו בקיאים בלשון הערב חתרו לתקן ההעתקה ההיא ולא השלימו מה שחתרו מזה, וכמעט אומר שתקונם לא היה ראוי לשים לב עליו. ונשאר הספר משובש ומבולבל, ונמנעה הבנתו ממנו, אנחנו קהל המעיינים.

ובראותי זה, עם מה ששערתי מעוצם התועלת המגיע מענינו, להיות[ו] קצר, קטן הכמות גדול האיכות להפליא: וזה לפי שבו הנבחר והיקר ממה שבספר אבונצר בהגיון אשר הוא גדול המעלה והשיעור מאד, כי אבן רשד בזה הספר לקט מאותם הפנינים אשר יסד עליהם אבונצר ספרו בהגיון מה שנראה אליו הכרחי אין מנוס מבלעדיו כפי כונתו, והלך בעקבותיו לא סר מהם ימין ושמאל. והנה לזאת הסבה דחקתי את עצמי, אעפ״י שהשעה דוחקת אותי והזמן ינהגני בכבדות, ואשוב להעתיקו מראש.

והמעיד האל אם כונתי לתפוש ולהשיג המעתיק הראשון והמתקנים אחריו ולשים לי שֵׁם במלאכה הזאת! אין דבר, חי יי, זולתי בקשת האמת המעיד לעצמו ומסכים מכל צד.

ומהאל הגומל אשאל גמול על טרחי. והשלמתי המלאכה הזאת עשרים טבת משנת תשעים לפרט האלף הששי ליצירה, במגדול טרשקו. ישתבח העוזר ויתעלה לנצח אמן:

Document 6.1.1.4. Samuel b. Judah of Marseilles, postface to his translation of Alexander of Aphrodisias's *De anima* (from the Arabic of Ishāq ibn Ḥunain).

Neubauer-Renan, *Écrivains*, 565–566; Steinschneider, 'The Translator's Epilogue at the End of Alexander of Aphrodisias' *De Anima*'.

נשלם מאמר האסכנדר **בנפש**, ונשלמה העתקתו ת״ל. וחובר בלשון יון, מדעתי העתיקו מהלשון ההוא ללשון ערבי אסחק בן חנין, כמו שאתה רואה בראש המאמר. אחר כן העתקתיו אני מלשון ערבי ללשון עברי.

והוא עמוק קשה להבין, כי המחבר כה משפטו בכל דבריו, כל שכן בחכמה הזאת אשר היא חכמת הנפש החמורה והעמוקה בטבעה. ומה שיוסיף על עמקו וקשיו הוא מה שספרנו מהיותו מועתק מלשון אל לשון פעמַים. אבל אעפ״י שהוא עמוק וקשה לשתי סבות—האחת

מצדו והשנית מצד העתקותיו כמו שתארנו—הנה מי שהורגל בחכמה הרגל נכון יכול להבינו בשלמות.

אשרי המחכה ויגיע להבין אמריו. כי המאמר הזה שלם בתכלית השלימות, נכבד ויקר, גדול מאד משישובח. וזה לשתי סבות: האחת כי מחברו הוא השלם שבחכמים, לא [קם] כמוהו גאון ונביא אמת בסיעת הפילוסופים כביכול. אפי׳ ארסט׳ו בן ניקומאכוס המפורסם בחכמה ביונים, אשר המציא החכמות כפי מה שאמרו והיה הראש הראשון—לא הגיע אל שעור האיש הזה בחכמה, ודבריו במאמר הזה וזולתו יעידו על אמתת דברי. כל שכן וקל וחומר כי באמת לא היה בכל הפלוסופים הבאים אחר ארסט׳ו עד היום הזה מאותם שהגיעונו דבריהם אחד שיגיע למדרגתו ולשעורו בחכמה. והסבה השנית: כי הוא כבר חובר בחכמה היקרה הגבוהה מכל החכמות, היא החכמה הטבעית, ובחלק היותר נכבד מכל חלקיה הוא חלק הנפש. כי החלק הזה הוא אינו טבעי לבד, אך יש לו מבא גדול באלהיות. גם החכמה הטבעית בשאר חלקיה יש לה שתוף עצום עם החכמה האלהית. והחלק הזה—רצוני חלק הנפש—עולה על כלם בשתוף והקירוב אשר בינו ובין החכמה האלהית.

הנה לשתי סבות אלו היה המאמר הזה הנכבד שבמאמרים. בחנהו דרך המחקר השלם והעיון התמים. כזהב יצא. [...]

Document 6.1.1.5. Samuel b. Judah of Marseilles, postface to his revision of Jacob b. Makhir's translation of Jabir ibn Aflaḥ's *Epitome of the Almagest.*

Neubauer-Renan, *Écrivains*, 562–563; Paris, BNF, MS héb. 1024, ff.107a–108a.

אמר שמואל בר׳ יהודה יא״ע [= יגן אלוהים עליו] בן משולם בן יצחק בן שלמה נ״ע הידוע בהמון מילש דמרשילייא. [...]

שמעה אזני ותבן רוממות הספר הזה אשר חבר בה אחד היה נשיא אלהים, חכם ישמעאלי ידוע ומכונה בהמון ג׳אבר בן אפלח האשבילי. ומכותלי בית הספר חבר בחכמה הזאת ג״כ החכם בן רשד אשר יצק מים על ידיו, הכרתי כי מושב אלהים ישב.

ובעת ההיא לא יכלה נפשי להשיגו כי לא נמצא מועתק זולתי לחכם ר׳ נתן המאתי נ״ע כפי מה שאחשוב. והיתה העתקתו מקולקלת מאד כפי מה שאמרו. גם לא היתה נמצאת כי אם ביד יחידים. וכביתי אש תשוקתי הגדולה הבוערת, החלפתי פרה בחמור, הוא **מג׳סטי** בן רשד אשר הזכרתי. כי הייתי חושב להיותו אחרון ומפורסם בחכמה שספרו מספיק לא יצטרך לחבור אחר. ומפני הסבות האלו לא דלקתי אז אחרי הספר הזה היקר. ואנחם ממנו בתמורתו. וכי לא היתה אז בחינתי שלמה להיותי מתחנך בחכמות ותחלת מעשי בהן—ערבו לי אז מאמרי בן רשד בספרו הנזכר מאד. אמרתי אין כמוהו בהפלגה ועלוי.

ויהי כי ארכו הימים ועברו שלשים משנותי השתוקקה נפשי לשוב בחכמה הזאת הגבוהה ואשים את פני אל הספר הראשון, החמור הגדול, אשר חבר בה הראש הראשון בטלמיוס אלפלודי הידוע **באלמג׳סטי**, והוא הספר אשר למדתי תחלה בנערותי וכי ידעתי עומק הספר כספתי על כל פנים לשוב לקוראו מראש. וקראתי אותו בעת ההיא עם החכם אחי אן בונדוי דמרשיליא השלם בכל מדע צעיר ממני לימים יאע״ו [= יגן אלהים עליו וינצרנו]. ועלתה לנו קריאתו בשלמות מלבד המאמר האחרון כי תכפונו המעיקים. גם עלה בידי אז פירוש שלשת המאמרים הראשונים ולא הוכן לי אז לפרש עוד שאר מאמרי הספר עד תומו כאשר חתרתי,

לעוצם הצרות התכופות השיגוני. והיה העיון הזה במגדול טרשקו, כי שם היה ביתנו כעת ההיא. והיו ימי כל״ה שנה:

ומאשר נפקחו עיני בימים ההם בחכמה הזאת הנכבדת ושאר החכמות כלן או רובן, בכולן שיערתי מדברי בן רשד בספרו בחכמה הזאת, כי הטוב הנמצא בהם היה מלוקט מהספר הזה היקר מזהב ומפז. ורבי הטובות המופלאות הנמצאות בספר הזה נשמטו מהספר ההוא, וכמעט שכל ספרו אין טוב בו זולתי מה שלקט מזה.

התחזקנו והתאמצנו לזה אני ואחי החכם הנזכר, ומעיל גבהות הנפש ויקרתה להשגת החכמה מעלינו הפשטנו, ונפן ונסע ללכת מעירנו טרנקטאלייש והשתדלנו עד שהגיע לידינו שם ערבי הספר הזה, מדוייק מאד כפי מנהג הספרים. והעתקנו ממנו בעת ההיא בחפזה, כלואים יומַים באחד מבתי דורשי החכמה אשר שם בלחם צר ובמים לחץ, כי נטרדנו להשיב הערבי לבעליו ולשוב אל ביתנו בגלגל העסקים הכלים הֱכרחנו, ולקטנו מהספר על זולת ההתדבקות המקומות שהיו יותר מסופקים אלינו, לא עלה מקובץ הנלקט לשמינית הספר. אמרתי בחפזי: זה ינחמנו.

ויהי היום ורוח גדולה וחזק מפרק כחותי ומשבר עצמותי טלטלני טלטלה הנידני מקני מלוני נשאתני פה, יגעתי ומצאתי הספר הזה מועתק לחכם ר׳ יעקב בר׳ מכיר נ״ע. לא נח לבי ושקט, הערמתי וגאון נפשי לפנים בגלגל הספר היקר השפלתי עד השגתי עם בעליו והרשני להעתיקו. וכתבתי אותו והיה מכתב הספר שהעתקתי ממנו מכתב החכם המעתיק הנזכר. והייתי משער תמיד בעת כתבי אותו כי רבו תעיותיו. אמרתי אז לשוא טרחתי ושקר נחלתי.

סוף דבר, הכל נשמע לא שלותי עד שנעזרתי עזר אלהי והגיע לידי היום פה הערבי הראשון אשר ממנו לקטתי ראשונה מה שלקטתי ודיקתי הספר הזה כולו פעמַים עמו וגם מצאתי בו נסחאות רבות לא היו בהעתקת החכם הנזכר. וגם מאותה ההעתקה חסרו דברי בעל הספר באלמג׳רה וכדור מוגשם, ותמונות יאותו בכוכבים הקיימים. וכל החסר השלמתי והמעוות תקנתי כיכולתי. וחי נפשי טרחתי בדבר הזה יותר ועיפה נפשי לו מאשר אם [בנדפס: עם] העתקתיו מהערבי בכללות, לעוצם השגיאות מצאתי בהעתקה. וספרו לי כי החכם ר׳ משה תבון ז״ל העתיקו ג״כ אך לא הגיע לידי ולא ראיתי אותה ההעתקה.

אלו מציאות מופלאות נשאות. ׳מי חכם ויבן אלה, נבון וידעם׳ [הושע יד י]. בית יעקב לכו ונלכה באור ה׳ הגדול והנורא הזה [עפ״י ישע׳ ב ה], כי לא היה כמו המחבר הזה לפניו מלך בחכמה הזאת ונביא אמת, ואחריו מי ימצאנו? הוא תפש הכהן הגדול בחכמה הזאת בטלמיוס אלפלודי. וזכר תפישות עצומות ופירש מאמריו וגלה סתריו וקצר אריכויותיו, הכל שריר וקיים כמו שתראו בספר אם תעיינו בו. ומי כמוהו מורה מושל מקשיב על [בנדפס: עד] דבר אמת.

עד הנה עזרנו ה׳ יהי שמו מבורך. היה זה בשנת שתים וארבעים משנותי א׳ טבת משנת שש ותשעים לאלף הששי לבריאת עולם, באייגש המדינה [...]

Section Seven. Routine Medical Translations: The Fourteenth Century

Part 2. Latin–Hebrew Translations

7.2.1. Leon Joseph of Carcassonne (1365, Carcassonne–1418, Perpignan)

Document 7.2.1.1. Leon Joseph, preface to his Hebrew translation of Gerard de Solo's *Pratica super nono Almansoris* (c. 1394).

Freudenthal, 'Hebrew Science in Medieval Midi', 39–52 (text edition with H. Yishai).

[1] אָמַר לֵיאוֹן יוֹסֵף הַמַּעְתִּיק, הַיּוֹשְׁבִי קַארְקָאשׁוֹנָה.

[2] הִנֵּה מֵאָז פָּקַחְתִּי עֵינַיִם לִדְרוֹשׁ וְלָתוּר בַּחָכְמוֹת הַחִיצוֹנִיּוֹת, אֲשֶׁר הֵם בַּמִּין וּבַמִּסְפָּר רַבּוֹת, כְּמִסְפַּר יְמֵי שָׁבוּעַ זֹאת, וְלָזֶה אֵין כּוּלָּם כְּאֶחָד נִגְדָּרוֹת, וְנִקְשְׁרָה תּוֹחַלְתִּי בָּהֶם לָדַעַת לַחְקוֹר מֵהֶם חֲקִירוֹת מִתְחַלְּפוֹת, גָּדְלוּ מַעֲלוֹת הַחָכְמוֹת הָהֵם בְּעֵינַי בעל [צ״ל: מֵעַל] כָּל הַתְּהִילּוֹת. וְזֶה, מֵרוֹב כִּסְפִּי וְתַאֲוָתִי לַעֲמוֹד עַל דְּרִישָׁתָם וְהִשְׁתּוֹקְקִי עַל חֲקִירָתָם, וְהָלַכְתִּי אַחֲרֵי עִקְּבוֹת חֲכָמֵינוּ, הַנִּמְצָאִים הַיּוֹם גַּם מִתְּמוֹל גַּם מִשִּׁלְשׁוֹם, לַחֲשׁוֹב מַחֲשָׁבוֹת יֵאוֹרוּ עֵינַי בָּהֶם, אוֹר הַשֵּׂכֶל וְהַלְּבָבוֹת. בְּאָמְרִי: הִגִּיעוּ הֵם בְּמַה שֶּׁהִגִּיעוּ הַשְּׁלֵמִים שֶׁכְּבָר חָלְפוּ וְהָלְכוּ לָהֶם זֶה יָמִים.

[3] וּמָצָאתִי הֱיוֹת הַחִסָּרוֹן הַנִּמְצָא בָּהֶם וְלִקְצָת אֻומָּתֵינוּ בַּדּוֹר הַזֶּה גָּדוֹל וְעָצוּם. וְדִבְרֵיהֶם בַּחָכְמוֹת הַחִיצוֹנִיּוֹת 'כְּדִבְרֵי הַסֵּפֶר הֶחָתוּם' [ישע' כט יא]. וְאָמַרְתִּי: אוּלַי כִּסְפִּי גָּדוֹל מֵהַשָּׂגָתִי, וְהַחִסָּרוֹן מַגִּיעַ מִצִּדִּי, לְעוֹמֶק הַמּוּשָּׂג וְקִצּוּר שִׂכְלִי וְהַשָּׂגָתִי וְדַלּוּתִי וְדַקּוּתִי וְסִכְלוּתִי מֵהַשִּׂיג עִנְיָנָם, לֹא מִסִּכְלוּתָם וְחֶסְרוֹן יְדִיעָתָם, וַיֶּחְכְּמוּ מִכָּל הָאָדָם [עפ״י מל״א ה יא], וְהִנְנִי מִכְּלָלָם שָׂרִיד. וְהָיִיתִי כְּמַחֲרִיד.

[6] גַּם קְצָת מֵהֶם, אֲשֶׁר הָאֵל חֲנָנָם לְעַיֵּין בָּהֶם, לִבְרוֹר הָאוֹכֶל מִתּוֹךְ הַפְּסוֹלֶת [עפ״י שבת עד ע״א], וְכָל אֶחָד לְשִׂכְלוֹ חָקַר, רִמּוֹן מָצָא, תּוֹכוֹ אָכַל קְלִיפָּתוֹ זָרַק [עפ״י חגיגה טו ע׳א], הוּכְרְחוּ לַעֲשׂוֹת זֶה בְּהֶסְתֵּר וּבְהֶצְנֵעַ, בַּחֲוָחִים וּבַמַּחֲבוֹאוֹת [שה״ש ב יד], לֹא הָיוּ רַשָּׁאִים לְהַרְבִּיץ הַחָכְמָה בַּשְּׁוָקִים וּבָרְחוֹבוֹת, וּלְהִתְוַכֵּחַ בָּהֶם לְהֵרָאוֹת פָּנִים לָהֶם, וּלְהוֹשִׁיב יְשִׁיבָה בָּרַבִּים, לְהַמְצִיא הָאֱמֶת עַל בּוּרְיוֹ, כִּי לֹא יוּדַע הָאֱמֶת כ״א בְּהֶפְכּוֹ.

[7] וְזֶה מְפַחְדָּם מִלְּשׁוֹן הֲמוֹן הַסְּכָלִים, כִּי הֵם מְתֵי מְעַט וְהֵמָּה הָרַבִּים. מְצוֹרָף לָזֶה מִיִּרְאַת קְצָת הַתּוֹרִיִּים, אֲשֶׁר מִיֶּתֶר הַחָכְמוֹת מוּפְשָׁטִים, אֲשֶׁר יָדָם תְּקֵפָה עַל הַמְעַיְּינִים—לֹא בְּכֹחָם וְרֹב יְדִיעָתָם, רַק לְחוֹזֶק יְדֵיהֶם, וְרוֹב נִכְלֵיהֶם, לִהְיוֹת עֲדַת הֶהָמוֹן הַהוּא נִשְׁמָע אֲלֵיהֶם, בְּחָשְׁבָם הֱיוֹת הַחָכְמוֹת הָהֵם וְהַמְעַיְּינִים בָּהֶם יוֹצְאִים מִכְּלַל בַּעֲלֵי הַתּוֹרָה. אֵלֶּה מִדּוֹת הַחֲסֵרִים וְהַמְדַבְּרִים עַל הַחֲכָמִים 'תּוֹעָה' [עפ״י ישע' לב ו]. וְהִנֵּה רָאִיתִי הַכָּתוּב עַל סֵפֶר הָרַמְבָּ״ם **מוֹרֵה הַנְּבוֹכִים** וְאֶת אֲשֶׁר עָשׂוּ לוֹ בַּיָּמִים הָרִאשׁוֹנִים, עִם הֱיוֹתוֹ גָּאוֹן וְנִסְמָךְ בְּחָכְמַת הַתּוֹרָה יוֹתֵר מֵהֶם לְאֶלֶף וְלִרְבָבָה, וְהַצֶּדֶק וְהָאֱמֶת אִתּוֹ, וּמַעֲשָׂיו מוֹכִיחִין אוֹתוֹ.

[8] וּמִפְּנֵי זֶה גָּדַל הַחִסָּרוֹן וְהִפְלִיג הֶעְדֵּר בִּקְצָת הָאֻומָּה, אַחַר הַהִשְׁתַּתְּפוּת [?]. בְּהוֹסִיף עַל זֶה סִבָּה: כִּי רוֹב סִפְרֵי הַחָכְמוֹת הָהֵם חִבְּרוּם חֲכָמִים מִבְּנֵי הָעַמִּים, כִּיְוָנִים, וּבְנֵי יִשְׁמָעֵאל, אוֹ הַנּוֹצְרִים וְיֶתֶר הַמַּאֲמִינִים אֲשֶׁר דָּתֵיהֶם שׁוֹנוֹת מִדַּת הַיְּהוּדִים, וְהִגִּיעוּ הַסְּפָרִים אֵלֵינוּ עַל יְדֵי הַמַּעְתִּיקִים. וְאוּלַי הֵם אוֹ קְצָתָם לֹא הָיוּ בְּקִיאִים בַּלָּשׁוֹן הַהוּא אוֹ הַלְּשׁוֹנוֹת עַל הַשְּׁלֵמוּת, וְהֶחֱלִיפוּ

מִלָּה אַחַת בְּאַחֶרֶת, בְּצֶלֶם אוֹ בִּדְמוּת; אוֹ חִסְּרוּ מֵהֶם אוֹ הוֹסִיפוּ, לְחֶסְרוֹן יְדִיעָתָם בַּלָּשׁוֹן הַהוּא יְדִיעָה בְּעַצְמוּת; אוֹ נָפַל בַּסְּפָרִים אֲשֶׁר הֶעְתִּיקוּ מֵהֶם טָעוּת, וּמִפְּנֵי זֶה נָפְלָה הַשְּׁגִיאָה בַּסְּפָרִים הָהֵם וְקִיצּוּר הַהַשָּׂגָה וְהַלֵּאוּת.

[12] וּבִרְאוֹתִי אֵלֶּה הַסִּבּוֹת הַמּוֹנְעוֹת, וְיָדַעְתִּי לִימּוּד הַחָכְמוֹת הַנִּזְכָּרוֹת אֲשֶׁר בֵּין הַנּוֹצְרִים יְדוּעוֹת נוֹרָא מְאֹד וְנִפְלָא—וּמִי יְכִילֶנּוּ, יַעֲמוֹד עַל עִנְיָנוֹ וְיַקִּיפֶנּוּ ?—אָמַרְתִּי בְּלִבִּי: אֶלְמַד מְעַט מִידִיעַת לְשׁוֹנָם, וְאֵשֵׁב בִּישִׁיבוֹתֵיהֶם וּבֵית עִיּוּנָם, וְאֵלֵךְ אַחֲרֵי עִקְּבוֹתָם, לְמַעַן אוּכַל לְהוֹעִיל לְעַצְמִי וּלְזוּלָתִי בְּהַשִּׂיגֵי דִּבְרֵיהֶם וּקְצָת מֵאֲמִתַּת סִפְרֵיהֶם. וְאַשְׁקִיף בְּמַה שֶּׁהִשְׁקִיפוּ הַשְּׁלֵמִים מֵהֶם, אִם הַזְּמַן יַסְפִּיק בְּזֶה וְלֹא יִמְנָעֵנִי מוֹנֵעַ וּמִקְרֶה, וְאִם הַשֵּׂכֶל יָכוֹל לְהַשְׁקִיף עָלָיו וְלֹא יַטְרִידֵהוּ הַמָּחוֹז [חסר?] מִמֶּנּוּ וְאֵלָיו.

[15] וְהִנֵּה נִמְנְעָה מִמֶּנּוּ יְדִיעַת הַחָכְמוֹת בְּכַיּוֹצֵא בָּזֶה, לֹא שֶׁיִּהְיֶה שִׂכְלֵנוּ לְמַטָּה מִשִּׂכְלֵיהֶם, כִּי גַּם לָנוּ לֵבָב כְּמוֹהֶם. אָמְנָם מְנָעוּנוּ הַסִּבּוֹת אֲשֶׁר קֶדֶם זִכְרוֹנָם, עִם גָּלוּתֵינוּ וּלְחָצֵנוּ, וְרוֹב תַּעֲנוּגָם, הוֹדָם, וַהֲמוֹנָם. וּמִפְּנֵי זֶה הֵמָּה עָלוּ מַעְלָה מַעְלָה בְּחָכְמוֹת הַחִיצוֹנִיּוֹת, וַאֲנַחְנוּ יָרַדְנוּ מַטָּה, לְרוֹב הַטְּרָדוֹת הַמְּעִיקוֹת.

⁘

[20] וּבְהַרְגִּישִׁי אֶת כָּל זֶה אָמַרְתִּי: ׳אֵין נָבוֹן וְחָכָם׳ [עפ״י בר׳ מא לט] בְּאִיקְלִימֵנוּ זֶה. הָלַכְתִּי אַחֲרֵי עִקְּבוֹת הַנּוֹצְרִים הַחֲכָמִים וְהַיְשִׁירוּנִי בַּדְּבָרִים, וְהֶרְאוּנִי הַדְּרָכִים, וְגִלּוּ אֵלַי הַסְּפָרִים, חֲדָשִׁים גַּם יְשָׁנִים. וְרָאִיתִי בֵּינֵיהֶם דִּבְרֵי הֶחָכָם הָרוֹפֵא גּוֹרְדוֹ אֲשֶׁר בְּרוֹב דְּבָרָיו תְּשׁוּבָתוֹ בְּצִדּוֹ. הִפְלִיג לַעֲשׂוֹת בְּעִיּוּן וּבִמְלָאכָה, וּלְכֶתֶר שֵׁם טוֹב זָכָה. רַבּוּ סְפָרָיו וְגָדְלוּ יְקָרוֹתָיו, נִתְעַלָּה עַל כָּל אֶחָיו.

[21] וְהִנֵּה הִגִּיעוּ אֵלֵינוּ קְצָת סְפָרָיו, וְהוּא: **שׁוֹשַׁן הָרְפוּאָה וְהַקְדָּמַת הַיְּדִיעָה**. אָמְנָם הִגִּיעוּ לָנוּ בְּהַעְתָּקָה מְאֹד חֲלוּשָׁה, סֻלָּם רָעוּעַ הוּא. הִנֵּה כִּי הַיְּהוּדִים אֲשֶׁר הֶעְתִּיקוּהוּ בָּאֲרָצוֹת הָאֵלֶּה לֹא הִשִּׂיגוּהוּ רַק בְּאֶמְצָעוּת הַלָּעוֹזוֹת—לָעֲזוּ הַלָּאטִין וְהוֹסִיפוּ בְּטָעֻיּוֹת. [...]

[22] וּבְחַפְּשִׂי אַמְתְּחוֹת סִפְרֵיהֶם וּמָצָאתִי בֵּינֵיהֶם שְׁנֵי סְפָרִים חֲדָשִׁים, לִ׳שְׁנֵי צַנְתְּרוֹת הַזָּהָב׳ [עפ״י זכר׳ ד יב] נִמְשָׁלִים, מַה מְּאֹד הֵם טוֹבִים וּרְאוּיִים, אֲשֶׁר מִבְּנֵי אֻמָּתֵנוּ נֶעְדָּרִים. הֵמָּה סֵפֶר הֶחָכָם גִּירָאבְט דִּישׁוּלְא, וְסֵפֶר הֶחָכָם מַאִישְׁטְרִי יוֹאַן דִּי טוֹרְנָאמִירָא, אֲשֶׁר זֶה עֶשֶׂר שָׁנִים שָׁמַעְתִּי שִׁמְעָם וְלֹא יָכוֹלְתִּי לְהַשִּׂיגָם—לֹא בָּהָר, וְאִם הוּא הַמָּקוֹם אֲשֶׁר מִמֶּנּוּ חֻצְּבוּ, וְלֹא בְּאוִינְיוֹן, וְלֹא בְּיֶתֶר הַמְּקוֹמוֹת הַטּוֹבִים. כִּי הִנֵּה בִּמְצִיאוּת מְעַטִּים, מְצוֹרָף לְזֶה שֶׁחַכְמֵי הָהָר הָיוּ מַחְרִימִים וּמְנַדִּים כָּל מִי שֶׁיִּמְכְּרֵם לְזוּלַת הַנּוֹצְרִים. וְהוֹצֵאתִי בְּהַשָּׂגָתָם מָמוֹנוֹת כְּפִי כֹּחִי, וְלֹא הִרְגַּשְׁתִּי בְּמִי וָמִי. וְהִגִּיעוּ לְיָדַי בְּזֹאת הַשָּׁנָה שֶׁהִיא שְׁנַת תִּשְׁעִים וְאַרְבַּע לְמִסְפַּר הַנּוֹצְרִים וְחֶשְׁבּוֹנָם. וְחֵי הָאֱמֶת קְנִיתִים בִּכְפָלַיִם מִשּׁוֹוְיָם, לְחִשְׁקִי בָּהֶם וּבְתַאֲוָתִי וּרְצוֹנִי בָּם.

[23] וְרָאִיתִי הַפְלָגַת טוּבָם, מַרְאִים פָּנִים לְכָל הַדְּבָרִים, וּמְשִׁיבִים כַּהֲלָכָה לְכָל הָעִנְיָנִים. וְהַפּוֹעֵל עִמָּהֶם יַעֲמוֹד נִשְׁעָן כְּעַמּוּד נִבְצָר, לֹא יִירָא מִדָּבָר, וּמֵרִבְבוֹת עֲדַת הָרוֹפְאִים, אִם יָבוֹאוּ אֵלָיו בְּנִסְיוֹנוֹת וּבְוִיכּוּחִים, לִהְיוֹתָם סוֹלֶת נְקִיָּיה לֹא קָדַם כְּמוֹתָם בִּזְכִירָה. וְאֶחְשׁוֹב כִּי לֹא יִתְאַחֲרוּ וְהַיּוֹתֵר טוֹב מֵהָרְפוּאָה בָּחֲרוּ.

⁘

[24] וְהִנֵּה מַתְחִיל אֲנִי בְּסֵפֶר **גיראבט** הַבָּנוּי עַל הַסֵּפֶר הַתְּשִׁיעִי מִסֵּפֶר **אל מנצור**, אֲשֶׁר הוּא מְעַט הַכַּמּוּת וְרַב הָאֵיכוּת, ׳חָזָק מִצֹּר׳ [יחז׳ ג ט], מֵשִׁיב כָּרָאוּי וּמַגִּיד כַּהֲלָכָה. וּמְלַאכְתּוֹ מְלָאכָה בְּטוּחָה, וְנוֹתֵן טַעַם לְכָל דְּבָרָיו, טַעַם לְשֶׁבַח. לֹא נֶעְלַם מִמֶּנּוּ דָּבָר וְלֹא נִשְׁכַּח. וְלָכֵן שָׁקְטָה הַהַסְכָּמָה לְהַתְחִיל בְּהַעְתָּקָתוֹ.

[25] וְכַאֲשֶׁר יִגְזוֹר הַשֵּׁם בְּהַשְׁלָמָתוֹ, אַתְחִיל אַחֲרָיו סֵפֶר **טורנא מירא**, אֲשֶׁר הָיָה בְּיָמָיו עַל כָּל חַכְמֵי הָהָר לְרֹאשׁ וְהֵם עָמְדוּ תַּחְתָּיו. רְאִיתִיו בְּעֵינַי וְדִבַּרְתִּי עִמּוֹ. חָסִיד הָיָה בְּדִבְרוֹ, לֹא הָיָה חֵקוֹ כְּחוֹק חַכְמֵי דּוֹרוֹ—לְבַזּוֹת הַיְּהוּדִים אֲשֶׁר בִּמְלֶאכֶת הָרְפוּאָה מִתְעַסְּקִים. יַעַן הָיָה מֵיִשְׁרָם בְּכֹחוֹ. בִּצְרוֹר חַיֵּי חֲסִידֵי הָאֻומּוֹת תִּהְיֶה צְרוּרָה נַפְשׁוֹ. וְסֵפֶר אַחֵר לֹא דָּמָה אֵלָיו בְּיוֹפִי חָכְמָתוֹ. דִּבֵּר בָּאֵיבָרִים וְכֹחוֹתֵיהֶם, וְהַחֳלָיִים הַמִּתְהַוִּים בָּהֶם. שׁוֹאֵל וּמֵשִׁיב, מְבָרֵר וּמְלַבֵּן, כְּאָב הַמְּלַמֵּד אֶת הַבֵּן. חִדֵּשׁ אֲמָרִים נְכוֹחִים וִישָׁרִים, יִמְתְּקוּ וְיֶעֶרְבוּ לַשּׁוֹמְעִים.

[26] וְהִנֵּה עָלָה עַל רוּחִי לְהַעְתִּיק הַסְּפָרִים הָהֵם—לֹא לְעַצְמִי אֲבָל לְזוּלָתִי מֵהָעָם, אֲשֶׁר עִמִּי הַיּוֹם, וַאֲשֶׁר יָבוֹאוּ אַחֲרַי, הַבִּלְתִּי יוֹדְעִים מְאוּמָה בִּלְשׁוֹנָם. כִּי בְּעָבְרָם עֲלֵיהֶם וּקְרָא[וּ]ם, וְיִרְאוּ שְׁלֵמוּתָם וְיוֹפִי סִידּוּרָם, יְבָרְכוּנִי בִּגְלָלָם וְיִזְכְּרוּנִי בְּלִבָּם לִהְיוֹתִי סִבַּת חִיּוּתָם, וְשֶׁיּוּכְלוּ עֲמוֹד נֶגֶד הָרוֹפְאִים הַנּוֹצְרִים בְּנִסְיוֹנוֹת וּבְוִיכּוּחִים. וְאֵלֶּה חַיֵּי שְׁנַיִים. יָדַעְתִּי כִּי עִם אֵלֶּה הַסְּפָרִים יָשִׂימוּ דִּבְרֵי הַנּוֹצְרִים לְאַיִן—אִם לֹא יִמְשְׁכוּ [!] אֶת בְּשָׂרָם בַּיַּיִן, לֶאֱכוֹל וְלִשְׁתּוֹת וְלָחוֹג חַגּוֹת, כִּי אָז לֹא יוֹעִילוּ לָהֶם הַהַעְתָּקוֹת, יַעַן יִהְיוּ לָהֶם כְּמוֹ שֶׁלֹּא הָועְתְּקוּ, וּמוֹרָשֵׁי לְבָבוֹתָם נִתְּקוּ. אָמְנָם אִם יִקְרְאוּ בָּהֶם פְּעָמִים רַבּוֹת, יוֹצִיאוּ מֵהֶם תּוֹעָלוֹת מִתְחַלְּפוֹת, וְיַעַמְדוּ ׳בְּהַשְׁקֵט וּבְבִטְחָה׳ [עפ״י ישע׳ ל טו], וְיִשְׁנוּ בְּמִטָּתָם בִּמְנוּחָה. לֹא יִפְחֲדוּ מִ׳קּוֹל נוֹגֵשׂ׳ [עפ״י איוב ג יח] בְּשֵׁבֶט הָרְפוּאָה יַכֶּם, כִּי יוּכְלוּ לִלְחוֹם מִלְחֲמוֹתֵיהֶם, לֹא יִירְאוּ מֵהֲמוֹן הָרוֹפְאִים, אַף כִּי יִהְיוּ מְחֻוכָּמִים. וְכָל שְׁאָר סִפְרֵי מְלֶאכֶת הָרְפוּאָה יוּכְלוּ לְהַסְתִּיר וּלְהַעֲלִים, וְעִם שְׁנֵי אֵלֶּה אִם יַחְרִימוּ תַּחְרִים. [...]

∴

[30] וְהִנֵּה אֶכְתּוֹב כָּל הָרְפוּאוֹת [הַ]מֻּורְכָּבוֹת בְּלָשׁוֹן לאטין אַחֲרֵי הַעְתִּיקִי אוֹתָם בִּלְשׁוֹנֵנוּ, בג״ה, לְמַעַן אוֹעִיל לַמַּתְחִילִים לְהִתְעַסֵּק בַּמְּלָאכָה, יוּכְלוּ לְסַדֵּר בְּלִי עָמָל וִיגִיעָה, יִרְאוּ עַצְמָם חֲכָמִים וּנְבוֹנִים, זִקְנֵי הַמְּלָאכָה, וּשְׁלֵמִים, גַּם כִּי יִהְיוּ מַתְחִילִים, לֹא רָאוּ מְאוֹרוֹת בַּמְּלָאכָה כָּל הַיָּמִים. לֹא יְבַזּוּ אוֹתָם הָעַטָּרִים וְלֹא יִצְטָרְכוּ לְעַיֵּין בְּסֵפֶר שְׁמוֹת הַסַּמִּים, אִם יֵדַע הָרְפוּאוֹת עַל פֶּה—כַּכָּתוּב ׳וְשָׁמַר רַגְלְךָ מִלָּכֶד׳ [עפ״י משלי ג כו]—וְיַחְזִיקוּהוּ לְרוֹפֵא חָשׁוּב.

Bibliography

Abbreviations

B Babylonian Talmud

HÜ Moritz Steinschneider, *Die hebraeischen Übersetzungen des Mittelalters und die Juden als Dolmetscher*, Berlin, 1893; reprinted Graz, 1956; translated as: Moritz Steinschneider, *The Hebrew Translations of the Middle Ages and the Jews as Transmitters*, ed. Charles H. Manekin, Y. Tzvi Langermann, and Hans Hinrich Biesterfeldt, Dordrecht: Springer, 2013–

Primary Sources and Their Translations

Aderet, Solomon, *Teshuvot ha-Rashba*, ed. H.Z. Dimitrovsky, 2 vols, Jerusalem: Mossad HaRav Kook, 1990

Alḥarizi, Judah, *Kitāb al-Durar: A Book in the Praise of God and the Israelite Communities*, ed. Joshua Blau, Paul Fenton, and Joseph Yahalom, Jerusalem: Ben-Zvi Institute, 2009 [Hebrew]

Averroes, *Drei Abhandlungen über die Conjunction des separaten Intellects mit dem Menschen von Averroes (Vater und Sohn), aus dem Arabischen übersetzt von Samuel Ibn Tibbon*, ed. and tr. J. Hercz, Berlin, 1869

Averroes, *Commentary on Plato's Republic*, ed. E.I.J. Rosenthal, Cambridge: Cambridge University Press, 1956

Averroes, 'The Hebrew Translation of Averroes' Prooemium to His "Long Commentary on Aristotle's Physics"', ed. and tr. S. Harvey, *Proceedings of the American Academy for Jewish Research*, 52 (1985), 55–84

Averroes, *Middle Commentary on Porphyry's 'Isagoge'* [...] *and on Aristotle's 'Categoriae' Translated from the Original Arabic and Hebrew and Latin versions* [...] *by Herbert A. Davidson*, Cambridge MA: The Medieval Academy of America, 1969

Baḥya b. Joseph Ibn Paqudah, *Sefer Ḥovot ha-levavot*, tr. R. Judah Ibn Tibbon, ed. A[braham] Zifroni, Jerusalem, 1928

Bar Ḥiyya, Abraham, *Chibbur ha-Meschicha we-ha-Tischboreth. Lehrbuch der Geometrie des Abraham bar Chija*, ed. and annotated by Michael Guttmann, Berlin: Meḳitse nirdamim, 1913

Bar Ḥiyya, Abraham, *La obra enciclopédica Yesodé ha-Tebuna u-Migdal ha-Emuna de R. Abraham bar Ḥiyya ha-Bargeloni*, ed. and tr. José-Maria Millás-Vallicrosa, Madrid/Barcelona: C.S.I.C., 1952

Bar Ḥiyya, Abraham, צורת הארץ ותבנית כדורי הרקיע וסדר מהלך כוכביהם, Offenbach, 1720; 2nd edn, ed. Rafael Lasri, Jerusalem: Ha-Makhon li-Tekhunah we-Qiddush ha-Hodesh, 2009

'Doeg the Edomite', 'Preface to his Collection of Translations', in 'Twelfth-Century Latin Medicine in Hebrew Garb: Doeg the Edomite as a Cultural Intermediary', ed. and tr. Gad Freudenthal, Michael McVaugh, and Katelyn Mesler, *Medieval Encounters*, 26/3 (2020), 226–284

Falaqera, Shem-Tov Ibn, *Moreh ha-moreh*, ed. Yair Schiffman, Jerusalem: World Union of Jewish Studies, 2001

Gundissalinus, Dominicus, 'The Medieval Hebrew Translations of Dominicus Gundissalinus', tr. Yossef Schwartz, in *Latin-into-Hebrew*, 2: *Texts and Studies*, ed. Alexander Fidora, Harvey J. Hames, and Yossef Schwartz, Leiden: Brill, 2013, 19–45

Gundissalinus, Dominicus, '*Sefer ha-nefeš* (*Tractatus de anima*)', ed. Yossef Schwartz, in *Latin-into-Hebrew, 2: Texts and Studies*, ed. Alexander Fidora, Harvey J. Hames, and Yossef Schwartz, Leiden: Brill, 2013, 225–279

Hallevi, Judah, *The Kuzari*, ed. and with commentary by Gedalia [Gideon] Brecher, vol. 2, Prague, 1839

Ḥayyuj, Judah ben David, *Grammatische Werke des R. Jehuda Chajjug* […] *enthaltend*: (1) *Von den ruhenden Buchstaben*. (2) *Von den Zeitwörtern mit doppelten Buchstaben*. (3) *Von der Punktuation*. [in Abraham Ibn Esra's translation], ed. Leopold Dukes, Stuttgart: Adolph Krabbe, 1844

Ḥayyuj, Judah ben David, ספר אותיות הנוח. ספר פעלי הכפל: *Two Treatises on Verbs Containing Feeble and Double Letters by R. Jehuda Hayug of Fez, Translated into Hebrew from the Original Arabic by R. Moses Giḳatilia of Cordova; to which is Added the Treatise on Punctuation by the Same Author Translated by Aben Ezra: Edited from Bodleian MSS with an English Translation*, ed. John W. Nutt, London/Berlin: Asher, 1870

Ibn al-Muṯannā, ספר בטעמי לוחות אלכוארזמי: *Ibn al-Muthannâ's Commentary on the Astronomical Tables of al-Khwârizmî*, ed. and tr. Bernard R. Goldstein, New Haven CT: Yale University Press, 1967

Ibn Janāḥ, Jonah, *Sefer ha-Shorashim*, tr. Judah Ibn Tibbon, ed. Wilhelm (Binyamin Zeev) Bacher, Berlin: Mekize Nirdamim, 1896

Ibn Janāḥ, Jonah, *Sefer ha-Riqmah*, tr. Judah Ibn Tibbon, ed. Michael Wilenski; 2nd edn, completed by David Tené and Zeev Ben-Hayim, 2 vols, Jerusalem: Israel Academy for Hebrew, 1964

Ibn Paqūda, Baḥyā ben Joseph, *Sefer Ḥovot ha-levavot*, Hebrew translation by Judah Ibn Tibbon, ed. A[braham] Zifroni, Tel Aviv: Maḥbarot le-sifrut/Mossad HaRav Kook, 1949

Ibn Tibbon, Judah, 'Ethical Will', in *Hebrew Ethical Wills*, ed. and tr. Israel Abrahams, 2 vols, Philadelphia PA: The Jewish Publication Society, 1926, I, 51–92

Ibn Tibbon, Judah, 'Ethical Will', in *Judah b. Sha'ul Ibn Tibbon's 'Iggeret ha-musar'*, ed. Pinḥas Qoraḥ, Modi'in, 2007 [Hebrew]

Ibn Tibbon, Moses, 'Ibn al-Djezzar's *Viaticum*, hebr. צידת דרכים. a) Vorrede des Moses ibn Tibbon, nach ms. Netter (Schönblum 1869), Uri 413, München 19', ed. M. Steinschneider, *Monatsschrift für Geschichte und Wissenschaft des Judentums*, 38. Jahrgang, 4 (1894), Miscelle 30, 180–181

Ibn Tibbon, Samuel, *Ma'amar Yiqqawu ha-mayim*, ed. M. Bisliches, Pressburg [Bratislava], 1837

Ibn Tibbon, Samuel, 'Samuel and Moses Ibn Tibbon on Maimonides' Theory of Providence', ed. Z. Diesendruck, *Hebrew Union College Annual*, 11 (1936), 341–366

Ibn Tibbon, Samuel, *Otot ha-Shamayim: Samuel Ibn Tibbon's Hebrew Version of Aristotle's* Meteorology, ed. and tr. Resianne Fontaine, Leiden: E.J. Brill, 1995

Ibn Tibbon, Samuel, 'Perush ha-millot ha-zarot [Glossary of Unfamiliar Terms]', in R. Moshe ben Maimon, *Moreh ha-Nevukhim*, ed. J. Even-Shmuel, Jerusalem: Mossad HaRav Kook, repr. 2000, 11–92

Ibn Tibbon, Samuel, 'Preface [to his Translation of the *Guide of the Perplexed*]', in R. Moshe ben Maimon, *Moreh ha-Nevukhim*, ed. J. Even-Shmuel, Jerusalem: Mossad HaRav Kook, repr. 2000, 117–122

Ibn Tibbon, Samuel, *Samuel Ibn Tibbon's Commentary on Ecclesiastes: The Book of the Soul of Man*, ed. James T. Robinson, Tübingen: Mohr Siebeck, 2007

Ibn Tibbon, Samuel, 'Samuel Ibn Tibbon's *Ma'amar Yiqqawu ha-Mayim*. A Philosophical and Exegetical Treatise', ed. Rebecca Kneller-Rowe, PhD diss., Tel-Aviv University, 2011 [Hebrew]

Ibn Tibbon, Samuel, *'The Book of the Soul of Man': Commentary on Ecclesiastes*, ed. James T. Robinson, Jerusalem: World Union of Jewish Studies, 2016 [Hebrew]

Jonathan of Lunel, 'Letter of R. Jonathan of Lunel to Maimonides', 1:33–35, in Salomon Aaron Wertheimer, *Sefer Ginzey Yerushalayim*, Jerusalem, 1906 [Hebrew]

Lichtenberg, A.L. (ed.), *Qovetz Teshuvot ha-Rambam we-igerotaw*, 3 vols, Leipzig, 1852

Maimonides, Moses, *Moreh ha-nevukhim*, tr. Judah Alḥarizi, ed. Aryeh Loeb Schlossberg, with notes by Simon Scheyer and Solomon Munk, London, 1851

Maimonides, Moses, *Qovetz Teshuvot ha-Rambam we-igerotaw*, ed. A.L. Lichtenberg, 3 vols, Leipzig, 1852

Maimonides, Moses, *Shemonah Peraqim le-ha-Rambam, 'im Be'ur ḥadash be-ṣeruf haqdamat R"S Ibn Tibbon z"l*, ed. David Slutzky, Warsaw, 1864; repr. without the name of the editor, Berlin: B. Cohen, 1927

Maimonides, Moses, *(Medical) Aphorisms of Moses*, Hebrew translation of R. Nathan Hameathi (1283), ed. S. Muntner, Jerusalem: Mossad HaRav Kook, 1961 [Hebrew]

Maimonides, Moses, *Haqdamot ha-Rambam le-ferush ha-Mishnah*, ed. Mordecai Dov Rabinovich, Jerusalem: Mossad HaRav Kook, 1962

Maimonides, Moses, *The Guide of the Perplexed*, tr. S. Pines, Chicago IL: University of Chicago Press, 1963

Maimonides, Moses, *A Maimonides Reader*, tr. Isadore Twersky, New York: Behrman House, 1972

Maimonides, Moses, *Letters of Maimonides*, tr. Leon D. Stitskin, New York: Yeshiva University Press, 1977

Maimonides, Moses, *Ma'amar Teḥiyyat ha-metim*, tr. Judah Alḥarizi, ed. Abraham Halkin, in *'Ma'amar Teḥiyyat ha-metim la-Rambam, betargumo shel R. Yehudah Alḥarizi, Koveẓ al yad*, 9/19 (1979), 129–150

Maimonides, Moses, *Iggerot ha-Rambam* [= *Letters and Essays of Maimonides*], 2 vols, ed. and tr. I. Shailat, Maaleh Adumim, 1988–1989 [Hebrew and Judeo-Arabic]

Nathan ha-Me'ati, 'Nathans Vorrede zur Übersetzung des *Kanons*', ed. M. Steinschneider, *Monatsschrift für Geschichte und Wissenschaft des Judentums*, 38. Jahrgang, 4 (1893–1894), Miscelle 29, 179–180

Nathan ha-Me'ati, 'The Preface by Nathan ha-Me'ati to his Hebrew Translation of Ibn Sīnā's *Canon* (1279)', ed. Joshua O. Leibowitz, *Koroth*, 7/1–2 (1976), Hebrew section, 1–7

Nathan ha-Me'ati, 'The Preface by Nathan ha-Me'ati', tr. Joshua O. Leibowitz [and Miriam Bodian], *Koroth*, 7/1–2 (April 1976), Hebrew section, I–VIII; repr. in *Koroth*, 14 (2000) (anniversary volume), 149–154

Samuel ben Judah, 'The Translator's Epilogue at the End of Alexander of Aphrodisias' *De Anima*', ed. Moritz Steinschneider, *Oṣar tov for the Year 5647* [= Hebrew supplement of *Magazin für die Wissenschaft des Judenthums*, 14 (1887)], 8–10

Samuel ben Solomon ben Nathan ha-Me'ati, 'Aus Samuel b. Salomo's Vorrede zu ibn Zohr', ed. M. Steinschneider, *Monatsschrift für Geschichte und Wissenschaft des Judentums*, 37. Jahrgang, 7 (1892–1893), Miscelle 17, 316–317

Secondary Literature

Baneth, David Z., 'Different Translation Methods in the Hebrew Versions of the *Kuzari*', *Leshonenu*, 21 (1957), 293–307 [Hebrew]

Beit-Arié, Malachi, *Hebrew Codicology: Historical and Comparative Typology of Hebrew Medieval Codices based on the Documentation of the Extant Dated Manuscripts Using a Quantitative Approach*, preprint internet English version 0.2 (April 2018) [Hebrew]

Berman, Lawrence V., 'Greek into Hebrew: Samuel ben Judah of Marseilles, 14th-Century Philosopher and Translator', in *Jewish Medieval and Renaissance Studies*, ed. Alexander Altmann, Cambridge MA: Harvard University Press, 1967, 289–320

Bos, Gerrit, 'Medical Terminology in the Hebrew Tradition: Nathan ben Eliezer ha-Me'ati', *Revue des Études Juives*, 172/3–4 (2013), 305–321

Bos, Gerrit (ed.), *Novel Medical and General Hebrew Terminology from the 13th Century*, Volume 2, Journal of Semitic Studies, Supplement Series 30, Oxford, 2013

Brague, Rémi, *Aristote et la question du monde*, Paris: Presses universitaires de France, 1988

Brener, Ann, 'Stealing Wisdom: A Story of Books (and Book-Thieves) from Immanuel of Rome's *Mahbarot*', *Prooftexts*, 28 (2008), 1–27

Brinken, Ann-Dorothee von den, 'Center of the Earth: Arin', in *Trade, Travel and Exploration in the Middle Ages: An Encyclopedia*, ed. by John Block Friedman and Kristen Mossler Figg, Oxford: Routledge, 2010, 103–104

Cohn, Willy, 'Jüdische Übersetzer am Hofe Karls I. von Anjou, König von Sizilien (1266–1285)', *Monatsschrift für Geschichte und Wissenschaft des Judenthums*, 79. Jahrgang, 3 (1935), 246–260; repr. in id., *Juden und Staufer in Unteritalien und Sizilien*, Aalen, 1978, 50–64

Davidson, Herbert A., 'The Study of Philosophy as a Religious Obligation', *Religion in a Religious Age*, ed. S.D. Goitein, Cambridge MA: Association for Jewish Studies, 1974, 53–68; revised version in id., *Maimonides the Rationalist*, Oxford: Littman Library of Jewish Civilization, 2011, 1–14

Delgado, J. Martinez, 'Moshe ben Shemuel ha-Kohen ibn Chiquitilla, el traductor', *Miscelánea de Estudios Árabes y Hebraicos. Sección hebrea*, 51 (2002), 119–157

Drory, Rina, *Models and Contacts: Arabic Literature and its Impact on Medieval Jewish Culture*, Leiden: Brill, 2000

Emery, Richard W., 'Jewish Physicians in Medieval Perpignan', *Michael*, 12 (1991), 113–134

Forte, Doron, 'Back to the Sources: Alternative Versions of Maimonides' Letter to Samuel Ibn Tibbon and Their Neglected Significance', *Jewish Studies Quarterly*, 23 (2016), 47–90

Fraenkel, Carlos, *From Maimonides to Samuel Ibn Tibbon: The Transformation of the* Dalālat al-Ḥā'irīn *into the* Moreh ha-Nevukhim, Jerusalem: Magnes Press, 2007 [Hebrew]

Freudenthal, Gad, '(Al-)Chemical Foundations for Cosmological Ideas: Ibn Sīnā on the Geology of an Eternal World', in *Physics, Cosmology and Astronomy, 1300–1700: Tension and Accommodation*, Boston Studies in the Philosophy of Science 126, ed. Sabetai Unguru, Dordrecht: Kluwer, 1991, 47–73; repr. in id., *Science in the Medieval Hebrew and Arabic Traditions* [Essay XII]

Freudenthal, Gad, *Aristotle's Theory of Material Substance: Form and Soul, Heat and Pneuma*, Oxford: Clarendon Press, 1995

Freudenthal, Gad, *Science in the Medieval Hebrew and Arabic Traditions*, Aldershot: Ashgate, 2005

Freudenthal, Gad, 'Samuel Ibn Tibbon's Avicennian Theory of an Eternal World', *Aleph: Historical Studies in Science and Judaism*, 8 (2008), 41–129

Freudenthal, Gad, 'Les controverses maïmonidiennes', *Cahiers du judaïsme*, 28 (2010), 38–52

Freudenthal, Gad, 'The Aim and Structure of Steinschneider's *Die hebraeischen Übersetzungen des Mittelalters*. The Historiographic Underpinnings of a Masterpiece and Their Untoward Consequences', in *Studies on Steinschneider: Moritz Steinschneider and the Emergence of the Science of Judaism in Nineteenth-Century Germany*, ed. Reimund Leicht and Gad Freudenthal, Boston: Brill, 2011, 191–211

Freudenthal, Gad, 'R. Judah Ibn Tibbon's Epistle to R. Asher b. Meshullam of Lunel', *Ta-Shma: Studies in Judaica in Memory of Israel M. Ta-Shma*, ed. R. Reiner et al., Alon Shevut: Tevunot, 2011, 651–672 [Hebrew]

Freudenthal, Gad, 'Abraham Ibn Ezra and Judah Ibn Tibbon as Cultural Intermediaries. Early Stages in the Introduction of Non-Rabbinic Learning into Provence in the Mid-Twelfth Century', in *Exchange and Transmission Across Cultural Boundaries: Philosophy, Mysticism and Science in the Mediterranean World*, ed. Sarah Stroumsa and Haggai Ben-Shammai, Jerusalem: The Israel Academy of Science and Humanities, 2013, 52–81

Freudenthal, Gad, 'Why did Hebrew Science in Medieval Midi not Take Off? The Insights of Leon Josef of Carcassonne (1394)', *Korot*, 23 (2016), 25–52 [Hebrew]

Freudenthal, Gad, 'The Brighter Side of Medieval Christian-Jewish Polemical Encounters: Transfer of Medical Knowledge in the Midi (Twelfth–Fourteenth Centuries)' *Medieval Encounters*, 24 (2018), 29–61

Freudenthal, Gad, 'The Medieval Hebrew Reception of Avicenna's Account of the Formation and Perseverance of Dry Land: Between Bold Naturalism and Fideist Literalism (Samuel and Moses Ibn Tibbon, Jacob ben Sheshet, Moses Melguiri, Gershon ben Solomon, Levi ben Gerson, Levi ben Abraham, Isaac and Ḥayyim Israeli, et al.)', in *The Arabic, Hebrew and Latin Reception of Avicenna's Physics and Cosmology*, ed. Dag Nikolaus Hasse and Amos Bertolacci, Berlin: Walter de Gruyter, 2018, 270–311

Freudenthal, Gad, 'Science and Medicine', in *The Cambridge History of Judaism*, VI: *The Middle Ages: The Christian World*, ed. Robert Chazan, Cambridge: Cambridge University Press, 2018, 702–741

Freudenthal, Gad (ed.), 'Arabic and Latin Cultures as Resources for the Hebrew Translation Movement: Comparative Considerations, Both Quantitative and Qualitative', in *Science in Medieval Jewish Cultures*, ed. Gad Freudenthal, Cambridge: Cambridge University Press, 2011, 74–105

Freudenthal, Gad (ed.), *Science in Medieval Jewish Cultures*, Cambridge: Cambridge University Press, 2011

Freudenthal, Gad (ed.), *Science and Philosophy in Ashkenazi Culture: Rejection, Toleration, and Accommodation*, Leipzig: Vandenhoeck and Ruprecht, 2009 [Jahrbuch des Simon-Dubnow-Instituts / Simon Dubnow Institute Yearbook 8 (for 2009)], 13–315

Freudenthal, Gad and Resianne Fontaine, 'Philosophy and Medicine in Jewish Provence, Anno 1199: Samuel Ibn Tibbon and Doeg the Edomite Translating Galen's *Tegni*', *Arabic Sciences and Philosophy*, 26 (2016), 1–26

Freudenthal, Gad and Ruth Glasner, 'Patterns of Medieval Translation Movements', in *De l'Antiquité tardive au moyen âge: Études de logique aristotélicienne et de philosophie grecque, syriaque, arabe et latine offertes à Henri Hugonnard-Roche*, ed. Elisa Coda and Cecilia Martini Bonadeo, Paris: Vrin, 2014, 245–252

Freudenthal, Gad and Mauro Zonta, 'Avicenna amongst Medieval Jews. The Reception of Avicenna's Philosophical, Scientific and Medical Writings in Jewish Cultures, East and West', *Arabic Sciences and Philosophy*, 22 (2012), 217–287

Gauthier, R[ené]-A[ntoine], 'Notes sur le début (1225–1240) du premier "averroïsme"', *Revue des sciences philosophiques et théologiques*, 66 (1982), 321–374

Goldstein, Bernard R., 'Astronomy and Astrology in the Works of Abraham Ibn Ezra', *Arabic Sciences and Philosophy*, 6 (1996), 9–21

Goshen-Gottstein, M.H., 'Translations and Translators in the Middle Ages. (iv): Samuel Ibn Tibbon's *Explanation of Unfamiliar Terms*', *Tarbiẓ*, 30 (1961), 385–395 [Hebrew]

Gutas, Dimitri, *Greek Thought, Arabic Culture: The Graeco-Arabic Translation Movement in Baghdad and Early 'Abbāsid Society (2nd–4th/5th–10th c.)*, London: Routledge, 1998

Guttmann, Julius, *Die Philosophie des Judentums*, Munich, 1933

Halkin, Abraham S., 'The Medieval Jewish Attitude Toward Hebrew', in *Biblical and Other Studies*, ed. Alexander Altmann, Cambridge MA: Harvard University Press, 1963, 233–248

Johnson, Aaron and Gad Freudenthal, 'A New Porphyry Fragment?', *The Classical Quarterly*, 70/1 (2020), 410–418

Kiener, Ronald C., 'The Hebrew Paraphrase of Saadiah Gaon's *Kitāb al-Amānāt wa'l-I'tiqādāt*', *AJS Review*, 11 (1986), 1–25

Kraemer, Joel L., *Maimonides: The Life and World of One of Civilization's Greatest Minds*, New York: Doubleday Religion, 2008

Krafft, Albrecht and Simon Deutsch (eds), *Die handschriftlichen hebräischen Werke der k. k. Hofbibliothek zu Wien*, Vienna, 1847

Langermann, Y. Tzvi, 'On the Beginnings of Hebrew Scientific Literature and on Studying Scientific Literature and on Studying History Through "*Maqbilot*" (Parallels)', *Aleph: Historical Studies in Science and Judaism*, 2 (2002), 169–189

Latin-into-Hebrew: Texts and Studies, Volume One: *Studies*, ed. Resianne Fontaine and Gad Freudenthal, Volume Two: *Texts in Contexts*, ed. Alexander Fidora, Harvey J. Hames, and Yossef Schwartz, Leiden: Brill, 2013

Marks, Richard G., 'Hindus and Hinduism in Medieval Jewish Literature', *Indo-Judaic Studies in the Twenty-First Century: A View from the Margin*, ed. Nathan Katz, New York: Palgrave Macmillan, 2007, 57–73

Marx, Alexander, 'The Correspondence between the Rabbis of Southern France and Maimonides, etc.', *Hebrew Union College Annual*, 3 (1926), 311–358

Marx, Alexander, 'Texts by and about Maimonides', *Jewish Quarterly Review*, n.s., 25 (1935), 371–428

Mascuch, Michael, Rudolf Dekker, and Arianne Baggerman, 'Egodocuments and History: A Short Account of the *Longue Durée*', *The Historian*, 78 (2016), 11–56

Melamed, Abraham, *The Myth of the Jewish Origins of Science and Philosophy*, Jerusalem: Magnes Press, 2010 [Hebrew]

Munk, Salomon, *Mélanges de philosophie juive et arabe*, Paris, 1859

Olszowy-Schlanger, Judith, 'The Science of Language Among Medieval Jews', *Science in Medieval Jewish Cultures*, ed. Gad Freudenthal, Cambridge: Cambridge University Press, 2011, 359–424

Pingree, David, 'The Indian and Pseudo-Indian Passages in Greek and Latin Astronomical and Astrological Texts', *Viator*, 7 (1976), 141–196

Pingree, David, *From Astral Omens to Astrology*, Rome: Istituto Italiano per l'Africa e l'Oriente, 1997

Rabin, Chaim. 'The History of the Translation of the *Canon* into Hebrew', *Melilah*, 3–4 (1950): 132–142; repr. in Hayyim Rabin, *Linguistic Studies: Collected Papers in Hebrew and Semitic Languages*, ed. Moshe Bar-Asher et al., Jerusalem: Israel Academy of Hebrew Language, 1999, 324–338 [Hebrew]

Ravitzky, Aviezer, 'Aristotle's *Meteorology* and the Maimonidean Modes of Interpreting the Account of Creation', *Aleph: Historical Studies in Science and Judaism*, 8 (2008), 361–400

Renan, Ernest [and Adolphe Neubauer], *Les Écrivains juifs français du XIVe siècle*, Paris, 1877 [= *Histoire littéraire de la France*, 27, 351–789]

Robinson, James T., 'The Ibn Tibbon Family: A Dynasty of Translators in Medieval Provence', in *Studies in Memory of Isadore Twersky*, ed. Jay Harris, Cambridge MA: Harvard University Press, 2005, 193–224

Robinson, James T., 'Samuel Ibn Tibbon', in *The Stanford Encyclopedia of Philosophy Archive: Summer 2014 Edition*, ed. Edward N. Zalta, https://plato.stanford.edu/archives/sum2014/entries/tibbon/ (accessed 1.8.21)

Roth, Norman, 'The "Theft of Philosophy" by the Greeks from the Jews', *Classical Folia*, 32 (1978), 53–67

Roth, Norman, 'Jewish Reactions to the *Arabiyya* and the Renaissance of Hebrew in Spain', *Journal of Semitic Studies*, 28 (1983), 63–84

Rothschild, Jean-Pierre, 'Motivations et méthodes des traductions en hébreu du milieu du XIIe à la fin du XVe siècle', in *Traduction et traducteurs au moyen âge*, ed. G. Contamine, Paris: Éditions du CNRS, 1989, 279–302

Sadan, Joseph, 'Un intellectuel juif au confluent de deux cultures: Yehuda al-Ḥarizi et sa biographie arabe', *Judíos y musulmanes en al-Andalus y al Magreb: Contactos intelectuales*, ed. M. Fierro, Madrid: Collection de la Casa de Velásquez, 2002, 105–151

Schacter, Jacob J. (ed.), *Judaism's Encounter with Other Cultures: Rejection or Integration?*, Northvale NJ: Jason Aronson, 1997

Schirmann, Jefim (Hayyim), *The History of Hebrew Poetry in Christian Spain and Southern France*, edited, expanded, and annotated by Ezra Fleischer, Jerusalem: The Magnes Press, 1997 [Hebrew]

Sela, Shlomo, *Abraham Ibn Ezra and the Rise of Medieval Hebrew Science*, Leiden: Brill, 2003

Sela, Shlomo, 'Queries on Astrology Sent from Southern France to Maimonides: Critical Edition of the Hebrew Text, Translation and Commentary', *Aleph: Historical Studies in Science and Judaism*, 4 (2004), 89–190

Sela, Shlomo and Gad Freudenthal, 'Abraham Ibn Ezra's Scholarly Writings: A Chronological Listing', *Aleph: Historical Studies in Science and Judaism*, 6 (2006), 13–55

Septimus, Bernard, *Hispano-Jewish Culture in Transition: The Career and Controversies of Ramah*, Cambridge MA: Harvard University Press, 1982

Shatzmiller, Joseph, 'Livres médicaux et éducation médicale: à propos d'un contrat de Marseille en 1316', *Mediaeval Studies*, 42 (1980), 463–470

Shatzmiller, Joseph, 'Au service de la Cour de Naples: Kalonymos d'Arles et Judah Romano', in *L'Écriture de l'histoire juive: Mélanges en l'honneur de Gérard Nahon*, ed. Danièle Iancu-Agou and Carol Iancu, Leuven, 2012, 163–169

Sirat, Colette, 'Les traducteurs juifs à la cour des rois de Sicile et de Naples', in *Traduction et traducteurs au moyen âge*, ed. Geneviève Contamine, Paris: CNRS, 1989, 169–191

Smith, David Eugene and Jekuthiel Ginsburg, 'Rabbi Ben Ezra and the Hindu-Arabic Problem', *The American Mathematical Monthly*, 25/3 (1918), 99–108

Sonne, I., 'Ein Beitrag zu der verschollenen Kuzariübersetzung des Jehuda ben Cardinal', *Monatsschrift für Geschichte und Wissenschaft des Judentums*, 72/1–2 (1928), 66–70

Steinschneider, Moritz, 'Zur Geschichte der Uebersetzungen aus dem Indischen in's Arabische und ihres Einflusses auf die arabische Literatur', *Zeitschrift der Deutschen Morgenländischen Gesellschaft*, 25/3 (1871), 378–428

Steinschneider, Moritz (ed.), 'The Translator's Preface to the *Antidotarium* [Hebrew]', *Jeschurun*, 7 (5631/1871), Hebrew section, 90–91 (corrections on 94)

Steinschneider, Moritz, 'Robert von Anjou und sein Verhältnis zu einigen gelehrten Juden', *Monatsschrift für Geschichte und Wissenschaft des Judentums*, 48 (1904), 713–717

Stern, Samuel M., 'Maimonides' Correspondence with the Scholars of Provence', *Zion*, 16 (1951), 18–29 [Hebrew]

Visi, Tamás, 'Berechiah ben Naṭronai ha-Naqdan's *Dodi ve-Neḵdi* and the Transfer of Scientific Knowledge from Latin to Hebrew in the Twelfth Century', *Aleph: Historical Studies in Science and Judaism*, 14/2 (2014), 9–73

Yoeli-Tlalim, Ronit, 'Exploring Persian Lore in the Hebrew *Book of Asaf*', *Aleph: Historical Studies in Science and Judaism*, 18 (2018), 1–23

Zonta, Mauro, 'La tradizione ebraica dell'Almagesto di Tolomeo', *Henoch*, 15 (1993), 325–350

Zonta, Mauro, 'Mineralogy, Botany and Zoology in Medieval Hebrew Encyclopaedias', *Arabic Sciences and Philosophy*, 6 (1996), 262–315

Zonta, Mauro, *Hebrew Scholasticism in the Fifteenth Century: A History and Source Book*, Dordrecht: Springer, 2006

Zonta, Mauro, 'Medieval Hebrew Translations of Philosophical and Scientific Texts: A Chronological Table', in *Science in Medieval Jewish Cultures*, ed. Gad Freudenthal, Cambridge: Cambridge University Press, 2011, 17–73

CHAPTER 12

Renaissance Scholars on Why They Translate Scientific and Philosophical Works from Arabic into Latin

Dag Nikolaus Hasse

Part One: Essay

A good number of texts provide explicit information about the motives and circumstances surrounding Arabic–(Hebrew)–Latin translations in the Renaissance, notably the many prefaces and dedicatory letters that accompany printed texts. Renaissance translations are successors to the medieval Arabic–Latin translations of scientific and philosophical texts, most of which had been produced in the twelfth to the thirteenth centuries. The new wave of translations stretches from about 1480, when Girolamo Ramusio in Damascus and Elia del Medigo in Italy started to translate Avicenna and Averroes, until 1549, the death of the last productive translator, Jacob Mantino.[1]

The documents presented here in English translation and in their original versions are arranged roughly in chronological order. Most of them are in Latin, the standard academic language of Christian Europe of the time, and were written by translators, revisers, editors of translations, or historians interested in translations; two documents are archival minutes of faculty meetings of Padua University (Texts 4.1 and 4.2). The translators included in this collection fall into two groups: those who translated from Arabic and worked in Arabic-speaking countries, like Girolamo Ramusio and Andrea Alpago, and those who translated from Hebrew versions of Arabic texts and worked in Northern Italy and Rome, like Elia del Medigo, Paolo Ricci, Abraham de Balmes, Calo Calonymos, and Jacob Mantino. The outcome of these seventy years of translation effort is impressive in size and quality: Avicenna's *Canon of Medicine* was made accessible in new and improved Latin versions, and many commentaries by Averroes then still unknown in Christian Europe could now be read in Latin, such as the *Long Commentary on the Posterior Analytics*, the Epitome of the *Metaphysics*, and the *Middle Commentary on De animalibus*.[2] But the translators also

1 For a comprehensive treatment of the transmission of Arabic sciences and philosophy in the Renaissance, see Hasse, *Success and Suppression*.

2 On Averroes translations in the Renaissance, see Tamani, 'Traduzioni ebraico-latine', 105–114;

 | DOI:10.1163/9789004472648_013

engaged with other Arabic scientific traditions, such as the medicine of Averroes, the astronomy of Alpetragius, and the astrology of Haly ibn Abenragel.

Patrons played an important role in many Renaissance translations, as the dedicatory letters presented below demonstrate. Three aristocratic patrons, in particular, were prominently involved in the Hebrew–Latin translations of Averroes: Giovanni Pico della Mirandola, Domenico Grimani, and Ercole Gonzaga. In the early 1480s, Elia del Medigo translated several texts for Giovanni Pico della Mirandola, who was continuing his studies at Padua University. Domenico Grimani, a young Venetian aristocrat and student at Padua University, was commissioning Averroes translations from Elia del Medigo in the 1480s, and later in life, when he became cardinal and patriarch of Aquila and famous for his humanist library, he ordered and financed Averroes translations by Abraham de Balmes, which were printed in 1523. In the 1520s, Ercole Gonzaga, bishop of Mantua and a student and admirer of the Aristotelian philosopher Pietro Pomponazzi, became the patron of some of Jacob Mantino's Averroes translations. While it is clear that Pico, Grimani, and Gonzaga actively commissioned and supported Hebrew–Latin translations, we do not know the degree to which other addressees were involved in the works dedicated to them. Addressees include Alberto Pio, prince of Carpi, a humanist Aristotelian; the Venetian doge Andrea Gritti, who had studied at Padua University; Gian Matteo Giberti, bishop of Verona; and Pope Leo X, Pope Paul III, and the papal military commander Guido Rangoni, all three of whom are dedicatees of translations by Jacob Mantino. It may well be the case that the dedicatee also supported the printing of the work.

The remarkable connection between the Averroes translations and the Italian high aristocracy is partly explainable by the fact that many of the Jewish translators of Averroes were physicians: Abraham de Balmes served as personal physician to Cardinal Domenico Grimani, and Jacob Mantino to Pope Paul III, while Calo Calonymos gave medical counsel to Andrea Gritti, the doge of Venice. But it is also explainable by common links to the University of Padua. Most of the dedicatees had studied in Padua, and some remained admirers of the Aristotelian philosophy taught there, in particular that of Pietro Pomponazzi and Averroes. The links hold firm, despite the facts that most of the translators and patrons involved lived in other cities for most of their lives, and that Padua University lost influence when it was closed down from 1509 to 1517. In the decades before and after the closure, the University of Padua was one of

---

Burnett, 'The Second Revelation', 185–198; Di Donato, 'Traduttori di Averroè', 25–49; and Hasse, *Success and Suppression*, chap. III and 341–357.

the biggest universities in Christian Europe, an intellectual hub that attracted students from many European countries, and which educated and hosted many scholars of the time.[3] Hence, while it is important not to overemphasize an alleged 'Padua School' of thought, the documents presented here underline the fact that there did exist in Padua an intellectual climate very sympathetic towards Averroes that radiated far beyond Padua. This climate was the social context for most Averroes translations of the Renaissance.

Remarkably enough, Padua is also the context for two translations of Avicenna's *Canon of Medicine* (see Texts 1 and 4).[4] The translators Girolamo Ramusio and Andrea Alpago studied philosophy and medicine in Padua and were later appointed physicians at the Venetian embassy in Damascus: Ramusio in 1483, Alpago (after Ramusio's death) in c. 1487. While Ramusio's attempt at translation was not completed and survives only as rough notes in a manuscript, Andrea Alpago's substantial corrections to the *Canon* became a success story. During his sojourn in Damascus, which lasted for three decades until 1517, Alpago learned much from reading Avicenna's medical and philosophical works with his teacher Ibn al-Makkī, a distinguished Arabic scholar of Damascus.[5] At the same time, Alpago remained in touch with the officials in Venice and Padua. After his return to the Veneto, he submitted his Arabic–Latin translations to the faculty (Collegio) of philosophers and physicians of Padua University. The Collegio officially recommended Alpago's corrections to the *Canon* in January 1521 (Texts 4.1 and 4.2), and the corrections, which are important philological accomplishments in themselves, were printed posthumously in 1527 by his nephew Paolo Alpago. From 1544 onwards, these corrections were often printed in early modern *Canon* editions, either in the margin or integrated into the text. Andrea Alpago is a remarkable figure, whose contacts bridge large distances and who had travelled 'in Cyprus, Syria, Egypt, and virtually the whole Orient' in search of trustworthy manuscripts, as his nephew Paolo Alpago relates (Text 4.3). Shortly before his death in 1522, Alpago was even offered a professorship in practical medicine at Padua University.[6]

3 Grendler, *The Universities of the Italian Renaissance*, 36–40.

4 On Renaissance translations of the *Canon of Medicine*, see Siraisi, *Avicenna in Renaissance Italy*, 133–143; and Hasse, *Success and Suppression*, 96–115. Cf. also d'Alverny, 'Avicenne et les médecins de Venise', 182–188.

5 On Andrea Alpago's mentor in Damascus, see Michot, 'A Mamlūk theologian's commentary', 195–198 ('Appendix I: *rays ebenmechi, praeceptor meus*'). See also Veit, 'Andrea Alpago und Schah Ismāʿīl', 457–465.

6 On Andrea Alpago, see the fundamental study by Lucchetta, *Il medico e filosofo*, and Hasse, *Success and Suppression*, 96–115, with further literature.

The existence of patrons, of an audience, and of an academic context therefore constituted an important foundation for Arabic–(Hebrew–)Latin translation activity in the Renaissance. The Collegio's adoption of Alpago's version of the *Canon* is motivated by another factor often mentioned in the documents below: the demand for and concern to have a correct text. The Paduan professors cannot imagine 'a more laudable and more necessary enterprise' than the correction of obscure and faulty passages in the medieval Latin version of Avicenna's *Canon*. In a culture where Avicenna and Averroes were prescribed reading in many Christian European universities, correct translations of these authors were an essential demand. The promise of uncorrupted texts that are more reliable than the medieval translations is proffered alike by Arabic–Latin translators, Hebrew–Latin translators, and humanist revisers. Dedicated humanists such as Paolo Ricci, Jean Bruyerin Champier, Alban Thorer, Miguel Jerónimo Lesdesma, and Antonius Stupa (Texts 3, 9–11, and 14) often add a polemical tone to this promise, by pointing to the barbarous style of the medieval translators and to the 'millions of mistakes' (Text 11) that distort these texts, as they claim. The Hebrew–Latin translators, in turn, argue that the Hebrew transmission of Averroes offers better and more correct texts than the Latin transmission, where most books by Averroes are 'corrupted', as Calo Calonymos puts it (Text 6.1). This is also because Jewish education emphasizes truth, whereas the Romans honour rhetoric first, as Abraham de Balmes argues (Text 5). Balmes even extends the hope for a correct text to the works of Aristotle himself, who was transmitted badly by the Greeks in antiquity, so that 'in remedying the defects of Aristotle's books one should trust the one Averroes more than all Greeks together' (Text 5). As is obvious here, the transmission of textual witnesses is an important topic for many translators: many of them claim to have found a particularly old or trustworthy manuscript that forms the basis of their translation or revision. The interest in transmission is also apparent, for example, in Grimani's inquiry into the tradition of Aristotle's *Metaphysics* and the different sequence of books in Greek, Arabic, and Latin. (Text 2).

The translators, however, are not only hunting for correct versions of existing texts, but also, and successfully, for new texts unknown to the Latins. The Paduan faculty of philosophers and physicians is most interested not only in Alpago's *Canon* translation, but also in 'the translation of some Arabic books not yet published' (Text 4.2). Six such translations, namely of philosophical works by Avicenna, were printed by Andrea Alpago's nephew in 1546, but Paolo attests that there existed further translations that do not seem to be extant today (Text 4.3). Jacob Mantino is 'touched by incredible joy' when he encounters a Hebrew version of Averroes's *Middle Commentary on De animalibus*, 'which had long been desired by the Latins' (Text 7.1), and Abraham de Balmes comments on the fact that his translation of the *Long Commentary on the Posterior Analytics* fills an important gap (Text 5). The greatest demand, apparently, is for new texts by Averroes, which are diligently collected by the editors of the monumental 1550/52 Giunta edition of the combined Aristotle and Averroes (Text 13).[7]

Intellectual interests are an important factor in all translation movements, and this is no less true in this case. The first translator chronologically, Elia del Medigo, already responds to the philosophical interests of his patron Domenico Grimani, who had discussed the problem of universals with him, defending an anti-realistic position. There is much evidence for a strong Renaissance interest in Averroes's commentaries on logic: Domenico Grimani commissions the translation of logical commentaries from Abraham de Balmes, who himself composes a *Liber de demonstratione*, which is printed together with his translations; Averroes's *Long Commentary on the Posterior Analytics* is translated three times, by Burana, Balmes, and Mantino; and the editors of the Giunta edition are fully abreast with the teaching of logic at Padua University, one of them (Marco degli Oddi) being a professor of logic himself. However, as one can see from the full list of Renaissance Arabic–(Hebrew–)Latin translations that I have published elsewhere,[8] other disciplines of philosophy are much in demand too, especially natural philosophy, psychology, and metaphysics. The translators themselves often share these interests, as is apparent from their own philosophical works, such as Elia del Medigo's Latin treatises on cosmology and psychology, and Calo Calonymos's work *On the Creation of the World Proved with Physical Reasons*.

In medicine, scholars were most interested in three areas: first, topics that were often taught at the universities, such as the principles of medicine, treatment methods, and fevers (see Text 7.3), which are covered for instance in Avicenna's *Canon of Medicine*, chapters I.1, I.4, and IV.1; second, the compatibility of philosophical and medical doctrines, and of Aristotle and Galen in particular, a topic that motivates Jean Bruyerin Champier to revise Averroes's *Colliget* (Text 9); and third, medical botany and pharmacology, a field to which the translator Andrea Alpago contributes much through his *Interpretatio Arabicorum nominum*, a Latin lexicon of Arabic proper names in the *Canon*, and of drug names in particular.

In astronomy and astrology, hardly any translations were produced, save for Calo Calonymos's translation of Alpetragius's astronomical *Theorica plane-*

7 On this important edition, see Burnett, 'Revisiting', 55–64.

8 Hasse, *Success and Suppression*, 72–75.

tarum and Antonius Stupa's revision of Haly filius Abenragel's astrological *De iudiciis astrorum*. The intellectual context of Calo's translation is noteworthy (Text 6.2): in the dedication, Calo praises the anti-Ptolemaic attitude of Averroes and Alpetragius, who had attacked Ptolemy's mathematical astronomy for not being in accordance with the physics of the universe. The physical problems of Ptolemy's astronomy are a vibrant topic for many intellectuals in the decades around 1500, among them Girolamo Fracastoro, Giovan Battista Amico, and Nicolaus Copernicus.[9] Calo clearly takes side with Alpetragius and Averroes against Ptolemy and laments the physical 'absurdities' that deter people from the discipline of astronomy as a whole.

Humanist motives play a role in the Renaissance translation movement in many different ways. Some have already been remarked upon, such as the humanist profile of patrons and editors; the search for reliable manuscripts; the comparison of textual witnesses; and the interest in correcting textual mistakes. Another important humanist topic, also in the documents below, is Latin style. Jacob Mantino's translations, for example, are written in a moderately classicizing Latin, which present Averroes and Avicenna in a form that is much more acceptable for readers with humanist education than the translations of the Middle Ages. Mantino puts this bluntly: the 'unkempt and mutilated' Latin style of the older translations is the reason 'why many today condemn the teaching of Averroes' (Text 7.2). A different and more polemical tone comes from full-blooded humanists such as Paolo Ricci, Miguel Ledesma, and Antonius Stupa, who deplore the 'barbarism and foolishness' of the medieval translators and commentators and produce texts in a thoroughly classicizing, and sometimes exquisite, Latin style.

The background to such remarks about 'barbarism' is the Greek/Arabic antagonism that looms large in the long-lasting Renaissance controversy over the value of Arabic sciences and philosophy.[10] This controversy centres on several main topics that reverberate in the documents below: the presumed corruption of medical knowledge in the Arabic–Latin tradition, the alleged irreligion of Averroes's philosophy, and the 'un-Ptolemaic' astrology of Arabic authorities like Albumasar. These accusations are usually uttered by humanists. It is all the more noteworthy that several humanists appear among the revisers and editors of Arabic–Latin translations and speak positively about Arabic science. The French humanist Bruyerin Champier defends Averroes as

9 On Amico and Fracastoro, see Di Bono, 'Copernicus, Amico, Fracastoro', 133–154. Cf. Hasse, 'Averroes' Critique of Ptolemy', 69–88.

10 On this controversy, see Klein-Franke, *Die klassische Antike in der Tradition des Islam*, 65–76; and Hasse, *Success and Suppression*, 138–139, 250–255, with further literature (459, n. 7).

'the greatest savourer of Galen's books' (Text 9). The Swiss humanist Thorer praises the Arabic physician Rhazes as being more useful for medical practice and less obscure than Hippocrates and Galen (Text 10). Miguel Ledesma assures his readers that there is nothing in his Avicenna edition that is not in full harmony with Galen (Text 11). But many humanists were scathing in their critique of Arabic–Latin scientific traditions and provoked the reaction of scholars such as Abraham de Balmes, who takes an openly anti-humanist stance. It is wrong, he claims, to aim for 'ornate eloquence' rather than for truth. And it is wrong to prefer the Greek commentators, and the Greeks in general, to Averroes, since philosophy did not originate with the Greeks, but with the barbarians, i.e., in the Near East, which is one reason why Averroes, as a barbarian, is more to be trusted on the interpretation of Aristotle than any Greek commentator (Text 5).

Historical arguments such as these, for which Abraham de Balmes draws on the Christian historian Eusebius (third century AD), are often employed by translators and editors. At the other end of the spectrum, one may locate the testimony of Girolamo Donzellini (Text 15), a humanist editor of Rhazes, whose preface to the edition, ironically, contains a hardliner humanist's view of history. Donzellini deplores the transmission of medicine from the Greeks to the Arabs, because as a result medicine was shipwrecked, and from the Arabs to the medieval Latins, because medicine remained there unproductive for a long time; only the Renaissance period under the Medici saw a resuscitation of medicine 'from the clear sources of the Greek'. Other historical accounts of the cross-cultural transmission of the sciences, such as those by Johannes Carion, Jean Bruyerin Champier, and Jacob Milich (Texts 8, 9, and 12), are less prejudiced than those of Balmes and Donzellini. It is in the Renaissance that a Western historiography of translation movements emerges. From the very beginning, this historiography, coloured by justificatory and polemical purposes, also serves to describe the author's own attitude towards the translation and cross-cultural transmission of the sciences and philosophy.

Part Two: Texts in Translation

Unless otherwise specified, translations are the author's own.

Text 1. Girolamo Ramusio

Girolamo Ramusio was the first Renaissance scholar who tried to improve the medieval Latin version of Avicenna's *Canon of Medicine* by comparing it to the Arabic. Like Andrea Alpago after him, Ramusio served as a physician at the Venetian embassy in Damascus. The result of his work on the *Canon* is extant in an Arabic manuscript that preserves Ramusio's Latin translation written between the lines of Arabic. This 'translation'—or, rather, rough notes for a translation, since Ramusio noted Latin equivalents for each Arabic term but did not produce an intelligible Latin syntax—must have been written between 1483, when Ramusio arrived in Damascus, and his premature death in 1486 in Beirut. Ramusio had studied philosophy and medicine at Padua University in the 1470s. The end of the manuscript preserves a letter to a friend called Flavius, in which Ramusio explains the circumstances and motives of his attempt at translation.[11]

Girolamo Ramusio, interlinear Arabic–Latin translation of Avicenna's *Canon*, Paris, BNF, MS arabe 2897, ff. 160v and 349r.

f.160v

I wanted to turn first to what is publicly lectured upon at the University of Padua. […]

f.349r

To Flavius by Ramusius. When writing in Arabic [i.e., copying an Arabic manuscript of the *Canon*], I have seen much that is in total conflict with the Latin of Avicenna [i.e., with Gerard of Cremona's translation of the *Canon*]. I have made notes of many things that the Latin ⟨translator⟩ had omitted or written in a distorted fashion; I have read much that I was not able to turn into Latin with precision because of my inexperience with the Arabic language. Hence,

11 On Girolamo Ramusio, see d'Alverny, 'Survivance et renaissance d'Avicenne', 95; and Hasse, *Success and Suppression*, 96.

learning from day to day, I correct and emend myself and my writing more successfully every day.

In copying this first ⟨book⟩ of Avicenna in Arabic, I have not retained any order, generally, for I have always started with the more useful things first, leaving the less useful for later. In order that you may understand the sequence, my best Flavius, listen: I have first turned to anatomy ⟨on leaf⟩ 1, ⟨then to the section⟩ on pulse, urine, and excretion from leaf 50; chapter 1.4 ⟨begins⟩ on leaf 107; the sixth subchapter on the faculties on leaf 160; the second fen on leaf 174; and the third fen on leaf 272. ⟨With respect to⟩ the rest of 1.4 and to anatomy, you will see that my Latin is corrected. But don't be astonished if not everything is as diligently corrected, since I understood everything ⟨only⟩ slowly, and in order to learn faster, I wrote in Arabic and impaired everything with ⟨my hastily written⟩ letters. I hope to return to everything next year, examine it with utmost care, and turn it into its best form. Hence I thank the mighty gods and Catta[?] etc.

Text 2. Elia del Medigo, *Domino Dominico Grimani*; ed. Kieszkowski, 76–77.

The Jewish scholar Elia del Medigo's translations are the starting-point of the Renaissance wave of Hebrew–Latin translations of Averroes's commentaries on Aristotle. Del Medigo was born in Candia on Crete (modern Heraklion), lived in Venice, Padua, Florence, Bassano, and Perugia from some time before 1480 until about 1490, and then returned to Crete, where he died in 1493. Del Medigo dedicated some of his translations to two young noblemen and students at Padua University, Domenico Grimani and Giovanni Pico della Mirandola, who would become famous also for their humanist interests, though it is their philosophical inclinations that, according to del Medigo, motivated them to order from him new translations of Averroes. Translating Averroes from Hebrew into Latin was a difficult undertaking for del Medigo, whose Latin was much less fluent than that of later translators such as Paolo Ricci or Jacob Mantino.[12]

Often in discussing philosophical matters, as is your habit [i.e., of the dedicatee, Domenico Grimani], you spoke to me about the order of the books of ⟨Aristotle's⟩ *Metaphysics*, ⟨i.e., about⟩ which one should be the first: either the book [*Alpha elatton*] that begins with ⟨the discussion of⟩ truth, as the ancient texts generally have it, or the following book [*Alpha*], as some new texts have it. For you argued that the book that is generally the first [*Alpha elatton*] is like a pref-

12 On Elia del Medigo, see Engel, *Elijah Del Medigo*; and Hasse, *Success and Suppression*, 77–78.

ace [...] to the entire *Metaphysics*, in particular according to the Arabic⟨–Latin⟩ translation, where the book [*Alpha*] that is found ⟨first⟩ among the Latins, is not the first. And because you have reasoned well, I wanted to translate for you the preface of the commentator [i.e., Averroes], which he wrote for book XII of the *Metaphysics*, where he explains the order of the books and many good things. And even though I have translated ⟨this text⟩ already for the most noble gentleman Giovanni ⟨Pico⟩, count of Mirandola, nevertheless I do not have this translation, and ⟨the two translations⟩ perhaps do not vary in any sentence. Often Your Magnificence also discussed the ⟨problem of⟩ universals, arguing that they do not exist outside the soul. Many Latins would disagree, even though in itself this is doubtless true. [...]

I do not want to say more about this issue for the moment, but rather I would like to translate what Averroes says about this in the above-mentioned passage [i.e., *Metaphysics*, VII.48 and VII.51], from which the truth of the matter appears. And even though these things are difficult, and many people may have erred ⟨on the matter⟩, nevertheless the truth will be plain to intelligent people like you, in particular because the difficult issues are explained by me through the words of the commentator in other places, as in the questions on the intellect [i.e., *the Long Commentary on De anima*, III.5], when speaking about the universal, and in other places. What ⟨Averroes⟩ says in the afore-mentioned preface contributes much to this issue, and in particular what I translated of his *Epitome* ⟨of the *Metaphysics*⟩. Because I know that you are able to know many good and difficult things by yourself, I have in this ⟨case⟩ preserved not only its meaning, but also the words in the best way possible. Farewell.

Text 3. Paolo Ricci

The Hebrew–Latin translator Paolo Ricci (d. 1541), a Christian convert from Judaism, teacher of philosophy and medicine, and, later in life, physician at the Habsburg court, stands out among the translators discussed in this chapter for his high-flown neo-classical Latin style. Ricci's humanist motives are apparent in the preface to his translation of Averroes's *Middle Commentary on De caelo*, published in 1511. He believes, as many humanist scholars of the time did, that the medieval Latin translations of Averroes are replete with errors. This is a topic he had discussed already with his 'companion philosophers' at Padua University, apparently in past years when he was still a student. The remedy he proposes is to translate anew from Jewish sources.[13]

---

13 On Paolo Ricci, see Roling, *Aristotelische Naturphilosophie*; and Hasse, *Success and Suppression*, 78–79.

Paolo Ricci, *Prefatio* (Milan, 1511), f.7r; English translation from Burnett, 'The Second Revelation', 193.

Not only these three prefaces, but the entire Latin version of Averroes, as I have once explained to some of our companion philosophers, abounds in many corruptions and errors. I wanted to warn you of this so that, when you see that some things in this preface [to the commentary on the *Physics*] and others which I have taken from the libraries of the Jews do not agree with the Latin version, you do not burst into a complaining and accusatory tone of voice. For everywhere there are Jews and some people skilled in the Arabic tongue whom, if you are a seeker after truth, you may consult and cross-examine as much as you can. In this way, you will recognize that everything has been transmitted carefully and precisely.

Text 4. Andrea Alpago

Andrea Alpago (d. 1522) is the most 'Oriental' of all translators of the Renaissance, since he lived for about thirty years in Damascus. A native of the city of Belluno, north of Venice, Alpago studied philosophy and medicine in Padua. In about 1487, he became the successor of Girolamo Ramusio as physician of the Venetian embassy in Damascus, and returned to Italy as late as December 1520. Alpago's corrections of the medieval Latin version of Avicenna's *Canon of Medicine* were a great success. They were officially embraced by the faculty of philosophers and physicians of the University of Padua in 1521, as the documents translated below show. They were printed by his nephew Paolo Alpago in 1527 and often integrated into *Canon* editions from 1544 onwards.[14]

Text 4.1. Collegio of the philosophers and physicians of the University of Padua, *Pro excellentissimo*; ed. Lucchetta, 88–89.

In the name of Christ Amen. In the year 1521 after Christ, indiction IX, on Monday, 21 January, at the 17th hour, in the church S. Urbano, where the Sacro Collegio of the most distinguished doctors of the arts and of medicine of Padua convened by order of Vincenzo Moscheta, the excellent doctor of the arts and medicine and Prior of his gentlemen counsellors, it was said and declared by the above-mentioned gentleman Prior:

14 On Andrea Alpago, see n. 6 above.

'Most excellent gentlemen doctors! The reason for this extraordinary convention of yours is the following: there is a distinguished doctor of the arts and medicine, the gentleman master Andrea ⟨Alpago⟩ of the city ⟨of Belluno⟩, who for many years lived in Oriental areas and in particular in Damascus, where he admirably performed the medical art and for a long time entertained familiarity and contacts with a certain excellent Arab physician, most expert in the Arabic language [i.e., Ibn al-Makkī]. From this man he learned much and thus became so brilliant that he corrected and translated certain passages of Avicenna. As a result, many passages of the same Avicenna, which had been obscure, were now made clear and obvious. And he also translated some further Arabic books into Latin. Surely in our time we could not have a more laudable and more necessary enterprise than this correction and translation, since through this the same passages that before, as was said above, had been obscure and ambiguous, were made clear and intelligible. And in fact this man deserves much praise, and it would be our task to recommend him to our most famous gentlemen Rectors ⟨of the city of Padua⟩ and to ask our magnificences that they deign to recommend this excellent man to the serene highness our Duke of Venice to the effect that his so precious works are made well known, and also that he has reason to proceed in this endeavour, namely in translating other passages of the same Avicenna and other books from Arabic into Latin.'

After this statement had been made, this man [i.e., Andrea Alpago] was much recommended by several most renowned doctors of the Collegio and various statements were made on the issue. Finally, the above-mentioned gentleman Prior suggested the following resolution: whether it would please them that the most renowned doctors of the arts and medicine, the gentlemen masters Nicolò da Genova, Ludovico Carensio, Girolamo Bagolino, and Giovanni Lorenzo da Sassoferrato, by the order of this Sacro Collegio, examine the works by the afore-mentioned most perfect doctor of the arts and medicine gentleman master Andrea of the city ⟨of Belluno⟩ and make a report to the Sacro Collegio.

Text 4.2. Collegio of the philosophers and physicians of the University of Padua, *Pro domino magistro Andrea de Cividale*; ed. Lucchetta, 90.

On Thursday, 24 January ⟨1521⟩, in the church S. Urbano, at the 22nd hour.

When the Sacro Collegio of the most distinguished doctors of the arts and medicine of Padua convened again, by order of the above-mentioned gentleman Prior, the excellent doctors who had before been chosen to examine the works of the mentioned gentleman, master Andrea of the city ⟨of Belluno⟩, spoke after they had inspected several works by the same master Andrea, translated from Arabic into Latin. They presented their report one by one and recommended them highly as useful and necessary.

When this report was made, the above-mentioned gentleman Prior suggested the following resolution, which is: in view of the excellent report made by the most distinguished doctors of the arts and medicine, gentlemen masters Nicolò da Genova, Ludovico Carensio, Girolamo Bagolino, and Giovanni Lorenzo da Sassoferrato, about the excellent translation and correction of passages in the books by Avicenna accomplished by the excellent doctor of the arts and medicine gentleman master Andrea of the city ⟨of Belluno⟩, and about the translation of some Arabic books not yet published, that the below-mentioned five most excellent doctors be elected by this Sacro Collegio to approach the magnificent gentlemen Rectors ⟨of the city⟩ and praise and favour this man because of these works in a way that befits this most sacred Collegio.

Text 4.3. Paolo Alpago, *Ter illustrissimo* (Venice, 1546), *1^{v}–*2^{v}, dedicatory letter to Cardinal Alessandro Farnese (d. 1589).

Paolo Alpago of Belluno, physician, wishes the best of fortune to the thrice famous and venerable gentleman Cardinal Alessandro Farnese, his most revered master.

Andrea Alpago of Belluno, my uncle, the renowned philosopher and physician, had translated—thrice famous and venerable master—the great book on medicine by Avicenna from Arabic into the Latin language, having redeemed it from innumerable faults, with great success, as very many people saw who have the book in their hands, and with great labour, expense, and risks, as I saw who, still a young man and his companion in writing and intensive study, did not leave his side a finger's breadth in the least when he, in his old age, was searching for the hiding places of the Arabic language and for trustworthy manuscripts in Cyprus, Syria, Egypt, and virtually the whole Orient. This book was first published some years ago [in 1527] by myself, but was published again recently [in 1544] in somewhat corrected form.

My uncle had translated some splendid treatises by the same Avicenna: the *Compendium of the Soul, On Destination, i.e., on the State of Souls after Separation from their Bodies, Aphorisms on the Soul, On Definitions, On Questions,* (and) *On Divisions of the Sciences.*[15] These were lying around with me for a longer time

15 These texts are: *Compendium on the Soul* (GP10), *The Immolation Destination* (GM2),

than was appropriate and than I wished, concealed by darkness and location, because I was hindered both by the laborious work on this greater edition [i.e., the *Canon* edition of 1544] and by private occupations. [...]

Hence, send this most splendid light to those who are in darkness; and see ⟨yourself⟩ what Avicenna thinks about the soul and how much he agrees with the Christians. And when we realize that we have done you a favour with this, most famous and venerable gentleman, you may expect some further ⟨translated texts⟩ of this kind, treatises not only by the same Avicenna, but also by other Arabs, which had been rendered into Latin by the same person, my uncle, and which are worthy, in my opinion, of having a place in the palaces of princes, namely, *On Poisons, On the Correction of Errors that Occur in the Regimen of Health, On Pleasant Medicaments for Princes, On Precious Stones*, and many other such texts.

Text 5. Abraham de Balmes

Abraham de Balmes (d. 1523), a Jewish scholar from Lecce, counts among the most productive Hebrew–Arabic translators of the Renaissance. His translations include works by Avempace (Ibn Bāǧǧa), Alfarabi (al-Fārābī), and Alhazen (Ibn al-Hayṯam), but the majority of them were concerned with Averroes and his works on logic. He shared this interest in logic with his patron Cardinal Domenico Grimani, whom Balmes served as personal physician. Grimani commissioned and financed the printing of seven Averroes translations by Balmes in 1523. Balmes's dedication to Grimani (who, three decades before, had been the patron of Elia del Medigo) is remarkable for its rejoinder to the anti-Arabic polemics of his time. Balmes answers with a cultural history of philosophy, which is based on Eusebius. The inventors of philosophy were barbarians, not the Greeks, Balmes claims, and Averroes is their worthy successor, who is a greater expert on Aristotle than the Greek commentators and fully to be trusted in matters of logic.[16]

Abraham de Balmes, *Divo Dominico Grimano* (Venice, 1523), sig. AA2r–v and sig. AA3r, dedicatory letter to Domenico Grimani concerning his Hebrew–Latin translation of Averroes.

extracts from *Notes* (GS12a), *Definitions* (GL6), *Answers to Ten Questions* (GP9), and *The Divisions of Philosophy* (GS1). Cf. d'Alverny, 'Andrea Alpago interprète', 2. The classificatory numbers refer to Gutas, *Avicenna and the Aristotelian Tradition*, second edition.

16 On Abraham de Balmes, see Giuliano Tamani, 'Le traduzioni ebraico-latine di Abraham', 613–635; and Hasse, *Success and Suppression*, 89–91.

Often, my most esteemed Sir, I have contemplated the fact that our hard work, which we have produced with your support and dedicated to your renowned name, will amount to nothing if it is not useful for the many people studying the arts. For this reason, we have tried with full zeal, with our own effort, and with the help of friends to have the money needed for the printing of a portion of our translations and editions, since my means do not suffice for ⟨printing⟩ everything at once. [...]

But perhaps a vain adversary might say: what have we to do with Averroes, a barbarian, given that our Peripatetic philosophy originated with the Greeks, not with the barbarians? These people should read, I beg, the tenth ⟨book⟩ of the *Preparation of the Gospel* by Eusebius [of Caesarea, third century AD], who says in the second chapter that all those among the Greeks who truly philosophized are recent ⟨thinkers⟩. [...]

Then, more than 600 years after the Trojan times, there flourished among the Greeks those who were called 'the wise', who invented nothing else but some short and pleasant maxims for the benefit of human life.[17] Then, after a long interval, it is reported that the genus of philosophers came into being—for Pythagoras was the first of all those who clearly invented also the term 'philosophy'—who only learned from the barbarians, but nothing from the Greeks, as someone dared to say. We therefore have to acknowledge that the entire philosophy of the Greeks originated with the barbarians. [...]

Given this, let us come to Aristotle's philosophy, which we cannot call 'correct' but rather 'corrupt' among the Greeks, for the reason that the descendants of Neleus, ⟨son of Coriscus,⟩ concealed the books of Aristotle and Theophrastus in some trench and finally handed them over, ruined by worms and humidity, to Apellicon of Teos [d. c. 84 BC]. Apellicon was more a lover of books than of wisdom and let the books be transcribed, wishing to amend the ruined passages, but the text was not added correctly and he produced the books replete with errors, as Strabo relates in book 13 ⟨chapter 1.54⟩. Who therefore will be more capable than Averroes properly to correct the books in accordance with Aristotle's intention? Averroes, as we have often read, wrenched his ⟨correct⟩ text out of a bad translation and wrote a commentary that conforms to the meaning of the Greek manuscript, which is why we think that in remedying the defects of Aristotle's books one should trust more the one Averroes than all Greeks together. But let us leave this aside and return to those worthless people who, blown up with flatulence rather than philosophy, polemicize against

17 The reference here is to the traditional seven sages of Greece, already referred to in Plato's *Protagoras* (342e–343a), Thales of Miletus, Pittacus of Mytilene, Bias of Priene, Solon of Athens, Cleobulus of Lindus, Myson of Chenae, and Chilon of Sparta.

our Averroes in order to boast about speaking Greek, saying that he did not understand Aristotle's intention because his doctrine often diverges from the Greek.

To these people one should answer that they may well correct their corrupt and ruined Greek manuscripts and align them with Averroes's interpretation, or they may restore for us a speaking and living Aristotle, who informs us about whether the emendations of Averroes are closer to the truth or those of the Greeks. But let us grant that Averroes did not agree with Aristotle, since they believe they have to censure him in such a way; ⟨still,⟩ Averroes needs to be praised and revered highly, because his doctrine, which they do not believe to be Aristotle's, is closer to truth and reason than the doctrine that these people claim and declare to be Aristotle's. For this reason, it would be more honourable for them to learn Averroes's doctrine from his good followers than shamefully to tear him to pieces with venomous teeth.

Let them be happy with the little wind they make about him; we shall imitate Averroes and believe in most of what he says, and in particular on logic, where there is no religious problem to be feared. We therefore offer these books, which are dedicated to you, divine Grimani, to our descendants, so that they realize that they have these books because of you, since we have published them upon your commission, dedicated them to your name, and published them with your support. For they have in this volume Averroes's *Long Commentary on the Posterior Analytics* of Aristotle, which had not been known among the Latins ⟨and⟩ which we have recently translated according to the intention of Averroes, adding separate comments. [...]

Let no one believe that we have written this to exhibit ornate eloquence and a pretence at oratorical skill; rather we have translated it by employing our commonly used words and the forms that appear everywhere, since I prefer to be criticized for the misuse of rhetoric rather than for changing the meaning or the wording of the author; first, because from my very first years when I was immersed in my Hebrew letters in my Talmudic schools, where truth was given priority over eloquence, which was despised, and secondly but more importantly, because the difference between languages usually gives rise to frequent breaches of the norms of speech. For the Romans thought that witty eloquence always deserves to be honoured first, the Hebrews, truth.

Text 6. Calo Calonymos ben David

Calo Calonymos ben David of Naples came from a Jewish family of southern Italy, but moved to Venice later in life, in 1517, where he continued to practise as a much-esteemed physician. Two Hebrew–Latin Averroes translations by Calo were printed in

Venice in 1527 and dedicated to Alberto Pio, prince of Carpi, a humanist Aristotelian who had been a student of the Paduan philosopher Pietro Pomponazzi. In his dedication, Calo emphasizes the great potential of Hebrew for correcting the Latin translations of Averroes. The second dedication accompanies the 1531 printing of Calo's translation of *On Astronomy* by Alpetragius (al-Biṭrūǧī), which was meant to replace the medieval translation by Michael Scot. Calo praises this work for its Aristotelian and anti-Ptolemaic astronomy, which avoids the physical absurdities, as he sees it, of Ptolemy's mathematical astronomy.[18]

Text 6.1. Calo Calonymos, *Illustri domino Alberto Pio* (Venice, 1527), verso of title-page, dedicatory letter to Alberto Pio, prince of Carpi, concerning his Hebrew–Latin translation of Averroes.

Excellent gentleman, ⟨Alberto Pio⟩! Averroes, the prince of the commentators, in different phases of his life composed three sorts of commentaries, i.e., long commentaries, paraphrases, and epitomes, on many logical, physical, metaphysical, and ethical volumes of Aristotle, the light of philosophy. In old age, Averroes did not cease to compose very many little books and letters in which he corrected himself with regard to many questions he had already addressed in those commentaries, and in which he expressed himself more clearly where in these texts he had proceeded inadequately. Almost all these books can be found among the Jews, and, in fact, in correct condition—not corrupted like most books ⟨by Averroes⟩ among the Latins—so that the intention of Averroes with reference to the intention of Aristotle in all his above-mentioned works is ascertained perfectly.

Even though I do not have the capacity to translate all those little books into Latin, which exceed my powers in quantity and quality, since I am impeded by my work as a physician in this famous city of Venice, I nevertheless decided to translate now into Latin the most refined little book or letter by Averroes, *About the Connection of the Abstract Intellect with the Human Being*, add it to our translation of Averroes's *Incoherence of the Incoherence* of the philosophy of Algazel[19] and to our own book *On the Creation of the World Proved with Physical Reasons*, and to print everything together. This is why, since I often consider

18 On Calo Calonymos, see Di Donato, 'Traduttori di Averroè', 38–40.

19 These two texts are: *Treatise on the Contact of the Separate Intellect with Man* (*Maqāla fī ttiṣāl al-ʿaql al-mufāriq bi-l-insān*), not extant in Arabic (no. 43.2), and the famous *Tahāfut al-Tahāfut* (no. 56). The classificatory numbers refer to Endress, 'Averrois Opera', 339–381.

in my mind that your famous Excellence flourishes in all genres of literature and especially in Peripatetic philosophy, that I dedicate this very useful little book to you as evidence of our service, which is growing every day, if God wills.

Text 6.2. Calo Calonymos, *Reverendissimo domino* [...] *Ioanni Mattheo de Gibertis episcopo* (Venice, 1531), sig. aa1v, dedicatory letter to Gian Matteo Giberti, bishop of Verona, concerning his Hebrew–Latin translation of Alpetragius.

When I fully devoted myself to ⟨studying⟩ the theory of Ptolemy for many years, I confess to immortal God that I perceived in this theory so many and such important points of conflict with physical reasons that I never felt satisfied and hence I almost retreated from it. In fact, some, even many, avoid the study of astronomy, because they see that in heaven there are things of such absurdity that nobody can observe.

This is why, awakened by a divine counsel, I began to study this little work together with Elia, my beloved son-in-law and a very intelligent man, and I was satisfied as I revisited from the top ⟨of the heavens⟩, as it were, the physical causes of the diversities in the heavenly spheres. And if the book I had in hand was faulty ⟨in any place⟩, I corrected it with another exemplar, paying much attention especially to the shape of the letters. Hence, intellectually weak though I might be, I published this little work and translated it into Latin with the same number of words as lucidly and plainly as I could, so that people will be able to reach this knowledge of the movements of the spheres in harmony with physical science, as a result of which the science of astronomy will be elucidated by being proven with true demonstrations. The astronomy of Ptolemy, however, lacks both demonstrations through effect [*quia*] and demonstrations through cause [*propter quid*], as Averroes testifies in the proem to Ptolemy's *Almagest*, because according to this doctrine astronomy itself can be demonstrated in the other liberal arts.

Even though this little book was transmitted to the Latins before, that previous translation [i.e., by Michael Scot] is so obscure that hardly anything good can be extracted from it, and occasionally the meaning of the author is completely eliminated, which is something you rarely hear about.

Text 7. Jacob Mantino

Jacob Mantino (d. 1549) stands out among all Hebrew–Latin translators in the Renaissance for his accomplished and moderately humanist Latin, for the high social standing he achieved in Bologna, Venice, and Rome, and for the appreciation he enjoyed among his contemporaries, such as the Giunta editors, who, after his death, praised the 'golden' translations of this 'most learned' scholar.[20] Mantino was a physician much respected among the high clergy, noble families, and ambassadors, and served as personal physician to Pope Paul III from 1533 onwards. The dedications of some of his translations to high-ranking patrons demonstrate his humanist motivation and high esteem for Averroes and Avicenna. Mantino and his patrons, Bishop Ercole Gonzaga and Doge Andrea Gritti of Venice, had studied in Padua or Bologna, where Mantino expected to find readers for his translations.[21]

Text 7.1. Jacob Mantino, *Leoni decimo pontifici maximo* (Rome, 1521), sig. A1v, dedicatory letter to Pope Leo X concerning his Hebrew–Latin translation of Averroes.

Many years later [after Aristotle], Averroes of Cordoba, who acquired the title 'commentator' with full right, expounded, as he had done with other books of Aristotle, the books entitled *On the Parts* and *On the Generation of Animals* in a brief, but clearly divine paraphrase, in which Averroes defends most vehemently the teaching of Aristotle against Galen, Avicenna, and many others. When I recently encountered this paraphrase, which had long been desired by the Latins, written in our, i.e., Hebrew, characters, I was touched by incredible joy. And I did not stop until I had rendered it, as best as I can, in Latin speech, convinced that what I have produced will be most welcome to those who teach the science dealing with nature. Because just as it is difficult to acquire knowledge of the natural world without Aristotle, likewise it is not recommendable, at least in my opinion, to teach Aristotle without Averroes.

Text 7.2. Jacob Mantino, *Herculi Consagae electo Mantuano domino* (1523, ed. Kaufmann, 222), dedicatory letter to Ercole Gonzaga, bishop of Mantua, concerning his Hebrew–Latin translation of Averroes.

There exist almost innumerable, as it were, interpreters and commentators of Aristotle, among whom only one, Averroes of Cordoba, a Muslim, comes closest to the sentences of Aristotle himself, as nobody in fact can doubt. [...]

20 Aristotle and Averroes, *Omnia* [...] *opera* (Venice, 1562), I, 2, f.319r: 'Hucusque doctissimi Mantini candide lector aurea super hoc primo Posteriorum pervenit translatio. Cetera vero morte preventus perficere haud potuit.'

21 On Jacob Mantino, the most informative study is still that by Kaufmann, 'Jacob Mantino', 30–60, 207–229. Cf. Hasse, *Success and Suppression*, 79–80.

And although in the books we are going to translate and in those we have translated already we do not claim for ourselves eloquence in Latin—for I admit that I have not attained it—we will nevertheless not imitate that translation [of the *Long Commentary on the Metaphysics*] that was produced for the Latins long ago [by Michael Scot] in an ugly and barbarous way, but we will, to the best of our ability, attempt to render the sentences of the author intact and intelligible. Therefore, I have decided to send this *Epitome of the Metaphysics* by Averroes to the printer, since much appears unkempt and mutilated, because of a distorted translation, in Averroes's *Long Commentary* which the Latins have—this, in fact, is a common feature of all old translations of Averroes and is the reason why many today condemn the teaching of Averroes.

Text 7.3. Jacob Mantino, *Andree Griti serenissimo ac excellentissimo Venetiarum principi* (Venice, 1530), sig. A1v–A2r, dedicatory letter to doge Andrea Gritti of Venice concerning his Hebrew–Latin translation of Avicenna's *Canon*.

But because Avicenna in writing had used a pagan and idiosyncratic Arabic language, which is not that easily acquired by Latin people, the translation of his works abounds in many and great errors, which Andrea of Belluno, the famous physician of our time who is equally learned in the Arabic and Latin languages, has laudably emended for the most part. He was not, however, able to purify the field entirely from foreign seeds; rather, many remain, which cloud the truth of reading as with some fog.

This is why I decided—because I always believed that nothing is more appropriate for humans than to contribute benefit to all mortals in any way whatsoever—to turn now to some commentaries by Avicenna, just as I rendered many ⟨texts⟩ in various disciplines from Hebrew into Latin, translate them into Latin purged and, as far as possible, fully cleaned from all blemishes, and offer them to readers as manifest evidence of my mind. When people compare my translation with those of others, they will see for themselves what they owe to me.

Because three parts especially of Avicenna's ⟨*Canon*⟩ are publicly lectured upon at the universities—namely the first part of the first book, which is called the first fen of the first book [on the definition and principles of medicine], the fourth part, which is named the fourth fen of the first book [on treatment methods], and the first part of the fourth book, which is entitled the first fen of the fourth book [on fevers]—I proposed to render all these ⟨parts⟩ into Latin speech and started with the fourth part of the first fen, since this part seems to be of greater usefulness for the general art of healing than the other parts.

Text 8. Johannes Carion

Johannes Carion (d. 1537), an influential German Protestant historian, represents an evolving historical interest in translation movements, and in Arabic–Latin translations in particular. His attribution of the *Almagest* translation to the circle of Frederick II Hohenstaufen, however, is not correct historically: this translation was made by Gerard of Cremona in twelfth-century Toledo.

Johannes Carion, *Chronica* (Wittenberg, 1533), s.v. 'Fridericus der ander, der xxiiii. Deudsche Keisar'.

This laudable emperor [i.e., Frederick II Hohenstaufen] I cannot lament enough, gifted as he was with so many virtues but afflicted so much by the popes that it is appalling. He knew many languages, Latin, German, Greek, and Saracen [i.e., Arabic], and fostered the arts. He had the *Almagest* of Ptolemy transferred for the first time from the Saracen language into Latin and thus renewed the fine art of astronomy, which nobody in the whole of Europe had been learning for a long time.

Text 9. Jean Bruyerin Champier

In 1537, the French physician and humanist Jean Bruyerin Champier published the allegedly 'first Latin version' of Averroes's medical magnum opus *Colliget* (*Kitāb al-Kullīyyāt fī l-ṭibb, Book on the Generalities of Medicine*), which in fact is a humanist revision of the medieval Latin translation by Bonacosa, produced in 1285. Bruyerin Champier was still alive in 1560, when he dedicated his own *De re cibaria libri XXII* to Michel de l'Hôpital, who had just been appointed chancellor of France. The Averroes edition of 1537 is accompanied by an eight-page prefatory epistle, written in the house of Bruyerin's uncle Symphorien Champier, and is addressed to two French physicians. Here Bruyerin not only justifies his 'translation' as being of benefit to medicine, but also explains, with remarkable accuracy, the historical migration of books and scientific knowledge, via translations, from Greek to Arabic and from Arabic to Latin. He also comments on the advantages of the work on medicine by Averroes, who succeeded in harmonizing Galen with Aristotle.[22]

Jean Bruyerin Champier, *Praefatio* (Lyon, 1537), sig. A3r–A4v.

22 On Jean Bruyerin Champier, see Burnett, 'The Second Revelation', 194–195; and Hasse, *Success and Suppression*, 6 and 115–121.

When I reflected upon the fact that almost all the works of our prince Galen were translated into Latin, together with the works of later Greek authors, due to the steady effort of learned men, it came to my mind that I too could help medicine if I were to treat as diligently as possible the matter I have undertaken, which was either not tried by others or greatly corrupted and distorted. In this matter I have spent enormous labour, because the manuscripts that have been printed are so highly corrupt that I could not derive any, or hardly any, help from them. But some months ago, a very old manuscript fell into my hands which contained three or four sections of these *Collectanea*, and which went back to the time when the philosophy and the medicine of the Arabs and Moors entered France. ⟨In view of⟩ these works, I think at this point it is my duty to teach how the learning of these barbarous nations came to us.

When the great flourishing of learning collapsed in Athens and Gothic barbarity invaded the Roman empire, some Greek books of both the philosophers and the physicians migrated to the Arabs and Moors—and to the Spanish, too, who were held prisoners under the yoke of the Arabs and used their language and laws—especially the books of Aristotle, as well as those of Galen; even Plato seems to have been read by Avicenna. Thus, it happened that they [the Arabs] translated many volumes of both authors from Greek into their own language. For it is known that this nation [the Arabs] was most zealous in the study of the humanities, with the result that Averroes, Alfarabi, Avicenna, and innumerable others of the same period philosophized on the basis of these books, which they thought alone sufficed for the investigation of truth; and they wholly concentrated on this effort and poured all the force of their intellect into writing interpretations and explanations of both authors [i.e., Aristotle and Galen].

When Alfonso, a man with the greatest desire for the sciences and especially for mathematics, reigned in Spain at the time when the Moorish still held Betica [i.e., the southern province of Spain], it easily happened that, partly because of the vicinity, partly because of the frequent commerce between the people, books written in the language of Averroes and the other Moors were transported into the north-eastern province of Spain. There they were somehow rendered into Latin by some Spaniard or, since the schools of philosophy and medicine were already flourishing in Paris, they were transported from Spain to France and were brought to Paris, just as the monk John Scotus [Eriugena] had translated the works of Dionysius [i.e., Pseudo-Dionysius Areopagita] into Latin when Charles the Great was emperor [*sic*: when Charles the Bald was king, around AD 860].

Moreover, the Jews did not wish at all to be debarred from these doctrines. Of these Moses Aegyptus [i.e., Maimonides]—the most penetrating follower of

Aristotle, who philosophized at the time of Averroes—was, as far as I know, the first to embark on this enterprise. Afterwards, others imitated him and brought to light many works belonging to every kind of teaching. [… Bruyerin Champier mentions that he was motivated by Jérôme Monteux to revise Averroes's *Colliget*.]

For ⟨Jérôme Monteux⟩ very often and frankly admitted that he had made more progress by reading our Averroes—in general, the greatest savourer of Galen's books—than by reading any of the late Greeks other ⟨than Galen⟩. This statement can be boldly embraced by anybody who has most diligently studied the commentaries of all of them. For he will see clearly that Averroes did not write with the purpose of censuring Galen, to whom he is so addicted and whom he praises so excellently, or defaming him, but that he was working with full zeal and in every way to free Aristotle—to whom he also dedicated himself very much—if possible, from the stain of significant error, and to harmonize one with the other [i.e., Galen with Aristotle].

Text 10. Alban Thorer

Alban Thorer (1489–1550) was a Swiss humanist and physician, professor of Latin and rhetoric at the University of Basel, and a Latin–German translator of Andreas Vesalius's anatomy. In 1544, Thorer published an edition of nine medical works, among them the *Liber ad Almansorem* by Rhazes (Abū Bakr ibn Zakariyā' al-Rāzī), which, rather than a translation proper, is a humanist revision of Gerard of Cremona's twelfth-century translation. The title-page and the preface comment programmatically on the value of Arabic medicine and on the deficiencies, as Thorer saw it, of medieval Arabic–Latin translations.

Alban Thorer, *Illustrissimo* […] *abbati* (Basel, 1544), title-page and preface to his humanist revision of Rhazes, sig. A2r.

Title-page

These are the works of the supreme physician Abubeter Rhazes Maomethus, ⟨which are⟩ rather excellent because of the great amount of practice and experience and because of the most certain information, drawn from logical demonstrations, on all non-natural conditions, and also because of the richest collection of medicaments. Nothing is more useful than these ⟨books⟩ for practice, because he explains everything obscure in Hippocrates or diffuse in Galen in the most faithful and learned way and illuminates it.

They have been translated into Latin by Gerard of Toledo, physician of Cremona, by Andreas Vesalius of Brussels, and by Alban Thorer of Winterthur, and have now for the first time been most rigorously and diligently collated with an old manuscript and ⟨thus⟩ restored, so that they can be understood by students of medicine. Nothing could be given to them that is more salubrious for the use of miserable mortal beings against so many kinds of illnesses.

Preface

That the Arabs seem to be, and in fact are, less pleasing and more confused and turbid than the Greeks, one should attribute less to the ⟨Arab⟩ authors themselves, whose language in general is considered inelegant, ungraceful, and rude, than to their most incompetent translators, who make everything terrible through unkempt style, false and obscure Arabic and barbaric terms, and utmost inelegance. Without doubt, if these authors were available in texts of their own language, and if this language were as known as Greek and Latin are, they would not cause such a nausea in most readers.

Text 11. Miguel Jerónimo Ledesma

Miguel Jerónimo Ledesma (d. 1547), professor of medicine in Valencia and one of the protagonists of humanism in Spain, published in 1547 a 'translation' of Avicenna's *Canon medicinae*, book 1, fen 1, which, in fact, is a much-abbreviated stylistic revision of Gerard of Cremona's translation of the *Canon*. In the dedication to Thomas de Villanueva, archbishop of Valencia, Ledesma offers a motivation for his work, claiming that Avicenna's medicine, which he finds to be in total accordance with Galen, was seriously distorted by 'barbarous' medieval translators and commentators.[23]

Miguel Jerónimo Ledesma, *Epistola nuncupatoria* (Valencia, 1547/1548), sig. A2v–A3r, preface to humanist revision of Avicenna.

Using a unique order and always behaving as a translator of Galen, Avicenna attempts in a very learned way to fulfil all duties of the best physician. But partly because he was ignorant of Greek and Latin, and partly because he was using incorrectly translated ⟨texts by⟩ Paul of Aegina and Galen, as will be apparent to whoever compares them with Avicenna, nobody should be astonished at the fact that Avicenna had occasionally made mistakes when recounting simple medicines, in the method of treatment, as well as in other

23 On Miguel Jerónimo Ledesma, see Ballester, 'The Circulation', 183–199; Siraisi, *Avicenna in Renaissance Italy*, 138–139; and Hasse, *Success and Suppression*, 99.

things that pertain to teaching the nature of medical subjects and principles. In addition to all this, the very deplorable calamity befell Avicenna that he had a barbarous translator and even more barbarous commentators. Hence, if someone sets out to consider the issue thoroughly and without any prejudice, he will realize that because of the corrupt translation of Galen, the lapse of time [i.e., from Avicenna until today], and the barbarism and foolishness of the commentators inevitably millions of mistakes affect the author himself.

In commiseration, I made a great effort to restore to the Arabic truth, if not the entire Avicenna, at least this one of his many books. It is not worth the trouble to notify where things are discarded, added, or changed, because everyone can use his eyes to examine the details at his own discretion. However, I do not want to leave unmentioned the following, that there is not one passage in Avicenna that we did not either confirm through the statement by Galen from where it is taken, or reject through a disagreeing statement by Galen. For this work, our very ancient Avicenna codex was at hand, a manuscript that deviates much from the vulgate ⟨text⟩. Likewise, at times even the recent translator Andrea ⟨Alpago⟩ of Belluno, rather than following the truth, actually follows the interpretation of Gentile ⟨da Foligno⟩ or Niccolò ⟨Bertruccio?⟩, or some other ⟨Latin commentator⟩. In addition to our own extraordinary efforts in discovering the peculiarities of languages, an associate was consulted who is skilled equally in the Arabic language and in medicine.

Text 12. Jacob Milich

Jacob Milich (1501–1559), the first anatomy professor of Wittenberg University, includes a succinct history of the Greek–Arabic–Latin transmission of medicine in his 'Speech about the Life of Avicenna', which encourages young people to study medicine.

Jacob Milich, *Oratio de Avicennae vita* (Wittenberg, 1550), sig. B7v.

Even though after the expulsion of the Greek language many ancient authors were also expelled, some people thirsty for knowledge were nevertheless looking for sources. For this reason, many writings of Hippocrates, Galen, and Ptolemy were translated by the Saracens into the Arabic language, from which not much later they were translated into Latin with the help of emperors Lothar and Frederick II.

Text 13. Giunta editors

The eleven-volume edition of the combined Aristotle and Averroes, which appeared in 1550–1552 at the Giunta press in Venice, counts among the most ambitious editorial projects in science and philosophy of the Renaissance. The edition was reprinted three times, with some modifications and additions: in 1560, 1562, and 1574/75. The first edition is accompanied by a long prefatory fascicle written by one of the editors, Marco degli Oddi, a teacher of logic at Padua University. From this preface we learn that Giovanni Battista Bagolino of Verona, himself the son of a Paduan logic professor, was the driving force behind the edition and did most of the editorial and philological work in preparation, but he died too early to see the volumes in print. Bagolino is praised in particular for his care in correcting medieval translations and for collecting all the Renaissance Hebrew–Latin translations of Averroes that he could find. The dedicatory letter by Tommaso Giunta contains a programmatic accolade of Averroes as a commentator superior to his Greek predecessors.[24]

Giunta editors, *Aristotelis Stagiritae* [...] *opera* (Venice, 1550/1552), title-page and I, f.2v; English translation from Burnett, 'Revisiting', 55 and 58.

Title-page

All the commentaries of Averroes of Cordoba on these works [i.e., of Aristotle] that have come down to us, and other books of his on logic, philosophy and medicine, of which some too, having escaped the notice of the Latins, have recently been translated by Jacob Mantino; others have been translated by the same scholar in a clearer and more faithful way than ever before, and the rest have been most diligently corrected in almost innumerable places from the manuscripts and the best printed books of the most celebrated philosophers of this time of ours, each having been adorned with a large number of marginal notes.

Dedicatory Letter by Tommaso Giunta

When Aristotle dealt with principles, methods, and general things in such a way that he left many things to be inspected and investigated more carefully by others, the Greeks made little—or rather no—effort in doing this. But the Arabs,

24 On the Giunta edition, see the classic article by Schmitt, 'Renaissance Averroism'; and Burnett, 'Revisiting'.

not content with mere translations, thought that the whole subject matter—i.e. the things themselves which had to be dealt with—should be investigated by them more carefully and fully. In this Averroes especially can be praised. His most solid teaching is not so much drawn from, as squeezed out of, the water-springs of the Greeks. He shone out so much that he alone rightly has claimed the name of 'Commentator' for himself. And now it should be clear amongst everybody who has practised philosophy in recent centuries that those parts of philosophy which had been omitted by Aristotle have been investigated more carefully by no other person, and no one has established them on more solid foundations.

Text 14. Antonius Stupa

Antonius Stupa was a Swiss humanist (d. 1551) who worked as a corrector at the Petri publishing house in Basel. He did not produce a translation proper, but a revision of a major work of Arabic astrology: the *De iudiciis astrorum* by Haly filius Abenragel ('Alī ibn abī l-Riǧāl). This text had been translated in the thirteenth century into Old Castilian and from Old Castilian into Latin by two Italian notaries at the court of King Alfonso x. Stupa's work was motivated by his disgust about the 'Spanish, French, and Italian phrases' in this medieval version.[25]

Antonius Stupa, *Epistola nuncupatoria* (Basel, 1551), sig. a3r–v, dedicatory letter to Haly filius Abenragel; English translation from Burnett, 'The Second Revelation', 190–191, slightly adapted.

Abenragel was an Arab, as has just been pointed out, and wrote in Arabic, and Jehuda son of Musca translated the work from Arabic into Spanish (as is clear from the translators' preface); and from this afterwards Egidio of Parma with Petrus Regius translated it into Latin, so that Spanish, French, and Italian phrases, rather than the meanings and expressions of the Latin language, were retained everywhere (something that was tolerated in those times). For what Latin writer has ever said 'charista baldi' for 'caritas annonae' or 'frumenti'? This author abounds in words like this, as anyone who compares our edition with the translation in the Venetian edition [first printed 1485] will easily discover. We do not make the comparison out of a desire to dash out of the hands of scholars that translation and replace it by our own, but so that they can agree to our purging and grant that a style that is barbarous and more French and

25 On Antonius Stupa, see Jenny, 'Antonius Stuppa', 49–83.

Spanish than Latin (as will be shown in our arguments) should be transcribed into the better diction of the Latin language, changed and made more pure, presuming that they have some desire for the purity of this diction.

Text 15. Girolamo Donzellini

A final example of the growing historical awareness for translation movements comes from the Italian physician Girolamo Donzellini (d. 1587), who served as the editor of the Basel 1564 edition of Rhazes's *Liber ad Almansorem*, book IX, together with Leonardo Giacchini's commentary. In a long preface of eighteen pages, Donzellini, in truly humanist manner, elaborates on the nobility of medicine and includes a short history of the discipline, which is clearly prejudiced against the merits of Arabic and medieval medicine.

Girolamo Donzellini, *Praefatio* (Basel, 1564), sig. B4r–v, preface to edition of Rhazes.

When the science of medicine was transported from the Greeks to the Arabs, it was shipwrecked, and when the Latins received it from the Arabs, they were very unproductively involved in it for a long time. God, having finally mercy on our fate, brought the sciences back to light, together with the competence in languages, and also illuminated this divine science [i.e., medicine]: a number of men were awakened who taught the science from the clear sources of the Greek. For when the Ottoman fury began to mistreat Greece with arms, the most learned scholars of the Greeks, unable to bear this barbarism, came to Italy, bringing with them treasures of books. They were received most respectfully by the family of the Medici and brought light to all disciplines that were for the most part corrupt. Since that time medicine began to flourish.

Part Three: Original Texts

Text 1. Girolamo Ramusio, *Ad Flavium*, Paris, BNF, MS arabe 2897, ff.160v and 349r.

f.160v

Volui prius videre que in gymnasio Patavino publice leguntur.

f.349r

Ad Flavium Ramusius. Inter scribendum arabice multa vidi que latino Abuali omnino repugnant, multa annotavi que latinus dimisit aut incorrepta scripsit, multa perlegi que non ita ad unguem latina facere potui ob imperitiam arabice lingue, quare in dies adiscens in dies melius corrigo meque ac scripta mea castigo; dum igitur hunc primum Abuali arabice scripsi nullum ordinem fere servavi, nam ad utiliora prius semper accessi posteriora relinquens minus utilia, et ut ordinem reperias, Flavi optime, aduerte: primo vidi usque ad anathomiam 1, de pulsu, urina et egestione a charta 50, quarta primi a charta 107, doctrina sexta de virtutibus a charta 160, fen secunda a charta 174, fen tertia a charta 272. Reliquum quarte primi et de anathomia invenies latinum meum correctum, nec admireris si omnia non ita dilligenter castigata sint, cum sensim omnia perceperim, et ut citius adiscerem arabice scripsi et litteris omnia labefeci. Spero inde futuro anno omnia repetere et dilligentissime omni studio perscrutari et in optimam formam redigere, quare diis magnis et Catte gratias ago etc.

Text 2. Elia del Medigo, *Domino Dominico Grimani*; ed. Kieszkowski, 76–77.

Sepe philosophando, ut tui moris est, dixisti mihi de ordine librorum methaphysice, quis debet esse primus, an liber incipiens de veritate, sicut invenitur communiter in libris antiquis, an sequens, sicut invenitur in quibusdam novis. Dicebas enim quod ille liber qui communiter est primus est tamquam prohemium [...] totius libri metaphysice, et maxime secundum translationem arabicam, in ipsa enim non est primus, qui invenitur apud latinos. Et quia bene ratiocinatus fuisti, volui transducere (tibi) prohemium Commentatoris, quod fecit in XIIo Metaphysice, in quo ponit ordinem librorum et multa bona. Et quamvis alias transduxi dignissimo domino Johanni comiti Mirandolano, tamen illam transductionem non habeo, et forte in nulla sententia variatur. Multotiens etiam tractabat magnificentia vestra de universalibus, dicens ipsa non esse extra animam, cui multi latini opponunt, quamvis in se sine dubio est verum. [...]

De hoc autem nolo ad presens prolongare, sed tamen volui transducere dicta Commentatoris in hoc in loco allegato, ex quibus verbis apparet veritas huius. Et quamvis difficilia sint, et forte multi errabunt, tamen intelligenti sicut vos apparebit veritas, et maxime quia id quod est difficile declaratum est a me per verba Commentatoris in aliis locis, in questionibus de intellectu, loquendo de universali, et in aliis locis. Id etiam quod dicit in prohemio predicto multum facit in hoc, et id quod transduxi de sua Summa de hoc maxime. Quia ego scio quod ex te potes multa bona et difficilia cognoscere, in hoc autem non solum sententiam suam servavi, ymmo et verba meliore modo, quo fieri possit. Valete.

Text 3. Paolo Ricci, *Prefatio* (Milan, 1511), f.7r.

Nec solum eiuscemodi tria prooemia, sed universa in Averoi latina editio, ut quandoque aliquibus ex conphilosophis nostris patefeci, crebris corruptellis erroribusque abundat. Haec admonuisse volui ne quando aliqua huius prooemii et aliorum subsequentium que de hebreorum bibliotheca excerpsi, a latina editione discriminari conspexeris, in querulam et criminatoriam vocem prorumpas. Ubique enim hebrei et nonnulli arabice lingue adsunt eruditi, quos si veri explorator fueris consule pro viribusque examina, unde omnia quidem sedulo ac adamussim tradita esse dignosces.

Text 4. Andrea Alpago

Text 4.1. Collegio of the philosophers and physicians of the University of Padua, *Pro excellentissimo*; ed. Lucchetta, 88–89.

In Christi nomine amen. Anno a nativitate eiusdem millesimo quingentesimo vigesimo primo, indictione nona, die lune vigesimo primo ianuarii in ecclesia S. Urbani hora decimaseptima, ubi, convocato Sacro Collegio clarissimorum artium et medicine doctorum Padue mandato excellentis artium et medicine doctoris domini Vincentii Moscheta, prioris, et dominorum consiliariorum suorum, per prefatum dominum priorem dictum et exspositum [*sic*] fuit:

'Excellentissimi domini doctores, causa convocationis extraordinarie vestrorum est ista, quia est quidam preclarus artium et medicine doctor dominus magister Andreas de Cividale qui per plures annos stetit in partibus orientalibus et maxime in Damascho ubi se mirabiliter gesit [*sic*] circa medicinam et longo tempore habuit familiaritatem et comercium cum quodam excellenti medico arabo et sapientissimo in lingua araba, a quo multa didicit et sic effectus est tam preclarus quod correxit et translatavit certa loca Avicene, propter quam correctionem et translatationem multa loca ipsius Avicene, que antea erant obscura, reddita sunt clara et manifesta, ac etiam certos alios libros arabos in latinum translatavit. Certe temporibus nostris rem magis laudabilem et necessariam habere non poteramus quam correctionem et traductionem ipsam, cum ipsa correctione et traductione loca ipsa que per prius, ut supra dictum est, erant obscura et ambigua, facta sint clara et intelligibilia. Et profecto vir iste meretur magnam laudem et nostrum esset ipsum comendare penes clarissimos dominos Rectores nostros et suas magnificentias rogare ut talem virum excellentem comendare dignarentur serenissimo domino Duci nostro Venetiarum ad hoc, ut opera sua tam digna manifestarentur et haberet etiam causam prosequendi in re ipsa, videlicet in traslatando alia loca ipsius Avicene et alios libros ex arabo in latinum.'

Qua expositione facta, talis vir plurimum commendatus fuit per nonnullos celeberrimos doctores in ipso Collegio existentes et fuerunt super tali re facti diversi sermones, tandem prefatus dominus Prior posuit hunc partitum, videlicet quibus placet quod excellentissimi artium et medicine doctores domini magistri Nicolaus de Ianua, Ludovius Carensius, Hieronimus Bagolinus, et Iohanne Laurentius de Sassoferrato auctoritate huius Sacri Collegii videant opera dicti consumatissimi artium et medicine doctoris domini magistri Andree de Cividale et refferrant huic Sacro Collegio.

Text 4.2. Collegio of the philosophers and physicians of the University of Padua, *Pro domino magistro Andrea de Cividale*; ed. Lucchetta, 90.

Die iovis XXIIII ianuarii, in ecclesia S. Urbani, hora XXII.

Convocato iterum Sacro Collegio famosissimorum artium et medicinae doctorum Padue mandato ultrascripti domini prioris ultrascripti excellentes doctores, ultra ellecti ad videndum opera prefati domini magistri Andree de Cividale, rettulerunt et relationem suam fecerunt singulatim cum vidissent nonnulla opera eiusdem magistri Andree, translatata ex arabo in latinum: illaque sumopere comendarunt tamquam utilia et necessaria.

Qua relatione habita, iamdictus dominus Prior posuit hoc partitum, videlicet: attenta optima relatione facta per clarissimos artium et medicine doctores dominos magistros Nicolaum de Ianua, Ludovicum Carensum, Hieronimum Bagolinum et Iohannem Laurentium de Sassoferrato de optima translactione [*sic*] et correctione locorum in libris Avicene facta per excellentem artium et medicine doctorem dominum magistrum Andream de Cividale nec non etiam de translactione quorundam librorum arabum nondum editorum, quod per istud Sacrum Collegium elligantur infrascripti quinque excellentissimi doctores qui accedant ad magnificos dominos Rectores et eo modo et tali talem virum propter huiusmodi opera laudent et approbent quo deceat hoc Sacratissimum Collegium.

Text 4.3. Paolo Alpago, *Ter illustrissimo* [...] (Venice, 1546), *1^{v}–*2^{v}.

Paulus Alpagus Bellunensis phisicus, ter Illustrissimo ac Reverendissimo Domino Alexandro Cardinali Farnesio, domino suo colendissimo, foelicitatem.

Verterat, ter Illustrissime ac Reverendissime Domine, ex Arabica in Latinam linguam innumeris mendis vindicatum Andreas Alpagus Bellunensis patruus meus philosophus ac medicus insignis maximum de re medica Avicennae volumen; quanta foelicitate videre quam plurimi quorum prae manibus est, quantis laboribus, sumptu ac periculis vidi ego qui adolescens adhuc, dum is iam longevus in Cypro, Syria, Aegypto, ac toto pene oriente, Arabicae linguae latebras, et codicum fidem perquireret, ab eius latere ne transversum quidem unguem discessi, styli etiam ac lucubrationum socius. Id est a me aliquot ab hinc annis primo in lucem editum, secundo vero nuperrime aliquanto emendatius.

Verterat idem patruus meus eiusdem Avicennae libellos aureos scilicet Compendium de anima, De Mahad idest de dispositione animarum post ipsarum separationem a corporibus, Aphorismos de anima, De diffinitionibus et quaesitis, De divisionibus scientiarum. Haec iacuere apud me diutius, quam par erat quamque ego cupiebam, tenebris ac situ obducta, cum ego essem et lucubrando maiore illo volumine, et privatis negociis impeditus. [...]

Illuminare igitur iubar fulgentissimum his, qui in tenebris sunt; et quid Avicenna de anima sentiat, quamque cum Christianis conveniat, vide. Ac si in hoc gratificatos nos esse non nihil Illustrissimae ac Reverendissimae D⟨ominae⟩ tuae intelligemus, nonnulla alia id genus, tum eiusdem Avicennae, tum aliorum Arabum opuscula ab eodem patruo meo latinitate donata, digna meo iudicio, quae in Principum aulis degant, scilicet De venenis, De correctione errorum quae accidunt in regimine sanitatis, De medicinis principum non horribilibus, De lapidibus pretiosis, et pleraque alia, expecta.

Text 5. Abraham de Balmes, *Divo Dominico Grimano* (Venice, 1523), sig. AA2r–v and sig. AA3r.

Saepe mecum estimavi, reverendissime Domine, nostras lucubrationes, quas te favente aedidimus et tuo inclyto nomini inscripsimus, nihil fore nisi multitudini studiosae bonarum artium prodessent. Quam ob rem omni conatu et propria industria et amicorum adminiculis pro aliqua portione nostrarum traductionum aeditionumque imprimenda exponendas pecunias habere curavimus, cum simul ad omnes non suppetant mihi vires. [...]

Sed dicet forsan ventosus adversarius: quid nobis cum Averroy barbaro viro? Cum nostra perypathetica philosophia a graecis non a barbaris ipsis ortum habuerit? Legant hi quaesumus graeci viri Eusebii Pamphili decimum de evangelica praeparatione, qui illius secundo capite ait quod iuniores omnes sunt qui apud graecos recte aliquid philosophati sunt. [...]

Deinde annis post troiana tempora pluribus quam sexcentis illi qui sapientes apud graecos vocati sunt floruerunt, qui nihil aliud quam sententias quasdam commode ac breviter dictas ad utilitatem humanae vitae invenerunt. Magno deinde spatio temporis interposito philosophorum genus incepisse fertur, Pythagoras enim primus omnium fuit quem philosophiae quoque nomen invenisse constat, qui non nisi a barbaris didicit, a graecis autem nihil eum habuisse ausus quispiam est dicere? Fateri ergo oportet a barbaris graeciae philosophiam omnem ortam esse. [...]

Sed dato hoc ad Aristotelis philosophiam veniamus, quam non possumus dicere correctam potius quam corruptam esse apud graecos, eo quod Aristotelis et Theophrasti libros in fossa quadam Nelei posteri occuluerunt, quos et tineis et humiditate labefactos tandem Apeliconi Teio tradiderunt, qui cum magis librorum esset studiosus quam sapientiae, volens corrosiones emendare eos transcribendos dedit, scriptura non recte suppleta, quapropter libros edidit erroribus plenos, ut Strabo libro tertiodecimo refert. Quis ergo valebit illos rectius ad mentem Aristotelis castigare quam unus Averroys? Quem saepius legimus extorsisse suum textum falso traductum et expositionem scripsisse sententiae codicis graeci conformem, quare in perficiendo defectus librorum Aristotelis magis uni Averroy quam omnibus graecis credendum censemus. Sed missa isthaec faciamus, ad hos futiles viros redeamus qui potius ventositate quam philosophia inflati ut graecizare se ostentent in nostrum Averroym oblatrant, dicentes quod Aristotelis mentem non habuerit, quia a graeca littera sua sententia saepe recedit.

Quibus dicendum est quod suos graecos codices corruptos et labefactos castigent et ad Averroys mentem reducant, vel Aristo⟨telem⟩ loquentem vivum nobis restituant, qui certos nos faciat, an Averroys vel graecorum castigationes sint veriores? Sed esto quod Averroys non scripserit ad mentem Aristotelis quando sic ipsum vituperare putant, nimis admodum magnificare et venerari eum censemus, dum Averroys sententia, quam isti negant fuisse Aristo⟨telis⟩, veritati et rationi conformior cernitur quam sit illa quam Aristote⟨lis⟩ illi fuisse affirmant et praedicant. Qua de re honoratius esset eis Averroys doctrinam a suis bonis imitatoribus audire quam viperino dente illum turpiter in cassum lacerare.

Gaudeant ergo ventulo illi suo, nos Averroym imitemur, Averroy credamus in maiore parte suorum dictorum praecipue in logica, ubi nullum fidei scandalum est metuendum. Hos ergo dive Grimane libros ita nostris posteris tibi inscriptos tradimus, ut per te illos habuisse censeant, quando te iubente eos aedidimus, tuoque nomini inscripsimus et te favente publicos facimus; habebunt enim in hoc volumine nondum apud latinos visa magna Averroys commentaria super libros posteriorum analiticorum Aristotelis iuxta litterae seriem, quam etiam ad mentem Averroys noviter traduximus et commentis singulis apposuimus. [...]

Nec credat aliquis quod eloquentiae ornatu et orationis fucu ea scripserimus, sed nostris communibus usitatis verbis et modis passim venientibus ipsa tradimus, cum maluerim de abusu eloquentiae quam de mutata sententia aut serie authoris increpandum esse, tum quia mea prima ineunte aetate meis hebraeis litteris meis talmudisticis gymnasiis immissus sum, quibus veritas praeponitur spreta eloquentia, eo magis quia linguarum disparitas contrarias eloquii normas saepe parire solet. Romani enim dicacitatem, hebraei vero veritatem semper praehonorandam censuerunt.

Text 6. Calo Calonymos ben David

Text 6.1. Calo Calonymos, *Illustri domino Alberto Pio* (Venice, 1527), verso of title-page.

Averoes expositorum princeps, Excellens Domine, in plurimis Aristotelis philosophie luminis voluminibus et logicis et physicis et metaphysicis et moralibus tria genera commentariorum, magna scilicet commentaria, paraphrases et epithomata in diversis suis etatibus edidit. Nec destitit in senectute multos et multos edere libellos et epistolas quibus quidem se castigavit in multis quesitis de eis que in predictis commentariis dixerat ac se clarius elucidat si diminute in eis se gesserat. Que omnia fere volumina apud hebreos reperiuntur et correcta quidem, non autem corrupta ut plurima que apud latinos, ex quibus perfecte mens Averrois ad mentem Aristotelis in omnibus suis operibus iam dictis elicitur.

Et si mihi exercitio medendi in hac Inclyta Venetiarum urbe impedito non sit facultas libellos illos omnes excedentes quidem et quantitate et qualitate vires meas in latinum transferendi, decrevi tamen in presentiarum excultissimum libellum seu epistolam ipsius Averrois De connexione intellectus abstracti cum homine latinitati donare eumque annectere nostre traductioni Destructionis Averrois destructionum philosophie Algazelis ac libro nostro De creatione mundi physicis rationibus probata omnesque una impressioni tradere. Quare sepius animo mecum ipse volutans tuam Illustrem Dominationem omni litterarum genere presertim philosophie peripatetice florere, hunc iccirco libellum perutilem tibi dedico pro inditio nostre servitutis in dies domino concedente amplius augende.

Text 6.2. Calo Calonymos, *Reverendissimo domino* [...] *Ioanni Mattheo de Gibertis episcopo Veronensi* (Venice, 1531), sig. aa1v.

Et cum per multos annos sane Ptolemaei incubuerim theorice, immortalem testor deum mihi ipsi nunquam satisfacere valui perspiciens aequidem in ea tot et tanta phisicis adversantia rationibus, itaque me poene ab ea removi; quinimmo nonnulli ac etiam complures preterea astrologiae incumbere pretereunt, videntes in coelo tam absurda inesse que contemplari potest nemo.

Quare divino excitatus consilio huic studium adhibens opusculo una cum Elia genero delecto viro quidem acuto mihi ipsi satisfeci tanquam repetens ab alto phisicas diversitatum in orbibus inventarum causas. Et si mihi mendosus accomodatus fuit liber cum alio tamen exemplo ac nimia vigilia precipue litteralibus figuris eum correxi, itaque ingenii imbecillitate hoc opusculum in lucem edidi et id in latinum totidem verbis lucide et plane ut potui transtuli, ut homines ad hanc motuum orbium cognitionem cum phisice scientiae consensu pervenire possint ac valeant et sic astrologiae scientia veris probata demonstrationibus illustrabitur, illa vero Ptolemaei teste Averroe in sui Almagesti prohemio et demonstratione quia et demonstratione propter quid orbata est. Quoniam ergo iuxta eam doctrinam in caeteris liberalibus artibus astrologia ipsa concludi potest.

Et si hic libellus ad latinos iam antea traditus est, illa tamen translatio ita involuta est, ut vix ab ea aliquid boni evelli possit, quinimmo interdum autoris penitus aufertur sensus, quod auditu difficile est.

Text 7. Jacob Mantino

Text 7.1. Jacob Mantino, *Leoni decimo pontifici maximo* (Rome, 1521), sig. A1v.

Multis post annis Averrois Cordubensis (qui unus interpretis nomen iure merito est assequtus) ut caeteros Aristotelis libros ita et eos qui de partibus et generatione animalium inscribuntur explicavit, conscripta brevi quadam sed certe divina paraphrasi, in qua Aristotelis sententiam adversus Galenum, Avicennam et alios complures accerrime defendit. Hanc quum diu a Latinis desyderatam nuper offendissem nostris, hoc est Haebraicis, exaratam litteris, incredibili gaudio sum affectus. Nec prius destiti quam eam pro virili meo in latinam converterem orationem; facturum me arbitratus gratissimum iis qui scientiam, quae de natura est, profitentur. Nam ut rerum naturalium cognitionem sine Aristotele adipisci difficile est, ita Aristotelem sine Averroi profiteri, meo quidem iudicio, non est valdae probandum.

Text 7.2. Jacob Mantino, *Herculi Consagae electo Mantuano domino*; ed. Kaufmann, 222.

Huius [sc. Aristotelis] equidem expositores atque commentatores innumerabiles prope dixerim extant, inter quos unus tantum Averroys Cordubensis machometanus sententiis ipsius Aristotelis maxime accedere nullus prorsus potest ambigere. [...]

Et quamvis in his libris vertendis et iam conversis a nobis latinam eloquentiam non profiteamur, fateor enim me eam non esse assecutum, illam tamen traductionem quae pridem fede et barbare latinis data fuit atque obscure non imitabimur, sed pro viribus conabimur sententiam integram authori reddere et intelligibilem. Quapropter hoc epithoma metaphisice Averroys prelo transferre decrevi, cum in longa commentatione ipsius quam latini habent multa inculta et mutiliata [*sic*] appareant, propter depravatam traductionem, idque profecto est familiare omnibus priscis traductionibus Averroys fuitque causa ut multi hac etate doctrinam Averroys damnent.

Text 7.3. Jacob Mantino, *Andree Griti serenissimo ac excellentissimo Venetiarum principi* (Venice, 1530), sig. A1v–A2r.

Sed quoniam Avicenna in scribendo gentilitio ac sibi peculiari Arabico idiomate usus est, quod a latinis hominibus non ita facile comparatur, eius operum traductio maximis ac multis erroribus scatet, quos Andreas Belunensis, etatis nostrae medicus insignis et Arabica latinaque lingua pariter eruditus, magna ex parte laudabiliter emendaverit, alieno tamen semine campum penitus expurgare non potuit, sed adhuc plurima relicta sunt quae veluti nebula quadam veritatem lectionis obducant.

Quamobrem ego qui nihil magis proprium hominis esse semper existimaverim quam quoquo modo universis utilitatem conferre mortalibus, quemadmodum pleraque in variis facultatibus ex hebraico in latinum sermonem converti, ita nunc aliquos Avicennae commentarios aggressus, eos maculis omnibus emendatos, ac quoad fieri possit, absolutissimos latinitate donatos, tanquam evidens animi mei indicium legentibus offerre decrevi; qui postquam meam cum aliorum interpretationibus comparaverint, quid mihi debeant ipsi viderint.

Cum autem tres potissimum Avicennae partes in gymnasiis publice legantur, videlicet prima pars primi libri, quae prima fen primi dicitur, et quarta, quae quarta fen primi nuncupatur, ac prima pars quarti libri, quae prima fen quarti appellatur, ea omnia in latinam orationem vertere proponens a quarta fen primi libri interpretari exorsus sum quia haec pars maiorem caeteris in universali medendi arte utilitatem afferre videtur.

Text 8. Johannes Carion, *Chronica* (Wittenberg, 1533), s.v. 'Fridericus der ander, der xxiiii. Deudsche Keisar'.

Diesen loeblichen Keisar kan ich nicht gnug klagen, der mit so viel schoenen tuegenden begnadet gewesen, Noch haben ihm die Baepst so hart zu gesetzt, das zu erbarmen ist. Er hat viel sprachen gekoent, Latin, Deudsch, Grekisch, und Sarracenisch, Hat auch die kuensten gefurdert. Das Almagestum Ptolemei hat er erstlich aus Sarracenischer sprach in Latin bringen lassen, und dadurch die schoene kunst Astronomia, die inn gantzem Europa kein mensch lange zeit gelernet hat, widder auff bracht.

Text 9. Jean Bruyerin Champier, *Praefatio* (Lyon, 1537), sig. A3r–A4v.

Consideranti nanque mihi Galeni principis nostri universa pene opera Latinitate donata, simul ac Grecorum aliorum posteriorum, diligentia virorum doctorum, venit in mentem me quoque posse medicinam adiuvare si hanc quam aggredior materiam, aliis vel intentatam vel corruptissime depravatam, quam possem diligentissime tractarem. Laboravi equidem hac in re vehementer quoniam codices quos typis excusos habemus depravatissimi essent, unde aut nihil aut parum adiumenti ab iis capere potuimus. Verum ab hinc menses aliquot inciderat in manus nostras codex vetustissimus, qui tres quatuorve sectiones horum Collectaneorum complectebatur quique referebat eam tempestatem qua in Gallias immigrarunt Arabum atque Mauritanorum tum philosophia, tum medicina. Porro mei officii esse puto, eadem opera docere, quo modo harum gentium barbararum doctrina ad nos deuolarit.

Postquam itaque Athenarum florentissimum studium corruit, Romanumque imperium Gotthica barbaries inuasit, migrarunt ad Arabas Mauritanosque, adde et Hispanos (qui sub Arabum iugo vincti tenebantur atque eorum lingua ac legibus utebantur), Graeci aliquot tum philosophorum, tum medicorum libri, potissimumque Aristotelis, simul ac Galeni, (quinetiam Platonem legisse uidetur Auicenna). Quo factum est ut complura utriusque volumina e Graeco in suam linguam conuerterint, constat nanque eam gentem bonarum scientiarum fuisse studiosissimam, quamobrem Auerrhous, Alpharabius, Auicenna, aliique innumeri eodem seculo iis philosophati sunt libris, quos solos ueritati indagandae sufficere arbitrati sunt; in hocque negocium toti incumbentes, uimque omnem ingenii effundentes utriusque interpretationes atque expositiones susceperunt.

Cum vero in Hispania imperaret Alphonsus literarum, et maxime mathematicarum, sitientissimus, Mauris adhuc Bethicam obtinentibus, facile fuit ut partim ob uiciniam, partim ob frequens populorum commercium comportarentur in Hispaniam citeriorem Auerrhoi aliorumque Maurorum lingua conscripti libri, ubi ab Hispano quopiam Latinitate utcumque donati sint, aut florentibus iam in Lutecia Parisiorum philosophiae medicinaeque studiis ex Hispania in Gallias deportati illic transferri potuerunt, quemadmodum imperium tenente Carolo Magno Ioannes Scotus monachus Dionysii opera converterat in Latinum.

Praeterea Haebrei horum placitorum exortes minime esse voluerunt, quorum primus (quantum intelligo) Moses Aegyptius, acerrimus Aristotelis sectator et Auerrhoi seculo philosophatus, negocium tale est aggressus, quem postea alii imitati multa in lucem emiserunt in omni doctrinae genere monumenta. [...]

Frequentissime etenim ingenueque confessus est [i.e., Hieronymus Montuus], se plus per Auerrhoi nostri lectionem profecisse (alioqui Galeni operum maximus helluo) quam per ullum alium posteriorum Graecorum. Cuius sententiae audacter subscribere poterit is qui in omnium commentariis diligentissime sit uersatus. Quippe liquido perspiciet non eo animo scripsisse Auerrhoum ut Galenum (cui tantum sit addictus, quemque tam egregie laudet) taxaret calumniareturque, uerum id omni studio modisque omnibus satagere ut Aristotelem (cui etiam summopere fauebat), si fieri posset, ab erroris insignis nota liberaret huncque cum illo conciliaret.

Text 10. Alban Thorer, *Illustrissimo* [...] *abbati* (Basel, 1544), title-page and preface, sig. A2r.

Title-page

Abubetri Rhazae Maomethi, ob usum experientiamque multiplicem et ob certissimas ex demonstrationibus logicis indicationes ad omnes praeter naturam affectus, atque etiam propter remediorum uberrimam materiam, summi medici opera exquisitiora, quibus nihil utilius ad actus practicos extat, omnia enim penitus quae habet aut Hippocrates obscuriora aut Galenus fusiora, fidelissime doctissimeque exponit et in lucem profert.

Per Gerarduum Toletanum medicum Cremonensem, Andream Vesalium Bruxellensem, Albanum Torinum Vitoduranum latinitate donata, ac iam primum quam castigatissime ad uetustum codicem summo studio collata et restaurata, sic ut a medicinae candidatis intelligi possint. Quibus nihil prorsus salutarius in miserorum mortalium usum aduersus tot morborum species conferri potuit.

Preface

Quod vero minus quam Graeci iucundi et turbulentiores squalidique magis videntur et revera existunt Arabes, non tam autoribus ipsis, quorum idioma in universum inelegans, inconcinnum et horridulum habetur, quam ineptissimis ipsis interpretibus adscribendum autumo, qui incomptissimo charactere, vitiatis et obscuris dictionibus Arabicis et Barbaris, plurimaque inconcinnitate omnia horrificarunt. Et procul dubio, si tales autores sua lingua scripti extarent, et eadem nobis tam nota atque Graeca et Latina foret, non tantam plerisque lectoribus nauseam mouerent.

Text 11. Miguel Jerónimo Ledesma, *Epistola nuncupatoria* (Valencia, 1547/1548), sig. A2v–A3r.

Avicenna vero erudite admodum et singulari ordine usus et Galeni se usquequaque faciens interpretem omnes medici optimi numeros absolvere conatur. Caeterum quia tum Graecitatis et Latinitatis ignarus erat, tum quod perperam translatum Paulum et Galenum habebat, quod intelliget qui hosce illi contulerit, mirari nemo debet quod aliquando tum in tradendis simplicibus medicamentis, tum in medendi methodo atque etiam in ceteris que ad rerum et principiorum medicine naturam perdiscendam vergunt, fuerit lapsus. Porro preter hec omnia illud satis lachrymabile damnum accidit Avicennae, quod nactus est barbarum interpretem barbarioresque multo enarratores. Unde si quis rem penitius et sine ullo affectu velit perpendere, intelliget ex corrupta Galeni tralatione, ex temporis diuturnitate, ex enarratorum barbarie et ineptiis, non potuisse nisi errorum myriadas in autorem ipsum dimanare.

Quod ego miseratus si non totum Avicennam certe hunc ex tot libris unum ad Arabicam veritatem enixus sum emendare. Ubi quae abiecta, addita aut commutata sint, non est opere precium referre, singula nanque quivis oculis poterit pro arbitrio perlustrare. Illud tamen non tacebo nullum esse Avicenne locum quem vel Galeni dicto, ex quo desumptus est, non confirmemus, vel eiusdem sententia, cum ab illo dissentit, antiquemus. Cui labori praesto fuit vetustissimus noster codex Avicennicus manu scriptus longe a vulgato dissidens. Item Andreas Bellunensis novus interpres atqui is aliquando Gentilis aut Nicoli aut alterius cuiuspiam sententiam verius sequitur quam veritatem, quibus praeter peculiaria nostra in vestigandis linguarum proprietatibus studia, adde consultum fuisse socium Arabicae linguae non minus quam rei medicae peritum.

Text 12. Jacob Milich, *Oratio de Avicennae vita* (Wittenberg, 1550), sig. B7v.

E⟨t⟩si enim pulsa lingua graeca, veteres autores etiam exulabant, tamen aliqui studiosiores requirebant fontes. Quare et in linguam Arabicam pleraque Hippocratis Galeni et Ptolemei opera a Sarracenis in linguam Arabicam sunt conversa, ex qua lingua non multo post nostrorum imperatorum beneficio Lotharii ac Friderici secundi in latinam linguam translata sunt.

Text 13. Giunta editors, *Aristotelis Stagiritae* [...] *opera* (Venice, 1550/1552), title-page and 1, f.2v (dedicatory letter).

Title-page

Averrois Cordubensis in ea opera omnes qui ad nos pervenere commentarii, aliique ipsius in logica, philosophia, et medicina libri: Quorum aliqui non amplius a Latinis visi, nuper a Iacob Mantino sunt conversi: Alii ab eodem clarius ac fidelius, quam unquam antea ab aliis, translati: Caeteri ex manuscriptis, optimisque codicibus Philosophorum hac nostra aetate celeberrimorum, innumeris pene locis diligentissime castigati: Singuli compluribus margineis scholiis exornati.

Dedicatory Letter

Sed cum Aristoteles principia, modos et quae generalia sunt ita tractasset ut aliis multa diligentius inspicienda ac contemplanda relinqueret, in eo Graeci parum admodum, ne dicam nihil, laboris sibi sumpserunt. At Arabes, non contenti nudis interpretationibus, materiam totam, hoc est res ipsas de quibus tractandum fuerat, multo diligentius ac fusius sibi inspiciendas putaverunt, idque vel praecipuum in Averroe laudatur, cuius solidissima doctrina de Graecorum fontibus non magis hausta quam expressa usque eo enituit ut solus 'commentatoris' nomen sibi iure vendicarit, ac iam constet inter omnes qui proximis seculis sunt philosophati, eas philosophiae partes quae ab Aristotele sunt omissae, ab alio hactenus nemine vel diligentius inspectas vel fundamentis solidioribus fuisse constitutas.

Text 14. Antonius Stupa, *Epistola nuncupatoria* (Basel, 1551), sig. a3r–v.

Fuit autem, ut modo dictum est, hic author Arabs et Arabice conscripsit, quem Yhuda filius Muscae (ut in interpretum prooemio patet) de Arabico in Hispanicum idioma transtulit atque ex hoc post in Latinum sermonem vertit Aegidius Parmensis una cum Petro Regio, sic ut phrases Hispanicae, Gallicae et Italicae ubique magis fuerint servatae (quod illa tempora tulerunt) quam proprietates et dictiones Latini sermonis. Quis enim Latinorum dixit unquam 'charistiam bladi' pro 'caritate annonae' seu 'frumenti', quibus locutionibus scatebat hic author, quod ille facile deprehendet qui illam Venetam translationem cum hac nostra conferet. Non hoc agimus ut velimus illam de manibus studiosorum excutere et nostram recipi, sed ut purgationem ferant concedantque barbaram et magis Gallicam et Hispanicam (quemadmodum in rationibus nostris patebit) quam Latinam versationem, in meliorem Latinae linguae sermonem transcribi, mutari et puriorem fieri, si puritatis huius sermonis eos detinet aliqua cupiditas.

Text 15. Girolamo Donzellini, *Praefatio* (Basel, 1564), sig. B4r–v.

A Graecis ad Arabas delata [sc. medicinae scientia], naufragium fecit, ac Latini ab Arabibus illam recipientes, diu admodum infoeliciter in illa versati sunt. Deus tandem nostram sortem miseratus, cum una cum linguarum peritia, scientias in lucem reuocaret, hanc etiam divinam artem illustravit, excitatis aliquot viris, qui e limpidis Graecorum fontibus illam docerent. Siquidem ex quo Othomanicus furor Graeciam armis coepit divexare, eius barbariem non ferentes Graecorum doctissimi in Italiam venerunt, librorum thesauros secum adferentes, qui a Medicea familia potissimum honorifice accepti, disciplinis omnibus magna ex parte corruptis, lumen attulerunt. Ex eo tempore medicina florere coepit.

Bibliography

Primary Sources

Alpago, Paolo, *Ter illustrissimo ac reverendissimo domino Alexandro Cardinali Farnesio*, in *Avicennae philosophi praeclarissimi ac medicorum principis Compendium de anima, De mahad* [...] *Aphorismi de anima, De diffinitionibus et quaesitis, De divisione scientiarum*, Venice: Giunta, 1546; repr. Farnborough: Gregg, 1969, *1v–*2v

Balmes, Abraham de, *Divo Dominico Grimano*, in Averroes, *Libri Posteriorum analiticorum*, Venice: per Io. Antonium & fratres de Sabio, 1523, sig. AA2r–AA3r

Bruyerin Champier, Jean, *Praefatio*, in Averroes, *Collectaneorum de re medica* [...] *sectiones tres*, Lyon: Gryphius, 1537, sig. A2r–B1v

Calonymos, Calo, *Illustri domino Alberto Pio Carpensium Domino*, in Averroes, *Subtilissimus liber Averois*, Venice: Pederzano, 1527, verso of title-page

Calonymos, Calo, *Reverendissimo domino domino Ioanni Mattheo de Gibertis episcopo Veronensi*, in Alpetragius, *Theorica planetarum*, Venice: Lucantonio Giunta, 1531, sig. aa1v

Carion, Johannes, *Chronica*, Wittenberg: Rhaw, 1533

Collegio of the philosophers and physicians of the University of Padua, *Pro excellentissimo artium et medicine doctore domino magistro Andrea de Cividale*, in Francesca Lucchetta, *Il medico e filosofo Bellunese Andrea Alpago (†1522), traduttore di Avicenna*, Padua: Antenore, 1964, 88–89

Collegio of the philosophers and physicians of the University of Padua, *Pro domino magistro Andrea de Cividale*, in Francesca Lucchetta, *Il medico e filosofo Bellunese Andrea Alpago (†1522), traduttore di Avicenna*, Padua: Antenore, 1964, 90

del Medigo, Elia, *Dignissimo nobili acutissimo philosopho domino Dominico Grimani*, in Bohdan Kieszkowski, 'Les rapports entre Elie de Medigo et Pic de la Mirandole (d'après le ms. lat. 6508 de la Bibliothèque nationale)', *Rinascimento*, 4 (1964), 76–77

Donzellini, Girolamo, *Praefatio*, in Leonardus Iacchinus (Leonardo Giacchini), *In nonum Rasis ad Almansorem*, Basel: Petrus Perna, 1564, sig. A2r–C2v

Giunta editors, *Aristotelis Stagiritae omnia quae extant opera*, Venice: Giunta, 1550/1552, ff.2r–4v

Ledesma, Miguel Jerónimo, *Epistola nuncupatoria*, in Avicenna, *Prima primi Canonis Avicenne sectio*, Valencia: Ioannes Mey, 1547/1548, sig. A2r–A3r

Mantino, Jacob, *Leoni decimo pontifici maximo*, in Averroes, *Paraphrasis* [...] *de partibus et generatione animalium*, Rome: Marcello Silber, 1521, sig. A1v–A2r

Mantino, Jacob, *Herculi Consagae electo Mantuano domino suo illustrissimo*, in Averroes, *Epitome Metaphysicae*, Bologna: heredes Benedicti Hectoris, 1523; ed. David Kaufmann, 'Jacob Mantino: Une page de l'histoire de la Renaissance', *Revue des études juives*, 26 (1893), 221–223

Mantino, Jacob, *Andree Griti serenissimo ac excellentissimo Venetiarum principi*, in Avicenna, *Quarta fen primi libri*, Venice: Lucantonio Giunta, 1530, sig. A1v–A2v

Milich, Jacob, *Oratio de Avicennae vita*, Wittenberg: Veit Kreutzer, 1550

Ramusio, Girolamo, *Ad Flavium*, in Avicenna, *Qānūn*, Paris, BNF, MS arabe 2897

Ricci, Paolo, *Prefatio*, in Averroes, *Hoc opere contenta* [...], Milan: Leonardo Vegio, 1511, f.7r

Stupa, Antonius, *Epistola nuncupatoria*, in Haly filius Abenragel, *Libri de iudiciis astrorum*, Basel: Petri, 1551, sig. a2r–b1v

Thorer, Alban, *Illustrissimo summaque veneratione dignissimo principi ac domino domino Ioanni Rodolpho Stoer Murbacensis ac Lutrensis collegiorum abbati*, in Rhazes, *Opera exquisitiora*, Basel: Petri, 1544, sig. a2r–a3v

Secondary Literature

Ballester, Luis García, 'The Circulation and Use of Medical Manuscripts in Arabic in Sixteenth-Century Spain', *Journal for the History of Arabic Science*, 3 (1979), 183–199

Burnett, Charles, 'The Second Revelation of Arabic Philosophy and Science: 1492–1575', in *Islam and the Italian Renaissance*, ed. Charles Burnett and Anna Contadini, London: The Warburg Institute, 1999, 185–198

Burnett, Charles, 'Revisiting the 1552–1550 and 1562 Aristotle-Averroes Edition', in *Renaissance Averroism and its Aftermath: Arabic Philosophy in Early Modern Europe*, ed. Anna Akasoy, Guido Giglioni, Dordrecht: Springer, 2013, 55–64

d'Alverny, Marie-Thérèse, 'Avicenne et les médecins de Venise', in Marie-Thérèse d'Alverny, *Avicenne en occident: Recueil d'articles de Marie-Thérèse d'Alverny réunis en hommage à l'auteur*, Paris: J. Vrin, 1993 [article XIII]

d'Alverny, Marie-Thérèse, 'Andrea Alpago interprète et commentateur d'Avicenne', in Marie-Thérèse d'Alverny, *Avicenne en occident: Recueil d'articles de Marie-Thérèse d'Alverny réunis en hommage à l'auteur*, Paris: J. Vrin, 1993 [article XIV]

d'Alverny, Marie-Thérèse, 'Survivance et renaissance d'Avicenne à Venise et Padoue', in Marie-Thérèse d'Alverny, *Avicenne en occident: Recueil d'articles de Marie-Thérèse d'Alverny réunis en hommage à l'auteur*, Paris: J. Vrin, 1993 [article XV]

Di Bono, Mario, 'Copernicus, Amico, Fracastoro and Ṭūṣī's Device: Observations on the Use and Transmission of a Model', *Journal for the History of Astronomy*, 26 (1995), 133–154

Di Donato, Silvia, 'Traduttori di Averroè e traduzioni ebraico-latine nel dibattito filosofico del XV e XVI secolo', in *L'averroismo in età moderna (1400–1700)*, ed. Giovanni Licata, Macerata: Quodlibet, 2013, 25–49

Endress, Gerhard, 'Averrois Opera: A Bibliography of Editions and Contributions to the Text', in *Averroes and the Aristotelian Tradition: Sources, Constitution and Reception of the Philosophy of Ibn Rushd*, ed. Gerhard Endress and Jan A. Aertsen with the assistance of Klaus Braun, Leiden et al.: Brill, 1998, 339–381

Engel, Michael, *Elijah Del Medigo and Paduan Aristotelianism: Investigating the Human Intellect*, London et al.: Bloomsbury, 2017

Jenny, Beat Rudolf, 'Antonius Stuppa—ein vergessener Humanist aus dem Bergell', *Bündner Monatsblatt*, 3/4 (1975), 49–83

Grendler, Paul F., *The Universities of the Italian Renaissance*, Baltimore MD/London: The Johns Hopkins University Press, 2002

Gutas, Dimitri, *Avicenna and the Aristotelian Tradition: Introduction to Reading Avicenna's Philosophical Works. Second, Revised and Enlarged Edition, Including an Inventory of Avicenna's Authentic Works*, Leiden: Brill, 2014

Hasse, Dag Nikolaus, 'Averroes' Critique of Ptolemy and its Reception by John of Jandun and Agostino Nifo', in *Averroes' Natural Philosophy and its Reception in the Latin West*, ed. Paul J.J.M. Bakker, Leuven: Leuven University Press, 2015, 69–88

Hasse, Dag Nikolaus, *Success and Suppression: Arabic Sciences and Philosophy in the Renaissance*, Cambridge MA/London: Harvard University Press, 2016

Kaufmann, David, 'Jacob Mantino: Une page de l'histoire de la Renaissance', *Revue des études juives*, 26 (1893), 30–60, 207–229

Kieszkowski, Bohdan, 'Les rapports entre Elie de Medigo et Pic de la Mirandole (d'après le ms. lat. 6508 de la Bibliothèque nationale)', *Rinascimento*, 4 (1964), 41–91

Klein-Franke, Felix, *Die klassische Antike in der Tradition des Islam*, Darmstadt: Wissenschaftliche Buchgesellschaft, 1980

Lucchetta, Francesca, *Il medico e filosofo Bellunese Andrea Alpago (†1522), traduttore di Avicenna*, Padua: Antenore, 1964

Michot, Yahya J., 'A Mamlūk theologian's commentary on Avicenna's *Risāla aḍḥawiyya* being a translation of a part of the *Dar' al-taʿāruḍ* of Ibn Taymiyya, with introduction, annotation, and appendices. Part I', *Journal of Islamic Studies*, 14/2 (2003), 149–203

Roling, Bernd, *Aristotelische Naturphilosophie und christliche Kabbalah im Werk des Paulus Ritius*, Tübingen: Niemeyer, 2007

Schmitt, Charles B., 'Renaissance Averroism Studied through the Venetian Editions of Aristotle-Averroes (with Particular Reference to the Giunta Edition of 1550–2)', in Charles B. Schmitt, *The Aristotelian Tradition and Renaissance Universities*, London: Variorum Reprints, 1984 [article VIII]

Siraisi, Nancy, *Avicenna in Renaissance Italy: The Canon and Medical Teaching in Italian Universities after 1500*, Princeton NJ: Princeton University Press, 1987

Tamani, Giuliano, 'Le traduzioni ebraico-latine di Abraham de Balmes', in *Biblische und Judaistische Studien: Festschrift für Paolo Sacchi*, ed. Angelo Vivian, Frankfurt am Main et al.: Lang, 1990, 613–635

Tamani, Giuliano, 'Traduzioni ebraico-latine di opere filosofiche e scientifiche', in *L'Hébreu au temps de la Renaissance: Ouvrage collectif recueilli et édité*, ed. Ilana Zinguer, Leiden et al.: Brill, 1992, 105–114

Veit, Raphaela, 'Andrea Alpago and Schah Ismāʿīl: Der Begründer der Safavidendynastie im Zeugnis eines Venezianer Gesandtschaftsarztes, Händlers und Informanten', in *Das Charisma: Funktionen und symbolische Repräsentationen*, ed. Pavlína Rychterová et al., Berlin: Akademie-Verlag, 2008, 457–465

Index of Personal Names

Entries are alphabetized up to the first comma (where present), ignoring spaces, diacritics, *hamza* ('), *'ayn* (ʿ), particles (such as de, del, etc.), and the prefixes al- and Pseudo-.

Index of Subjects and Placenames

Entries are alphabetized up to the first comma (where present), ignoring spaces, diacritics, *hamza* ('), *'ayn* ('), particles (such as de, del, etc.), and the prefix al-.

Index of Sources

Entries are alphabetized to the first comma (where present), ignoring spaces, diacritics, *hamza* ('), *'ayn* ('), particles (such as de, del, etc.), and the prefixes al- and Pseudo-.

www.ingramcontent.com/pod-product-compliance
Lightning Source LLC
Chambersburg PA
CBHW031318230926
51848CB00001B/3
* 9 7 8 9 0 0 4 7 6 0 8 9 9 *